T0281544

Accelerator Physics

Fourth Edition

Accelerator Physics

Fourth Edition

S. Y. Lee
Indiana University, USA

NEW JERSEY • LONDON • SINGAPORE • BEIJING • SHANGHAI • HONG KONG • TAIPEI • CHENNAI • TOKYO

Published by

World Scientific Publishing Co. Pte. Ltd.
5 Toh Tuck Link, Singapore 596224
USA office: 27 Warren Street, Suite 401-402, Hackensack, NJ 07601
UK office: 57 Shelton Street, Covent Garden, London WC2H 9HE

British Library Cataloguing-in-Publication Data
A catalogue record for this book is available from the British Library.

ACCELERATOR PHYSICS
4th Edition

ISBN 978-981-3274-67-9
ISBN 978-981-3274-78-5 (pbk)

For any available supplementary material, please visit
https://www.worldscientific.com/worldscibooks/10.1142/11111#t=suppl

Printed in Singapore

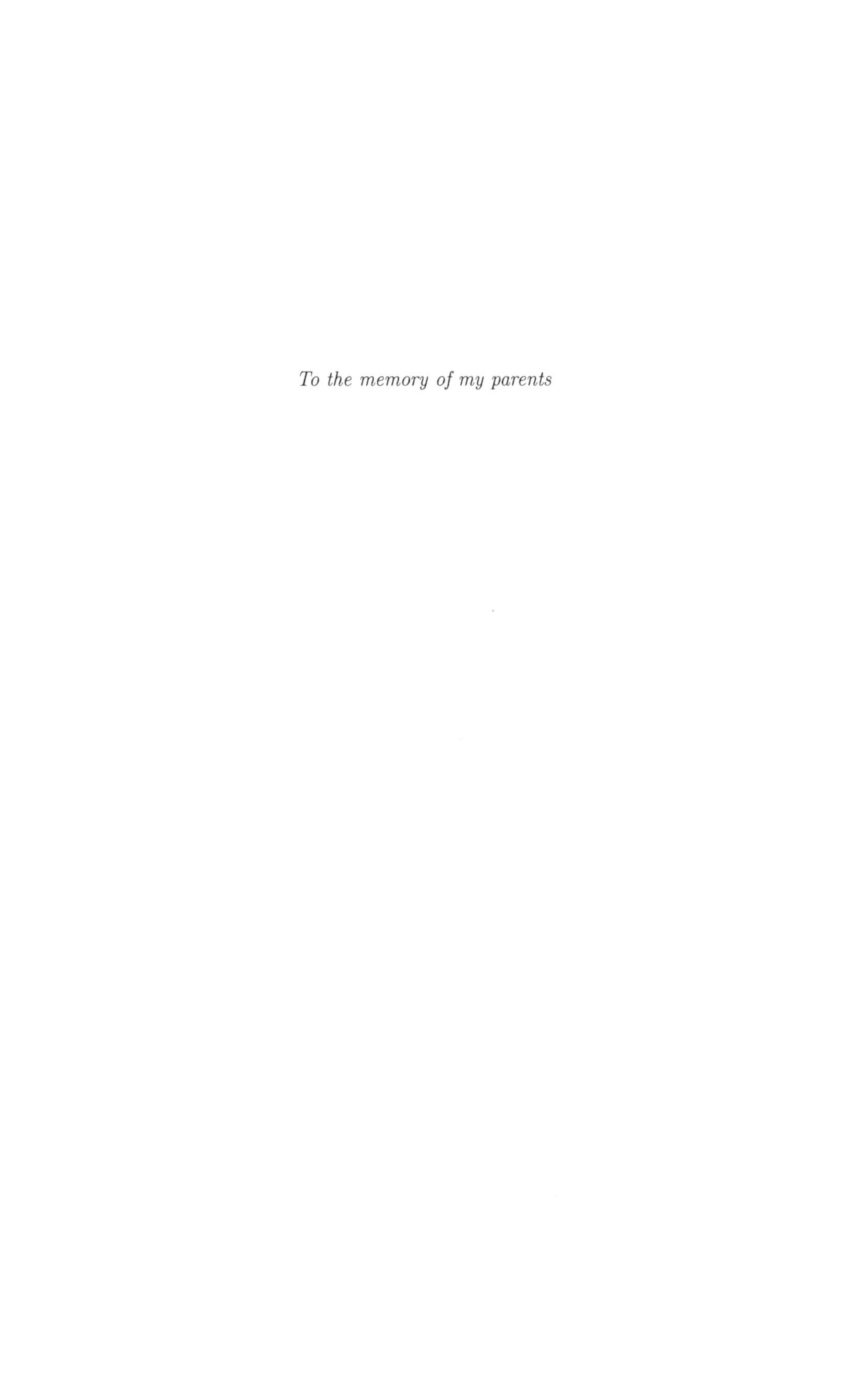

To the memory of my parents

Preface

Accelerator science research and development has been very active for more than 100 years. High energy particle accelerators have provided many scientific discoveries in the 20th century, and will continue to pave the scientific discoveries in the 21st century. The confirmation of the God particle or the Higgs boson discovery on July 4, 2012 is a clear example. These discoveries led to international efforts for the design and construction of Higgs-factories and a very large hadron collider (VLHC). Accelerator research and development will continue to power innovations in the 21st century.

The success of high brilliance ultrafast femtosecond X-ray laser from the linac-coherent-light-source (LCLS) project drives the construction of the LCLS-II and the X-FEL based on the superconducting linac technology. This also provides incentive for diffraction limited light sources, such as energy recovery linacs, and upgrade of storage ring based light sources.

High power hadron beams have been used in the production of neutron, meson, muon, neutrino beams, that find applications in condensed matter physics, high energy and nuclear physics, and possibly the nuclear transmutation of the nuclear wastes or the "energy amplifiers." Recent efforts on the Fixed-Field Alternating-Gradient accelerators are of great interests to future high power accelerators.

There are more than 5000 X-ray radiation therapy accelerators in the world. Hundred years of radiation biology research paves the efforts of recent particle (hadron) beam radiotherapy. As of 2017, there are 79 hadron medical centers for particle radiotherapy. There are more than 45 hadron medical centers in the planing stage.

This fourth edition keeps the structure of the previous editions. The design concepts of recent diffraction limited electron storage ring are discussed in Chapter 4.3. I expand the coverage of the non-linear beam dynamics in Chapter 2. I revise some homework problems, and correct mis-prints in earlier Editions. However, for beginners in accelerator physics, one should study Secs. II-IV in Chapter 2, and Secs. I-II in Chapter 3. Your comments and criticisms to this revised edition are appreciated.

S.Y. Lee
Bloomington, Indiana, U.S.A.
June 10, 2018

Preface to Third Edition

Accelerator science took off in the 20th century. Accelerator scientists invent many innovative technologies to produce and manipulate high energy and high quality beams that are instrumental to progresses in natural sciences. Many kinds of accelerators serve the need of research in natural and biomedical sciences, and the demand of applications in industry.

In the 21st century, accelerators will become even more important in applications that include industrial processing and imaging, biomedical research, nuclear medicine, medical imaging, cancer therapy, energy research, etc. Accelerator research aims to produce beams in high power, high energy, and high brilliance frontiers. These beams addresses the needs of fundamental science research in particle and nuclear physics, condensed matter and biomedical sciences. High power beams may ignite many applications in industrial processing, energy production, and national security.

Accelerator Physics studies the interaction between the charged particles and electromagnetic field. Research topics in accelerator science include generation of electromagnetic fields, material science, plasma, ion source, secondary beam production, nonlinear dynamics, collective instabilities, beam cooling, beam instrumentation, detection and analysis, beam manipulation, etc.

The textbook is intended for graduate students who have completed their graduate core-courses including classical mechanics, electrodynamics, quantum mechanics, and statistical mechanics. I have tried to emphasize the fundamental physics behind each innovative idea with least amount of mathematical complication. The textbook may also be used by advanced undergraduate seniors who have completed courses on classical mechanics and electromagnetism. For beginners in accelerator physics, one begins with Secs. 2.I–2.IV in Chapter 2, and follows by Secs. 3.I–3.II in Chapter 3 for the basic betatron and synchrotron motion. The study continues onto Secs. 2.V, 2.VIII, and 3.VII for chromatic aberration and collective beam instabilities. After these basic topics, the rf technology and basic physics of linac are covered in Secs. 3.V, 3.VI, 3.VIII in Chapter 3. The basic accelerator physics course ends with physics of electron storage rings in Chapter 4, and some advanced topics of free electron laser and beam-beam interaction in Chapter 5.

For beginners, one should pay great attention to the Floquet transformation of Sec. 2.II that can be used to solve Hill's equation with perturbations. Similarly, some scaling properties of bunch longitudinal distribution in Sec. 3.II are handy for beam manipulation, data analysis, and machine design. The Hamiltonian formalism and canonical transformation, often used to solve particle motion in this book, can provide a better physics picture in beam dynamics.

In this revised edition, I include some recently published information on beam manipulation and detection methods, advanced data analysis. I revise some homework problems, and correct mis-prints in the second edition. The homework is designed to solve a particular problem by providing step-by-step procedures to minimize frustration. The answer is usually listed at the end of each homework problem so that the result can be used in practical design of accelerator systems. I take this opportunity to enhance the content of Sec. 2.VII. Your comments and criticisms on this revised edition are appreciated.

S.Y. Lee, Bloomington, Indiana, U.S.A.
June 10, 2011

Preface to Second Edition

Since the appearance of the first edition in 1999, this book has been used as a textbook or reference for graduate-level "Accelerator Physics" courses. I have benefited from questions, criticism and suggestions from colleagues and students. As a response to these suggestions, the revised edition is intended to provide easier learning explanations and illustrations.

Accelerator Physics studies the interaction between the charged particles and electromagnetic field. The applications of accelerators include all branches of sciences and technologies, medical treatment, and industrial processing. Accelerator scientists invent many innovative technologies to produce beams with qualities required for each application.

This textbook is intended for graduate students who have completed their graduate core-courses including classical mechanics, electrodynamics, quantum mechanics, and statistical mechanics. I have tried to emphasize the fundamental physics behind each innovative idea with least amount of mathematical complication. The textbook may also be used by undergraduate seniors who have completed courses on classical mechanics and electromagnetism. For beginners in accelerator physics, one begins with Secs. 2.1–2.4 in Chapter 2, and follows by Secs. 3.1–3.2 for the basic betatron and synchrotron motion. The study continues onto Secs. 2.5, 2.8, and 3.7 for chromatic aberration and collective beam instabilities. After these basic topics, the rf technology and basic physics of linac are covered in Secs. 3.5, 3.6, 3.8 in Chapter 3. The basic accelerator physics course ends with physics of electron storage rings in Chapter 4.

I have chosen the Frenet-Serret coordinate-system of $(\hat{x}, \hat{s}, \hat{z})$ for the transverse radially-outward, longitudinally-forward, and vertical unit base-vectors with the right-hand rule: $\hat{z} = \hat{x} \times \hat{s}$. I have also chosen positive-charge to derive the equations of betatron motion for all sections of the Chapter 2, except a discussion of $\pm$-signs in Eq. (2.22). The sign of some terms in Hill's equation should be reversed if you solve the equation of motion for electrons in accelerators.

The convention of the rf-phase differs in linac and synchrotron communities by $\phi_{\rm linac} = \phi_{\rm synchrotron} - (\pi/2)$. To be consistent with the synchrotron motion in Chapter 3, I have chosen the rf-phase convention of the synchrotron community to describe the synchrotron equation of motion for linac in Sec. 3.8.

In this revised edition, I include two special topics: free electron laser (FEL) and beam-beam interaction in Chapter 5. In 2000, several self-amplified spontaneous emission (SASE) FEL experiments have been successfully demonstrated. Many light source laboratories are proposing the fourth generation light source using high gain FEL based on the concept of SASE and high-gain harmonic generation (HGHG). Similarly, the success of high luminosity B-factories indicates that beam-beam interaction remains very important to the basic accelerator physics. These activities justify the addition of two introductory topics to the accelerator physics text.

Finally, the homework is designed to solve a particular problem by providing step-by-step procedures to minimize frustrations. The answer is usually listed at the end of each homework problem so that the result can be used in practical design of accelerator systems. I would appreciate very much to receive comments and criticism to this revised edition.

S.Y. Lee, Bloomington, Indiana, U.S.A.
November, 2004

Preface to First Edition

The development of high energy accelerators began in 1911 when Rutherford discovered the atomic nuclei inside the atom. Since then, high voltage DC and rf accelerators have been developed, high-field magnets with excellent field quality have been achieved, transverse and longitudinal beam focusing principles have been discovered, high power rf sources have been invented, high vacuum technology has been improved, high brightness (polarized/unpolarized) electron/ion sources have been attained, and beam dynamics and beam manipulation schemes such as beam injection, accumulation, slow and fast extraction, beam damping and beam cooling, instability feedback, etc. have been advanced. The impacts of the accelerator development are evidenced by many ground-breaking discoveries in particle and nuclear physics, atomic and molecular physics, condensed-matter physics, biomedical physics, medicine, biology, and industrial processing.

Accelerator physics and technology is an evolving branch of science. As the technology progresses, research in the physics of beams propels advancement in accelerator performance. The advancement in type II superconducting material led to the development of high-field magnets. The invention of the collider concept initiated research and development in single and multi-particle beam dynamics. Accelerator development has been impressive. High energy was measured in MeV's in the 1930's, GeV's in the 1950's, and multi-TeV's in the 1990's. In the coming decades, the center of mass energy will reach 10-100 TeV. High intensity was 10^9 particles per pulse in the 1950's. Now, the AGS has achieved 6×10^{13} protons per pulse. We are looking for 10^{14} protons per bunch for many applications. The brilliance of synchrotron radiation was about 10^{12} [photons/s mm^2 mrad^2 0.1% $(\Delta\lambda/\lambda)$] from the first-generation light sources in the 1970's. Now, it reaches 10^{21}, and efforts are being made to reach a brilliance of $10^{29} - 10^{34}$ in many FEL research projects.

This textbook deals with basic accelerator physics. It is based on my lecture notes for the accelerator physics graduate course at Indiana University and two courses in the U.S. Particle Accelerator School. It has been used as preparatory course material for graduate accelerator physics students doing thesis research at Indiana University. The book has four chapters. The first describes historical accelerator development. The second deals with transverse betatron motion. The third chapter concerns synchrotron motion and provides an introduction to linear accelerators. The fourth deals with synchrotron radiation phenomena and the basic design principles of low-emittance electron storage rings. Since this is a textbook on basic accelerator physics, topics such as nonlinear beam dynamics, collective beam instabilities, etc., are mentioned only briefly, in Chapters 2 and 3.

Attention is paid to deriving the action-angle variables of the phase space coordinates because the transformation is basic and the concept is important in understanding the phenomena of collective instability and nonlinear beam dynamics. In the design of synchrotrons, the dispersion function plays an important role in particle stability, beam performance, and beam transport. An extensive section on the dispersion function is provided in Chapter 2. This function is also important in the design of low-emittance electron storage ring lattices.

The SI units are used throughout this book. I have also chosen the engineer's convention of $j = -i$ for the imaginary number. The exercises in each section are designed to have the student apply a specific technique in solving an accelerator physics problem. By following the steps provided in the homework, each exercise can be easily solved.

The field of accelerator physics and technology is multi-disciplinary. Many related subjects are not extensively discussed in this book: linear accelerators, induction linacs, high brightness beams, collective instabilities, nonlinear dynamics, beam cooling physics and technology, linear collider physics, free-electron lasers, electron and ion sources, neutron spallation sources, muon colliders, high intensity beams, vacuum technology, superconductivity, magnet technology, instrumentation, etc. Nevertheless, the book should provide the understanding of basic accelerator physics that is indispensable in accelerator physics and technology research.

S.Y. Lee, Bloomington, Indiana, U.S.A.
January, 1998

Acknowledgments

I would like to thank colleagues, particularly D. Li, H. Huang, X. Kang, M. Ellison, K.M. Fung, M. Bai, A. Riabko, S. Cousineau, C. Beltran, W. Guo, X. Huang, V. Ranjbar, Y. Zhang, X. Pang, L. Yang, Y. Jing and T. Luo, who made many useful suggestions to this revised edition. During the course of this work, I have benefited greatly from the collaboration with Drs. David Caussyn, Y. Wang, D. Jeon, K.Y. Ng, Y. Yan and Prof. A. Chao. I owe special thanks to Margaret Dienes for editing the first edition of this book. Supports from the National Science Foundation and Department of Energy to our graduate students and postdocs are indispensible to research carried out in past years. Results of these research are included in this latest edition. The responsibility for all errors lies with me. Your comments and corrections will be highly appreciated.

Contents

Preface vii

Preface to Third Edition ix

Preface to Second Edition xi

Preface to First Edition xiii

Acknowledgments xv

Symbols and Notations xxix

List of Tables xxxiii

1 Introduction 1

I Historical Developments 4
I.1 Natural Accelerators 5
I.2 Electrostatic Accelerators 5
I.3 Induction Accelerators 6
I.4 Radio-Frequency (RF) Accelerators 8
I.5 Colliders and Storage Rings 16
I.6 Synchrotron Radiation Storage Rings 18
II Layout and Components of Accelerators 18
II.1 Acceleration Cavities 19
II.2 Accelerator Magnets 20
II.3 Other Important Components 22
III Accelerator Applications 22
III.1 High Energy and Nuclear Physics 22
III.2 Solid-State and Condensed-Matter Physics 23
III.3 Other Applications 23
Exercise 24

2 Transverse Motion **33**
I Hamiltonian for Particle Motion in Accelerators 34
I.1 Hamiltonian in Frenet-Serret Coordinate System 35
I.2 Magnetic Field in Frenet-Serret Coordinate System 37
I.3 Equation of Betatron Motion 38
I.4 Particle Motion in Dipole and Quadrupole Magnets 39
Exercise 40
II Linear Betatron Motion 44
II.1 Transfer Matrix and Stability of Betatron Motion 44
II.2 Courant–Snyder Parametrization 47
II.3 Floquet Transformation 48
A. Betatron tune 50
B. FODO cell 50
C. Doublet cells 52
II.4 Action-Angle Variable and Floquet Transformation 53
II.5 Courant–Snyder Invariant and Emittance 55
A. Emittance of a beam 56
B. The σ-matrix 57
C. Emittance measurement 57
D. Gaussian distribution function 59
E. Adiabatic damping and the normalized emittance 60
II.6 Stability of Betatron Motion: A FODO Cell Example 60
II.7 Symplectic Condition 61
II.8 Effect of Space-Charge Force on Betatron Motion 62
A. The Kapchinskij-Vladimirskij (KV) distribution 62
B. The space charge force 63
C. The envelope equation for a space charge dominated beam . 65
D. A uniform focusing paraxial system 66
E. Space-charge force for gaussian distribution 67
Exercise 69
III Effect of Linear Magnet Imperfections 80
III.1 Closed-Orbit in the Presence of Dipole Field Error 80
A. The perturbed closed orbit and Green's function 80
B. Distributed dipole field error 82
C. The integer stopband integrals 82
D. Statistical estimation of closed-orbit errors 83
E. Closed-orbit correction 83
F. Effects of dipole field error on orbit length 84
III.2 Extended Matrix Method for the Closed Orbit 86
III.3 Application of Dipole Field Error 86
A. Orbit bumps 86
B. Fast kick for beam extraction 87

		C. Effects of rf dipole field, rf knock-out	89
		D. Orbit response matrix and accelerator modeling	91
		E. Model Independent Analysis	94
	III.4	Quadrupole Field (Gradient) Errors	95
		A. Betatron tune shift	95
		B. Betatron amplitude function modulation (beta-beat)	96
		C. The half-integer stopband integrals	96
		D. Example of one quadrupole error in FODO cell lattice	98
		E. Statistical estimation of stopband integrals	98
		F. Effect of a zero tune shift π-doublet quadrupole pair	98
	III.5	Basic Beam Observation of Transverse Motion	99
		A. Beam position monitor (BPM)	99
		B. Measurements of betatron tune and phase-space ellipse	100
	III.6	Application of Quadrupole Field Error	102
		A. β-function measurement	102
		B. Tune jump	102
	III.7	Beam Spectra	103
		A. Transverse spectra of a particle	103
		B. Fourier spectra of a single beam with finite time span	105
		C. Fourier spectra of many particles and Schottky noise	106
	III.8	Beam Injection and Extraction	108
		A. Beam injection and extraction	108
		B. Beam extraction	109
	III.9	Mechanisms of Emittance Dilution and Diffusion	110
		A. Emittance diffusion due to random scattering processes	110
		B. Space charge effects	111
		C. Emittance evolution measurements and modeling	114
	Exercise		115
IV	Off-Momentum Orbit		122
	IV.1	Dispersion Function	122
		A. FODO cell	124
		B. Dispersion function in terms of transfer matrix	125
		C. Effect of dipole and quadrupole error on dispersion function	126
	IV.2	$\mathcal{H}$-Function, Action, and Integral Representation	126
	IV.3	Momentum Compaction Factor	128
		A. Transition energy and phase-slip factor	129
		B. Phase stability of synchrotron motion	130
		C. Effect of dispersion on the response matrix of the ORM	131
	IV.4	Dispersion Suppression and Dispersion Matching	132
	IV.5	Achromat Transport Systems	134
	IV.6	Transport Notation	136
	IV.7	Experimental Measurements of Dispersion Function	137

	IV.8	Transition Energy Manipulation	138
		A. γ_T jump schemes	139
		B. Flexible momentum compaction (FMC) lattices	141
	IV.9	Minimum $\langle\mathcal{H}\rangle$ Modules	146
	Exercise		150
V	Chromatic Aberration		158
	V.1	Chromaticity Measurement and Correction	159
		A. Chromaticity measurement	159
		B. Chromatic correction	160
		C. Nonlinear modeling from chromaticity measurement	162
	V.2	Nonlinear Effects of Chromatic Sextupoles	163
	V.3	Chromatic Aberration and Correction	163
		A. Systematic chromatic half-integer stopband width	164
		B. Chromatic stopband integrals of FODO cells	165
		C. The chromatic stopband integral of insertions	166
		D. Effect of the chromatic stopbands on chromaticity	166
		E. Effect of sextupoles on the chromatic stopband integrals	167
	V.4	Lattice Design Strategy	168
	Exercise		169
VI	Linear Coupling		171
	VI.1	The Linear Coupling Hamiltonian	171
	VI.2	Effects of an Isolated Linear Coupling Resonance	173
		A. Normal modes at a single linear coupling resonance	174
		B. Resonance precessing frame and Poincaré surface of section	174
		C. Initial horizontal orbit	175
		D. General linear coupling solution	176
	VI.3	Experimental Measurement of Linear Coupling	177
	VI.4	Linear Coupling Correction with Skew Quadrupoles	180
	VI.5	Linear Coupling Using Transfer Matrix Formalism	181
	Exercise		181
VII	Nonlinear Resonances		186
	VII.1	Nonlinear Resonances Driven by Sextupoles	186
		A. Tracking methods	186
		B. The leading order resonances driven by sextupoles	187
		C. The third order resonance at $3\nu_x = \ell$	189
		D. Experimental measurement of a $3\nu_x = \ell$ resonance	191
		E. Other 3rd-order resonances driven by sextupoles	192
	VII.2	Higher-Order Resonances	193
	VII.3	Nonlinear Detuning from Sextupoles and Octupoles	196
	VII.4	Betatron Tunes and Nonlinear Resonances	197
		A. Emittance growth, beam loss and dynamic aperture	198
		B. Tune diffusion rate and dynamic aperture	199

C. Space charge effects . . . 201
Exercise . . . 203
VIII Collective Instability and Landau Damping . . . 210
VIII.1 Impedance . . . 210
A. Resistive wall impedance . . . 210
B. Space-charge impedance . . . 211
C. Broad-band impedance . . . 212
D. Narrow-band impedance . . . 212
E. Properties of the transverse impedance . . . 212
VIII.2 Transverse Wave Modes . . . 213
VIII.3 Effect of Wakefield on Transverse Wave . . . 214
A. Beam with zero frequency spread . . . 216
B. Beam with finite frequency spread . . . 216
C. A model of collective motion . . . 217
VIII.4 Frequency Spread and Landau Damping . . . 218
A. Landau damping . . . 218
B. Solutions of dispersion integral with Gaussian distribution . 220
Exercise . . . 221
IX Synchro-Betatron Hamiltonian . . . 224
Exercise . . . 228

3 Synchrotron Motion **229**
I Longitudinal Equation of Motion . . . 230
I.1 The Synchrotron Hamiltonian . . . 233
I.2 The Synchrotron Mapping Equation . . . 235
I.3 Evolution of Synchrotron Phase-Space Ellipses . . . 236
I.4 Some Practical Examples . . . 237
I.5 Summary of Synchrotron Equations of Motion . . . 237
A. Using t as independent variable . . . 237
B. Using longitudinal distance s as independent variable . . . 238
Exercise . . . 238
II Adiabatic Synchrotron Motion . . . 241
II.1 Fixed Points . . . 241
II.2 Bucket Area . . . 242
II.3 Small-Amplitude Oscillations and Bunch Area . . . 244
A. Gaussian beam distribution . . . 244
B. Synchrotron motion in reference time coordinates . . . 245
C. Approximate action-angle variables . . . 246
II.4 Small-Amplitude Synchrotron Motion at the UFP . . . 247
II.5 Synchrotron Motion for Large-Amplitude Particles . . . 247
A. Stationary synchrotron motion . . . 248
B. Synchrotron tune . . . 248

II.6 Experimental Tracking of Synchrotron Motion 249
Exercise 251
III RF Phase and Voltage Modulations 256
III.1 Normalized Phase-Space Coordinates 256
III.2 RF Phase Modulation and Parametric Resonances 259
A. Effective Hamiltonian near a parametric resonance 260
B. Dipole mode 260
C. Island tune 262
D. Separatrix of resonant islands 263
III.3 Measurements of Synchrotron Phase Modulation 264
A. Sinusoidal rf phase modulation 264
B. Action angle derived from measurements 265
C. Poincaré surface of section 266
III.4 Effects of Dipole Field Modulation 267
A. Chaotic nature of parametric resonances 269
B. Observation of attractors 270
C. The hysteretic phenomena of attractors 272
D. Systematic property of parametric resonances 273
III.5 RF Voltage Modulation 275
A. The equation of motion with rf voltage modulation 275
B. The perturbed Hamiltonian 276
C. Parametric resonances 277
D. Quadrupole mode 277
E. The separatrix 279
F. The amplitude dependent island tune of 2:1 parametric resonance 279
III.6 Measurement of RF Voltage Modulation 280
A. Voltage modulation control loop 280
B. Observations of the island structure 281
Exercise 282
IV Nonadiabatic and Nonlinear Synchrotron Motion 285
IV.1 Linear Synchrotron Motion Near Transition Energy 286
A. The asymptotic properties of the phase space ellipse 288
B. The Gaussian distribution function at transition energy . . 289
IV.2 Nonlinear Synchrotron Motion at $\gamma \approx \gamma_T$ 289
IV.3 Beam Manipulation Near Transition Energy 292
A. Transition energy jump 292
B. Momentum aperture for faster beam acceleration 292
C. Flatten the rf wave near transition energy 292
IV.4 Synchrotron Motion with Nonlinear Phase Slip Factor 293
IV.5 The QI Dynamical Systems 295
Exercise 299

V Beam Manipulation in Synchrotron Phase Space . . . 300
V.1 RF Frequency Requirements . . . 301
A. The choice of harmonic number . . . 302
B. The choice of rf voltage . . . 302
V.2 Capture and Acceleration of Proton and Ion Beams . . . 303
A. Adiabatic capture . . . 303
B. Non-adiabatic capture . . . 303
C. Chopped beam at the source . . . 305
V.3 Bunch Compression and Rotation . . . 305
A. Bunch compression by rf voltage manipulation . . . 306
B. Bunch compression using unstable fixed point . . . 307
C. Bunch rotation using buncher/debuncher cavity . . . 308
V.4 Debunching . . . 309
V.5 Beam Stacking and Phase Displacement Acceleration . . . 309
V.6 Double rf Systems . . . 310
A. Synchrotron equation of motion in a double rf system . . . 311
B. Action and synchrotron tune . . . 312
C. The $r \leq 0.5$ case . . . 312
D. The $r > 0.5$ case . . . 313
E. Action-angle coordinates . . . 314
F. Small amplitude approximation . . . 316
G. Sum rule theorem and collective instabilities . . . 316
V.7 The Barrier RF Bucket . . . 317
A. Equation of motion in a barrier bucket . . . 318
B. Synchrotron Hamiltonian for general rf wave form . . . 319
C. Square wave barrier bucket . . . 319
D. Hamiltonian formalism . . . 321
E. Action-angle coordinates . . . 322
V.8 Beam-stacking in Longitudinal Phase space . . . 323
Exercise . . . 326
VI Fundamentals of RF Systems . . . 330
VI.1 Pillbox Cavity . . . 330
VI.2 Low Frequency Coaxial Cavities . . . 332
A. Shunt impedance and Q-factor . . . 334
B. Filling time . . . 336
C. Qualitative feature of rf cavities . . . 336
D. The rf cavity of the IUCF cooler injector synchrotron . . . 337
E. Wake-function and impedance of an RLC resonator model . 339
VI.3 Beam Loading . . . 339
A. Phasor . . . 340
B. Fundamental theorem of beam loading . . . 340
C. Steady state solution of multiple bunch passage . . . 341

VI.4 Beam Loading Compensation and Robinson Instability 342
A. Robinson dipole mode instability 343
B. Qualitative feature of Robinson instability 344
Exercise 345
VII Longitudinal Collective Instabilities 348
VII.1 Beam Spectra of Synchrotron Motion 349
A. Coherent synchrotron modes 349
B. Coherent synchrotron modes of a kicked beam 351
C. Measurements of coherent synchrotron modes 352
VII.2 Collective Microwave Instability in Coasting Beams 354
VII.3 Longitudinal Impedance 355
A. Space-charge impedance 355
B. Resistive wall impedance 357
C. Narrowband and broadband impedance 358
VII.4 Single Bunch Microwave Instability 358
A. Negative mass instability without momentum spread 358
B. Landau damping with finite frequency spread 359
C. Keil-Schnell criterion 360
D. Microwave instability near transition energy 362
E. Microwave instability and bunch lengthening 363
F. Microwave instability induced by narrowband resonances . . 364
Exercise 365
VIII Introduction to Linear Accelerators 367
VIII.1 Historical Milestones 367
VIII.2 Fundamental Properties of Accelerating Structures 370
A. Transit time factor 370
B. Shunt impedance 371
C. The quality factor Q 371
VIII.3 Particle Acceleration by EM Waves 372
A. EM waves in a cylindrical wave guide 373
B. Phase velocity and group velocity 374
C. TM modes in a cylindrical pillbox cavity 375
D. Alvarez structure 377
E. Loaded wave guide chain and the space harmonics 378
F. Standing wave, traveling wave, and coupled cavity linacs . . 381
G. High Order Modes (HOMs) 383
VIII.4 Longitudinal Particle Dynamics in a Linac 383
A. The capture condition in an electron linac with $v_{\rm p} = c$ . . . 384
B. Energy spread of the beam 385
C. Synchrotron motion in proton linacs 386
VIII.5 Transverse Beam Dynamics in a Linac 387
Exercise 390

4 **Physics of Electron Storage Rings** **397**
I Fields of a Moving Charged Particle 401
I.1 Non-relativistic Reduction 403
I.2 Radiation Field for Particles at Relativistic Velocities 403
Example 1: linac 404
Example 2: Radiation from circular motion 404
I.3 Frequency and Angular Distribution 405
A. Frequency spectrum of synchrotron radiation 407
B. Asymptotic property of the radiation 409
C. Angular distribution in the orbital plane 409
D. Angular distribution for the integrated energy spectrum . . 409
E. Frequency spectrum of radiated energy flux 410
I.4 Quantum Fluctuation 411
Exercise 413
II Radiation Damping and Excitation 415
II.1 Damping of Synchrotron Motion 415
II.2 Damping of Betatron Motion 419
A. Transverse (vertical) betatron motion 419
B. Horizontal betatron motion 420
II.3 Damping Rate Adjustment 422
A. Increase U to increase damping rate (damping wiggler) . . 422
B. Change $\mathcal{D}$ to re-partition the partition number 422
C. Robinson wiggler 424
II.4 Radiation Excitation and Equilibrium Energy Spread 425
A. Effects of quantum excitation 425
B. Equilibrium rms energy spread 426
C. Adjustment of rms momentum spread 428
D. Beam distribution function in momentum 428
II.5 Radial Bunch Width and Distribution Function 429
II.6 Vertical Beam Width 431
II.7 Beam Lifetime 432
A. Quantum lifetime 432
B. Touschek lifetime 434
II.8 Summary: Radiation Integrals 437
Exercise 438
III Emittance in Electron Storage Rings 443
III.1 Emittance of Synchrotron Radiation Lattices 443
A. FODO cell lattice 444
B. Double-bend achromat (Chasman-Green lattice) 447
C. Theoretical Minimum Emittance (TME) lattice 450
D. Three-bend achromat 451
E. Summary of Lattice Properties and QBA 452

F. Design concepts of recent light source upgrades 454
III.2 Insertion Devices 456
A. Ideal helical undulators or wigglers 457
B. Characteristics of radiation from undulators 460
III.3 Effect of IDs on beam dynamics 461
A. Effect of IDs on beam emittances 462
B. Effect of IDs on momentum spread 463
C. Effect of ID induced dispersion functions 463
D. Effect of IDs on the betatron tunes 465
III.4 Beam Physics of High Brightness Storage Rings 466
Exercise 469

5 Special Topics in Beam Physics 475
I Free Electron Laser (FEL) 476
I.1 Small Signal Regime 478
A. Vlasov equation in longitudinal phase-space coordinates .. 479
B. The free electron laser gain 481
I.2 Interaction of the Radiation Field with the Beam 483
A. Perturbation solution of the Maxwell-Vlasov equations ... 483
B. High gain regime 484
I.3 High Gain FEL Facilities 486
Exercise 486
II Beam-Beam Interaction 488
II.1 The Beam-Beam Force in Round Beam Geometry 488
A. The beam-beam potential 489
B. Dynamics betatron amplitude functions 489
C. Disruption factor 490
II.2 The Coherent Beam-Beam Effects 491
II.3 Nonlinear Beam-Beam Effects 492
II.4 Experimental Observations and Numerical Simulations 493
II.5 Beam-Beam Interaction in Linear Colliders 497
Exercise 498

A Classical Mechanics and Analysis 501
I Hamiltonian Dynamics 501
I.1 Canonical Transformations 501
I.2 Fixed Points 502
I.3 Poisson Bracket 502
I.4 Liouville Theorem 502
I.5 Floquet Theorem 503
II Stochastic Beam Dynamics 504
II.1 Central Limit Theorem 504

II.2 Langevin Equation of Motion 505
A. Random walk method 505
B. Other stochastic integration methods 506
II.3 Fokker-Planck Equation 507
III Methods of Data Analysis in Beam Physics 508

B Numerical Methods and Physical Constants 511
I Fourier Transform 511
I.1 Nyquist Sampling Theorem 511
I.2 Discrete Fourier Transform 512
I.3 Digital Filtering 513
I.4 Some Simple Fourier Transforms 514
II Cauchy Theorem and the Dispersion Relation 514
II.1 Cauchy Integral Formula 514
II.2 Dispersion Relation 515
III Useful Handy Formulas 515
III.1 Generating Functions for Bessel Functions 515
III.2 The Hankel Transform 516
III.3 The Complex Error Function [30] 516
III.4 A Multipole Expansion Formula 516
III.5 Cylindrical Coordinates 516
III.6 Gauss' and Stokes' Theorems 517
III.7 Vector Operation 517
III.8 2D Magnetic Field in Multipole Expansion 518
IV Maxwell's Equations 518
IV.1 Lorentz Transformation of EM Fields 519
IV.2 Cylindrical Waveguides 519
A. TM modes: $H_s = 0$ 519
B. TE modes: $E_s = 0$ 520
IV.3 Voltage Standing Wave Ratio 521
V Physical Properties and Constants 521

Bibliography 525

Index 527

Symbols and Notations

- α, phase space damping rate
- $\alpha_{\rm ad}$, the adiabaticity coefficient of synchrotron motion
- $\alpha_{\rm b}(\phi_{\rm s})$, running bucket phase space area reduction factor
- $\alpha_{\rm c}$, momentum compaction factor
- $\alpha_x = -\beta_x'/2, \;\; \alpha_z = -\beta_z'/2$
- $\alpha_{xx} = \partial Q_x/\partial J_x$ nonlinear betatron detuning parameter
- $\alpha_{xz} = \partial Q_x/\partial J_z$ nonlinear betatron detuning parameter
- $\alpha_{zz} = \partial Q_z/\partial J_z$ nonlinear betatron detuning parameter
- a, b, the horizontal and the vertical envelope radii in KV equation
- $A_x = \beta_x/\sigma_s$, $A_z = \beta_z/\sigma_s$, hour-glass scaling factors for luminosity
- $\vec{A}$, vector potential
- $\mathcal{A}$, longitudinal phase space area of one bunch
- $\tilde{\mathcal{A}}$, longitudinal phase space area of all bunches in the ring
- $\mathcal{A}_{\rm B}$, longitudinal bucket area
- B or n_b, the number of bunches in a storage ring
- $\mathbf{B}$, betatron amplitude matrix
- $B_1 = \partial B_z/\partial x$, gradient function of a quadrupole magnet
- $B_n = \partial^n B_z/\partial x^n$, $2(n+1)$th multipole of a magnet
- b_n, a_n, multipole expansion coefficients of magnetic fields
- $B\rho = p_0/e$, momentum rigidity of the beam
- $B_c = m^2c^2/(e\hbar) \approx 4.4 \times 10^9$ T, Schwinger critical field
- β_x, β_z, betatron amplitude functions, or called the Courant-Synder parameter, or the Twiss parameter
- $\gamma_x = (1+\alpha_x^2)/\beta_x$, $\gamma_z = (1+\alpha_z^2)/\beta_z$
- γ, β, Lorentz's relativistic factors
- $\gamma_{\rm T}$, transition energy $\gamma_{\rm T} mc^2$
- C, circumference of the machine
- $C_\gamma = 4\pi r_e/3(mc^2)^3 = \begin{cases} 8.846 \times 10^{-5}\ \text{m/(GeV)}^3 & \text{for electrons} \\ 4.840 \times 10^{-14}\ \text{m/(GeV)}^3 & \text{for muons} \\ 7.783 \times 10^{-18}\ \text{m/(GeV)}^3 & \text{for protons} \end{cases}$
- $C_q = 55\hbar/32\sqrt{3}mc = 3.83 \times 10^{-13}$ m quantum fluctuation coefficient (electron)
- $C_x = \partial Q_x/\partial\delta$, $C_z = \partial Q_z/\partial\delta$, chromaticities
- C_y stands for either C_x or C_z
- $c = 299792458$ m/s, speed of light
- D or D_x, horizontal dispersion function
- D_z, vertical dispersion function
- $\mathcal{D}$, damping re-partition number

- $\mathcal{D} = \sigma_s / f_{\rm bb}$, beam-beam disruption parameter
- $\delta = \Delta p/p_0$, fractional momentum deviation
- $\hat{\delta}$, the maximum fractional momentum spread of a beam
- δ or δ_1, the resonance proximity parameter
- $\delta_{\rm skin} = \sqrt{2/\mu\sigma_c\omega}$, skin depth of conductors
- ϵ_0, permittivity of the vacuum
- $\epsilon_x, \epsilon_z, \epsilon_\perp$, transverse emittances
- $\epsilon_{{\rm n},x} = \beta\gamma\epsilon_x$, $\epsilon_{{\rm n},z} = \beta\gamma\epsilon_z$, normalized emittances
- $\mathcal{E}$ electric field across a cavity gap
- $\mathcal{E}_0$ the amplitude of the electric field across a cavity gap
- $\mathcal{F}$, the emittance dependent factor for electron storage rings, i.e. $\epsilon_x = \mathcal{F} C_q \gamma^2 \theta^3$
- $F_{\rm B} = 2\pi R_0/\sqrt{2\pi}\sigma_s$, bunching factor
- f_0 revolution frequency
- $f = B\rho/B_1\ell$ focal length of a quadrupole
- $f_{\rm bb}$, focal length of beam-beam interaction
- (ϕ, δ), synchrotron phase space coordinates with $\delta = \Delta p/p_0$
- $(\phi, \Delta E/\omega_0)$, synchrotron phase space coordinates
- $(\frac{R}{h}\phi, -\delta)$, synchrotron phase space coordinates
- $(\phi, \mathcal{P})$, normalized synchrotron phase space coordinates with $\mathcal{P} = -(h|\eta|/\nu_s)\delta$
- $\Phi, \Phi_x, \Phi_z, \Phi_\perp$, transverse phase advance per cell or per period
- $g = 1 + 2\ln(b/a)$, geometric factor of electromagnetic wave in a wave guide
- h, harmonic number of the rf frequency, $f_{\rm rf} = hf_0$
- H, Hamiltonian
- $\mathcal{H} = \gamma_x D^2 + 2\alpha_x DD' + \beta_x D'^2$, dispersion $\mathcal{H}$-function
- $I_{\rm d}$ or $J_{\rm d}$, the dispersion action
- I_x, I_z or J_x, J_z, horizontal and vertical betatron actions
- I_s or J_s, the longitudinal action
- $\mathcal{J}_x, \mathcal{J}_E, \mathcal{J}_z$, damping partition numbers
- I_i's $(i = 1, 2, 3 \ldots)$, radiation integrals
  $I_1 = \int(D/\rho)ds$; $\quad I_2 = \int(1/\rho^2)ds$
  $I_3 = \int(1/|\rho|^3)ds$; $\quad I_{3a} = \int(1/\rho^3)ds$
  $I_4 = \int(D/\rho)[(1/\rho^2) + 2K]ds$; $\quad I_5 = \int(\mathcal{H}/|\rho|^3)ds$
- $k = \omega/c$, wave number
- $K(s) = B_1/B\rho$, gradient function of a magnet
- $K_x(s) = 1/\rho^2 - K(s)$, horizontal focusing function
- $K_z(s) = K(s)$, vertical focusing function
- $K_{\rm sc} = 2Nr_0/\beta^2\gamma^3$, space charge perveance
- $\kappa = K_{\rm sc}L/2\epsilon_\perp\Phi_\perp$, effective space charge perveance parameter
- $K_{\rm w} = eB_{\rm w}\lambda_{\rm w}/2\pi mc$, wiggler or undulator parameter

- $K_{\rm w,rms} = K_{\rm w}$, for helical wiggler or undulator $K_{\rm w,rms} = K_{\rm w}/\sqrt{2}$, for planar wiggler or undulator
- L, length of a periodic cell or superperiod
- $\lambda_{\rm C} = h/(m_e c) = 2.426 \times 10^{-12}$ m, Compton wavelength
- $\lambda_{\rm w}$, the wiggler period
- $\mathcal{L}$, luminosity
- $\mu_0 = 4\pi \times 10^{-7}$ Tm/A, permeability of the vacuum
- $\mu_{\rm c}$, permeability of a conducting medium
- μ, permeability of a medium
- $\vec{\mu}$, magnetic dipole moment
- $M(s_2|s_1)$, (2×2, 3×3, or 4×4) transfer matrix for linear betatron motion
- $\mathbf{M}(s)$, betatron transfer matrix of a periodic beam transport section
- $\nu_{x,z}$, betatron tunes
- $\nu_{\rm s}$, synchrotron tune for $\phi_{\rm s} = 0$ at zero synchrotron amplitude
- $\Delta\nu_{\rm sc} = K_{\rm sc}L/4\pi\epsilon$, Laslett space charge tune shift
- N, number of particle per unit length, for a Gaussian bunch: $\hat{N} = N_{\rm B}/\sqrt{2\pi}\sigma_s$
- $N_{\rm B}$, number of particles per bunch
- n, field gradient index, focusing index
- ω_0, revolution angular frequency
- $\omega_{\rm c}$, critical angular frequency
- ω_β, angular frequency of betatron motion
- $\omega_{\rm r}$ resonance frequency of an rf cavity
- ω, angular frequency of electromagnetic waves
- P, superperiod
- $P_{\rm d}$, power dissipation in rf cavity
- $P_{\rm ST} = -8/5\sqrt{3}$, Sokolov-Ternov radiative polarization
- $Q_{x,z}$, (nonlinear) betatron tunes
- $Q_{\rm s} = \nu_{\rm s}\sqrt{|\cos\phi_{\rm s}|}$, synchrotron tune
- $\tilde{Q}_{\rm s}$, the amplitude dependent synchrotron tune
- Q–factor, quality factor of rf cavity
- ρ, bending radius of a dipole magnet
- $\rho_{\rm c} = 1/\sigma_{\rm c}$, resistivity of a conductor
- $\rho_{\rm fel} = \mu_0 n_0 e^2 \lambda_{\rm w}^2 K_{\rm w}^2/(4\pi^2\gamma_r^3 m)$, FEL or Pierce parameter
- $\rho(x, s, z)$, distribution function
- $R(A_x, A_z)$, hour-glass reduction factor for the luminosity
- R or R_0, average radius of a synchrotron
- $R_{\rm c} = Z_{\rm c} = \sqrt{L/C}$, characteristic impedance of a transmission line
- $R_{\rm s} = 1/\sigma_{\rm c}\delta_{\rm skin} = \sqrt{\mu_0\omega/2\sigma_{\rm c}}$ surface resistance of a conductor
- $R_{ij}, T_{ijk}, U_{ijkl}$, transport matrices

- R_x, R_z, the horizontal and the vertical envelope radii
- $R_{\rm sh}$, shunt impedance of an rf cavity
- $r_{\rm sh}$, shunt impedance of rf cavities per unit length
- RRR$=\rho(273{\rm K}, 0\,{\rm Tesla})/\rho(10{\rm K}, 0\,{\rm Tesla})$, residual resistance of a conducting wire
- $r_0 = e^2/4\pi\epsilon_0 mc^2$ classical radius of the particle with mass m
- $\vec{r}_0(s)$, a reference orbit in an accelerator or a transport line
- $r_{\rm c}$, bunch compression ratio
- $\sigma_{\rm c} = 1/\rho_{\rm c}$, conductivity
- σ–matrix
- σ_x, σ_z, rms bunch bunch widths
- σ_s or σ_ℓ, rms bunch bunch length
- $\tau_{\rm ad}$ adiabatic time
- $\tau_{\rm nl}$, nonlinear time
- (τ, δ), synchrotron phase space coordinates
- $(\tau, \dot{\tau}/\omega_0)$, normalized synchrotron phase space coordinates
- T_0 and T, revolution periods for a reference particle and other particles
- $T_{\rm s}$, period of synchrotron motion
- U_0, energy loss per revolution due to synchrotron radiation in dipoles
- $U_{\rm w}$, total synchrotron radiation-energy loss per revolution including wigglers
- V, V_0, $V_{\rm rf}$, rf voltage
- $v_{\rm p}$, phase velocity
- $v_{\rm g}$, group velocity
- $W_{\rm st}$, stored energy in rf cavity
- $w_{\rm st}$, stored energy per unit length in rf cavity
- ξ, ξ_x, ξ_z, linear beam-beam tune shift parameter
- $\xi_{\rm sc}, \xi_{x,{\rm sc}}, \xi_{z,{\rm sc}}$, linear space charge tune shift parameter
- $(\hat{x}, \hat{s}, \hat{z})$, Frenet-Serret coordinate system defined by a reference orbit $\vec{r}_0(s)$
- (x, x'), horizontal betatron phase space coordinates
- $(x, \mathcal{P}_x)$, horizontal normalized phase space coordinates
- (y, y'), either x or z betatron phase space coordinates
- $(y, \mathcal{P}_y)$ either x or z normalized phase space coordinates
- η, phase slip factor
- Υ beamstrahlung parameter
- (z, z'), vertical betatron phase space coordinates
- $(z, \mathcal{P}_z)$, vertical normalized phase space coordinates
- $Z_{\rm sh}$, shunt impedance
- $Z_0 = \mu_0 c = 1/\epsilon_0 c \approx 377\,\Omega$, vacuum impedance
- $Z_{\rm sc}$, space charge impedance
- $\zeta_{_N}(w) = \sin Nw\pi/\sin w\pi$, the enhancement function

List of Tables

1. Table 1.1: Induction linac and achievements (p. 7)
2. Table 2.1: Percentage of particles in the confined phase space volume (p. 59)
3. Table 2.2: Linear coupling resonances and their driving terms (p. 172)
4. Table 2.3: Resonances due to sextupole and their driving terms (p. 188)
5. Table 3.1: Bucket length, bucket height, and bucket area factors (p. 242)
6. Table 3.2: Formula for bucket area in conjugate phase space variables (p. 242)
7. Table 3.3: The adiabatic and nonlinear times of some proton synchrotrons (p. 285)
8. Table 3.4: RF parameters of some proton synchrotrons (p. 300)
9. Table 3.5: SFP and UFP of a double rf system (p. 311)
10. Table 3.6: Some characteristic properties of RF cavities (p. 333)
11. Table 3.7: Typical space-charge impedance at $\gamma = \gamma_T$ (p. 356)
12. Table 3.8: Characteristic behavior of collective instability without landau damping (p. 358)
13. Table 3.9: Parametric dependence of the SLAC cavity geometry (p. 377)
14. Table 3.10: Some parameters of basic cylindrical cavity cells (p. 378)
15. Table 3.11: Properties of rf bucket in conjugate phase space variables (p. 386)
16. Table 4.1: Properties of some electron storage rings (p. 400)
17. Table 4.2: Properties of some high energy storage rings (p. 412)
18. Table 4.3: parameters of some undulators and wigglers (p. 457)
19. Table 5.1: Parameter list of high luminosity e^+e^- colliders (p. 494)
20. Table B.1: Zeroes of Bessel function for TM and TE modes (p. 523)

Chapter 1

Introduction

The first accelerator dates back to prehistoric-historic times, when men built bows and arrows for hunting. The race to build modern particle accelerators began in 1911 when Rutherford discovered the atomic nucleus by scattering α-particles off gold foil. These activities produced a series of innovative ideas such as the voltage rectifier (Cockcroft-Walton) and the Van de Graaff DC accelerators, the rf linac accelerators, the classic cyclotrons, the betatrons, the separate sector cyclotrons, the synchrotrons, and eventually storage rings and colliding beams.

The physics and technology of accelerators and storage rings involves many branches of science, including electromagnetism, solid-state properties of materials, atomic physics, superconductivity,nonlinear mechanics, spin dynamics, plasma physics, quantum physics, radiofrequency, and vacuum technology. Accelerators have found many applications: they are used in nuclear and particle physics research, in industrial applications such as ion implantation and lithography, in biological and medical research with synchrotron light sources, in material science and medical research with spallation neutron sources, etc. Accelerators have also been used for radiotherapy, food sterilization, waste treatment, etc.

A major application of particle accelerators is experimental nuclear and particle physics research. Advances in technology have allowed remarkable increases in energy and luminosity[1] for fundamental physics research. High energy was measured in MeV's in the 1930's, and is measured in TeV's in the 1990s. The beam intensity was about 10^9 particles per pulse (ppp) in the 1950's, and is about 10^{14} ppp in the 1990s. Since 1970, high energy and high luminosity colliders have become basic tools in nuclear and particle physics research. As physicists probe deeper into the inner structure of matter, high energy provides new territory for potential discoveries, and indeed new energy frontiers usually lead to new physics discoveries. The evolution of

[1]The luminosity $\mathcal{L}$ is defined as the rate of particle encountering per unit area in a collision process (see Exercise 1.7). The commonly used dimension is [cm^{-2} s^{-1}]. The counting rate in a detector is $\mathcal{L}\sigma$, where σ is the cross-section of a reaction process.

accelerator development can be summarized by the Livingston chart shown in Fig. 1.1, where the equivalent proton kinetic energy for a fixed target experiment is plotted as a function of time. The total center of mass energy for fixed target proton-proton collision is $\sqrt{2mc^2(\text{KE} + 2mc^2)}$, where m is the proton mass and KE is the kinetic energy of the moving particle in the Laboratory frame (see Exercise 1.6).

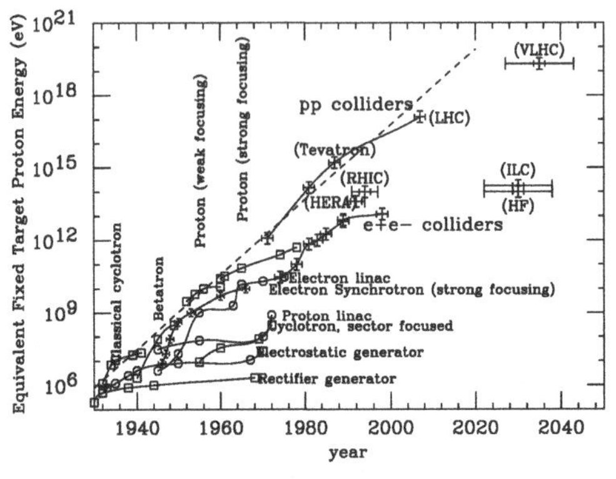

Figure 1.1: The Livingston Chart: The equivalent fixed target proton beam energy versus time in years. Note that innovative ideas provide substantial jump in beam energy. The dashed line, drawn to guide the trend, corresponds to beam energy doubling in every two years. The discovery of Higgs boson in 2014 at the LHC at 125 GeV/c^2 re-ignites efforts to build a Higgs factory in the form of e$^+$e$^-$ circular collider besides the efforts of the very large hadron collider (VLHC) and the international linear collider (ILC).

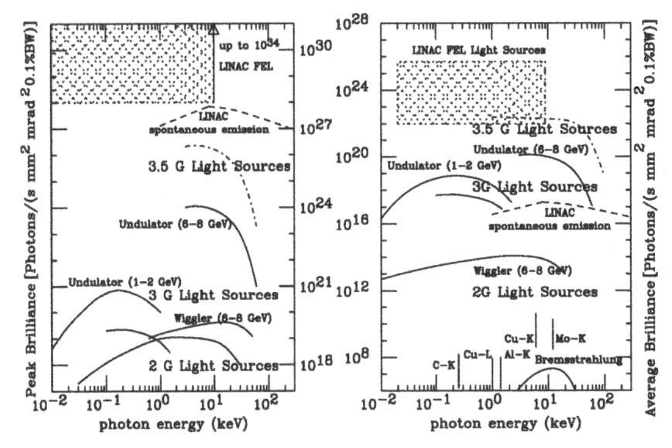

Figure 1.2: The peak and average Photon Brilliance, defined as photons/(mm^{2}mrad^{2}s (0.1%$\Delta\omega/\omega$)), vs. photon beam energy generated by high quality electron beams in storage rings and in linac FELs. Recent progresses in linac light sources push the brilliance to the level of 10^{34} photons/(mm^{2}mrad^{2}s (0.1%$\Delta\omega/\omega$)). A new Generation of storage ring light sources Can achieve about the same brilliance.

In the 1940's, scientists discovered that high energy electron beams in synchrotron could generate high energy Photon beams. With its flexible photon energy and high brilliance, photon sources produced by high-brightness electron beams have surpass the conventional optical sources. Applications of high energy photon beams include atomic physics, biology, chemistry, condensed matter physics and medicine. There are tremendous progresses in building high energy and high brightness electron sources and special insertion devices for photon production. Figure 1.2 shows the peak photon brilliance, defined as the photon beam intensity divided by its phase-space volume or in [number of photons/(mm^2 mrad^2 s (0.1% $\Delta\omega/\omega$))] as a function of photon beam energy from storage rings and linacs.

High power proton beams can be used to produce high flux neutron beam for fundamental material science research. The high power proton beams have also been actively considered to incinerate nuclear waste and provide energy amplification for future global energy needs. The secondary beams from high power hadron beams provides neutrino beams for research in high energy nuclear physics, High intensity heavy-ion beams had also been actively pursued for inertial fusion evaluation.

Frontiers in accelerator physics and technology research

Accelerator physics is a branch of applied science. Innovations in technology give rise to new frontiers in beam physics research. Since higher energy leads to new discoveries, and higher luminosity leads to higher precision in experimental results, the frontiers of accelerator physics research are classified into the frontiers of high energy and high brightness. Some of these topics in beam physics are as follows.

- High energy: For high energy hadron accelerators, research topics cover high field superconducting magnets, the stability of high-brightness beams, emittance preservation, and nonlinear dynamics, etc. For lepton colliders, research topics include high acceleration gradient structures, wakefields and emittance preservation, high power rf sources, linear collider technologies, etc.
- High luminosity: To provide a detailed understanding of CP violation and other fundamental symmetry principles of interactions, dedicated meson (Φ, τ-charm, B) factories were constructed in 1990s. Since the neutron flux from spallation neutron sources is proportional to the proton beam power, physics and technology for high-intensity low-loss proton sources are important (See e.g., the *National Spallation Neutron Source Design Report* (Oak Ridge, 1997)). Furthermore, a high-intensity proton source can be used to drive secondary beams such as kaons, pions, and muons. With high-intensity μ beams, $\mu^+\mu^-$ collider studies are also of current interest.
- High-brightness beams: Beam-cooling techniques have been extensively used in attaining high-brightness hadron beams. Stochastic cooling has been successfully applied to accumulate anti-protons. This led to the discovery of W and Z bosons, and b and t quarks. Electron cooling and laser cooling have been applied to many low energy storage rings used in atomic and nuclear physics research. Ionization cooling is needed for muon beams in $\mu^+\mu^-$ colliders. Taking advantage of radiation cooling, synchrotron light sources with high-brightness electron beams are used in medical, biological, and condensed-matter physics research. Sub-picosecond photon beams would be important to time-resolved experiments. A high power tunable free-electron laser would be useful for chemical and technical applications. The linac light source and very low emittance storage ring projects will provide very high brilliance X-ray sources for scientific applications.

- Accelerator applications: The medical use of accelerators for radiation treatment,[2] isotope production, sterilization of medical tools, etc., requires safety, reliability, and ease in operation. Higher beam power density with minimum beam loss can optimize safety in industrial applications such as ion implantation, electron-beam welding, material testing, food sterilization, etc.

Research topics in accelerator physics include beam cooling, nonlinear beam dynamics, beam-beam interactions, collective beam instability, beam manipulation techniques, ion sources, space-charge effects, beam instrumentation development, novel acceleration techniques, etc. Accelerator technology research areas include superconducting materials, high power rf sources, high gradient accelerating structures, etc. This book deals only with the fundamental aspects of accelerator physics. It serves as an introduction to more advanced topics such as collective beam instabilities, nonlinear beam dynamics, beam-cooling physics and technologies, rf physics and technology, magnet technology, etc. First, the technical achievements in accelerator physics of past decades will be described.

I Historical Developments

A charged particle with charge q and velocity $\vec{v}$ in the electromagnetic fields $(\vec{\mathcal{E}}, \vec{B})$ is exerted by the Lorentz's force $\vec{F}$:

$$\vec{F} = q(\vec{\mathcal{E}} + \vec{v} \times \vec{B}). \tag{1.1}$$

The charge particle can only gain or lose its energy by its interaction with the electric field $\vec{\mathcal{E}}$. Since the magnetic force is perpendicular to both $\vec{v}$ and $\vec{B}$, the charged particle will move on a circular arc. In particular, when the magnetic flux density is perpendicular to $\vec{v}$, the momentum and the **momentum rigidity** $B\rho$ of the charged particle are

$$\begin{aligned} p &= mv = |q|B\rho, \\ B\rho\ [\mathrm{T\!-\!m}] &= \frac{p}{|q|} = \frac{A}{Z} \times 3.33564 \times p\ [\mathrm{GeV/c/u}], \end{aligned} \tag{1.2}$$

where ρ is the bending radius m is the mass of the particle, $B\rho$ is measured in Tesla-meter, and the momentum is measured in GeV/c per amu, and A and Ze are the atomic mass number and charge of the particle.

Accelerators are composed of ion sources, cavity and magnet components that can generate and maintain electromagnetic fields for beam acceleration and manipulation, devices to detect beam motion, high vacuum components for attaining excellent beam

[2]See e.g., P.L. Petti and A.J. Lennox, ARNS **44**, 155 (1994).

lifetime, undulators and wigglers to produce high brilliance photon beam, targets for producing secondary beams, etc. Accelerators can be classified as linear or circular, electrostatic or radio frequency, continuous (CW, DC or coasting beam) or bunched and pulsed. They are designed to accelerate electrons (leptons) or hadrons, stable or radioactive ions. Accelerators are classified as follows, in no specific chronological order.

I.1 Natural Accelerators

Radioactive accelerators

In 1911, Rutherford, with Hans Geiger and Ernest Marsden, employed α particles escaping the Coulomb barrier of Ra and Th nuclei to investigate the inner structure of atoms.[3] He demonstrated the existence of a positively charged nucleus with a diameter less than 10^{-13} m. This led to the introduction of Bohr's atomic model, and the revolution of quantum mechanics in the early 20th century. In 1919, Rutherford also used α particles to induce the first artificial nuclear reaction, $\alpha + {}_{14}\mathrm{N} \rightarrow {}_{17}\mathrm{O} + \mathrm{H}$. This discovery created an era of search for high-voltage sources for particle acceleration that can produce high-intensity high-energy particles for the study of nuclear transmutation.

Cosmic rays

Cosmic rays arise from galactic source accelerators. Nuclei range from n and H to Ni; heavy elements have been measured with energies up to 3×10^{20} eV.[4] Muons were discovered in cosmic-ray emulsion experiments in 1936 by C.D. Anderson, S.H. Neddermeyer, and others. Pions were discovered in 1947 in emulsion experiments. Interest in the relativistic heavy ion collider (RHIC) was amplified by the cosmic ray emulsion experiments.

I.2 Electrostatic Accelerators

X-ray tubes

William David Coolidge in 1926 achieved 900-keV electron beam energy by using three X-ray tubes in series. Such a cascade type structure is called the Coolidge tube.

[3]The kinetic energy of α particles that tunnel through the Coulomb barrier to escape the nuclear force is typically about 6 MeV.

[4]See J.A. Simpson, *Ann. Rev. Nucl. Sci.* **33**, 323 (1983) and R. Barnett *et al.*, *Phys. Rev.* (Particle Data Group) **D54**, 1 (1996). An event with energy 3×10^{20} eV had been recorded in 1991 by the Fly's Eye atmospheric-fluorescence detector in Utah (see *Physics Today*, p. 19, Feb. 1997; p. 31, Jan. 1998).

Cockcroft-Walton electrostatic accelerator

In 1930, John Douglas Cockcroft and Ernst Thomas Sinton Walton developed a high-voltage source by using high-voltage rectifier units. In 1932, they reached 400-kV terminal voltage to achieve the first man-made nuclear transmutation:[5] p+Li→2He. The maximum achievable voltage was limited to about 1 MV because of sparking in air. Since then, Cockcroft-Walton accelerators have been widely used as the first-stage ion-beam accelerator. Recently, they are being replaced by more compact, economical, and reliable radio frequency quadrupole (RFQ) accelerators.

Van de Graaff and tandem accelerators

In 1931, R.J. Van de Graaff developed the electrostatic charging accelerator.[6] In the Van de Graaff accelerator, the rectifier units are replaced by an electrostatic charging belt, and the high-voltage terminal and the acceleration tube are placed in a common tank with compressed gas for insulation, which increases the peak acceleration voltage. Placement of the high-voltage terminal at the center of the tank, the charge-exchange process on a negatively charged atomic beams can provide tandem acceleration to the stripped positively-charged nuclei.[7] Today the voltage attained in tandem accelerators is about 25 MV. When the Van de Graaff accelerator is used for electron acceleration, it has the brand name Pelletron.

I.3 Induction Accelerators

According to Faraday's law of induction, when the magnetic flux changes, it induces electric field along the path that encompasses the magnetic flux:

$$\oint_C \mathcal{E}\cdot d\vec{s} = \dot{\Phi}, \qquad \Phi = \int_S \vec{B}\cdot d\vec{S}. \tag{1.3}$$

Here $\mathcal{E}$ is the induced electric field, Φ is the total magnetic flux, $d\vec{s}$ is the differential of the line integral that surrounds the surface area, $d\vec{S}$ is the differential of the surface integral, and $\vec{B}$ is the "magnetic field"[8] enclosed by the contour C. The induced electric field can be used for beam acceleration.

[5] J.D. Cockcroft and E.T.S. Walton, *Proc. Roy. Soc.* A**136**, 619 (1932); A**137**, 229 (1932); A**144**, 704 (1934). Cockcroft and Walton shared 1951 Nobel Prize in physics. See also Brian Cathcart, *The Fly in the Cathedral,* (Farrar, Straus and Giroux; NY, 2005)

[6] R.J. Van de Graaff, J.G. Trump, and W.W. Buechner, *Rep. Prog. Phys.* **11**, 1 (1946).

[7] R.J. Van de Graaff, *Nucl. Inst. Methods* **8**, 195 (1960).

[8] We will use "magnetic field" as a synonym for "magnetic flux density."

A: Induction linac

The induction linac was invented by N.C. Christofilos in the 50's for the acceleration of high-intensity beams.[9] A linear induction accelerator (LIA) employs a ferrite core arranged in a cylindrical symmetric configuration to produce an inductive load to a voltage gap. Each LIA module can be viewed as a low-Q 1:1 pulse transformer. When an external current source is discharged through the circuit, the electric field at the voltage gap along the beam axis is used to accelerate the beam. A properly pulsed stack of LIA modules can be used to accelerate high-intensity short-pulse beams with a gradient of about 1 MV/m and a power efficiency of about 50%.[10] Table 1.1 lists the achievements of some LIA projects.

Table 1.1: Induction linac projects and achievements

Project	Laboratory	I (kA)	E (MeV)	Beam width (ns)	Repetition rate (Hz)
ETA II	LLNL	3	70	50	1
ETA III	LLNL	2	6	50	2000
ATA	LLNL	10	50	50	1000

B: Betatron

Let ρ be the mean radius of the beam pipe in a basic magnet configuration of a betatron. If the total magnetic flux enclosed by the beam circumference is ramped up by a time-dependent magnetic flux density, the induced electric field along the beam axis and the particle momentum are

$$\oint \mathcal{E}\cdot ds = 2\pi\rho\mathcal{E} = \pi\rho^2\dot{B}_{\rm av}, \qquad \mathcal{E} = \frac{1}{2}\dot{B}_{\rm av}\rho, \qquad \dot{p} = e\mathcal{E} = \frac{1}{2}e\dot{B}_{\rm av}\rho,$$
$$p = \frac{1}{2}eB_{\rm av}\rho = eB_{\rm g}\rho, \qquad \text{or} \qquad B_{\rm g} = \frac{1}{2}B_{\rm av}, \tag{1.4}$$

where $\mathcal{E}$ is the induced electric field, $B_{\rm av}$ is the average magnetic flux density inside the circumference of the beam radius. We obtain the betatron principle: *the guide field $B_{\rm g}$ is equal to 1/2 of the average field $B_{\rm av}$*, first stated by R. Wideröe in 1928.[11]

[9]See e.g., J.W. Beal, N.C. Christofilos and R.E. Hester, *IEEE Trans. Nucl. Sci.* NS **16**, 294 (1958) and references therein; Simon Yu, Review of new developments in the field of induction accelerators, in *Proc. LINAC96* (1996).

[10]See e.g., R.B. Miller, in *Proc. NATO ASI on High Brightness Transport in Linear Induction Accelerators*, A.K. Hyder, M.F. Rose, and A.H. Guenther, Eds. (Plenum Press, 1988); R.J. Briggs, *Phys. Rev. Lett.* **54**, 2588 (1985); D.S. Prono, *IEEE Trans. Nucl. Sci.* NS**32**, 3144 (1985); G.J. Caporaso, *et al., Phys. Rev. Lett.* **57**, 1591 (1986); R.B. Miller, *IEEE Trans. Nucl. Sci.* NS**32**, 3149 (1985); G.J. Caporaso, W.A. Barletta, and V.K. Neil, *Part. Accel.* **11**, 71 (1980).

[11]In 1922, Joseph Slepian patented the principle of applying induction electric field for electron beam acceleration in the U.S. patent 1645304.

Figure 1.3 is a schematic drawing of a betatron, where particles circulate in the vacuum chamber with a guide field B_g, which is equal to half of the average flux density B_{av} enclosed by the orbiting particle.

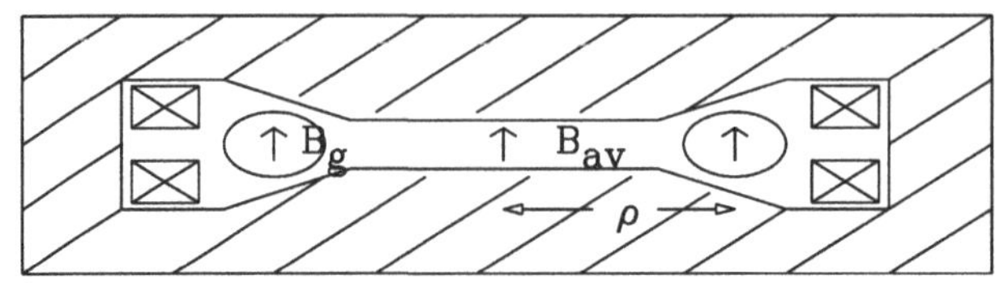

Figure 1.3: Schematic drawing of a betatron. The guide field for beam particles B_g must equal to the average flux density B_{av} enclosed by the orbiting path.

It took many years to understand the stability of transverse motion. This problem was solved in 1941 by D. Kerst and R. Serber.[12] We design a magnet so that the magnetic field is

$$B_z = B_0 \left(\frac{R}{r}\right)^n, \qquad \text{with} \qquad n = -\frac{R}{B_0}\left(\frac{dB_z}{dr}\right)_{r=R}, \tag{1.5}$$

where R is the orbit radius of a reference particle, r is a radius with small deviation from R, and n is the *focusing index*. Let $x = r - R$ and z be small radial and vertical displacements from a reference orbit, then the equations of motion become

$$\frac{d^2z}{dt^2} + \omega^2 nz = 0, \qquad \frac{d^2x}{dt^2} + \omega^2(1-n)x = 0. \tag{1.6}$$

The motion is stable and simple harmonic if $0 \le n \le 1$ (see Exercise 1.14). The resulting frequencies of harmonic oscillations are $f_x = f_0\sqrt{1-n}$ and $f_z = f_0\sqrt{n}$, where $f_0 = \omega/2\pi = v/(2\pi R)$ is the revolution frequency, and v is the speed of the particle.

In 1940 D. Kerst built and operated the first betatron achieving 2.3 MeV at University of Illinois. In 1949 he constructed a 315-MeV betatron[13] at the University of Chicago with parameters $\rho = 1.22$ m, $B_g = 9.2$ kG, $E_{\rm inj} = 80-135$ keV, $I_{\rm inj} = 1-3$ A. The magnet weighed about 275 tons and the repetition rate was about 6 Hz. The limitations of the betatron principle are (1) synchrotron radiation loss (see Chapter 4) and (2) the transverse beam size limit due to its intrinsic weak-focusing force.

I.4 Radio-Frequency (RF) Accelerators

Since the high-voltage source can induce arcs and corona discharges, it is difficult to attain very high voltage in a single acceleration gap. It would be more economical to

[12] D. Kerst and R. Serber, *Phys. Rev.* **60**, 53 (1941). See also Exercise 1.14. Since then, the transverse particle motion in all types of accelerators has been called *betatron motion.*

[13] D.W. Kerst *et. al.*, Phys. Rev. **78**, 297 (1950).

make the charged particles pass through the acceleration gap many times. This concept leads to many different rf accelerators,[14] which can be classified as linear (RFQ, linac) and cyclic (cyclotron, microtron, and synchrotron). An important milestone in rf acceleration is the discovery of the phase-focusing principle by E. M. McMillan and V. Veksler in 1945 (see Ref. [21] and Chap. 2, Sec. IV.3). Accelerators using an rf field for particle acceleration are described in the following subsections.

A. LINAC

In 1925 G. Ising pointed out that particle acceleration can be achieved by using an alternating radio-frequency field. In 1928 R. Wideröe reported the first working rf accelerator, using a 1-MHz, 25-kV oscillator to produce 50-kV potassium ions shown in the top plot of Fig. 1.4. In 1931 D.H. Sloan and E.O. Lawrence built a linear accelerator using a 10-MHz, 45 kV oscillator to produce 1.26 MV Hg^+ ion.[15]

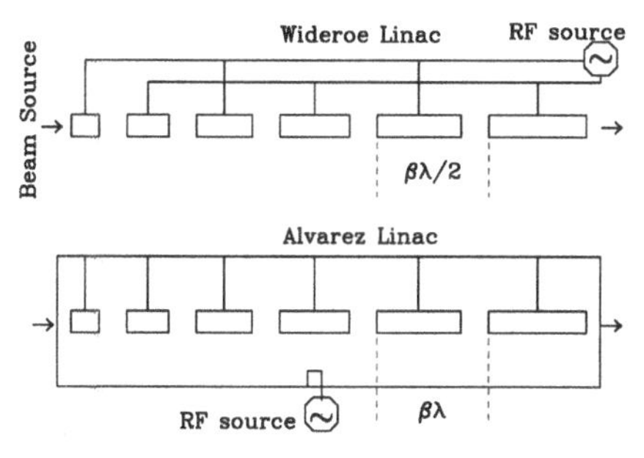

Figure 1.4: Top: schematic drawing of the Wideröe rf LINAC, where drift tubes shield particles from the decelerating rf electric field. Wideröe used a 1-MHz, 25-kV oscillator to produce 50-kV potassium ions. Bottom: enclosing the drift tubes in a metallic cylinder, the capacitance of the gap can be coupled to the inductance for a resonance cavity to achieve a higher efficiency in acceleration gradient. This cavity invented by Alvarez is called Alvarez linac or drift-tube linac (DTL).

Since the distance between adjacent drift tubes is $\beta\lambda/2 = \beta c/(2f_{\rm rf})$, it would save space by employing higher frequency rf sources. However, the problem associated with a high frequency structure is that it radiates rf energy at a rate of $P = \omega_{\rm rf} C V_{\rm rf}^2$, where $\omega_{\rm rf}$ is the rf frequency, C is the gap capacitance, and $V_{\rm rf}$ is the rf voltage. The rf radiation power loss increases with the rf frequency. To eliminate rf power loss, the drift tube can be placed in a cavity so that the electromagnetic energy is stored in the form of a magnetic field (inductive load). At the same time, the resonant frequency of the cavity can be tuned to coincide with that of the accelerating field.[16]

In 1948 Louis Alvarez and W.K.H. Panofsky constructed the first 32-MV drift-tube proton linac (DTL or Alvarez linac) shown schematically in the bottom plot of Fig. 1.4. Drift-tube linacs have been used as injectors for high energy accelerators at

[14]The rf sources are classified into VHF, UHF, microwave, and millimeter waves bands. The microwave bands are classified as follows: L band, 1.12-1.7 GHz; S band, 2.6-3.95 GHz; C band, 3.95-5.85 GHz; X band, 8.2-12.4 GHz; K band, 18.0-26.5 GHz; millimeter wave band, 30-300 GHz. See also Exercise 1.2.

[15]D.H. Sloan and E.O. Lawrence, *Phys. Rev.* **38**, 2021 (1931).

[16]L. Alvarez, *Phys. Rev.* **70**, 799 (1946).

BNL, KEK, Fermilab, SNS, and CERN. In the 1970's Los Alamos constructed the first side-coupled cavity linac (CCL), reaching 800 MeV. Fermilab upgraded part of its linac with the CCL to reach 400 MeV kinetic energy in 1995.

After World War II, rf technology had advanced far enough to make magnetron and klystron amplifiers that could provide MW rf power at 3 GHz (S band).[17] Today, the highest energy linac has achieved 50-GeV electron energy operating at S band (around 2.856 GHz) at SLAC, and has achieved an acceleration gradient of about 20 MV/m, fed by klystrons with a peak power higher than 40 MW in a 1-μs pulse length. To achieve 100 MV/m, about 25 times the rf power would be needed. The peak power is further enhanced by pulse compression schemes.

Superconducting cavities have substantially advanced in recent years. At the Continuous Electron Beam Accelerator Facility (CEBAF) at the Thomas Jefferson National Accelerator Laboratory (JLAB) in Virginia, about 160 m of superconducting cavity was installed for attaining a beam energy up to 6 GeV in 5 paths using 338 five-kW CW klystrons. During the LEP-II upgrade more than 300 m of superconducting rf cavity was installed for attaining more than 100-GeV beam energy. Many accelerator laboratories, such as KEK in Japan, Cornell and Fermilab in the U.S. and DESY in Germany, are collaborating in the effort to achieve a high-gradient superconducting cavity for a linear collider design called the International Linear Collider (ILC). Normally, a superconducting cavity operates at about 5–10 MV/m. After extensive cavity wall conditioning, single-cell cavities have reached far beyond 25 MV/m.[18]

B: RFQ

In 1970, I.M. Kapchinskij and V.A. Teplyakov invented a low energy radio-frequency quadrupole (RFQ) accelerator – a new type of low energy accelerator. Applying an rf electric field to the four-vane quadrupole-like longitudinally modulated structure, a longitudinal rf electric field for particle acceleration and a transverse quadrupole field for focusing can be generated simultaneously. Thus the RFQs are especially

[17] The klystron, invented by Varian brothers in 1937, is a narrow-band high-gain rf amplifier. The operation of a high power klystron is as follows. A beam of electrons is drawn by the induced voltage across the cathode and anode by a modulator. The electrons are accelerated to about 400 kV with a current of about 500 A. As the beam enters the input cavity, a small amount of rf power (< 1 kW) is applied to modulate the beam. The subsequent gain cavities resonantly excite and induce micro-bunching of the electron beam. The subsequent drift region and penultimate cavity are designed to produce highly bunched electrons. The rf energy is then extracted at the output cavity, which is designed to decelerate the beam. The rf power is then transported by rf waveguides. The wasted electrons are collected at a water-cooled collector. If the efficiency were 50%, a klystron with the above parameters would produce 100 MW of rf power. See also E.L. Ginzton, "The $100 idea", *IEEE Spectrum*, **12**, 30 (1975).

[18] See e.g., J. Garber, *Proc. PAC95*, p. 1478 (IEEE, New York 1996). Single-cell cavities routinely reach 30 MV/m and beyond.

useful for accelerating high-current low-energy beams. Since then many laboratories, particularly Los Alamos National Laboratory (LANL), Lawrence Berkeley National Laboratory (LBNL), and CERN, have perfected the design and construction of RFQ's, which are replacing Cockcroft-Walton accelerators as injectors to linac and cyclic accelerators.[19]

C: Cyclotron

The *synchrotron frequency* for a non-relativistic particle in a constant magnetic field is nearly independent of the particle velocity, i.e.,

$$\omega_{\rm syn} = \frac{eB_0}{\gamma m} \quad \xrightarrow{\gamma \approx 1} \quad \omega_{\rm cyc} = \frac{eB_0}{m}, \tag{1.7}$$

where B_0 is the magnetic field, and m is the particle mass. In 1929 E.O. Lawrence combined the idea of a nearly constant revolution frequency and Ising's idea of the rf accelerator (see Sec. I.4A of Wiederöe linac), he invented the *cyclotron.*[20] Historical remarks in E.O. Lawrence's Nobel lecture are reproduced below:

> One evening early in 1929 as I was glancing over current periodicals in the University library, I came across an article in a German electrical engineering journal by Wideröe on the multiple acceleration of positive ions. ... This new idea immediately impressed me as the real answer which I had been looking for to the technical problem of accelerating positive ions, ... Again a little analysis of the problem showed that a uniform magnetic field had just the right properties – that the angular velocity of the ion circulating in the field would be independent of their energy so that they would circulate back and forth between suitable hollow electrodes in resonance with an oscillating electric field of a certain frequency which has come to be known as the *cyclotron frequency.*

If two D plates (dees) in a constant magnetic field are connected to an rf electric voltage generator, particles can be accelerated by repeated passage through the rf gap, provided that the rf frequency is an integer multiple of the cyclotron frequency, $\omega_{\rm rf} = h\omega_0$. On January 2, 1931 M.S. Livingston demonstrated the cyclotron principle by accelerating protons to 80 keV in a 4.5-inch cyclotron, where the rf potential applied across the the accelerating gap was only 1000 V. In 1932 Lawrence's 11-inch cyclotron reached 1.25-MeV proton kinetic energy that was used to split atoms, just a few months after this was accomplished by the Cockcroft-Walton electrostatic

[19]See e.g. A. Pisent, *HIGH POWER RFQS*, Proc. of PAC09, 75 (2009)

[20]E.O. Lawrence and N.E. Edlefsen, *Science*, **72**, 376 (1930). See e.g. E.M. McMillan, *Early Days in the Lawrence Laboratory (1931-1940)*, in *New directions in physics*, eds. N. Metropolis, D.M. Kerr, Gian-Carlo Rota, (Academic Press, Inc., New York, 1987). The *cyclotron* was coined by Malcolm Henderson, popularized by newspaper reporters; see M.S. Livingston, *Particle Accelerators: A Brief History*, (Harvard, 1969).

accelerator. Since then, many cyclotrons were designed and built in Universities.[21] Figure 1.5 shows a schematic drawing of a classical cyclotron.

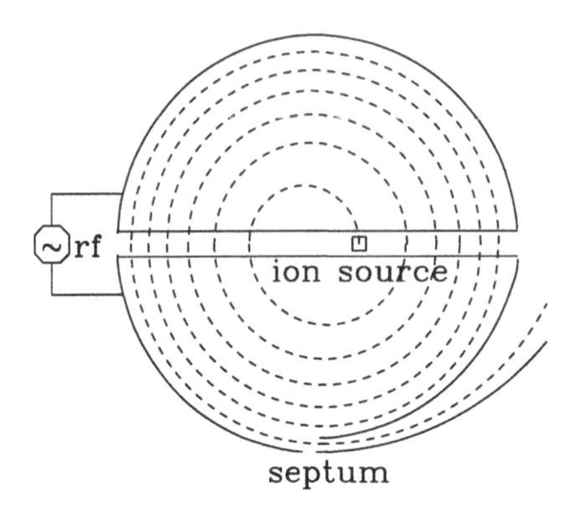

Figure 1.5: Schematic drawing of a classical cyclotron. Note that the radial distance between adjacent revolutions becomes smaller as the turn number increases [see Eq. (1.9)]. A septum is a device that can kick the beam into an external beam transport line.

The momentum p and kinetic energy T of the extracted particle are $p = m\gamma\beta c$ and $T = mc^2(\gamma - 1) = p^2/[(\gamma+1)m]$. Using Eq. (1.2), we obtain the kinetic energy per amu as

$$\frac{T}{A} = \frac{e^2 B_0^2 R_0^2}{(\gamma+1)m_u}\left(\frac{Z}{A}\right)^2 \equiv K\left(\frac{Z}{A}\right)^2, \tag{1.8}$$

where $B_0 R_0 = B\rho$ is the magnetic rigidity, Z and A are the charge and atomic mass numbers of the particle, m_u is the atomic mass unit, and K is called the *K-value or bending limit* of a cyclotron. In the non-relativistic limit, the K-value is equal to the proton kinetic energy T in MeV, e.g. K200 cyclotron can deliver protons with 200 MeV kinetic energy.

The iron saturates at a field of about 1.8 T (depending slightly on the quality of iron and magnet design). The total volume of iron-core is proportional to the cubic power of the beam rigidity $B\rho$. Thus the weight of iron-core increases rapidly with its K-value: Weight of iron $= W \sim (B\rho)^3 \sim K^{1.5}$, where $B\rho$ is the beam rigidity. Typically, the magnet for a K-100 cyclotron weighs about 160 tons. The weight problem can be alleviated by using superconducting cyclotrons.[22]

Beam extraction systems in cyclotrons is challenging. Let V_0 be the energy gain per revolution. The kinetic energy at N revolutions is $K_N = eNV_0 = e^2B^2r^2/2m$, where e is the charge, m is the mass, B is the magnetic field, and r is the beam radius at the N-th revolution. The radius r of the beam at the N-th revolution becomes

$$r = \frac{1}{B}\left(\frac{2mV_0}{e}\right)^{1/2} N^{1/2}, \tag{1.9}$$

[21] M.S. Livingston, *J. Appl. Phys*, **15**, 2 (1944); **15**, 128 (1944); W.B. Mann, *The Cyclotron*, (Wiley, 1953); M.E. Rose, *Phys. Rev.*, **53**, 392 (1938); R.R. Wilson, *Phys. Rev.*, **53**, 408 (1938); *Am. J. Phys.*, **11**, 781 (1940); B.L. Cohen, *Rev. Sci. Instr.*, **25**, 562 (1954).

[22] See H. Blosser, in *Proc. 9th Int. Conf. on Cyclotrons and Applications*, p. 147 (1985).

i.e. the orbiting radius increases with the square root of the revolution number N. The beam orbit separation in successive revolutions may becomes small, and thus the septum thickness becomes a challenging design problem.

Two key difficulties associated with classical cyclotrons are the orbit stability and the relativistic mass effect. The orbit stability problem was partially solved in 1945 by D. Kerst and R. Serber (see Exercise 1.14). The maximum kinetic energy was limited by the kinetic mass effect. Because the relativistic mass effect can destroy particle synchronism [see Eq. (1.7)], the upper limit of proton kinetic energy attainable in a cyclotron is about 12 MeV (See Exercise 1.4).[23] Two ideas proposed to solve the dilemma are the isochronous cyclotron and the synchrocyclotron.

Isochronous cyclotron

The radius of an orbiting particle and the magnetic field that maintain isochronism with a constant ω are

$$\rho = \frac{p}{eB} = \frac{p}{\gamma m\omega} = \frac{c}{\omega}\left(1 - \frac{E_0^2}{E^2}\right)^{1/2},$$

$$B_z = \frac{\gamma m\omega}{e} = \frac{\omega}{ec^2}E(\rho) = \frac{\omega E_0}{ec^2}\left[1 - \left(\frac{\omega\rho}{c}\right)^2\right]^{-1/2}. \tag{1.10}$$

where $E_0 = mc^2$ is the rest energy, ω is the angular revolution frequency, and B_z or B is the guide field. When the magnetic flux density is shaped according to Eq. (1.10), the focusing index becomes $n < 0$, and the vertical motion is unstable.

In 1938 L.H. Thomas pointed out that, by using an azimuthal varying field, the orbit stability can be retained while maintaining the isochronism. The isochronous cyclotron is also called the azimuthal varying field (AVF) cyclotron. Orbit stability can be restored by shaping the magnetic pole plates with hills and valleys.[24] The success of sector-focused cyclotron constructed by J.R. Richardson *et al.* led to the proliferation of the separate sector cyclotron, or ring cyclotron in the 1960's.[25] It gives stronger "edge" focusing for attaining vertical orbit stability. Ring cyclotrons are composed of three, four, or many sectors. Many universities and laboratories built ring cyclotrons in the 1960's.

Synchrocyclotron

Alternatively, synchronization between cyclotron frequency and rf frequency can be achieved by using rf frequency modulation (FM). FM cyclotrons can reach 1-GeV

[23] H. Bethe and M. Rose, *Phys. Rev.* **52**, 1254 (1937).

[24] L.H. Thomas, *Phys. Rev.* **54**, 580 (1938).

[25] H.A. Willax, *Proc. Int. Cyclotron Conf.* 386 (1963); Design and operation aspects of a 1.3 MW high power proton ring cyclotron at the PSI by M. Seidel *et. al.* is available at http://accelconf.web.cern.ch/AccelConf/IPAC10/papers/tuyra03.pdf

proton kinetic energy.[26] The synchrocyclotron uses the same magnet geometry as the weak-focusing cyclotrons. Synchronism between the particle and the rf accelerating voltage is achieved by ramping the rf frequency. Because the rf field is cycled, i.e. the rf frequency synchronizes with the revolution frequency as the energy is varied, synchrocyclotrons generate pulsed beam bunches. Thus the average intensity is low. The synchrocyclotron is limited by the rf frequency detuning range, the strength of the magnet flux density, etc. Currently two synchrocyclotrons are in operation, at CERN and at LBL.

D: Microtron

The accelerating rf cavities are expensive, it would be economical to use the rf structure repetitively. The microtron, originally proposed by V. Veksler in 1944, is designed to do this. Repetitive use requires synchronization between the orbiting and the rf periods. For example, if the energy gain per turn is exactly equal to the rest mass of the electron, the energy at the $n-1$ passage is $\gamma m_0 c^2 = n m_0 c^2$. the orbit period is an integral multiple of the fundamental cyclotron period: $T = n\frac{2\pi m}{eB}$. Thus, if the rf frequency $\omega_{\rm rf}$ is an integral multiple of the fundamental cyclotron frequency, the particle acceleration will be synchronized. Such a scheme or its variation was invented by V. Veksler in 1945.

The synchronization concept can be generalized to include many variations of magnet layout, e.g. the race track microtron (RTM), the bicyclotron, and the hexatron. The resonance condition for the RTM with electrons traveling at the speed of light is

$$n\lambda_{\rm rf} = \beta c \Delta t = 2\pi\beta\frac{\Delta E}{ecB}, \tag{1.11}$$

where ΔE is the energy gain per passage through the rf cavity, B is the bending dipole field, $\lambda_{\rm rf}$ is the rf wavelength, and n is an integer. This resonance condition simply states that the increase in path length is an integral multiple of the rf wavelength.

Some operational microtrons are the three-stage MAMI microtron at Mainz, Germany,[27] and the 175-MeV microtron at Moscow State University. Several commercial models have been designed and built by DanFisik.The weight of the microtron also increases with the cubic power of beam energy.

E: Synchrotrons, weak and strong focusing

After E.M. McMillan and V. Veksler discovered the phase focusing principle of the rf acceleration field in 1945, a natural evolution of the cyclotron was to confine the particle orbit in a well-defined path while tuning the rf system and magnetic field

[26] For a review, see R. Richardson, *Proc. 10th Int. Conf. on Cyclotrons and Their Applications*, IEEE CH-1996-3, p. 617 (1984).

[27] See e.g., H. Herminghaus, in *Proc. 1992 EPAC*, p. 247 (Editions Frontières, 1992).

to synchronize particle revolution frequency.[28] The first weak-focusing proton synchrotron, with focusing index $0 < n < 1$, was the 3-GeV Cosmotron in 1952 at BNL. A 6-GeV Bevatron constructed at LBNL in 1954, led to the discovery of antiprotons in 1955.

An important breakthrough in the design of synchrotron came in 1952 with the discovery of the strong-focusing or the alternating-gradient (AG) focusing principle by E.D. Courant, H.S. Snyder and M.S. Livingston.[29] Immediately, J. Blewett invented the electric quadrupole and applied the alternating-gradient-focusing concept to linac[30] solving difficult beam focusing problems in early day rf linacs. Here is "*Some Recollection on the Early History of Strong Focusing*" in the publication BNL 51377 (1980) by E.D. Courant:

> Came the summer of 1952. We have succeeded in building the Cosmotron, the world's first accelerator above one billion volts. We heard that a group of European countries were contemplating a new high-energy physics lab with a Cosmotron-like accelerator (only bigger) as its centerpiece, and that some physicists would come to visit us to learn more about the Cosmotron. ...
>
> Stan (Livingston) suggested one particular improvement: In the Cosmotron, the magnets all faced outward. ... Why not have some magnets face inward so that positive secondaries could have a clear path to experimental apparatus inside the ring?
>
> ... I did the calculation and found to my surprise that the focusing would be strengthened simultaneously for both vertical and horizontal motion. ... Soon we tried to make the gradients stronger and saw that there was no theoretical limit – provided the alterations were made more frequent as the gradient went up. Thus it seemed that aperture could be made as small as one or two inches – against 8×24 inches in the Cosmotron, 12×48 in the Bevatron, and even bigger energy machines as we then imagined them. With these slimmer magnets, it seemed one could now afford to string them out over a much bigger circles, and thus go to 30 or even 100 billion volts.

The first strong-focusing 1.2 GeV electron accelerator was built by R. Wilson at Cornell University. Two strong-focusing or alternating-gradient (AG) proton synchrotrons, the 28-GeV CERN PS (CPS) and the 33-GeV BNL AGS, were completed in 1959 and 1960 respectively. The early strong-focusing accelerators used combined-function magnets, i.e., the pole-tips of dipoles were shaped to attain a strong quadrupole field. For example, the bending radius and quadrupole field gradients of AGS magnets are respectively $\rho = 85.4$ m, and $B_1 = (\partial B/\partial x) = \pm 4.75$ T/m

[28] Frank Goward and D.E. Barnes converted a betatron at Telecommunication Research Laboratory into a synchrotron in August 1946. A few months earlier, J.R. Richardson, K. MacKenzie, B. Peters, F. Schmidt, and B. Wright modified the fixed frequency 37-inch cyclotron at Berkeley to a synchro-cyclotron for a proof of synchrotron principle. A research team at General Electric Co. at Schenectady built a 70 MeV electron synchrotron to observe synchrotron radiation in October 1946. See also E.J.N. Wilson, *50 years of synchrotrons*, Proc. of the EPAC96 (1996).

[29] E.D. Courant, H.S. Snyder and M.S. Livingston, *Phys. Rev.* **88**, 1188 (1952).

[30] J. Blewett, *Phys. Rev.* **88**, 1197 (1952).

at $B = 1.15$ T. This corresponds to a focusing index of $n = \pm 352$. The strengths of a string of alternating focusing and defocusing lenses were adjusted to produce net strong focusing effects in both planes.

The strong focusing idea was patented by a U.S. engineer, N.C. Christofilos,[31] living in Athens, Greece. Since then, the alternating-gradient (AG or strong-focusing) principle and a cascade of AG synchrotrons, proposed by M. Sands,[32] has become a standard design concept of high energy accelerators.

Since the saturation properties of quadrupole and dipole fields in a combined function magnet are different, there is advantage in machine tuning with separate quadrupole and dipole magnets. The Fermilab Main Ring was the first separate function accelerator.[33] Most present-day accelerators are separate-function machines. For conventional magnets, the maximum dipole field strength is about 1.5 T and the maximum field gradient is approximately $1/a$ [T/m] (see Exercise 1.12), where a is the aperture of the quadrupole in meters. For superconducting magnets, the maximum field and field gradient depends on superconducting coil geometry, superconducting coil material, and magnet aperture.

I.5 Colliders and Storage Rings

The total center-of-mass energy obtainable by having a high energy particle smash onto a stationary particle is limited by the kinematic transformation (see Exercise 1.6). To boost the available center-of-mass energy, two beams are accelerated to high energy and made to collide at interaction points.[34] Since the lifetime of a particle beam depends on the vacuum pressure in the beam pipe, stability of the power supply, intrabeam Coulomb scattering, Touschek scattering, quantum fluctuations, collective instabilities, nonlinear resonances, etc., accelerator physics issues have to be evaluated in the design, construction, and operation of colliders. Beam manipulation techniques such as beam stacking, bunch rotation, invention of beam cooling including stochastic beam cooling, invented by S. Van de Meer, and electron beam cooling, invented by Budker in 1966, etc., are essential in making the collider a reality.[35]

[31]N.C. Christofilos, *Focusing system for ions and electrons*, U.S. Patent No. 2736799 (issued 1956). Reprinted in *The Development of High Energy Accelerators,* M.S. Livingston, ed. (Dover, New York, 1966).

[32]M. Sands, *A proton synchrotron for 300 GeV*, MURA Report 465 (1959). MURA stands for Mid-Western University Research Association.

[33]The Fermi National Accelerator Laboratory was established in 1967. The design team adopted a cascade of accelerators including proton linac, rapid cycling booster synchrotron, and a separate function Main Ring.

[34]A.M. Sessler, *The Development of Colliders*, LBNL-40116, (1997). The collider concept was patented by R. Wideröe in 1943. The first collider concept based on "storage rings" was proposed by G.K. O'Neill in Phys. Rev. Lett. **102**, 1418 (1956).

[35]S. Van de Meer, *Stochastic Damping of Betatron Oscillations in the ISR*, CERN internal report CERN/ISR-PO/72-31 (1972); H. Poth, *Phys. Rep.* **196**, 135 (1990) and references therein.

The first proton-proton collider was the intersecting storage rings (ISR) at CERN completed in 1969. ISR was the test bed for physics ideas such as stochastic beam cooling, high vacuum, collective instabilities, beam stacking, phase displacement acceleration, nonlinear beam-beam force, etc. It reached 57 A of single beam current at 30 GeV. It stopped operation in 1981.

The first electron storage ring (200 MeV) was built by B. Touschek *et al.* in 1960 in Rome. It had only one beam line and an internal target to produce positrons, and it was necessary to flip the entire ring by 180° to fill both beams. Since the Laboratoire de l'Accelerateur Lineaire (LAL) in Orsay had a linac, the storage ring was transported to Orsay in 1961 to become the first e^+e^- collider. The Stanford-Princeton electron-electron storage ring was proposed in 1956 but completed only in 1966. An e^-e^- collider moved from Moscow to Novosibirsk in 1962 began its beam collision in 1965.

Since the 1960's, many e^+e^- colliders have been built. Experience in the operation of high energy colliders has led to an understanding of beam dynamics problems such as beam-beam interactions, nonlinear resonances, collective (coherent) beam instability, wakefield and impedance, intrabeam scattering, etc. Many e^+e^- colliders such as CESR at Cornell, SLC and PEP at SLAC, PETRA and DORIS at DESY, VEPP's at Novosibirsk, TRISTAN at KEK, and LEP at CERN led to major discoveries in particle physics. The drive to reach higher beam energy provided incentives for research on high power klystrons, new acceleration methods, etc. High energy lepton colliders such as NLC, JLC, and CLIC are expanding linear accelerator technology. On the luminosity frontier, the Φ-factory at Frascati, J/Ψ-factories at the PEPC in Beijing, B-factories such as PEP-II at SLAC and TRISTAN-II at KEK, and the SUPER-B at Frascati aim to reach 10^{33-35} cm^{-2} s^{-1}.

Proton-antiproton colliders such as the Tevatron at Fermilab and Sp$\bar{\text{p}}$S at CERN had made many discoveries. The discovery of type-II superconductors[36] led to the successful development of superconducting magnets, which have been successfully used in the Tevatron to attain 2-TeV c.m. energy, and in HERA to attain 820-GeV proton beam energy. At present, the CERN LHC (14-TeV c.m. energy) and the BNL RHIC (200-GeV/u heavy ion c.m. energy) are operational. The (40-TeV) SSC proton collider in Texas was canceled in October 1993. Physicists are contemplating a very large hadron collider with about 60–100 TeV beam energy and possible muon-muon colliders.

[36]Type II superconductors allow partial magnetic flux penetration into the superconducting material so that they have two critical fields $B_{c1}(T)$ and $B_{c2}(T)$ in the phase transition, where T is the temperature. The high critical field makes them useful for technical applications. Most type II superconductors are compounds or alloys of niobium; commonly used alloys are NbTi and Nb_3Sn.

I.6 Synchrotron Radiation Storage Rings

Since the discovery of synchrotron light from a then high energy (80-MeV) electron synchrotron in 1947, the synchrotron light source has become an indispensable tool in basic atomic and molecular physics, condensed-matter physics, material science, biological, chemical, and medical research, and material processing. Worldwide, about 70 light sources are in operation or being designed or built.

Specially designed high-brightness synchrotron radiation storage rings are classified into generations. Those in the first generation operate in the parasitic mode from existing high energy e^+e^- colliders. The second generation comprises dedicated low-emittance light sources. Third-generation light sources produce high-brilliance photon beams from insertion devices using dedicated high-brightness electron beams. Many 3rd Generation light storage rings at 1-6 GeV have been operational since 1990. They serve scientific communities in the world. Efforts in research on the fourth generation (coherent) light sources based on linac free electron laser from a long undulator have been very successful. This leads to construction of linac light sources in the US, Japan, Europe, and China. Similarly, the emittance of storage ring light sources have been pushed to pico-meter level. These upgraded light source will serve scientific communities.

The back-scattering of laser beam ($\lambda \sim 1$ μm) by a high energy electron beam ($E \sim 20 - 60$ MeV) can produce tunable high brightness hard-X-ray. The small foot-print of this device can be used in many university laboratories for biomedical and material science research. This inverse Compton or Thompson X-ray (ICX, or TX) source is also of great interest in accelerator physics research.

II Layout and Components of Accelerators

A high energy accelerator complex is composed of ion sources, buncher/debuncher, chopper, pre-accelerators such as the high-voltage source or RFQ, drift-tube linac (DTL), booster synchrotrons, storage rings, and colliders. Figure 1.6 shows the drawing of a small accelerator complex, now decommissioned, at Indiana University Cyclotron Facility. Particle beams are produced from ion sources, where charged ions are extracted by a high-voltage source to form a beam. The beam pulse is usually prebunched and chopped into appropriate sizes, accelerated by a DC accelerator or RFQ to attain the proper velocity needed for a drift-tube linac. The beams can be injected into a chain of synchrotrons to reach high energy. Some basic accelerator components are described in the following subsections.

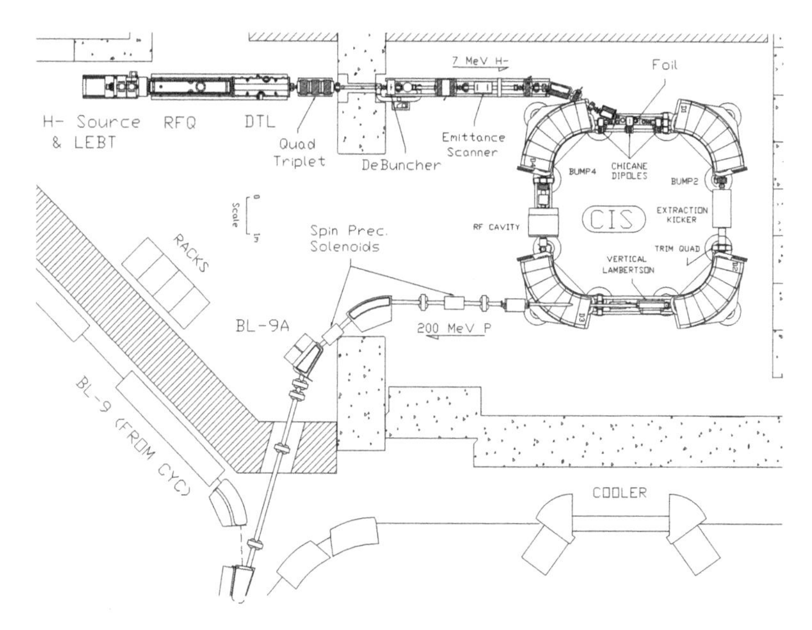

Figure 1.6: A small accelerator, the Cooler Injector Synchrotron (CIS) at the Indiana University Cyclotron Facility. The source, RFQ, DTL, debuncher, chopper, the CIS synchrotron with 4 dipoles, and a transfer line are shown to illustrate the basic structure of an accelerator system. The circumference is 17.36 m.

II.1 Acceleration Cavities

The electric fields used for beam acceleration are of two types: the DC acceleration column and the rf cavity.[37] The DC acceleration column is usually used in low energy accelerators such as the Cockcroft-Walton, Van de Graaff, etc.

The rf acceleration cavity provides high longitudinal electric field at an rf frequency that ranges from a few hundred kHz to 10–30 GHz. For a particle with charge e, the energy gain/loss per passage through a cavity gap is

$$\Delta E = e\Delta V, \tag{1.12}$$

where $\Delta V = V_0 \sin(\omega_{\rm rf} t + \phi)$ is the effective gap voltage, $\omega_{\rm rf}$ is the rf frequency, V_0 is the effective peak accelerating voltage, and ϕ is the phase angle. Low frequency rf cavities are usually used to accelerate hadron beams, and high frequency rf cavities to accelerate electron beams.

[37] In recent years, new acceleration schemes such as inverse free-electron laser acceleration, laser plasma wakefield acceleration, etc., have been proposed for high-gradient accelerators. See e.g., *Advanced Accelerator Concepts*, AIP Conf. Proc. No. 398, S. Chattopadhyay, *et al.*, Eds. (1996) and reference therein.

Acceleration of the bunch of charged particles to high energies requires synchronization and phase focusing. The synchronization is achieved by matching the rf frequency with particle velocity, and the phase focusing is achieved by choosing a proper phase angle between the rf wave and the beam bunch.

II.2 Accelerator Magnets

Accelerator magnets requires a stringent field uniformity condition in order to minimize un-controllable beam orbit distortion and beam loss. Accelerator magnets are classified into field type of dipole magnets for beam orbit control, quadrupole magnets for beam size control, sextupole and higher-order multipole magnets for the control of chromatic and geometric aberrations.

Accelerator magnets are also classified into conventional iron magnets and superconducting magnets. The conventional magnets employ iron or silicon-steel with OFHC copper conductors. The superconducting magnets employ superconducting coils to produce high field magnets.

Dipoles

Dipole magnets are used to guide charged particle beams along a desired orbit. From the Lorentz force law, the bending angle θ in passing through a dipole, and the integrated dipole field in a ring are

$$\theta = \frac{e}{p_0}\int_{s_1}^{s_2} Bdl = \frac{1}{B\rho}\int_{s_1}^{s_2} Bdl, \qquad \oint Bdl = 2\pi p_0/e = 2\pi B\rho. \tag{1.13}$$

where p_0 is the momentum of the beam, and $B\rho = p_0/e$ is the momentum rigidity of the beam, and the total bending angle for a circular accelerator is 2π.

The conventional dipole magnets are made of laminated silicon-steel plates for the return magnetic flux to minimize eddy current and hysteresis loss. Solid block of high permeability soft-iron can also be used for magnets in the transport line or cyclotrons, that requires DC magnetic field. A gap between the iron yoke is used to shape dipole field. The iron plate can be C-shaped for a C-dipole (see Exercise 1.10 and the left plot of Fig. 1.7), or H-shaped for H-dipole. Since iron saturates at about 1.7 T magnet flux, the maximum attainable field for iron magnet is about 1.8 T. To attain a higher dipole field, superconducting coils can be used. These magnets are called superconducting magnets.

Superconducting magnets that use iron to enhance the attainable magnetic field is also called superferric magnets. For high field magnets, e.g. 5-12 T, blocks of superconducting coils are used to simulate the cosine-theta current distribution (see Exercise 1.9). The right plot of Fig. 1.7 shows the cross-section of the high field SSC dipole magnets.

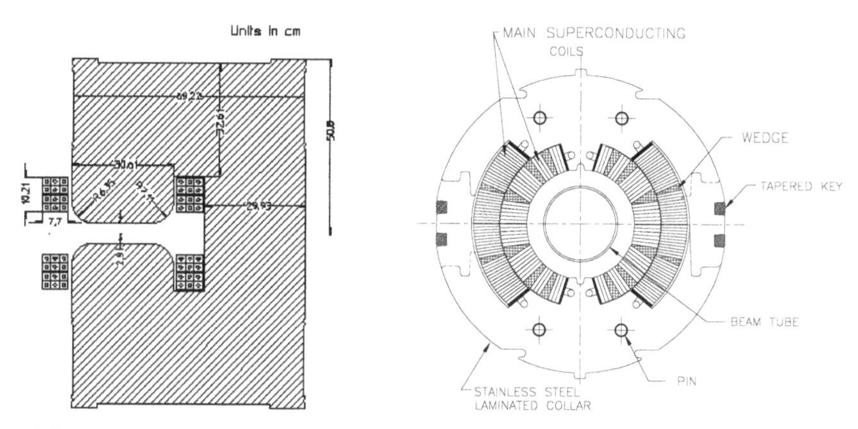

Figure 1.7: The cross-sections of a C-shaped conventional dipole magnet (left, courtesy of G. Berg at IUCF) and an SSC superconducting dipole magnet (right, courtesy of R. Gupta at LBNL). For conventional magnets, the pole shape is designed to attain uniform field in the gap. The rectangular blocks shown in the left plot are oxygen free high conductivity (OFHC) copper coils. For superconducting magnets, the superconducting coils are arranged to simulate the cosine-theta like distribution.

Quadrupoles

A stack of laminated iron plates with a hyperbolic profile can be used to produce quadrupole magnet (see Exercise 1.12), where the magnetic field of an ideal quadrupole is

$$\vec{B} = B_1(z\hat{x} + x\hat{z}), \tag{1.14}$$

where $B_1 = \partial B_z/\partial x$ evaluated at the center of the quadrupole, and $\hat{x}, \hat{s}$, and $\hat{z}$ are the unit vectors in the horizontal, azimuthal, and vertical directions. For a charged particle passing through the center of a quadrupole, the magnetic field and the Lorentz force are zero. At a displacement (x, z) from the center, the Lorentz force for a particle with charge e and velocity v along $\hat{s}$ direction is

$$\vec{F} = evB_1\hat{s} \times (z\hat{x} + x\hat{z}) = -evB_1 z\hat{z} + evB_1 x\hat{x}. \tag{1.15}$$

The equations of motion become

$$\frac{1}{v^2}\frac{d^2x}{dt^2} = \frac{eB_1}{\gamma mv}x, \qquad \frac{1}{v^2}\frac{d^2z}{dt^2} = -\frac{eB_1}{\gamma mv}z. \tag{1.16}$$

Thus a focusing quadrupole in the horizontal plane is also a defocusing quadrupole in the vertical plane and vice versa. Defining the focusing index as

$$n = R^2\frac{eB_1}{\gamma mv} = \frac{R^2}{B\rho}\frac{\partial B_z}{\partial x}, \tag{1.17}$$

we obtain

$$\frac{d^2x}{ds^2} = \frac{n}{R^2}x, \qquad \frac{d^2z}{ds^2} = -\frac{n}{R^2}z \tag{1.18}$$

in a quadrupole, where $s = vt$ is the longitudinal distance along the $\hat{s}$ direction.

II.3 Other Important Components

Other important components in accelerators are ion sources;[38] monitors for beam position, beam current and beam loss; beam dump; emittance meters; vacuum ports and pumps; beam orbit and stopband correctors; skew quadrupoles, sextupoles, octupoles, and other nonlinear magnets for nonlinear stopband correction; orbit bumps, kickers, and septum; power supplies, etc. For high energy experiments, sophisticated particle detectors are the essential sources of discovery. For synchrotron radiation applications in electron storage rings, wigglers and undulators are used to enhance the photon beam quality.

The timing and operation of all accelerator components (including experimental devices) are controlled by computers. Computer control software retrieves beam signals, and sets proper operational conditions for accelerator components. The advance in computer hardware and software provides advanced beam manipulation schemes such as slow beam extraction, beam stacking accumulation, stochastic beam cooling, etc.

III Accelerator Applications

III.1 High Energy and Nuclear Physics

To probe into the inner structure of the fundamental constituents of particles, high energy accelerators are needed. Historical advancement in particle and nuclear physics has always been linked to advancement in accelerators. High energy accelerators have provided essential tools in the discovery of $\bar{p}$, Ω, $J/\Psi, Z^0, W^\pm \cdots$, etc. Observation of a parton-like structure inside a proton provided proof of the existence of elementary constituents known as quarks. The IUCF cyclotron had been used to understand the giant M1 resonances in nuclei. The Tevatron at Fermilab facilitated the discovery of the top quark in 1995. Radioactive beams may provide nuclear reactions that will provide information on nucleo-synthesis of elements in the early universe.

High energy colliders such as HERA (30-GeV electrons and 820-GeV protons), Tevatron (1-TeV on 1-TeV proton-antiproton collider), SLC (50-GeV on 50-GeV e^+e^- collider), LEP (50-100-GeV on 50-100-GeV e^+e^- collider) led the way in high energy physics in the 1990s. High luminosity colliders, such as the B-factories at SLAC

[38] B. Wolf, ed., *Handbook of Ion Sources* (CRC Press, New York, 1995); R. Geller, *Electron Cyclotron Resonance Ion Sources and ECR Plasma* (Inst. of Phys. Pub., Bristol, 1996).

and KEK and the Φ-factory at DAΦNE, provided dedicated experiments for understanding the symmetry of the fundamental interactions. The RHIC (100-GeV/u on 100-GeV/u heavy ion collider) provided important information on the phase transitions of quark-gluon plasma. The JLAB 4-12-GeV continuous electron beams allow high resolution (0.1 fm) studies of the electromagnetic properties of nuclei. The LHC (7-TeV on 7-TeV proton-proton collider) at CERN will lead high energy physics research at the beginning of the 21st century.

High power accelerators are also considered to be particle sources for mesons, muons, neutrinos. The fixed-field alternating-gradient (FFA) accelerator, sprung out of the discovery of the alternating-gradient (or strong focusing) principle, was proposed by the Midwestern University Research Association (MURA) group in the 1950's. Since the magnetic field in FFA is stationary, particle-beams can be accelerated rapidly. There are renewed efforts in exploring the feasibility of achieving high power beam, rapid acceleration of muon beams, etc.[39] in recent years. High power particle sources are also important in accelerator physics.

III.2 Solid-State and Condensed-Matter Physics

Ion implantation,[40] synchrotron radiation sources, and neutron back-scattering have provided important tools for solid-state and condensed-matter physics research. Neutron sources have been important sources for research aimed at understanding the properties of metals, semiconductors, and insulators. Free-electron lasers with short pulses and high brightness in a wide spectrum of frequency ranges have been used extensively in medical physics, solid-state physics, biology, and biochemistry.

III.3 Other Applications

Electron beams can be used to preserve and sterilize agricultural products. Beam lithography is used in industrial processing. Radiation has been used in the manufacture of polymers, radiation hardening for material processing, etc. Particle beams have been used to detect defects and metal fatigue of airplanes, ships, and strategic equipment.

Since the discovery of X-ray in 1895, radiation has been used in medical imaging, diagnosis, and radiation treatments. Radiation can be used to terminate unwanted tumor growth with electron, proton, or ion beams. In particular, proton and heavy-ion beams have become popular in cancer radiation therapy because these beam particles deposit most of their energy near the end of their path. By controlling the beam energy, most of the beam energy can be deposited in the cancerous tumor with

[39]See S. Machida, *et. al.*, *Nature Physics*, **8**, 243 (2012) and references therein.

[40]The ion implantation, invented by W. Shockley in 1954 (U.S. Patent 2787564), has become an indispensable tool in the semiconductor industry.

less damage to surrounding healthy cells. Beams have also been used in radiation sterilization, isotope production for radionuclide therapy,[41] etc.

Exercise 1: Basics

1. Show that the magnet rigidity $B\rho$ is related to the particle momentum p by

$$B\rho\ [\mathrm{Tm}] = \frac{p}{Ze} = \begin{cases} 3.3357\ p & [\mathrm{GeV/c}] \quad \text{for singly charged particles} \\ \frac{3.3357}{Z}\ p & [\mathrm{GeV/c}] \quad \text{for particles with charge } Ze \end{cases},$$

   where B is the magnetic flux density, ρ is the bending radius, p is the beam momentum, and Ze is the charge of the particle.

   (a) (1) Calculate the magnetic rigidity of proton beams at the IUCF Cooler Ring (kinetic energy 500 MeV), RHIC (momentum 250 GeV/c), Tevatron (momentum 1 TeV/c) SSC (momentum 20 TeV/c). (2) Calculate the momentum rigidity of 19.7 TeV Au-ion beam in RHIC, and 287 TeV Pb-ion beam in LHC.

   (b) If the maximum magnetic flux density for a conventional dipole is 1.7 Tesla, what is the total length of dipole needed for each of these accelerators? What is the total length of dipoles needed in each accelerator if superconducting magnets are used with magnetic fields $B = 3.5$ T (RHIC), 5 T (Tevatron) and 6.6 T (SSC)?

2. The resonance frequency of a LC circuit is $f_r = 1/2\pi\sqrt{LC}$. Assuming that you can build a capacitor with a minimum capacitance of $C = 1$ pF, what value of inductance L is needed to attain 3 GHz resonance frequency? What is your conclusion from this exercise? Can you use a conventional LC circuit for microwave tuning?[42]

3. Consider a uniform cylindrical beam with N particles per unit length in a beam of radius a; show that a test charged particle traveling along at the same velocity as the beam, v, experiences a repulsive *space-charge* force,

$$\vec{F} = \begin{cases} \dfrac{e^2 N}{2\pi\epsilon_0 a^2 \gamma^2}\vec{r} & r \le a \\ \dfrac{e^2 N}{2\pi\epsilon_0 r^2 \gamma^2}\vec{r} & r > a \end{cases}$$

   where $\gamma = 1/\sqrt{1-\beta^2}$ and e is the charge of the beam particle.

   (a) Estimate the space-charge force for the SSC low energy booster at injection with kinetic energy 800 MeV and $N_{\mathrm{B}} = 10^{10}$ particles per bunch. We assume a Gaussian distribution with rms bunch length $\sigma_s = 2$ m and beam diameter 4 mm. For a Gaussian beam, we use the maximum N given by $N_{\mathrm{B}}/(\sqrt{2\pi}\sigma_s)$ at the center of the bunch.

---

[41]See e.g. Bert M. Coursey and R. Nath, *Phys. Today*, p. 25, April 2000.

[42]See V.F.C. Veley, *Modern Microwave Technology* (Prentice Hall, Englewood Cliffs, NJ, 1987). For an order of magnitude estimation, a 5-cm-radius single loop with wire 0.5 mm thick will yield an inductance of about 3×10^{-7} H.

(b) What happens if the test charged particle travels in the opposite direction in the head-on collision process? Estimate the space-charge force for the e^+e^- colliding beam at SLC, where the beam parameters are $E = 47$ GeV and $N_B = 2 \times 10^{10}$, the rms bunch length is $\sigma_s = 0.6$ mm, and the beam size is 3 μm. If this force is exerted by a quadrupole, what is the equivalent field gradient?

4. In a cyclotron, the *synchronous* frequency is $\omega = eB/\gamma m = \omega_0/\gamma$, where $\omega_0 = eB/m$ is the cyclotron frequency, and γ is the relativistic energy factor. Use the following steps, in the uniform acceleration approximation, to prove that, if a sinusoidal voltage $V_{\rm rf} = V \cos \omega_0 t$ is applied to the dees, the maximum attainable kinetic energy is $\sqrt{2eVmc^2/\pi}$, where e and m are the charge and mass of the particle.

   (a) Let ψ be the rf phase of the particle. Show that the equation of motion in a uniform acceleration approximation is $d\psi/dt = (\gamma^{-1} - 1)\omega_0, \quad d\gamma^2/dt = a \cos \psi$, where $a = 2\omega_0 eV/\pi mc^2$.

   (b) Defining a variable $q = a \cos \psi$, show that the equation of motion becomes $(q/\sqrt{a^2 - q^2})dq = (2\gamma - 2)\omega_0 \, d\gamma$. Integrate this equation and show that the maximum kinetic energy attainable is $\sqrt{2eVmc^2/\pi}$.

5. The total power radiated by an accelerated charged particle is given by Larmor's formula:

$$P = \frac{1}{4\pi\epsilon_0}\frac{2e^2\dot{v}^2}{3c^3} = \frac{1}{4\pi\epsilon_0}\frac{2e^2}{3m^2c^3}\left(\frac{dp_\mu}{d\tau} \cdot \frac{dp_\mu}{d\tau}\right)$$

where $d\tau = dt/\gamma$ is the proper time and p_μ is the four-momentum.[43]

   (a) In a linear accelerator, the motion is along a straight path. The power radiated is

$$P = \frac{1}{4\pi\epsilon_0}\frac{2e^2}{3m^2c^3}\left(\frac{dp}{dt}\right)^2 = \frac{1}{4\pi\epsilon_0}\frac{2e^2}{3m^2c^3}\left(\frac{dE}{dx}\right)^2,$$

   where dE/dx is the rate of energy change per unit distance. The ratio of radiation power loss to power supply from an external accelerating source is

$$\frac{P}{dE/dt} = \frac{1}{4\pi\epsilon_0}\frac{2e^2}{3m^2c^3v}\left(\frac{dE}{dx}\right) \approx \frac{2}{3}\frac{r_e}{mc^2}\left(\frac{dE}{dx}\right),$$

   where $r_e = 2.82 \times 10^{-15}$ m is the classical radius of the electron. Assuming that electrons gain energy from 1 GeV to 47 GeV in 3 km at SLC, what is the ratio of power loss to power supply? In the Next Linear Collider (NLC), the gradient of the accelerating cavities will increase by a factor of 10. What will be the ratio of radiation power loss to power supply? What is your conclusion from this exercise?

   (b) In a circular accelerator, $\vec{p}$ changes direction while the change in energy per revolution is small, i.e.

$$\frac{d\vec{p}}{d\tau} = \gamma\omega|\vec{p}| \gg \frac{1}{c}\frac{dE}{d\tau},$$

[43] See J.D. Jackson, *Classical Electrodynamics*, 2nd ed., p. 468 (1975).

where $\omega = \beta c/\rho$, and ρ is the bending radius. The radiated power becomes

$$P = \frac{2}{3}\frac{r_0}{mc}\gamma^2\omega^2|\vec{p}|^2 = \frac{\beta^4 c C_\gamma}{2\pi}\frac{E^4}{\rho^2},$$

$$C_\gamma = \frac{4\pi r_0}{3(mc^2)^3} = \begin{cases} 8.85\times 10^{-5}[\mathrm{m}/(\mathrm{GeV})^3] & \text{for electrons,} \\ 7.78\times 10^{-18}[\mathrm{m}/(\mathrm{GeV})^3] & \text{for protons,} \end{cases}$$

where r_0 is the classical radius of the particle, m is the mass, and The radiative energy loss per revolution of an isomagnetic storage ring becomes

$$U_0 = \beta^3 C_\gamma E^4/\rho.$$

i. Calculate the energy dissipation per revolution for electrons at energy $E =$ 50 GeV and 100 GeV in LEP, where $\rho = 3096.175$ m and the circumference is 26658.853 m.

ii. Find the energy loss per turn for protons in SSC, where the magnetic field is 6.6 Tesla at 20 TeV, the circumference is 87120 m, and the bending radius is 10187 m.

iii. Show that the power radiated per unit length in dipoles for a beam is

$$P(\mathrm{W/m}) = 10^6 \cdot \frac{U_0[\mathrm{MeV}]}{2\pi\rho\ [\mathrm{m}]} I\ [\mathrm{A}],$$

where U_0 is the energy loss per revolution, I is the total beam current, and ρ is the radius of curvature in the dipole. Find the synchrotron radiation power loss per unit length in LEP and SSC, where the circulating beam currents are respectively 3 mA and 70 mA.

6. The center of mass (c.m.) energy of two particles with mass m and energy $\gamma_{\rm cm} mc^2$ colliding head on has a total c.m. energy of $2\gamma_{\rm cm} mc^2$. (1) Show that the total c.m. energy for this collision is equivalent to a fixed target collision at the energy of $\gamma_{\rm FT} mc^2$ with another particle with mass m at rest if $\gamma_{\rm FT} = 2\gamma_{\rm cm}^2 - 1$. (2) What is the equivalent center of mass energy for the collision of a cosmic ray proton at the energy of 3×10^{20} eV with another proton? (3) In December 2010, LHC had Pb_{82}^{208} on Pb_{82}^{208} collision at 287 TeV per beam. If this were a fixed target experiment, what energy of the lead ion should be in order to achieve the same center of mass energy?

7. The luminosity, $\mathcal{L}$ [cm^{-2} s^{-1}], is a measure of the probability (rate) of particle encounters per unit area in a collision process. Thus the total counting rate of a physics event is $R = \sigma_{\rm phys}\mathcal{L}$, where $\sigma_{\rm phys}$ is the cross-section of a physics process.

   (a) In fixed target experiments, the luminosity is given by $\mathcal{L} = (dN_{\rm B}/dt)n_{\rm target}$, where $dN_{\rm B}/dt$ is the number of beam particles per second on target, and $n_{\rm target}$ is the target thickness measuring the number of atoms per cm^2. The average luminosity is given by $\langle\mathcal{L}\rangle = \langle dN_{\rm B}/dt\rangle n_{\rm target}$, where $\langle dN_{\rm B}/dt\rangle = N_{\rm B} f$. Here $N_{\rm B}$ is the number of particles per pulse (bunch) and f is the pulse repetition rate. Consider a fixed target experiment, where the beam repetition rate is 0.4 Hz, beam particle per pulse is 10^{13}, the beam pulse length is 150 ns, and the target thickness is 4 mg/cm^2 Au foil. Find the instantaneous and average luminosities of the fixed target experiment. What is the advantage of stretching the beam pulse length to 1 s in this experiment?

(b) When two beams collide head-on, the luminosity is

$$\mathcal{L} = 2\, f\, N_1\, N_2 \int \rho_1(x,z,s_1)\rho_2(x,z,s_2)dxdzdsd(\beta ct),$$

where $s_1 = s + \beta ct, s_2 = s - \beta ct$, f is the encountering frequency, N_1 and N_2 are the numbers of particles, and ρ_1 and ρ_2 are the normalized distribution functions for these two bunches. Using a Gaussian bunch distribution,

$$\rho(x,z,s) = \frac{1}{(2\pi)^{3/2}\sigma_x\sigma_z\sigma_s} \exp\left\{-\frac{x^2}{2\sigma_x^2} - \frac{z^2}{2\sigma_z^2} - \frac{s^2}{2\sigma_s^2}\right\},$$

where σ_x, σ_z, and σ_s are respectively the horizontal and vertical rms bunch widths and the rms bunch length, show that the luminosity for two bunches with identical distribution profiles is

$$\mathcal{L} = \frac{fN_1N_2}{4\pi\sigma_x\sigma_z}.$$

Show that when two beams are offset by a horizontal distance b, the luminosity is reduced by a factor $\exp\{-b^2/4\sigma_x^2\}$.

8. Show that the magnetic field on the axis of a circular cylindrical winding of uniform cross-section is

$$B_{\parallel}(s) = \frac{\mu_0 J}{2}\left\{(\ell - s)\ln\frac{b + (b^2 + [\ell - s]^2)^{1/2}}{a + (a^2 + [\ell - s]^2)^{1/2}} + s\ln\frac{b + (b^2 + s^2)^{1/2}}{a + (a^2 + s^2)^{1/2}}\right\}$$

where ℓ is the length of the solenoid, J is the current density, a, b are the inner and outer cylindrical radii respectively, and s is the distance from one end of the solenoid. For an ideal solenoid, set $s = \ell/2, b \to a$, show that the magnetic field and the inductance are

$$B_{\parallel} = \mu_0 nI,$$
$$L = \mu_0 n^2 \ell S = \mu_0 n^2 \times \text{ volume of the solenoid},$$

where n is the number of turns per unit length, I is the current in each turn, and S is the cross-section area of the solenoid. Note that the total energy stored in the magnet is given by the magnetic energy.

9. From elementary physics, the field at a distance r from a long straight wire carrying current I is

$$B = \mu_0 I/2\pi r$$

along a direction tangential to a circle with radius r around the wire.

(a) Show that the 2D magnetic field at location $y = x + jz$ for a long straight wire is

$$B_z(x,z) + jB_x(x,z) = \frac{\mu_0 I}{2\pi(y - y_0)},$$

where j is an imaginary number, the current I is positive if it points out of paper, and $y_0 = x_0 + jz_0$ is the position of the current filament.[44]

[44]See R.A. Beth, *J. Appl. Phys.* **37**, 2568 (1966); **38**, 4689 (1967).

(b) If the current per unit area of an infinitely long circular current sheet is

$$\lambda(r, \phi) = (I_1/2a)\cos\phi\ \delta(r-a),$$

where I_1 is the total dipole current and (r, ϕ) are the cylindrical coordinates with $x = r\cos\phi$ and $z = r\sin\phi$, show that the magnetic field inside the current sheet is

$$B_z = -\mu_0 I_1/4a, \quad B_x = 0.$$

This is the cosine-theta current distribution for a dipole. High-field superconducting dipoles are normally made of current blocks that simulate the cosine-theta distribution.

(c) The Beth current sheet theorem states that the magnetic fields in the immediate neighborhood of a two-dimensional current sheet are

$$B(y_+) - B(y_-) = j\mu_0(dI/dy),$$

where y_+ and y_- are the complex coordinates $y = x + jz$ at an infinitesimal distance from the current sheet, and dI/dy is the current per unit length. Apply this theorem to show that the cosine-theta current distribution on a circular cylinder gives rise to a pure dipole field inside the cylinder.

10. Show that the magnetic field at the coordinate $y = x + jz$, due to a thin current wire located at coordinates $y_0 = x_0 + jz_0$, between two sheets of parallel plates with infinite permeability is[45]

$$B_z + jB_x = \frac{\mu_0 I}{4g}\left[\tanh\frac{\pi(y-y_0^*)}{2g} + \coth\frac{\pi(y-y_0)}{2g}\right],$$

where g is the gap between two parallel plates. The current flows in the $\hat{x}\times\hat{z}$ direction.

11. Show that the dipole field and the inductance of a window-frame dipole with two sheets of parallel plates having infinite permeability are

$$B = \mu_0 NI/g = \mu_0 nI,$$
$$L = \mu_0 N^2 \ell w/g = \mu_0 n^2 \times \text{ volume of the dipole},$$

where N is the number of turns, I is the current in each turn, g is the gap between two iron plates, and $n = N/g$ is the number of turns per unit gap length, ℓ and w are the length and width of the dipole. The total power dissipation is $P = (NI)^2R$, where $R = \rho\ell/A$ is the resistance, A as the cross-sectional area of the conductor, and ρ is the resistivity of the coil.

12. Following Maxwell's equation, $\nabla\times\vec{B} = 0$ in the current-free region, and the magnetic field can be derived from a magnetic potential, $\Phi_{\rm m}$, with $\vec{B} = -\nabla\Phi_{\rm m}$. For a quadrupole field with $B_z = Kx, B_x = Kz$, show that the magnetic potential is

[45] S.Y. Lee, *Nucl. Inst. Meth.* **A300**, 151 (1991). Use the following identities:

$$\tanh\frac{\pi y}{2} = \frac{4y}{\pi}\sum_{k=1}^{\infty}\frac{1}{(2k-1)^2+y^2}, \quad \coth\frac{\pi y}{2} = \frac{2}{\pi y} + \frac{4y}{\pi}\sum_{k=1}^{\infty}\frac{1}{(2k)^2+y^2}.$$

$\Phi_{\rm m} = -Kxz$. The equipotential curve is $xz =$ constant. Thus the pole shapes of quadrupoles are hyperbolic curves with $xz = a^2/2$. The pole-tip field is $B_{\rm pole\ tip} = Ka$. To avoid the magnetic field saturation in iron, the pole-tip field in a quadrupole is normally designed to be less than 0.9 Tesla. The achievable gradient is $B_1 = B_{\rm pole\ tip}/a$. Show that the gradient field is

$$B_1 = 2\mu_0 NI/a^2,$$

where NI is the number of ampere-turns per pole, and a is the half-aperture of the quadrupole. The inductance in an ideal quadrupole is

$$L = \frac{8\mu_0 N^2 \ell}{a^2}(x_c^2 - \frac{a^4}{12x_c^2}) \approx \frac{8\mu_0 N^2 \ell}{a^2} x_c^2,$$

where x_c is the distance of the conductor from the center of the quadrupole. In reality, x_c^2 should be replaced by $x_c^2 + x_c w_c$, where w_c is the width of the pole.

13. Consider a pair of conductors with cross-sections independent of the azimuthal coordinate s, and surrounded by isotropic and homogeneous medium with permittivity ϵ and permeability μ. Maxwell's equations are

$$\nabla \cdot (\epsilon \vec{E}) = 0, \qquad \nabla \times \vec{E} = -\tfrac{\partial \vec{B}}{\partial t}, \quad \nabla \cdot \vec{B} = 0, \qquad \nabla \times \vec{H} = \frac{\partial \epsilon \vec{E}}{\partial t},$$

where the external charge and current are zero. Let $\hat{x}, \hat{z}$ and $\hat{s}$ form the basis of an orthonormal coordinate system. For a transverse guided field propagating in the $+\hat{s}$ direction, we assume

$$\vec{E}(\vec{r}, t) = \vec{E}_\perp(x, z) e^{-j(ks-\omega t)},$$
$$\vec{H}(\vec{r}, t) = \vec{H}_\perp(x, z) e^{-j(ks-\omega t)}, \quad \vec{B}_\perp = \mu \vec{H}_\perp,$$

where fields are all transverse with phase velocity $v_{\rm p} = \omega/k$.

(a) Show that the frequency ω and the wave number k of the electromagnetic wave satisfy the dispersion relation $\omega = k/\sqrt{\epsilon\mu}$. Show that the transverse electromagnetic fields satisfy the static electromagnetic field equation,

$$\left(\frac{\partial^2}{\partial x^2} + \frac{\partial^2}{\partial z^2}\right) \vec{E}_\perp(x, z) = 0, \quad \left(\frac{\partial^2}{\partial x^2} + \frac{\partial^2}{\partial z^2}\right) \vec{H}_\perp(x, z) = 0,$$

and the transverse plane wave obeys the relation $\vec{H} = \frac{1}{Z}\, \hat{s} \times \vec{E}_\perp$, where $Z = \sqrt{\mu/\epsilon}$ is the intrinsic impedance of the medium.

(b) Show that, because of the transverse nature of the electromagnetic field, the electric field can be represented by

$$\vec{E}(x, z) = -\nabla_\perp \phi(x, z),$$

where ϕ is the electric potential, and $\nabla_\perp$ is the transverse gradient with respect to the transverse coordinates. By definition, the capacitance per unit length is $C = \lambda/V$, where $V = \phi_1 - \phi_2$ is the potential difference between two conductors, and λ is the charge per unit length on conductors. Using Ampere's law, show that

$$\oint \vec{H} \cdot d\vec{r} = \lambda/\epsilon Z = \lambda v_{\rm p},$$

where $I = \lambda v_{\rm p}$ is the current per unit length, and $d\vec{r} = dx\hat{x} + dz\hat{z}$.

(c) Similarly, the inductance per unit length is

$$L = \frac{1}{I}\frac{d\Phi}{ds} = \frac{\mu}{I}\int_{c1}^{c2} \vec{H} \cdot (\hat{s} \times d\vec{r}),$$

where the integral is carried out between two conductors. Show that there is a general relation:

$$C\,L = \mu\epsilon = 1/v_{\rm p}^2.$$

The characteristic impedance of the transmission line is given by $R_c = \sqrt{L/C} = V/I$, where C and L are the capacitance and the inductance per unit length.

(d) Show that the capacitance and the inductance per unit length of a coaxial cable with inner and outer radii r_1 and r_2 are

$$C = \frac{2\pi\epsilon}{\ln(r_2/r_1)}, \qquad L = \frac{\mu}{2\pi}\ln\frac{r_2}{r_1}.$$

Fill out the following table for some commonly used coaxial cables.

Type	Diameter [cm]	Capacitance [pF/m]	Inductance [μH/m]	R_c [Ω]	Delay time [ns/m]
RG58/U	0.307	93.5		50	
RG174/U	0.152	98.4		50	
RG218/U	1.73	96.8		50	

14. Derive the transverse equations of motion for electrons in a betatron[46] by the following procedures. In the cylindrical coordinate system, the equation of motion for electrons is

$$\frac{dp_r}{dt} - \gamma m r\dot{\theta}^2 = -er\dot{\theta}B_z, \qquad \frac{dp_z}{dt} = er\dot{\theta}B_r,$$

where $\hat{r}, \hat{z}$ are respectively the radially outward and vertically upward directions, B_r, B_z are the radial and vertical components of the magnetic flux density, θ is the azimuthal angle, and $\dot{\theta} = v/r$ is the angular velocity. If the vertical component of the magnetic flux density is

$$B_z = B_0\left(\frac{r}{R}\right)^{-n} \approx B_0\left(1 - n\frac{r-R}{R} + \cdots\right),$$

where n is the field index. Then the radial magnetic field with $B_r = 0$ at $z = 0$ is

$$B_r = z\left(\frac{\partial B_z}{\partial r}\right)_{r=R} = -\frac{nB_0}{R}z + \cdots.$$

Show that the equations of motion become

$$\ddot{\xi} + \omega_0^2(1-n)\xi = 0, \quad \ddot{\zeta} + \omega_0^2 n\zeta = 0,$$

where $\xi = (r-R)/R$, $\zeta = z/R$, and $\omega_0 = v/R = eB_0/\gamma m$ is the angular velocity of the orbiting particle. Show that the stability of betatron motion requires $0 \le n \le 1$.

[46]See D. Kerst and R. Serber, *Phys. Rev.* **60**, 53 (1941). Because of this seminal work, the transverse oscillations of charged particles in linear or circular accelerators are generally called *betatron oscillations*.

15. Ion sources are indispensable to all applications in accelerators. For electron beams, there are thermionic sources, rf gun sources, laser-driven electron sources, etc. For charged ion beams, there are many different configurations for generating plasma sources for beam extraction.[47] Charged ion beams are usually drawn from a space-charge ion source at zero initial velocity. The flow of charged ions is assumed to be laminar. In the space-charge dominated limit, the electric field between the anode and the cathode is maximally shielded by the beam charge. The maximum beam current occurs when the electric field becomes zero at the emitter. Assume a simplified geometry of two infinite parallel plates so that the the motion of ions is one-dimensional. Let s be the distance coordinate between the parallel plates with $s = 0$ at the emitter, and $s = a$ at the anode. The Poisson equation becomes

$$\frac{d^2V}{ds^2} = -\frac{\rho}{\epsilon_0},$$

where V is the electric potential, ρ is the ion density in the parallel plate, and ϵ_0 is the permittivity.

   (a) In the non-relativistic limit with laminar flow, show that the Poisson equation becomes

$$\frac{d^2V}{ds^2} = \frac{J}{\epsilon_0}\left(\frac{m}{2e}\right)^{1/2} V^{-1/2},$$

   where $J = \rho v$ is the current density, e and m are the charge and mass of the ion, and v is the velocity of the ion.

   (b) For a space-charge dominated beam, the condition of maximum space-charge shielding is equivalent to $V = 0$ and $dV/ds = 0$ at $s = 0$. Show that the maximum current is

$$J = \chi\frac{V_0^{3/2}}{a^2}, \quad \chi = \frac{4\epsilon_0}{9}\left(\frac{2e}{m}\right)^{1/2},$$

   where V_0 is the extraction voltage at the anode, and χ is the perveance of the ion source. The relation of the current to the extraction voltage is called Child's law.[48]

   (c) Show that the space-charge perveance parameters for electron, proton, deuteron, He^+, N^+, and Ar^+ ion sources are given by the following table. Here the microperveance is defined as $1\ \mu P = 1 \times 10^{-6}\ A/V^{3/2}$.

	e	p	D^+	He^+	N^+	A^+
χ (μP)	2.334	0.0545	0.0385	0.0272	0.0146	0.00861

16. **The Paraxial Ray Equation**: In the free space, the electric potential obeys the Laplace equation $\nabla^2 V = 0$. Using the basis vectors $(\hat{r}, \hat{\phi}, \hat{s})$ for the cylindrical coordinates in paraxial geometry, where r is the radial distance from the axis of symmetry, ϕ is the azimuthal coordinate, and s is the longitudinal coordinate, we expand the

[47] See e.g. *Proc. Int. Symp. on Electron Beam Ion Sources, AIP Conf. Proc.* No. 188 (1988); *Production and Neutralization of Negative Ions and Beams, AIP Conf. Proc.* No. 210 (1990).

[48] C.D. Child, *Phys. Rev.* **32**, 492 (1911); I. Langmuir, *Phys. Rev.* **32**, 450 (1913). See also A.T. Forrester, *Large Ion Beams* (Wiley, New York, 1988).

position vector as $\vec{R} = r\hat{r} + s\hat{s}$. Let $V_0(s)$ be the electric potential on the axis of symmetry. Show that the electric potential $V(r,s)$ and the electric field $\vec{E} = E_r\hat{r} + E_s\hat{s}$ are

$$V(r,s) = V_0(s) - \frac{V_0^{(2)}}{4}r^2 + \frac{V_0^{(4)}}{64}r^4 + \cdots,$$
$$E_r = \frac{V_0^{(2)}}{2}r - \frac{V_0^{(4)}}{16}r^3 + \cdots,$$
$$E_s = -V_0^{(1)} + \frac{V_0^{(3)}}{4}r^2 - \frac{V_0^{(5)}}{64}r^4 + \cdots,$$

where $V_0^{(n)}$ correspond to nth-derivative of V_0 with respect to s. The equation of motion for a non-relativistic particle in the electric field is $m\ddot{\vec{R}} = e\vec{E}$, where the overdot represents the time derivative. Show that the equation of motion for the radial coordinate, known as *the paraxial ray equation*, becomes

$$Vr'' + \frac{1}{2}V'r' + \frac{1}{4}V''r = 0,$$

where V replaces V_0 for simplicity and the prime is the derivative with respective to s. The paraxial ray equation can be used to analyze the beam envelope in electrostatic accelerators.[49]

17. Consider a line charge inside an infinitely long circular conducting cylinder with radius b. The line-charge density per unit length is λ, and the coordinates of the line charge are $\vec{a} = (a\cos\phi, a\sin\phi)$, where a is the distance from the center of the cylinder, and ϕ is the phase angle with respect to the $\hat{x}$ axis. Show that the induced surface charge density on the cylinder is[50]

$$\sigma(b,\phi_{\rm w}) = -\frac{\lambda}{2\pi b}\frac{b^2 - a^2}{b^2 + a^2 - 2ba\cos(\phi_{\rm w} - \phi)}$$
$$= -\frac{\lambda}{2\pi b}\left[1 + 2\sum_{i=1}^{\infty}\left(\frac{a}{b}\right)^n \cos n(\phi_{\rm w} - \phi)\right].$$

where $\phi_{\rm w}$ is the angular coordinate of the cylindrical wall surface. This result is the basis of beam position monitor design.

[49] V.K. Zworykin *et al.*, *Electron Optics and the Electron Microscope*, (Wiley, 1945); J.R. Pierce, *Theory and Design of Electron Beams*, (Van Nostrand, 1949); V.E. Cosslett, *Introduction to Electron Optics*, (Oxford, 1950); F. Terman, *Radio Engineers' Handbook*, (McGraw-Hill, 1943).

[50] Let the image charge be located at $\vec{c} = (c\cos\phi,\ c\sin\phi)$, then the electric potential for infinite line charges at $\vec{r}$ is

$$\Phi(r) = \frac{\lambda}{2\pi\epsilon_0}\ln|\vec{r} - \vec{a}| + \frac{\lambda_{\rm i}}{2\pi\epsilon_0}\ln|\vec{r} - \vec{c}|.$$

The electric field is $E = -\nabla\Phi$. Using the condition $E_\phi = 0$ on the conducting wall surface in the cylindrical coordinate, we obtain $c = b^2/a$ and $\lambda_{\rm i} = -\lambda$. The induced surface charge density is $\sigma = \epsilon_0 E_r$. The multipole expansion can be obtained by using the identity $\cos n\theta + j\sin n\theta = e^{jn\theta}$.

Chapter 2

Transverse Motion

The transverse particle motion in an accelerator is divided into a closed orbit and a small-amplitude betatron motion around the closed orbit, where the closed orbit in a synchrotron is defined as a particle trajectory that closes onto itself after a complete revolution, a closed orbit in a linac or cyclotron is the orbit with zero betatron oscillation amplitude. Particle motion with a small deviation from the closed orbit will oscillate around the closed orbit. The terminology of betatron motion is derived from the seminal work of D. Kerst and R. Serber on the stability of transverse particle motion in a betatron. It is now used for transverse motion in all types of accelerators.

In synchrotrons, bending magnets are needed to provide complete revolution of the particle beam. This defines a closed orbit. Betatron motion around the closed orbit is determined by an arrangement of quadrupoles, called accelerator *lattice*. In actual accelerators, magnetic field errors are unavoidable, the closed orbit and the betatron motion will be perturbed. Lattice design has to take these field errors into account. Since the bending angle of a dipole depends on the particle momentum, the resulting closed orbit is a function of the particle momentum. In the first-order approximation, the deviation of the closed orbit is proportional to the fractional off-momentum deviation $(p - p_0)/p_0$, where p_0 is the momentum of a reference particle. The dispersion function, defined as the derivative of the closed orbit with respect to the fractional off-momentum variable, and the chromatic aberration of the betatron motion play a major role in the accelerator's performance. Furthermore, careful correction of linear and nonlinear resonances and feedback of collective beam instabilities are important for high-intensity and high-brightness beams.

Various aspects of transverse particle motion will be discussed in this chapter. In principle, the method discussed in this chapter can also be applied to a linac or a transport line, where the betatron motion is equivalent to an initial value problem. In Sec. I, we derive the particle Hamiltonian in the Frenet-Serret coordinate system. For accelerator practitioners, who are not familiar with Hamiltonian dynamics, can skip the formal formulation of Sec. I and jump right on to the Hill's equations: Eq. (2.22) of Sec. II, where we examine the properties of linear betatron motion. We discuss the Floquet transformation to action-angle variables, beam distribution, beam emittance, and properties of the envelope function. In Sec. III, we study the effects of linear magnetic imperfections (dipole and quadrupole field errors) and their roles in beam manipulation. Section IV deals with the off-momentum closed orbit and its implications for longitudinal synchrotron motion, and also with the lattice design strategies for variable $\gamma_{\rm T}$ and minimum dispersion action. Section V describes the chromatic aberration and its correction, and Section VI describes linear betatron coupling. In Sec. VII, we examine the effects of low-order nonlinear resonances. Section VIII introduces the basic concept of transverse collective instabilities and Landau damping. Section IX lays out a general framework for the synchrotron-betatron coupling Hamiltonian.

I Hamiltonian for Particle Motion in Accelerators

The motion of a charged particle in electromagnetic field $\vec{\mathcal{E}}$ and $\vec{B}$ is governed by the Lorentz force,

$$\frac{d\vec{p}}{dt} = \vec{F} = e(\vec{\mathcal{E}} + \vec{v} \times \vec{B}), \tag{2.1}$$

where $\vec{p} = \gamma m\vec{v}$ is the mechanical momentum, $\vec{v} = d\vec{r}/dt$ is the velocity, m is the mass, e is the charge, and $\gamma = 1/\sqrt{1 - v^2/c^2}$ is the relativistic Lorentz factor. The energy and momentum of the particle are $E = \gamma mc^2 = mc^2 dt/d\tau$ and $\vec{p} = m\gamma\vec{v} = md\vec{r}/d\tau$, where τ is the proper time with $dt/d\tau = \gamma$. The electric and magnetic fields are related to the vector potential $\vec{A}$ and scalar potential Φ via $\vec{\mathcal{E}} = -\nabla\Phi - \partial\vec{A}/\partial t$, and $\vec{B} = \nabla \times \vec{A}$. With the Lagrangian: $L = -mc^2\sqrt{1 - v^2/c^2} - e\Phi + e\vec{v}\cdot\vec{A}$, Equation (2.1) can be derived from Lagrange's equation

$$\frac{d}{dt}\left(\frac{\partial L}{\partial \vec{v}}\right) - \frac{\partial L}{\partial \vec{r}} = 0. \tag{2.2}$$

The canonical momentum, the Hamiltonian, and Hamilton's equations of motion are

$$\vec{P} = \frac{\partial L}{\partial \vec{v}} = \vec{p} + e\vec{A}, \tag{2.3}$$

$$H = \vec{P}\cdot\vec{v} - L = c[m^2c^2 + (\vec{P} - e\vec{A})^2]^{1/2} + e\Phi, \tag{2.4}$$

$$\dot{x} = \frac{\partial H}{\partial P_x}, \quad \dot{P}_x = -\frac{\partial H}{\partial x}, \quad \text{etc.,} \tag{2.5}$$

where the overdot is the derivative with respect to time t, and $(x, P_x), \cdots$ pairs are conjugate phase-space coordinates with respect to any reference point in space. Particle motion in accelerators is usually confined to small deviations from a well-defined reference orbit. The phase space coordinate system around this reference orbit is called the Frenet-Serret coordinate system shown in Fig. 2.1.

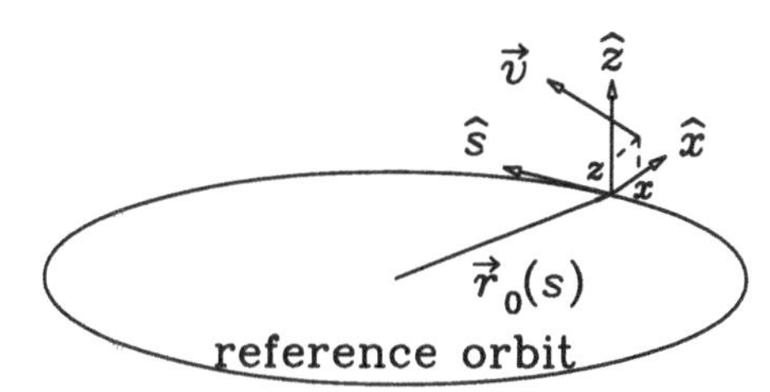

Figure 2.1: Curvilinear coordinate system for particle motion in synchrotrons. $\vec{r}_0(s)$ is the reference orbit, $\hat{x}$, $\hat{s}$ and $\hat{z}$ form the basis of the curvilinear coordinate system. Any point in the phase space can be expressed by $\vec{r} = \vec{r}_0 + x\hat{x} + z\hat{z}$. Here x and z are betatron coordinates.

I.1 Hamiltonian in Frenet-Serret Coordinate System

Let $\vec{r}_0(s)$ be the reference orbit (see Fig. 2.1), where s is the path length measured along the reference orbit from an initial point. The tangent unit vector to the reference orbit, the unit vector perpendicular to the tangent vector and on the tangential plane,[1] and the unit vector orthogonal to the tangential plane are

$$\hat{s}(s) = \frac{d\vec{r}_0(s)}{ds}, \qquad \hat{x}(s) = -\rho(s)\frac{d\hat{s}(s)}{ds}, \qquad \hat{z}(s) = \hat{x}(s) \times \hat{s}(s)\ , \tag{2.6}$$

where $\rho(s)$ defines the radius of curvature. The vectors $(\hat{x}, \hat{s}, \hat{z})$ form the orthonormal basis for the right-handed Frenet-Serret curvilinear coordinate system with

$$\hat{x}'(s) = \frac{1}{\rho(s)}\hat{s}(s) + \tau(s)\hat{z}(s), \quad \hat{z}'(s) = -\tau(s)\hat{x}(s)\ , \tag{2.7}$$

where the prime denotes differentiation with respect to s, and $\tau(s)$ is the torsion of the curve. For simplicity, we discuss only planar geometry, where $\tau(s) = 0$. The particle trajectory around the reference orbit can be expressed as

$$\vec{r}(s) = \vec{r}_0(s) + x\hat{x}(s) + z\hat{z}(s)\ . \tag{2.8}$$

To express the equation of motion in terms of the reference orbit coordinate system (x, s, z), we perform a canonical transformation by using the generating function

$$F_3(\vec{P}; x, s, z) = -\vec{P}\cdot[\vec{r}_0(s) + x\hat{x}(s) + z\hat{z}(s)]\ , \tag{2.9}$$

where $\vec{P}$ is the momentum in the Cartesian coordinate system. The conjugate momenta for the coordinates (x, s, z) (see Appendix A) and the field momentum vector in the curvilinear coordinate system A_s, A_x and A_z are

$$p_s = -\frac{\partial F_3}{\partial s} = \left(1 + \frac{x}{\rho}\right)\vec{P}\cdot\hat{s}, \quad p_x = -\frac{\partial F_3}{\partial x} = \vec{P}\cdot\hat{x}, \quad p_z = -\frac{\partial F_3}{\partial z} = \vec{P}\cdot\hat{z}.$$
$$A_s = \left(1 + \frac{x}{\rho}\right)\vec{A}\cdot\hat{s}, \quad A_x = \vec{A}\cdot\hat{x}, \quad A_z = \vec{A}\cdot\hat{z}\ .$$

[1] Using Eq. (2.6), we find $a_{\rm centripetal} = |d^2\vec{r}_0/dt^2| = (ds/dt)^2|(d/ds)(d\vec{r}_0/ds)| = v^2|(d\hat{s}/ds)|$, where $v = ds/dt$ is the tangential velocity. The magnitude of the bending radius is $\rho = v^2/a_{\rm centripetal} = |ds/d\hat{s}|$.

The new Hamiltonian becomes

$$H = e\Phi + c\left\{m^2c^2 + \frac{(p_s - eA_s)^2}{(1+x/\rho)^2} + (p_x - eA_x)^2 + (p_z - eA_z)^2\right\}^{1/2}, \quad (2.10)$$

Note that A_s and p_s are not simply the projections of vectors $\vec{A}$ and $\vec{P}$ in the $\hat{s}$ direction. In the new coordinate system, Hamilton's equation becomes

$$\dot{s} = \frac{\partial H}{\partial p_s},\ \dot{p}_s = -\frac{\partial H}{\partial s};\quad \dot{x} = \frac{\partial H}{\partial p_x},\ \dot{p}_x = -\frac{\partial H}{\partial x};\quad \dot{z} = \frac{\partial H}{\partial p_z},\ \dot{p}_z = -\frac{\partial H}{\partial z}. \quad (2.11)$$

The next step is to use s as the independent variable instead of time t [20]. With the relation $dH = (\partial H/\partial p_x)dp_x + (\partial H/\partial p_s)dp_s = 0$ or

$$x' = \frac{dx}{ds} = \frac{\dot{x}}{\dot{s}} = \left(\frac{\partial H}{\partial p_x}\right)\left(\frac{\partial H}{\partial p_s}\right)^{-1} = \frac{\partial(-p_s)}{\partial p_x}, \quad \text{etc.}, \quad (2.12)$$

where the prime denotes differentiation with respect to s, Hamilton's equations of motion become

$$t' = \frac{\partial p_s}{\partial H},\ H' = -\frac{\partial p_s}{\partial t};\quad x' = -\frac{\partial p_s}{\partial p_x},\ p_x' = \frac{\partial p_s}{\partial x};\quad z' = -\frac{\partial p_s}{\partial p_z},\ p_z' = \frac{\partial p_s}{\partial z}. \quad (2.13)$$

Here s as the independent variable, $-p_s$ is the new Hamiltonian, and the conjugate phase-space coordinates are $(x, p_x;\ z, p_z;\ t, -H)$.

Since the scalar and vector potentials Φ and $\vec{A}$ depend on position, the new Hamiltonian is a function of the independent coordinate s. However, the repetitive nature of the accelerator components, the dependence of the new Hamiltonian on s is *periodic*. The periodic nature of the new Hamiltonian can be fruitfully exploited in the analysis of linear and nonlinear betatron motion with the Floquet theorem. The new Hamiltonian $\tilde{H} = -p_s$ becomes

$$\tilde{H} = -\left(1+\frac{x}{\rho}\right)\left[\frac{(H-e\Phi)^2}{c^2} - m^2c^2 - (p_x - eA_x)^2 - (p_z - eA_z)^2\right]^{1/2} - eA_s, \quad (2.14)$$

with $(x, p_x, z, p_z, t, -H)$ as the phase-space coordinates. The energy and momentum of a particle are given by $E = H - e\Phi$ and $p = \sqrt{E^2/c^2 - m^2c^2}$. Since the transverse momenta p_x and p_z are much smaller than the total momentum p, we expand the Hamiltonian up to second order in p_x and p_z

$$\tilde{H} \approx -p\left(1+\frac{x}{\rho}\right) + \frac{1+x/\rho}{2p}\left[(p_x - eA_x)^2 + (p_z - eA_z)^2\right] - eA_s. \quad (2.15)$$

I.2 Magnetic Field in Frenet-Serret Coordinate System

The scale factors for the Frenet-Serret coordinate system are $h_x = 1, h_s = 1 + \frac{x}{\rho}, h_z = 1$. The differential path length is $d\ell^2 = h_x^2 dx^2 + h_s^2 ds^2 + h_z^2 dz^2$, and the differential operators are

$$\begin{aligned}
\nabla\Phi &= \frac{\partial\Phi}{\partial x}\hat{x} + \frac{1}{h_s}\frac{\partial\Phi}{\partial s}\hat{s} + \frac{\partial\Phi}{\partial z}\hat{z}, \\
\nabla\cdot\vec{A} &= \frac{1}{h_s}\left[\frac{\partial(h_s A_1)}{\partial x} + \frac{\partial A_2}{\partial s} + \frac{\partial(h_s A_3)}{\partial z}\right], \\
\nabla\times\vec{A} &= \frac{1}{h_s}\left[\frac{\partial A_3}{\partial s} - \frac{\partial(h_s A_2)}{\partial z}\right]\hat{x} + \left[\frac{\partial A_1}{\partial z} - \frac{\partial A_3}{\partial x}\right]\hat{s} + \frac{1}{h_s}\left[\frac{\partial(h_s A_2)}{\partial x} - \frac{\partial A_1}{\partial s}\right]\hat{z}, \\
\nabla^2\Phi &= \frac{1}{h_s}\left[\frac{\partial}{\partial x}h_s\frac{\partial\Phi}{\partial x} + \frac{\partial}{\partial s}\frac{1}{h_s}\frac{\partial\Phi}{\partial s} + \frac{\partial}{\partial z}h_s\frac{\partial\Phi}{\partial z}\right],
\end{aligned}$$

where $A_1 = \vec{A}\cdot\hat{x}$, $A_2 = \vec{A}\cdot\hat{s}$, and $A_3 = \vec{A}\cdot\hat{z}$. In particle accelerators, we consider only the case with zero electric potential with $\Phi = 0$, furthermore, for an accelerator with transverse magnetic fields, we can assume $A_x = A_z = 0$. The two-dimensional magnetic field can be expressed as

$$\begin{aligned}
&\vec{B} = B_x(x,z)\hat{x} + B_z(x,z)\hat{z}, \\
&B_x = -\frac{1}{h_s}\frac{\partial(h_s A_2)}{\partial z} = -\frac{1}{h_s}\frac{\partial A_s}{\partial z}, \quad B_z = \frac{1}{h_s}\frac{\partial(h_s A_2)}{\partial x} = \frac{1}{h_s}\frac{\partial A_s}{\partial x},
\end{aligned} \tag{2.16}$$

with $A_s = h_s A_2$. Using Maxwell's equation $\nabla\times\vec{B} = 0$, we have

$$\frac{\partial}{\partial z}\frac{1}{h_s}\frac{\partial A_s}{\partial z} + \frac{\partial}{\partial x}\frac{1}{h_s}\frac{\partial A_s}{\partial x} = 0. \tag{2.17}$$

General solutions of B_x, B_z and A_s can be obtained through power series expansion (see Exercise 2.1.3).

In the rectangular coordinate system with $h_s = 1$ or $\rho = \infty$, we have $\nabla_\perp^2 A_s = 0$, and A_s can be expanded in power series as

$$A_s = B_0\Re\left[\sum_{n=0}^{\infty}\frac{b_n + ja_n}{n+1}(x+jz)^{n+1}\right], \tag{2.18}$$

where j is the imaginary number, $\Re[...]$ represents the real part, and $B_z = \frac{\partial A_s}{\partial x}$ and $B_x = -\frac{\partial A_s}{\partial z}$. Normally the normalization constant B_0 is chosen as the main dipole field strength such that $b_0 = 1$, i.e. $B_0 b_0 = -[B\rho]/\rho$, where $B\rho$ is the momentum rigidity of the beam, ρ is the bending radius. The resulting magnetic flux density and the effective multipole field on the beam particles become[2]

$$B_z + jB_x = B_0 \sum_{n=0}^{\infty} (b_n + ja_n)(x + jz)^n \tag{2.19}$$

$$\frac{1}{B\rho}(B_z + jB_x) = \frac{1}{\rho} \sum_{n=0}^{\infty} (b_n + ja_n)(x + jz)^n,$$

$$b_n = \frac{1}{B_0 n!} \frac{\partial^n B_z}{\partial x^n}\Big|_{x=z=0}, \qquad a_n = \frac{1}{B_0 n!} \frac{\partial^n B_x}{\partial x^n}\Big|_{x=z=0},$$

where b_n, a_n are called $2(n+1)$th multipole coefficients with dipole b_0, dipole roll a_0, quadrupole b_1, skew quadrupole a_1, sextupole b_2, skew sextupole a_2, etc.[3] Here after, we use the notation: $B_n = \frac{\partial^n B_z}{\partial x^n}\Big|_{x=z=0}$. The complex 2D magnetic field representation in $B_z + jB_x$ is called the Beth representation (see Exercise 1.10).

Since $\nabla \times \vec{B} = 0$ in the current free region, the magnetic field can also be derived from a scalar magnetic potential $\Phi_{\rm m}$, i.e. $\vec{B} = -\nabla\Phi_{\rm m}$ (see Exercise 2.1.3). The scalar magnetic potential is

$$\Phi_{\rm m} = -B_0 \Im \left[\sum_{n=0}^{\infty} \frac{b_n + ja_n}{n+1} (x + jz)^{n+1} \right], \tag{2.20}$$

where $\Im[...]$ represents the imaginary part of the expression.

I.3 Equation of Betatron Motion

Disregarding the effect of synchrotron motion (see Sec. IX Chap. 2), Hamilton's equations of betatron motion are

$$x' = \frac{\partial \tilde{H}}{\partial p_x}, \quad p_x' = -\frac{\partial \tilde{H}}{\partial x}, \qquad z' = \frac{\partial \tilde{H}}{\partial p_z}, \quad p_z' = -\frac{\partial \tilde{H}}{\partial z}. \tag{2.21}$$

With the transverse magnetic fields of Eq. (2.16), the betatron equations of motion become

$$x'' - \frac{\rho + x}{\rho^2} = \pm \frac{B_z}{B\rho} \frac{p_0}{p} \left(1 + \frac{x}{\rho}\right)^2, \qquad z'' = \mp \frac{B_x}{B\rho} \frac{p_0}{p} \left(1 + \frac{x}{\rho}\right)^2, \tag{2.22}$$

---

[2]The multipole expansion of the magnetic field is usually re-scaled to obtain

$$B_z + jB_x = B_0 \sum_{n=0}^{\infty} (b_n + ja_n)(\frac{x + jz}{r_{\rm b}})^n,$$

where $r_{\rm b}$ is a reference radius. The reference radius for the multipole expansion of superconducting magnets is often chosen to be 1/2 or 2/3 of the inner coil radius, e.g. $r_{\rm b} = 1$ cm for the SSC and LHC, and $r_{\rm b} = 2.54$ cm for RHIC and Tevatron. The resulting b_n and a_n coefficients are dimensionless.

[3]Note that the multipole convention used in Europe differs from that in the U.S. In Europe, physicists use b_1, a_1 for dipole and dipole roll, b_2, a_2 for quadrupole and skew quadrupole, etc.

where we neglect higher-order terms, the upper and lower signs correspond to the positive and negative charged particle respectively, p is the momentum of the particle, p_0 is the momentum of a reference particle, $B\rho = p_0/e$ is the *magnetic rigidity*, and e is the charge of a particle. The sign convention is chosen such that $B\rho$ is positive. Alternatively, Eq. (2.22) can be derived through Newton's law of acceleration (see Exercise 2.1.2). The equations of motion are given by

$$\ddot{r} - r\dot{\theta}^2 = \frac{ev_s B_z}{\gamma m} = \pm\frac{v_s^2 B_z}{B\rho}, \qquad \ddot{z} = \mp\frac{v_s^2 B_x}{B\rho},$$

which can be transformed into Eq. (2.22) by changing the time variable to the coordinate of orbital distance s, i.e. $x'' = \ddot{r}/v_s^2$.

I.4 Particle Motion in Dipole and Quadrupole Magnets

We consider a on-momentum particle with $p = p_0$, expand the magnetic field up to first order in x and z, i.e.

$$B_z = \mp B_0 + \frac{\partial B_z}{\partial x}x = \mp B_0 + B_1 x, \qquad B_x = \frac{\partial B_z}{\partial x}z = B_1 z, \tag{2.23}$$

where $B_0/B\rho = 1/\rho$ signifies the dipole field in defining a closed orbit, and the quadrupole gradient function $B_1 = \partial B_z/\partial x$ is evaluated at the closed orbit. The betatron equations of motion, Eq. (2.22), become Hill's equation:

$$\begin{aligned} &x'' + K_x(s)x = 0, \qquad &z'' + K_z(s)z = 0, \\ &K_x = 1/\rho^2 + K_1(s), \qquad &K_z = -K_1(s), \end{aligned} \tag{2.24}$$

where $K_1(s) = \mp B_1(s)/B\rho$ is the effective focusing function with dimension $[\mathrm{m}^{-2}]$. Here the upper and lower signs correspond respectively to the positive and negative charged particles. The sign-convention is $K_1 > 0$ for horizontal focusing, and thus vertical defocusing. The focusing index is defined as $n(s) = \pm\rho^2 K_1(s)$, or $K_x = \frac{1}{\rho^2}(1 - n)$ and $K_z = \frac{1}{\rho^2}n$. A weak focusing accelerator requires $0 \le n(s) \le 1$, while a strong-focusing accelerator, $|n| \gg 1$, e.g. $n(s) \approx \pm 350$ for the AGS. Some observations about the linearized betatron equations (2.24) are given below.

- In a quadrupole, where $1/\rho = 0$, we have $K_x = -K_z$, i.e. a horizontally focusing quadrupole is also a vertically defocusing quadrupole and vice versa.
- A horizontal bending dipole has a focusing function $K_x = 1/\rho^2$, and $K_z = 0$. A dipole with entrance and exit angles perpendicular to the edge of the dipole field is called a *sector dipole* (see Fig. 2.2a). The entrance and exit angles of particle trajectories in non-sector type dipoles are not perpendicular to the dipole edge. There is an edge focusing/defocusing effect (see Exercise 2.2.2) on all dipoles.

- The focusing functions K_x, K_z is periodic functions of the longitudinal coordinate s in one revolution. One can design an accelerator lattice with many identical focusing periods. The number of identical building blocks is called the *superperiod* P. The solution of periodic Hill's equation satisfies the *Floquet theorem*.

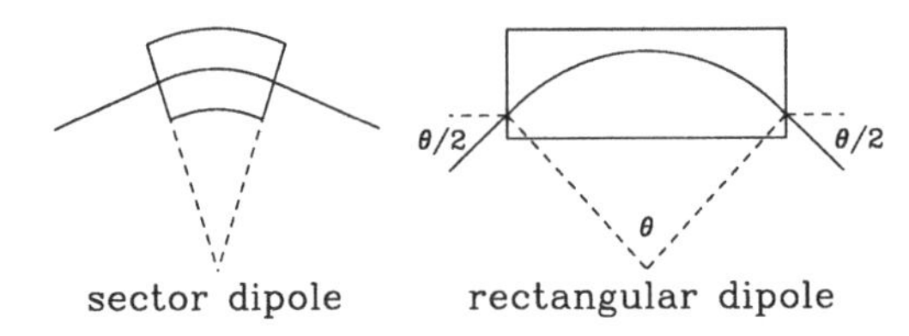

Figure 2.2: Schematic drawing of the particle trajectory in a *sector dipole* and in a *rectangular dipole*. Note that the particle orbit is perpendicular to the pole-faces of the sector dipole magnet, and makes an angle $\theta/2$ with the pole-faces in the rectangular dipole.

Exercise 2.1

1. In the Frenet-Serret coordinate system $(\hat{x}, \hat{s}, \hat{z})$, transverse magnetic fields are

$$B_x = -\frac{1}{1+x/\rho}\frac{\partial A_s}{\partial z}, \quad B_z = \frac{1}{1+x/\rho}\frac{\partial A_s}{\partial x}.$$

   Derive Eq. (2.22) from the Hamiltonian of Eq. (2.15).

2. Derive Eq. (2.22) through the following geometric argument. Let $(\hat{x}, \hat{s}, \hat{z})$ be local polar coordinates inside a dipole. The particle coordinate is

$$\vec{r} = (\rho + x)\hat{x} + z\hat{z},$$

   where ρ is the bending radius. The momentum of the particle is $\vec{p} = \gamma m\dot{\vec{r}}$, where γ is constant in the static magnetic field, and the overdot corresponds to the derivative with respect to time t. Similarly, $d\vec{p}/dt = \gamma m\ddot{\vec{r}}$.

   (a) Using Eq.(2.7), show that

$$\dot{\vec{r}} = \dot{x}\hat{x} + (\rho + x)\dot{\theta}\hat{s} + \dot{z}\hat{z},$$
$$\ddot{\vec{r}} = [\ddot{x} - (\rho + x)\dot{\theta}^2]\hat{x} + [2\dot{x}\dot{\theta} + (\rho + x)\ddot{\theta}]\hat{s} + \ddot{z}\hat{z},$$

   where $\theta = s/\rho$ is the angle associated with the reference orbit, i.e. $ds = \rho d\theta$.

   (b) Using $d\vec{p}/dt = e\vec{v} \times \vec{B}$, with $\vec{B} = B_x\hat{x} + B_z\hat{z}$, show that

$$\ddot{x} - (\rho + x)\dot{\theta}^2 = \frac{v_s^2 B_z}{B\rho}, \quad \ddot{z} = -\frac{v_s^2 B_x}{B\rho},$$

   where $B\rho = \gamma m v_s/e$ is the momentum rigidity and v_s is the longitudinal velocity.

(c) Transform the time coordinate to the longitudinal distance s with $ds = \rho d\theta$, where $d\theta = v_s dt/(\rho + x)$, and show that

$$x'' - \frac{\rho + x}{\rho^2} = \frac{B_z}{B\rho}(1 + \frac{x}{\rho})^2, \qquad z'' = -\frac{B_x}{B\rho}(1 + \frac{x}{\rho})^2,$$

where the prime is the derivative with respect to s.

3. Inside the vacuum chamber of an accelerator, we have $\nabla \times \vec{B} = 0$ and $\nabla \times \vec{E} = 0$. Thus the electric field and magnetic field can be expanded by scalar potentials with $\vec{B} = -\nabla\Phi_{\rm m}, \vec{E} = -\nabla\Phi_{\rm e}$, where both scalar potentials satisfy the Laplace equation with $\nabla^2\Phi = 0$, where Φ stands for either $\Phi_{\rm m}$ or $\Phi_{\rm e}$. In the curvilinear coordinates (x, s, z), we then have

$$\nabla^2\Phi = \frac{1}{1 + hx}\frac{\partial}{\partial x}([1 + hx]\frac{\partial\Phi}{\partial x}) + \frac{\partial^2\Phi}{\partial z^2} + \frac{1}{1 + hx}\frac{\partial}{\partial s}(\frac{1}{1 + hx}\frac{\partial\Phi}{\partial s}) = 0,$$

where $h = 1/\rho$, and ρ is the radius of curvature. Expressing the scalar potential in power series of particle coordinates,

$$\Phi = \sum_{i,j>0} A_{ij}\frac{x^i}{i!}\frac{z^j}{j!},$$

show that A_{ij} satisfies the following iteration relation:

$$\begin{aligned} A_{i,j+2} &= -A''_{i,j} - ihA''_{i-1,j} + ih'A'_{i-1,j} - A_{i+2,j} - (3i+1)hA_{i+1,j} \\ &\quad -3ihA_{i-1,j+2} - i(3i-1)h^2A_{i,j} - 3i(i-1)h^2A_{i-2,j+2} \\ &\quad -i(i-1)^2h^3A_{i-1,j} - i(i-1)(i-2)h^3A_{i-3,j+2}, \end{aligned}$$

where the prime is the derivative with respect to s. Assuming $A_{00} = 0, A_{10} = 0$, and $A_{01} = -B_{00}$ in a rectangular coordinate system with $h = h' = 0$, show that the magnetic potential, up to the fourth order with $i + j \le 4$, is[4]

$$\begin{aligned} \Phi &= -B_{00}z + \frac{1}{2}A_{20}(x^2 - z^2) + A_{11}xz + \frac{1}{6}A_{30}(x^3 - 3xz^2) + \frac{1}{2}B''_{00}xz^2 \\ &\quad +\frac{1}{2}A_{21}x^2z + \frac{1}{6}(B''_{00} - A_{21})z^3 + \frac{1}{24}A_{40}(x^4 - 6x^2z^2 + z^4) \\ &\quad +\frac{1}{12}A''_{20}(-3x^2z^2 + z^4) + \frac{1}{6}A_{31}(x^3z - xz^3) - \frac{1}{6}A''_{11}xz^3. \end{aligned}$$

4. The field components in the current-free region of an axial symmetric solenoid are

$$B_x = x\sum_{k=0}^{\infty} b_{2k+1}(x^2 + z^2)^k, \quad B_z = z\sum_{k=0}^{\infty} b_{2k+1}(x^2 + z^2)^k, \quad B_s = \sum_{k=0}^{\infty} b_{2k}(x^2 + z^2)^k.$$

[4]A word of caution: the magnetic potential obtained here can not be used as the potential in the Hamiltonian of Eq. (2.14). In particular, the potential for a quadrupole is given by the A_{11} term and the skew quadrupole arises from the A_{20} term, etc. However, this serves as a general method for deriving the magnetic field map.

(a) Show that the coefficients are

$$b_{2k+1} = \frac{-1}{2(k+1)} b'_{2k}, \quad b_{2k+2} = \frac{1}{2(k+1)} b'_{2k+1},$$

where the prime is the derivative with respect to s. Show that the vector potential is

$$A_x = z \sum_{k=0}^{\infty} \frac{b_{2k}}{2(k+1)} (x^2 + z^2)^k, \quad A_z = -x \sum_{k=0}^{\infty} \frac{b_{2k}}{2(k+1)} (x^2 + z^2)^k, \quad A_s = 0.$$

In a cylindrical coordinate system, where $\vec{r} = x\hat{x} + z\hat{z}$, $r = \sqrt{x^2 + z^2}$, and $\hat{\phi} = (-z\hat{x} + x\hat{z})/r$, show that the vector potential can be expressed as

$$\vec{A} = -\left[\frac{1}{2} r\, b_0(s) - \frac{1}{16} r^3\, b''_0(s) + \cdots\right] \hat{\phi}.$$

(b) The Hamiltonian of Eq. (2.15) for the particle motion in the solenoid is

$$H = -p + \frac{1}{2p}[(p_x - eA_x)^2 + (p_z - eA_z)^2].$$

Show that the linearized equation of motion is (see also Exercise 2.6.2)

$$x'' + 2gz' + g'z = 0, \qquad z'' - 2gx' - g'x = 0,$$

where $g = eb_0/2p = eB_{\parallel}/2p$ is the strength of the solenoid. The linearized equation can be solved analytically. Letting $y = x + jz$, show that the coupled equation of motion becomes

$$y'' - j2gy' - jg'y = 0.$$

Transforming the coordinates into the rotating frame, show that the system is decoupled, i.e.

$$\bar{y} = ye^{-j\theta(s)}, \quad \text{where} \quad \theta = \int_0^s g ds,$$

$$\bar{y}'' + g^2\bar{y} = 0.$$

Thus the solenoidal field, in the rotating frame, provides both horizontal and vertical focusing, independent of the direction of the solenoidal field. Note also that the effects of the ends of a solenoid, included in the g' terms, have been included to obtain this Hill's equation in the rotating frame.

(c) Up to third order, show that the equation of motion is

$$\begin{aligned} x'' + 2gz' + g'z &= \frac{g''}{2} z'(x^2 + z^2) + \frac{g'''}{8} z(x^2 + z^2), \\ z'' - 2gx' - g'x &= -\frac{g''}{2} x'(x^2 + z^2) - \frac{g'''}{8} x(x^2 + z^2). \end{aligned}$$

5. Consider the transverse magnetic field in the Frenet-Serret coordinate system.[5] For normal multipoles with mid-plane symmetry with

$$B_z(z) = B_z(-z), \quad B_x(z) = -B_x(-z), \quad B_s(z) = -B_s(-z),$$

the most general form of expansion is

$$B_z = \sum_{i,k=0}^{\infty} b_{i,k} x^i z^{2k}, \quad B_x = z\sum_{i,k=0}^{\infty} a_{i,k} x^i z^{2k}, \quad B_s = z\sum_{i,k=0}^{\infty} c_{i,k} x^i z^{2k},$$

where a, b, c can be determined from Maxwell's equations: $\nabla \times \vec{B} = 0$ and $\nabla \cdot \vec{B} = 0$. Show that Maxwell's equations give the following relations:

$$a_{i,k} = \frac{i+1}{2k+1} b_{i+1,k}, \qquad c_{i,k} + \frac{1}{\rho} c_{i-1,k} = \frac{1}{2k+1} b'_{i,k},$$

$$c'_{i,k} + \frac{(i+1)^2}{\rho(2k+1)} b_{i+1,k} + \frac{(i+1)(i+2)}{2k+1} b_{i+2,k} + \frac{2(k+1)}{\rho} b_{i-1,k+1} + 2(k+1) b_{i,k+1} = 0,$$

where the prime is the derivative with respect to s. Assuming that we can measure the B_z at the mid-plane as a function of x, s, i.e.

$$B_z(z=0) = B_{0,0} + B_{1,0}x + B_{2,0}x^2 + B_{3,0}x^3 + \cdots,$$

where $B_{i,0}$ are functions of s, show that the field map is

$$\begin{aligned}
B_z &= B_{0,0} + B_{1,0}x + B_{2,0}x^2 - (B_{2,0} + \frac{B''_{0,0}}{2} + \frac{B_{1,0}}{2\rho})z^2 + B_{3,0}x^3 \\
&\quad -\{3B_{3,0} + \frac{2B_{2,0}}{\rho} - \frac{1}{\rho}(B_{2,0} + \frac{B_{1,0}}{2\rho} + \frac{B''_{0,0}}{2}) + \frac{1}{2}[B''_{1,0} - (\frac{B'_{0,0}}{\rho})']\}xz^2 + \cdots, \\
B_x &= B_{1,0}z + 2B_{2,0}xz + 3B_{3,0}x^2z - \frac{1}{3}\{3B_{3,0} + \frac{2B_{2,0}}{\rho} \\
&\quad -\frac{1}{\rho}(B_{2,0} + \frac{B_{1,0}}{2\rho} + \frac{B''_{0,0}}{2}) + \frac{1}{2}[B''_{1,0} - (\frac{B'_{0,0}}{\rho})']\}z^3 + \cdots, \\
B_s &= B'_{0,0}z + (B'_{1,0} - \frac{B'_{0,0}}{\rho})xz + (B'_{2,0} - \frac{B'_{1,0}}{\rho} + \frac{B'_{0,0}}{\rho^2})x^2z \\
&\quad -\frac{1}{3}(B'_{2,0} + (\frac{B_{1,0}}{2\rho})' + \frac{B'''_{0,0}}{2})z^3 + \cdots.
\end{aligned}$$

Show that in a pure multipole magnet, where $\rho \to \infty$, the magnetic field can be expanded as

$$B_z + jB_x = \sum_{n=0} B_{n,0}(x+jz)^n - \frac{B''_{0,0}}{2}z^2 + \frac{B''_{1,0}}{2}(x+jz)z^2 + \cdots,$$

where j is the complex number. Thus for a finite length quadrupole with $B'_{1,0} \neq 0$, the end field has an octupole-like magnetic multipole field.

---

[5]See K. Steffen, CERN **85-19**, p. 25 (1985).

II Linear Betatron Motion

Transverse particle motion around a closed orbit is called betatron motion. Since the amplitude of betatron motion is normally small, we study the linearized Hill's equation: Eq. (2.24). The focusing functions are normally arranged to be periodic with $K_{x,z}(s+L) = K_{x,z}(s)$, where L is the length of a periodic structure in an accelerator. For example, Fig. 2.3 shows a schematic drawing of the Fermilab booster lattice, where four combined function magnets are arranged to form a basic focusing-defocusing periodic (FODO) cell. Exploiting the periodic nature, we apply the Floquet theorem (see Appendix A, Sec. I.5) to facilitate the design of an accelerator lattice, In this section we study linear betatron motion. betatron tune, envelope equation. Floquet transformation, the action and Courant–Snyder invariant, σ-matrix, beam distribution and emittance.

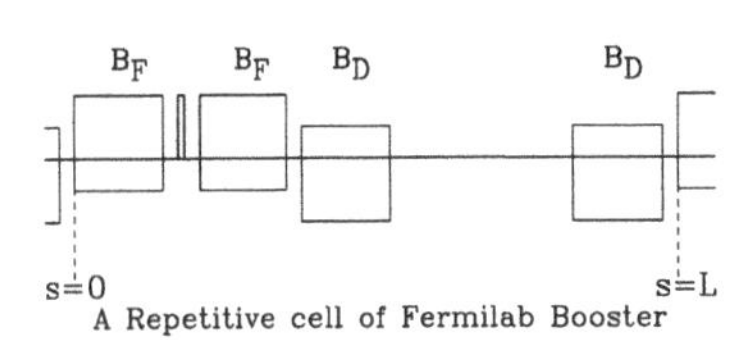

Figure 2.3: A schematic drawing of the Fermilab booster lattice, made of 24 FODO cells with cell-length 19.7588 m. Each period consists of four combined-function magnets of length 2.8896 m and focusing function $K_{\rm F} = 0.02448$ m^{-2} and $K_{\rm D} = -0.02082$ m^{-2}. A small trim focusing quadrupole is used to change the betatron tune. The nominal betatron tunes are $\nu_x = 6.7$ and $\nu_z = 6.8$.

II.1 Transfer Matrix and Stability of Betatron Motion

Because accelerator components usually have uniform or nearly uniform magnetic fields, the focusing functions $K_{x,z}(s)$ are essentially piecewise constant. Let y, y' represent either horizontal or vertical phase-space coordinates, then Eq. (2.24) becomes

$$y'' + K_y(s)y = 0, \tag{2.25}$$

with the periodic condition $K_y(s+L) = K_y(s)$. The solution (y, y') is continuous for a finite K_y. We neglect the subscript y hereafter for simplify our notation. With a constant K, the solution is

$$y(s) = \begin{cases} a\ \cos(\sqrt{K}s + b), & K > 0, \\ as + b, & K = 0, \\ a\ \cosh(\sqrt{-K}s + b), & K < 0. \end{cases} \tag{2.26}$$

The integration constants a and b are determined by the initial values of y_0 and y'_0. We define the betatron state-vector and obtain the betatron transfer matrix $M(s|s_0)$ as

$$\mathbf{y}(s) = \begin{pmatrix} y(s) \\ y'(s) \end{pmatrix}; \qquad \mathbf{y}(s) = M(s|s_0)\mathbf{y}(s_0). \tag{2.27}$$

Here we use (x, x') as the betatron state vector, but bear in mind that the conjugate phase space coordinates are (x, p_x) of the Hamiltonian (2.14), and the phase space evolution should be described by Hamilton's equations (2.21). The transfer matrix for a constant focusing function K is

$$M(s|s_0) = \begin{cases} \begin{pmatrix} \cos\sqrt{K}\ell & \frac{1}{\sqrt{K}}\sin\sqrt{K}\ell \\ -\sqrt{K}\sin\sqrt{K}\ell & \cos\sqrt{K}\ell \end{pmatrix} & K > 0\text{: focusing quad.} \\ \begin{pmatrix} 1 & \ell \\ 0 & 1 \end{pmatrix} & K = 0\text{: drift space} \\ \begin{pmatrix} \cosh\sqrt{|K|}\ell & \frac{1}{\sqrt{|K|}}\sinh\sqrt{|K|}\ell \\ \sqrt{|K|}\sinh\sqrt{|K|}\ell & \cosh\sqrt{|K|}\ell \end{pmatrix} & K < 0\text{: defocusing quad.} \end{cases}$$

where $\ell = s - s_0$. In thin-lens approximation with $\ell \to 0$, the transfer matrix for a quadrupole reduces to

$$M_{\text{focusing}} = \begin{pmatrix} 1 & 0 \\ -1/f & 1 \end{pmatrix}, \quad M_{\text{defocusing}} = \begin{pmatrix} 1 & 0 \\ 1/f & 1 \end{pmatrix}, \tag{2.28}$$

where f is the focal length given by[6] $f = \lim_{\ell\to 0} \frac{1}{|K|\ell}$. Similarly, the transfer matrix for a pure sector dipole with $K_x = 1/\rho^2$ is

$$M_x(s, s_0) = \begin{pmatrix} \cos\theta & \rho\sin\theta \\ -\frac{1}{\rho}\sin\theta & \cos\theta \end{pmatrix} \xrightarrow{(\theta\ll 1)} \begin{pmatrix} 1 & \ell \\ 0 & 1 \end{pmatrix}, \tag{2.29}$$

where $\theta = \ell/\rho$ is the orbiting angle and ρ is the bending radius, and ℓ is the length of the dipole. In small-angle approximation, the transfer matrix of a dipole is equivalent to that of a drift space.

The transfer matrix for any intervals made up of subintervals is just the product of the transfer matrices of these subintervals, e.g. $M(s_2|s_0) = M(s_2|s_1)M(s_1|s_0)$. Using these matrices, particle motion can be tracked through accelerator elements. Combining all segments, the solution of a second-order differential equation can be expressed as

$$y(s) = C(s, s_0)y_0 + S(s, s_0)y_0', \quad y'(s) = C'(s, s_0)y_0 + S'(s, s_0)y_0',$$

where C' and S' are the derivatives of C and S with respect to s, and y_0 and y_0' are the initial phase-space coordinates at s_0. The solutions $C(s, s_0)$ and $S(s, s_0)$ are respectively called the cosine-like and sine-like solutions with boundary conditions

$$C(s_0, s_0) = 1, \; S(s_0, s_0) = 0, \quad C'(s_0, s_0) = 0, \; S'(s_0, s_0) = 1.$$

[6]The convention for the transfer matrix of a thin-lens quadrupole is $M_{\text{quad}} = \begin{pmatrix} 1 & 0 \\ -1/f & 1 \end{pmatrix}$, where $f > 0$ for a focusing quadrupole and $f < 0$ for a defocusing quadrupole. In this case, $f = \lim_{\ell\to 0} 1/(K\ell)$.

The solution of Eq. (2.24) can be expressed in terms of the transfer matrix as[7]

$$\mathbf{y}(s) = M(s|s_0)\mathbf{y}(s_0), \quad M(s|s_0) = \begin{pmatrix} C(s,s_0) & S(s,s_0) \\ C'(s,s_0) & S'(s,s_0) \end{pmatrix},$$

where (y_0, y_0') and (y, y') are the particle phase-space coordinates at the entrance and exit of accelerator elements. For any two linearly independent solutions y_1, y_2 of Hill's equation, the Wronskian is independent of time, i.e.

$$W(y_1, y_2, s) \equiv y_1 y_2' - y_1' y_2, \quad \frac{dW}{ds} = 0. \tag{2.31}$$

The Wronskian obeys $W(s) = [\det M]W(s_0)$. Thus $\det M = 1$, or $\det M(s|s_0) = W(C, S, s) = 1$.

An accelerator is usually constructed with repetitive modules. Let L be the length of a module with $K(s + L) = K(s)$. The number of identical modules that form a complete accelerator is called the *superperiod* P. For example, $P = 24$ for the Fermilab booster shown in Fig. 2.3, $P = 12$ for the AGS at BNL, and $P = 8$ for the bare lattice of LEP at CERN. The transfer matrix M of one repetitive period composed of n elements is a periodic function of s with a period L, i.e.

$$\mathbf{M}(s) \equiv M(s + L|s) = M_n \cdots M_2 M_1,$$

where the M_i's are the transfer matrices of the constituent elements. Using the periodicity condition, we find

$$\begin{aligned} M(s_2 + L|s_1) &= \mathbf{M}(s_2)M(s_2|s_1) = M(s_2|s_1)\mathbf{M}(s_1), \\ \mathbf{M}(s_2) &= M(s_2|s_1)\mathbf{M}(s_1)[M(s_2|s_1)]^{-1}. \end{aligned} \tag{2.32}$$

Since $\mathbf{M}(s_2)$ and $\mathbf{M}(s_1)$ are related by a similarity transformation, the periodic transfer matrix has identical eigenvalues. The transfer matrix for passing through P superperiods is $M(s + PL|s) = [\mathbf{M}(s)]^P$, and for passing through m revolutions becomes $[\mathbf{M}(s)]^{mP}$.

The necessary and sufficient condition for stable orbital motion is that all matrix elements of the matrix $[\mathbf{M}(s)]^{mP}$ remain bounded as m increases. Let λ_1, λ_2 be the eigenvalues and v_1, v_2 be the corresponding eigenvectors of the matrix $\mathbf{M}$. Since $\mathbf{M}$ has

[7]The transfer matrix for the uncoupled betatron motion can be expressed as

$$\begin{pmatrix} x \\ x' \\ z \\ z' \end{pmatrix}_2 = \begin{pmatrix} M_x(s_2|s_1) & 0 \\ 0 & M_z(s_2|s_1) \end{pmatrix} \begin{pmatrix} x \\ x' \\ z \\ z' \end{pmatrix}_1, \tag{2.30}$$

where the M's are the 2×2 transfer matrices.

a unit determinant, the eigenvalues are the reciprocals of each other, i.e. $\lambda_1 = 1/\lambda_2$, and $\lambda_1 + \lambda_2 = \text{Trace}(\mathbf{M})$. The eigenvalue satisfies the equation

$$\lambda^2 - \text{Trace}(\mathbf{M})\lambda + 1 = 0.$$

Let $\text{Trace}(\mathbf{M}) = 2\cos(\Phi)$. We find that Φ is real if $\text{Trace}(\mathbf{M}) \le 2$, and Φ is complex if $\text{Trace}(\mathbf{M}) > 2$. The eigenvalues are $\lambda_1 = e^{j\Phi}$ and $\lambda_2 = e^{-j\Phi}$, where Φ is the betatron phase advance of a periodic cell.

Expressing the initial condition of beam coordinates (y_0, y_0') as a linear superposition of the eigenvectors, i.e. $\begin{pmatrix} y_0 \\ y_0' \end{pmatrix} = av_1 + bv_2$, where v_1 and v_2 are the eigenvectors associated with eigenvalues λ_1 and λ_2 respectively, we find that the particle coordinate after the mth revolution becomes

$$\begin{pmatrix} y_m \\ y_m' \end{pmatrix} = \mathbf{M}^m \begin{pmatrix} y_0 \\ y_0' \end{pmatrix} = a\lambda_1^m v_1 + b\lambda_2^m v_2.$$

The stability of particle motion requires that λ_1^m and λ_2^m not grow with m. Thus a necessary condition for orbit stability is to have a real betatron phase advance Φ, or

$$|\text{Trace}(\mathbf{M})| \le 2. \tag{2.33}$$

II.2 Courant–Snyder Parametrization

The most general form for matrix $\mathbf{M}$ with *unit modulus* can be parametrized as

$$\mathbf{M} = \begin{pmatrix} \cos\Phi + \alpha\sin\Phi & \beta\sin\Phi \\ -\gamma\sin\Phi & \cos\Phi - \alpha\sin\Phi \end{pmatrix} = \mathbf{I}\cos\Phi + \mathbf{J}\sin\Phi, \tag{2.34}$$

where α, β and γ are Courant–Snyder parameters,[8] Φ is the phase advance, $\mathbf{I}$ is the unit matrix, and

$$\mathbf{J} = \begin{pmatrix} \alpha & \beta \\ -\gamma & -\alpha \end{pmatrix}, \quad \text{with} \quad \text{Trace}(\mathbf{J}) = 0, \quad \mathbf{J}^2 = -\mathbf{I} \text{ or } \beta\gamma = 1 + \alpha^2. \tag{2.35}$$

Similarity transformation of the matrix $\mathbf{M}$ can also be parametrized as that of Eq. (2.34). The ambiguity in the sign of $\sin\Phi$ can be resolved by requiring β to be a positive definite number if $|\text{Trace}(\mathbf{M})| \le 2$, and by requiring $\text{Im}(\sin\Phi)>0$ if $|\text{Trace}(\mathbf{M})| > 2$. The definition of the phase factor Φ is still ambiguous up to an integral multiple of 2π. This ambiguity will be resolved when the matrix is tracked along the accelerator elements. Using the property of matrix $\mathbf{J}$, we obtain the De Moivere's theorem:

$$\mathbf{M}^k = (\mathbf{I}\cos\Phi + \mathbf{J}\sin\Phi)^k = \mathbf{I}\cos k\Phi + \mathbf{J}\sin k\Phi,$$
$$\mathbf{M}^{-1} = \mathbf{I}\cos\Phi - \mathbf{J}\sin\Phi.$$

[8]The α, β, and γ parameters have nothing to do with the relativistic Lorentz factor.

With the similarity transformation of Eq. (2.32), the values of the Courant–Snyder parameters $\alpha_2, \beta_2, \gamma_2$ at s_2 are related to $\alpha_1, \beta_1, \gamma_1$ at s_1 by (see Exercise 2.2.8)

$$\begin{pmatrix} \beta \\ \alpha \\ \gamma \end{pmatrix}_2 = \begin{pmatrix} M_{11}^2 & -2M_{11}M_{12} & M_{12}^2 \\ -M_{11}M_{21} & M_{11}M_{22} + M_{12}M_{21} & -M_{12}M_{22} \\ M_{21}^2 & -2M_{21}M_{22} & M_{22}^2 \end{pmatrix} \begin{pmatrix} \beta \\ \alpha \\ \gamma \end{pmatrix}_1, \tag{2.36}$$

where M_{ij} are the matrix elements of $M(s_2|s_1)$.

1. The evolution of the betatron amplitude function in a drift space is

$$\beta_2 = \frac{1}{\gamma_1} + \gamma_1 \left(s - \frac{\alpha_1}{\gamma_1}\right)^2 = \beta^* + \frac{(s - s^*)^2}{\beta^*},$$
$$\alpha_2 = \alpha_1 - \gamma_1 s = -\frac{(s - s^*)}{\beta^*}, \qquad \gamma_2 = \gamma_1 = \frac{1}{\beta^*}.$$

Note that γ is constant in a drift space, and $s^* = \alpha_1/\gamma_1$ is the location for an extremum of the betatron amplitude function with $\alpha(s^*) = 0$.

2. Passing through a thin-lens quadrupole, the evolution of betatron function is given by

$$\beta_2 = \beta_1, \qquad \alpha_2 = \alpha_1 + \frac{\beta_1}{f}, \qquad \gamma_2 = \gamma_1 + \frac{2\alpha_1}{f} + \frac{\beta_1}{f^2},$$

where f is the focal length of the quadrupole. Thus a thin-lens quadrupole gives rise to an angular kick to the betatron amplitude function without changing its magnitude.

II.3 Floquet Transformation

Since the focusing function $K(s)$ is a periodic function, Eq. (2.25) can be solved by using the Floquet transformation:

$$y(s) = aw(s)e^{j\psi(s)}, \quad y^*(s) = aw(s)e^{-j\psi(s)}, \tag{2.37}$$

where a is a constant, and w and ψ are the amplitude and phase functions. Since $K(s)$ is real, the amplitude and phase functions satisfy

$$w'' + Kw - \frac{1}{w^3} = 0, \qquad \psi' = \frac{1}{w^2}. \tag{2.38}$$

They are the *betatron envelope* and phase equations. The integration constant in the phase equation is chosen to be 0 so that the w^2 is exactly the Courant-Snyder β-function in Eq. (2.40).

Any solution of Eq. (2.25) is a linear superposition of the linearly independent solutions y and y^*. The mapping matrix $M(s_2|s_1)$ is

$$M(s_2|s_1)=\begin{pmatrix} \frac{w_2}{w_1}\cos\psi - w_2 w_1' \sin\psi & w_1 w_2 \sin\psi \\ -\frac{(1+w_1w_1'w_2w_2')}{w_1w_2}\sin\psi - (\frac{w_1'}{w_2}-\frac{w_2'}{w_1})\cos\psi & \frac{w_1}{w_2}\cos\psi + w_1 w_2'\sin\psi \end{pmatrix}, \tag{2.39}$$

where $w_1 = w(s_1), w_2 = w(s_2), \psi = \psi(s_2) - \psi(s_1)$, $w_1' = w'(s_1), w_2' = w'(s_2)$, and the prime is the derivative with respect to s.

Let $s_2 - s_1 = L$ be the length of a periodic beam line, i.e. the focusing function $K(s)$ satisfies $K(s) = K(s+L)$. Using the Floquet theorem (see Appendix A, Sec I.5) with the periodic boundary conditions to the amplitude and phase functions, and equating the matrix $\mathbf{M}$ of a complete period in Eq. (2.39) to Eq. (2.34), we obtain

$$\begin{aligned} &w_1 = w_2 = w, \quad w_1' = w_2' = w', \quad \psi(s_1+L) - \psi(s_1) = \Phi. \\ &w^2 = \beta, \qquad \alpha = -ww' = -\beta'/2. \end{aligned} \tag{2.40}$$

With the integration constant of Eq. (2.38), the amplitude of the betatron motion is exactly equal to the square root of the Courant–Snyder parameter $\beta(s)$, which will be referred to as the *betatron amplitude function.* The Courant–Snyder parameter α is related to the slope of the betatron amplitude function. The betatron phase advance of one period is $\Phi = \int_0^L \frac{ds}{\beta(s)}$. In the smooth approximation, we have $\Phi = L/\langle\beta\rangle$, or $\langle\beta\rangle = L/\Phi$. The betatron wavelength is $\lambda_\beta = 2\pi\langle\beta\rangle$. Substituting $\beta = w^2$ back into Eq. (2.38), we obtain

$$\frac{1}{2}\beta'' + K\beta - \frac{1}{\beta}\left[1 + (\frac{\beta'}{2})^2\right] = 0, \quad \text{or} \quad \alpha' = K\beta - \frac{1}{\beta}(1+\alpha^2). \tag{2.41}$$

The transfer matrix of Eq. (2.39) from s_1 to s_2 in any beam transport line becomes

$$\begin{aligned} M(s_2|s_1) &= \begin{pmatrix} \sqrt{\frac{\beta_2}{\beta_1}}(\cos\psi + \alpha_1\sin\psi) & \sqrt{\beta_1\beta_2}\sin\psi \\ -\frac{1+\alpha_1\alpha_2}{\sqrt{\beta_1\beta_2}}\sin\psi + \frac{\alpha_1-\alpha_2}{\sqrt{\beta_1\beta_2}}\cos\psi & \sqrt{\frac{\beta_1}{\beta_2}}(\cos\psi - \alpha_2\sin\psi) \end{pmatrix} \\ &= \begin{pmatrix} \sqrt{\beta_2} & 0 \\ -\frac{\alpha_2}{\sqrt{\beta_2}} & \frac{1}{\sqrt{\beta_2}} \end{pmatrix} \begin{pmatrix} \cos\psi & \sin\psi \\ -\sin\psi & \cos\psi \end{pmatrix} \begin{pmatrix} \frac{1}{\sqrt{\beta_1}} & 0 \\ \frac{\alpha_1}{\sqrt{\beta_1}} & \sqrt{\beta_1} \end{pmatrix} \\ &\equiv \mathbf{B}(s_2)\begin{pmatrix} \cos\psi & \sin\psi \\ -\sin\psi & \cos\psi \end{pmatrix}\mathbf{B}^{-1}(s_1), \end{aligned} \tag{2.42}$$

where $\beta_1, \alpha_1, \gamma_1$, and $\beta_2, \alpha_2, \gamma_2$ are values of betatron amplitude functions at s_1 and s_2 respectively, $\psi = \psi(s_2) - \psi(s_1)$, and we have defined the betatron amplitude matrix $\mathbf{B}(s)$ and its inverse as

$$\mathbf{B}(s) = \begin{pmatrix} \sqrt{\beta(s)} & 0 \\ -\frac{\alpha(s)}{\sqrt{\beta(s)}} & \frac{1}{\sqrt{\beta(s)}} \end{pmatrix} \quad \text{and} \quad \mathbf{B}^{-1}(s) = \begin{pmatrix} \frac{1}{\sqrt{\beta(s)}} & 0 \\ \frac{\alpha(s)}{\sqrt{\beta(s)}} & \sqrt{\beta(s)} \end{pmatrix}. \tag{2.43}$$

We note, from Eq. (2.42), that the linear betatron motion becomes coordinate rotation after the normalization of the phase-space coordinates with the $\mathbf{B}^{-1}$ matrix. Applying Floquet theorem to a repetitive period, where $s_2 = s_1 + L$ with $K(s_2) = K(s_1)$, we obtain $\beta_1 = \beta_2$, $\alpha_1 = \alpha_2$, and the transfer matrix of Eq. (2.42) reduces to the Courant–Snyder parametrization of Eq. (2.34).

A. Betatron tune (number of betatron oscillations in one revolution):

We consider an accelerator of circumference $C = PL$ with P identical superperiods. The phase change per revolution is $P\Phi$. The betatron tune ν_y, or Q_y, defined as the *number of betatron oscillations in one revolution*, is

$$Q_y = \nu_y = \frac{P\Phi_y}{2\pi} = \frac{1}{2\pi}\int_s^{s+C} \frac{ds}{\beta_y(s)}. \tag{2.44}$$

The betatron oscillation frequency is $\nu_y f_0$, where f_0 is the revolution frequency. The general solution of Eq. (2.25) becomes

$$y(s) = a\sqrt{\beta_y(s)}\cos\left[\psi_y(s) + \xi_y\right] \quad \text{with} \quad \psi_y(s) = \int_0^s \frac{ds}{\beta_y(s)}, \tag{2.45}$$

where a, ξ_y are constants to be determined from initial conditions. This is a pseudo-harmonic oscillation with varying amplitude $\beta_y^{1/2}(s)$. The local betatron wavelength is $\lambda = 2\pi\beta_y(s)$.

Introducing the coordinate $\eta(\phi_y)$, and "time" coordinate to ϕ_y, Hill's equation becomes

$$\eta = \frac{y}{\sqrt{\beta_y}}, \quad \phi_y(s) = \frac{1}{\nu_y}\int^s \frac{ds}{\beta_y(s)}, \qquad \frac{d^2\eta}{d\phi_y^2} + \nu_y^2\eta = 0. \tag{2.46}$$

The phase function ("time variable") ϕ_y increases by 2π in one revolution. The linear betatron motion is simple harmonic.

Example 1: FODO cell in thin-lens approximation

A FODO cell (Fig. 2.4) is made of a pair of focusing and defocusing quadrupoles with or without dipoles in between:

$$\left\{\frac{1}{2}\text{QF} \quad \text{O} \quad \text{QD} \quad \text{O} \quad \frac{1}{2}\text{QF}\right\}$$

where O stands for either a dipole or a drift space: FODO cells are often used in beam transport in arcs and transport lines.[9]

[9]The accelerator lattice is usually divided into arcs and insertions. Arcs are curved sections that transport beams for a complete revolution. Insertions (or straight sections) are usually used for physics experiments, rf cavities, injection and extraction systems, etc.

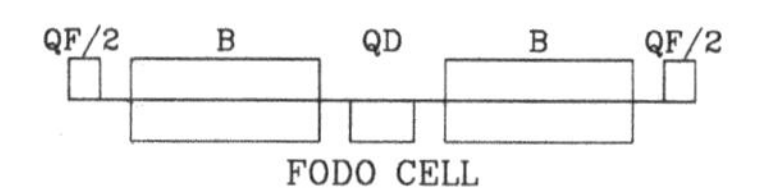

Figure 2.4: A schematic plot of a FODO cell, where the transfer matrices for dipoles (B) can be approximated by drift spaces, and QF and QD are the focusing and defocusing quadrupoles.

The transfer matrix in the thin-lens approximation, is[10]

$$\begin{aligned} \mathbf{M} &= \begin{pmatrix} 1 & 0 \\ -\frac{1}{2f} & 1 \end{pmatrix} \begin{pmatrix} 1 & L_1 \\ 0 & 1 \end{pmatrix} \begin{pmatrix} 1 & 0 \\ \frac{1}{f} & 1 \end{pmatrix} \begin{pmatrix} 1 & L_1 \\ 0 & 1 \end{pmatrix} \begin{pmatrix} 1 & 0 \\ -\frac{1}{2f} & 1 \end{pmatrix} \\ &= \begin{pmatrix} 1 - \frac{L_1^2}{2f^2} & 2L_1(1 + \frac{L_1}{2f}) \\ -\frac{L_1}{2f^2}(1 - \frac{L_1}{2f}) & 1 - \frac{L_1^2}{2f^2} \end{pmatrix} \end{aligned}$$

where the focusing and defocusing quadrupoles have focal lengths $\pm f$ respectively and L_1 is the drift length between quadrupoles. Because of the repetitive nature of FODO cells, the transfer matrix can be identified with the Courant–Snyder parametrization of Eq. (2.34):

$$\begin{aligned} \cos\Phi &= \frac{1}{2}\mathrm{Trace}(\mathbf{M}) = 1 - \frac{L_1^2}{2f^2} \qquad \text{or} \qquad \sin\frac{\Phi}{2} = \frac{L_1}{2f}, \\ \beta_{\mathrm{F}} &= \frac{2L_1\,(1 + \sin(\Phi/2))}{\sin\Phi}, \qquad \alpha_{\mathrm{F}} = 0. \end{aligned}$$

The parameter Φ is the phase advance per cell, and β_{F} and α_{F} are values of the betatron amplitude functions at the center of the focusing quadrupole. The betatron tune for a machine with N FODO cells is $\nu = N\Phi/2\pi$. The above procedure can be performed at any position of the FODO cell, and the corresponding Courant–Snyder parameters are values of the betatron amplitude functions at that position. For example, we find

$$\begin{aligned} \beta_{\mathrm{D}} &= \frac{2L_1\,(1 - \sin(\Phi/2))}{\sin\Phi}, \qquad \alpha_{\mathrm{D}} = 0, \\ \beta_{\text{mid. point}} &= \frac{L_1}{\sin\Phi}\left(2 - \sin^2\frac{\Phi}{2}\right), \qquad \alpha_{\text{mid. point}} = \pm\frac{1}{\cos(\Phi/2)} \end{aligned}$$

at the center of the defocusing quadrupole, and at the midpoint between the QF and the QD respectively. We can also use the transfer matrix of Eq. (2.42) to find the betatron amplitude functions at other locations (see Exercise 2.2.8).

The solid and dashed lines in the upper plot of Fig. 2.5 show the betatron amplitude functions $\beta_x(s)$ and $\beta_z(s)$ for the AGS. The middle plot shows the dispersion

[10]The transfer matrices of dipoles are represented by those of drift spaces, where we neglect the effect of $1/\rho^2$ focusing and edge focusing. The transfer matrix for vertical motion can be obtained by reversing focusing and defocusing elements.

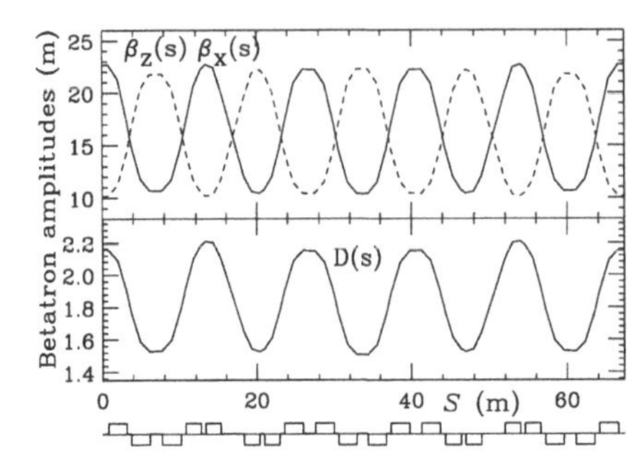

Figure 2.5: The betatron amplitude functions for one superperiod of the AGS lattice, which made of 20 combined-function magnets. The upper plot shows β_x (solid line) and β_z (dashed line). The middle plot shows the dispersion function D_x. The lower plot shows schematically the placement of combined-function magnets. Note that the superperiod can be well approximated by five regular FODO cells. The phase advance of each FODO cell is about 52.8°.

function $D(s)$, to be discussed in Sec. IV. The AGS lattice has 12 superperiods, each composed of 20 combined-function dipoles, shown schematically in the bottom plot of Fig. 2.5. The AGS lattice can be well approximated by 60 FODO cells with a phase advance of 52.8° for a betatron tune of 8.8, and a half-cell length of $L_1 = 6.726$ m for a complete circumference of 807.12 m.

Example 2: Doublet cells

The values of the horizontal and vertical betatron functions in FODO cells alternate in magnitude, i.e.

$$\frac{\beta_x}{\beta_z} \approx \frac{1+\sin\Phi/2}{1-\sin\Phi/2}, \quad \text{and} \quad \frac{1-\sin\Phi/2}{1+\sin\Phi/2},$$

at the focusing and defocusing quadrupoles respectively. The beam size variation increases with the phase advance of the FODO cell. In some applications, a paraxial beam transport system provides a simpler geometrical beam matching solution. Some examples of paraxial beam transport beam lines are the doublet, the triplet, and the solenoidal transport systems. In the following example, we consider a doublet beam line, shown schematically in Fig. 2.6.

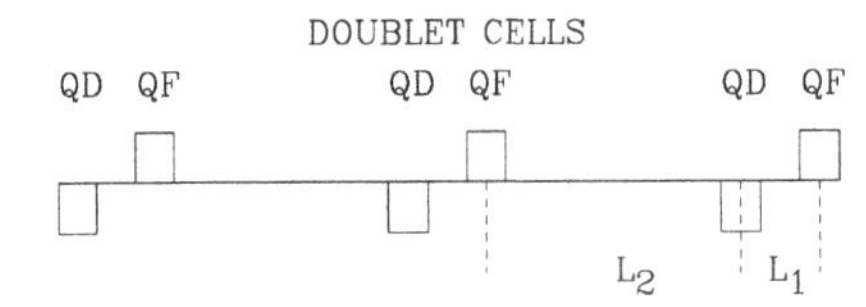

Figure 2.6: A schematic plot of a doublet transport line, where two quadrupoles are separated by a distance L_1, and the long drift space L_2 between two quadrupoles can be filled with dipoles.

The phase advance of a doublet cell, in thin-lens approximation, and the maximum and minimum values of the betatron amplitude function are (see Exercise 2.2.13)

$$\sin\frac{\Phi}{2} = \frac{\sqrt{L_1 L_2}}{2f}, \qquad \beta_{\max} = \frac{L_1 + L_2 + L_1 L_2/f}{\sin\Phi}, \qquad \beta_{\min} = \frac{f^2}{L_2}\sin\Phi.$$

where we have assumed equal focusing strength for the focusing and the defocusing quadrupoles, f is the focal length of the quadrupoles, and L_1 and L_2 are the lengths of the drift spaces shown in Fig. 2.6. If $L_1 \ll L_2$, the horizontal and vertical betatron amplitude functions are nearly identical along the transport line. Thus the doublet can be considered as an example of the paraxial transport system. Other paraxial transport systems are triplets and solenoidal focusing channels (see Exercise 2.2.12).

II.4 Action-Angle Variable and Floquet Transformation

The Hill equation, $y'' + K(s)y = 0$, can be derived from a pseudo-Hamiltonian

$$H = \frac{1}{2}y'^2 + \frac{1}{2}K(s)y^2, \tag{2.47}$$

where (y, y') are conjugate phase-space coordinates. We want to transform (y, y') to the action-angle coordinates, where ψ in Eq. (2.45) serves as the angle-coordinate. There are two possible generating functions: either $F_1(y, \psi)$ or $F_3(y', \psi)$ (see Exercise 2.21). Using Eq. (2.45) we find F_1, and obtain the action-coordinate as:

$$\begin{aligned} y' &= -\frac{y}{\beta}(\tan\psi - \frac{\beta'}{2}), \iff F_1(y, \psi) = \int_0^y y'dy = -\frac{y^2}{2\beta}(\tan\psi - \frac{\beta'}{2}), \\ J &= -\frac{\partial F_1}{\partial\psi} = \frac{y^2}{2\beta}\sec^2\psi = \frac{1}{2\beta}[y^2 + (\beta y' + \alpha y)^2]. \end{aligned} \tag{2.48}$$

where (ψ, J) are the angle (betatron phase) and action coordinates, and $y' = \partial F_1/\partial y$ is verified easily from the generating function $F_1(y, \psi)$. With the canonical transformation, the new Hamiltonian becomes

$$\tilde{H} = H + \frac{\partial F_1}{\partial s} = \frac{J}{\beta}. \tag{2.49}$$

Hamilton's equation gives $\psi' = \partial\tilde{H}/\partial J = 1/\beta(s)$, which recovers Eq. (2.38). Since the new Hamiltonian is independent of the phase coordinate ψ, the action J is invariant, i.e. $\frac{dJ}{ds} = -\frac{\partial\tilde{H}}{\partial\psi} = 0$. Using Eq. (2.48), we obtain

$$y = \sqrt{2\beta J}\,\cos\psi, \qquad y' = -\sqrt{\frac{2J}{\beta}}\,[\sin\psi + \alpha\cos\psi], \tag{2.50}$$

where $\alpha = -\beta'/2$. The action J is the phase space area enclosed by the invariant torus:

$$J = \frac{1}{2\pi}\int_{\text{torus}} dy'dy = \frac{1}{2\pi}\oint y'dy. \tag{2.51}$$

The Jacobian of the transformation from (y, y') to (J, ψ) is equal to 1. A word of convention: The area of the phase space ellipse is $2\pi J$, where we usually use (πJ) as action in unit of [π-mm-mrad] or [$\pi\mu$m], or [πnm], or [πpm]. Thus sometimes

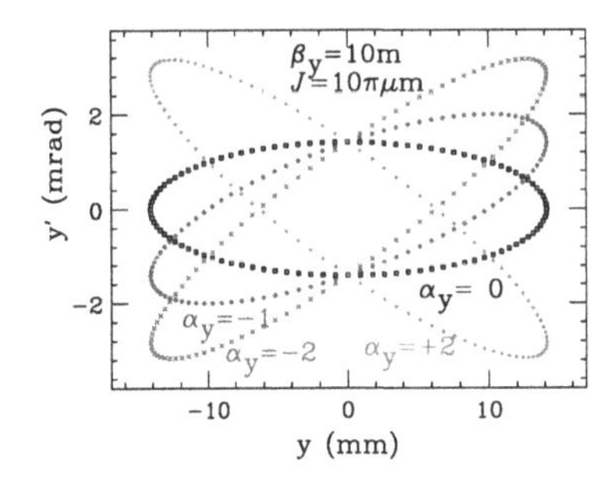

Figure 2.7: The betatron phase space ellipses of a particle with actions $J = 10\pi$ mm mrad. The betatron parameters are $\beta_y = 10$m, and α_y shown by each curve. The scale for the ordinate y is mm, and y' in mrad. The betatron parameters for each ellipse are marked on the graph. All ellipses has the maximum y coordinate at $\sqrt{2\beta_y J}$. The maximum angular coordinate y' is $\sqrt{2(1+\alpha_y^2)J/\beta_y}$. All ellipses have the same phase space area of $2\pi J$.

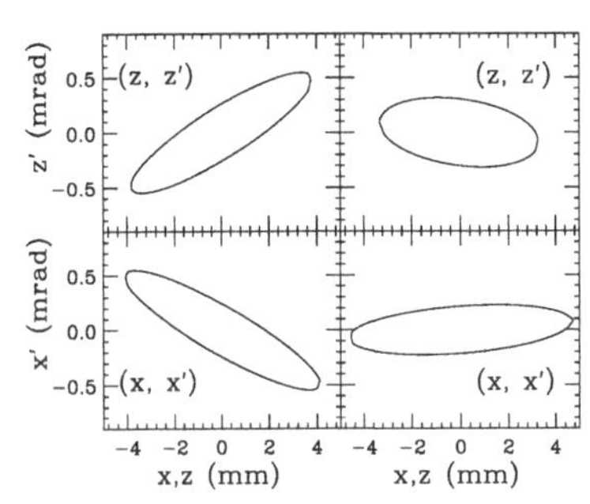

Figure 2.8: The horizontal and vertical betatron ellipses for a particle with actions $J_x = J_z = 0.5\pi$ mm-mrad at the end of the first dipole (left plots) and the end of the fourth dipole of the AGS lattice (see Fig. 2.5). The scale for the ordinate x or z is in mm, and that for the coordinate x' or z' is in mrad. Left plots: $\beta_x = 17.0$ m, $\alpha_x = 2.02$, $\beta_z = 14.7$ m, and $\alpha_z = -1.84$. Right plots: $\beta_x = 21.7$ m, $\alpha_x = -0.33$, $\beta_z = 10.9$ m, and $\alpha_z = 0.29$.

one writes Eq. (2.50) as $y = \sqrt{\frac{2\beta J}{\pi}}\cos\psi$, and similarly for y'. The factor π cancels the unit of π in action. Figure 2.7 shows the phase ellipses with identical action of $J = 10\pi$ mm-mrad.

Figure 2.8 shows the phase-space ellipses (x, x') and (z, z') for a particle with actions $J_x = J_z = 0.5\pi$ mm-mrad at the ends of the first and the fourth dipoles of the AGS lattice (see Fig. 2.5). Such a phase-space ellipse is also called the Poincaré map, where the particle phase-space coordinates are plotted in each revolution. The consecutive phase-space points can be obtained by multiplying the transfer matrices, i.e.

$$\begin{pmatrix} x \\ x' \end{pmatrix}_{n+1} = \mathbf{M}_x \begin{pmatrix} x \\ x' \end{pmatrix}_n, \qquad \begin{pmatrix} z \\ z' \end{pmatrix}_{n+1} = \mathbf{M}_z \begin{pmatrix} z \\ z' \end{pmatrix}_n,$$

where $\mathbf{M}_x$ and $\mathbf{M}_z$ are the transfer matrices of one complete revolution. The Poincaré map of betatron motion at a fixed azimuth s is also called the Poincaré *surface of section*. If the betatron tune is not a rational number, the consecutive phase-space points of the particle trajectory will trace out the entire ellipse. The areas enclosed by the horizontal and vertical ellipses are equal to $2\pi J_x$ and $2\pi J_z$ respectively. As the particle travels in an accelerator, the shape of the phase-space ellipse may vary but the area enclosed by the ellipse is invariant.

A. Normalized phase space coordinates

We define the normalized conjugate phase-space coordinate $\mathcal{P}_y$ as

$$\mathcal{P}_y = \beta y' + \alpha y = -\sqrt{2\beta J}\sin\psi. \tag{2.52}$$

A particle trajectory in the normalized phase-space coordinates $(y, \mathcal{P}_y)$ moves clockwise on a circle with radius $\sqrt{2\beta J}$ as phase advance ψ increases. In terms of the betatron amplitude matrix of Eq. (2.43), the normalized phase space coordinates are expressed as

$$\begin{pmatrix} y \\ \mathcal{P}_y \end{pmatrix} = \sqrt{\beta}\,\mathbf{B}^{-1}\begin{pmatrix} y \\ y' \end{pmatrix} \quad \text{and} \quad \begin{pmatrix} y \\ y' \end{pmatrix} = \frac{1}{\sqrt{\beta}}\mathbf{B}\begin{pmatrix} y \\ \mathcal{P}_y \end{pmatrix}. \tag{2.53}$$

B. Using the orbital angle θ as the independent variable

The Hamiltonian $\tilde{H}$ of Eq. (2.49) depends on the independent variable s. Because $\beta(s)$ is not a constant, the phase advance is modulated along the accelerator orbital trajectory. Sometimes it is useful to use the orbiting angle as "time" coordinate in order to obtain a global Fourier expansion of particle motion. We use the generating function for coordinate transformation:

$$F_2(\psi, \bar{J}) = \left(\psi - \int_0^s \frac{ds}{\beta} + \nu\theta\right)\bar{J} \quad \Longrightarrow \quad \bar{\psi} = \psi - \int_0^s \frac{ds}{\beta} + \nu\theta, \quad \bar{J} = J,$$

Here $\theta = s/R$ is the orbiting angle of the reference orbit and R is the mean radius of an accelerator. The transformation compensates the modulated phase-advance function with conjugate coordinates $(\bar{\psi}, \bar{J})$ and new Hamiltonian $\tilde{H} = \nu\bar{J}/R$. Scaling the "time coordinate" from s to θ, the re-scaled new Hamiltonian and the corresponding coordinate-transformation are

$$\bar{H} = R\tilde{H} = \nu\bar{J}. \tag{2.54}$$

$$y = \sqrt{2\beta\bar{J}}\cos(\bar{\psi} + \chi(s) - \nu\theta), \tag{2.55}$$

$$\mathcal{P}_y = \beta y' + \alpha y = -\sqrt{2\beta\bar{J}}\sin(\bar{\psi} + \chi(s) - \nu\theta),$$

where $\chi(s) = \int_0^s ds/\beta$. The transformation is useful in expressing a general betatron Hamiltonian in action-angle coordinates for obtaining a global Fourier expansion in the nonlinear resonance analysis. Hereafter, the notation $(\bar{\psi}, \bar{J})$ is simplified to (ψ, J).

II.5 Courant–Snyder Invariant and Emittance

Using the general solution $y(s)$ of Eq. (2.45), we obtain $\beta y' + \alpha y = -a\beta^{1/2}(s)\sin(\nu\phi(s) + \delta)$. The Courant–Snyder invariant defined by

$$C(y, y') = \frac{1}{\beta}\left[y^2 + (\alpha y + \beta y')^2\right] = \gamma y^2 + 2\alpha y y' + \beta y'^2 \tag{2.56}$$

is equal to *twice the action.* The trajectory of particle motion with initial condition (y_0, y_0') follows an ellipse described by $C(y, y') = \epsilon$. The phase space enclosed by (y, y') of Eq. (2.56) is equal to $\pi\epsilon$ (see Fig. 2.9). The quantity "$\pi\epsilon$" is identify as emittance in the unit of [πmm-mrad] or [$\pi\mu$m] or [πnm] or simply [μ-meter] or [μm], [nano-meter] or [nm], [pico-meter] or [pm], etc.. In each unit, the factor π is implied, explicitly stated or not. The maximum betatron amplitude is $\sqrt{\beta\epsilon}$, where π is ignored in the calculation, or sometimes explicitly expressed as $\sqrt{\beta\epsilon/\pi}$ to cancel the π is the unit of emittance.

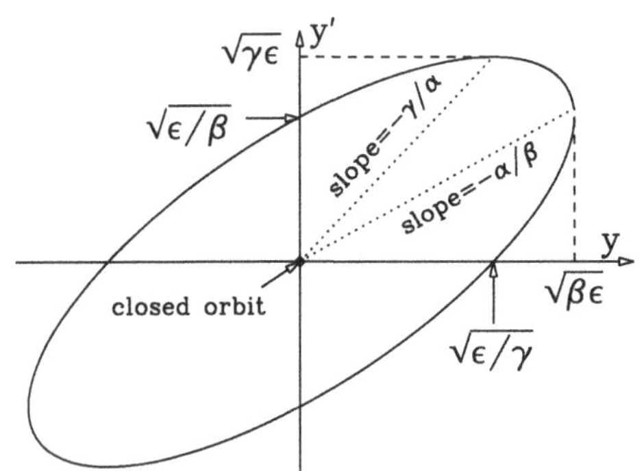

Figure 2.9: The Courant-Snyder invariant ellipse. The area enclosed by the ellipse is equal to $\pi\epsilon$, where ϵ is twice the betatron action; β is the betatron amplitude functions, and $\alpha = -\frac{1}{2}\beta'$, $\gamma = (1 + \alpha^2)/\beta$. The maximum amplitude of betatron motion is $\sqrt{\beta\epsilon}$, and the maximum divergence (angle) of the betatron motion is $\sqrt{\gamma\epsilon}$ (see the footnote in Eq. (2.50) for the convention of the emittance and action. The center of the ellipse is the reference orbit or closed orbit (c.o.)).

A. The emittance of a beam

A beam is composed of particles distributed in phase space. Depending on the initial beam preparation, we approximate a realistic beam distribution function by some simple analytic formula. Neglecting dissipation and diffusion processes, each particle in the distribution function has its invariant Courant–Snyder ellipse.

Given a normalized distribution function $\rho(y, y')$ with $\int \rho(y, y')dydy' = 1$, the moments of the beam distribution are

$$\langle y\rangle = \int y\rho(y, y')dydy', \quad \langle y'\rangle = \int y'\rho(y, y')dydy',$$
$$\sigma_y^2 = \int (y - \langle y\rangle)^2\rho(y, y')dydy', \quad \sigma_{y'}^2 = \int (y' - \langle y'\rangle)^2\rho(y, y')dydy',$$
$$\sigma_{yy'} = \int (y - \langle y\rangle)(y' - \langle y'\rangle)\rho(y, y')dydy' = r\sigma_y\sigma_{y'},$$

where σ_y and $\sigma_{y'}$ are the rms beam widths, $\sigma_{yy'}$ is the correlation, and r is the correlation coefficient. The rms beam emittance is defined as

$$\epsilon_{\rm rms} \equiv \sqrt{\sigma_y^2\sigma_{y'}^2 - \sigma_{yy'}^2} = \sigma_y\sigma_{y'}\sqrt{1 - r^2}. \tag{2.57}$$

The rms emittance of a ring beam in phase space, i.e. particles uniformly distributed in phase coordinate ψ at a fixed action J of Eq. (2.50), is $\epsilon_{\rm rms} = J$. If the accelerator is composed of linear elements such as dipoles and quadrupoles, the emittance defined in Eq. (2.57) is invariant. The rms emittance is equal to the phase-space area enclosed by the Courant–Snyder ellipse of the rms particle (see Exercise 2.2.14).

Although incorrect, the term "emittance" is often loosely used as twice the action variable of betatron oscillations. The betatron oscillations of "a particle" with an "emittance" ϵ is

$$y(s) = \sqrt{\beta\epsilon}\,\cos[\nu\phi(s) + \delta]. \tag{2.58}$$

Figure 2.9 shows a Courant–Snyder invariant ellipse with phase space area $\pi\epsilon$,[11] the rms beam width is $\sqrt{\beta(s)\epsilon}$, and the beam rms divergence y' is $\sqrt{\gamma(s)\epsilon}$. Since $\gamma = (1+\alpha^2)/\beta$, the transverse beam divergence is smaller at a location with a large $\beta(s)$ value, i.e. all particles travel in parallel paths. In accelerator design, a proper $\beta(s)$ value is therefore important for achieving many desirable properties.

B. The σ-matrix

The σ-matrix of a beam distribution is defined as

$$\sigma = \begin{pmatrix} \sigma_{11} & \sigma_{12} \\ \sigma_{12} & \sigma_{22} \end{pmatrix} = \begin{pmatrix} \sigma_y^2 & \sigma_{yy'} \\ \sigma_{yy'} & \sigma_{y'}^2 \end{pmatrix} = \langle (\mathbf{y} - \langle \mathbf{y}\rangle)(\mathbf{y} - \langle \mathbf{y}\rangle)^\dagger \rangle,$$
$$\sigma(s_2) = M(s_2|s_1)\sigma(s_1)M(s_2|s_1)^\dagger. \tag{2.59}$$

where $\mathbf{y}$ is the betatron state-vector of Eq. (2.27), $\mathbf{y}^\dagger = (y, y')$ is the transpose of $\mathbf{y}$, and $\langle \mathbf{y}\rangle$ is the first moment. The rms emittance defined by Eq. (2.57) is the determinant of the σ-matrix, i.e. $\epsilon_{\rm rms} = \sqrt{\det\sigma}$ (see also Exercise 2.2.14). It is easy to verify that $\mathbf{y}^\dagger\sigma^{-1}\mathbf{y}$ is invariant under linear betatron motion. An invariant beam distribution is

$$\rho(y, y') = \rho(\mathbf{y}^\dagger\sigma^{-1}\mathbf{y}). \tag{2.60}$$

C. Emittance measurement

The emittance can be obtained by measuring the σ-matrix. The beam profile of protons and ions is usually measured by using wire scanners or ionization profile monitors. Synchrotron light monitors are commonly used in electron storage rings. More recently, laser light has been used to measure electron beam size in the submicron range. Using the rms beam width and Courant–Snyder parameters, we can

[11] The accelerator scientists commonly use π-mm-mrad as the unit of emittance. However, the factor π is also often omitted. In beam width calculation, we get $\sigma_y = \sqrt{\pi\epsilon_y\beta_y/\pi}$. The synchrotron light source community also uses nano-meter (nm) as the unit for emittance. In fact, the factor π is implied and omitted in the literature.

deduce the emittance of the beam. Two methods commonly used to measure the rms emittance are discussed below.

C1. Quadrupole tuning method

Using Eq. (2.59), we find

$$\sigma_{11}(s_2) = \sigma_{11}(s_1)\left[M_{11} + \frac{\sigma_{12}(s_1)}{\sigma_{11}(s_1)}M_{12}\right]^2 + \frac{\epsilon_{\rm rms}^2}{\sigma_{11}(s_1)}M_{12}^2, \tag{2.61}$$

where $\sigma_{ij}(s_1)$'s are elements of the σ matrix at the entrance of the quadrupole with $\epsilon_{\rm rms}^2 = \sigma_{11}\sigma_{22} - \sigma_{12}^2$, and $\sigma_{11}(s_2)$ is the 11-element of the σ-matrix at the profile monitor location s_2. For a setup of a quadrupole and a drift space, we find $M_{12} = (1/\sqrt{K})\sin(\sqrt{K}\ell_{\rm q}) + L\cos(\sqrt{K}\ell_{\rm q})$ and $M_{11} = \cos(\sqrt{K}\ell_{\rm q}) - \sqrt{K}L\sin(\sqrt{K}\ell_{\rm q})$, where $K = B_1/B\rho$ is the focusing function, $\ell_{\rm q}$ is the length of the quadrupole, and L is the distance between the quadrupole and the beam profile monitor. In thin lens approximation, we find $M_{12} \to (L + \frac{\ell_{\rm q}}{2})$ $M_{11} \to 1 - (L + \frac{\ell_{\rm q}}{2})g$, and

$$\sigma_{11}(s_2) \approx \sigma_{11}(s_1)\left(1 - (L + \frac{\ell_{\rm q}}{2})g + \frac{\sigma_{12}(s_1)}{\sigma_{11}(s_1)}(L + \frac{\ell_{\rm q}}{2})\right)^2 + \frac{\epsilon_{\rm rms}^2}{\sigma_{11}(s_1)}(L + \frac{\ell_{\rm q}}{2})^2,$$

where $g = Kl_{\rm q}$ is the effective quadrupole strength.

The $\sigma_{11}(s_2)$ data by varying quadrupole strength g can be used to fit a parabola. The rms emittance $\epsilon_{\rm rms}$ can be obtained from the fitted parameters. This method is commonly used at the end of a transport line, where a fluorescence screen or a wire detector (harp) is used to measure the rms beam size.

C2. Moving screen method

Using a movable fluorescence screen, the beam size at three spots can be used to determine the emittance. Employing the transfer matrix of drift space, the rms beam widths at the second and third positions are

$$\begin{cases} R_2^2 = \sigma_{11} + 2L_1\sigma_{12} + L_1^2\sigma_{22}, \\ R_3^2 = \sigma_{11} + 2(L_1 + L_2)\sigma_{12} + (L_1 + L_2)^2\sigma_{22}, \end{cases} \tag{2.62}$$

where $\sigma_{11} = R_1^2$, σ_{12} and σ_{22} are elements of the σ matrix at the first screen location, and L_1 and L_2 are respectively drift distances between screens 1 and 2 and between screens 2 and 3. The solution σ_{12} and σ_{22} of Eq. (2.62) can be used to obtain the rms beam emittance: $\epsilon_{\rm rms} = \sqrt{\sigma_{11}\sigma_{22} - \sigma_{12}^2}$.

If screen 2 is located at the waist, i.e. $dR_2^2/dL_1 = 0$, then the emittance can be determined from rms beam size measurements of screens 1 and 2 alone. The resulting emittance is

$$\epsilon^2 = \left(R_1^2R_2^2 - R_2^4\right)/L_1^2.$$

This method is commonly used to measure the electron emittance in a transport line.

D. The Gaussian distribution function

The equilibrium beam distribution in the linearized betatron phase space may be any function of the invariant action. However, the Gaussian distribution function

$$\rho(y, y') = \mathcal{N} \exp\left(\frac{-1}{2\det\sigma}(\sigma_{22}y^2 - 2\sigma_{12}yy' + \sigma_{11}y'^2)\right) \tag{2.63}$$

is commonly used to evaluate the beam properties. Expressing the normalized Gaussian distribution in the normalized phase space, we obtain

$$\rho(y, \mathcal{P}_y) = \frac{1}{2\pi\sigma_y^2} e^{-(y^2+\mathcal{P}_y^2)/2\sigma_y^2}, \tag{2.64}$$

where $\langle y^2 \rangle = \langle p_y^2 \rangle = \sigma_y^2 = \beta_y \epsilon_{\rm rms}$ with an rms emittance $\epsilon_{\rm rms}$. Transforming $(y, \mathcal{P}_y)$ into the action-angle variables (J, ψ) with

$$y = \sqrt{2\beta_y J}\cos\psi, \qquad \mathcal{P}_y = -\sqrt{2\beta_y J}\sin\psi.$$

The Jacobian of the transformation is $\frac{\partial(y,\mathcal{P}_y)}{\partial(\psi,J)} = \beta_y$, and the distribution function becomes

$$\rho(J) = \frac{1}{\epsilon_{\rm rms}} e^{-J/\epsilon_{\rm rms}}, \quad \rho(\epsilon) = \frac{1}{2\epsilon_{\rm rms}} e^{-\epsilon/2\epsilon_{\rm rms}}, \tag{2.65}$$

where $\epsilon = 2J$. The percentage of particles contained within $\epsilon = n\epsilon_{\rm rms}$ is $1 - e^{-n/2}$, shown in Table 2.1.

Table 2.1: Percentage of particles in the confined phase-space volume

$\epsilon/\epsilon_{\rm rms}$	2	4	6	8
Percentage in 1D [%]	63	86	95	98
Percentage in 2D [%]	40	74	90	96

The maximum phase-space area that particles can survive in an accelerator is called the *admittance*, or the *dynamic aperture*. The admittance is determined by the vacuum chamber size, the kicker aperture, and nonlinear magnetic fields. To achieve good performance of an accelerator, the emittance should be kept much smaller than the admittance. Note that some publications assume 95% emittance, i.e. the phase-space area contains 95% of the beam particles, $\epsilon_{95\%} \approx 6\,\epsilon_{\rm rms}$ for a Gaussian distribution. For superconducting accelerators, a dynamic aperture of 6σ or more is normally assumed for magnet quench protection. For electron storage rings, quantum fluctuations due to synchrotron radiation are important; the machine acceptance usually requires about 10σ for good quantum lifetime.

Accelerator scientists in Europe use $\epsilon = 4\epsilon_{\rm rms}$ to define the beam emittance. This convention arises from the fact that the rms beam emittance of a KV beam is equal to 1/4 of the full KV beam emittance [see Eq. (2.73)]. A uniform phase space distribution in an ellipse $y^2/a^2 + y'^2/b^2 = 1$ has an rms emittance equal to $\pi ab/4$.

E. Adiabatic damping and the normalized emittance: $\epsilon_{\rm n} = \beta\gamma\epsilon$

The Courant–Snyder invariant of Eq. (2.56), derived from the phase-space coordinate y, y', is not invariant when the energy is changed. To obtain the Liouville invariant phase-space area, we should use the conjugate phase-space coordinates (y, p_y) of the Hamiltonian in Eq. (2.14). Since $p_y = py' = mc\beta\gamma y'$, where m is the particle's mass, p is its momentum, and $\beta\gamma$ is the Lorentz relativistic factor, the *normalized emittance* defined by $\epsilon_{\rm n} = \beta\gamma\epsilon$ is invariant. The beam emittance decreases with increasing beam momentum, i.e. $\epsilon = \epsilon_{\rm n}/\beta\gamma$. This is called *adiabatic damping.* The adiabatic phase-space damping of the beam can be visualized as follows. The transverse velocity of a particle does not change during acceleration, while the transverse angle $y' = p_y/p$ becomes smaller as the particle momentum increases, and thus the beam emittance $\epsilon = \epsilon_{\rm n}/\beta\gamma$ becomes smaller. The adiabatic damping also applies to beam emittance in proton or electron linacs.

On the other hand, the beam emittance in electron storage rings increases with energy ($\sim \gamma^2$) resulting from the quantum fluctuation (see Chap. 4). The corresponding normalized emittance is proportional to γ^3, where γ is the relativistic Lorentz factor.

II.6 Stability of Betatron Motion: A FODO Cell Example

In this section, we illustrate the stability of betatron motion using a FODO cell example. We consider a FODO cell with quadrupole focal length f_1 and $-f_2$, where the $\pm$ signs designates the focusing and defocusing quadrupoles respectively. The transfer matrix of $\{\frac{1}{2}{\rm QF}_1 \ {\rm O}\ {\rm QD}_2\ {\rm O}\ \frac{1}{2}{\rm QF}_1\}$ is

$$\begin{aligned}
\mathbf{M} &= \begin{pmatrix} 1 & 0 \\ -\frac{1}{2f_1} & 1 \end{pmatrix}\begin{pmatrix} 1 & L_1 \\ 0 & 1 \end{pmatrix}\begin{pmatrix} 1 & 0 \\ \frac{1}{f_2} & 1 \end{pmatrix}\begin{pmatrix} 1 & L_1 \\ 0 & 1 \end{pmatrix}\begin{pmatrix} 1 & 0 \\ -\frac{1}{2f_1} & 1 \end{pmatrix} \\
&= \begin{pmatrix} 1 + \frac{L_1}{f_2} - \frac{L_1}{f_1} - \frac{L_1^2}{2f_1f_2} & 2L_1(1 + \frac{L_1}{2f_2}) \\ \frac{1}{f_2} - \frac{1}{f_1} - \frac{L_1}{f_1f_2} + \frac{L_1}{2f_1^2} + \frac{L_1^2}{4f_1^2f_2}) & 1 + \frac{L_1}{f_2} - \frac{L_1}{f_1} - \frac{L_1^2}{2f_1f_2} \end{pmatrix},
\end{aligned}$$

where L_1 is the drift length between quadrupoles. Identifying the transfer matrix with the Courant–Snyder parametrization, we obtain

$$\cos\Phi_x = 1 + \frac{L_1}{f_2} - \frac{L_1}{f_1} - \frac{L_1^2}{2f_1f_2}, \qquad \cos\Phi_z = 1 - \frac{L_1}{f_2} + \frac{L_1}{f_1} - \frac{L_1^2}{2f_1f_2}.$$

The stability condition, Eq. (2.33), of the betatron motion is equivalent to the following conditions:

$$|1 + 2X_2 - 2X_1 - 2X_1X_2| \le 1 \quad \text{and} \quad |1 - 2X_2 + 2X_1 - 2X_1X_2| \le 1, \tag{2.66}$$

where $X_1 = L_1/2f_1$ and $X_2 = L_1/2f_2$. The solution of Eq. (2.66) is shown in Fig. 2.10, which is usually called the necktie diagram. The lower and the upper boundaries of the shaded area correspond to $\Phi_{x,z} = 0$ or π respectively. Since the stable region is limited by $X_{1,2} \le 1$, the focal length should be larger than one-fourth of the *full cell length.*

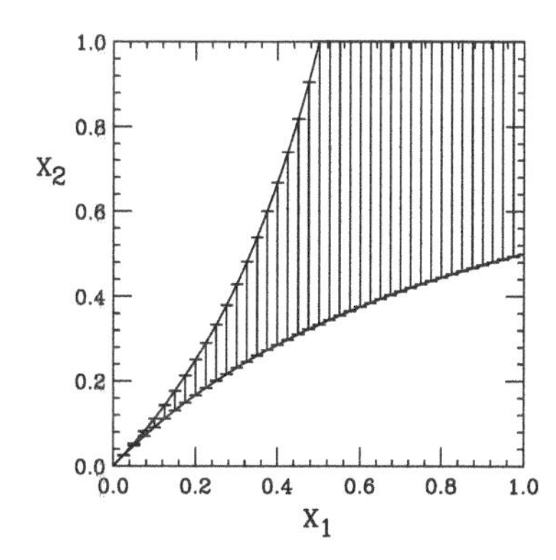

Figure 2.10: The "necktie diagram" for the stable region of a FODO cell lattice shown in the shaded area of focusing strengths $X_1 = L_1/2f_1$ vs $X_2 = L_1/2f_2$. where L_1 is the half cell length, f's are focal lengths. The lower and upper boundaries correspond to $\Phi_{x,z} = 0$ or $180°$ respectively. When X_1 is at the lower part of stability boundary, the phase advance of the FODO cell is $\Phi_x = 0$. At the boundary of the stability $X_1 = 1$, the phase advance $\Phi_x = \pi$.

The phase advances Φ_x and Φ_z of repetitive FODO cells should be less than π. The phase advances of a complex repetitive lattice-module with more than 2 quadrupoles can be larger than π. For example, the phase advance of a flexible momentum compaction (FMC) module is about $3\pi/2$ (see Sec. IV.8 and Exercise 2.4.17) and the phase advance of a minimum emittance double-bend achromat module is about 2.4π (see Sec. III.1; Chap. 4). In general, the stability of betatron motion is described by $|\cos\Phi_x| \le 1$ and $|\cos\Phi_z| \le 1$ for any type of accelerator lattice or repetitive transport line.

II.7 Symplectic Condition

The 2×2 transfer matrix M with $\det M = 1$ satisfies $\tilde{M}JM = J$, where $\tilde{M}$ is the transpose of the matrix M, and $J = \begin{pmatrix} 0 & 1 \\ -1 & 0 \end{pmatrix}$. In general, the transfer matrix of a Hamiltonian flow of n degrees of freedom satisfies

$$\tilde{M}JM = J, \qquad J = \begin{pmatrix} 0 & I \\ -I & 0 \end{pmatrix}, \tag{2.67}$$

where $\tilde{M}$ is the transpose of the transfer matrix M, and $J^2 = -I$, $\tilde{J} = -J$, $J^{-1} = -J$ with I as the $n \times n$ unit matrix. A $2n \times 2n$ matrix, M, is said to be Symplectic if it satisfies Eq. (2.67).[12] The matrices I and J are symplectic.

If the matrix M is symplectic, then M^{-1} is also symplectic and $\det M = 1$. If M and N are symplectic, then MN is also symplectic. Since the set of symplectic matrices satisfies the properties that (1) the unit matrix I is symplectic, (2) if M is symplectic then M^{-1} is symplectic, and (3) if M and N are symplectic, then MN is also symplectic, the set of symplectic matrices form a group denoted by $Sp(2n)$. The properties of real symplectic matrices are described below.

- The eigenvalues of symplectic matrix M must be real or must occur in complex conjugate pairs, i.e. λ and λ^*. The eigenvalues of a real matrix M or the roots of the characteristic polynomial $P(\lambda) = |M - \lambda I| = 0$ have real coefficients.

- Since $|M| = 1$, zero can not be an eigenvalue of a symplectic matrix.

- If λ is an eigenvalue of a real symplectic matrix M, then $1/\lambda$ must also be an eigenvalue. They should occur at the same multiplicity. Thus eigenvalues of a symplectic matrix are pairs of reciprocal numbers. For a symplectic matrix, we have $K^{-1}(\tilde{M} - \lambda I)K = M^{-1} - \lambda I = -\lambda M^{-1}(M - \lambda^{-1} I)$ or $P(\lambda) = \lambda^{2n} P(\frac{1}{\lambda})$. If we define $Q(\lambda) = \lambda^{-n} P(\lambda)$, then $Q(\lambda) = Q(\frac{1}{\lambda})$.

II.8 Effect of Space-Charge Force on Betatron Motion

The betatron amplitude function $w = \sqrt{\beta_y}$ of the Floquet transformation satisfies Eq. (2.38). Defining the envelope radius of a beam as $R_y = \sqrt{\beta_y \epsilon_y}$, where ϵ_y is the emittance, the envelope equation becomes

$$R_y'' + K_y R_y - \frac{\epsilon_y^2}{R_y^3} = 0, \tag{2.68}$$

where the prime corresponds to the derivative with respect to s. Based on the Floquet theorem, we can impose a periodic condition, $R_y(s) = R_y(s + L)$ to the envelope equation, if K_y is a periodic function of s, i.e. $K_y(s) = K_y(s+L)$, with L as the length of a repetitive period. The periodic envelope solution, aside from a multiplicative constant, is equal to the betatron amplitude function. The envelope function of an *emittance dominated beam is* $R_y = \sqrt{\beta_y \epsilon_y}$. *When the self-induced space-charge force is included in the betatron motion, what happens to the beam envelope?*

[12]The transfer matrix M expressed in this form corresponds to the transfer matrix for phase space coordinates $(q_1, q_2, \cdots, q_n; p_1, p_2, \cdots, p_n)$. If we choose the phase space coordinates as $(q_1, p_1, q_2, p_2, \cdots$, the J matrix will be defined slightly differently.

A. The Kapchinskij-Vladimirskij distribution

It is known that the Coulomb mean-field from an arbitrary beam distribution is likely to be nonlinear. For example, The Exercise 5.2.1 shows the Coulomb mean field of a Gaussian beam distribution. In 1959, Kapchinskij and Vladimirskij (KV) discovered an ellipsoid beam distribution that leads to a perfect linear space-charge force within the beam radius. This distribution function is called the KV distribution.[13]

Particles, in the KV distribution, are uniformly distributed on a constant total emittance surface of the 4-dimensional phase space, i.e.

$$\rho(x, \mathcal{P}_x, z, \mathcal{P}_z) = \frac{Ne}{\pi^2 a^2 b^2}\delta\left(\frac{1}{a^2}\left(x^2+\mathcal{P}_x^2\right)+\frac{1}{b^2}\left(z^2+\mathcal{P}_z^2\right)-1\right), \tag{2.69}$$

$$\rho(J_x, J_z) = \frac{4Ne}{\epsilon_x\epsilon_z}\delta\left(\frac{2J_x}{\epsilon_x}+\frac{2J_z}{\epsilon_z}-1\right). \tag{2.70}$$

where N is the number of particles per unit length, e is the particle's charge, a and b are envelope radii of the beam, x and z are the transverse phase-space coordinates, $\mathcal{P}_x = R'_x$, and $\mathcal{P}_z = R'_z$ are the corresponding normalized conjugate phase-space coordinates, ϵ_x and ϵ_z are the horizontal and vertical emittances, and the envelope radii are $a = \sqrt{\beta_x\epsilon_x}$ and $b = \sqrt{\beta_z\epsilon_z}$. Thus beam particles are uniformly distributed along an action line

$$\frac{J_x}{\epsilon_x}+\frac{J_z}{\epsilon_z}=\frac{1}{2}, \tag{2.71}$$

Some properties of the KV distribution are as follows.

1. Integrating the conjugate momenta, the distribution function becomes

$$\rho(x,z) = \frac{Ne}{\pi ab}\,\Theta\left(1-\frac{x^2}{a^2}-\frac{z^2}{b^2}\right) \tag{2.72}$$

   where the $\Theta(\xi)$ function is equal to 1 if $\xi \geq 0$, and 0 if $\xi < 0$. In fact, the KV particles are uniformly distributed in any two-dimensional projection of the four-dimensional phase space.

2. The rms emittances of the KV beam are

$$\epsilon_{x,\rm rms} = \frac{\langle x^2\rangle}{\beta_x} = \frac{\epsilon_x}{4}, \quad \epsilon_{z,\rm rms} = \frac{\langle z^2\rangle}{\beta_z} = \frac{\epsilon_z}{4}. \tag{2.73}$$

   Thus the rms envelope radii are equal to half of the beam radii in the KV beam.

[13] I.M. Kapchinskij and V.V. Vladimirskij, *Proc. Int. Conf. on High Energy Accelerators*, p. 274 (CERN, Geneva, 1959).

B. The Coulomb mean-field due to all beam particles

The next task is to calculate the effect of the average space-charge force. Neglecting the longitudinal variations, beam particles can be viewed as a charge distribution in an infinite long wire with a line-charge density given by Eq. (2.72). The electric field at the spatial point (x, z) is

$$\begin{aligned}\vec{E}(x,z) &= \frac{Ne}{2\pi\epsilon_0\ \pi ab}\int\!\!\int dx'dz'\ \Theta(1-\frac{x'^2}{a^2}-\frac{z'^2}{b^2})\frac{(x-x')\hat{x}+(z-z')\hat{z}}{(x-x')^2+(z-z')^2}\\ &= \frac{2Ne}{2\pi\epsilon_0}\left(\frac{x}{a(a+b)}\hat{x}+\frac{z}{b(a+b)}\hat{z}\right),\end{aligned} \tag{2.74}$$

where ϵ_0 is the vacuum permittivity. A noteworthy feature of the KV distribution function is that the resulting mean-field inside the beam envelope radii is linear! If the external focusing force is also linear, the KV distribution is a self-consistent distribution function. Including the mean-magnetic-field, the force on the particle at (x, z) is

$$\vec{F}(x,z) = \frac{2Ne^2}{2\pi\epsilon_0\gamma^2}\left(\frac{x}{a(a+b)}\hat{x}+\frac{z}{b(a+b)}\hat{z}\right), \tag{2.75}$$

where γ is the relativistic energy factor. Hill's equations of KV beams of motion become

$$x'' + \left(K_x(s) - \frac{2K_{\rm sc}}{a(a+b)}\right)x = 0, \qquad z'' + \left(K_z(s) - \frac{2K_{\rm sc}}{b(a+b)}\right)z = 0, \tag{2.76}$$

where the prime is a derivative with respect to the longitudinal coordinate s, and $K_{\rm sc}$ is the "normalized" space-charge *perveance* parameter given by

$$K_{\rm sc} = \frac{2Nr_0}{\beta^2\gamma^3}, \tag{2.77}$$

where $r_0 = e^2/4\pi\epsilon_0 mc^2$ is the classical radius of the particle, and N is the number of particles per unit length. Performing Floquet transformation of the linear KV-Hill equation $x = w_x e^{j\psi_x}$ and $z = w_z e^{j\psi_z}$, we obtain

$$w_x'' + \left(K_x - \frac{2K_{\rm sc}}{a(a+b)}\right)w_x + \frac{1}{w_x^3} = 0, \qquad \psi_x'' = \frac{1}{w_x^2}, \tag{2.78}$$

$$w_z'' + \left(K_z - \frac{2K_{\rm sc}}{b(a+b)}\right)w_z + \frac{1}{w_z^3} = 0, \qquad \psi_z'' = \frac{1}{w_z^2}. \tag{2.79}$$

Multiplying Eq. (2.78) by $\sqrt{\epsilon_x}$ and Eq. (2.79) by $\sqrt{\epsilon_z}$, and identifying $a = w_x\sqrt{\epsilon_x}$ and $b = w_z\sqrt{\epsilon_z}$, we obtain the KV envelope equations, or simply the KV equations:

$$a'' + K_x a - \frac{2K_{\rm sc}}{a+b} - \frac{\epsilon_x^2}{a^3} = 0, \qquad b'' + K_z b - \frac{2K_{\rm sc}}{a+b} - \frac{\epsilon_x^2}{b^3} = 0. \tag{2.80}$$

Solving the KV envelope equation is equivalent to finding the betatron amplitude function in the presence of the space-charge force. The usefulness of the KV equation has been further extended to arbitrary ellipsoid distribution functions provided that the envelope functions a and b are equal to *twice the rms envelope radii*, and the emittances ϵ_x and ϵ_z are equal to four times the rms emittances.[14]

If the external force is periodic, i.e. $K_x(s) = K_x(s+L)$, the KV equation can be solved by imposing the periodic boundary (closed orbit) condition (Floquet theorem)

$$a(s) = a(s+L), \quad b(s) = b(s+L). \tag{2.81}$$

A numerical integrator or differential equation solvers can be used to find the envelope function of the space-charge dominated beams. The matched beam envelope solution can be obtained by a proper closed orbit condition of Eq. (2.81).

For beams with an initial mismatched envelope, the envelope equation can be solved by using the initial value problem to find the behavior of the mismatched beams. For space-charge dominated beams, the envelope solution can vary widely depending on the external focusing function, the space-charge parameter, and the beam emittance. To understand the physics of the mismatched envelope, it is advantageous to extend the envelope equation to Hamiltonian dynamics as discussed below.

C. Hamiltonian formalism of the envelope equation

Introducing the pseudo-envelope momenta as $p_a = a'$ and $p_b = b'$, we can derive the KV equations (2.80) from the envelope Hamiltonian:

$$\begin{aligned} H_{\rm env} &= \frac{1}{2}\left(p_a^2 + p_b^2\right) + V_{\rm env}(a,b) \\ V_{\rm env}(a,b) &= \frac{1}{2}(K_x a^2 + K_z b^2) - 2K_{\rm sc}\ln(a+b) + \frac{\epsilon_x^2}{2a^2} + \frac{\epsilon_z^2}{2b^2}, \end{aligned} \tag{2.82}$$

where $V_{\rm env}(a,b)$ is the envelope potential. The matched beam envelope is the equilibrium solution (the betatron amplitude function) of the envelope Hamiltonian. For example, if we start from the condition with envelope momenta $p_a = p_b = 0$, the matched envelope radii are located at the minimum potential energy location, i.e.

$$\frac{\partial V_{\rm env}}{\partial a}(a_{\rm m}, b_{\rm m}) = \frac{\partial V_{\rm env}}{\partial b}(a_{\rm m}, b_{\rm m}) = 0,$$

where $a_{\rm m}$ and $b_{\rm m}$ are the matched envelope radii. The envelope oscillations of a mismatched beam can be determined by the perturbation around the matched solution

$$V_{\rm env} = \frac{1}{2}\frac{\partial^2 V_{\rm env}}{\partial a^2}(a - a_{\rm m})^2 + \frac{1}{2}\frac{\partial^2 V_{\rm env}}{\partial b^2}(b - b_{\rm m})^2 + \cdots.$$

[14] P.M. Lapostolle, *IEEE Trans. Nucl. Sci.* **NS-18**, 1101 (1971); F.J. Sacherer, *ibid.* 1105 (1971); J.D. Lawson, P.M. Lapostolle, and R.L. Gluckstern, *Part. Accel.* **5**, 61 (1973); E.P. Lee and R.K. Cooper, *ibid.* **7**, 83 (1976).

Using the second-order derivatives, we can obtain the envelope tune, which is equal to twice betatron tune at $K_{\rm sc} = 0$.

D. An example of a uniform focusing paraxial system

First we consider a beam in a uniform paraxial focusing system, where the focusing function is

$$K_x = (2\pi/L)^2 .$$

Here L is the betatron wavelength, and the betatron amplitude function is $\beta_{x0} = L/2\pi$. With $a = b$ in Eq. (2.80), the envelope Hamiltonian is

$$H_{\rm env} = \frac{1}{2}p_a^2 + V_{\rm env}(a); \qquad V_{\rm env}(a) = \frac{1}{2}\left(\frac{2\pi}{L}\right)^2 a^2 - K_{\rm sc}\ln a + \frac{\epsilon_x^2}{2a^2}.$$

When the space-charge force is negligible, we find that the matched envelope radius is $a_{\rm m0} = \sqrt{\epsilon_x L/2\pi} = \sqrt{\epsilon_x\beta_x}$, and the second-order derivative at the matched envelope radius

$$\left(\frac{d^2V_{\rm env}}{da^2}\right)^{1/2} = 2(\frac{2\pi}{L}),$$

which is twice the betatron tune (see also Exercise 2.2.15) and is independent of the envelope-oscillation amplitude.

Now, we consider the effect of space charge on the envelope function. The matched envelope radius is obtained from the solution of $dV_{\rm env}/da = 0$, i.e.

$$a_{\rm m}^2 = \epsilon_x\beta_x = \epsilon_x\left(\frac{L}{2\pi}\right)\left[\kappa + \sqrt{\kappa^2+1}\right], \qquad \kappa = \frac{K_{\rm sc}L}{2\epsilon_x 2\pi} = \frac{K_{\rm sc}L_{\rm tot}}{2\epsilon_x\Phi_{\rm tot}} \tag{2.83}$$

where κ is the effective space-charge parameter, and $L_{\rm tot}$ and $\Phi_{\rm tot}$ are the total length and total phase advance of a transport system.[15] Equation (2.83) indicates that the betatron amplitude function increases by a factor $\kappa+\sqrt{\kappa^2+1}$ due to the space-charge force. The second-order derivative of the potential at the matched radius is

$$\left(\frac{d^2V_{\rm env}}{da^2}\right)^{1/2} = 2\frac{2\pi}{L}\left(1 - \frac{\kappa}{(\kappa+\sqrt{\kappa^2+1})}\right)^{1/2},$$

which is the phase advance per unit length of small amplitude envelope oscillation in the presence of the Coulomb potential. When the space-charge perveance parameter is zero, the phase advance of the envelope oscillation is twice of that of the betatron oscillation, and when the space-charge force is large, as $\kappa \to \infty$, the phase advance of the small-amplitude envelope oscillations can maximally be depressed to $\sqrt{2}\,(2\pi/L)$.

[15] The Laslett (linear) space-charge tune shift is related to the space-charge perveance parameter by $\xi_{\rm sc} \equiv \Delta\nu_{\rm sc} = K_{\rm sc}L_{\rm tot}/4\pi\epsilon_x = \kappa\nu$, where ν is the tune.

There is a large envelope detuning from 2μ to $\sqrt{2}\mu$, where μ is the betatron phase advance. A nonlinear envelope resonance can be excited when perturbation exists and a resonance condition is satisfied.[16]

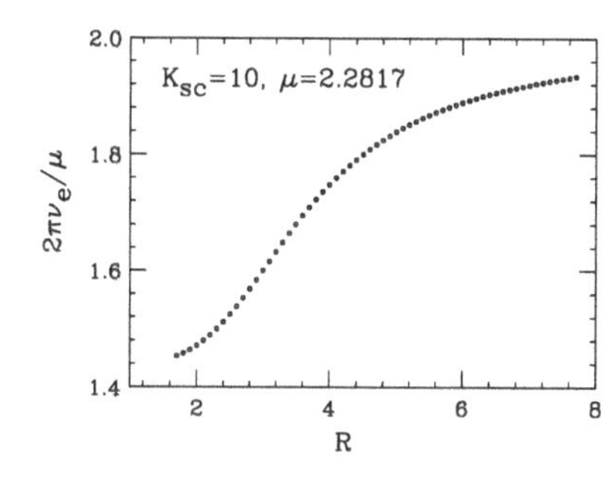

Figure 2.11: The phase advance of the envelope oscillations divided by the original betatron phase advance for a high space charge beam with $K_{\rm sc} = 10$, $\mu = 2.28175$. The matched radius is $R_0 = a_{\rm m}\sqrt{2\pi/(\mu\epsilon_x L)} = 1.4199$ in this example. See Eq. (2.83) for the matched envelope radius. When the envelope radius is mismatched from R_0, the envelope radius oscillates around R_0 at an envelope tune depending on its maximum radius oscillation amplitude. The ordinate R is the normalized maximum envelope radius of the beam.

Figure 2.11 shows the envelope tune of a space charge dominated beam with $K_{\rm sc} = 10$ and a phase advance of $\mu = 2.2817$ radian (or $\nu = \mu/2\pi$ for the unperturbed betatron tune) as a function of the maximum amplitude of the envelope oscillation. At a large envelope amplitude, the envelope tune approaches twice the unperturbed betatron tune. Near the matched envelope radius (or small amplitude envelope oscillations), the envelope tune approaches $\sqrt{2}$ times the unperturbed betatron tune.

The single particle betatron phase advance per unit length is obtained by substituting Eq. (2.83) into Eq. (2.76), i.e. $\Phi_x = \frac{2\pi}{L}(\sqrt{\kappa^2+1} - \kappa)$. When the space charge parameter κ is small, the incoherent space-charge (Laslett) tune shift is equal to $\Delta\nu_{\rm sc} = \xi_{\rm sc} = \kappa$. When the space charge parameter κ is large, the betatron tune can be depressed to zero.

E. Space-charge force for Gaussian distribution

Since the emittance growth rate is usually much faster than a synchrotron period, this justifies the performance of only 2D simulation for a slice of the beam at the longitudinal bunch center. For a beam with linear particle density N and bi-Gaussian charge distribution

$$\rho(x,z) = \frac{Ne}{2\pi\sigma_x\sigma_z} e^{-x^2/2\sigma_x^2 \, -z^2/2\sigma_z^2}, \tag{2.84}$$

[16] S.Y. Lee and A. Riabko, *Phys. Rev.* E **51**, 1609 (1995); A. Riabko *et al.*, *Phys. Rev.* E **51**, 3529 (1995); C. Chen and R.C. Davidson, *Phys. Rev.* E**49**, 5679 (1994); *Phys. Rev. Lett.* **72**, 2195 (1994). See also Ref. [8] for an exploration of the space-charge dynamics.

with $\sigma_{x,z}$ being the rms horizontal and vertical beam radii including contribution coming from momentum dispersion, the transverse 2D space-charge potential is

$$V_{\rm sc}(x,z) = \frac{K_{\rm sc}}{2}\int_0^\infty \frac{\exp\left[-\frac{x^2}{2\sigma_x^2+t}-\frac{z^2}{2\sigma_z^2+t}\right]-1}{\sqrt{(2\sigma_x^2+t)(2\sigma_z^2+t)}}\,dt, \tag{2.85}$$

where $K_{\rm sc}$ is the space-charge perveance of Eq. (2.77), r_0 is the particle classical radius, and β and γ are the relativistic parameters. In the simulation, we set the bunch intensity with $N_{\rm B}$ particles and an rms bunch-length σ_s to obtain $N = N_{\rm B}/\sqrt{2\pi}\sigma_s$. The space-charge force on each particle is obtained by Hamilton's equation. Thus each beam particle passing through a length Δs experiences a space-charge kick

$$\frac{\Delta x'}{\Delta s} = -\frac{\partial V_{\rm sc}}{\partial x}, \qquad \frac{\Delta z'}{\Delta s} = -\frac{\partial V_{\rm sc}}{z}. \tag{2.86}$$

We expand the space-charge potential in Taylor series in order to study the systematic space charge resonances:

$$\begin{aligned} V_{\rm sc}(x,z) \;=\; & -\frac{K_{\rm sc}}{2}\left\{\left[\frac{x^2}{\sigma_x(\sigma_x+\sigma_z)}+\frac{z^2}{\sigma_z(\sigma_x+\sigma_z)}\right]-\frac{1}{4\sigma_x^2(\sigma_x+\sigma_z)^2}\left[\frac{2+r}{3}x^4+\right.\right. \\ & \left. +\frac{2}{r}x^2z^2+\frac{1+2r}{3r^3}z^4\right]+\frac{1}{72\sigma_x^3(\sigma_x+\sigma_z)^3}\left[\frac{8+9r+3r^2}{5}x^6+\right. \\ & \left.\left. +\frac{3(3+r)}{r}x^4z^2+\frac{3(3r+1)}{r^3}x^2z^4++\frac{8r^2+9r+3}{5r^5}z^6\right]+\cdots\right\}, \end{aligned} \tag{2.87}$$

with $r = \sigma_z/\sigma_x$. The first term inside the curly brackets represents the linear force, which gives rise to linear space charge (Laslett) tune shift. The second and the third terms drive the 4th and 6th order resonances.

The linear space charge tune shift parameters become

$$\xi_{\rm sc,x/z} \equiv |\Delta\nu_{\rm sc,x/z}| = \begin{cases} \dfrac{K_{\rm sc}}{4\pi}\displaystyle\oint \frac{1}{\sigma_x(\sigma_x+\sigma_z)}\beta_x ds \\ \dfrac{K_{\rm sc}}{4\pi}\displaystyle\oint \frac{1}{\sigma_z(\sigma_x+\sigma_z)}\beta_z ds \end{cases} \xrightarrow{\text{roundbeam}} \frac{2\pi R K_{\rm sc}}{8\pi\epsilon_{\rm rms}}. \tag{2.88}$$

Particles at the center of the beam has a betatron tune shift $-\xi_{\rm sc,x/z}$, and large betatron amplitude particles have small betatron tune shift. Since particles at different betatron amplitudes have different betatron tune shift, the space charge force produces an *incoherent* tune spread $\xi_{\rm sc}$. The space charge parameter of the KV distribution in Eq. (2.83) is

$$\xi_{\rm KV,sc} = \frac{2\pi R K_{\rm sc}}{16\pi\epsilon_{\rm rms}}.$$

The space charge parameter of Gaussian distribution is a factor of 2 larger than that of a beam with uniform distribution. The space charge tune shift of all particles in the KV beam is identically $\xi_{\rm KV,sc}$. It is still *incoherent*.

Exercise 2.2

1. The focusing function $K(s)$ for most accelerator magnets can be assumed to be piece-wise constant. Show that

$$K(s) = 0, \qquad M(s_2|s_1) = \begin{pmatrix} 1 & s \\ 0 & 1 \end{pmatrix},$$

$$K(s) = K \geq 0, \quad M(s_2|s_1) = \begin{pmatrix} \cos\sqrt{K}s & \frac{1}{\sqrt{K}}\sin\sqrt{K}s \\ -\sqrt{K}\sin\sqrt{K}s & \cos\sqrt{K}s \end{pmatrix},$$

$$K(s) = K < 0, \quad M(s_2|s_1) = \begin{pmatrix} \cosh\sqrt{|K|}s & \frac{1}{\sqrt{|K|}}\sinh\sqrt{|K|}s \\ \sqrt{|K|}\sinh\sqrt{|K|}s & \cosh\sqrt{|K|}s \end{pmatrix},$$

with $s = s_2 - s_1$. Show that the mapping matrix M for a short quadrupole of length ℓ, in the thin-lens approximation, is

$$M = \begin{pmatrix} 1 & 0 \\ -\frac{1}{f} & 1 \end{pmatrix}$$

where $f = \lim_{\ell\to 0}(K\ell)^{-1}$, is the focal length of a quadrupole. For a focusing quad, $f > 0$; and for a defocusing quad, $f < 0$.

2. When a particle enters a dipole at an angle δ with respect to the normal edge of a dipole (see drawing below), there is a quadrupole effect. This phenomenon is usually referred to as edge focusing. Using edge focusing, the zero-gradient synchrotron (ZGS) was designed and constructed in the 1960's at Argonne National Laboratory. The ZGS was made of 8 dipoles with a circumference of 172 m attaining the energy of 12.5 GeV. Its first proton beam was commissioned on Sept. 18, 1963. See L. Greenbaum, *A Special Interest* (Univ. of Michigan Press, Ann Arbor, 1971). We use the convention that $\delta > 0$ if the particle trajectory is closer to the center of the bending radius. Show that the transfer matrices for the horizontal and vertical betatron motion due to the edge focusing are

$$M_x = \begin{pmatrix} 1 & 0 \\ \frac{\tan\delta}{\rho} & 1 \end{pmatrix} \quad M_z = \begin{pmatrix} 1 & 0 \\ -\frac{\tan\delta}{\rho} & 1 \end{pmatrix}$$

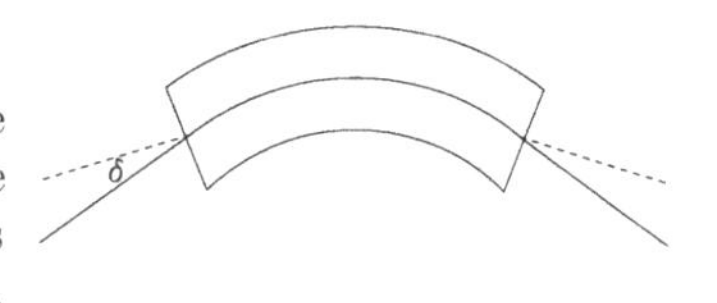

where δ is the entrance or the exit angle of the particle with respect to the normal direction of the dipole edge. Thus the edge effect with $\delta > 0$ gives rise to horizontal defocusing and vertical focusing.

3. The particle orbit enters and exits a *sector dipole* magnet perpendicular to the dipole edges. Assuming that the gradient function of the dipole is zero, i.e. $\partial B_z/\partial x = 0$, show that the transfer matrix is

$$M_x = \begin{pmatrix} \cos\theta & \rho\sin\theta \\ -\frac{\sin\theta}{\rho} & \cos\theta \end{pmatrix}, \quad M_z = \begin{pmatrix} 1 & \ell \\ 0 & 1 \end{pmatrix}$$

where θ is the bending angle, ρ is the bending radius, and ℓ is the length of the dipole. Note that a sector magnet gives rise to horizontal focusing.

4. The entrance and exit edge angles of a rectangular dipole are $\delta_1 = \theta/2$ and $\delta_2 = \theta/2$, where θ is the bending angle. Find the horizontal and vertical transfer matrices for a rectangular dipole (Fig. 2.2b).

5. For a weak-focusing accelerator, $K_z(s) = n/\rho^2 = $ constant and $K_x = (1-n)/\rho^2$, where ρ is the radius of the accelerator. The focusing index n is

$$n(s) = \frac{\rho(s)}{B_z(s,0,0)} \frac{\partial B_z(s,x,0)}{\partial x}\Big|_{x=0},$$

where we have chosen the coordinate system shown in Fig. 2.1. Solve the following problems by using the uniform focusing approximation with constant n.

   (a) Show that the horizontal and vertical transfer matrices are

$$M_x = \begin{pmatrix} \cos(\sqrt{1-n}\ s/\rho) & (\rho/\sqrt{1-n})\sin(\sqrt{1-n}\ s/\rho) \\ -(\sqrt{1-n}/\rho)\sin(\sqrt{1-n}\ s/\rho) & \cos(\sqrt{1-n}\ s/\rho) \end{pmatrix},$$

$$M_z = \begin{pmatrix} \cos(\sqrt{n}\ s/\rho) & (\rho/\sqrt{n})\sin(\sqrt{n}\ s/\rho) \\ -(\sqrt{n}/\rho)\sin(\sqrt{1-n}\ s/\rho) & \cos(\sqrt{n}\ s/\rho) \end{pmatrix}.$$

   (b) Show that the betatron tunes are $\nu_x = (1-n)^{1/2}$ and $\nu_z = n^{1/2}$, and the stability condition is $0 \le n \le 1$.

   (c) If N equally spaced straight sections, with $K_x = K_z = 0$, are introduced into the accelerator lattice adjacent to each combined-function dipole, calculate the mapping matrix for the basic period and discuss the stability condition.

6. The path length for a particle orbit in an accelerator is

$$C = \oint \sqrt{[1+(x/\rho)]^2 + x'^2 + z'^2}\, ds.$$

Show that the average orbit length of the particle with a vertical betatron action J_z is longer by

$$\frac{\Delta C}{C} = \frac{1}{2}\langle\frac{1+\alpha_z^2}{\beta_z}\rangle J_z,$$

where α_z and β_z are betatron amplitude functions. In the smooth approximation, the betatron amplitude function is approximated by $\langle\beta_z\rangle = R/\nu_z$, and the betatron oscillations can be expressed as

$$z = \hat{z}\cos\left(\frac{s}{\langle\beta_z\rangle} + \chi_z\right),$$

where R, ν_z and $\hat{z}$ are the average radius, the vertical betatron tune, and the vertical betatron amplitude respectively, and χ_z is an arbitrary betatron phase angle of the particle. Show that the average orbit length of a particle executing betatron oscillations is longer by

$$\frac{\Delta C}{C} = \frac{\nu_z^2}{4R^2}\hat{z}^2.$$

Thus the orbit length depends quadratically on the betatron amplitude.

7. In a strong-focusing synchrotron, the art (or science) of magnet arrangement is called lattice design. The basic building blocks of a lattice are usually FODO cells. A FODO cell is composed of QF OO QD OO, where QF is a focusing quadrupole, OO represents either a drift space or bending dipoles of length L_1, and QD is a defocusing quadrupole. The length of a FODO cell is $L = 2L_1$. Using the thin-lens approximation,

(a) Find the mapping matrix and the phase advance of the FODO cell and discuss the stability condition.

(b) Find the parameters β, α at the quadrupoles and at the center of the drift space as a function of L_1 and Φ. Find the phase advance Φ that minimizes the betatron amplitude function at the focusing quadrupole location.

8. Using Eq. (2.41), show that $\beta''' + 4\beta' K + 2\beta K' = 0$. Solve this equation for a drift space and a quadrupole respectively, and show that the solution of this equation must be one of the following forms:

$$\begin{cases} \beta = a + bs + cs^2, & \text{drift space} \\ \beta = a\cos 2\sqrt{K}s + b\sin 2\sqrt{K}s + c, & \text{focusing quadrupole} \\ \beta = a\cosh 2\sqrt{|K|}s + b\sinh 2\sqrt{|K|}s + c, & \text{defocusing quadrupole.} \end{cases}$$

(a) In a drift space, where there are no quadrupoles, Show also that the betatron amplitude function is given by

$$\beta(s) = \beta_1 - 2\alpha_1 s + \gamma_1 s^2 = \beta^* + \frac{(s - s^*)^2}{\beta^*},$$

where the parameters α_1, β_1 and γ_1 at betatron function values at the beginning of the element, or $s = 0$; and β^* is the betatron function at the symmetry point $s = s^*$ with $\beta' = 0$. This means that $s^* = \beta_1/(\alpha_1 + 1/\alpha_1)$, and $\beta^* = \beta_1/(1+\alpha_1^2) = 1/\gamma^* = 1/\gamma_1$. Note that $\gamma_1 = (1 + \alpha_1^2)/\beta_1 = 1/\beta^*$, i.e. γ_1 is constant in a drift space.

(b) Using the similarity transformation Eq. (2.32), show that the Courant–Snyder parameters $\alpha_2, \beta_2, \gamma_2$ at s_2 are related to $\alpha_1, \beta_1, \gamma_1$ at s_1 by

$$\begin{pmatrix} \beta_2 \\ \alpha_2 \\ \gamma_2 \end{pmatrix} = \begin{pmatrix} M_{11}^2 & -2M_{11}M_{12} & M_{12}^2 \\ -M_{11}M_{21} & M_{11}M_{22} + M_{12}M_{21} & -M_{12}M_{22} \\ M_{21}^2 & -2M_{21}M_{22} & M_{22}^2 \end{pmatrix} \begin{pmatrix} \beta_1 \\ \alpha_1 \\ \gamma_1 \end{pmatrix},$$

where M_{ij} are the matrix elements of $M(s_2|s_1)$. Use these equations to verify your solution to part (a). Similarly, the betatron function inside the focusing and defocusing quadrupole are respectively given by

$$\begin{aligned} \beta_{\text{focusing}}(s) &= \frac{1}{2}(\beta_1 - \frac{\gamma_1}{K})\cos 2\sqrt{K}s - \frac{\alpha_1}{\sqrt{K}}\sin 2\sqrt{K}s + \frac{1}{2}(\beta_1 + \frac{\gamma_1}{K}), \\ \beta_{\text{defocusing}}(s) &= \frac{1}{2}(\beta_1 + \frac{\gamma_1}{K})\cosh 2\sqrt{K}s - \frac{\alpha_1}{\sqrt{K}}\sinh 2\sqrt{K}s + \frac{1}{2}(\beta_1 - \frac{\gamma_1}{K}), \end{aligned}$$

9. Use the transfer matrix $M(s_2|s_1)$ of Eq. (2.42) to show that, when a particle is kicked at s_1 by an angle θ, the displacement at a downstream location is

$$\Delta x_2 = \theta\sqrt{\beta_1\beta_2}\sin\psi,$$

where β_1 and β_2 are values of betatron functions at s_1 and s_2 respectively, and $\psi = \psi(s_2) - \psi(s_1)$ is the betatron phase advance between s_1 and s_2. The quantity $\sqrt{\beta_1\beta_2}\sin\psi$ is usually called the kicker arm. To minimize the kicker magnet strength θ, the injection or extraction kickers are located at a high β locations with a 90° phase advance.

10. Transforming the betatron phase-space coordinates onto the normalized coordinates with
$$Y = \frac{1}{\sqrt{\beta}} y, \quad \mathcal{P}_Y = \frac{1}{\sqrt{\beta}}(\alpha y + \beta y'),$$
or
$$\begin{pmatrix} Y \\ \mathcal{P}_Y \end{pmatrix} = \mathbf{B}^{-1}\begin{pmatrix} y \\ y' \end{pmatrix} \quad \text{with} \quad \mathbf{B}^{-1} = \begin{pmatrix} 1/\sqrt{\beta} & 0 \\ \alpha/\sqrt{\beta} & \sqrt{\beta} \end{pmatrix} \quad \text{and} \quad \mathbf{B} = \begin{pmatrix} \sqrt{\beta} & 0 \\ -\alpha/\sqrt{\beta} & \sqrt{\beta} \end{pmatrix},$$
show that the betatron transfer matrix in normalized coordinates becomes
$$\tilde{M}(s_2|s_1) = \begin{pmatrix} \cos\psi & \sin\psi \\ -\sin\psi & \cos\psi \end{pmatrix},$$
i.e. the betatron transfer matrix becomes coordinate rotation with rotation angle equal to the betatron phase advance. Show that the transfer matrix of Eq. (2.42) becomes $M(s_2|s_1) = \mathbf{B}_2\tilde{M}\mathbf{B}_1^{-1}$, where $\mathbf{B}_2$ and $\mathbf{B}_1$ are the betatron amplitude matrices at $s = s_2$ and s_1 respectively.

11. Show that the Floquet transformation of Eq. (2.55) transforms the Hamiltonian of Eq. (2.47) into Eq. (2.54).

12. Often a solenoidal field has been used to provide both the horizontal and the vertical beam focusing for the production of secondary beams from a target (see Exercise 2.1.4). The focusing channel can be considered as a focusing-focusing (FOFO) channel. We consider a FOFO focusing channel where the focusing elements are separated by a distance L. Use the thin-lens approximation to evaluate beam transport properties of a periodic FOFO channel.

    (a) Show that the phase advance of a FOFO cell is
    $$\sin\frac{\Phi}{2} = \frac{1}{2}\sqrt{\frac{L}{f}},$$
    where f is the focal length given by $f^{-1} = g^2\ell = \Theta^2/\ell$, ℓ is the length of the solenoid, $g = B_\parallel/2B\rho$ is the effective solenoid strength, $B_\parallel$ is the solenoid field, and $\Theta = g\ell$ is the solenoid rotation angle.

    (b) Show that the maximum and minimum values of the betatron amplitude function are
    $$\beta_{\max} = L/\sin\Phi, \quad \beta_{\min} = f\sin\Phi.$$

13. The doublet configuration consists of a pair focusing and defocusing quadrupoles with equal focusing strength separated by a small distance L_1 as a beam focusing unit. The doublet pairs are repeated at intervals $L_2 \gg L_1$ for beam transport (Fig. 2.6). These quadrupole doublets can be used to maintain round beam configuration during beam transport. Using the thin-lens approximation with equal focal length for the focusing and defocusing quadrupoles, describe the properties of betatron motion in a doublet transport line.

(a) Show that the betatron phase advance in a doublet cell is

$$\psi = \psi_{x,z} = 2\arcsin\left(\sqrt{L_1 L_2}/2f\right),$$

where f is the focal length of the quadrupoles.

(b) Show that the maximum betatron amplitude function is approximately

$$\beta_{\rm max} = (L_1 + L_2 + L_1 L_2/f)/\sin\psi.$$

(c) Show that the minimum betatron amplitude function is

$$\beta^* = \sqrt{L_1\left(4f^2 - L_1 L_2\right)/4L_2}.$$

(d) Sketch the betatron amplitude functions and compare your results with that of the FODO cell transport line.

14. *Statistical definition of beam emittance:*[17] We consider a statistical distribution of N non-interacting particles in phase space (x, x'). Let $\rho(x, x')$ be the distribution function with

$$\int \rho(x, x')dxdx' = 1.$$

The first and second moments of beam distribution are

$$\langle x\rangle = \frac{1}{N}\sum x_i = \int x\rho(x, x')dxdx', \quad \langle x'\rangle = \frac{1}{N}\sum x'_i = \int x'\rho(x, x')dxdx',$$

$$\sigma_x^2 = \frac{1}{N}\sum(x_i - \langle x\rangle)^2, \quad \sigma_{x'}^2 = \frac{1}{N}\sum(x'_i - \langle x'\rangle)^2,$$

$$\sigma_{xx'} = \frac{1}{N}\sum(x_i - \langle x\rangle)(x'_i - \langle x'\rangle) = r\sigma_x\sigma_{x'}.$$

Here σ_x and $\sigma_{x'}$ are rms beam widths, and r is the correlation coefficient. The rms emittance is defined as

$$\epsilon_{\rm rms} = \sigma_x\sigma_{x'}\sqrt{1-r^2}.$$

(a) Assuming that particles are uniformly distributed in an ellipse

$$x^2/a^2 + x'^2/b^2 = 1,$$

show that the total phase-space area is $\mathcal{A} = \pi ab = 4\pi\epsilon_{\rm rms}$. The factor 4 has often been used in the definition of the full emittance, i.e. $\epsilon = 4\epsilon_{\rm rms}$, to ensure that the phase-space area of such an ellipse is $\pi\epsilon$.

(b) Show that the rms emittance defined above is invariant under a coordinate rotation

$$X = x\cos\theta + x'\sin\theta, \quad X' = -x\sin\theta + x'\cos\theta,$$

and show that the correlation coefficient $R = \sigma_{XX'}/\sigma_X\sigma_{X'}$ is zero if we choose the rotation angle to be

$$\tan 2\theta = \frac{2\sigma_x\sigma_{x'}r}{\sigma_x^2 - \sigma_{x'}^2}.$$

Show that σ_X and $\sigma_{X'}$ reach extrema at this rotation angle.

[17]See P. Lapostolle, *IEEE Trans. Nucl. Sci.* **NS-18**, 1101 (1971), and J. Buon, CERN **91-04**, 30 (1991). The statistical definition of beam emittance is applicable to all phase space coordinates.

(c) In accelerators, particles are distributed in the Courant-Snyder ellipse:

$$I(x,x') = \gamma x^2 + 2\alpha x x' + \beta x'^2,$$

where α, β, γ are betatron amplitude functions. If the beam distribution function is a function of $I(x,x')$, show that

$$\epsilon_{\rm rms} = \frac{\sigma_x^2}{\beta} = \frac{\sigma_{x'}^2}{\gamma}, \quad r = -\frac{\alpha}{\sqrt{\beta\gamma}}, \quad \text{or} \quad \begin{pmatrix} \sigma_x^2 & \sigma_{xx'} \\ \sigma_{xx'} & \sigma_{x'}^2 \end{pmatrix} = \epsilon_{\rm rms} \begin{pmatrix} \beta & -\alpha \\ -\alpha & \gamma \end{pmatrix}.$$

and

$$\mathbf{x}^\dagger \sigma^{-1} \mathbf{x} = \frac{1}{\epsilon_{\rm rms}}(\gamma x^2 + 2\alpha x x' + \beta x'^2).$$

(d) Show that the σ matrix is transformed, in the linear betatron motion, according to

$$\begin{pmatrix} \sigma_x^2 & \sigma_{xx'} \\ \sigma_{xx'} & \sigma_{x'}^2 \end{pmatrix}_2 = M(s_2|s_1) \begin{pmatrix} \sigma_x^2 & \sigma_{xx'} \\ \sigma_{xx'} & \sigma_{x'}^2 \end{pmatrix}_1 M(s_2|s_1)^\dagger,$$

where $M^\dagger$ is the transpose of the matrix M. Use this result to show that $\mathbf{x}^\dagger \sigma^{-1} \mathbf{x}$ of Eq. (2.60) is invariant under betatron motion and thus an invariant beam distribution function is a function of $\mathbf{x}^\dagger \sigma^{-1} \mathbf{x}$. The transport equation for the σ-matrix can be used to measure the σ-matrix elements and derive the rms beam emittance. For a thick quadrupole lens, show that Eq. (2.61) becomes

$$\begin{aligned} \sigma_x^2(s_2) &= \sigma_x^2(s_1) \Big\{ \cos\sqrt{K}\ell_{\rm q} - L\sqrt{K}\sin\sqrt{K}\ell_{\rm q} + \frac{\sigma_{xx'}(s_1)}{\sigma_x^2(s_1)} \Big(\frac{1}{\sqrt{K}} \sin\sqrt{K}\ell_{\rm q} \\ &\quad + L\cos\sqrt{K}\ell_{\rm q} \Big) \Big\}^2 + \frac{\epsilon_{\rm rms}^2}{\sigma_x^2(s_1)} \left[\frac{1}{\sqrt{K}} \sin\sqrt{K}\ell_{\rm q} + L\cos\sqrt{K}\ell_{\rm q} \right]^2, \end{aligned}$$

where $K = B_1/B\rho$ and $\ell_{\rm q}$ and the focusing function and the length of the quadrupole, and L is the length of the drift space between the quadrupole and the profile monitor.

(e) Particle motion in synchrotrons obeys Hamiltonian dynamics with

$$x' = \frac{dx}{ds}, \quad \frac{dx'}{ds} = -\frac{\partial H}{\partial x}.$$

Show that

$$\frac{d\epsilon^2}{ds} = -2\sigma_x^2(\langle x' \frac{\partial H}{\partial x}\rangle - \langle x'\rangle\langle \frac{\partial H}{\partial x}\rangle) + 2\sigma_{xx'}(\langle x \frac{\partial H}{\partial x}\rangle - \langle x\rangle\langle \frac{\partial H}{\partial x}\rangle).$$

For a linear Hamiltonian, we have $\partial H/\partial x = Kx$, where $K(s)$ is the focusing function. Show that the rms emittance is conserved. What would your conclusion be if the Hamiltonian were nonlinear?

15. Consider a beam of noninteracting particles in an accelerator with focusing function $K_y(s)$, where the particle betatron coordinate obeys Hill's equation

$$y'' + K_y(s) y = 0.$$

Let Y be the envelope radius of the beam with emittance ϵ, i.e. $Y(s) = \sqrt{\beta(s)\epsilon}$.

(a) Show that the envelope equation of motion is

$$Y'' + K_y(s)Y - \frac{\epsilon^2}{Y^3} = 0.$$

(b) Show that the envelope equation can be derived from the envelope Hamiltonian[18]

$$H_{\rm env} = \frac{1}{2}P^2 + V_{\rm env}, \qquad V_{\rm env} = \frac{1}{2}K_yY^2 + \frac{\epsilon^2}{2Y^3},$$

where (P, Y) are conjugate envelope phase-space coordinates with $P = Y'$, and $V_{\rm env}$ is the "potential energy". In a smooth focusing approximation, $K_y(s) = (2\pi/L)^2$, where L is the wavelength of the betatron oscillations. Using the smooth approximation, we have $\langle K_y \rangle = (2\pi Q_y/C)^2$ obtained from Floquet transformation to Hill's equation, where C is the circumference, and Q_y is the betatron tune. The corresponding average betatron wavelength is C/Q_y, and the average betatron amplitude function is $\langle \beta_y \rangle = R/Q_y$, where R is the average radius. The equivalent betatron amplitude function is $\beta_y = L/2\pi$. The matched beam radius is given by $dV_{\rm env}/dY = 0$, i.e. $Y_{\rm m} = \sqrt{L\epsilon/2\pi}$. Show that the solution of the betatron motion and the solution of the envelope equation are

$$y = \sqrt{\frac{L\epsilon}{2\pi}}\cos(\frac{2\pi s}{L} + \chi_\beta),$$
$$Y^2 = \sqrt{A^2 + \left(\frac{L\epsilon}{2\pi}\right)^2} + A\cos(2\frac{2\pi s}{L} + \chi),$$

where the parameters A and χ are determined by the initial beam conditions. Thus the envelope of a mis-injected beam bunch will oscillate at twice the betatron oscillation frequency (the quadrupole mode).

(c) Let us make Floquet transformation to the envelope equation in part (a) with

$$R = \frac{Y}{\sqrt{\beta\epsilon}}, \qquad \phi = \frac{1}{\nu}\int_0^s \frac{ds}{\beta(s)},$$

where β is the betatron amplitude function, and ν is the betatron tune. Show that the normalized envelope R satisfies the equation:

$$\frac{d^2}{d\phi^2}R + \nu^2 R - \frac{\nu^2}{R^3} = 0.$$

Using $(R, P_R = dR/d\phi)$ as the conjugate phase space coordinates, we obtain the envelope Hamiltonian as $H = \frac{1}{2}P_R^2 + V_{\rm env}(R)$, where the envelope potential is

$$V_{\rm env} = \frac{1}{2}\nu^2R^2 + \frac{\nu^2}{2R^2}.$$

Show that the exact solution of the envelope equation is

$$R^2 = \sqrt{1 + a^2} + a\cos(2\nu\phi + \chi),$$

[18]See S.Y. Lee and A. Riabko, *Phys. Rev.* **E51**, 1609 (1995); ibid. *Phys. Rev.* **E51**, 3529 (1995).

where a is the envelope mismatch amplitude. Note: if the *square* of the rms beam width is plotted as a function of revolution turns, the resulting oscillation will be sinusoidal. The envelope Hamiltonian is, in fact, linear.

16. The Courant–Snyder phase-space ellipse of a synchrotron is $\gamma y^2 + 2\alpha y y' + \beta y'^2 = \epsilon$, where α, β and γ are the Courant–Snyder parameters. If the injection optics is mismatched with $\gamma_1 y^2 + 2\alpha_1 y y' + \beta_1 y'^2 = \epsilon$, find the emittance growth factor (note that the easiest way to estimate the emittance growth is to transform the injection ellipse into the normalized coordinates of the ring optics. The deviation of the injection ellipse from a circle in the normalized phase space corresponds to the emittance growth).

    (a) Transform the injection ellipse into the normalized coordinates of the ring lattice, and show that the injection ellipse becomes

$$\left(\frac{\beta}{\beta_1} + \frac{(\alpha_1\beta - \beta_1\alpha)^2}{\beta\beta_1}\right) Y^2 + 2\frac{\alpha_1\beta - \alpha\beta_1}{\beta} YP + \frac{\beta_1}{\beta} P^2 = \epsilon,$$
$$Y = \frac{1}{\sqrt{\beta}}\, y, \quad P = \frac{1}{\sqrt{\beta}}(\alpha y + \beta y').$$

    (b) Transform the ellipse to the upright orientation, and show that the major and minor axes of the ellipse are

$$F_+ = \left(X_{\rm mm} + \sqrt{X_{\rm mm}^2 - 1}\right)^{1/2}, \quad F_- = \left(X_{\rm mm} - \sqrt{X_{\rm mm}^2 - 1}\right)^{1/2},$$

    where the mismatch factor $X_{\rm mm}$ is (see Exercise 2.2.14)

$$X_{\rm mm} = \frac{1}{2}(\gamma_1\beta + \beta_1\gamma - 2\alpha_1\alpha) = \frac{1}{2\epsilon_{\rm rms}}(\beta\sigma_{x'}^2 + \gamma\sigma_x^2 + 2\alpha\sigma_{xx'}).$$

    Note that the rms quantities $\sigma_x, \sigma_{x'}$ and $\sigma_{xx'}$ can be measured from the injected beam. What happens to the beam if the beam is injected into a perfect linear machine where there is no betatron tune spread? Show that the tune of the envelope oscillations is twice the betatron tune (see Exercise 2.2.15).

    (c) In general, nonlinear betatron detuning arises from space-charge forces, nonlinear magnetic fields, chromaticities, etc. Because the betatron tune depends on the betatron amplitude, the phase-space area of the mis-injected beam will decohere and grow. Show that the emittance growth factor is

$$F_+^2 = \left(X_{\rm mm} + \sqrt{X_{\rm mm}^2 - 1}\right).$$

    (d) Let the betatron amplitude function at the injection point be $\beta_x = 17.0$ m and $\alpha_x = 2.02$. The injection ellipse of a beam with emittance 5π mm-mrad is given by $x^2/a^2 + x'^2/b^2 = 1$, where $a = 5.00$ mm and $b = 1.00$ mrad. Find the final beam emittance after nonlinear decoherence.

17. At an interaction point (IP) of a collider, or at a symmetry point in a storage ring, the lattice betatron functions are usually designed to an appropriate $\beta_{x,z}^*$ value with symmetry condition: $\alpha_{x,z}^* = 0$. The resulting betatron amplitude functions in the straight section become $\beta_{x,z} = \beta_{x,z}^* + s^2/\beta_{x,z}^*$ (see Exercise 2.2.8). The luminosity, $\mathcal{L}$,

measuring the probability of particle encounters in a head on collision of two beams, is

$$\mathcal{L} = 2fN_1N_2 \int \rho_1(x, z, s_1)\rho_2(x, z, s_2)\, dx\, dz\, ds\, d(\beta ct),$$

where $s_1 = s + \beta ct$ and $s_2 = s - \beta ct$.

(a) Assuming Gaussian bunch distribution with

$$\rho(x, z, s) = \frac{1}{(2\pi)^{3/2}\sigma_x\sigma_z\sigma_s} \exp\left\{-\frac{x^2}{2\sigma_x^2} - \frac{z^2}{2\sigma_z^2} - \frac{s^2}{2\sigma_s^2}\right\},$$

where $\sigma_x = \sqrt{\beta_x\epsilon_x}$, $\sigma_z = \sqrt{\beta_z\epsilon_z}$, σ_s are respectively the rms beam sizes in x, z, s directions, show that the luminosity, in a short bunch condition with $\sigma_s \ll \beta^*_{x,z}$, is

$$\mathcal{L} = R(A)\frac{fN_1N_2}{4\pi\sigma_x^*\sigma_z^*},$$

where R is the reduction factor, and $\sigma_x^* = \sqrt{\beta_x^*\epsilon_x}$ and $\sigma_z^* = \sqrt{\beta_z^*\epsilon_z}$ are rms beam size at the IP.

(b) Because of finite bunch-length σ_s, show that the luminosity reduction factor for two identical Gaussian distributions is

$$R(A_x, A_z) = \frac{2}{\sqrt{\pi}} \int \frac{e^{-\zeta^2} d\zeta}{\sqrt{(1 + (\zeta^2/A_x^2))(1 + (\zeta^2/A_z^2))}}$$

where $A_{x,z} = \beta^*_{x,z}/\sigma_s$ is a measure of the betatron amplitude variation at the interaction point. In a short bunch approximation with $A_x \gg 1$ and $A_z \gg 1$, we obtain $R(A_x, A_z) \approx 1$. Most colliders operate at a condition $A_{x,z} \approx 1$. The luminosity reduction due to finite bunch-length is called the hour-glass effect.

(c) For a round beam with $A = A_x = A_z$, show that (see Section 7.1.3. in Ref. [30])

$$R(A) = \sqrt{\pi}Ae^{A^2}\,\mathrm{erfc}(A) \approx \frac{\sqrt{\pi}A(1 + 0.2836A + 0.07703A^2)}{(1 + 0.47047A)^3},$$

where the latter approximate identity is valid up to about $A \le 2.5$. Asymptotically, we have $R(A) \to 1$ for $A \to \infty$. Plot $R(A)$ as a function of A and show that the actual luminosity is

$$\mathcal{L} = R(A)\mathcal{L}_0 = \frac{fN_1N_2}{4\pi\epsilon_\perp\sigma_s}\sqrt{\pi}e^{A^2}\,\mathrm{erfc}(A)$$

for a given σ_s, where $\epsilon_\perp = \epsilon_x = \epsilon_z$. Plot $\mathcal{L}$ as a function of A. Does the luminosity decrease at $A < 1$?

(d) For a flat beam with $\beta_x^* \gg \sigma_s$, i.e. $A_x \gg 1$, show that the reduction factor becomes

$$R = \frac{2A_z}{\sqrt{\pi}} \int \frac{e^{-A_z^2\zeta^2} d\zeta}{\sqrt{1+\zeta^2}} = \frac{A_z}{\sqrt{\pi}} e^{\frac{A_z^2}{2}} K_0(\frac{A_z^2}{2}),$$

where K_0 is the modified Bessel function. Calculate the reduction factor as a function of A_z and show that the luminosity is (use 3.364.3 of Ref. [31])

$$\mathcal{L} = R(A_z)\mathcal{L}_0 = \frac{fN_1N_2}{4\pi\sqrt{\beta_x\epsilon_x\epsilon_z\beta_z}} \frac{1}{\sqrt{\pi}} \sqrt{A_z} e^{\frac{A_z^2}{2}} K_0(\frac{A_z^2}{2}).$$

18. **Focusing of atomic beams:**[19] There are now two types of polarized ion sources: the atomic beam polarized ion source (ABS), and the optically pumped polarized ion source (OPPIS) producing mainly hydrogen and deuterium ions. The ABS has produced polarized H^- ions with about 75% polarization at a peak current of 150 μA with 100 μs duration and 5 Hz repetition rate. Similarly, OPPIS has been able to produce a polarized H^- ion source up to 400 μA with 80% polarization at a normalized emittance near 1 π mm-mrad. The principle of the ABS is to form atomic beams in a discharge tube called a *dissociator*. As the beam travels through the beam tube, the spin states of the atoms are selected in a *separation magnet*, which is a quadrupole or a sextupole.[20] The non-uniform magnetic field preferentially selects one spin state (Stern-Gerlach effect). This exercise illustrates the focusing effect due to a sextupole field. Let $\vec{\mu} = \tilde{g}\mu_B\vec{J}$ be the magnetic moment of the atomic beam, where $\tilde{g}$ is the Landé g-factor, $\mu_B = 5.788 \times 10^{-5}$eV/T is the Bohr magneton, and $\vec{J}$ is the angular momentum of the atom. The magnetic energy of the atomic beam in the magnetic field $\vec{B}$ and the force acting on the hydrogen atom are

$$W = -\vec{\mu}\cdot\vec{B}. \qquad\qquad \vec{F} = \nabla(\vec{\mu}\cdot\vec{B}) = \pm\mu_a\nabla|\vec{B}|$$

for two quantized spin 1/2 states of the hydrogen atom, i.e. the electron spin is quantized along the $\vec{B}$ direction, and $\mu_a \approx \mu_B$ for the hydrogen-like atom.

(a) Show that the sextupole field focuses the spin state of the atomic beam with lower magnetic dipole energy; in other words, it defocuses the spin state with higher magnetic dipole energy. The atoms not contained in the beam pipe will be pumped away. It is worth pointing out that there is no preferred direction of the spin projection inside the sextupole. The electron spin is quantized with respect to the magnetic field. The selected atoms, which have a preferential one-spin state, will pass through the *transition* region. Here the the magnetic field is slowly changed to align all atomic polarization into the uniform field *ionizer* region, where in the high-field regime the nuclear spin can be flipped by rf field, the polarized ions are formed by the bombardment of electron beams, and the polarized ions are drawn by the electric field to form a polarized ion beam.

(b) When a quadrupole is used to replace the sextupole magnet, show that the effective force on the atom is a dipole field.

(c) If the temperature of the dissociator is 60 K, what is the velocity spread of the atomic beam? Discuss the effect of velocity spread of the atomic beam.

19. **A paraxial focusing system (lithium lens):** A strong paraxial focusing system can greatly increase the yield of the secondary beams. To this end, the lithium lens or a strong solenoid has been used. The Li lens was first used at Novosibirsk for focusing the e^+e^- beams. It became the essential tool for anti-proton collection at Fermilab.[21] A cylindrical lithium rod carrying a uniform current pulse can create a large magnetic field. The magnetic flux density is

$$B_\theta(r) = \frac{\mu I r}{2\pi r_0^2},$$

[19] See e.g., W. Haberli, *Ann. Rev. Nucl. Sci.*, **17**, 373 (1967).

[20] H. Friedburg, and W. Paul, *Naturwiss.* **38**, 159 (1951); H.G. Bennewitz and W. Paul, *Z. Phys.* **139**, 489 (1954).

[21] B.F. Bayanov, *et al.*, *Nucl. Inst. Methods.* **190**, 9 (1981).

where I is the current, r is the distance from the center of the rod, r_0 is the radius of the Li conductor, and $\mu \approx \mu_0 = 4\pi \times 10^{-7}$ Tm/A is the permeability.

(a) Find the focusing function for the 8-GeV kinetic energy antiprotons if $I = 500$ kA, $r_0 = 10$ mm, and the length is 15 cm. What is the focal length?

(b) The total nuclear reaction cross-section between the antiprotons and the Li nucleus is given by the geometric cross-section, i.e. $\sigma_{\rm p,A} = \pi(r_{\rm p}^2 + R_{\rm A}^2)$, where $r_p = 0.8$ fm, $R_{\rm A} = 1.3 \times A^{1/3}$ fm, and A is the atomic mass number. The atomic weight is 6.941 g, and the density is 0.5 g/cm^3, show that the nuclear reaction length is about 1 m. To minimize the beam loss, choose the length of the Li lens to be less than 10% of the nuclear reaction length.

(c) Find the magnetic pressure $P = B^2/2\mu_0$ that acts to compress the Li cylinder in units of atmospheric pressure (1 atm $= 1.013 \times 10^5$ N/m).

20. Low energy synchrotrons often rely on the bending radius $K_x = 1/\rho^2$ for horizontal focusing and edge angles in dipoles for vertical focusing. Find the lattice property of the low energy synchrotron described by the following input data file (MAD). What is the effects of changing the edge angle and dipole length? Discuss the stability limit of the lattice.

```
TITLE,"CIS BOOSTER (1/5 Cooler), (90degDIP)"
! CIS =1/5 of Cooler circumference =86.82m / 5 =17.364m
! It accelerates protons from 7 MeV to 200 MeV in 1-5 Hz.
LCELL:=4.341          ! cell length 17.364m/4
L1:= 2.0              ! dipole length
L2:=LCELL-L1          ! straight section length
RHO:=1.27324
EANG:=12.*TWOPI/360 ! use rad. for edge angle
ANG := TWOPI/4
OO : DRIFT,L=L2
BD : SBEND,L=L1, ANGLE=ANG, E1=EANG,E2=EANG, K2=0.
SUP: LINE=(BD,OO)     ! a superperiod
USE,SUP,SUPER=4
PRINT,#S/E
TWISS,DELTAP=0.0,TAPE
STOP
```

21. The action angle coordinate transformation of Eq. (2.48) can also be carried out by using the generating function $F_3(y', \psi)$. Show that the generating function is

$$F_3(y', \psi) = \frac{\beta y'^2}{2\left(\tan\psi - \frac{\beta'}{2}\right)}$$

and show that the new Hamiltonian is also Eq. (2.49).

III Effect of Linear Magnet Imperfections

In the presence of magnetic field errors, Hill's equations (2.22) are

$$x'' + K_x(s)x = \frac{\Delta B_z}{B\rho}, \qquad z'' + K_z(s)z = -\frac{\Delta B_x}{B\rho}, \tag{2.89}$$

$$\Delta B_z + j\Delta B_x = B_0 \sum_{n=0}^{\infty} (b_n + ja_n)\,(x + jz)^n.$$

where ΔB_z and ΔB_x are the perturbing fields, B_0 is the main dipole field, b_0 and b_1 are respectively the dipole and quadrupole field errors, b_2 is the sextupole field error, etc. The a's are skew magnetic field errors. This section addresses the linear betatron perturbation resulting from the dipole (b_0) and quadrupole (b_1) field errors, and illustrates possible beam manipulation by using the perturbing fields.

Based on our study of the betatron motion in Sec. II, we will show that linear magnet imperfections have two major effects: (1) closed-orbit distortion due to dipole field error, and (2) betatron amplitude function perturbation due to quadrupole field error. The effect of linear betatron coupling due to the skew quadrupole term, a_1, and the solenoid will be discussed in Sec. VI.

III.1 Closed-Orbit in the Presence of Dipole Field Error

Up to now, we have assumed perfect dipole magnets with an ideal reference closed orbit that passes through the center of all quadrupoles. In reality, dipole field errors may arise from errors in dipole length or power supply, dipole roll giving rise to a horizontal dipole field, a closed orbit not centered in the quadrupoles, and feed-down from higher-order multipoles.

A. The perturbed closed orbit and Green's function

First, we consider a single thin dipole field error at a location $s = s_0$ with a kick-angle $\theta = \Delta B dt/B\rho$ in an otherwise ideal accelerator, where ΔB is the dipole field error, dt is the dipole thickness (length), $\Delta B dt$ is the integrated dipole field error and $B\rho = p_0/e$ is the momentum rigidity of the beam. Let

$$\mathbf{y}_- = \begin{pmatrix} y_0 \\ y_0' - \theta \end{pmatrix}, \quad \mathbf{y}_+ = \begin{pmatrix} y_0 \\ y_0' \end{pmatrix}$$

be the phase-space closed-orbit state-vectors just before and just after the kick element located at s_0. The closed-orbit condition is

$$\mathbf{M}\begin{pmatrix} y_0 \\ y_0' \end{pmatrix} = \begin{pmatrix} y_0 \\ y_0' - \theta \end{pmatrix}, \tag{2.90}$$

where $\mathbf{M}$ is the one-turn transfer matrix of Eq. (2.34) for an ideal accelerator. The resulting closed orbit at s_0 is

$$y_0 = \frac{\beta_0\theta}{2\sin\pi\nu}\cos\pi\nu, \qquad y_0' = \frac{\theta}{2\sin\pi\nu}(\sin\pi\nu - \alpha_0\cos\pi\nu),$$

where ν is the betatron tune and α_0, β_0 are the values of the betatron amplitude functions at kick dipole location s_0.

The closed orbit at other location s in the accelerator can be obtained from the propagation of betatron oscillations Eq. (2.30). Using the transfer matrix of Eq. (2.42), we obtain

$$\begin{pmatrix} y(s) \\ y'(s) \end{pmatrix}_{\rm co} = M(s|s_0)\begin{pmatrix} y_0 \\ y_0' \end{pmatrix}, \qquad \Longrightarrow \qquad y_{\rm co}(s) = G(s,s_0)\theta(s_0), \quad (2.91)$$

$$G(s,s_0) = \frac{\sqrt{\beta(s)\beta(s_0)}}{2\sin\pi\nu}\cos(\pi\nu - |\psi(s) - \psi(s_0)|),$$

where $G(s,s_0)$ is the Green function of Hill's equation. The presence of $\sin\pi\nu$ in the denominator of Green's function shows that the closed orbit may not exist if the betatron tune is an integer. The orbit response arising from a dipole field error is given by the product of the Green function and the kick angle. The right plot of Fig. 2.12 shows the closed-orbit perturbation in the AGS booster due to a dipole field error of 6.82 mr. The left plot is a schematic drawing of the resulting closed orbit around an ideal orbit. Since the betatron tune of the AGS booster is 4.82, the closed-orbit perturbation is dominated by the $n = 5$ harmonic, showing 5 complete oscillations in Fig. 2.12.

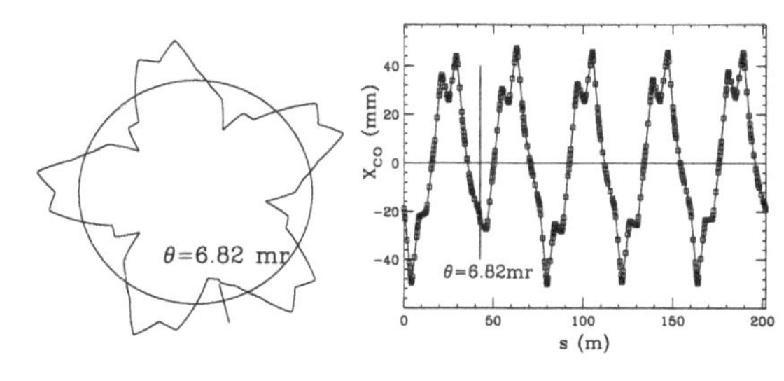

Figure 2.12: Left: schematic plot of the closed-orbit of the AGS booster resulting from a horizontal kicker with kick-angle $\theta = 6.82$ mr at the location marked by a straight line. Since the betatron tune of 4.82 is close to the integer 5, the closed orbit is dominated by the fifth error harmonic. Right: the closed orbit as a function of the longitudinal distance.

Equation (2.91) shows that the closed orbit becomes infinite when the condition $\sin\pi\nu = 0$ is encountered. The orbit kicks in every turn due to a dipole error coherently add up, making the closed orbit unstable. The left plot of Fig. 2.13 shows schematically the evolution of a phase-space trajectory (y, y') in the presence of a dipole error when the betatron tune is an integer. Since the angular kick $\Delta y' = \theta$, where θ is the kick angle of the error dipole, is in the same direction in each revolution, the closed orbit does not exist. The betatron tunes are chosen to avoid integers. In other words, if the betatron tune is near an integer, the closed orbit becomes very sensitive to dipole field error.

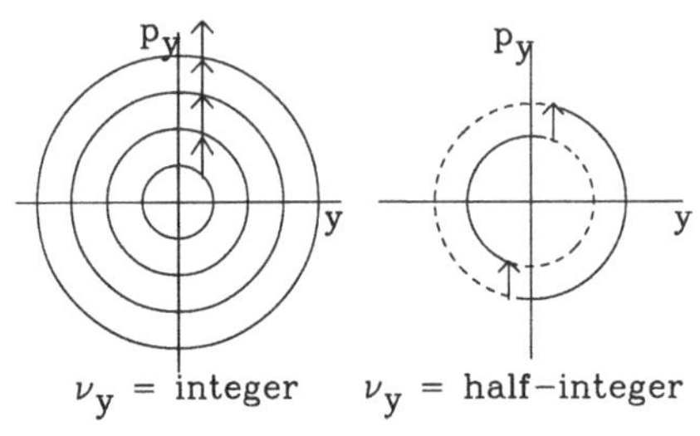

Figure 2.13: Left, a schematic plot of the closed-orbit perturbation due to an error dipole kick when the betatron tune is an integer. Here $\Delta p_y = \beta_y \Delta y' = \beta_y \theta$, where θ is the dipole kick angle and β_y is the betatron amplitude function value at the dipole. Right, a schematic plot of the particle trajectory resulting from a dipole kick when the betatron tune is a half-integer; here the angular kicks from two consecutive orbital revolutions cancel each other.

If the betatron tune is a half-integer, the angular kicks of two consecutive revolutions cancel each other (see right plot of Fig. 2.13). For the closed orbit, it is better to choose a betatron tune closer to a half-integer. However, we will show later that the quadrupole field error will produce betatron amplitude-function instability at a half integer tune. Thus the betatron tunes should also avoid half-integers.

B. Distributed dipole field error

In reality, dipole field errors are distributed around the accelerator. Since Hill's equation with distributed dipole field errors $\Delta\theta(t) = (\Delta B(t)/B\rho)dt$ is linear, the closed orbit can be obtained by a linear superposition of dipole kicks, i.e.

$$\begin{aligned} y_{\rm co}(s) &= \int_s^{s+C} G(s,t)\frac{\Delta B(t)}{B\rho}dt = \frac{\sqrt{\beta(s)}}{2\sin\pi\nu}\int_s^{s+C}\sqrt{\beta(t)}\frac{\Delta B(t)}{B\rho}\cos(\pi\nu+\psi(s)-\psi(t))dt \\ &= \frac{\nu\sqrt{\beta(s)}}{2\sin\pi\nu}\int_\phi^{\phi+2\pi}\left[\beta^{3/2}(\varphi)\frac{\Delta B(\varphi)}{B\rho}\right]\cos\nu(\pi+\phi-\varphi)d\varphi, \end{aligned} \tag{2.92}$$

where $\phi(s) = (1/\nu)\int_0^s dt/\beta(t)$, and $\psi(s) = \nu\phi(s)$. It is easy to verify that Eq. (2.92) is the closed-orbit solution of the inhomogeneous Hill equation

$$\frac{d^2y}{ds^2} + K_y(s)y = \frac{\Delta B}{B\rho}, \tag{2.93}$$

where $\Delta B = \pm\Delta B_z$ for horizontal motion and $\Delta B = \mp\Delta B_x$ for vertical motion (the upper sign for positive charge, and the lower sign for negative charge).

C. The integer stopband integrals

Since square bracketed term in the integrand of Eq. (2.92) is a periodic function of 2π, it can be expressed in a Fourier series and obtain the closed orbit:

$$f(\phi) \equiv \nu\beta^{3/2}(\phi)\frac{\Delta B(\phi)}{B\rho} = \sum_{k=-\infty}^{\infty} f_k\, e^{jk\phi}, \tag{2.94}$$

$$f_k = \frac{1}{2\pi}\oint \nu\beta^{3/2}\frac{\Delta B}{B\rho}\, e^{-jk\phi}\, d\phi = \frac{1}{2\pi}\oint \sqrt{\beta}\,\frac{\Delta B}{B\rho}\, e^{-jk\phi}\, ds \tag{2.95}$$

$$y_{\rm co}(s) = \sqrt{\beta(s)}\sum_{k=-\infty}^{\infty}\frac{\nu f_k}{\nu^2-k^2}e^{jk\phi} \approx \frac{\sqrt{\beta(s)}\,|f_{[\nu]}|\,\cos([\nu]\phi+\chi)}{(\nu-[\nu])}, \tag{2.96}$$

where the Fourier amplitude f_k is called the *integer stopband integral* with $f_{-k} = f_k^*$. The closed orbit has simple poles at integer harmonics. The simple pole structure in Eq. (2.96) indicates that the closed orbit is most sensitive to the error harmonics closest to the betatron tune. The resulting closed orbit is usually dominated by a few harmonics near $[\nu]$, an integer nearest the betatron tune.

D. Statistical estimation of closed-orbit errors

In practice, the perturbing field error $\Delta B/B\rho$, due mainly to random construction errors in the dipole magnets and misalignment errors in the quadrupoles, is not known a priori. During the design stage of an accelerator, a statistical argument is usually used to estimate the rms closed orbit,

$$y_{\rm co,rms} \approx \frac{\beta_{\rm av}}{2\sqrt{2}|\sin\pi\nu|}\sqrt{N}\theta_{\rm rms}, \tag{2.97}$$

where $\beta_{\rm av}$, N, and $\theta_{\rm rms}$ are respectively the average β-function, the number of dipoles with field errors, and the rms angular kick angle.

Now we consider the dipole field error generated by quadrupole misalignment. When quadrupole magnets are misaligned by a distance Δy, the effective angular kick and the resulting closed orbit error are

$$\theta = \frac{B_1\ell}{B\rho}\Delta y = \frac{\Delta y}{f},$$

$$y_{\rm co,rms} \approx \left\{\frac{\beta_{\rm av}}{2\sqrt{2}f_{\rm av}|\sin\pi\nu|}\sqrt{N_{\rm q}}\right\}\Delta y_{\rm rms}, \tag{2.98}$$

where $B_1 = \partial B_z/\partial x$ is the quadrupole gradient, f is the focal length, $N_{\rm q}$ is the number of quadrupoles and $f_{\rm av}$ is the average focal length. The coefficient in curly brackets is called a *sensitivity factor* for quadrupole misalignment. For example, if the sensitivity factor is 20, an rms quadrupole misalignment of 0.1 mm will result in a rms closed-orbit distortion of 2 mm. The sensitivity factor increases with the size of an accelerator.

E. Closed-orbit correction

Closed-orbit correction is an important task in accelerator commissioning. If the closed orbit is large, the beam lifetime and dynamical aperture can be severely reduced. First, any major sources of dipole error should be corrected. The remaining closed orbit can generally be corrected by either the stopband correction scheme, or harmonic correction scheme, or χ^2-minimization method.

With a few dipole correctors, the stopband near $k = [\nu]$ is

$$f_{[\nu]} = \frac{1}{2\pi\nu} \sum_i \sqrt{\beta_i}\, \theta_i\, e^{-j[\nu]\phi_i}, \tag{2.99}$$

where θ_i is the angular kick of the ith corrector. Placing these correctors at high-β locations with a phase advance between correctors of $[\nu]\phi_i \approx \pi/2$, one can adjust the real and imaginary parts independently.

The harmonic closed-orbit correction method uses distributed dipole correctors powered with a few harmonics nearest the betatron tune to minimize a set of stopband integrals f_k. For example, if $N_{\rm c}$ dipole correctors are powered with

$$\theta_i = \frac{1}{\sqrt{\beta_i}}(a_k \cos k\phi_i + b_k \sin k\phi_i), \quad (i = 1, \cdots, N_{\rm c}),$$

where β_i and ϕ_i are the betatron amplitude function and the betatron phase at the ith kicker location, the kth stopband can be corrected by adjusting the a_k and b_k coefficients. A few harmonics can be superimposed to eliminate all dangerous stopbands.

Another orbit correction method is the χ^2-minimization procedure. Let $N_{\rm m}$ be the number of BPMs and $N_{\rm c}$ the number of correctors. Let $y_{i,{\rm co}}$ and Δ_i be the closed-orbit deviation and BPM resolution of the ith BPM.[22] The aim is to minimize χ^2 of closed orbit error by varying $\theta_1, \theta_2, \ldots$ of $N_{\rm c}$ correctors, where

$$\chi^2 = \sum_{i=1}^{N_{\rm m}} \frac{|y_{i,{\rm co}}|^2}{\Delta_i^2}.$$

All orbit correction schemes minimize only error-harmonics nearest the betatron tune. Because the closed orbit is not sensitive to error-harmonics far from the betatron tune, these harmonics can hardly be changed by closed-orbit correction schemes.

In many beam manipulation applications such as injection, extraction, manipulation with an internal target, etc., local closed-orbit bumps are often used. Possible schemes of local orbit bumps are the "four-bump method" discussed in Sec. III.3 and the "three-bump method" (see Exercise 2.3.4).

[22]The BPM resolution depends on the stability of the machine and on the number of bits and the effective width of the pickup electrode (PUE). For example, the BPM resolution for the data acquisition system with a 12-bit ADC and a 40-mm effective width PUE is about 10 μm. If an 8-bit ADC is used, the resolution is worsened by a factor of 16. The BPM resolution for proton storage rings is about 10 to 100 μm.

F. Effects of dipole field error on orbit length

The path length of a circulating particle in the Frenet-Serret coordinate system is

$$C = \oint \sqrt{(1+x/\rho)^2 + x'^2 + z'^2}\, ds \approx C_0 + \oint \frac{x}{\rho}\, ds + \cdots, \tag{2.100}$$

where C_0 is the orbit length of the unperturbed orbit, and higher order terms associated with betatron motion are neglected. Since a dipole field error gives rise to a closed-orbit distortion, the circumference of the closed orbit may be changed as well. We consider the closed-orbit change due to a single dipole kick at $s = s_0$ with kick angle θ_0. Using Eq. (2.92), we find the change in circumference as

$$\Delta C = C - C_0 = \theta_0 \oint \frac{G_x(s, s_0)}{\rho}\, ds = D(s_0)\, \theta_0, \tag{2.101}$$

$$D(s_0) = \oint \frac{G_x(s, s_0)}{\rho} ds = \frac{\sqrt{\beta_x(s_0)}}{2\sin \pi\nu_x} \oint \frac{\sqrt{\beta_x(s)}}{\rho} \cos(\pi\nu_x - |\psi_x(s) - \psi(s_0)|) ds.$$

Here $D(s_0)$ is the value of the dispersion function at s_0 (see Sec. IV). The change in orbit length due to a dipole field error is equal to the dispersion function times the orbital kick angle. When dipole field errors are distributed in a ring, the change in the total path length becomes

$$\Delta C = \oint D(s) \frac{\Delta B_z(s)}{B\rho} ds. \tag{2.102}$$

In many cases, the dipole field errors are generated by power supply ripple, ground vibration, traffic and mechanical vibration, tidal action, etc., and thus the circumference is modulated at some modulation frequencies. The modulation frequency from ground vibration is typically less than 10 Hz. The power supply ripple can produce modulation frequency at some harmonics of 50 or 60 Hz, and the frequency generated by mechanical vibrations is usually of the order of kHz. Normally, particle motion in an accelerator can tolerate small-amplitude modulation provided that modulation frequencies do not induce betatron or synchrotron resonances. However, if a modulation frequency is equal to the betatron or synchrotron frequency, particle motion will be strongly perturbed. For example, an rf dipole field operating at a betatron sideband[23] can kick the beam out of the vacuum chamber; this is called *rf knock-out*. This method can be used to measure the betatron tune.

[23]The FFT spectra of a transverse phase-space coordinate display rotational harmonics at integer multiples of the revolution frequency and the betatron lines next to the rotation harmonics. These betatron frequency lines are called the betatron sidebands. See Sec. III.7 for details.

III.2 Extended Matrix Method for the Closed Orbit

The inhomogeneous differential equation (2.89) for the closed orbit of the betatron oscillation can be solved by the extended 3×3 transfer matrix method. For example, the equation of motion for a dipole field error in a combined function magnet is $x'' + Kx = \frac{\Delta B_z}{B\rho}$. The betatron phase-space coordinates before and after the combined function quadrupole is given by the extended transfer matrix

$$\begin{pmatrix} x \\ x' \\ 1 \end{pmatrix}_2 = \begin{pmatrix} \cos\sqrt{K}\ell & \frac{1}{\sqrt{K}}\sin\sqrt{K}\ell & \frac{\Delta B_z}{B\rho K}(1-\cos\sqrt{K}\ell) \\ -\sqrt{K}\sin\sqrt{K}\ell & \cos\sqrt{K}\ell & \frac{\Delta B_z}{B\rho\sqrt{K}}\sin\sqrt{K}\ell \\ 0 & 0 & 1 \end{pmatrix} \begin{pmatrix} x \\ x' \\ 1 \end{pmatrix}_1, \quad (2.103)$$

where $\ell = s_2 - s_1$. In thin lens approximation, the transfer matrix of Eq. (2.103) becomes

$$M(s_2|s_1) = \begin{pmatrix} 1 & 0 & 0 \\ -1/f & 1 & \theta \\ 0 & 0 & 1 \end{pmatrix},$$

where $\theta = \Delta B_z \ell / B\rho$ and $f = 1/K\ell$ are respectively the dipole kick angle and the focal length of the perturbing element. Dipole field error can also arise from quadrupole misalignment. Let $\Delta y_{\rm q}$ be the quadrupole misalignment. The resulting extended transfer matrix in the thin-lens approximation is

$$M_{\rm quad} = \begin{pmatrix} 1 & 0 & 0 \\ -1/f & 1 & -\Delta y_{\rm q}/f \\ 0 & 0 & 1 \end{pmatrix}.$$

The 3×3 extended transfer matrix can be used to obtain the closed orbit of betatron motion. For example, the closed-orbit equation (2.90) is equivalent to

$$\begin{pmatrix} y_0 \\ y_0' \\ 1 \end{pmatrix} = \begin{pmatrix} 1 & 0 & 0 \\ 0 & 1 & \theta \\ 0 & 0 & 1 \end{pmatrix} \begin{pmatrix} M_{11} & M_{12} & 0 \\ M_{21} & M_{22} & 0 \\ 0 & 0 & 1 \end{pmatrix} \begin{pmatrix} y_0 \\ y_0' \\ 1 \end{pmatrix}, \quad (2.104)$$

where M's are matrix elements of 2×2 one-turn transfer matrix for an ideal machine, and θ is the dipole kick angle. Similarly, the 3×3 extended transfer matrix can be used to analyze the sensitivity of the closed orbit to quadrupole misalignment by multiplying the extended matrices along the transport line.[24]

III.3 Application of Dipole Field Error

Sometime, we create imperfections in an otherwise perfect accelerator for beam manipulation. Examples are the local-orbit bump, one-turn kicker for fast extraction, rf knock-out, etc.

---

[24] S.Y. Lee, S. Tepikian, *Proc. IEEE PAC Conf.*, p. 1639, (IEEE, Piscataway, N.J., 1991).

A. Orbit bumps

To facilitate injection, extraction, or special-purpose beam manipulation,[25] the orbit of beams can be bumped to a desired transverse position at specified locations. In this example, we discuss the four-bump method facilitated by four thin dipoles with kick angles θ_i ($i = 1, 2, 3, 4$). Using Eq. (2.92), we obtain

$$y_{\rm co}(s) = \frac{\sqrt{\beta(s)}}{2\sin\pi\nu}\sum_{i=1}^{4}\sqrt{\beta_i}\,\theta_i\ \cos(\pi\nu - |\psi - \psi_i|),$$

where $\theta_i = (\Delta B \Delta s)_i / B\rho$ and $(\Delta B \Delta s)_i$ are the kick-angle and the integrated dipole field strength of the i-th kicker. The conditions that the closed orbit is zero outside these four dipoles are $y_{\rm co}(s_4) = 0$, $y'_{\rm co}(s_4) = 0$, or

$$\sqrt{\beta_1}\theta_1\cos[\pi\nu-\psi_{41}] + \sqrt{\beta_2}\theta_2\cos[\pi\nu-\psi_{42}] + \sqrt{\beta_3}\theta_3\cos[\pi\nu-\psi_{43}] + \sqrt{\beta_4}\theta_4\cos\pi\nu = 0,$$
$$\sqrt{\beta_1}\theta_1\sin[\pi\nu-\psi_{41}] + \sqrt{\beta_2}\theta_2\sin[\pi\nu-\psi_{42}] + \sqrt{\beta_3}\theta_3\sin[\pi\nu-\psi_{43}] + \sqrt{\beta_4}\theta_4\sin\pi\nu = 0,$$

where $\psi_{ji} = \psi_j - \psi_i$ is the phase advance from s_i to s_j. Expressing θ_3 and θ_4 in terms of θ_1 and θ_2, we obtain

$$\begin{cases} \sqrt{\beta_3}\theta_3 = -(\sqrt{\beta_1}\theta_1\sin\psi_{41} + \sqrt{\beta_2}\theta_2\sin\psi_{42})/\sin\psi_{43}, \\ \sqrt{\beta_4}\theta_4 = (\sqrt{\beta_1}\theta_1\sin\psi_{31} + \sqrt{\beta_2}\theta_2\sin\psi_{32})/\sin\psi_{43}. \end{cases} \tag{2.105}$$

The orbit displacement inside the region of orbit bumps can be obtained by applying the transfer matrix to the initial coordinates. Using four bumps, we can adjust the orbit displacement and the orbit angle to facilitate ease of injection and extraction, to avoid unwanted collisions, and to avoid limiting-aperture in accelerator.

The three-bump method (see Exercise 2.3.4) has also been used for local orbit bumps. Although the slope of the bumped particle orbit can not be controlled in the three-bump method, this method is usually used for the local orbit correction because of its simplicity. Occasionally, two bumps can be used at favorable phase-advance locations in accelerators. Figure 2.14 shows an example of a local orbit bump using three dipoles. Since the two outer bumps happen to be nearly 180° apart in the betatron phase advance, the middle bump dipole has negligible field strength.

B. Fast kick for beam extraction

To extract a beam bunch from accelerator, a fast kicker magnet is usually powered in about 10–100 ns rise and fall times in order to bump beam bunches into the extraction

[25] Other examples are orbit bump at the aperture restricted area, internal target area, avoiding unwanted collisions in colliders, etc. For example, the counter-circulating e^+ and e^- beams, or the $\bar{p}$ and p beams in a collider can be made to avoid crossing each other in a common vacuum chamber with electrostatic separators.

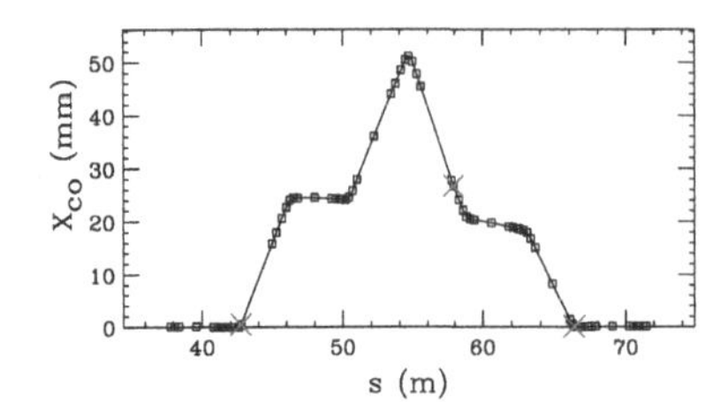

Figure 2.14: A simple orbit bump produced by three dipole kickers marked with symbol X in the AGS booster lattice. Since the first and third kickers are nearly 180° apart in the betatron phase advance, the local orbit bump is essentially accomplished with these outer two kickers. In this example, there are 3 focusing and 2 defocusing quadrupoles between two outer bump dipoles.

channel, where a septum is located.[26] With the transfer matrix of Eq. (2.42), the transverse displacement of the beam is

$$\Delta x_{\rm co}(s) = \left\{\sqrt{\beta_x(s_{\rm k})\beta_x(s)}\sin(\Delta\phi_x(s))\right\}\theta_{\rm k}, \tag{2.106}$$

where $\theta_{\rm k} = \int B_{\rm k} ds/B\rho$ is the kicker strength (angle), $B_{\rm k}$ is the kicker dipole field, $\beta_x(s_{\rm k})$ is the betatron amplitude function evaluated at the kicker location, $\beta_x(s)$ is the amplitude function at location s, and $\Delta\phi_x(s)$ is the phase advance from $s_{\rm k}$ of the kicker to location s. The quantity in curly brackets in Eq. (2.106) is called the *kicker lever arm.*

To achieve a minimum kicker angle, the septum is located about 90° phase advance from the kicker, and the values of the betatron amplitude function at the septum and kicker locations are also optimized to obtain the largest kicker lever arm. Similar constraints apply to the kicker in the transverse feedback system, the kicker array for stochastic cooling, etc.

Figure 2.15 shows a schematic drawing of the cross-section of a Lambertson septum magnet. A beam is bumped from the center orbit $x_{\rm c}$ to a bumped orbit $x_{\rm b}$. At the time of fast extraction, a kicker kicks the beam from the bumped orbit to the the extraction channel at $x_{\rm k}$, where the uniform dipole field bends the beam into the extraction channel. The iron in the Lambertson magnet is shaped to minimize the field leakage into the field-free region and the septum thickness, that is of the order of 4-10 mm depending on the required magnetic field strength.

[26]The **kicker** is an electric or magnetic device that provides an angular deflection to charged particle beam at a fast rise and fall times so that it can selectively deflect some beam bunches without affecting others. The electric kicker applies the traveling wave to a stripline type waveguide. The magnetic kicker employs ferrite material to minimize eddy-current effects. The rise and fall times of the kickers range from 10 ns to 100's ns. The **septum** is a device with an aperture divided into a field-free region and a uniform-field region, where the former will not affect the circulating beams, and the latter can direct the beam into an extraction or injection channels. Depending on the application, one can choose among different types of septum, such as wire septum, current sheet septum, Lambertson septum, etc.

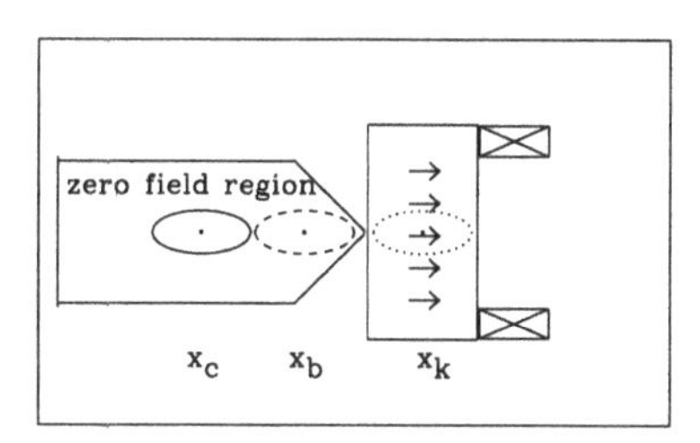

Figure 2.15: A schematic drawing of the central orbit x_c, bumped orbit x_b, and kicked orbit x_k in a Lambertson septum magnet. The blocks marked with X are conductor-coils, The ellipses marked beam ellipses with closed orbits x_c, x_b, and x_k. The arrows indicated a possible magnetic field direction for directing the kicked beams downward or upward in the extraction channel.

C. Effects of rf dipole field, rf knock-out

In the presence of a localized rf dipole, Hill's equation is

$$\frac{d^2y}{ds^2} + K(s)y = \theta_a \sin\omega_{\rm m}t \sum_{n=-\infty}^{\infty} \delta(s - nC), \tag{2.107}$$

where $\theta_a = \Delta B\ell/B\rho$ and $\omega_{\rm m}$ are respectively the kick angle and the angular frequency of the rf dipole, C is the circumference, and $t = s/\beta c$ is the time coordinate. The periodic delta function reflects the fact that beam particles encounter the kicker field only once per revolution.

With coordinate transformation: $\eta = y/\sqrt{\beta}$, $\phi = \frac{1}{\nu}\int_0^s ds/\beta$, Eq. (2.107) becomes

$$\frac{d^2\eta}{d\phi^2} + \nu^2\eta = \frac{\nu\sqrt{\beta_0}\theta_a}{2\pi} \sum_{n=-\infty}^{\infty} \sin(n + \nu_{\rm m})\phi,$$

where $\nu_{\rm m} = \omega_{\rm m}/\omega_0$ is the modulation tune, β_0 is the value of the betatron amplitude function at the rf dipole location, ω_0 is the orbital angular frequency, and we use $\delta(s - nC) = \frac{1}{|ds/d\phi|}\delta(\phi - 2\pi n)$. The solution of the inhomogeneous Hill's equation is $\eta = A\cos\nu\phi + B\sin\nu\phi + \eta_{\rm co}$, where A and B are the amplitude of betatron motion determined by the initial conditions, and the particular solution $\eta_{\rm co}$ is the coherent time dependent closed orbit,

$$\eta_{\rm co} = \sum_{n=-\infty}^{\infty} \frac{\nu\sqrt{\beta_0}\theta_a}{2\pi[\nu^2 - (n + \nu_{\rm m})^2]} \sin(n + \nu_{\rm m})\phi. \tag{2.108}$$

The discrete nature of the localized kicker generates error harmonics $n + \nu_{\rm m}$ for all $n \in (-\infty, \infty)$. For example, if the betatron tune is 8.8, large betatron oscillations can be generated by an rf dipole at any of the following modulation tunes: $\nu_{\rm m} = 0.2, 0.8, 1.2, 1.8, \ldots$. The coherent betatron motion of the beam in the presence of an rf dipole at $\nu_{\rm m} \approx \nu$ (modulo 1) with initial condition $y = y' = 0$ is

$$\begin{aligned} y(s) &= \frac{\sqrt{\beta(s)\beta_0}\,\theta_a}{2\pi} \sum_{n=-\infty}^{\infty} \frac{1}{\nu^2 - (n + \nu_{\rm m})^2} [\nu\sin(n + \nu_{\rm m})\phi - (n + \nu_{\rm m})\sin\nu\phi] \\ &\approx -\left[\frac{\sqrt{\beta(s)\beta_0}\,\theta_a}{4\pi}\frac{s}{R}\right] \cos\frac{\nu s}{R} + \ldots, \end{aligned} \tag{2.109}$$

where the last approximate identity is obtained by expanding the term in the sum with $n + \nu_m \approx \nu$, and retaining only the dominant term. Equation (2.109) indicates that the beam is driven coherently by the rf dipole, and the amplitude of betatron motion grows linearly with time.The coherent growth time of the betatron oscillation is inversely proportional to $|\nu_{\rm m} - \nu|$ (mod 1). Beyond the coherent time, the beam motion is out of phase with the external force and leads to damping. This process is related to the Landau damping to be discussed in Sec. VIII.4.

Figure 2.16 shows the measured betatron coordinate (lower curve) at a beam position monitor (BPM) after applying rf knock-out kicks to the beam in the IUCF Cooler Ring, and the fractional part of the betatron tune (upper curve), that, in this experiment, is equal to the knockout tune. The rf dipole was on from 1024 to 1536 revolutions starting from the triggering time. At revolution number 2048, the beam was imparted a transverse kick. Note the linear growth of the betatron amplitude during the rf dipole-on time. Had the rf dipole stayed on longer, the beam would have been driven out of the vacuum chamber, called the "*rf knock out.*"

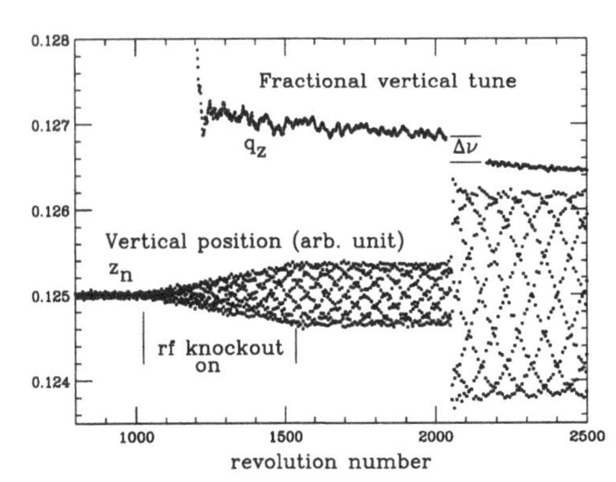

Figure 2.16: The lower curve shows the measured vertical betatron oscillations at one BPM in the IUCF Cooler resulting from an rf dipole kicker at the betatron frequency. The rf dipole was turned on for 512 revolutions, and the beam was imparted by a one-turn kicker after another 512 revolutions. The betatron amplitude grew linearly during the rf knockout-on time. The upper curve shows the fractional part of the betatron tune obtained by counting the phase advance in the phase-space map using data of two BPMs.

The fractional betatron tune, shown in the upper trace, is measured by averaging the phase advance from the Poincaré map (see Sec. III.5), where data from two BPMs are used. This two-kick method can be used to provide a more accurate measurement of the dependence of the betatron tune on the betatron amplitude. The power supply ripple at the IUCF cooler ring gives rise to a betatron tune modulation of the order of 2×10^{-3} at 60 Hz and its harmonics. On the other hand, the dependence of the betatron tune on the betatron action is typically 10^{-4} per 1π mm-mrad. To measure this small effect in the environment of the existing power supply ripple, the two-kick method was used to measure the instantaneous betatron tune change at the moment of the second kick.[27]

[27]See M. Ellison *et al.*, *Phys. Rev. E* **50**, 4051 (1994).

The rf dipole can be adiabatically turned on to induce coherent betatron oscillations for betatron tune measurement without causing serious emittance dilution.[28] Figure 2.17 shows the vertical beam profile measured at the AGS during the adiabatic turn-on/off of an rf dipole. When the rf dipole was on, the beam profile became larger because the beam was executing coherent betatron oscillations, and the profile was obtained from the integration of many coherent betatron oscillations. As the rf dipole is adiabatically turned off, the beam profile restored to its original shape.

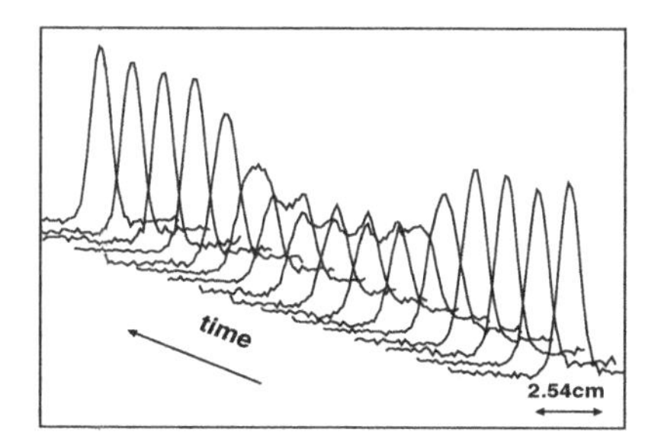

Figure 2.17: The beam profile measured from an ionization profile monitor (IPM) at the AGS during the adiabatic turn-on/off of an rf dipole. The beam profile appeared to be much larger during the time that the rf dipole was on because the profile was an integration of many coherent synchrotron oscillations. After the rf dipole was adiabatically turned off, the beam profile restored back to its original shape (Graph courtesy of M. Bai at BNL).

The induced coherent betatron motion can be used to overcome the intrinsic spin resonances during polarized beam acceleration. Furthermore, the measurement of the coherent betatron tune shift as a function of the beam current can be used to measure the real and imaginary parts of the transverse impedance (see Sec. VIII). This method is usually referred to as the beam transfer function (BTF).

D. Orbit response matrix and accelerator modeling

Equation (2.92) shows that the beam closed orbit in a synchrotron is equal to the propagation of the dipole field error through Green's function of Hill's equation. If the closed-orbit response to a small dipole field perturbation can be accurately measured, Green's function of Hill's equation can be modeled. The orbit response matrix (ORM) method measures the closed-orbit response induced by a known dipole field perturbation. The resulting response functions can be used to calibrate quadrupole strengths, BPM gains, quadrupole misalignment, quadrupole roll, dipole field integral, sextupole field strength, etc. The ORM method has been successfully used to model many electron storage rings.[29]

We consider a set of small dipole perturbation given by θ_j, $j = 1, ..., N_{\rm b}$, where $N_{\rm b}$ is the number of dipole kickers. The measured closed orbit y_i at the ith beam

[28] M. Bai *et al.*, Phys. Rev. **E56**, 6002 (1997); Phys. Rev. Lett. **80**, 4673 (1998); Ph.D. Thesis, Indiana University (1999); see also S.Y. Lee, PRSTAB **9**, 074001 (2006).

[29] See J. Safranek and M.J. Lee, *Proc. Orbit Correction and Analysis in Circular Accelerators*, AIP Conf. Proc. No. **315**, 128 (1994); J. Safranek and M.J. Lee, *Proc. 1994 European Part. Accel. Conf.* 1027 (1997). J. Safranek, *Proc. 1995 IEEE Part. Accel. Conf.*, 2817 (1995).

position monitors from a dipole perturbation is

$$y_i = \mathbf{R}_{ij}\theta_j, \quad j = 1, ..., N_{\rm b} \quad i = 1, ..., N_{\rm m}. \tag{2.110}$$

($N_{\rm b}$ can differs from $N_{\rm m}$). The response matrix $\mathbf{R}$ is equal to the Green's function G_y of Eq. (2.91) and another term resulting from the orbit length change due to the dipole kick to be discussed in Sec. IV.3.C.

Experimentally, we measure $\mathbf{R}_{ij}$ ($i = 1, \cdots, N_{\rm m}$) vs the dipole kick at θ_j ($j = 1, \cdots, N_{\rm b}$). The full set of the measured response matrix $\mathbf{R}$ can be employed to model the dipole and quadrupole field errors, the calibration of the BPM gain factor, sextupole misalignment, etc. The outcome of response matrix modeling depends on the BPM resolution, the number of BPMs and kickers, and the machine stability during the experimental measurement.

The ORM method minimizes the difference between the measured and model matrices $\mathbf{R}_{\rm exp}$ and $\mathbf{R}_{\rm model}$. Let

$$\mathbf{W}_k = \frac{|\mathbf{R}_{{\rm model},ij} - \mathbf{R}_{{\rm exp},ij}|}{\sigma_i} \tag{2.111}$$

be the difference between the closed-orbit data measured and those derived from a model, where σ_i is the rms error of ith measurements. Here the number of index k is $N_{\rm b} \times N_{\rm m}$, and the model response matrix can be calculated from MAD[23], SYNCH[24], or COMFORT[26] programs. The measured response matrix needs calibration in the kicker angle and BPM gain, i.e.

$$\mathbf{R}_{{\rm exp},ij} = \frac{\mathbf{R}_{{\rm data},ij}}{f_j g_i},$$

where f_j is the calibration factor of the jth kicker, and g_i is the gain factor of the ith BPM. The ORM accelerator modeling is to minimize the error of the vector $\mathbf{W}$ by minimizing the χ-square (χ^2) defined as

$$\chi^2 = \frac{1}{N_b \cdot N_m} \sum_k \mathbf{W}_k^2.$$

We consider sets of parameters w_m's that are relevant to accelerator model and orbit measurement. Some of these parameters are kicker angle calibration factor, the BPM gain factor, the dipole angle and dipole roll, the quadrupole strength and roll, sextupole strength, etc. The ORM modeling is to find a new set of w_m-parameters such that

$$||\mathbf{W}(w_m)|| = 0. \tag{2.112}$$

First, we begin with parameters w_m and evaluate $\mathbf{W}(w_m)$. The idea is to find a new set of parameters $w_m + \Delta w_m$ that satisfies Eq. (2.112), i.e.

$$\mathbf{W}_k(w_m + \Delta w_m) \approx \mathbf{W}_k(w_m) + \frac{d\mathbf{W}_k}{dw_m}\Delta w_m = 0. \tag{2.113}$$

To evaluate Δw_m, we invert matrix $\mathcal{W} \equiv \frac{d\mathbf{W}_k}{dw_m}$, which has the dimension of is $(N_b \cdot N_m) \times N_p$. Here, N_p is the number of parameters. In our application to accelerator physics, $(N_b \cdot N_m) \gg N_p$. The singular value decomposition (SVD) algorithm decomposes the matrix $\mathcal{W}$ into

$$\mathcal{W} = \frac{\mathbf{dW_k}}{\mathbf{dw_m}} = \mathbf{U\Lambda V}^{\mathrm{T}}, \tag{2.114}$$

where $\mathbf{V}^{\mathrm{T}}$ is a real orthonormal $N_p \times N_p$ matrix with $\mathbf{VV}^{\mathrm{T}} = \mathbf{V}^{\mathrm{T}}\mathbf{V} = 1$, $\mathbf{\Lambda}$ is a diagonal $N_p \times N_p$ matrix with elements $\mathbf{\Lambda}_{11} = \sqrt{\lambda_1} \geq \mathbf{\Lambda}_{22} = \sqrt{\lambda_2} \cdots \geq 0$, and $\mathbf{U} = \mathbf{AV\Lambda}^{-1}$ is a $(N_m \cdot N_b) \times N_p$ matrix with $\mathbf{U}^{\mathrm{T}}\mathbf{U} = \mathbf{1}$.[30] Here $\lambda_1, \lambda_2, \cdots$ are eigenvalues of the matrix $\mathcal{W}^{\mathrm{T}}\mathcal{W}$, and $\mathbf{V}$ is composed of orthonormal eigenvectors of $\mathcal{W}^{\mathrm{T}}\mathcal{W}$, i.e. $\mathcal{W}^{\mathrm{T}}\mathcal{W} = \mathbf{V\Lambda^2V}^{\mathrm{T}}$. The SVD-method sets all eigenvalues $\lambda_i \leq \lambda_c$, $(i > r)$ to $\lambda_i = 0$, $(i > r)$, where λ_c is called the tolerance level and r is called the *rank* of the matrix $\mathcal{W}$. Setting all $\lambda_i = 0$ $(i > r)$ is equivalent setting $\Delta w_i = 0$ for $i > r$. This means that these dynamical parameters have no relevance to the measured data. Once the SVD of matrix $\mathcal{W}$ is obtained, one finds Δw_m as

$$\Delta w_m = -\left(\mathbf{V\Lambda}^{-1}\mathbf{U}^{\mathrm{T}}\right)\mathbf{W}(w_m),$$

where $\mathbf{\Lambda}^{-1}$ is a diagonal matrix with $\mathbf{\Lambda}_{11}^{-1} = 1/\sqrt{\lambda_1}, \cdots, \mathbf{\Lambda}_{rr}^{-1} = 1/\sqrt{\lambda_r}$ and 0 for all remaining diagonal elements with $i > r$. The iterative procedure continues until $|\Delta w_m|$ or the change of χ^2 are small.

The response matrix modeling has been successfully implemented in many electron storage rings, where the BPM resolution is about 1~10 μm. The method has been used to calibrate kicker angle, BPM gain, quadrupole strength and roll, sextupole misalignment, dipole and quadrupole power supplies, etc. The method is also applicable to proton synchrotrons, where the BPM resolution is usually of the order of 100 μm.

In accelerator modeling, the dimension of the matrix $\mathcal{W}$, $(N_m \cdot N_b) \times N_p$, can be large. The inversion of a very large matrix may become time consuming. It is advantageous to model accelerator parameters in sequences, e.g., (1) kicker angle calibration f_j, (2) BPM gain g_i, (3) quadrupole strength ΔK_i, (4) dipole angle calibration, (5) dipole roll, etc. These steps are sometimes essential in attaining a reliable set of model parameters.

For high-power synchrotrons, beam particles are injected, accelerated and extracted in a short time duration. For example, the proton storage ring (PSR) at Los Alamos National Laboratory accumulates protons for 3000 turns and the beam bunch is extracted after accumulation for high-intensity short-pulse neutron production. The closed orbit data can be obtained by averaging betatron oscillations in a

[30]The SVD decomposition of a $m \times n$ matrix $\mathcal{W}$ in Eq. (2.114) can also be carried out in such a way that $\mathbf{U}$ and $\mathbf{V}$ are respectively orthonormal real $m \times m$ and $n \times n$ matrices with $\mathbf{U^TU} = \mathbf{UU^T} = \mathbf{1}$ and $\mathbf{V^TV} = \mathbf{VV^T} = \mathbf{1}$, and $\mathbf{\Lambda}$ is a $m \times n$ diagonal matrix.

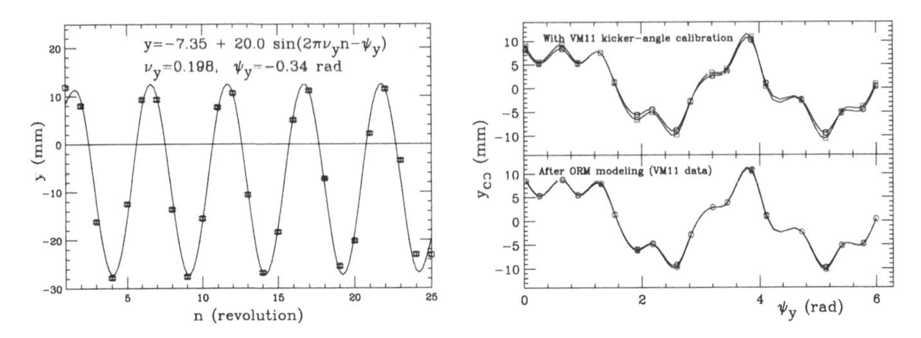

Figure 2.18: Left, digitized betatron oscillation data of one BPM are used to derive betatron amplitude, phase and tune, and closed orbit offset. Right, top and bottom plots show the closed orbit data compared with Green's function of Eq. (2.91) at a calibrated vertical steerer angle before and after ORM modeling.

single turn injection. The betatron oscillations of each BPM can be used to obtain the betatron amplitude, phase and tune, and the closed orbit (see the left plot of Fig. 2.18). These information can be used in the ORM analysis for accelerator modeling.[31] The right plots of Fig. 2.18 shows an example of typical fit in ORM modeling. The success of accelerator modeling depends critically on the orbit and tune stability, the number of BPMs and orbit steerers, proper set of experimental data for attaining relevant parameters.

E. Model Independent Analysis

Using turn-by-turn BPM data excited by resonant pinger discussed in Sec. III.3C, one can also carry out response matrix analysis for accelerator modeling, called Model Independent Analysis (MIA). This method has been successfully applied to SLC linac, PEP-II and Advanced Photon Source.[32] For the application of MIA in a storage ring, one uses an rf dipole pinger to excite coherent betatron oscillation and measures the response function with turn-by-turn BPM digitizing system (See Sec. III, in Appendix A, where we introduce the independent component analysis (ICA) for beam measurements).

[31]X. Huang *et al.*, *Analysis of the Orbit Response Matrix Measurement for PSR*, Technote: PSR-03-001 (2003).

[32]J. Irwin, C.X. Wang, and Y.T. Yan, Phys. Rev. Lett. **82**, 1684 (1999); C.X. Wang, Ph.D. Thesis, Stanford University (1999); J. Irwin and Y.T. Yan, Proceedings of EPAC 2000, p. 151 (2000); C.X. Wang, Vadim Sajaev, and C.Y. Yao, Phys. Rev. ST Accel. Beams **6**, 104001 (2003).

III.4 Quadrupole Field (Gradient) Errors

The betatron amplitude function discussed in Sec. II depends on the distribution of quadrupole strengths. What happens to the betatron motion if some quadrupole strengths deviate from their ideal design values? We found in Sec. III.1 that the effect of dipole field error on the closed orbit would be minimized if the betatron tune was a half-integer. Why don't we choose a half-integer betatron tune?

This section addresses effects of quadrupole field error that can arise from variation in the lengths of quadrupoles, errors in quadrupole power supply, horizontal closed-orbit deviation in sextupoles,[33] etc. These errors correspond to the b_1 term in Eq. (2.19).

A. Betatron tune shift

Including the gradient error, Hill's equation for the perturbed betatron motion about a closed orbit is

$$\frac{d^2y}{ds^2} + [K_0(s) + k(s)]\, y = 0, \tag{2.115}$$

where $K_0(s)$ is the focusing function of the ideal machine discussed in Sec. II, and $k(s)$ is a small perturbation. The perturbed focusing function $K(s) = K_0(s)+k(s)$ satisfies a weaker superperiod condition $K(s+C) = K(s)$, where C is the circumference. Let $\mathbf{M}_0$ be the one-turn transfer matrix of the ideal machine, i.e.

$$\mathbf{M}_0(s) = I\cos\Phi_0 + J\sin\Phi_0, \qquad I = \begin{pmatrix} 1 & 0 \\ 0 & 1 \end{pmatrix}, \quad J(s) = \begin{pmatrix} \alpha(s) & \beta(s) \\ -\gamma(s) & -\alpha(s) \end{pmatrix},$$

where $\Phi_0 = 2\pi\nu_0$ is the unperturbed betatron phase advance in one revolution, ν_0 is the unperturbed betatron tune, and $\alpha(s)$, $\beta(s)$, and $\gamma(s)$ are betatron amplitude functions of the unperturbed machine.

The transfer matrix of an infinitesimal localized gradient perturbation with length ds_1 is

$$m(s_1) = \begin{pmatrix} 1 & 0 \\ -k(s_1)ds_1 & 1 \end{pmatrix}. \tag{2.116}$$

The one-turn transfer matrix becomes $\mathbf{M}(s_1) = \mathbf{M}_0(s_1)m(s_1)$:

$$\mathbf{M}(s_1) = \begin{pmatrix} \cos\Phi_0 + \alpha_1\sin\Phi_0 - \beta_1 k(s_1)ds_1\sin\Phi_0 & \beta_1\sin\Phi_0 \\ -\gamma_1\sin\Phi_0 - [\cos\Phi_0 + \alpha_1\sin\Phi_0]k(s_1)ds_1 & \cos\Phi_0 - \alpha_1\sin\Phi_0 \end{pmatrix},$$

[33] Substituting $x = x_{\rm co} + x_\beta$ and $z = z_\beta$ into the sextupole field of Eq. (2.19), we obtain

$$\Delta B_z = \frac{1}{2}B_2(x_{\rm co}^2 + 2x_{\rm co}x_\beta + x_\beta^2 - z_\beta^2), \qquad \Delta B_x = B_2(x_{\rm co}z_\beta + x_\beta z_\beta),$$

where $B_2 = \partial^2 B_z/\partial x^2$. Thus an off-center horizontal orbit in a sextupole generates a dipole field $\frac{1}{2}B_2 x_{\rm co}^2$, and a quadrupole field gradient $B_2 x_{\rm co}$. This process is called *feed-down*.

where $\alpha_1 = \alpha(s_1)$, $\beta_1 = \beta(s_1)$, and $\gamma_1 = \gamma(s_1)$. The phase advance of the perturbed machine can be obtained from the trace of $\mathbf{M}$, i.e.

$$\cos\Phi - \cos\Phi_0 = -\frac{1}{2}\beta(s_1)k(s_1)ds_1\sin\Phi_0, \quad \text{or} \quad \Delta\Phi \approx \frac{1}{2}\beta(s_1)k(s_1)ds_1,$$

where $\Delta\Phi = \Phi - \Phi_0$, and the betatron tune shift is $\Delta\nu - \frac{1}{4\pi}\beta(s_1)k(s_1)ds_1$. Here the betatron tune shift depends on the product of the gradient error and the betatron amplitude function at the error quadrupole; it is positive for a focusing quadrupole, and negative for a defocusing quadrupole. For a distributed gradient error, the tune shift is

$$\Delta\nu = \frac{1}{4\pi}\oint\beta(s_1)k(s_1)ds_1 = \frac{1}{4\pi}\oint\beta(s_1)\frac{\Delta B_1(s_1)}{B\rho}ds_1. \tag{2.117}$$

The betatron tunes are particularly sensitive to gradient errors at high-β locations. Thus the power supply for high-β quadrupoles should be properly regulated in high energy colliders, and high-brightness storage rings.

B. Betatron amplitude function modulation (beta-beat)

To evaluate the effect of the gradient error on the betatron amplitude function, we again consider an infinitesimal quadrupole kick at s_1 of Eq. (2.116). The one-turn transfer matrix at s_2 and the change of the off-diagonal matrix element are

$$\mathbf{M}(s_2) = M(s_2 + C|s_1)\ m(s_1)\ M(s_1|s_2),$$
$$\Delta[\mathbf{M}(s_2)]_{12} = -k_1 ds_1\beta_1\beta_2\ \sin[\nu_0(\phi_1 - \phi_2)]\ \sin[\nu_0(2\pi + \phi_2 - \phi_1)],$$

where $\beta_1 = \beta(s_1), \beta_2 = \beta(s_2)$ $\phi_1 = \psi(s_1)/\nu_0$, and $\phi_2 = \psi(s_2)/\nu_0$ are the values of the unperturbed betatron functions. Since $\mathbf{M}_{12} = \tilde{\beta}_2\sin\Phi$, where $\tilde{\beta}_2 = \tilde{\beta}(s_2)$ and Φ are the the perturbed betatron function and betatron phase advance, we find

$$(\Delta\beta_2)\sin\Phi_0 = \Delta\mathbf{M}_{12} - \beta_2\cos\Phi_0\Delta\Phi = -\frac{1}{2}k_1 ds_1\beta_1\beta_2\cos[2\nu_0(\pi - \phi_1 + \phi_2)],$$

where $\Delta\beta_2 = \tilde{\beta}_2 - \beta_2$ and $\Delta\Phi = \Phi - \Phi_0$. Removing the subscript 2 and integrating over the distributed gradient errors, we obtain

$$\begin{aligned} \frac{\Delta\beta(s)}{\beta(s)} &= -\frac{1}{2\sin\Phi_0}\int_s^{s+C} k(s_1)\beta(s_1)\cos[2\nu_0(\pi + \phi - \phi_1)]\,ds_1 \\ &= -\frac{\nu_0}{2\sin\Phi_0}\int_\phi^{\phi+2\pi} k(\phi_1)\beta^2(\phi_1)\cos[2\nu_0(\pi + \phi - \phi_1)]\,d\phi_1, \end{aligned} \tag{2.118}$$

where $\phi = (1/\nu_0)\int_0^s ds/\beta$. The factor $\sin 2\pi\nu_0$ in the denominator of Eq. (2.118) implies that the betatron amplitude function diverges when ν_0 is a half-integer. We can also verify that $\Delta\beta/\beta$ satisfies (see Exercise 2.3.10)

$$\frac{d^2}{d\phi^2}\left[\frac{\Delta\beta(s)}{\beta(s)}\right] + 4\nu_0^2\left[\frac{\Delta\beta(s)}{\beta(s)}\right] = -2\nu_0^2\beta^2 k(s). \tag{2.119}$$

C. The half-integer stopband integrals

In a manner similar to the closed-orbit analysis, we expand the gradient error function $\nu_0\beta^2 k(s)$, which is a periodic function of s, in a Fourier series and obtain

$$\nu_0\beta^2 k(s) = \sum_{p=-\infty}^{\infty} J_p\, e^{jp\phi}, \qquad J_p = \frac{1}{2\pi}\oint \beta\, k(s)\, e^{-jp\phi}\, ds. \tag{2.120}$$

$$\frac{\Delta\beta(s)}{\beta(s)} = -\frac{\nu_0}{2}\sum_{p=-\infty}^{\infty}\frac{J_p e^{jp\phi}}{\nu_0^2-(p/2)^2} \approx -\frac{J_{p0}\cos(p_0\phi+\chi_0)}{(2\nu_0-p_0)}, \tag{2.121}$$

where J_p is the pth harmonic half-integer stopband integral. The tune shift of Eq. (2.117) is $\Delta\nu = J_0/2$. The perturbation $\Delta\beta(s)/\beta(s)$ can usually be approximated by one or two leading harmonics, where p_0 is the integer nearest $2\nu_0$, and χ_0 is the phase angle of J_{p0}. The betatron amplitude function is most sensitive to error harmonics of $\beta^2 k(s)$ nearest $2\nu_0$. The amplitude function becomes infinite when $2\nu_0$ approaches an integer; this is called a half-integer resonance. When the betatron tune is a half-integer, a quadrupole error can generate coherent additive phase-space kicks every revolution. The "closed orbit" of the betatron amplitude function will cease to exist, as shown in the left plot of Fig. 2.19. The betatron tune should avoid all half-integers.

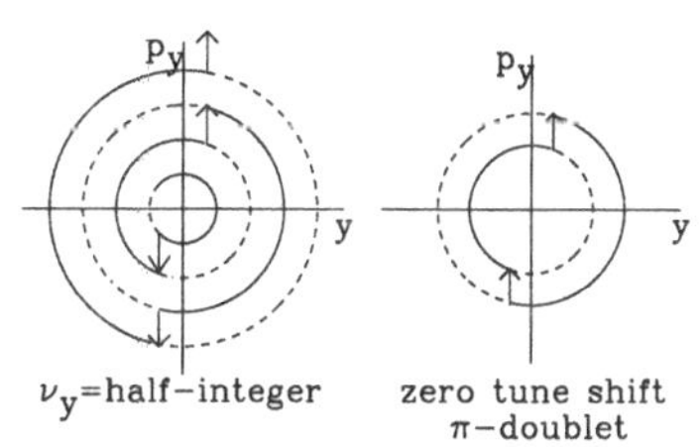

Figure 2.19: Left, schematic plot of a particle trajectory at a half-integer betatron tune resulting from an error quadrupole kick $\Delta p_y = \beta_y\Delta y' = -\beta_y y/f$, where f is the focal length, y is the displacement from the quadrupole center, and β_y is the betatron amplitude function at the quadrupole. The quadrupole kick is proportional to the displacement y. At a half-integer betatron tune, y changes sign in each consecutive revolution and the kick angles coherently add in each revolution to produce unstable particle motion. Right: cancellation of two kicks by zero tune shift π-doublets, that produce only local perturbation to betatron motion.

The evolution of phase-space coordinates resulting from a quadrupole kick is [using Eq. (2.116)] $\Delta y = 0$ and $\Delta y' = -k(s_1)y\, ds_1 = -y/f$, where $f = 1/(k_1 ds_1)$ is the focal length of the error quadrupole, and y is the displacement from the center of the closed orbit. The change of the slope y' is proportional to the displacement y. The left plot of Fig. 2.19 shows the behavior of a quadrupole kick at a half-integer tune, where the quadrupole kicks are coherently additive. This will lead to an ever increasing betatron amplitude. Thus the half-integer stopband gives rise to unstable betatron motion. When the betatron tune is an integer, quadrupole kicks will resemble the left plot in Fig. 2.13. Thus an integer betatron tune is also a half-integer resonance.

The right plot shows the effect of zero tune shift π-doublets, which give rise only to local betatron perturbation.

The stopband width is defined by $\delta\nu_p \equiv |J_p|$ such that $|\Delta\beta(s)/\beta(s)|_{\max} \approx 1$ at $\nu_0 \approx \frac{p}{2} \pm \frac{1}{2}\delta\nu_p$. This means that the betatron tune should differ from a half-integer by at least the stopband width. When the betatron tunes are inside the stopband, the beam size will increase by at least a factor of $\sqrt{2}$, and beam loss may occur.

D. Example of one quadrupole error in FODO cell lattice

We consider a simple accelerator lattice made of 18 FODO cells with half cell length 10-m, and dipole length 8 m bending angle 10°. The betatron tunes are set at $\nu_x = 4.79302$ and $\nu_z = 4.78298$ by quadrupoles. Now, consider an 1% decrease in focusing quadrupole strength at the end of the 10th cell. The top 2 plots of Fig. 2.20 show $\Delta\beta/\beta$, a pure sinusoidal function oscillating at twice betatron tune agreeing with Eq. (2.118). The regular oscillation of $\Delta\beta/\beta$ gives a beating of the betatron amplitude function, called *beta-beat*, dominated by the 9th and 10th harmonics. At the error quadrupole location, the kick in $\Delta\beta/\beta$ is equal to $-2\nu_0^2\beta k ds$. This particular property can be used to model the accelerator and discover mis-behaved quadrupoles.

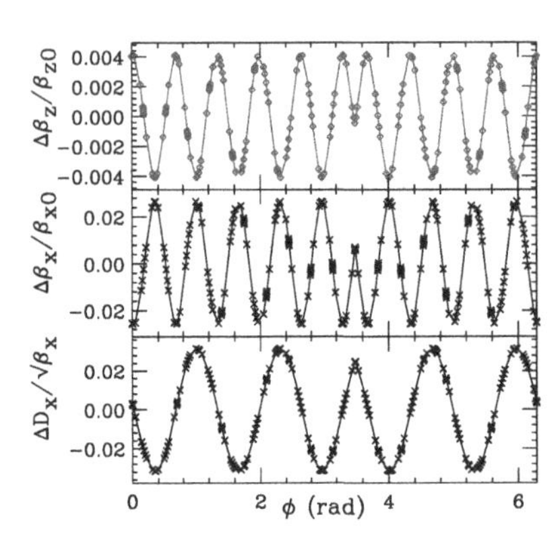

Figure 2.20: Perturbation of betatron amplitude functions vs ϕ (either $\phi_x = \frac{1}{\nu_x}\int_0^s ds/\beta_x$ or $\phi_z = \frac{1}{\nu_z}\int_0^s ds/\beta_z$) resulting from 1% decrease in gradient strength of the 10th focusing quadrupole. The betatron amplitude function perturbation is dominated by harmonics nearest $[2\nu_x]$ and $[2\nu_z]$. Since $\beta_x/\beta_z \sim 6.37$ at the focusing quadrupole location, the resulting error $\Delta\beta_x/\beta_x$ is about $6.37\Delta\beta_z/\beta_z$. A single kick at the error quadrupole location can be identified in the top 2 plots. The bottom plot shows the effect of quadrupole error on dispersion function shown as $\Delta D_x/\sqrt{\beta_x}$ vs $\phi = \phi_x$. A single kick at the error quadrupole location is visible to the dispersion closed orbit.

E. Statistical estimation of stopband integrals

Again, in the design stage of an accelerator, if we do not know a priori the gradient error, the stopband integral can be estimated by statistical argument as

$$J_p \approx \frac{1}{4\pi f_{\rm av}}\beta_{\rm av}\sqrt{N_{\rm q}}\left(\frac{\Delta K}{K}\right)_{\rm rms},$$

where $\beta_{\rm av}, f_{\rm av}, N_{\rm q}$, and $(\Delta K/K)_{\rm rms}$ are respectively the average β value, the average focal length, the number of quadrupoles, and the rms relative gradient error.

F. Effect of a zero tune shift π-doublet quadrupole pair

A zero tune shift π-doublet (or the zero tune shift half-wave doublet) is composed of two quadrupoles separated by 180° in betatron phase advance with zero tune shift. Using the zero tune shift condition, we obtain

$$\beta_1 \Delta K_1 \Delta L_1 + \beta_2 \Delta K_2 \Delta L_2 = 0, \tag{2.122}$$

where $\beta_{1,2}$ are betatron amplitude functions at quadrupole pair locations, and $\Delta K_i \Delta L_i$ is the integrated field strength of the ith quadrupole. The zero tune shift condition of Eq. (2.122) also produces a zero stopband integral at $p = [2\nu]$, i.e. $J_{[2\nu]} = 0$. Since the stopband integral $J_{[2\nu]}$ of a zero tune shift π-doublet is zero, the doublet has little effect on the global betatron perturbation shown in the right plot of Fig. 2.19, where the betatron perturbation due to the first quadrupole is canceled by the second quadrupole. Zero tune shift π-doublets can be used to change the dispersion function and the transition energy (to be discussed in Sec. IV.8). We find that the zero tune shift π-doublet produces a zero stopband width. On the other hand, a zero tune shift *quarter-wave* quadrupole pair produces a maximum contribution to the half-integer stopband. Employing such modules, we can correct half-integer betatron stopbands.

III.5 Basic Beam Observation of Transverse Motion

Measurements of beam properties are important in improving the performance of a synchrotron. In this section we discuss some basic beam diagnosis tools. Further detailed discussions can be found in the literature.[34]

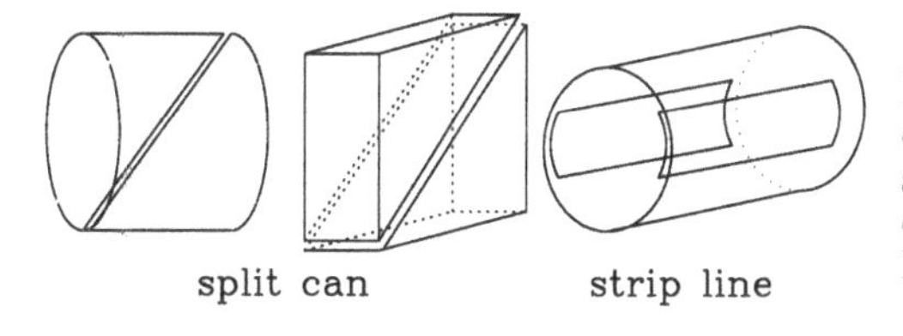

Figure 2.21: A schematic drawing of electric beam position monitors. The split-can type BPM has the advantage of linear response; the strip-line type has a larger transfer function.

A. Beam position monitor (BPM)

Transverse beam position monitors (BPMs) or pickup electrodes (PUEs) are usually composed of two or four conductor plates or various button-like geometries. Figure 2.21 shows a sketch of some simple electric BPM geometries used mainly in

[34]R. Littauer, AIP Conf. Proc. **105**, 869 (1983); R. Shafer, *IEEE Trans. Nucl. Sci.* NS**32**, 1933 (1985); J.L. Pellegrin, SLAC PUB-2522 (1980), and *Proc. 11th Int. Conf. on High Energy Accelerators*, p. 459 (1980); H. Koziol, CERN **89-05**, p. 63 (1989); M. Serio, CERN **91-04**, p. 136 (1991); P. Strehl, CERN **87-10**, p. 99 (1987); J. Boer, R. Jung, CERN **84-15**, p. 385 (1984).

proton synchrotrons. The button BPMs are used mainly in electron storage rings, where the bunch length is small. As the beam passes by, the induced image electric charges on the plates can be transmitted into a low impedance circuit, or the induced voltage can be measured on a high impedance port such as the capacitance between the electrode and the surrounding vacuum chamber.

The BPM can have an electrostatic, e.g. split electrodes and buttons, or a magneto-static, e.g. small secondary loop winding, configurations. An electrostatic monitor is equivalent to a current generator, where the image charge is detected by the shunt capacitance of the electrode to ground. Similarly, a magnetic loop monitor is equivalent to a voltage generator with a series inductor, which is the self-inductance of the loop. The voltage is proportional to the rate of variation of the magnetic flux associated with the beam current linked to the loop.

In general, the beam position is

$$y \approx \frac{w}{2}\frac{U_+ - U_-}{U_+ + U_-} = \frac{w}{2}\frac{\Delta}{\Sigma}, \tag{2.123}$$

where U_+ and U_- are either the current or the voltage signals from the right (up) and left (down) plates, $\Delta = U_+ - U_-$ is called the difference signal or the Δ-signal, $\Sigma = U_+ + U_-$ is the sum signal, and $w/2$ is the effective width of the PUE. Depending on the geometry of the PUE, the relation Eq. (2.123) may require nonlinear calibration. Measurements of the normalized difference signal with proper calibration provide information about the beam transverse coordinates. If we digitize beam centroid positions turn by turn, we can measure the betatron motion. On the other hand, sampling the position data at a slower rate, we can obtain the closed-orbit information from the DC component.

B. Measurements of betatron tune and phase-space ellipse

If the betatron oscillations from the BPM systems can be digitized turn by turn, the betatron tune can be determined from the FFT of the transverse oscillations (see Appendix B). Figure 2.22 shows the data for the horizontal betatron oscillation of a beam after a transverse kick at the IUCF cooler ring. The top plot shows the digitized data at two BPM positions (x_1 and x_2). The lower plot shows the FFT spectrum of the x_1 data. From the FFT spectrum, we find that the horizontal and vertical tunes of this experiment were $\nu_x = 3.758$ and $\nu_z = 4.683$ respectively.

The phase-space trajectory can be optimally derived from the measured betatron coordinates at two locations with a phase advance of an odd multiple of 90°. With Eq. (2.39), we obtain x_1' and the invariant phase-space ellipse;

$$x_1' = \frac{\csc\psi_{21}}{\sqrt{\beta_1\beta_2}}\,x_2 - \frac{(\cot\psi_{21} + \alpha_1)}{\beta_1}\,x_1,$$

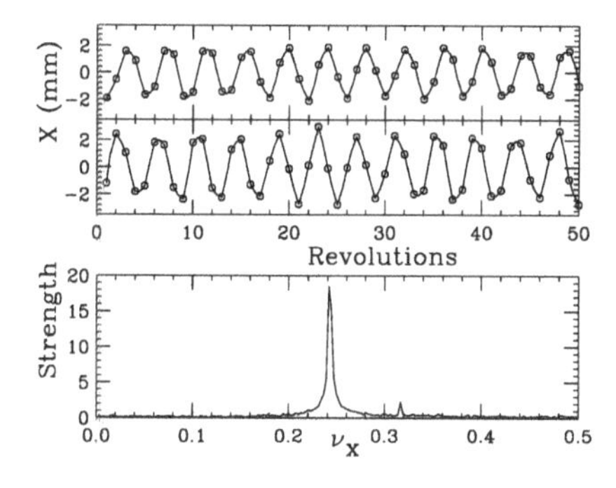

Figure 2.22: The measured betatron coordinates at two horizontal BPMs, after the beam is imparted a magnetic kick, vs the revolution number. The solid line is drawn to guide the eye. A total of 385 data points are used to obtain the FFT spectrum showing the fractional part of the horizontal betatron tune. The fractional part of the horizontal betatron tune is $\nu_x = 0.242 \pm 0.002$. The observed vertical betatron tune at $\nu_z = 0.317$ may result from linear coupling or from a tilted horizontal kicker.

$$x_1^2 + \left(\sqrt{\frac{\beta_1}{\beta_2}} \csc \psi_{21}\, x_2 - \cot \psi_{21}\, x_1 \right)^2 = 2\beta_1 J, \tag{2.124}$$

where $\psi_{21} = \psi_2 - \psi_1$ is the betatron phase advance between two BPMs, β_1 and β_2 are the values of betatron amplitude function at two BPMs, and $\alpha_1 = -\beta_1'/2$ at the first BPM. The area enclosed by the (x_2, x_1) ellipse is $2\pi\sqrt{\beta_1\beta_2}\,|\sin\psi_{21}|J$, and J is the betatron action. Dots in the left plot of Fig. 2.23 shows the measured (x_2, x_1) ellipse, where the solid line is obtained from Eq. (2.124) by fitting $\sqrt{\beta_1/\beta_2}$ and ψ_{21} parameters for the orientation, and $\beta_1 J_1$ for the size of the ellipse. If the betatron amplitude function β_1 is independently measured, the action of the ellipse can be determined.

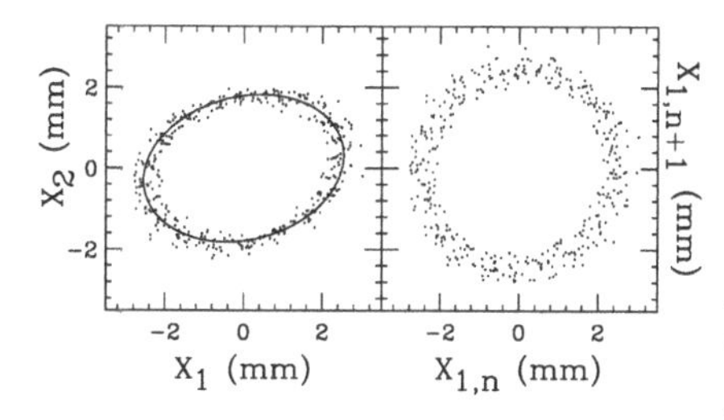

Figure 2.23: Left, the phase-space ellipse (x_2, x_1) of Fig. 2.22. The solid line shows the ellipse of Eq. (2.124) with $\sqrt{\beta_2/\beta_1} = 1.4$, $\psi_{21} = 80°$, and $2\beta_1 J = 8 \times 10^{-6}$ m^2. If β_1 is independently measured, the action of the betatron orbit can be obtained. Right, phase-space ellipse of single digitized BPM data in successive revolutions. Because the betatron tune is nearly 3.75, the phase space is an upright circle.

Two BPMs separated by about 90° in phase advance are useful for obtaining a nearly upright transverse phase-space ellipse. The turn by turn digitized data require a high bandwidth digitizer and a large memory transient recorder. However, if available hardware is limited, the phase-space ellipse can be obtained by using digitized data of successive turns of a single BPM. The right plot of Fig. 2.23 shows $x_{1,n+1}$ vs $x_{1,n}$. Because the horizontal tune in this example is 3.758 (see Fig. 2.22), the phase-space ellipse of $(x_{1,n}, x_{1,n+1})$ is nearly a circle. The area enclosed is $2\pi\beta_1|\sin 2\pi\nu_x|J$.

III.6 Application of Quadrupole Field Error

By using the quadrupole field error, the optical properties of the lattice can be altered, measured, or manipulated. Examples of β-function measurement and the betatron tune jump are described below. Other applications, such as π-doublets for dispersion function manipulation, will be discussed in Sec. IV.8.

A. β-function measurement

Using Eq. (2.117), we can derive the betatron amplitude function by measuring the betatron tune as a function of the quadrupole strength. The average betatron amplitude function $\langle\beta_{x,z}\rangle$ at a quadrupole becomes

$$\langle\beta_{x,z}\rangle = 4\pi\frac{\Delta\nu_{x,z}}{\Delta K\ell},$$

where $\Delta K\ell$ is the change in the integrated quadrupole strength. Figure 2.24 shows an example of the measured fractional part of the betatron tune vs the strength of a quadrupole at the IUCF cooler ring. The "average" betatron amplitude function at the quadrupole can be derived from the slopes of the betatron tunes. In this example, the slope of the horizontal betatron tune is larger than that of the vertical, and thus the horizontal betatron function is larger than the vertical one. Since the fractional part of the horizontal tune increases with the defocusing quadrupole strength, the actual horizontal tune is below an integer. For the IUCF cooler ring, we have $\nu_x = 4 - q_x$ and $\nu_z = 5 - q_z$, where q_x and q_z are the fractional parts of betatron tunes.

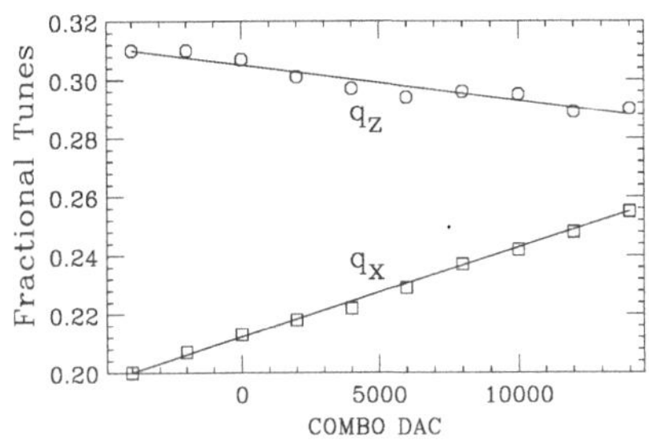

Figure 2.24: The horizontal and vertical tunes, determined by the FFT spectrum of the betatron oscillations, vs quadrupole field strength. The slope of the betatron tune vs quadrupole field can be used to determine the average betatron amplitude function at the quadrupole location. Because the fractional parts of betatron tunes were $q_x = 4 - \nu_x$ and $q_z = 5 - \nu_z$, the fractional horizontal tune appeared to "increase" with the strength of the horizontal defocusing quadrupole.

B. Tune jump

The vector polarization of a polarized beam is defined as the percent of particles whose spins lie along a quantization axis, e.g. the polarization of a proton beam is $P = (N_+ - N_-)/(N_+ + N_-)$, where $N_\pm$ are the numbers of particles with their spin projection lying along and against the quantization axis. For polarized beams in a planar accelerator, the quantization axis can be conveniently chosen to lie along the vertical direction that coincides with the vertical guide field.

According to the Thomas-BMT equation, the polarization vector precesses about the vertical axis at $G\gamma$ turns per revolution, where $G = (g-2)/2$ is the anomalous g-factor and γ is the Lorentz relativistic factor.[35] Thus $G\gamma$ is called the spin tune. Since the spin tune increases with the beam energy, acceleration of a polarized beam may encounter spin depolarization resonances [27], where the "imperfection resonance" arises from the vertical closed-orbit error, and the "intrinsic resonance" is produced by the vertical betatron motion. The imperfection resonance can be corrected by vertical orbit correctors to achieve proper spin harmonic matching. The AGS had 96 closed-orbit correctors for imperfection resonance harmonics.[36]

The intrinsic resonance in low/medium energy synchrotrons can be overcome by the tune jump method. When the $G\gamma$ value reaches an intrinsic spin resonance, the betatron tune is suddenly changed to avoid the resonance. This betatron tune jump can be achieved by using a set of ferrite quadrupoles with a very fast rise time. The AGS had 10 fast ferrite quadrupoles to produce a tune jump of about 0.3 in about 2.5 μs.

The amount of tune change is given by Eq. (2.117), where ΔB_1 is the quadrupole gradient of tune jump quadrupoles. Because of the integer and half-integer stopbands, the magnitude of tune jump is limited to about $\Delta\nu_z \approx 0.3$. With a large tune jump, beam dynamics issues such as non-adiabatic betatron amplitude function mismatch, linear betatron coupling, nonlinear resonances, non-adiabatic closed-orbit distortion, etc., should be carefully evaluated. Since the betatron tune of AGS is about 8.8, the important half-integer stopbands are located at $p = 17$ and 18. Placement of tune jump quadrupoles to minimize the stopband integral can reduce non-adiabatic perturbation to the betatron motion. Similarly, the non-adiabatic closed-orbit perturbation due to the misalignment of tune jump quadrupoles can also be analyzed.[37]

III.7 Beam Spectra

A. Transverse spectra of a particle

A circulating particle passes through the pickup electrode (PUE) at fixed time intervals T_0, where $T_0 = 2\pi R/\beta c$ is the revolution period, R is the average radius, and βc

[35] $G = 1.79284739$ for protons, and 0.0011596522 for electrons.

[36] At AGS, a 5% partial snake has recently been used to overcome all imperfection resonances. See, e.g., H. Huang *et al., Phys. Rev. Lett.* **73**, 2982 (1994).

[37] An rf dipole has recently been used to overcome these intrinsic spin resonances. See, e.g., M. Bai *et al., Phys. Rev. Lett.* **80**, 4673 (1998). A 20 G-m rf dipole was used to replace 10 ferrite quadrupoles with an integrated field strength of $\int B_1 ds = 15$ T.

is the speed. The current of the orbiting charged particle observed at the PUE[38] is

$$I(t) = e\sum_{-\infty}^{\infty}\delta(t - nT_0) = \frac{e}{T_0}\sum_{n=-\infty}^{\infty} e^{jn\omega_0 t} = \frac{e}{T_0} + 2\frac{e}{T_0}\sum_{n=1}^{\infty}\cos n\omega_0 t, \tag{2.125}$$

where e is the charge of the particle, $\delta(t)$ is the Dirac δ-function, j is the complex number, and $\omega_0 = \beta c/R = 2\pi f_0$ is the angular frequency. Note that the periodic occurrence of current pulses is equivalent to equally spaced Fourier harmonics.

The top plot of Fig. 2.25 shows the periodic time domain current pulses. The middle plot shows the frequency spectra of the particles occurring at all "rotation harmonics." Passing the signal into a spectrum analyzer for fast Fourier transform (FFT), we observe a series of power spectra at integral multiples of the revolution frequency nf_0, shown in the bottom plot. The DC current is e/T_0, and the rf current is $2e/T_0$. Because the negative frequency components are added to their corresponding positive frequency components, the rf current is twice the DC current.

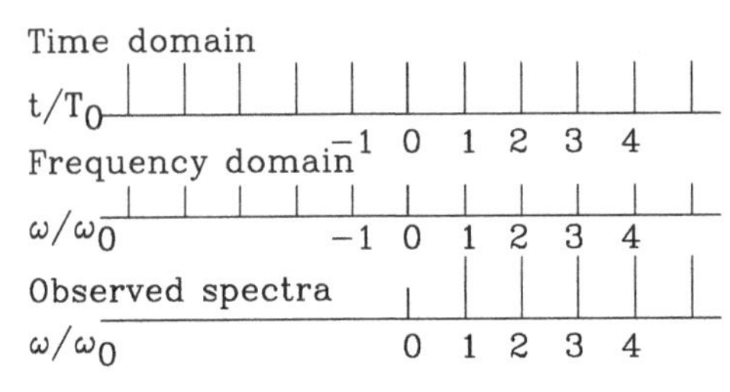

Figure 2.25: A schematic drawing of current pulses in the time domain (upper plot), in the frequency domain (middle plot), and observed in a spectrum analyzer (bottom plot). The rf current is twice the DC current because negative frequency components are added to their corresponding positive frequency components.

If we apply a transverse impulse (kick) to the beam bunch, the beam will begin betatron oscillations about the closed orbit. The BPM measures the transverse coordinates of the centroid of the beam charge distribution (dipole moment).

$$\begin{aligned}
d(t) &= I(t)y(t) = I(t)\left[y_0 + \hat{y}\cos(\omega_\beta t)\right], \\
\langle d(t)\rangle &= \langle I(t)\rangle y_0, \\
d_\beta &= \frac{e\hat{y}}{2T_0}\sum_{n=-\infty}^{\infty}\left(e^{j(n+Q_y)\omega_0 t} + e^{j(n-Q_y)\omega_0 t}\right) = \frac{e\hat{y}}{T_0}\sum_{n=-\infty}^{\infty}\cos\left[(n+Q_y)\omega_0 t\right].
\end{aligned}$$

where y_0 is the offset due to the closed-orbit error or the BPM misalignment, $\hat{y}$ is the amplitude of the betatron oscillation, and $\omega_\beta = Q_y\omega_0$ is the betatron angular frequency with betatron tune Q_y. The DC component $\langle d(t)\rangle$ of the dipole moment can be obtained by applying a low-pass filter to the measured BPM signal. Employing a band-pass filter, the betatron oscillation can be measured, that can be expanded in Fourier harmonics. In the frequency domain, the BPM signals of the betatron

[38]We assume that the bandwidth of PUEs is much larger than the revolution frequency. In fact, the bandwidth of PUEs is normally from 100's MHz to a few GHz.

oscillation contain sidebands at the revolution frequency lines, i.e. $f = (n + Q_y)f_0$ with integer n. Figure 2.26 shows a SPEAR3 spectrum around the rf frequency of 476.312 MHz. The harmonic number is 372, and thus the upper betatron sideband corresponds to $n = 367$, and the lower betatron sideband corresponds to $n = -377$. The vertical betatron tune is 6.18. The betatron sidebands are classified into fast wave, backward wave, and slow wave to be discussed briefly in Sec. VIII.

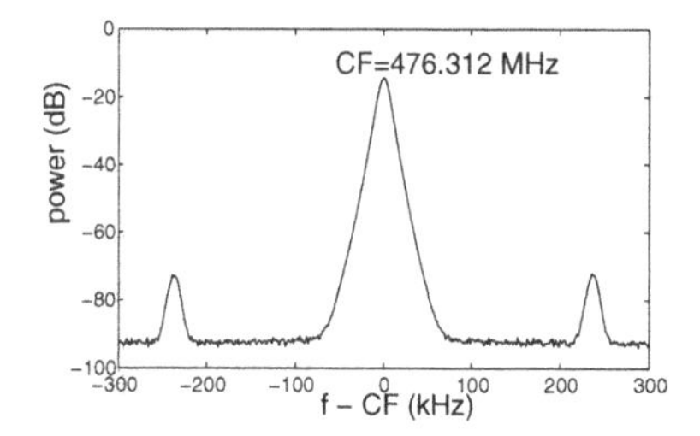

Figure 2.26: The spectrum around the rf frequency of 476.312 MHz for the SPEAR3. The upper betatron sideband corresponds to $n = 366$, and the lower betatron sideband corresponds to $n = -378$. The corresponding vertical betatron tune is about 6.18. The betatron sidebands are invisible during production runs. The sidebands appear only when the vertical chromaticity is set to 0. (Graph: courtesy of Xiaobiao Huang).

B. Fourier spectra of a single beam with finite time span

We note that a periodic δ-function current pulse in time gives rise to equally spaced Fourier spectra at all revolution harmonics. We ask what happens if the beam distribution has a finite time span with

$$I(t) = N_B e \sum_{n=-\infty}^{\infty} \rho(t - nT_0), \qquad \int_{-T_0/2}^{T_0/2} \rho(t - nT_0)dt = 1, \tag{2.126}$$

where the density distribution is normalized. There are many possible forms of beam distribution. We discuss two simple examples as follows.

1. If the beam is confined by a barrier rf wave or a double rf system, the beam distribution, approximated by the rectangular distribution, and its Fourier transform are

$$\rho(t - nT_0) = \frac{1}{2\Delta}\Theta(\Delta - |t - nT_0|); \tag{2.127}$$

$$I(\omega) = \frac{1}{2\pi}\int_{-\infty}^{\infty} I(t)e^{-j\omega t}dt = \frac{N_B e\omega_0}{2\pi}\left[\frac{\sin\omega\Delta}{\omega\Delta}\right]\sum_{n=-\infty}^{\infty}\delta(\omega - n\omega_0).$$

where $\Delta = \sqrt{3}\sigma_t$ is the bunch width in time. If the beam is confined by a sinusoidal rf cavity, the distribution can be a cosine-like function, parabolic function, or other distributions. The Fourier spectrum form-factor of a parabolic distribution $\rho(t) = 2\sqrt{\hat{\tau}^2 - t^2}/\pi\tau^2$ with $\hat{\tau} = 2\sigma_t$ is $J_0(\omega\tau) + J_2(\omega\tau)$, where J's are Bessel's function. The form-factor of a triangular distribution function $\rho(t) = [(1 - |t|/a)/a]\Theta(a - |t|)$ with $a = \sqrt{6}\sigma_t$ is $2(1 - \cos(\omega a))/(\omega a)^2$.

2. The beam distribution for electrons in storage rings is usually described by a Gaussian distribution due to the quantum fluctuation. The current pulse and its Fourier transform are

$$\rho(t - nT_0) = \frac{1}{\sqrt{2\pi}\sigma_t} e^{-(t-nT_0)^2/2\sigma_t^2}, \tag{2.128}$$

$$I(\omega) = \frac{N_B e\omega_0}{2\pi}\left[e^{-\omega^2\sigma_t^2/2}\right]\sum_{n=-\infty}^{\infty}\delta(\omega - n\omega_0). \tag{2.129}$$

Figure 2.27 shows the spectra form factor for Gaussian, rectangular and uniform elliptical beam distributions with an rms bunch length of 1 ns. In the frequency domain, the spectrum of the beam pulse is truncated by a form factor that depends on the time domain distribution function. In general, the Fourier spectrum of a beam with time width σ_t extends to about $1/\sigma_t$, e.g. a 6 dB roll-off frequency is $f_{\rm roll\ off} \sim 0.187/\sigma_t$ for a Gaussian beam. The frequency spectra of a long bunch, e.g. 1 m bunch length (3.3 ns), will have coherent spectra limited by a few hundred MHz. Since all particles in the bunch are assumed to have an identical revolution frequency, the Fourier spectra of Eqs. (2.127) and (2.129) are δ-function pulses bounded by the envelope factors. If there is a revolution frequency spread, the δ-function pulses are replaced by pulses with finite frequency width.

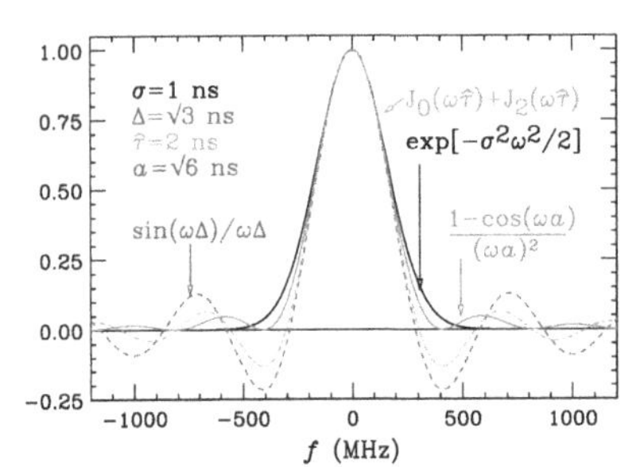

Figure 2.27: The form factors for Fourier spectra of a beam with Gaussian, rectangular, elliptical uniform, and triangular distributions with rms bunch length $\sigma_t = 1$ ns. The form factor serves as the envelope of the revolution comb lines shown in Fig. 2.25. All distributions with same rms bunch length has the same 6dB roll-off frequency, $f_{\rm roll-off} = 0.187/\sigma_t$. The coherent signal of a rectangular bunch can extend beyond the roll-off frequency.

C. Fourier spectra of many particles and Schottky noise

We consider N charged particles evenly distributed in the ring. The beam current observed at a PUE is

$$I(t) = e\sum_{n=-\infty}^{\infty}\delta(t - n\frac{T_0}{N}) = \frac{Ne}{T_0}\sum_{n=-\infty}^{\infty} e^{jn(N\omega_0)t}. \tag{2.130}$$

Note that the first Fourier harmonic is located at $N\omega_0$, and the spacing of Fourier harmonics is also $N\omega_0$.[39] If the number of particles is large, e.g. $N > 10^8$, the

[39]The analysis of equally spaced short bunches in the ring has identical Fourier spectra, i.e. if there are B bunches in the ring, the first coherent Fourier harmonic is $B\omega_0$.

frequency spectrum is practically outside the bandwidth of PUEs, and the spectrum is simply invisible. This means that the beam appears to have no rf signal. The beam that fills the accelerator is called a "DC beam," or a "coasting beam."

Similarly, the frequency spectra of the transverse dipole moment of N equally spaced particles give rise to a betatron sideband around the coherent orbital harmonics $N\omega_0 \pm \omega_\beta$. When N is a very large number, e.g. $N > 10^8$, the coherent betatron frequency becomes too high to be visible to PUEs.

It is important to realize that particles are not uniformly distributed in a circular accelerator. The longitudinal signal of N particles in a PUE is

$$\begin{aligned} I(t) &= e\sum_{i=1}^{N}\sum_{n=-\infty}^{\infty}\delta(t-t_i-nT_i) = 2\pi e\sum_{n=-\infty}^{\infty}\sum_{i=1}^{N}\omega_i e^{jn\omega_i(t-t_i)} \\ &\approx Nef_0 + 2ef_0\sum_{n=1}^{\infty}\sum_{i=1}^{N}\cos(n\omega_0 t + \Delta\phi_i(t)). \end{aligned} \tag{2.131}$$

The beam signal arising from random phase in charged particle distribution is called the Schottky noise. The power spectrum at each revolution harmonic from a low noise PUE is proportional to the number of particles.

Similarly, the dipole moments of the ith particle and the beam are respectively

$$d_i(t) = e\hat{y}_i\cos(\omega_{\beta i}t+\chi_i)\sum_{n=-\infty}^{\infty}\delta(t-nT_i-t_{0i}) = \frac{e\hat{y}_i}{T_i}\cos(\omega_{\beta i}t+\chi_i)\sum_{n=-\infty}^{\infty}e^{jn\omega_i(t-t_{0i})}$$

$$d(t) = \sum_{i=1}^{N}\frac{e\hat{y}_i}{T_i}\cos(\omega_{\beta i}t+\chi_i)\sum_{n=-\infty}^{\infty}e^{jn\omega_i(t-t_{0i})}.$$

Normally, the coherent betatron sidebands of a nearly uniform distribution are beyond the bandwidth of PUEs. However, the average power of the dipole moment can be measured. This is called the Schottky noise signal. If the particles are randomly distributed, the average power of the dipole moment is

$$P_{\rm av} = \frac{1}{2T}\int_{-T}^{T}|d^2(t)|dt, \tag{2.132}$$

where $2T$ is the sampling time. For practical consideration, T is of the order of minutes, this means that dipole moments of particles with frequencies within $T^{-1} \approx 10^{-2}$ Hz may interfere with one another. The resulting Schottky power can be contaminated by particle-to-particle correlation. Measurements with varying sampling times can be used to minimize the effect of particle correlation.

Since the phases $\omega_i t_{0i}$ and χ_i are random and uncorrelated, the Schottky power is proportional to the number of particles, i.e.

$$P_{\rm av} = \sum_{i=1}^{N}\frac{e^2\hat{y}_i^2}{4T_i^2} \quad \text{at} \quad \omega = n\langle\omega_0\rangle \pm \langle\omega_\beta\rangle. \tag{2.133}$$

The power spectrum resembles the single-particle frequency spectra located at $n\omega_0 \pm \omega_\beta$, i.e. betatron sidebands around all rotation harmonics. The Schottky signal can be used to monitor betatron and synchrotron tunes, frequency and phase space distributions, etc. It is the essential tool used for stochastic beam cooling.

III.8 Beam Injection and Extraction

A. Beam injection and extraction

Electrons generated from a thermionic gun or photocathode are accelerated by a high voltage gap to form a beam. The beam is captured in a linac or a microtron and accelerated to a higher energy for injection into other machines. Similarly, ions are produced from a source, e.g. a duoplasmatron, and extracted by a voltage gap to form a beam. The beam is accelerated by a DC accelerator or an RFQ for injection into a linac (DTL). The medium energy beam is then injected into various stage of synchrotrons.

A1. The strip or charge-exchange injection scheme

There are many schemes for beam injection into a synchrotron. The charge exchange injection involves H^- or H_2^+ ions, where a stripping foil with a thickness of a few μg/cm^2 to a few mg/cm^2 is used to strip electrons. The injection procedure is as follows. The closed orbit of the circulating beam is bumped onto the injection orbit of the H^- or H_2^+ beam by a closed-orbit bump and a set of *chicane magnets*, as shown in Fig. 2.28a.[40] Since the injection orbit coincides with the closed orbit of the circulating protons without violating Liouville's theorem, the resulting phase-space area will be minimized. The injected beam can be painted in phase space by changing the closed orbit during the injection. The injection efficiency for this injection scheme is high, except that we must take into account the effect of emittance blow-up through multiple Coulomb scattering due to the stripping foil (see Exercise 2.3.12).

Although the intensity of the H^- source is an order of magnitude lower than that of the H^+ source, a higher capture efficiency and a simpler injection scenario more than compensate the loss in source intensity. Most modern booster synchrotrons and some cyclotrons employ a H^- source. However, since the last electron in H^- has only about 0.7 eV binding energy, it can easily be stripped by a strong magnetic field at high energy.

[40]The chicane magnet may sometimes be replaced by punching a hole through the iron of a main dipole magnet provided that the saturation effect at high field is properly compensated.

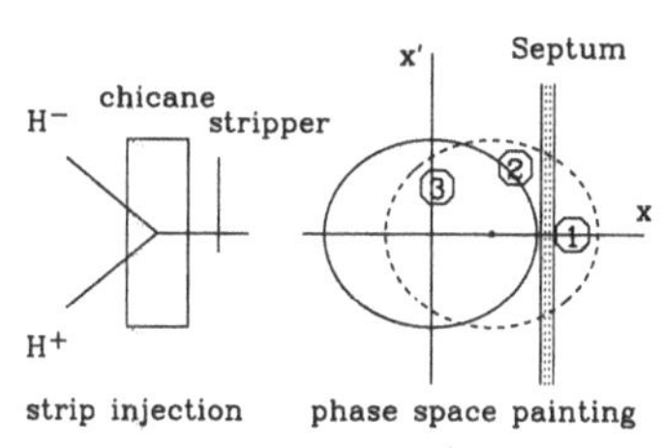

Figure 2.28: Schematic drawings of (a) a chicane magnet that merges the H^- and H^+ orbits onto the stripper, (b) the process of betatron phase-space painting. During the injection, the closed orbit is bumped near the septum (dashed ellipse) so that the injected beam marked (1) is captured within the dynamical aperture. Because of the betatron motion, the injected beam can avoid the septum in the succeeding revolutions marked (2) and (3). As the bumped orbit collapses during the injection time, the phase space is painted for beam accumulation.

A2. Betatron phase-space painting, cooling, radiation damping

The injection of protons or heavy ions into a synchrotron needs careful phase-space manipulation. The procedure is to bump the closed orbit of the circulating beam near the injection septum. The stable phase-space ellipse is shown as the dashed line in Fig. 2.28b. If the betatron tune and the orbit bump amplitude are properly adjusted, the particle distribution in betatron phase space can be optimized. This procedure is called phase-space painting. The injection efficiency is usually lower. The efficiency may be enhanced by employing betatron resonances.

Injection of the electron beam is similar to that of proton or ion beams except that the injected electrons damp to the center of the phase space because of the synchrotron radiation damping (see Chap. 4). At the time of injection pulse, the closed-orbit of circulating beams is bumped (kicked) close to a septum magnet so that the injection beam bunch is within the dynamical aperture of the synchrotron. At the completion of the injection procedure, the bump is removed, and the injected beamlet will damp and merge with the circulating beam bunch. The combination of phase-space painting and damping accumulation can be used to provide high-brightness electron beams in storage rings.

A3. Other injection methods

A method that has been successfully applied at the ISR is momentum phase-space stacking. This requires an understanding of the momentum closed orbit or the dispersion function, and of rf phase displacement acceleration. The method will be discussed in Chap. 3. This method is also commonly used in low energy cooler rings for cooling, and stacking accumulation of proton or polarized proton beams.

B. Beam extraction

B1. Fast single turn extraction and box-car injection

When a beam bunch is ready for extraction, orbit bump is usually excited. A fast kicker is fired to take the beam into the extraction channel of a septum magnet (see Sec. III.3). The extracted beam can be delivered to experimental areas, or be transferred into a another synchrotrons, called the box-car injection scheme.

B2. Slow extraction

Slow (beam) extraction by peeling-off high intensity beams can provide a higher duty factor[41] for many applications such as high reaction rate nuclear and particle physics experiments, medical treatment, etc. Slow extraction employs nonlinear magnets to drive a small fraction of the beam particles onto a betatron resonance.

The slow extraction using the third order resonance will be discussed in Sec. VII.1. Similarly, beam particles can be slowly extracted by employing the half integer resonance (see Exercise 3.3.3). Large-amplitude particles moving along the separatrix are intercepted by a thin (wire) septum that takes the particles to another septum on the extraction channel. The efficiency depends on the thin-septum thickness, the value of the betatron amplitude function at the septum location, betatron phase advance between the nonlinear magnet and the septum location, etc. More recently, efforts have been made to improve the uniformity of the extracted beam by stochastic excitation of the beam with noise; this is called stochastic slow extraction.

III.9 Mechanisms of Emittance Dilution and Diffusion

A. Emittance diffusion due to random scattering processes

In actual accelerators, noise may arise from various sources such as power supply ripple, ground vibration, intrabeam scattering, residual gas scattering, etc. This can induce emittance dilution and beam lifetime degradation. Our understanding of betatron motion provides us with a tool to evaluate the effect of noise on emittance dilution.

If the betatron angle y' is instantaneously changed by an angular kick θ, the resulting change in the betatron action is

$$\Delta I = I(y, y' + \theta) - I(y, y') = \theta(\alpha y + \beta y') + \frac{1}{2}\beta\theta^2. \tag{2.134}$$

If the angular kicks are uncorrelated, and the beam is composed of particles with many different betatron phases, the increase in emittance due to the random scattering processes is obtained by averaging betatron oscillations and kick angles, i.e.

$$\Delta\epsilon_{\rm rms} = 2\langle\Delta I\rangle = \langle\beta\theta^2\rangle \approx \langle\beta_\perp\rangle\langle\theta^2\rangle. \tag{2.135}$$

Random angular kicks to the beam particles arise from dipole field errors, non-resonant and non-adiabatic ground vibration, injection and extraction kicker noises,

[41]The duty factor is defined as the ratio of beam usage time to cycle time.

intrabeam Coulomb scattering, and multiple Coulomb scattering from gas molecules. Multiple scattering from gas molecules inside the vacuum chamber can cause beam emittance dilution, particularly at high $\beta_\perp$-function locations. This effect can also be important in the strip-injection of the H^- and H_2^+ ion sources from the stripping foil. The emittance growth rate can be obtained from the well-known multiple Coulomb scattering. Exercise 2.3.12 gives an example of estimating the emittance growth rate.

Other effects are due to the angular kicks from synchrotron radiation, quantum fluctuation resulting from energy loss, diffusion processes caused by rf noise, etc.; these will be addressed in Chaps. 3 and 4.

A1. Beam Lifetime

The single beam lifetime is determined by nuclear scattering on residual gas in the beam pipe, multiple scattering on the residual gas, ion or electron trapping due to residual gas scattering, photo desorption, intrabeam Coulomb scattering, Touschek effect (to be discussed in Chap.4 II.7), lifetime effect due to nonlinear resonances (see Sec. VII), etc. In a collider, beam lifetime is further reduced by beam-beam effects, particle loss due to beam-beam collisions, beam aperture limitation, etc.

B. Space charge effects

The repulsive Coulomb mean-field field of a beam can generate defocusing force to reduce the effective external focusing. The space-charge effect is characterized by an incoherent Laslett tune shift parameter $\xi_{\rm sc} = \Delta\nu_{\rm sc}$ (see Exercise 2.3.2). The tune shift parameter for low energy linacs at the ion source can be large, i.e. the betatron tune can be detuned to a value nearly 0 (see Sec. II.8). The incoherent space-charge tune shift for low energy synchrotron has a typical value of 0.2 – 0.6, which is about 10% or less of the betatron tunes. Yet, almost all low energy synchrotrons suffer space-charge induced emittance growth. We try to illustrate possible mechanisms.

B1. The coherent envelope oscillations due to space-charge force

We consider the effect of coherent envelope oscillations, pioneered by Sacherer,[42] with a simple KV model of 1D paraxial system (see Sec. II.8, where Hill's and the envelope equations are

$$\begin{cases} y'' + \left(k(s) - \dfrac{K_{\rm sc}}{R_{\rm b}^2(s)}\right) y = 0, & |y| \le R_{\rm b}(s) \\ y'' + k(s)y - \dfrac{K_{\rm sc}}{y} = 0, & |y| > R_{\rm b}(s) \end{cases} \tag{2.136}$$

[42] See F.J. Sacherer, *Transverse Space-Charge Effects in Circular Accelerators*, Ph.D. Thesis, UC Berkeley [Report No. UCRL-18454, UC Berkeley, 1968].

$$R_{\rm b}'' + k(s)R_{\rm b} - \frac{\epsilon^2}{R_{\rm b}^3} - \frac{K_{\rm sc}}{R_{\rm b}} = 0. \tag{2.137}$$

Here y stands for either the particle's horizontal or vertical betatron coordinate, $k(s)$ is the focusing function, $K_{\rm sc}$ is the space-charge perveance parameter, defined in Eq. (2.77), $\epsilon = 4\epsilon_{\rm rms}$ is the KV beam emittance, $R_{\rm b} = \sqrt{\beta(s)\epsilon}$ is the KV beam envelope radius, and $\beta(s)$ is the betatron amplitude function. For a KV beam, all particles are within the envelope radius. Making Floquet transformation with

$$R = \frac{R_{\rm b}}{\sqrt{\beta(s)\epsilon}}, \quad \eta = \frac{y}{\sqrt{\beta(s)}}, \qquad \phi = \frac{1}{\nu}\int_0^s \frac{ds}{\beta(s)}, \tag{2.138}$$

we transform Hill's and envelope equations into

$$\ddot{\eta} + \nu^2\eta - \frac{\nu^2\beta(s)K_{\rm sc}}{\epsilon R^2}\eta = 0 \qquad (y \le R_{\rm b}), \tag{2.139}$$

$$\ddot{R} + \nu^2 R - \frac{\nu^2}{R^3} - \frac{\nu^2\beta(s)K_{\rm sc}}{\epsilon R} = 0, \tag{2.140}$$

where ν is the betatron tune, and the over-dots are derivative with respect to the independent variable (time-coordinate) ϕ.

For synchrotrons, the space-charge terms in Hill's and envelope equations can be considered as a small perturbation unless a resonance condition is encountered. We expand the envelope radius around the unperturbed closed orbit with $R = 1 + r + \Delta$, where Δ is a ϕ-independent constant shift in the equilibrium radius and r is the ϕ-dependent term depending on the dynamics of the machine. We expand the space-charge factor:

$$\frac{\nu\beta(s)K_{\rm sc}}{2\epsilon} = \xi_{\rm sc}\left(1 + \sum_{n=1}^{\infty} q_n \cos(n\phi + \chi_n)\right) \tag{2.141}$$

in Fourier series, where

$$\xi_{\rm sc} = \frac{1}{2\pi}\oint \frac{\nu\beta(s)K_{sc}}{2\epsilon}d\phi = \frac{1}{4\pi}\oint \frac{\beta(s)K_{sc}}{[R_{\rm b}(s)]^2}ds, \tag{2.142}$$

$$\xi_{\rm sc}q_n = \frac{1}{\pi}\oint \frac{\nu\beta(s)K_{sc}}{2\epsilon}\cos(n\phi + \chi_n)d\phi. \tag{2.143}$$

The parameter $\xi_{\rm sc}$ is the Laslett (incoherent) linear space-charge tune shift parameter and $\xi_{\rm sc}q_n$ and χ_n are the Fourier amplitude and phase of the n-th harmonic. Substituting Eq. (2.141) into Eq. (2.140), we obtain $\Delta = \xi_{\rm sc}/2\nu$ and

$$\ddot{r} + (4\nu^2 - 4\nu\xi_{\rm sc})r \approx 2\nu\xi_{\rm sc}\sum_{n=1}^{\infty} q_n \cos(n\phi + \chi_n). \tag{2.144}$$

The space-charge force plays two roles in the envelope equation. It decreases the envelope tune from $\nu_{\rm env} = 2\nu$ to $\nu_{\rm env} = 2\nu - \xi_{\rm sc}$, and it generates a perturbation term, where the Fourier harmonic in the intrinsic betatron amplitude function serves as a harmonic perturbation to the envelope equation. The envelope radius, or the *perturbed betatron amplitude function*, is resonantly excited by the harmonic $n \approx \nu_{\rm env}$ with

$$r \approx \frac{2\nu\xi_{\rm sc}q_n}{-n^2 + (4\nu^2 - 4\nu\xi_{\rm sc})} \cos(n\phi + \chi_n). \tag{2.145}$$

Figure 2.29 shows the space-charge perturbed vertical betatron amplitude function (solid line), the original betatron amplitude function (dashed line), and the normalized envelope radius R (dotted line), obtained from a PIC simulation calculation for the Proton Storage Ring (PSR) at Los Alamos National Laboratory.[43] The PSR is a fixed energy synchrotron with 90.26 m circumference, $\nu_x = 3.19$ and $\nu_z = 2.19$. It serves as a compressor to compress 1.16 ms (3214 turns in PSR) of proton pulse from the 800 MeV Linac into a high intensity proton pulse of about 180 ns. Since the vertical betatron tune of the PSR is about 2.19, the dominant perturbing harmonic in the envelope equation is 4. The reduced envelope radius R shown in Fig. 2.29 clearly shows 4 oscillations in one circumference.

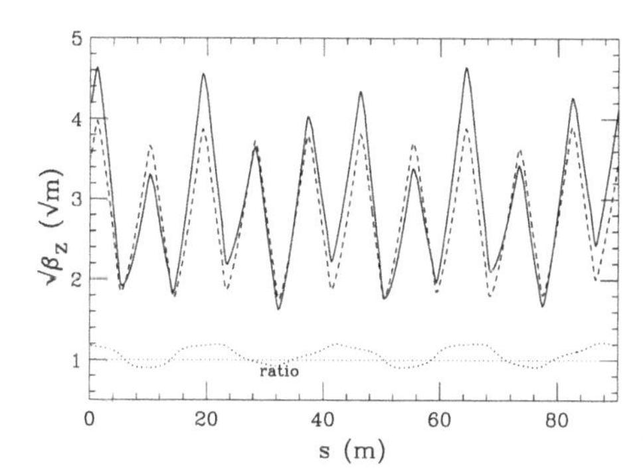

Figure 2.29: The square root of the perturbed vertical betatron amplitude function (solid line), for a beam with high intensity (4.37×10^{13} particles) in the PSR at LANL, is compared with the square root of the intrinsic betatron amplitude function (dashed line). The ratio of these two betatron amplitude function, shown as dotted line, is the reduced envelope radius R defined in Eq. (2.138). Note that the average of R is slightly larger than 1.

Substituting Eq. (2.141) into Hill's equation (2.139), we obtain

$$\ddot{\eta} + \nu^2\eta - \frac{2\nu\xi_{\rm sc}}{R^2}\left(1 + \sum_{n=1}^{\infty} q_n \cos(n\phi + \chi_n)\right)\eta = 0. \tag{2.146}$$

The particle tune is $\nu_{\rm p} \approx \nu - \xi_{\rm sc}$. Rightfully, $\xi_{\rm sc}$ is called the linear space-charge tune shift parameter. One speculates that a large envelope oscillation shown in Eq. (2.145) may cause a large particle oscillation at $n = 2(\nu - \xi_{\rm sc})$ for the Mathieu instability. However, after a closer inspection by substituting $R = 1 + \Delta + r$ into Eq. (2.146),

[43] S. Cousineau, Ph.D. thesis, Indiana University, 2002.

we find that the resonance strength is actually zero, i.e. *the envelope oscillation of a beam can not affect particle motion inside the envelope.* If particle motion inside the beam core is not affected by the envelope oscillation, what is the mechanism for emittance dilution?

C. Emittance evolution measurements and modeling

Measurement and modeling of the emittance evolution at the Fermilab Booster provided a very interesting revelation of the essential emittance growth mechanism. Fermilab Booster is a rapid cycling accelerator at 15 Hz from 400 MeV to 8 GeV (see Fig. 2.3). It can deliver 100 kW beam power. An Ionization Profile Monitor (IPM) can be used to measure the beam profile by averaging 50 bunches turn-by turn! The profile monitor is located at the center of a long straight section, where $\beta_z \gg \beta_x$. The emittances can be derived from the profile data and accelerator modeling (See Sec. III, in Appendix A).

Space-charge effects play an important role. Figure 2.30 shows the normalized vertical rms emittance in the first 4000 revolutions for intensity varying from 8.4×10^{11} (2-turn injection) to 7.5×10^{12} particles per pulse. Note that the vertical emittance grew rapidly and beam loss occurred in the first few hundreds of revolutions. On the other hand, the horizontal normalized emittance was found to be nearly independent of the beam intensity. The question is how the space charge affects the emittance growth for the high intensity beams in the Fermilab Booster?

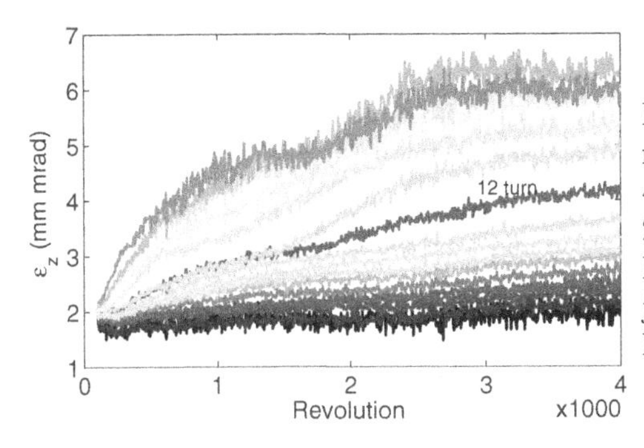

Figure 2.30: The normalized vertical rms emittance from 70-revolution to 4000-revolution for all data sets with 2-turn injection to 18-turn injection. Note the RED curve is for 12-turn injection which marks the border of two kinds of emittance growth behavior.

Fermilab Booster is a rapid cycling synchrotron (RCS) with a ramp rate of 15 Hz. Possible emittance growth mechanisms include (1) horizontal betatron motion excitation due to localized energy gain in 15 rf cavities distributed in half of the ring, (2) injection closed orbit error, (3) half-integer stopband discussed in the previous Section, (4) the Montague resonance at $2\nu_z - 2\nu_z = 0$ driven by the space charge potential, (5) effects of skew quadrupoles, and (6) effects of higher order multipoles such as the sextupole or octupoles. We set up to model the turn-by-turn emittance data including all effects listed above. Our results showed that the *sum and difference resonances induced by random skew-quadrupoles* were the main sources of emittance growth at the Fermilab Booster.[44] Our data analysis indicated that the effect of the envelope instability and the half-integer stopband was negligible. Efforts in using trim-quadrupole families to correct stopband integrals were *not* able to reduce beam emittance growth. However, it is possible that different machine operational conditions can have different space-charge emittance-growth mechanisms. Detailed experiments and modeling are essential in providing further insight on this important topic.

A type of RCS is the fixed-field-alternating-gradient (FFA) accelerator invented in the 1960s.[45] With fixed field accelerator magnets, the beam ramp rate and the beam power can be increased. The scaling FFA accelerators maintain constant betatron tunes without crossing linear betatron resonances. On the other hand, the non-scaling FFA accelerators that can reduce the aperture of accelerator magnets allow betatron tunes to run through many integers as the energy is increased. Intrinsic systematic resonances induced by the space charge force can become very important (see Sec. VII.4).

Exercise 2.3

1. Particle motion in the presence of magnetic field errors is (Sect. II.2)

$$y'' + K(s)y = \frac{\Delta B}{B\rho},$$

where y stands for either x or z, and $\Delta B = -\Delta B_x$ for z motion and $\Delta B = \Delta B_z$ for x motion. Define a new coordinate $\eta = y/\sqrt{\beta}$ with $\phi = (1/\nu)\int_0^s ds/\beta$ as the independent variable, the equation of motion becomes

$$\ddot{\eta} + \nu^2\eta = \nu^2\beta^{3/2}\frac{\Delta B}{B\rho},$$

where the overdot is the derivative with respect to the independent variable ϕ.

(a) For the dipole field error, where $\Delta B/B\rho$ is only a function of the independent variable s, show that Eq. (2.92) is a solution of the above equation. Show that the Green function $G(\phi, \phi_1) = [\cos\nu(\pi - |\phi - \phi_1|)]/2\nu\sin\pi\nu$ satisfies the following equation:

$$\left(\frac{d^2}{d\phi^2} + \nu^2\right)G(\phi, \phi_1) = \delta(\phi - \phi_1)$$

[44] X. Huang, S.Y. Lee, K.Y. Ng, and Y. Su, *PRST Accelerators and Beams*, **9**, 014202 (2006); For a Gaussian beam distribution, the space charge potential of Exercise 5.2.1 can be used in space charge force calculation.

[45] F.T. Cole, "*O Camelot! A MURA memoir*", in *Proc. of Cyclotron Conference 2001* (http://www.jacow.org/); see also Mills F. E., *Early FFA Development*, *ibid.*, p. 195 (2001).

Use the Green function to verify the solution given by Eq. (2.92). Use Eq. (2.94) for the Fourier expansion of the dipole field error $\beta^{3/2}\Delta B/B\rho$ show that the closed orbit arising from the dipole error is

$$y_{\rm co}(s) = \sqrt{\beta(s)} \sum_{k=-\infty}^{\infty} \frac{\nu^2 f_k}{\nu^2 - k^2} e^{jk\phi},$$

where f_k are integer stopband integrals. Using a single stopband approximation. and limiting the closed-orbit deviation to less than 20% of the rms beam size, show that the integer stopband width $\Gamma_{[\nu]}$ is given by $\Gamma_{[\nu]} \approx 5\nu|f_{[\nu]}|/\sqrt{\epsilon_{\rm rms}}$, where $[\nu]$ is the integer nearest the betatron tune.

(b) For the quadrupole field error, $\Delta B/B\rho = -k(s)y$, where $k(s)$ is the error gradient function, the Floquet-transformed equation becomes $\ddot{\eta} + \nu^2\eta = -\nu^2\beta^2 k(s)\eta$. Using the harmonic expansion of Eq. (2.120), we find

$$\ddot{\eta} + \nu^2\eta = -\nu\left(\sum J_p e^{+jp\phi}\right)\eta.$$

The dynamics of particle motion is dominated by harmonics $p \approx 2\nu$, and thus we can approximate the equation as

$$\ddot{\eta} + (\nu^2 + 2\nu g_p \cos(p\phi + \chi_p))\eta = 0,$$

where $g_p = |J_p|$, and χ_p is the phase of J_p. Use Mathieu instability condition to find the half integer stopband width.

The above equation can be derived from the Hamiltonian:

$$H = \frac{1}{2}\dot{\eta}^2 + \frac{1}{2}\left\{\nu^2 + \nu g_p \cos(p\phi + \chi_p)\right\}\eta^2.$$

Let $\eta = \sqrt{2J/\nu}\cos\psi$, where J and ψ are action and angle variables, show that the Hamiltonian at $\nu \approx p/2$ becomes

$$H \approx \nu J + \frac{1}{2} J g_p \cos(2\psi - p\phi - \chi_p).$$

Now, use the generating function $F_2 = (\psi - \frac{p}{2}\phi - \frac{1}{2}\chi_p)I$, show that the new Hamiltonian becomes

$$H = \delta I + \frac{1}{2} I g_p \cos 2\Psi,$$

where $I = J$ is the new action, $\Psi = \psi - \frac{p}{2}\phi - \frac{1}{2}\chi_p$ is the new angle variable, and $\delta = \nu - \frac{p}{2}$ is the resonance proximity parameter. Show that the equation of motion for the action is

$$\ddot{I} = g_p^2 I + 2\delta g_p \cos 2\Psi I.$$

Show that the action for a particle sitting on resonance, i.e. $\delta = 0$, becomes

$$I = ae^{g_p\phi} + be^{-g_p\phi} = ae^{2\pi g_p n} + be^{-2\pi g_p n},$$

where n is the revolution number, i.e. the action the particle will exponentially increase with the revolution number.

2. Effect of space charge force on beam emittance growth is very complicated as shown in Fig. 2.30 in Chapter 2, Sec. III.9. This homework is intended to derive a space-charge tune-shift parameter used to characterize high intensity beams. From Exercise 1.3, we find that a particle at a distance $r \ll a$ from a uniformly distributed paraxial beam bunch experiences a space-charge defocusing force

$$\vec{F} = \frac{2mc^2 N r_0}{a^2\gamma^2}\vec{r},$$

where γ is the relativistic Lorentz factor, a is the beam radius, r_0 is the classical radius of the particle, $\vec{r} = x\hat{x} + z\hat{z}$. and N is the number of particles per unit length. For a longitudinal Gaussian bunch distribution, $N = N_{\rm B}/\sqrt{2\pi}\sigma_s$, where $N_{\rm B}$ is the number of particles per bunch and σ_s is the bunch length.

(a) Using Eq. (2.76) for a round beam, show that the space-charge force is equivalent to a defocusing quadrupole with strength

$$K(s) = -\frac{2Nr_0}{a^2\beta^2\gamma^3} = -\frac{K_{\rm sc}}{a^2},$$

where $K_{\rm sc} = 2Nr_0/(\beta^2\gamma^3)$ is the normalized space-charge perveance parameter used frequently in the transport of space-charge-dominated beams in linacs, and β and γ are Lorentz's relativistic factors.

(b) The rms beam radius is $a^2 = \beta_y\epsilon$, where β_y is the betatron amplitude function and $\epsilon = 4\epsilon_{\rm rms}$ is the KV-type emittance. Show that the betatron (Laslett) tune shift of the particle $r \ll a$ induced by the space-charge force is given by[46]

$$\Delta\nu_{\rm sc} = \frac{2\pi R K_{\rm sc}}{4\pi\epsilon} = \frac{F_{\rm B} N_{\rm B} r_0}{2\pi\epsilon_{\rm N}\beta\gamma^2},$$

where $F_{\rm B} = \frac{2\pi R}{\sqrt{2\pi}\sigma_\ell}$ is the bunching factor and $\epsilon_{\rm N} = \beta\gamma\epsilon$ is the normalized beam emittance.

3. Using $\eta = y/\sqrt{\beta_y}$ and $p_\eta = d\eta/d\phi$ with $\phi = (1/\nu_y)\int_0^s ds/\beta_y$ as conjugate phase-space coordinates, and ϕ as time variable, show that Eq. (2.93) can be derived from the Hamiltonian

$$H = \frac{1}{2}(p_\eta^2 + \nu^2\eta^2) + \Delta H(\eta),$$
$$\Delta H = -\nu^2\int_0^\eta \beta^{3/2} g(\eta')d\eta', \quad g(\eta) = \frac{\Delta B}{B\rho}.$$

[46]This formula, derived based on the uniform round beam distribution. is called the Laslett space-charge tune shift. For a KV type round beam, the tune shift is independent of the particle's transverse coordinate r. For other beam distribution functions, a smaller betatron-amplitude particle may have a larger tune shift. Since the space-charge tune shift depends on the particle's betatron amplitude, the Laslett space-charge tune shift is also known as the incoherent space-charge tune spread. The small amplitude space-charge tune shift parameter is usually used as one of the design criteria in high power beam accelerators. Typical space charge tune-shift parameter is 0.25 to 0.5.

(a) Letting $\eta = \sqrt{2J/\nu}\cos\psi$ and $d\psi/d\phi = \nu$, show that the Hamiltonian becomes

$$H = \nu J + \langle \Delta H(J,\psi)\rangle_\psi + [\Delta H - \langle \Delta H(J,\psi)\rangle_\psi],$$

where

$$\langle \Delta H(J,\psi)\rangle_\psi = -\frac{\nu^2}{2\pi}\int_0^{2\pi}\left[\int_0^\eta \beta^{3/2} g(\eta')d\eta'\right]d\psi.$$

Show that the betatron tune is shifted by the perturbation $\Delta\nu = \partial\langle \Delta H\rangle_\psi/\partial J$.

(b) We consider a cylindrical Gaussian bunch distribution

$$\rho(x,s,z) = N_{\rm B}\rho(r)\rho(s),$$
$$\rho(r) = \frac{1}{2\pi\sigma_r^2}e^{-(x^2+z^2)/2\sigma_r^2}, \qquad \rho(s) = \frac{1}{\sqrt{2\pi}\sigma_s}e^{-s^2/2\sigma_s^2},$$

where $N_{\rm B}$ is the number of particles in a bunch, $\rho(r)$ and $\rho(s)$ are respectively the transverse and longitudinal Gaussian distributions with rms width σ_r and σ_s. Assuming $\sigma_s \gg \sigma_r$, show that the Lorentz force for a particle at distance r from the center of the bunch is

$$g(r,s) = \frac{2N_{\rm B}r_0}{\beta^2\gamma^3 r}\left(1 - e^{-r^2/2\sigma_r^2}\right)\rho(s),$$

where $r_0 = e^2/4\pi\epsilon_0 mc^2$ is the classical radius of the particle. Replacing r by $\sqrt{2\beta J/\nu}\,\cos\psi$, evaluate the space-charge tune shift as a function of the amplitude r.

4. The closed orbit can be locally corrected by using steering dipoles. A commonly used algorithm is based on the "three-bumps" method, where three steering dipoles are used to adjust local-orbit distortion. Let θ_1, θ_2, and θ_3 be the three bump angles. Show that these angles must be related by

$$\theta_2 = -\theta_1\sqrt{\frac{\beta_1}{\beta_2}}\frac{\sin\psi_{31}}{\sin\psi_{32}}, \qquad \theta_3 = \theta_1\sqrt{\frac{\beta_1}{\beta_3}}\frac{\sin\psi_{21}}{\sin\psi_{32}},$$

where β_i is the β-function at the ith steering magnet, $\psi_{ji} = \psi_j - \psi_i$ is the phase advance from the ith to the jth steering dipoles, and the orbit distortion is localized between the first and third steering magnets. Obviously, a local orbit bump can be attained by two steering dipoles θ_1 and θ_3 if and only if $\psi_{31} = n\pi$, where the phase advance is an integer multiple of π.

5. The AGS is composed of 12 superperiods with 5 nearly identical FODO cells per superperiod, i.e. it can be considered as a lattice made of 60 FODO cells with betatron tunes $\nu_z = 8.8$ and $\nu_x = 8.6$. The circumference is 807.12 m.

   (a) Estimate the closed orbit sensitivity factor of Eq. (2.98).

   (b) Estimate the the rms half-integer stopband width of Eq. (2.121) for the AGS.

   (c) During the polarized beam acceleration at AGS, a set of 10 ferrite quadrupoles located at high-β_z locations are powered to change the vertical tune by $\Delta\nu_z = -0.25$ in about 2.5 μs. This means that each quadrupole changes the betatron tune by -0.025.

i. What is the effect of these tune jump quadrupoles on the horizontal tune?
ii. What are the stopband integrals due to these tune jump quadrupoles?
iii. What are the favorable configurations for these quadrupoles from the beam dynamics point of view?
iv. Are there advantages to installing 12 quadrupoles? What are they?

6. In the H^- or H_2^+ strip injection process, the closed orbit is bumped onto the stripper location during the injection pulse. The injection beam and the circulating beam merge at the same phase-space point. We assume that the values of the betatron amplitude functions are $\beta_x = \beta_z = 10$ m, the emittances are $\epsilon_x = \epsilon_z = 2.5\pi$ mm-mrad for the injection beam, and $\epsilon_x = \epsilon_z = 40\pi$ mm-mrad for the circulating beam. Where should the stripper be located with respect to the center of the circulating beam? What is the minimum width of the stripper? Sketch a possible injection system scenario including local orbit bumps.

7. Multi-turn injection of heavy ion beams requires intricate phase-space painting techniques. The injection beam arrives through the center of a septum while the circulating beam closed orbit is bumped near the septum position. During the beam accumulation process, the orbit bump is reduced to avoid beam loss through the septum. We assume that the 95% emittances are 50 π mm-mrad for the stored beam and 2.5 π mm-mrad for the injection beam, the betatron amplitude functions are $\beta_x = \beta_z = 10$ m, and the thickness of the wire septum is 1 mm. How far from the closed orbit of the circulating beam should the septum be located? What effect, if any, does the betatron tune have on the beam-accumulation efficiency?

8. At extraction, the 95% emittance of the beam is adiabatically damped to 5 π mm-mrad at $B\rho = 10$ Tm. The extraction septum is located 40 mm from the center of the closed orbit of the circulating beam. At the septum location, the betatron functions are $\beta_x = 10$ m, $\beta_z = 8$ m. The septum (current sheet) thickness is 7 mm. A ferrite one-turn kicker is located upstream with $\beta_x = 10$ m and $\beta_z = 8$ m. The phase advance between the septum and the kicker is 60°. Discuss a scenario for efficient single-bunch extraction. What is the kicker angle required for single-turn extraction? Assuming that the maximum magnetic flux density for a kicker is 0.1 T, what is the minimum length of the kicker? What advantage, if any, does an orbit bump provide?

9. Particle motion in the presence of closed-orbit error is $x = x_{\rm co} + x_\beta$, where $x_{\rm co}$ is the closed orbit and x_β is the betatron displacement.

 (a) Show that an off-center horizontal closed orbit in quadrupoles gives rise to vertical dipole field error, and a vertical one to horizontal dipole field error.
 (b) The magnetic field of a nonlinear sextupole is

$$\Delta B_z + j\Delta B_x = \frac{B_2}{2}(x^2 - z^2 + 2jxz),$$

 where $B_2 = \partial^2 B_z/\partial x^2|_{x=0,z=0}$. Show that a horizontal closed-orbit error in a normal sextupole produces quadrupole field error. Show that the effective quadrupole gradient is $\partial B_z/\partial x|_{\rm eff} = x_{\rm co} B_2$.

10. In the presence of gradient error, the betatron amplitude functions and the betatron tunes are modified. This exercise provides an alternative derivation of Eq. (2.119).[47]

[47]See also H. Zgngier, LAL report 77-35, 1977; B.W. Montague, CERN **87-03**, 75-90 (1987).

We define the betatron amplitude deviation functions A and B as

$$A = \frac{\alpha_1\beta_0 - \alpha_0\beta_1}{\sqrt{\beta_0\beta_1}}, \quad B = \frac{\beta_1 - \beta_0}{\sqrt{\beta_0\beta_1}},$$

where β_0 and β_1 are respectively the unperturbed and the perturbed betatron amplitude functions associated with the gradient functions K_0 and K_1, and α_0 and α_1 are related to the derivatives of the betatron amplitude functions. Thus β_0 and β_1 satisfy the Floquet equation:

$$\begin{aligned} \beta_0' &= -2\alpha_0, \quad \alpha_0' = K_0\beta_0 - \gamma_0, \quad d\psi_0/ds = 1/\beta_0, \\ \beta_1' &= -2\alpha_1, \quad \alpha_1' = K_1\beta_1 - \gamma_1, \quad d\psi_1/ds = 1/\beta_1, \end{aligned}$$

where ψ_0 and ψ_1 are the unperturbed and the perturbed betatron phase functions.

(a) Show that

$$\frac{dB}{ds} = -A\left(\frac{1}{\beta_0} + \frac{1}{\beta_1}\right), \quad \frac{dA}{ds} = +B\left(\frac{1}{\beta_0} + \frac{1}{\beta_1}\right) + \sqrt{\beta_0\beta_1}\,\Delta K,$$

where $\Delta K = K_1 - K_0$.

(b) In a region with no gradient error, show that $A^2 + B^2 =$ constant, i.e. the phase-space trajectory of A vs B is a circle.

(c) Show that the change of A at a quadrupole with gradient error is

$$\Delta A = \int \sqrt{\beta_0\beta_1}\,\Delta K\, ds \approx \langle\beta_0\rangle g$$

in thin-lens approximation, where $g = +\int \Delta K\, ds$ is the integrated gradient strength of the error quadrupole, and $\langle\beta_0\rangle$ is the averaged value of betatron function in the quadrupole.

(d) In thin-lens approximation, show that the change of A in a sextupole is

$$\Delta A \approx \beta_0 g_{\text{eff}},$$

where $g_{\text{eff}} = (B_2\Delta s/B\rho)x_{\text{co}}$ is the effective quadrupole strength, $(B_2\Delta s/B\rho)$ is the integrated sextupole strength, and x_{co} is the closed-orbit deviation from the center of the sextupole.

(e) If we define the average betatron phase function as

$$\bar{\phi} = \frac{1}{2\bar{\nu}}\int_{s_0}^{s}\left(\frac{1}{\beta_0} + \frac{1}{\beta_1}\right) ds, \qquad \text{with} \qquad \bar{\nu} = \frac{1}{4\pi}\int_{s_0}^{s_0+C}\left(\frac{1}{\beta_0} + \frac{1}{\beta_1}\right) ds,$$

show that the function B satisfies

$$\frac{d^2B}{d\bar{\phi}^2} + 4\bar{\nu}^2 B = -4\bar{\nu}^2\frac{(\beta_0\beta_1)^{3/2}}{\beta_0 + \beta_1}\Delta K.$$

Show that this equation reduces to Eq. (2.119) in the limit of small gradient error.

11. Show that the half-integer stopband integral J_p is *approximately* zero at $p = [2\nu]$ for two quadrupole kickers separated by 180° in betatron phase advance with zero betatron tune shift. Such a *zero tune shift π-doublet* can be used to change γ_T with minimum effects on betatron motion (see Sec. IV.8.A.3).

12. Multiple scattering from gas molecules inside the vacuum chamber can cause beam emittance dilution, particularly at high-β locations. This effect can also be important in the strip-injection process. This exercise estimates the emittance dilution rate based on the multiple scattering formula (see the particle properties data) for the rms scattering angle

$$\theta^2 = 2\theta_0^2 \approx 2\left(\frac{13.6[\mathrm{MeV}]z_\mathrm{p}}{\beta cp}\right)^2 \frac{x}{X_0},$$

where p, βc and z_p are momentum, velocity, and charge number of the beam particles, X_0 is the radiation length, and x is the target thickness. The radiation length is

$$X_0 = \frac{716.4A}{Z(Z+1)\ln(287/\sqrt{Z})} \quad [\mathrm{g/cm^2}]$$

where Z and A are the atomic charge and the mass number of the medium.

(a) Using the ideal gas law, $PV = nRT$, where P is the pressure, V is the volume, n is the number of moles, T is the temperature, and $R = 8.314$ [J (°K mol)$^{-1}$], show that the equivalent target thickness in [g/cm^2/s] at room temperature is

$$x = 1.641 \times 10^{-6}\beta P_\mathrm{g}[\mathrm{ntorr}]A_\mathrm{g} \quad [\mathrm{g/cm^2/s}],$$

where βc is the velocity of the beam, P_g is the equivalent partial pressure of a gas at room temperature $T = 293$ °K, and A_g is the gram molecular weight of a gas. Show that the emittance growth rate is

$$\frac{1}{\tau_\epsilon} = \frac{1}{\epsilon}\frac{d\epsilon}{dt} = 2.345\frac{\gamma\langle\beta_\perp\rangle\ [\mathrm{m}]}{\beta\epsilon_\mathrm{N}\ [\pi\ \mathrm{mm\ mrad}]}\left(\frac{z_\mathrm{p}}{pc\ [\mathrm{GeV}]}\right)^2 \frac{P_\mathrm{g}\ [\mathrm{nTorr}]A_\mathrm{g}}{X_{0g}\ [\mathrm{g/cm^2}]} \quad [\mathrm{h^{-1}}],$$

where $\langle\beta_\perp\rangle$ is the average transverse betatron amplitude function in the accelerator, X_{0g} is the radiation length of the gas, γ is the Lorentz relativistic factor, z_p is the charge of the projectile, and p is the momentum of the beam. Because the emittance growth is proportional to the betatron function, better vacuum at high-$\beta_\perp$ location is useful in minimizing the multiple scattering effects.

(b) During the H$^-$ strip-injection process, the H$^-$ passes through a thin foil of thickness t_foil [μg/cm^2]. Show that the emittance growth per passage is

$$\Delta\epsilon = 117.8\,\frac{\beta_{\perp,\mathrm{foil}}}{\beta^2(pc[\mathrm{MeV}])^2}\,\frac{t_\mathrm{foil}[\mu\mathrm{g/cm^2}]}{X_0[\mathrm{g/cm^2}]} \quad [\pi\,\mathrm{mm\,mrad}],$$

where $\beta_{\perp,\mathrm{foil}}$ is the betatron amplitude function at the stripper location, p is the momentum of the injected beam, βc is the velocity of the beam, and X_0 is the radiation length. Estimate the emittance growth rate per passage through carbon foil with H$^-$ beams at an injection energy of 7 MeV if $\beta_{\perp,\mathrm{foil}} = 2$ m and $t_\mathrm{foil} = 4$ [μg/cm^2].[48]

[48] If the stripping foil is too thin, the efficiency of charge exchange is small, and the proton yield is little. If the foil is too thick, the beam emittance will increase because of multiple Coulomb scattering. A compromise between various processes is needed in the design of accelerator components.

IV Off-Momentum Orbit

In Sec. III, we discussed the closed orbit in the presence of dipole field errors and quadrupole misalignment for a reference particle with momentum p_0. Using closed-orbit correctors, we can achieve an optimized closed orbit that essentially passes through the center of all accelerator components, particularly quadrupoles and sextupoles. This closed orbit is called the "golden orbit," and a particle with momentum p_0 is called a *synchronous* particle.[49]

However, a beam is made of particles with momenta distributed around the synchronous momentum p_0. What happens to particles with momenta different from p_0? Here we study the effect of off-momentum on the closed orbit. Its effect on betatron motion will be addressed in Sec V.

For a particle with momentum p, the momentum deviation is $\Delta p = p - p_0$ and the fractional momentum deviation is $\delta = \Delta p/p_0$. The fractional momentum deviation $\delta = \Delta p/p$ is typically small, e.g. $|\delta| \leq 10^{-4}$ for SSC, $\leq 5\cdot 10^{-3}$ for RHIC, $\leq 3\cdot 10^{-2}$ for anti-proton accumulators, $\leq 10^{-4}$ for the IUCF Cooler Ring, and $\leq 2\cdot 10^{-2}$ for typical electron storage rings. Since δ is small, we can study the motion of off-momentum particles perturbatively. In Sec. IV.1, we will find that the off-momentum closed orbit is proportional to δ in the first-order approximation, and the dispersion function is defined as the derivative of the off-momentum closed orbit with respect to δ. We will discuss the properties of the dispersion function; in particular, the integral representation, the dispersion action, and the $\mathcal{H}$-function will be introduced in Sec. IV.2. The momentum compaction factor and transition energy are discussed in Sec. IV.3, where we introduce the phase focusing principle of synchrotron motion. In Sec. IV.4, we examine the method of dispersion suppression in a beam line. In Sec. IV.5 we discuss the achromat transport system, and in Sec. IV.6 we introduce the standard transport notation. In Sec. IV.7 we describe methods of dispersion measurements and correction, and in Sec. IV.8 methods of transition energy manipulation. Minimum $\langle\mathcal{H}\rangle$ lattices are discussed in Sec. IV.9 and Sec. III in Chapter 4.

IV.1 Dispersion Function

Expanding Eq. (2.22) to first order in x, we obtain

$$x'' + \left(\frac{1-\delta}{\rho^2(1+\delta)} - \frac{K(s)}{(1+\delta)}\right)x = \frac{\delta}{\rho(1+\delta)}, \tag{2.147}$$

[49]The revolution frequency of a synchronous particle is defined as the revolution frequency of the beam. The frequency of the radio-frequency (rf) cavities has to be an integer multiple of the revolution frequency of the beam, i.e. a synchronous particle synchronizes with the rf electric field. The name "synchrotron" for circular accelerators is derived from the synchronism between the orbiting particles and the rf field.

where $K(s) = B_1/B\rho$ is the quadrupole gradient function with $B_1 = \partial B_z/\partial x$ evaluated at the closed orbit. Solutions of Eq. (2.147) for $\delta = 0$ were discussed in Sec. II. For an off-momentum particle with $\delta \neq 0$, the solution of the linearized inhomogeneous equation (2.147) can be expressed as a linear superposition of the particular solution and the solution of the homogeneous equation: $x = x_\beta(s) + D(s)\delta$,[50] where $x_\beta(s)$ and $D(s)$ satisfy the equations

$$x_\beta'' + (K_x(s) + \Delta K_x)x_\beta = 0, \tag{2.148}$$

$$D'' + (K_x(s) + \Delta K_x)D = \frac{1}{\rho} + O(\delta), \tag{2.149}$$

$$K_x = \frac{1}{\rho^2} - K(s), \quad \Delta K_x = \left[-\frac{2}{\rho^2} + K(s)\right]\delta + O(\delta^2).$$

In this section, we will neglect the chromatic perturbation term $\Delta K_x(s)$. The solution of the inhomogeneous equation is called the dispersion function, where $D(s)\delta$ is the *off-momentum closed orbit.* Aside from the chromatic perturbation ΔK_x, the solution of the homogeneous equation x_β is the betatron motion around the off-momentum closed orbit. To the lowest order in δ, the dispersion function obeys the inhomogeneous equation

$$D'' + K_x(s)D = 1/\rho\,. \tag{2.150}$$

If $K_x(s)$ and $\rho(s)$ are periodic functions of s with period L, we can impose the periodic closed-orbit condition on the dispersion function[51]

$$D(s + L) = D(s), \quad D'(s + L) = D'(s). \tag{2.151}$$

Since $K_x(s)$ and $\rho(s)$ are usually piecewise constant for accelerator components, the inhomogeneous equation can easily be solved by the matrix method. The solution of a linear inhomogeneous dispersion equation is a of the particular solution and the solution of the homogeneous equation:

$$\begin{pmatrix} D(s_2) \\ D'(s_2) \end{pmatrix} = M(s_2|s_1)\begin{pmatrix} D(s_1) \\ D'(s_1) \end{pmatrix} + \begin{pmatrix} d \\ d' \end{pmatrix}, \tag{2.152}$$

where the 2×2 matrix $M(s_2|s_1)$ is the transfer matrix for the homogeneous equation, and d and d' are the particular solution. Let $\bar{d}$ be shorthand notation for the *two-component dispersion vector* with transpose vector $(\bar{d})^\dagger = (d, d')$. The transfer matrix

[50]Including the dipole field error, the displacement x is $x = x_{\rm co}(s) + x_\beta(s) + D(s)\delta$, where $x_{\rm co}$ is the closed-orbit error discussed in Sec. III. A beam is composed of particles with different momenta. The normalized Gaussian distribution function of the beam is $\rho(\delta) = \frac{1}{\sqrt{2\pi}\sigma_\delta}\exp\{-\delta^2/2\sigma_\delta^2\}$, where σ_δ is the rms fractional off-momentum width.

[51]The closed-orbit condition for the dispersion function is strictly required only for one complete revolution $D(s) = D(s + C)$ and $D'(s) = D'(s + C)$, where C is the circumference. The local closed-orbit condition of Eq. (2.151) for repetitive cells is not a necessary condition. However, this local periodic closed-orbit condition facilitates accelerator lattice design.

in Eq. (2.152) can be expressed by the 3×3 matrix

$$\begin{pmatrix} D(s_2) \\ D'(s_2) \\ 1 \end{pmatrix} = \begin{pmatrix} M(s_2|s_1) & \bar{d} \\ 0 & 1 \end{pmatrix} \begin{pmatrix} D(s_1) \\ D'(s_1) \\ 1 \end{pmatrix}. \tag{2.153}$$

with

$$\bar{d} = \begin{cases} \begin{pmatrix} 0 \\ 0 \end{pmatrix} & \text{for drift space and quadrupole} \\ \begin{pmatrix} \frac{1}{\rho K_x}(1-\cos\sqrt{K_x}s) \\ \frac{1}{\rho\sqrt{K_x}}\sin\sqrt{K_x}s \end{pmatrix} & \text{for dipole with } K_x > 0, \\ \begin{pmatrix} \frac{1}{\rho|K_x|}(-1+\cosh\sqrt{|K_x|}s) \\ \frac{1}{\rho\sqrt{|K_x|}}\sinh\sqrt{|K_x|}s \end{pmatrix} & \text{for dipole with } K_x < 0. \end{cases}$$

The transfer matrix for a pure *sector* dipole, where $K_x = 1/\rho^2$ with ρ the bending radius θ the bend angle and $\ell = \rho\theta$ the length of the dipole, is

$$M = \begin{pmatrix} \cos\theta & \rho\sin\theta & \rho(1-\cos\theta) \\ -(1/\rho)\sin\theta & \cos\theta & \sin\theta \\ 0 & 0 & 1 \end{pmatrix} \overset{\theta\ll 1}{\Longrightarrow} \begin{pmatrix} 1 & \ell & \frac{1}{2}\ell\theta \\ 0 & 1 & \theta \\ 0 & 0 & 1 \end{pmatrix}. \tag{2.154}$$

A. Dispersion function of a FODO cell in thin-lens approximation

A FODO cell with dipole, as shown in Fig. 2.4, is represented by

$$\mathrm{C} = \{\frac{1}{2}\mathrm{QF} \quad \mathrm{B} \quad \mathrm{QD} \quad \mathrm{B} \quad \frac{1}{2}\mathrm{QF}\},$$

where QF and QD are focusing and defocusing quadrupoles, and B represents bending dipole(s). Using thin-lens approximation, we obtain

$$\mathbf{M} = \begin{pmatrix} 1 & 0 & 0 \\ -\frac{1}{2f} & 1 & 0 \\ 0 & 0 & 1 \end{pmatrix} \begin{pmatrix} 1 & L_1 & \frac{1}{2}L_1\theta \\ 0 & 1 & \theta \\ 0 & 0 & 1 \end{pmatrix} \begin{pmatrix} 1 & 0 & 0 \\ \frac{1}{f} & 1 & 0 \\ 0 & 0 & 1 \end{pmatrix} \begin{pmatrix} 1 & L_1 & \frac{1}{2}L_1\theta \\ 0 & 1 & \theta \\ 0 & 0 & 1 \end{pmatrix} \begin{pmatrix} 1 & 0 & 0 \\ -\frac{1}{2f} & 1 & 0 \\ 0 & 0 & 1 \end{pmatrix},$$

where L_1 is the half cell length, θ is the bending angle of a half cell, and f is the focal length of the quadrupoles. The closed-orbit condition of Eq. (2.151) becomes

$$\begin{pmatrix} D \\ D' \\ 1 \end{pmatrix}_{\mathrm{F}} = \begin{pmatrix} 1-\frac{L_1^2}{2f^2} & 2L_1(1+\frac{L_1}{2f}) & 2L_1\theta(1+\frac{L_1}{4f}) \\ -\frac{L_1}{2f^2}+\frac{L_1^2}{4f^3} & 1-\frac{L_1^2}{2f^2} & 2\theta(1-\frac{L_1}{4f}-\frac{L_1^2}{8f^2}) \\ 0 & 0 & 1 \end{pmatrix} \begin{pmatrix} D \\ D' \\ 1 \end{pmatrix}_{\mathrm{F}}. \tag{2.155}$$

The D_F and D'_F in Eq. (2.155) are values of the dispersion function and its derivative at the focusing quadrupole location. Using the Courant-Snyder parametrization for the 2×2 matrix, we obtain

$$\sin\frac{\Phi}{2}=\frac{L_1}{2f},\ \beta_F=\frac{2L_1(1+\sin(\Phi/2))}{\sin\Phi},\ \alpha_F=0;\quad D_F=\frac{L_1\theta(1+\frac{1}{2}\sin(\Phi/2))}{\sin^2(\Phi/2)},\ D'_F=0.$$

where Φ is the phase advance per cell. The dispersion function at other locations in the accelerator can be obtained by the matrix propagation method, Eq. (2.153). The dispersion function at the defocusing quadrupole location is

$$D_D=\frac{L_1\theta(1-\frac{1}{2}\sin(\Phi/2))}{\sin^2(\Phi/2)},\quad D'_D=0.$$

The middle plot of Fig. 2.5 shows the dispersion function of the AGS lattice, which can be approximated by a lattice made of 60 FODO cells. Some characteristic properties of the dispersion function of FODO cells are listed as follows:

- The dispersion function at the focusing quadrupole is larger than that at the defocusing quadrupole by a factor $(2+\sin(\Phi/2))/(2-\sin(\Phi/2))$, which is about 2 at $\Phi\approx 90°$.
- The dispersion function is proportional to $L_1\theta$, proportional to the product of the cell length and the bending angle of a FODO cell. For given L_1, θ, and Φ, the dispersion function of a FODO cell is nearly independent of the dipole length $\ell=\rho\theta$. When the phase advance is small, $\Phi<\frac{\pi}{2}$, the dispersion function is $D\sim\frac{L_1\theta}{\sin^2(\Phi/2)}\sim\frac{4L_1\theta}{\Phi^2}$.
- Missing dipole FODO (MD-FODO) cells are commonly used in accelerator design for its drift spaces for injection, extraction, and rf cavities. The dispersion function at the center of the focusing quadrupole of the MD-FODO cell is equal to that of the regular FODO cell at the same phase advance and total bending angle per cell in thin lens approximation.

B. Dispersion function in terms of transfer matrix

In general, the transfer matrix of a periodic cell can be expressed as

$$\mathbf{M}=\begin{pmatrix}M_{11}&M_{12}&M_{13}\\M_{21}&M_{22}&M_{23}\\0&0&1\end{pmatrix},\tag{2.156}$$

where M_{11}, M_{12}, M_{21} and M_{22} are given by Eq. (2.34). Using the closed-orbit condition of Eq. (2.151), we obtain

$$D=\frac{M_{13}(1-M_{22})+M_{12}M_{23}}{2-M_{11}-M_{22}}=\frac{M_{13}(1-\cos\Phi+\alpha\sin\Phi)+M_{23}\beta\sin\Phi}{2(1-\cos\Phi)},$$

$$D'=\frac{M_{13}M_{21}+(1-M_{11})M_{23}}{2-M_{11}-M_{22}}=\frac{-M_{13}\gamma\sin\Phi+M_{23}(1-\cos\Phi-\alpha\sin\Phi)}{2(1-\cos\Phi)},$$

where Φ is the horizontal betatron phase advance of the periodic cell, α, β and $\gamma = (1+\alpha^2)/\beta$ are the Courant–Snyder parameters for the horizontal betatron motion at a periodic-cell location s, and D and D' are the value of the dispersion function and its derivative at the same location. Solving M_{13} and M_{23} as functions of D and D', the 3×3 transfer matrix is

$$\mathbf{M} = \begin{pmatrix} \cos\Phi+\alpha\sin\Phi & \beta\sin\Phi & (1-\cos\Phi-\alpha\sin\Phi)D-\beta D'\sin\Phi \\ -\gamma\sin\Phi & \cos\Phi-\alpha\sin\Phi & \gamma D\sin\Phi+(1-\cos\Phi+\alpha\sin\Phi)D' \\ 0 & 0 & 1 \end{pmatrix} \qquad (2.157)$$

This representation of the transfer matrix is sometimes useful in studying the general properties of repetitive accelerator sections.

C. Effect of dipole or quadrupole field error on dispersion function

In the presence of dipole and quadrupole field error, perturbation to the dispersion function $\Delta D(s) = D(s) - D_0(s)$ obeys

$$(\Delta D)'' + [K_0(s) + k(s)]\Delta D(s) = \left[\frac{1}{\rho} - \frac{1}{\rho_0}\right] - k(s)D_0(s), \qquad (2.158)$$

where $D_0(s)$ is the unperturbed dispersion function, $K_0(s)$ and $\rho_0(s)$ are the unperturbed dipole and focusing functions, and $k(s)$ is the quadrupole field error. The inhomogeneous equation can be solved by employing Floquet transformation by neglecting the higher order term $k(s)\Delta D(s)$. The bottom plot of Fig. 2.20 shows $\Delta D(s)/\sqrt{\beta_x}$ induced by the gradient error of a single focusing quadrupole with 1% increase in strength. Outside the quadruple kick location, we find a pure sinusoidal betatron oscillation. Quadrupoles in dispersive locations can be used to produce a local dispersion bump resembling that of closed orbit bump in Sec. III.3.

It appears that the factor $\frac{1}{\rho} - \frac{1}{\rho_0}$ in Equation (2.158) can be large if length of all dipoles in a lattice is shortened or lengthened, e.g. ρ can be a factor of 2 larger or smaller than ρ_0 in if the dipole length is shortened or lengthened by a factor 2. Can the change of dipole length causes a large perturbation to the dispersion function? We have previously stated that the dispersion function of a FODO cell is essentially a function of $L_1\theta$, nearly independent of the dipole length. Here L_1 and θ are the length and bending angle in each half cell. This paradox is resolved by the fact that $\frac{1}{\rho} - \frac{1}{\rho_0}$ has small stopband integrals at harmonics near the betatron tune ν_x. On the other hand, if $\frac{1}{\rho(s)} - \frac{1}{\rho_0}$ has a large stopband near $[\nu_x]$, the perturbation to the dispersion function will be substantial.

IV.2 $\mathcal{H}$-Function, Action, and Integral Representation

The dispersion $\mathcal{H}$-function is defined as

$$\mathcal{H}(D, D') = \gamma_x D^2 + 2\alpha_x DD' + \beta_x D'^2 = \frac{1}{\beta_x}[D^2 + (\beta_x D' + \alpha_x D)^2]. \qquad (2.159)$$

Since the dispersion function satisfies the homogeneous betatron equation of motion in regions with no dipole ($1/\rho = 0$), the $\mathcal{H}$-function is invariant. In regions with dipoles, the $\mathcal{H}$-function is not invariant. For a FODO cell, the dispersion $\mathcal{H}$-function at the defocusing quadrupole is larger than that at the focusing quadrupole, i.e. $\mathcal{H}_{\rm F} \le \mathcal{H}_{\rm D}$, where

$$\mathcal{H}_{\rm F} = \frac{L_1\theta^2 \sin\Phi(1+\frac{1}{2}\sin\frac{\Phi}{2})^2}{2(1+\sin\frac{\Phi}{2})\sin^4\frac{\Phi}{2}}, \quad \mathcal{H}_{\rm D} = \frac{L_1\theta^2 \sin\Phi(1-\frac{1}{2}\sin\frac{\Phi}{2})^2}{2(1-\sin\frac{\Phi}{2})\sin^4\frac{\Phi}{2}}, \tag{2.160}$$

and the dispersion $\mathcal{H}$-function is proportional to the inverse cubic power of the phase advance.

Now we define the *normalized* dispersion phase-space coordinates as

$$\begin{cases} X_{\rm d} = \dfrac{1}{\sqrt{\beta_x}} D = \sqrt{2J_{\rm d}}\cos\Phi_{\rm d}, \\ P_{\rm d} = \sqrt{\beta_x} D' + \dfrac{\alpha_x}{\sqrt{\beta_x}} D = -\sqrt{2J_{\rm d}}\sin\Phi_{\rm d}, \end{cases} \tag{2.161}$$

where the dispersion action is $J_{\rm d} = \frac{1}{2}\mathcal{H}(D, D')$. In a straight section, $J_{\rm d}$ is invariant and $\Phi_{\rm d}$, aside from a constant, is identical to the betatron phase advance. In a region with dipoles, $J_{\rm d}$ is not constant. The change of the dispersion function across a *thin* dipole is $\Delta D = 0$ and $\Delta D' = \theta$, i.e. $\Delta X_{\rm d} = 0$, $\Delta P_{\rm d} = \sqrt{\beta_x}\,\Delta D' = \sqrt{\beta_x}\,\theta$, where θ is the bending angle of the dipole. The change in dispersion action is $\Delta J_{\rm d} = (\beta_x D' + \alpha_x D)\theta$. For FODO-cell lattice shown in Sec. IV.1.A the normalized dispersion coordinate $X_{\rm d}$ is nearly constant, i.e. $D \sim \sqrt{\beta_x}$, and $P_{\rm d}$ is small. Figure 2.31 shows the normalized dispersion phase-space coordinates in one superperiod of AGS lattice (see Fig. 2.5) that is approximately made of 5 FODO cells.

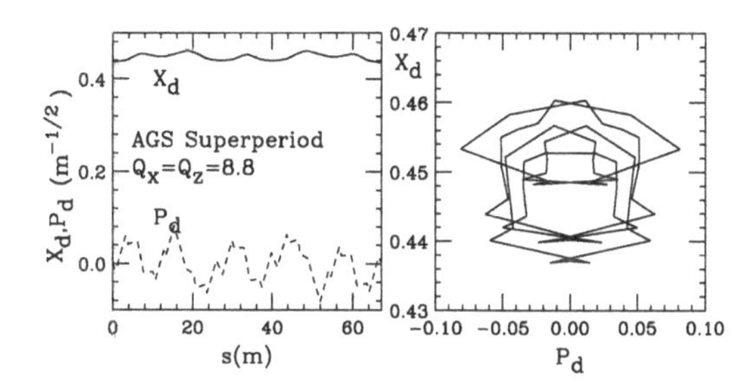

Figure 2.31: Left: Normalized dispersion phase-space coordinates $X_{\rm d}$ and $P_{\rm d}$ are plotted in a superperiod of the AGS lattice. Right: the coordinates are shown in $X_{\rm d}$ vs $P_{\rm d}$. The scales for both $X_{\rm d}$ and $P_{\rm d}$ are $\rm m^{1/2}$. Note that $X_{\rm d}$ is indeed nearly constant, and $(P_{\rm d}, X_{\rm d})$ propagate in a very small region of the dispersion phase-space (see also Fig. 2.38).

In contrast, the normalized dispersion phase-space coordinates for a double-bend achromat (DBA) lattice (see Sec. IV.5.A) shows different behavior. Figure 2.32 shows the normalized dispersion coordinates for the IUCF Cooler Ring, which is composed of 3 achromat straight-sections for electron cooling, rf cavities, etc., and 3 dispersive-sections for injection, momentum stacking, etc. The achromat sections are described by a single point at origin: $X_{\rm d} = P_{\rm d} = 0$. Inside dipoles, the normalized dispersion coordinates increase in magnitude. In dispersion matching sections, the normalized dispersion coordinates are located on invariant circles, that are nearly half-circles as shown in Fig. 2.32, i.e. the dispersion phase advance $\Phi_{\rm d}$ is nearly π in the dispersion matching section. Since the dispersion phase-advance is equal to the horizontal betatron phase-advance in a straight section, the horizontal betatron phase-advance is also nearly π.

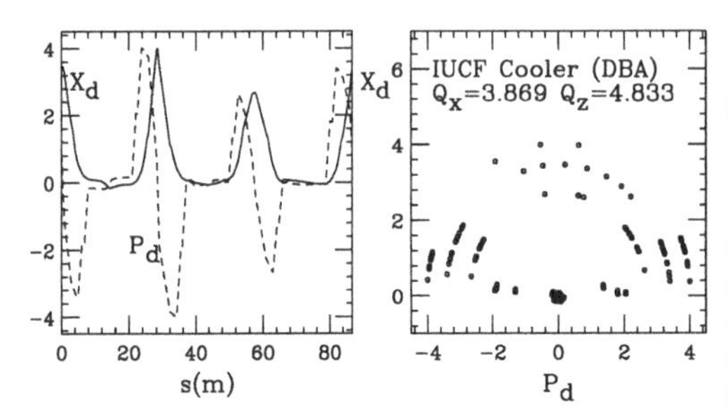

Figure 2.32: Left: Normalized dispersion phase-space coordinates $X_{\rm d}$ and $P_{\rm d}$ of the IUCF Cooler lattice are plotted. Right: The coordinates are shown in $X_{\rm d}$ vs $P_{\rm d}$ at the end of each lattice elements. The accelerator is made of six 60°-bends forming a 3 double-bend achromat modules, where the dispersion function is shown in Fig. 2.35. The scales for both $X_{\rm d}$ and $P_{\rm d}$ are $\rm m^{1/2}$.

The lattice function and the dispersion phase-space coordinates of the IUCF Cooler Ring differ substantially from the low emittance DBA lattice to be shown in Figs. 2.41. For an ion storage ring, the minimization of $\langle \mathcal{H} \rangle$ plays no important role in beam dynamics. Instead, a minimum β_z inside dipole will provide a criterion for the magnet gap g. Since the power required in the operation of a storage ring is proportional to g^2, it is preferable to design a machine with a minimum β_z inside dipoles, and the corresponding β_x will be large in dipole. The resulting dispersion phase-space coordinates in dipoles are much larger than those of minimum emittance DBA lattices shown in Fig. 2.41.

The dispersion function can also be derived from the dipole field error resulting from the momentum deviation. The angular kick due to the off-momentum deviation is $\theta = \frac{\Delta p}{p_0}\frac{1}{\rho}ds$, where ρ is the bending radius and ds is the differential length of the dipole. The corresponding dipole field error is $\frac{\Delta B}{B\rho} = \frac{\Delta p}{p_0}\frac{1}{\rho}$, Substituting the "dipole field error" in Eq. (2.92), we obtain the dispersion function of Eq. (2.101). The integral representation of the normalized dispersion functions is

$$\begin{cases} X_{\rm d}(s) = \dfrac{1}{2\sin\pi\nu_x}\displaystyle\int_s^{s+C} \dfrac{\sqrt{\beta_x(t)}}{\rho}\cos(\psi_x(t) - \psi_x(s) - \pi\nu_x)dt \\ P_{\rm d}(s) = \dfrac{-1}{2\sin\pi\nu_x}\displaystyle\int_s^{s+C} \dfrac{\sqrt{\beta_x(t)}}{\rho}\sin(\psi_x(t) - \psi_x(s) - \pi\nu_x)dt. \end{cases} \tag{2.162}$$

IV.3 Momentum Compaction Factor

Since the synchronization of particle motion in a synchrotron depends critically on the total path length, it is important to evaluate the effect of the off-momentum closed orbit on path length. Since the change in path length due to betatron motion is proportional to the square of the betatron amplitude [see Eq. (2.100)], the effect is small. The orbit deviation from a reference orbit of an off-momentum particle is linearly proportional to the product of the fractional off-momentum parameter δ and the dispersion function $D(s)$. The total path length will depend on the off-momentum parameter. The path difference and the "momentum compaction factor" are

$$\Delta C = \left[\oint \frac{D(s)}{\rho} ds\right] \delta, \qquad \alpha_c \equiv \frac{1}{C}\frac{d\Delta C}{d\delta} = \frac{1}{C}\oint \frac{D(s)ds}{\rho} \tag{2.163}$$

where $\langle D\rangle_i$ and θ_i are the average dispersion function and the bending angle of the ith dipole, and the last approximate identity uses thin-lens approximation. Since $D(s)$ is normally positive, the total path length for a higher momentum particle is longer. For example, the momentum compaction factor for a FODO lattice is

$$\alpha_c \approx \frac{(D_{\rm F} + D_{\rm D})\theta}{2L_1} \approx \frac{\theta^2}{\sin^2(\Phi/2)} \approx \frac{1}{\nu_x^2},$$

where L_1 and θ are the length and the bending angle of one half-cell, Φ is the phase advance of a FODO cell, and ν_x is the betatron tune (see Exercise 2.4.2).

A. Transition energy and the phase-slip factor

The importance of the momentum compaction factor will be fully realized when we discuss synchrotron motion in Chap. 3. In the meantime, we discuss the phase stability of synchrotron motion discovered by McMillan and Veksler [21].

Particles with different momenta travel along different paths in an accelerator. Since the revolution period is $T = 1/f = C/v$, where C is the circumference, and v is the speed of the circulating particle, the fractional difference of the revolution periods between the off-momentum and on-momentum particles and the "phase-slip-factor" are

$$\frac{\Delta f}{f_0} = -\frac{\Delta T}{T_0} = -\frac{\Delta C}{C} + \frac{\Delta v}{v} = -\left(\alpha_c - \frac{1}{\gamma^2}\right)\frac{\Delta p}{p_0} = -\eta\delta, \tag{2.164}$$

$$\eta = \alpha_c - \frac{1}{\gamma^2} = \frac{1}{\gamma_{\rm T}^2} - \frac{1}{\gamma^2}, \tag{2.165}$$

where $T_0 = 1/f_0$ is the revolution period of a synchronous particle, $\delta = \Delta p/p_0$ is the fractional momentum deviation, $\gamma_{\rm T} \equiv \sqrt{1/\alpha_c}$ is called the transition-γ, and $\gamma_{\rm T}mc^2$ or simply $\gamma_{\rm T}$ is the transition energy. For FODO cell lattices, $\gamma_{\rm T} \approx \nu_x$.

Below the transition energy, with $\gamma < \gamma_{\rm T}$ and $\eta < 0$, a higher momentum particle will have a revolution period shorter than that of the synchronous particle. Because a high energy particle travels faster, its speed compensates its longer path length in the accelerator, so that a higher energy particle will arrive at a fixed location earlier than a synchronous particle. Above the transition energy, with $\gamma > \gamma_{\rm T}$, the converse is true. Without a longitudinal electric field, the time slippage between a higher or lower energy particle and a synchronous particle is $T_0\eta\delta$ per revolution.

At $\gamma = \gamma_{\rm T}$ the revolution period is independent of the particle momentum. All particles at different momenta travel rigidly around the accelerator with equal revolution frequencies. This is the isochronous condition, which is the operating principle of AVF isochronous cyclotrons.

B. Phase stability of the bunched beam acceleration

Let $V(t) = V_0 \sin(h\omega_0 t + \phi)$ be the gap voltage of the rf cavity (see Fig. 2.33), where V_0 is the amplitude, ϕ is an arbitrary phase angle, h is an integer called the harmonic number, $\omega_0 = 2\pi f_0$ is the angular revolution frequency, and f_0 is the revolution frequency of a synchronous particle. A *synchronous* particle is defined as an ideal particle that arrives at the rf cavity at a constant phase angle $\phi = \phi_{\rm s}$, where $\phi_{\rm s}$ is the synchronous phase angle. The acceleration voltage at the rf gap and the acceleration rate for a synchronous particle are respectively given by

$$V_{\rm s} = V_0 \sin\phi_{\rm s}, \quad \dot{E}_0 = f_0 e V_0 \sin\phi_{\rm s}, \tag{2.166}$$

where e is the charge, E_0 is the energy of the synchronous particle, and the overdot indicates the derivative with respect to time t.

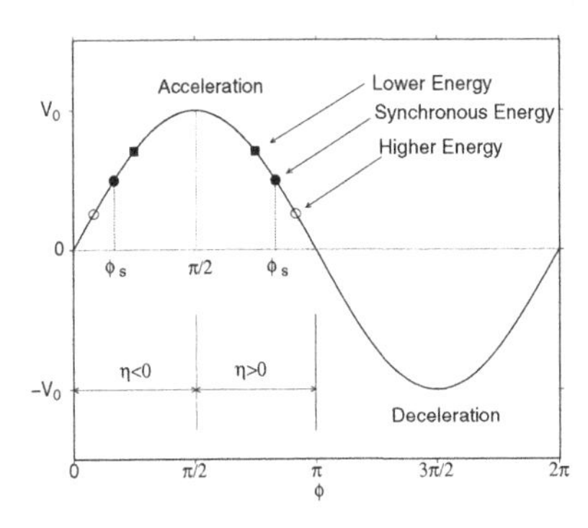

Figure 2.33: Schematic drawing of an rf wave, and the rf phase angles for a synchronous, a higher, and a lower energy particles (Graph courtesy of D. Li, LBNL). For a stable synchrotron motion, the phase focusing principle requires $0 < \phi_{\rm s} \le \pi/2$ for $\eta < 0$, and $\pi/2 < \phi_{\rm s} \le \pi$ for $\eta > 0$. Below the transition energy, with $0 \le \phi_{\rm s} < \pi/2$, a higher energy particle arrives at the rf gap earlier and receives less energy from the cavity. Thus the energy of the particle will becomes smaller than that of the synchronous particle. On the other hand, a lower energy particle arrives later and gains more energy from the cavity. This process gives rise to the phase stability of synchrotron motion.

A non-synchronous particle arriving at the rf cavity gap has a phase angle ϕ with respect to the rf field. The phase ϕ varies with time, and the acceleration rate is $\dot{E} = feV_0 \sin\phi$, where f is the revolution frequency. Combining with Eq. (2.166), we find the rate of change of the energy deviation is (see also Chap. 3, Sec. I)

$$\frac{d}{dt}\left(\frac{\Delta E}{\omega_0}\right) = \frac{1}{2\pi} e V_0 (\sin\phi - \sin\phi_{\rm s}), \tag{2.167}$$

where $\Delta E = E - E_0$ is the energy difference between the non-synchronous and the synchronous particles. Similarly, the equation of motion for the rf phase angle $\phi = -h\theta$, where θ is the actual angular position of the particle in a synchrotron, is

$$\frac{d}{dt}(\phi - \phi_{\rm s}) = -h\Delta\omega = h\omega_0\frac{\Delta T}{T_0} = h\eta\omega_0\frac{\Delta p}{p_0} = \frac{\eta h\omega_0^2}{\beta^2 E_0}\frac{\Delta E}{\omega_0}. \tag{2.168}$$

Equations (2.167) and (2.168) form the basic synchrotron equation of motion for conjugate phase-space coordinates ϕ and $\Delta E/\omega_0$. This is the equation of motion for a biased physical pendulum system, called synchrotron motion. The differential equation for the small amplitude phase oscillation is

$$\frac{d^2(\phi-\phi_{\rm s})}{dt^2} = \frac{\eta h\omega_0^2 eV_0}{2\pi\beta^2 E_0}(\sin\phi - \sin\phi_{\rm s}) \approx \frac{\eta\cos\phi_{\rm s} h\omega_0^2 eV_0}{2\pi\beta^2 E_0}(\phi-\phi_{\rm s}) = \omega_{\rm syn}^2(\phi-\phi_{\rm s});$$

$$\omega_{\rm syn} = \omega_0\sqrt{\frac{heV_0|\eta\cos\phi_{\rm s}|}{2\pi\beta^2 E_0}}. \qquad \begin{cases} 0 \le \phi_{\rm s} \le \pi/2 & \text{if } \gamma < \gamma_{\rm T} \text{ or } \eta < 0, \\ \pi/2 \le \phi_{\rm s} \le \pi & \text{if } \gamma > \gamma_{\rm T} \text{ or } \eta > 0. \end{cases} \tag{2.169}$$

where $\omega_{\rm syn}$ is the small-amplitude angular synchrotron frequency. The phase stability condition is $\eta\cos\phi_{\rm s} < 0$. Below the transition energy, with $0 \le \phi_{\rm s} < \pi/2$, a higher energy particle arrives at the rf gap earlier and receives less energy from the rf cavity (see Fig. 2.33). Thus the energy of the particle will gradually becomes smaller than that of the synchronous particle. On the other hand, a lower energy particle arrives later and gains more energy from the cavity. This process gives rise to the phase stability of synchrotron motion. Similarly, the synchronous phase angle should be $\pi/2 < \phi_{\rm s} \le \pi$ at $\gamma > \gamma_{\rm T}$.

Particles are accelerated through the transition energy in many medium energy synchrotrons such as the AGS, the Fermilab booster and main injector, the CERN PS, and the KEK PS. The synchronous angle has to be shifted from $\phi_{\rm s}$ to $\pi - \phi_{\rm s}$ across the transition energy within 10 to 100 μs. Fortunately, synchrotron motion around the transition energy region is very slow, i.e. $\omega_{\rm syn} \to 0$ at $\gamma \sim \gamma_{\rm T}$. A sudden change in the synchronous phase angle of the rf wave will not cause much beam dilution.

However, when the beam is accelerated through the transition energy, beam loss and serious beam phase-space dilution can result from space-charge-induced mismatch, nonlinear synchrotron motion, microwave instability due to wakefields, etc. An accelerator lattice with a negative momentum compaction factor, where the transition $\gamma_{\rm T}$ is an imaginary number, offers an attractive solution to these problems. Such a lattice is called an imaginary $\gamma_{\rm T}$ lattice. Particle motion in an imaginary $\gamma_{\rm T}$ lattice is always below transition energy, thus the transition energy problems can be eliminated. Attaining an imaginary $\gamma_{\rm T}$ lattice requires a negative horizontal dispersion in most dipoles, i.e. $\sum_i \langle D\rangle_i\theta_i < 0$. Methods of achieving a negative compaction lattice will be addressed in Sec. IV.8.

C: Effect of the dispersion function on orbit response matrix (ORM)

A dipole-kick θ_j at position s_j changes the closed orbit by $G(s, s_j)\theta_j$ and the circumference by $\Delta C = D(s_j)\theta_j$. The response matrix of the ORM experiment depends on the method of measurement:

1. Constant momentum: The change of the revolution period is $\Delta T = \Delta C/\beta c = D(s_j)\theta_j/\beta c$ at a constant velocity. Similarly the rf frequency *must* be adjusted according to $\Delta f/f = -\Delta T/T$ in order to maintain a constant momentum, The beam motion at this new rf frequency is on-momentum, i.e. $\delta = 0$ and the closed orbit is

$$x_{\rm co}(s_i) = G(s_i, s_j)\theta_j \tag{2.170}$$

   or the response matrix is $R_{i,j} = G(s_i, s_j)$ of Eq. (2.110). Sometimes, the rf cavity is turned off during the ORM measurement in proton accelerators. The beam, at a constant injection momentum, is "on-momentum" and the response matrix is $R_{i,j} = G(s_i, s_j)$.[52]

2. Constant path length: Some ORM experiments carry out at a constant rf frequency, i.e. the path length is constant. To maintain a constant pathlength, the beam has to orbit at an equivalent off-momentum "δ" $= \frac{1}{\alpha_c}\frac{\Delta C}{C_0}$ to compensate path length change by the dipole bump. Thus the corresponding closed orbit is

$$x_{\rm co}(s_i) = G(s_i, s_j)\theta_j + D(s_i)\delta = \left\{G(s_i, s_j) + \frac{D(s_i)D(s_j)}{2\pi R\alpha_c}\right\}\theta_j, \tag{2.171}$$

   where α_c is the momentum compaction factor, $D(s)$ is the dispersion function, and R is the mean radius of the accelerator. The response matrix becomes $R_{i,j} = G(s_i, s_j) + \frac{D(s_i)D(s_j)}{2\pi R\alpha_c}$.

IV.4 Dispersion Suppression and Dispersion Matching

Since bending dipoles are needed for beam transport in arc sections, the dispersion function can not be zero there. If the arc is composed of modular cells, such as FODO cells, etc., the dispersion function is usually constrained by the periodicity condition, Eq. (2.151), which simplifies lattice design. In many applications, the dispersion function should be properly matched in straight sections for optimal accelerator operation.[53] If the betatron and synchrotron motions are independent of each other,

[52] J. Kolski, Ph.D. Thesis (Indiana University, 2010) for ORM at PSR; Z. Liu, Ph.D. Thesis (Indiana University, 2011) for ORM at SNS.

[53] The curved transport line is usually called the arc, and the straight section that connects arcs is usually called the insertion, needed for injection, extraction, rf cavities, internal targets, insertion devices, and interaction regions for colliders.

the rms horizontal beam size is $\sigma_x^2(s) = \beta_x(s)\epsilon_{x,\rm rms} + D^2(s)\,\langle(\Delta p/p_0)^2\rangle$, where $\epsilon_{x,\rm rms}$ is the rms emittance. Thus the beam size of a collider at the interaction point can be minimized by designing a zero dispersion straight section. A zero dispersion function in the rf cavity region can be important to minimize the effect of synchro-betatron coupling resonances. We discuss here the general strategy for dispersion suppression.

First-order achromat theorem

The first-order achromat theorem states that a lattice of n repetitive cells is achromatic to first order if and only if $M^n = I$ or each cell is achromatic.[54] Here M is the 2×2 transfer matrix of each cell, and I is a 2×2 unit matrix. Let the 3×3 transfer matrix of a basic cell be

$$R = \begin{pmatrix} M & \bar{d} \\ 0 & 1 \end{pmatrix}, \quad \bar{d} = \begin{pmatrix} d \\ d' \end{pmatrix}, \tag{2.172}$$

where M is the 2×2 transfer matrix for betatron motion, and $\bar{d}$ is the dispersion vector. The transfer matrix of n cells is

$$R^n = \begin{pmatrix} M^n & (M^{n-1} + M^{n-2} + \cdot + 1)\bar{d} \\ 0 & 1 \end{pmatrix} = \begin{pmatrix} M^n & \bar{w} \\ 0 & 1 \end{pmatrix}, \tag{2.173}$$

where $\bar{w} = (M^n - I)(M - I)^{-1}\bar{d}$. Thus the achromat condition $\bar{w} = 0$ can be attained if and only if $M^n = I$ or $\bar{d} = 0$. An achromat section matches any zero dispersion function modules. A unit matrix achromat works like a transparent transport section for any dispersion functions.

Dispersion suppression

Applying the first-order achromat theorem, a strategy for dispersion function suppression can be derived. We consider a curved (dipole) achromatic section such that $M^n = I$. We note that one half of this achromatic section can generally be expressed as

$$R = \begin{pmatrix} -I & \bar{d} \\ 0 & 1 \end{pmatrix}, \quad \bar{d} = \begin{pmatrix} d \\ d' \end{pmatrix}. \tag{2.174}$$

Using the closed-orbit condition, Eq. (2.151), the dispersion function of the repetitive half achromat is $D = d/2,\ D' = d'/2$. If the dipole bending strength of the adjoining $-I$ section is halved, the transfer matrix and the dispersion function will be matched to zero value in the straight section, i.e.

$$R_{1/2} = \begin{pmatrix} -I & \frac{1}{2}\bar{d} \\ 0 & 1 \end{pmatrix} \quad \longrightarrow \quad \begin{pmatrix} d/2 \\ d'/2 \\ 1 \end{pmatrix} = \begin{pmatrix} -I & \frac{1}{2}\bar{d} \\ 0 & 1 \end{pmatrix} \begin{pmatrix} 0 \\ 0 \\ 1 \end{pmatrix}.$$

[54]See K. Brown and R Servranckx, p. 121 in Ref. [16].

Thus the zero dispersion section is matched to the arc by the dispersion suppression section.

When edge focusing is included, a small modification in the quadrupole strengths is needed for dispersion suppression. This is usually called the missing dipole dispersion suppressor (see Exercise 2.4.3c). The reduced bending strength scheme for dispersion suppression is usually expensive because of the wasted space in the cells. A possible variant uses $-I$ sections with full bending angles for dispersion suppression by varying the quadrupole strengths in the $-I$ sections. With use of computer programs such as MAD and SYNCH, the fitting procedure is straightforward.

Is the dispersion function unique?

A trivial corollary of the first-order achromat theorem is that a dispersion function of arbitrary value can be transported through a unit achromat transfer matrix, i.e. a 3×3 unit matrix.

Now we consider the case of an accelerator or transport line with many repetitive modules, which however do not form a unit transfer matrix. Is the dispersion function obtained unique? This question is easily answered by the closed-orbit condition Eq. (2.151) for the entire ring. The transport matrix of n identical modules is

$$R^n = \begin{pmatrix} M^n & (M^n - I)(M - I)^{-1}\bar{d} \\ 0 & 1 \end{pmatrix}, \tag{2.175}$$

where M is the transfer matrix of the basic module with dispersion vector $\bar{d}$. Using the closed-orbit condition, Eq. (2.151), we easily find that the dispersion function of the transport channel is uniquely determined by the basic module unless the transport matrix is a unit matrix, i.e. $M^n = I$. In the case of unit transport, any arbitrary value of dispersion function can be matched in the unit achromat. Since the machine tune can not be an integer because of the integer stopbands, the dispersion function of an accelerator lattice is uniquely determined.

IV.5 Achromat Transport Systems

If the dispersion function is not zero in a transport line, the beam closed orbit depends on particle momentum. However, it is possible to design a transport system such that the beam positions do not depend on beam momentum at both ends of the transport line. Such a beam transport system is called an achromat. The achromat theorem of Sec. IV.4 offers an example of an achromat.

A. The double-bend achromat

A double-bend achromat (DBA) or Chasman-Green lattice is a basic lattice cell frequently used in the design of low emittance synchrotron radiation storage rings. A

DBA cell consists of two dipoles and a dispersion-matching section such that the dispersion function outside the DBA cell is zero. It is represented schematically by

$$[\mathrm{OO}] \quad \mathrm{B} \quad \{\mathrm{O} \quad \mathrm{QF} \quad \mathrm{O}\} \quad \mathrm{B} \quad [\mathrm{OO}],$$

where [OO] is the zero dispersion straight section and {O QF O} is the dispersion matching section. The top plot of Fig. 2.34 shows a basic DBA cell.

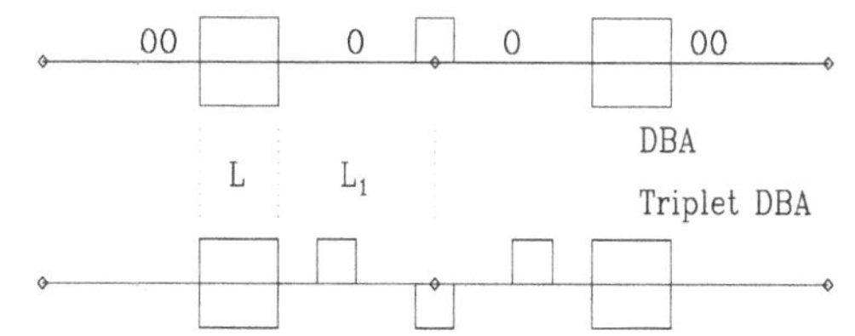

Figure 2.34: Schematic plots of DBA cells. Upper plot: standard DBA cell, where O and OO can contain doublets or triplets for optical match. Lower plot: triplet DBA, where the quadrupole triplet is arranged to attain betatron and dispersion function match of the entire module.

We consider a simple DBA cell with a single quadrupole in the middle. In thin-lens approximation, the dispersion matching condition is

$$\begin{pmatrix} D_{\rm c} \\ 0 \\ 1 \end{pmatrix} = \begin{pmatrix} 1 & 0 & 0 \\ -1/(2f) & 1 & 0 \\ 0 & 0 & 1 \end{pmatrix} \begin{pmatrix} 1 & L_1 & 0 \\ 0 & 1 & 0 \\ 0 & 0 & 1 \end{pmatrix} \begin{pmatrix} 1 & L & L\theta/2 \\ 0 & 1 & \theta \\ 0 & 0 & 1 \end{pmatrix} \begin{pmatrix} 0 \\ 0 \\ 1 \end{pmatrix}, \tag{2.176}$$

where f is the focal length of the quadrupole, θ and L are the bending angle and length of the dipole, and L_1 is the distance from the end of the dipole to the center of the quadrupole. The zero dispersion value at the entrance to the dipole is matched to a *symmetric* condition $D'_{\rm c} = 0$ at the center of the focusing quadrupole. The required focal length and the resulting dispersion function become

$$f = \frac{1}{2}\left(L_1 + \frac{1}{2}L\right), \quad D_{\rm c} = \left(L_1 + \frac{1}{2}L\right)\theta. \tag{2.177}$$

Note that the focal length needed in the dispersion function matching condition is independent of the dipole bending angle in thin-lens approximation, and it can easily be obtained from the geometric argument. The dispersion function at the symmetry point is proportional to the product of the effective length of the DBA cell and the bending angle.

Although this simple example shows that a single focusing quadrupole can attain dispersion matching, the betatron function depends on the magnet arrangement in the [OO] section, and possible other quadrupoles in the dispersion matching section. The dispersion matching condition of Eq. (2.177) renders a horizontal betatron phase advance Φ_x larger than π in the dispersion matching section (from the beginning of the dipole to the other end of the other dipole). The stability condition of betatron motion (see Sec. II.6) indicates that betatron function matching section [OO] can not be made of a simple defocusing quadrupole. A quadrupole doublet, or a triplet, is usually used in the [OO] section. Such DBA lattice modules have been widely applied in the design of electron storage rings.

A simple DBA cell is the triplet DBA (lower plot of Fig. 2.34), where a quadrupole triplet is located symmetrically inside two dipoles. This compact lattice was used for the SOR ring in Tokyo. Some properties of the triplet DBA storage ring can be found in Exercise 4.3.6.

B. Other achromat modules

The beam transport system in a synchrotron or a storage ring requires proper dispersion function matching. The design strategy is to use achromatic subsystems. An example of achromatic subsystem is the unit matrix module (see Sec. IV.4 on the first order achromat theorem). A unit matrix module can be made of FODO or other basic cells such that the total phase advance of the entire module is equal to an integer multiple of 2π. Achromatic modules can be optically matched with straight sections to form an accelerator lattice.

The achromatic transport modules are also important in the transport beamlines (see Exercises 2.4.12 to 2.4.15). The achromatic transport system find applications in high energy and nuclear physics experiments, medical radiation treatment, and other beam delivering systems.

IV.6 Transport Notation

In many applications, the particle coordinates in an accelerator can be characterized by a state vector $\vec{W}$, where the transpose is

$$\vec{W}^T = (\,W_1,\ \ W_2,\ \ W_3,\ \ W_4,\ \ W_5,\ \ W_6\,) = (\,x,\ \ x',\ \ z,\ \ z',\ \ \beta c\Delta t,\ \ \delta\,),$$

where βc is the speed of the particle, $\beta c\Delta t$ is the path length difference with respect to the reference orbit, and $\delta = \Delta p/p_0$ is the fractional momentum deviation of a particle. The transport of the state vector in linear approximation is

$$W_i(s_2) = \sum_{j=1}^{6} R_{ij}(s_2|s_1)\, W_j(s_1), \quad (i,j = 1,\cdots,6). \tag{2.178}$$

Note that the 2×2 diagonal matrices for the indices 1,2, and 3,4 are respectively the horizontal and vertical M matrices. The R_{13}, R_{23} R_{14}, R_{24} elements describe the linear betatron coupling. The R_{16}, R_{26} elements are the dispersion vector $\vec{d}$ of Eq. (2.172). Without synchrotron motion, we have $R_{55} = R_{66} = 1$. All other elements of the R matrix are zero.

In general, the nonlinear dependence of the state vector can be expanded as

$$\begin{aligned} W_i(s_2) &= \sum_{j=1}^{6} R_{ij} W_j(s_1) + \sum_{j=1}^{6}\sum_{k=1}^{6} T_{ijk} W_j(s_1) W_k(s_1) \\ &+ \sum_{j=1}^{6}\sum_{k=1}^{6}\sum_{l=1}^{6} U_{ijkl} W_j(s_1) W_k(s_1) W_l(s_1) + \cdots . \end{aligned} \tag{2.179}$$

For example, particle transport through a thin quadrupole is

$$\Delta x' = -\frac{x}{f(1+\delta)} = -\frac{x}{f} + \frac{x\delta}{f} - \frac{x\delta^2}{f} + \cdots,$$
$$\Delta z' = \frac{z}{f(1+\delta)} = \frac{z}{f} - \frac{z\delta}{f} + \frac{z\delta^2}{f} + \cdots,$$

and we obtain

$$R_{21} = -\frac{1}{f}, \quad T_{216} = +\frac{1}{f}, \quad U_{2166} = -\frac{1}{f}, \quad R_{43} = +\frac{1}{f}, \quad T_{436} = -\frac{1}{f}, \quad U_{4366} = +\frac{1}{f}.$$

Similarly, particle transport through a thin sextupole gives

$$\Delta x' = -\frac{S}{2(1+\delta)}(x^2 - z^2), \qquad \Delta z' = \frac{S}{1+\delta} xz,$$
$$T_{211} = -\frac{S}{2}, \quad T_{233} = \frac{S}{2}, \quad T_{413} = S, \quad \cdots .$$

where $S = -B_2\ell/B\rho$ is the integrated sextupole strength. Here we used the convention that $S > 0$ corresponds to a focusing sextupole. Tracing the transport in one complete revolution, we get the momentum compaction factor as $\alpha_c = R_{56}$. The program TRANSPORT[55] has often been used to calculate the transport coefficients in transport lines.

IV.7 Experimental Measurements of Dispersion Function

Digitized BPM turn by turn data can be used to measure the betatron motion. On the other hand, if the BPM signals are sampled at a longer time scale, the fast betatron oscillations are averaged to zero. The DC output provides the closed orbit of the beam. The dispersion function can be measured from the derivative of the closed orbit with respect to the off-momentum of the beam, i.e.

$$D = \frac{dx_{co}}{d(\Delta p/p_0)} = -\eta f_0 \frac{dx_{co}}{df_0}, \tag{2.180}$$

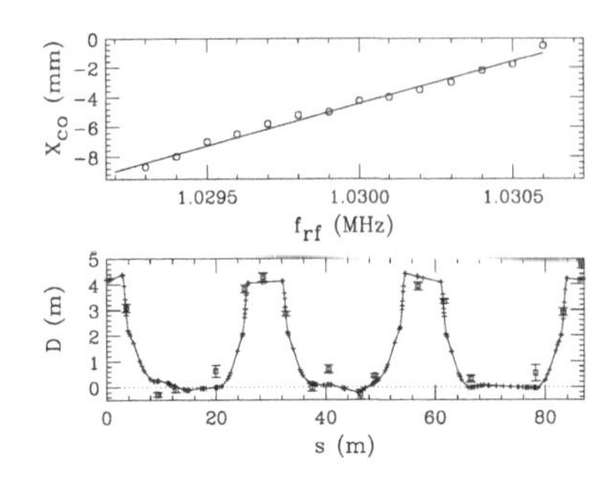

Figure 2.35: The upper plot shows the closed orbit at a BPM vs the rf frequency for the IUCF Cooler Ring. The slope of this measurement is used to obtain the "measured" dispersion function. The lower plot compares the measured dispersion function (rectangles) with that obtained from the MAD program (solid line).

where $x_{\rm co}$ is the closed orbit, f_0 is the revolution frequency, η is the phase-slip factor, and the momentum of the beam is varied by changing the rf frequency.

The upper plot of Fig. 2.35 shows the closed orbit at a BPM location vs the rf frequency at the IUCF Cooler Ring. Using Eq. (2.180), we can deduce the dispersion function at the BPM location. In the lower plot of Fig. 2.35 the "measured" dispersion functions of the IUCF Cooler Ring is compared with that obtained from the MAD program [23].[56] The accuracy of the dispersion function measurement depends on the precision of the BPM system, and also on the effects of power supply ripple. To improve the accuracy of the dispersion function measurement, we can induce frequency modulation to the rf frequency shift. The resulting closed orbit will have the characteristic modulation frequency. Fitting the resulting closed orbit with the known modulation frequency, we can determine the dispersion function more accurately.

IV.8 Transition Energy Manipulation

Medium energy accelerators often encounter problems during transition energy crossing, such as longitudinal microwave instability and nonlinear synchrotron motion. These problems can be avoided by an accelerator having a negative momentum compaction factor. The revolution period deviation ΔT for an off-momentum particle $\Delta p = p - p_0$ is given by Eq. (2.164). The accelerator becomes isochronous at the transition energy ($\gamma = \gamma_{\rm T}$).

There are many unfavorable effects on the particle motion near the transition energy. For example, the momentum spread of a bunch around transition can become so large that it exceeds the available momentum aperture, causing beam loss (see Chap. 3, Sec. IV). Since the frequency spread of the beam $\Delta\omega = -\eta\omega(\Delta p/p_0)$ vanishes at the transition energy, there is little or no Landau damping of microwave

[55]K.L. Brown, D.C. Carey, Ch. Iselin and F. Rothacker, CERN 80-04 (1980); D.C. Carey, FNAL Report TM-1046 (1981); D.C. Carey *et al.*, SLAC-R-530, Fermilab-Pub-98-310 (1998).

[56]The IUCF Cooler Ring lattice belongs generally to the class of double-bend achromats (see Exercise 2.4.12). A high dispersion straight section is used for momentum-stacking injection and zero dispersion straight sections are used for rf and electron cooling.

instability near transition (see Chap. 3, Sec. VII). As a result, the bunch area may grow because of collective instabilities. Furthermore, particles with different momenta may cross transition at different times, which leads to unstable longitudinal motion resulting in serious beam loss.

To avoid all the above unfavorable effects, it is appealing to eliminate transition crossing. The γ_T jump schemes have been used successfully to ease beam dynamics problems associated with the transition energy crossing; in these schemes some quadrupoles are pulsed so that the transition energy is lowered or raised in order to enhance the acceleration rate at the transition energy crossing. This has become a routine operation at the CERN Proton Synchrotron (PS).

Alternatively, one can design an accelerator lattice such that the momentum compaction factor α_c is negative, and thus the beam never encounters transition energy. This is called the "negative momentum compaction" or the "imaginary γ_T" lattice. All modern medium energy synchrotrons can be designed this way to avoid transition energy. We discuss below the methods of α_c manipulation, the transition energy jump scheme, and the design principle of the imaginary γ_T lattices.

A. γ_T jump schemes

In many existing low to medium energy synchrotrons, particle acceleration through the transition energy is unavoidable. Finding a suitable γ_T jump scheme can provide beam acceleration through transition energy without much emittance dilution and beam loss. Here we examine the strategy of γ_T jump schemes pioneered by the CERN PS group.[57]

In the presence of dipole field error, the closed orbit is given by Eq. (2.92). Substituting the dipole field error resulting from the off-momentum of a particle $\Delta B_z/B\rho = \frac{1}{\rho}\delta$ into Eq. (2.92), we obtain the horizontal off-momentum closed orbit $x_{co} = D(s)\delta$. Thus the dispersion function is

$$D(s_0) = \int_0^C \frac{G_x(s, s_0)}{\rho(s)} ds \approx \sum_i G_x(s_i, s_0)\theta_i, \tag{2.181}$$

where $\theta_{k,i}$ is the dipole angular kick in thin-lens approximation, $\rho(s)$ is the bending radius of dipoles, and the Green's function $G_x(s, s_0)$ is given in Eq. (2.91).

A.1 The effect of quadrupole field errors on the closed orbit

Consider N quadrupoles for the γ_T jump. We would like to evaluate the change of orbit length for off-momentum particles due to the γ_T jump quadrupoles. From Hill's

[57]W. Hardt, *Proc. 9th Int. Conf. on High Energy Accelerators* (USAEC, Washington, DC, 1974). See also T. Risselada, *Proc. CERN Accelerator School*, CERN-91-04, p. 161, 1991.

equation, the angular kick resulting from the ith $\gamma_{\rm T}$ jump quadrupole is

$$\theta_i = -K_i\ [x_{\rm co}(s_i) + D_i^*\delta], \tag{2.182}$$

where $K_i = -B_1\ell/B\rho$ is the strength of the ith $\gamma_{\rm T}$ jump quadrupole, assumed positive for a focusing quadrupole, and D_i^* is the perturbed dispersion function at $s = s_i$. Thus the change of the orbit length for the off-momentum particle is

$$\Delta C \approx \sum_i D_i\theta_i \approx -\left(\sum_i K_i D_i^* D_i\right)\delta, \tag{2.183}$$

where D_i is the unperturbed dispersion function, and we neglect higher-order terms in δ. Equation (2.183) indicates that quadrupoles at nonzero dispersion locations can be used to adjust the momentum compaction factor.

If N $\gamma_{\rm T}$-jump quadrupoles are used to change the momentum compaction factor, we obtain

$$C_0\Delta\alpha_{\rm c} = -\sum_{i=1}^{N} K_i D_i^* D_i. \tag{2.184}$$

The change in momentum compaction (called $\gamma_{\rm T}$ jump) depends on the unperturbed and perturbed dispersion functions at kick-quadrupole locations. An important constraint is that the betatron tunes should be maintained constant during the $\gamma_{\rm T}$ jump in order to avoid nonlinear betatron resonances, i.e.

$$\Delta Q_x = \frac{1}{4\pi}\sum_{i=1}^{N}\beta_{x,i}K_i = 0, \quad \Delta Q_z = -\frac{1}{4\pi}\sum_{i=1}^{N}\beta_{z,i}K_i = 0. \tag{2.185}$$

Thus we usually employ zero tune shift quadrupole pairs for the $\gamma_{\rm T}$ jump.

A.2 The perturbed dispersion function

The change in the closed orbit resulting from the quadrupole kicks can be obtained by substituting Eq. (2.182) into Eq. (2.181) to obtain the closed orbit solution:

$$\begin{aligned} [D^*(s) - D(s)]\,\delta &= -\sum_i G_x(s, s_i) K_i D_i^*\,\delta, \\ D_j^* &= D_j + \sum_i F_{ji} D_i^*, \end{aligned} \tag{2.186}$$

where $F_{ji} = -G_x(s_j, s_i)K_i$. The perturbed dispersion function at these quadrupole locations and the resulting change in momentum compaction become

$$\begin{aligned} \vec{D}^* &= (1-F)^{-1}\vec{D} = (1 + F + F^2 + F^3 + ...)\vec{D}. \\ \Delta\alpha_{\rm c} &= -\frac{1}{C_0}\sum_{ij}^{N} K_i(1 + F + F^2 + ...)_{ij} D_j D_i. \end{aligned}$$

A.3 $\gamma_{\rm T}$ jump using zero tune shift π-doublets

When zero tune shift pairs of quadrupoles separated by π in the betatron phase advance are used to produce a $\gamma_{\rm T}$ jump, the matrix F satisfies

$$F^n = 0 \quad \text{for} \quad n \geq 2. \tag{2.187}$$

This result can be easily proved by using the zero tune shift condition: $\beta_{x,k}K_k + \beta_{x,k+1}K_{k+1} = 0$ and the π phase advance condition:

$$\cos(\pi\nu_x - |\psi_k - \psi_j|) = -\cos(\pi\nu_x - |\psi_{k+1} - \psi_j|),$$
$$\cos(\pi\nu_x - |\psi_i - \psi_k|) = -\cos(\pi\nu_x - |\psi_i - \psi_{k+1}|).$$

Using the π-doublets, the perturbed dispersion function and the change in the momentum compaction factor become

$$D_i^* = (1+F)_{ij}D_j, \quad \text{or} \quad \Delta D_i = D_i^* - D_i = -G_x(s_i, s_j)K_jD_j,$$
$$\Delta\alpha_{\rm c} = -\frac{1}{C_0}\sum_i K_iD_i^2 + \sum_{ij} K_iK_jG_x(s_i, s_j)D_iD_j.$$

The change in the dispersion function is linear in K. The change in the momentum compaction factor contains a linear and a quadratic term in K. If the $\gamma_{\rm T}$ jump quadrupole pairs are located in the arc, where the unperturbed dispersion function is dominated by the zeroth harmonic in the Fourier decomposition, the term linear in K_i vanishes because of the zero tune shift condition.[58] The resulting change in the momentum compaction factor is a quadratic function of K_i.

Since the stopband integral of Eq. (2.120) at $p = [2\nu_x]$ due to the tune jump quadrupole pair is zero because of the zero tune shift condition, the π-doublet does not produce a large perturbation in the betatron amplitude function.

Thus if all quadrupoles used for $\gamma_{\rm T}$ jump are located in FODO cells, the amount of tune jump is second order in the quadrupole strength. On the other hand, $\gamma_{\rm T}$ jump using quadrupoles in straight sections can be made linear in quadrupole strength.

B. Flexible momentum compaction (FMC) lattices

Alternatively, a lattice having a *very small* or even *negative* momentum compaction factor can also be designed. Vladimirskij and Tarasov[59] introduced reverse bends in an accelerator lattice and succeeded in getting a negative orbit-length increase with momentum, thus making a negative momentum compaction factor. Another method

[58]This statement can be expressed mathematically as follows. If the zeroth harmonic term dominates, we have $D_i^2 \propto \beta_i$, and thus $\sum_i K_iD_i^2 \propto \sum_i K_i\beta_i = 0$ because of the zero tune shift condition.

[59]V.V. Vladimirski and E.K. Tarasov, *Theoretical Problems of the Ring Accelerators* (USSR Academy of Sciences, Moscow, 1955).

of designing an FMC lattice is called the *harmonic approach.*[60] In this method a systematic closed-orbit stopband is created near the betatron tune to induce dispersion-wave oscillations resulting in a high $\gamma_{\rm T}$ or an imaginary $\gamma_{\rm T}$. However, the resulting lattice is less tunable and the dispersion functions can be large. Thus the dynamical aperture may be reduced accordingly.

In 1972, Teng proposed an innovative scheme using negative dispersion at dipole locations, where the dispersion function can be matched by a straight section with a phase advance of π to yield little or no contribution to positive orbit-length increment.[61] This concept is the basis for *flexible momentum compaction* (FMC) lattices, which require negative dispersion functions at some sections of dipoles.

Trbojevic *et al.*[62] re-introduced a *modular approach* for the FMC lattice with a prescribed dispersion function. The dispersion phase-space maps are carefully matched to attain a lattice with a pre-assigned $\gamma_{\rm T}$ value. The module forms the basic building blocks for a ring with a negative momentum compaction factor or an imaginary $\gamma_{\rm T}$. The module can be made very compact without much unwanted empty space and, at the same time, the maximum value of the dispersion function can be optimized to less than that of the FODO lattice.

For attaining proper dispersion function matching, the normalized dispersion coordinates $X_{\rm d}$ and $P_{\rm d}$ of Eq. (2.161) are handy, i.e.

$$X = \frac{1}{\sqrt{\beta_x}} D = \sqrt{2J_{\rm d}} \cos\psi_{\rm d}, \qquad P = \sqrt{\beta_x} D' + \frac{\alpha_x}{\sqrt{\beta_x}} D = -\sqrt{2J_{\rm d}} \sin\psi_{\rm d}.$$

In the thin-element approximation, Eq. (2.150) indicates that $\Delta D = 0$ and $\Delta D' = \theta$ in passing through a thin dipole with bending angle θ. Therefore, in normalized $P_{\rm d}$-$X_{\rm d}$ space, the normalized dispersion vector changes by $\Delta P = \sqrt{\beta_x}\theta$ and $\Delta X = 0$. Outside the dipole ($\rho = \infty$), the dispersion function satisfies the homogeneous equation, and the dispersion action $J_{\rm d}$ is invariant, i.e. $P_{\rm d}$ and $X_{\rm d}$ lie on a circle $P^2 + X^2 = 2J_{\rm d}$. The phase angle $\psi_{\rm d}$ of the normalized coordinates is equal to the betatron phase advance. This dispersion phase-space plot can be helpful in the design of lattices and beam-transfer lines. It has also been used to lower the dispersion excursion during a fast $\gamma_{\rm T}$ jump at RHIC.[63]

---

[60] R. Gupta and J.I.M. Botman, *IEEE Trans. Nucl. Sci.* **NS-32**, 2308 (1985); T. Collins, *Beta Theory*, Technical Memo, Fermilab (1988); G. Guignard, *Proc. 1989 IEEE PAC*, p. 915 (1989); E.D. Courant, A.A. Garren, and U. Wienands, *Proc. 1991 IEEE PAC*, p. 2829 (1991).

[61] L.C. Teng, *Part. Accel.* **4**, 81 (1972).

[62] D. Trbojevic, D. Finley, R. Gerig, and S. Holmes, *Proc. 1990 EPAC*, p. 1536 (1990); K.Y. Ng, D. Trbojevic, and S.Y. Lee, *Proc. 1991 PAC*, p. 159 (1991); S.Y. Lee, K.Y. Ng, and D. Trbojevic, *Phys. Rev.* E**48**, 3040 (1993).

[63] D. Trbojevic, S. Peggs, and S. Tepikian, *Proc. 1993 IEEE Part. Accel. Conf.* p. 168 (1993).

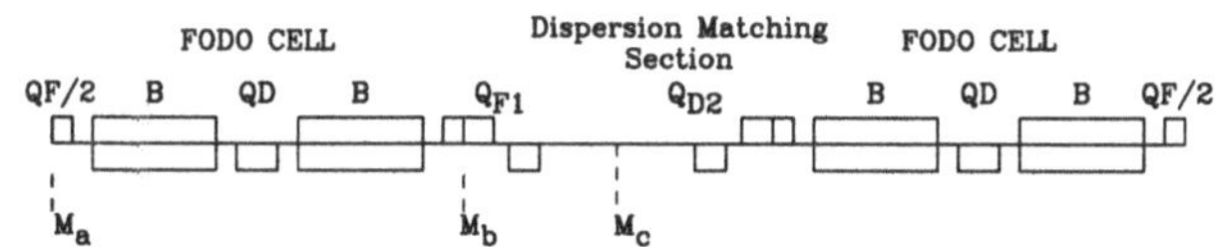

Figure 2.36: A schematic drawing of a basic module made of two FODO cells and an optical matching section.

B.1 The basic module and design strategy

A basic FMC module has two parts: (1) the FODO or DOFO cell, where the negative dispersion function in dipoles provides a negative momentum compaction factor, and (2) a matching section that matches the optical functions. We also assume reflection symmetry for all Courant-Snyder functions at symmetric points of the module. Although not strictly necessary, reflection symmetry considerably simplifies the analysis and optical matching procedure. For example, we consider a basic module composed of two FODO cells and a dispersion matching section shown schematically in Fig. 2.36:

$$\mathrm{M_a}\left\{\frac{1}{2}\mathrm{QF}\,B\,\mathrm{QD}\,B\,\frac{1}{2}\mathrm{QF}\right\}\,\mathrm{M_b}\,\{\mathrm{Q_{F_1}}\,\mathrm{O_1}\,\mathrm{Q_{D_2}}\,\mathrm{O_2}\}\,\mathrm{M_c}\,+\text{ reflection symmetry },$$

where $\mathrm{M_{a,b,c}}$ are marker locations, Q's are quadrupoles, O's are drift spaces, and B's stand for dipoles. The horizontal betatron transfer matrix of the FODO cell from the marker $\mathrm{M_a}$ to the marker $\mathrm{M_b}$ is [see Eq. (2.157) and Exercise 2.4.3]

$$M_{\rm FODO}=\begin{pmatrix}\cos\Phi & \beta_{\rm F}\sin\Phi & D_{\rm F}(1-\cos\Phi)\\ -\frac{1}{\beta_{\rm F}}\sin\Phi & \cos\Phi & \frac{D_{\rm F}}{\beta_{\rm F}}\sin\Phi\\ 0 & 0 & 1\end{pmatrix}, \tag{2.188}$$

where, for simplicity, we have chosen $\Phi_x=\Phi_z=\Phi$ for the betatron phase advance of the FODO cell, $\beta_{\rm F}$ and $D_{\rm F}$ are respectively the betatron amplitude and dispersion functions at the center of the focusing quadrupole for the regular FODO cell, and a symmetry condition $\beta'_{\rm F}=0$ and $D'_{\rm F}=0$ is assumed to simplify our transfer matrix in Eq. (2.188).

The procedure of optical function matching is (1) choose the desired value D_a of the dispersion function at the marker M_a; the dispersion function is propagated through FODO cell to obtained a dispersion vector at the marker M_b; (2) the optical functions are matched in the matching section. Figure 2.37 shows an example of the betatron amplitude functions for a matched FMC basic module with an added dipole in dispersion matching section in order to increase the packing factor. The J-PARC Main Ring in Japan, the PS2 design in CERN and the the Jlab EIC ring design employ various variations of the negative momentum compaction designs.[64]

[64]The packing factor and magnitude of the γ_T are optimized in the design process, see J-PARC design report, JAERI-Tech-2003-44/KEK-Report-2OO2-13; Y. Papaphilippou, et al., PAC09, 3805 (2009), and https://yannis.web.cern.ch/yannis/talks/PS2opticsLIS.pdf; S.A. Bogacz, in Proceedings of IPAC2017, 3350 (2017). Here, the packing factor is defined as the fraction of the circumference of an accelerator that is occupied by dipole magnets.

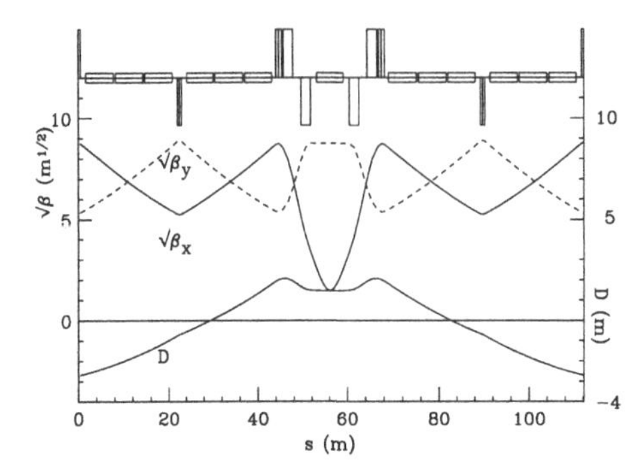

Figure 2.37: The lattice function of an FMC basic module. In this example, a dipole is added in the middle of dispersion matching section in order to increase the machine packing factor. Although the dipole in the matching section will contribute a positive value to the momentum compaction. The overall compaction factor can still be adjusted by a properly chosen D_a.

The dispersion function inside dipoles in FODO cells of an FMC module is mostly negative, and the resulting momentum compaction factor can become negative. Adjusting the initial dispersion function value D_a, the momentum compaction factor of the accelerator can be varied. Figure 2.38 shows an example of dispersion function matching for an FMC module by plotting the normalized dispersion phase-space coordinates $X_{\rm d}$ vs $P_{\rm d}$. A negative-momentum-compaction module requires $X_{\rm d} < 0$ in dipoles as demonstrated in the left plot of Fig. 2.38. The right plot shows a similar plot in thin-lens approximation. Although they look slightly different, the thin-element approximation can provide essential insight in the preliminary design where dispersion matching is required. Since there is no dipole in this example shown in Fig. 2.36, the dispersion phase-space coordinates are located on a circle as shown in Fig. 2.38, where the normalized dispersion phase-space coordinates $X_{\rm d} = D/\sqrt{\beta_x}$ vs $P_{\rm d} = (\alpha_x/\sqrt{\beta_x})D + \sqrt{\beta_x}D'$ are shown. If the lattice has a reflection symmetry at the marker $\rm M_c$, the matched dispersion phase-space coordinate is $P_{\rm d} = 0$.

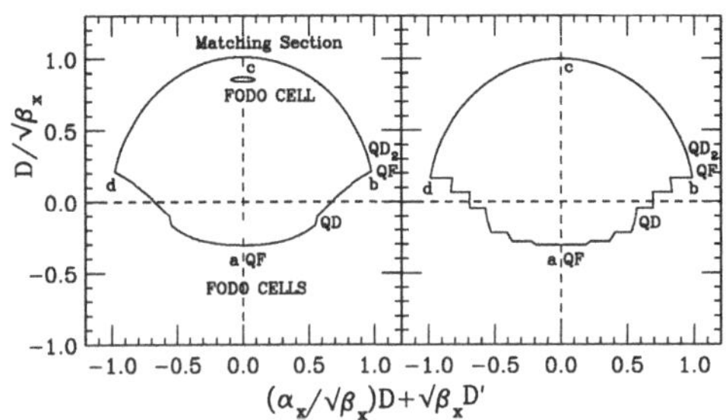

Figure 2.38: Left: An example of dispersion matching for a basic FMC module. The normalized dispersion phase-space coordinates for periodic FODO cells are marked "FODO CELLS." Right: The FMC cell in thin-lens approximation. Each dipole is divided into 3 segments. Each dipole segment changes only the coordinate P. There is no dipole in the matching section, thus the normalized phase-space torus is an arc of a circle.

B.2 Dispersion matching

The dispersion function at the beginning of the FODO cell is *prescribed* to have a negative value of $D_{\rm a}$ with $D'_{\rm a} = 0$. As we shall see, the choice of $D_{\rm a}$ essentially determines the dispersion excursion and the $\gamma_{\rm T}$ value of the module. Using the transfer matrix in Eq. (2.188), we find the dispersion function at marker $\rm M_b$ to be

$$D_{\rm b} = D_{\rm F} - (D_{\rm F} - D_{\rm a})\cos\Phi, \quad D'_{\rm b} = \frac{D_{\rm F} - D_{\rm a}}{\beta_{\rm b}}\sin\Phi\,, \tag{2.189}$$

where $\beta_{\rm b}$ is the betatron amplitude function at marker $\rm M_b$ with $\beta_{\rm b} = \beta_{\rm F}$. Now we assume that there is no dipole in the matching section, and the dispersion action is invariant in this region, i.e.

$$J_{\rm d,c} = J_{\rm d,b} = \frac{1}{2}\left[\frac{D_{\rm b}^2}{\beta_{\rm b}} + \beta_{\rm b} D_{\rm b}'^2\right] = J_{\rm d,F}[1 - 2(1-\zeta)\cos\Phi + (1-\zeta)^2]\,,$$

where $\zeta = D_{\rm a}/D_{\rm F}$ is the ratio of the desired dispersion at marker $\rm M_a$ to the dispersion function of the regular FODO cell, $J_{\rm d,b}, J_{\rm d,c}$ are dispersion actions at markers $\rm M_b$ and $\rm M_c$, and $J_{\rm d,F}$ is the dispersion action of the regular FODO cell at the focusing quadrupole location, given in thin-lens approximation by

$$J_{\rm d,F} = \frac{1}{2}\left[L_1\theta^2\frac{\cos\frac{\Phi}{2}(1+\frac{1}{2}\sin\frac{\Phi}{2})^2}{\sin^3\frac{\Phi}{2}(1+\sin\frac{\Phi}{2})}\right]. \tag{2.190}$$

B3. Other similar FMC modules

The above analysis can be applied to a basic FMC module composed of two DOFO cells and a dispersion matching section. Because the dispersion value at the defocusing quadrupole location is smaller than that at the focusing quadrupole location, a slightly smaller $|\zeta|$ can be used to minimize the magnitude of the dispersion function in the module.

To design a lattice with a higher packing factor, defined as the ratio of the total dipole length to the circumference, one may use a DOFODO in place of the FODO cell, i.e. three FODO cells instead of two are placed inside a basic module. The betatron transfer matrix in the DOFODO cell becomes

$$M_{a\to b} = \begin{pmatrix} \sqrt{\frac{\beta_{\rm F}}{\beta_{\rm D}}}\cos\frac{3}{2}\Phi & \sqrt{\beta_{\rm F}\beta_{\rm D}}\sin\frac{3}{2}\Phi & D_{\rm F} - D_{\rm D}\sqrt{\frac{\beta_{\rm F}}{\beta_{\rm D}}}\cos\frac{3}{2}\Phi \\ -\frac{1}{\sqrt{\beta_{\rm F}\beta_{\rm D}}}\sin\frac{3}{2}\Phi & \sqrt{\frac{\beta_{\rm D}}{\beta_{\rm F}}}\cos\frac{3}{2}\Phi & \frac{D_{\rm D}}{\sqrt{\beta_{\rm F}\beta_{\rm D}}}\sin\frac{3}{2}\Phi \\ 0 & 0 & 1 \end{pmatrix}, \tag{2.191}$$

where Φ is the phase advance of a FODO cell, and $\beta_{\rm F}, \beta_{\rm D}, D_{\rm F}, D_{\rm D}$ are the betatron amplitudes and dispersion values at the focusing and defocusing quadrupoles of the FODO cell. A similar analysis with a different number of FODO cells can be easily done. In general, the result will be a larger total dispersion value.

C. Reverse Bend and nsFFA accelerators

Consider a simple triplet cell made of combined function magnets, shown in the top plot of Fig. 2.39. The bottom plot shows the maximum dispersion function and the compaction vs the relative distribution of the bending angle for $\theta_D + \theta_F = 18^\circ$. When $(\theta_D - \theta_F)/(\theta_D + \theta_F) = 0$, both magnets have equal bending angle; if it is ± 1, one of the magnet is a pure quadrupole. When $|(\theta_D - \theta_F)/(\theta_D + \theta_F)| > 1$, one of the magnet is reverse bend. In particular, if the reverse bend is a focusing quadrupole, where the value of the dispersion function is maximum, the compaction factor can become 0. The non-scaling-Fixed-Field-Alternating-gradient (nsFFA) accelerator employs this property to achieve small enough phase slip factor for particles within a range of beams momentum rigidity using a fixed-frequency rf system. Similar concept works for the simple FODO cells.[65]

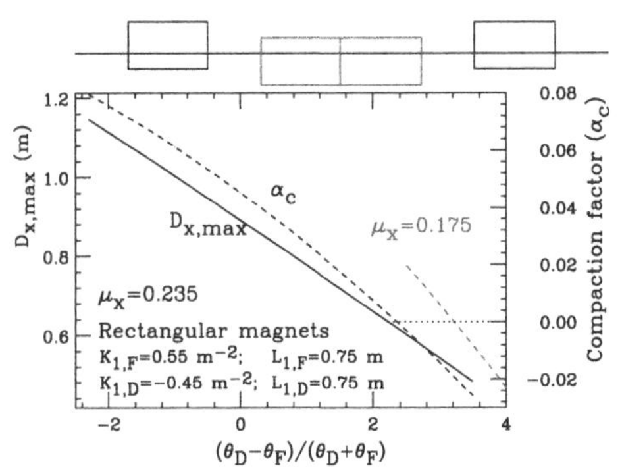

Figure 2.39: Top: An example of a triplet basic cell made of combined function magnets (cell length 5 m, combined function magnets 1.5 m, at a total bending angle $\theta_D + \theta_F = 18^\circ$). Bottom: Dispersion function (solid line) and the compaction factor (dashed line) vs the relative angular bend angle of the focusing and defocusing dipoles at $\mu_x = 0.235$. The compaction factor (dashed line) for $\mu_x = 0.175$ is shown for comparison.

IV.9 Minimum $\langle \mathcal{H} \rangle$ Modules

In electron storage rings, the natural (horizontal) emittance of the beam is determined by the average of the $\mathcal{H}$-function in the dipoles (see Chap. 4). A double-bend module (Fig. 2.34), also called Chasman-Green lattice, is made of two dipoles located reflection-symmetrically with respect to the center of the basic module:

$$\mathrm{M_a} \begin{pmatrix} \text{Triplet} \\ \text{or} \\ \text{Doublet} \end{pmatrix} \mathrm{B}\ \{\text{dispersion matching section}\}\ \mathrm{M_c} + \{\text{reflection symmetry}\}.$$

A quadrupole triplet or doublet matching section on the outside of the dipole B is the betatron amplitude matching section. If the achromat condition is imposed, the module is called a double-bend achromat (DBA). The zero dispersion region is

[65]See e.g. J.S. Berg, *Nucl. Instrum. Methods*, **A596**, 276 (2008).

usually used for insertion devices such as the undulator, the wiggler, and rf cavities. The dispersion matching section on the right side of the dipole can be made of a single quadrupole, a doublet, or a triplet. In this section, the strategy of minimizing $\langle\mathcal{H}\rangle$ inside dipoles will be discussed. To simplify our discussion, we will consider a single dipole lattice unit where the dispersion and betatron amplitude functions can be independently controlled.

The evolution of the $\mathcal{H}$-function in a sector dipole is (see Exercise 2.4.11)

$$\begin{aligned}\mathcal{H} &= \mathcal{H}_0 + 2(\alpha_0 D_0 + \beta_0 D_0')\sin\phi - 2(\gamma_0 D_0 + \alpha_0 D_0')\rho(1-\cos\phi)\\ &\quad +\beta_0\sin^2\phi + \gamma_0\rho^2(1-\cos\phi)^2 - 2\alpha_0\rho\sin\phi(1-\cos\phi),\end{aligned} \tag{2.192}$$

where $\mathcal{H}_0 = \gamma_0 D_0^2 + 2\alpha_0 D_0 D_0' + \beta_0 D_0'^2$; $\alpha_0, \beta_0, \gamma_0$, D_0 and D_0' are the Courant-Snyder parameters and dispersion functions at $s=0$; and $\phi = s/\rho$ is the coordinate of the bending angle inside the dipole. The average $\mathcal{H}$-function in the dipole becomes

$$\begin{aligned}\langle\mathcal{H}\rangle &= \mathcal{H}_0 + (\alpha_0 D_0 + \beta_0 D_0')\theta^2 E(\theta) - \frac{1}{3}(\gamma_0 D_0 + \alpha_0 D_0')\rho\theta^2 F(\theta)\\ &\quad +\frac{\beta_0}{3}\theta^2 A(\theta) - \frac{\alpha_0}{4}\rho\theta^3 B(\theta) + \frac{\gamma_0}{20}\rho^2\theta^4 C(\theta),\end{aligned} \tag{2.193}$$

$$E(\theta) = \frac{2(1-\cos\theta)}{\theta^2},\quad F(\theta) = \frac{6(\theta-\sin\theta)}{\theta^3},\quad A(\theta) = \frac{6\theta - 3\sin 2\theta}{4\theta^3},$$

$$B(\theta) = \frac{6 - 8\cos\theta + 2\cos 2\theta}{\theta^4},\quad C(\theta) = \frac{30\theta - 40\sin\theta + 5\sin 2\theta}{\theta^5},$$

where θ is the bend angle of the dipole. In the small-angle limit, $A\to 1, B\to 1, C\to 1, E\to 1$, and $F\to 1$. With the normalized scaling parameters $d_0 = \frac{D_0}{L\theta}$, $d_0' = \frac{D_0'}{\theta}$, $\tilde\beta_0 = \frac{\beta_0}{L}$, $\tilde\gamma_0 = \gamma_0 L$, $\tilde\alpha_0 = \alpha_0$, where $L = \rho\theta$ is the length of the dipole, the average $\mathcal{H}$-function becomes

$$\begin{aligned}\langle\mathcal{H}\rangle &= \rho\theta^3\Big\{\tilde\gamma_0 d_0^2 + 2\tilde\alpha_0 d_0 d_0' + \tilde\beta_0 d_0'^2 + (\tilde\alpha_0 E - \frac{\tilde\gamma_0}{3}F)d_0\\ &\quad +(\tilde\beta_0 E - \frac{\tilde\alpha_0}{3}F)d_0' + \frac{\tilde\beta_0}{3}A - \frac{\tilde\alpha_0}{4}B + \frac{\tilde\gamma_0}{20}C\Big\}.\end{aligned}$$

A. Minimum $\langle\mathcal{H}\rangle$-function with achromat condition

In the special case with the achromat condition $d_0 = 0$ and $d_0' = 0$, the average $\mathcal{H}$-function and its minimum value are

$$\langle\mathcal{H}\rangle = \rho\theta^3\left\{\frac{\tilde\beta_0}{3}A - \frac{\tilde\alpha_0}{4}B + \frac{\tilde\gamma_0}{20}C\right\} \quad\Longrightarrow\quad \langle\mathcal{H}\rangle_{\rm min,A} = \frac{G}{4\sqrt{15}}\rho\theta^3, \tag{2.194}$$

where we use the condition $\tilde\beta_0\tilde\gamma_0 = (1+\tilde\alpha_0^2)$ to obtain $\tilde\beta_0 = \frac{\sqrt{12}C}{\sqrt{5}G}$, $\tilde\alpha_0 = \frac{\sqrt{15}B}{G}$, and $G = \sqrt{16AC - 15B^2}$. The G-function decreases slowly with the dipole bending angle

θ shown in Fig. 2.40. The evolution of the betatron amplitude function in the dipole can be obtained from Eq. (2.36). In the small-angle approximation, the minimum betatron amplitude and its location are respectively $\beta^*_{\rm min,A} = \frac{3}{4\sqrt{60}}L$ and $s^*_{\rm min,A} = \frac{3}{8}L$.

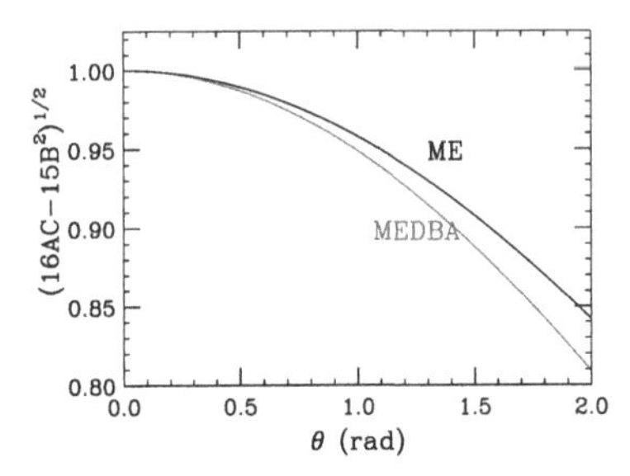

Figure 2.40: The minimum $\langle\mathcal{H}\rangle$ factors $G = \sqrt{16AC - 15B^2}$ for the DBA (lower curve) and $\tilde{G} = \sqrt{16\tilde{A}\tilde{C} - 15\tilde{B}^2}$ for the ME (upper curve) lattices are plotted as a function of the bending angle θ. The ME lattice data are for minimum $\langle\mathcal{H}\rangle$ without the achromat constraint. Note that $\langle\mathcal{H}\rangle$ is slightly smaller in a long dipole because of the $1/\rho^2$ focusing effect of the sector dipole.

It is difficult to design a lattice that can reach the theoretical minimum emittance. The typical emittance attained is about 2-4 times of the minimum emittance. The low-emittance DBA-lattice at the advanced photon source (APS) in Argonne National Laboratory is shown in Fig. 2.41, where the left-plot shows the optical functions with minimum β_x inside dipoles in order to minimize $\langle\mathcal{H}\rangle$. The middle and right plots show the normalized dispersion coordinates $(P_{\rm d}, X_{\rm d})$. The dispersion matching quadrupole at the center is split into two in order to leave space for a sextupole. Since the lattice is designed to minimize $\langle\mathcal{H}\rangle$ inside dipole, the normalized dispersion coordinates are small to be compared with those shown in Fig. 2.32. The entire achromat section of the DBA lattice is located at the origin $X_{\rm d} = P_{\rm d} = 0$. In the dispersion matching straight section, the normalized dispersion phase-space coordinates are located on a circle with the center at the origin. Many third generation high-brilliance light sources employ low emittance DBA-lattice for their storage ring. The details of emittance minimization procedure will be addressed in Chapter 4, Sec. III.

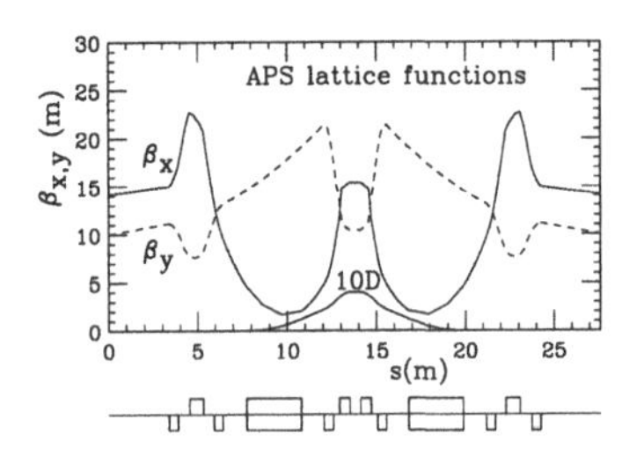

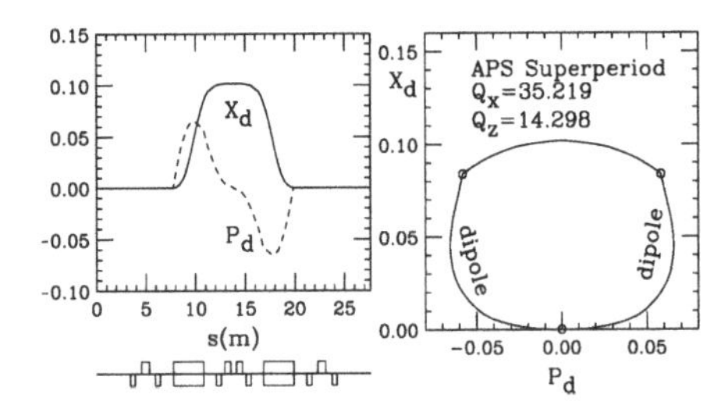

Figure 2.41: Left: The low emittance lattice functions for a superperiod of APS. The APS lattice has 40 superperiods so that the circumference is 1104 m. The tunes of this lattice are $Q_x = 35.219$, $Q_z = 14.298$. The momentum compaction factor is $\alpha_{\rm c} = 2.28 \times 10^{-4}$ in agreement with that of Eq. (2.197). Middle and Right: The normalized dispersion coordinates for the low emittance APS lattice is shown in one superperiod.

B. Minimum $\langle\mathcal{H}\rangle$ without achromat constraint

Without the achromat constraint, minimization of the $\mathcal{H}$-function can be achieved through the following steps. The minimum of $\langle\mathcal{H}\rangle$ is obtained by solving $\frac{\partial\langle\mathcal{H}\rangle}{\partial d_0} = 0$ and $\frac{\partial\langle\mathcal{H}\rangle}{\partial d_0'} = 0$ to obtain $d_{0,\min} = \frac{1}{6}F$ and $d'_{0,\min} = -\frac{1}{2}E$. The resulting $\langle\mathcal{H}\rangle$ becomes

$$\langle\mathcal{H}\rangle = \frac{1}{12}\rho\theta^3\left(\tilde{\beta}_0\tilde{A} - \tilde{\alpha}_0\tilde{B} + \frac{4\tilde{\gamma}_0}{15}\tilde{C}\right),$$

where $\tilde{A} = 4A - 3E^2, \tilde{B} = 3B - 2EF, \tilde{C} = \frac{9}{4}C - \frac{5}{4}F^2$. Using the relation $\tilde{\beta}_0\tilde{\gamma}_0 = 1 + \tilde{\alpha}_0^2$, we obtain $\langle\mathcal{H}\rangle_{\min} = \frac{\tilde{G}}{12\sqrt{15}}\rho\theta^3$, where $\tilde{G} = \sqrt{16\tilde{A}\tilde{C} - 15\tilde{B}^2}$ (see Fig. 2.40). Thus the minimum $\langle\mathcal{H}\rangle$ without achromatic constraint is a factor of 3 smaller than that with the achromat condition. A lattice designed with the constraint of minimum $\langle\mathcal{H}\rangle$ is called a theoretical minimum emittance (TME) lattice. The betatron amplitude function at the minimum $\langle\mathcal{H}\rangle$ is $\tilde{\beta}_0 = \frac{8}{\sqrt{15}\tilde{G}}\tilde{C}$, $\tilde{\alpha}_0 = \frac{\sqrt{15}}{\tilde{G}}\tilde{B}$, $\tilde{\gamma}_0 = \frac{2\sqrt{15}}{\tilde{G}}\tilde{A}$.The waist of the optimal betatron amplitude function for the minimum $\langle\mathcal{H}\rangle$ is located at the middle of the dipole, i.e. $s^* = L/2$. The corresponding minimum betatron amplitude function at the waist location is $\beta^*_{\min} = L/\sqrt{60}$ in small-angle approximation with $\theta \ll 1$.

Even though the minimum $\langle\mathcal{H}\rangle$ is one third of that with the achromat condition, the required minimum betatron amplitude function is $\beta^*_{\min} = \frac{4}{3}\beta^*_{\min,\mathrm{A}}$. The corresponding maximum betatron amplitude function will be reduced accordingly. We have discussed the minimum $\langle\mathcal{H}\rangle$ only in sector dipoles. In actual machine design, combined-function magnets with defocusing field may be used (see Chap. 4, Sec.III), where we will find that $\langle\mathcal{H}\rangle_{\min}$ is actually larger than for a separate function lattice.

C. Compaction factor in double-bend (DB) lattices

The dispersion function inside a sector dipole is

$$D(s) = \rho(1 - \cos\phi) + D_0\cos\phi + \rho D_0'\sin\phi, \tag{2.195}$$

$$D'(s) = \sin\phi - \frac{D_0}{\rho}\sin\phi + D_0'\cos\phi, \tag{2.196}$$

where ρ is the bending radius of the dipole, $\phi = s/\rho$ is the bend angle, and D_0 and D_0' are respectively the values of the dispersion function and its derivative at $s = 0$. For a matched double-bend module, the momentum compaction is

$$\alpha_{\mathrm{c}} = \frac{\rho}{L_{\mathrm{m}}}\left[\theta - \sin\theta + \frac{D_0}{\rho}\sin\theta + D_0'(1 - \cos\theta)\right] \approx \frac{\rho\theta^3}{L_{\mathrm{m}}}\left(\frac{1}{6} + \frac{D_0}{L\theta} + \frac{D_0'}{2\theta}\right) + \cdots,$$

where L_{m} is the length of one half of the double-bend module, $\theta = L/\rho$, and L is the length of the dipole. The momentum compaction factor depends on the initial dispersion function at the entrance of the dipole.

In small-angle approximation, the condition for negative momentum compaction is $6d_0 + 3d_0' \le -1$, where $d_0 = D_0/L\theta$, and $d_0' = D_0'/\theta$. The dispersion function in the rest of the module can be matched by quadrupole settings.

The momentum compaction in small dipole angle approximation for the isomagnetic DBA module with $D_0 = 0$ and $D_0' = 0$ and for the TME condition are

$$\alpha_{\rm c,DBA} \approx \frac{\rho\theta^2}{6R} \qquad \text{and} \qquad \alpha_{\rm c,TME} \approx \frac{\rho\theta^2}{12R}, \tag{2.197}$$

where R is the average radius of the storage ring and θ are the bending radius and the bending angle of each dipole. The momentum compaction factor of a DBA lattice is independent of the betatron tune. Finally, a reverse-bend dipole placed at the high dispersion straight section can also be used to adjust the momentum compaction factor of a DBA lattice. Such a lattice can provide a small-emittance negative momentum compaction lattice for synchrotron radiation sources.

Exercise 2.4

1. The dispersion function in a dipole satisfies the equation $D'' + K_x D = 1/\rho$. Let D_0 and D_0' be the dispersion function and its derivative at $s = 0$.

$$\begin{pmatrix} D(s) \\ D'(s) \\ 1 \end{pmatrix} = M \begin{pmatrix} D_0 \\ D_0' \\ 1 \end{pmatrix},$$

where M is the transfer matrix.

(a) Show that the transfer matrix for $K_x = K > 0$, and $K < 0$ are respectively

$$M = \begin{pmatrix} \cos\sqrt{K}s & \frac{1}{\sqrt{K}}\sin\sqrt{K}s & \frac{1}{\rho K}(1-\cos\sqrt{K}s) \\ -\sqrt{K}\sin\sqrt{K}s & \cos\sqrt{K}s & \frac{1}{\rho\sqrt{K}}\sin\sqrt{K}s \\ 0 & 0 & 1 \end{pmatrix}$$

$$M = \begin{pmatrix} \cosh\sqrt{|K|}s & \frac{1}{\sqrt{|K|}}\sinh\sqrt{|K|}s & \frac{1}{\rho|K|}(-1+\cosh\sqrt{|K|}s) \\ \sqrt{|K|}\sinh\sqrt{|K|}s & \cosh\sqrt{|K|}s & \frac{1}{\rho\sqrt{|K|}}\sinh\sqrt{|K|}s \\ 0 & 0 & 1 \end{pmatrix}$$

(b) Show that the transfer matrix of a sector magnet is given by Eq. (2.154).

(c) For a rectangular magnet, show that the horizontal transfer matrix is (see Exercise 2.2.3)

$$M_{\text{rectangular dipole}} = \begin{pmatrix} 1 & \rho\sin\theta & \rho(1-\cos\theta) \\ 0 & 1 & 2\tan(\theta/2) \\ 0 & 0 & 1 \end{pmatrix},$$

where ρ and θ are the bending radius and the bending angle.

(d) In thin-lens (small-angle) approximation, show that the transfer matrices M for quadrupoles and dipoles become

$$M_{\rm quad} = \begin{pmatrix} 1 & 0 & 0 \\ -1/f & 1 & 0 \\ 0 & 0 & 1 \end{pmatrix}, \quad M_{\rm dipole} = \begin{pmatrix} 1 & \ell & \ell\theta/2 \\ 0 & 1 & \theta \\ 0 & 0 & 1 \end{pmatrix},$$

where f is the focal length, and ℓ and θ are the length and bending angle of the dipole.

2. The bending arc of an accelerator lattice is usually composed of FODO cells. Each FODO cell is $[\frac{1}{2}$QF B QD B $\frac{1}{2}$QF$]$, where QF and QD are the focusing and defocusing quadrupoles with focal length f_1 and $-f_2$ respectively, and B is a dipole with bending angle θ. Let L be the half cell length.

(a) Using thin-lens approximation, show that the dispersion function and the betatron amplitude functions are

$$D_{\rm F} = \frac{L\theta}{\sin^2(\Phi_x/2)}(1+\frac{1}{4}T_-), \quad D_{\rm D} = \frac{L\theta}{\sin^2(\Phi_x/2)}(1-\frac{1}{4}T_+),$$

$$\beta_{x,{\rm F}} = \frac{L}{\sin(\Phi_x/2)}\sqrt{\frac{1+T_-}{1-T_+}}, \quad \beta_{x,{\rm D}} = \frac{L}{\sin(\Phi_x/2)}\sqrt{\frac{1-T_+}{1+T_-}},$$

where

$$S_\pm = \sin^2\frac{\Phi_x}{2} \pm \sin^2\frac{\Phi_z}{2}, \quad T_\pm = \frac{1}{4}\left(\sqrt{S_-^2+8S_+} \pm S_-\right),$$

and Φ_x and Φ_z are the horizontal and vertical betatron phase advance per cell.

(b) Simplify your result in part (a) with $\Phi_x = \Phi_z = \Phi$ and calculate the dispersion actions $J_{\rm d}({\rm QF}), J_{\rm d}({\rm QD})$ as a function of the phase advance per cell Φ. Plot $J_{\rm d}({\rm QF})/J_{\rm d}({\rm QD})$ as a function of Φ.

(c) Use the data in the table below to estimate the dispersion function of AGS, RHIC and SSC lattices in thin-lens approximation. Estimate the momentum beam size vs the betatron beam size in the arc.

	AGS	RHIC	Tevatron	SSC	LHC
$L_{\rm cell}$ (m)	13.45	29.6	59.5	180	97.96
Φ (deg)	52.5	90	75	90	90
Energy (GeV)	25	250	1000	20000	8000
$\epsilon_N\,(\pi\mu{\rm m})$	30	30	30	10	15
$(\Delta p/p_0)_{\rm rms}$	.005	0.003	0.001	0.0001	0.0001

(d) A collider lattice is usually made of arcs and insertions. The arc section is composed of regular FODO cells with bends, and the straight insertion section is composed of quadrupoles without dipoles. The dispersion suppressor matches the dispersion function in the arc to a zero dispersion value in the straight section. Show that the momentum compaction factor of such a lattice is

$$\alpha_{\rm c} \approx \frac{1}{\nu_{\rm arc}^2(1+L_{\rm s}/L_{\rm a})},$$

where $2\pi\nu_{\rm arc}$ is the total accumulated phase advance in the arcs, and $L_{\rm s}$ and $L_{\rm a}$ are the length of the straight section and the arc.

3. Show that the 3×3 transfer matrix of a repetitive cell is generally given by Eq. (2.157). Show that the transfer matrix of repetitive FODO cell is

$$M = \begin{pmatrix} \cos\Phi & \beta_{\rm F}\sin\Phi & 2D_{\rm F}\sin^2(\Phi/2) \\ -\gamma_{\rm F}\sin\Phi & \cos\Phi & \gamma_{\rm F}D_{\rm F}\sin\Phi \\ 0 & 0 & 1 \end{pmatrix},$$

where the symmetry conditions $\alpha_{\rm F} = 0$ and $D'_{\rm F} = 0$ are used, Φ is the phase advance per cell, $\beta_{\rm F}$ and $\gamma_{\rm F}$ are the Courant–Snyder parameters, and $D_{\rm F}$ is the dispersion function at the center of the quadrupole.

(a) Show that

$$M^2_{\Phi=90^\circ} = \begin{pmatrix} -1 & 0 & 2D_{\rm F} \\ 0 & -1 & 0 \\ 0 & 0 & 1 \end{pmatrix}, \quad M^4_{\Phi=90^\circ} = \begin{pmatrix} 1 & 0 & 0 \\ 0 & 1 & 0 \\ 0 & 0 & 1 \end{pmatrix}.$$

(b) Show that two FODO cells, each with 90° phase advance, match a zero dispersion region to a final dispersion of $D = 2D_{\rm F}$ and $D' = 0$.

(c) To match the dispersion function from a regular FODO cell in the arc to a zero value at the straight section, we need a dispersion suppressor. Adjoining the regular arc, the dispersion suppressor is composed of two reduced bending FODO cells, with bending angle θ_2 and θ_1 for each dipole.[66] Show that the conditions for zero dispersion after the dispersion suppressor are

$$\frac{\theta_1}{\theta} = \frac{1}{2(1-\cos\Phi)}, \quad \text{and} \quad \theta_1 + \theta_2 = \theta,$$

where θ is the bending angle of each dipole in the regular cell, and Φ is the phase advance of the FODO cell. At $\Phi = \pi/2$, these two FODO cells form the $-I$ unit. The theorem of dispersion suppression of Section IV.4 is verified.

(d) This exercise shows the effect of dispersion mismatch. Assuming that the accelerator lattice is made of n FODO cells, where $(n-1)$ FODO cells are [$\frac{1}{2}$QF B QD B $\frac{1}{2}$QF] with dipoles, and the bending magnets in the last FODO cell are replaced by drift spaces, show that the dispersion function at the entrance of the first FODO cell with a dipole is

$$D_1 = \frac{1-\cos n\Phi + \cos\Phi - \cos(n-1)\Phi}{2(1-\cos n\Phi)} D_{\rm F},$$

$$D'_1 = -\frac{\sin\Phi - \sin n\Phi + \sin(n-1)\Phi}{2(1-\cos n\Phi)} \gamma D_{\rm F},$$

where $D_{\rm F}$ is the dispersion function of the regular FODO cell at the center of the focusing quadrupole and Φ is the phase advance per cell. The resulting mismatched dispersion function can be very large at $n\Phi \approx 0 \pmod{2\pi}$, which is related to the integer stopband.

[66] A reduced bending cell can be represented by the following matrix with $\xi_1 = \theta_1/\theta$:

$$\begin{pmatrix} \cos\Phi & \beta_F\sin\Phi & \xi_1 D_F(1-\cos\Phi) \\ -\gamma_F\sin\Phi & \cos\Phi & \xi_1\gamma_F D_F\sin\Phi \\ 0 & 0 & 1 \end{pmatrix}.$$

4. Using thin-lens approximation, show that the momentum compaction factor α_c of an accelerator made of N FODO cells is

$$\alpha_c = \frac{1}{2\pi R}\oint \frac{D_x}{\rho} ds = \left(\frac{2\pi}{2N\sin\frac{\Phi}{2}}\right)^2 \approx \frac{1}{\nu_x^2},$$

where R is the average radius of the accelerator, Φ is the phase advance per cell, and ν_x is the horizontal betatron tune.

5. Consider a weak-focusing synchrotron (Exercise 2.2.5) with a constant focusing index $0 < n < 1$. Show that the lattice and dispersion functions are $\beta_x = \rho/\sqrt{1-n}$, $\beta_z = \rho/\sqrt{n}$, $D = \rho/(1-n)$, and the transition energy is $\gamma_T = \sqrt{1-n}$.

6. With the Floquet transformation, Eq. (2.150) can be transformed to

$$\frac{d^2\tilde{X}}{d\phi^2} + \nu^2\tilde{X} = \frac{\nu^2\beta^{3/2}}{\rho},$$

where $\tilde{X} = D/\sqrt{\beta}$, and $\phi = \int_0^s ds/\nu\beta$. Show that the solution of the above equation is

$$\frac{\beta^{3/2}}{\rho} = \sum_{k=-\infty}^{\infty} a_k e^{ik\phi}, \qquad a_k = \frac{1}{2\pi}\int_0^{2\pi}\frac{\beta^{3/2}}{\rho}e^{-ik\phi}d\phi,$$

$$D(s) = \nu^2\sqrt{\beta(s)}\sum_{k=-\infty}^{\infty}\frac{a_k e^{ik\phi(s)}}{\nu^2 - k^2}.$$

$$\alpha_c = \frac{\nu^3}{R}\sum_{k=-\infty}^{\infty}\frac{|a_k|^2}{\nu^2-k^2},$$

where $R = C/2\pi$ is the mean radius. In most accelerator design, the a_0 harmonic dominates, the dispersion function $D(s)$ is approximately $a_0\sqrt{\beta(s)}$. If $\rho \approx$ constant along the circumference, we have $a_0 = \frac{1}{2\pi\nu}\oint\frac{\beta^{1/2}}{\rho}ds \approx \frac{\langle\sqrt{\beta}\rangle}{\nu}$. Since $\nu = \oint ds/2\pi\beta \approx R/\langle\beta\rangle$, we find $\alpha_c \approx 1/\nu^2$.

7. Show that the integral representation of the dispersion function in Eq. (2.101) satisfies Eq. (2.149). Substituting the betatron coordinate into Eq. (2.163), show that the path-length change due to the betatron motion is

$$\begin{aligned}\Delta L &= \int_s^{s+C}\frac{x}{\rho}ds \\ &= [\sin 2\pi\nu_x X_d - (1-\cos 2\pi\nu_x)P_d]X_\beta + [(1-\cos 2\pi\nu_x)X_d + \sin 2\pi\nu_x P_d]P_\beta\end{aligned}$$

where $X_\beta = x/\sqrt{\beta_x}$ and $P_\beta = (\alpha_x x + \beta_x x')/\sqrt{\beta_x}$ are normalized betatron coordinates, and X_d, P_d are the normalized dispersion function phase-space coordinates of Eq. (2.162). Since the time average of the betatron motion is zero, $\langle X_\beta\rangle = \langle P_\beta\rangle = 0$, the path length depends on the betatron amplitude quadratically.

8. Show that orbit length change due to dipole field error is the product of the dipole kick angle and the dispersion function at the kicker location, i.e. $\Delta C = \sum_i D(s_i)\theta_i$, where $D(s_i)$ is the dispersion function at the dipole error location, and θ_i is the dipole field error.

9. The equation of motion for the vertical coordinate is

$$z'' + \frac{K_z(s)}{1+\delta} z = -\frac{\Delta B_x}{B\rho(1+\delta)}; \qquad \frac{\Delta B_x}{B\rho} = \frac{1}{\rho}(a_0 + b_1 z + a_1 x + 2b_2 xz + \cdots),$$

where $\delta = (p - p_0)/p_0$ is the fractional off-momentum deviation, $K_z(s)$ is the focusing function, $B\rho$ is the momentum-rigidity of the on-momentum particle. Here a_0 arises essentially from the dipole roll, b_1 is the gradient error, a_1 is the skew quadrupole field, and b_2 is the sextupole field. Substituting $x = x_{\rm co} + D_x\delta + x_\beta$, we obtain

$$z'' + \frac{\tilde{K}_z(s)}{1+\delta} z = -\frac{1}{(1+\delta)\rho}[a_0 + a_1(x_{\rm co} + x_\beta + D_x\delta) + 2b_2(x_\beta + D_x\delta)z],$$

where the effective focusing function is $\tilde{K}_z(s) = K_z(s) + b_1/\rho + 2b_2 x_{\rm co}$.

(a) Expand the vertical coordinate in $z = z_{\rm co} + D_z\delta + z_\beta$ and show that

$$\begin{aligned} z''_{\rm co} + \tilde{K}_z(s) z_{\rm co} &= -\frac{a_0}{\rho} - \frac{a_1}{\rho} x_{\rm co} \\ z_\beta + \tilde{K}_z(s) z_\beta &= -\frac{a_1 + 2b_2 z_{\rm co}}{\rho} x_\beta - \frac{2b_2}{\rho} x_\beta z_\beta \\ D''_z + \tilde{K}_z(s) D_z &= h_z(s) \\ h_z(s) &= +\tilde{K}_z(s) z_{\rm co} + \frac{a_0 + a_1 x_{\rm co}}{\rho} - \frac{a_1 + 2b_2 z_{\rm co}}{\rho} D_x(s). \end{aligned}$$

Here, both the closed orbit and the betatron coordinates are expanded in power series of the fractional off-momentum variable. Note that the horizontal and vertical closed orbit, betatron functions, and dispersion functions are all coupled.

(b) When coupling is small and the horizontal betatron tune is sufficiently far away from an integer, the vertical dispersion function is given by the solution of the dispersion function equation in (a). Show that the normalized dispersion functions are

$$\begin{aligned} \frac{D_z(s)}{\sqrt{\beta_z(s)}} &= \frac{1}{2\sin\pi\nu_z} \int_s^{s+C} \sqrt{\beta_z(t)} h_z(s) \cos(\pi\nu_z + \psi_z(s) - \psi_z(t)) dt, \\ \frac{\alpha_z}{\sqrt{\beta_z}} D_z + \sqrt{\beta_z} D'_z &= \frac{-1}{2\sin\pi\nu_z} \int_s^{s+C} \sqrt{\beta_z(t)} h_z(s) \sin(\pi\nu_z + \psi_z(s) - \psi_z(t)) dt, \end{aligned}$$

where ν_z, β_z, $\alpha_z = -\beta'_z/2$, ψ_z are the vertical betatron tune, the vertical betatron functions, and the vertical betatron phase function respectively.

(c) Carry out Floquet transformation by changing the independent variable from s to $\phi = \frac{1}{\nu_z}\int_0^s \frac{1}{\beta_z} ds$, and define the Fourier harmonics f_k of the perturbation as

$$f_k = \frac{1}{2\pi} \int_0^{2\pi} \sqrt{\beta_z} h_z(s) e^{-ik\phi} ds,$$

where k is an integer, show that

$$D_z(s) = \nu_z\sqrt{\beta_z(s)}\sum_{k=-\infty}^{\infty}\frac{f_k e^{ik\phi(s)}}{\nu_z^2 - k^2} \approx \sqrt{\beta_z(s)}\frac{|f_n|\cos(n\phi(s)+\xi_n)}{\nu_z - n},$$

where the second identity approximate the vertical dispersion function by a simple pole at $n = [\nu_z]$, the integer nearest the vertical betatron tune, and ξ_n is the phase of f_n. Estimate relative importance of various terms in $h_z(s)$ for a realistic accelerator.

10. In a straight section of an accelerator, $M_{13} = 0$ and $M_{23} = 0$. The values of the dispersion function at two locations in the beam line are related by

$$D_2 = M_{11}D_1 + M_{12}D_1', \quad D_2' = M_{21}D_1 + M_{22}D_1'.$$

Show that $\mathcal{H} = \gamma D^2 + 2\alpha DD' + \beta D'^2$ is invariant in the straight section.

11. In general, the dispersion function transfer matrix is given by Eq. (2.156). Show that the evolution of the $\mathcal{H}$-function is

$$\begin{aligned}\mathcal{H} &= \mathcal{H}_0 + 2(\alpha_0 D_0 + \beta_0 D_0')[M_{23}M_{11} - M_{13}M_{21}] \\ &\quad + 2(\gamma_0 D_0 + \alpha_0 D_0')[M_{13}M_{22} - M_{23}M_{12}] + \beta_0[M_{13}M_{21} - M_{23}M_{11}]^2 \\ &\quad + \gamma_0[M_{13}M_{22} - M_{23}M_{12}]^2 - 2\alpha_0[M_{13}M_{21} - M_{23}M_{11}][M_{13}M_{22} - M_{23}M_{12}]\end{aligned}$$

where $\mathcal{H}_0 = \gamma_0 D_0^2 + 2\alpha_0 D_0 D_0' + \beta_0 D_0'^2$; M_{ij} is a matrix element of the transfer matrix; and α_0, β_0, and γ_0 are Courant-Snyder parameters at the initial location.

   (a) Using the M_{ij} of Eq. (2.154), show that $\mathcal{H}$ in a sector dipole is given by Eq. (2.192).

   (b) Find $\mathcal{H}$ in a rectangular dipole (use the result of Exercise 2.4.1).

12. **Double-Bend Achromat:** Consider an achromatic bending system with two sector magnets and a focusing quadrupole midway between two dipoles, i.e.

$$\mathrm{B}[\rho,\theta] \quad \mathrm{O}[l] \quad \mathrm{Q_F}[K, l_\mathrm{q}] \quad \mathrm{O}[l] \quad \mathrm{B}[\rho,\theta].$$

Here K and l_q represent the focusing strength function and the length of the quadrupole. Show that the dispersion matching condition is

$$\rho\tan\frac{\theta}{2} + l = \frac{1}{\sqrt{K}}\cot\frac{\sqrt{K}l_\mathrm{q}}{2},$$

and that, in thin-lens approximation, the matching condition reduces to Eq. (2.177). The dispersion matching condition for DBA cell with rectangular dipoles is $\rho\sin\frac{\theta}{2}\cos\frac{\theta}{2} + l = \frac{1}{\sqrt{K}}\cot\frac{\sqrt{K}l_\mathrm{q}}{2}$. This basic achromat is also called a Chasman-Green lattice cell. The double-bend achromat (DBA) is commonly used in the design of low emittance storage rings, where quadrupole configurations are arranged to minimize $\langle\mathcal{H}\rangle$ in the dipole. Other achromat modules are

(a) the triple-bend achromat (TBA)

$$\mathrm{B}[\rho,\theta_\mathrm{o}]\quad \mathrm{O}[l_1]\quad \mathrm{Q_F}[K,l_\mathrm{q}]\quad \mathrm{O}[l_2]\quad \mathrm{B}[\rho,\theta_\mathrm{c}]\quad \mathrm{O}[l_2]\quad \mathrm{Q_F}[K,l_\mathrm{q}]\quad \mathrm{O}[l]\quad \mathrm{B}[\rho,\theta_\mathrm{o}]$$

which has been used in many synchrotron radiation light sources such as the ALS (Berkeley), TLS (Taiwan), KLS (Korea), and BESSY (Berlin), and

(b) the reverse-bend DBA

$$\mathrm{B}[\rho,\theta]\quad \mathrm{O}[l_1]\quad \mathrm{Q_F}[K,l_\mathrm{q}]\quad \mathrm{O}[l_2]\quad \mathrm{B}[-\rho,-\theta_\mathrm{r}]\quad \mathrm{O}[l_2]\quad \mathrm{Q_F}[K,l_\mathrm{q}]\quad \mathrm{O}[l]\quad \mathrm{B}[\rho,\theta]$$

where the reverse bend angle $\theta_\mathrm{r} \ll \theta$ can be used to adjust the desired momentum compaction factor.

13. **Achromatic translating system:** Show that the transport line with two sector dipoles $\mathrm{B}[\rho,\theta]\ \mathrm{O}[l_1]\ \mathrm{Q_F}[K,l_\mathrm{q}]\ \mathrm{O}[l_\mathrm{c}]\ \mathrm{Q_F}[K,l_\mathrm{q}]\ \mathrm{O}[l_1]\ \mathrm{B}[-\rho,-\theta]$ is achromatic if the following condition is satisfied:

$$\rho\sin\frac{\theta}{2}+l_1=\frac{l_\mathrm{c}\cos\sqrt{K}l_\mathrm{q}+\frac{2}{\sqrt{K}}\sin\sqrt{K}l_\mathrm{q}}{l_\mathrm{c}\sqrt{K}\sin\sqrt{K}l_\mathrm{q}-2\cos\sqrt{K}l_\mathrm{q}}.$$

Show that, in thin-lens approximation, $f_\mathrm{q}=[\ell_\mathrm{c}(\ell_1+\ell_\mathrm{B})]/[\ell_\mathrm{c}+2\ell_1+\ell_\mathrm{B}]$, where f_q is the focal length of the quadrupole and ℓ_B is the length of the dipole. Two quadrupoles are needed to provide dispersion matching.

14. Show the three sector dipole system $\mathrm{B}[\rho,\theta]\ \mathrm{O}[l]\ \mathrm{B}[\rho,\theta]\ \mathrm{O}[l]\ \mathrm{B}[\rho,\theta]$ is achromatic if the following condition is satisfied:

$$\frac{l}{\rho}=\frac{2\cos\theta+1}{\sin\theta}.$$

15. A set of four rectangular dipoles with zero net bending angle

$$\mathrm{B}[\rho,\theta]\ \mathrm{O}[l_1]\ \mathrm{B}[-\rho,-\theta]\ \mathrm{O}[l_2]\ \mathrm{B}[-\rho,-\theta]\ \mathrm{O}[l_1]\ \mathrm{B}[\rho,\theta]$$

has many applications. It can be used as a beam translation (chicane) unit to facilitate injection, extraction, internal target operation, etc. It can also be used as one unit of the wiggler magnet for modifying electron beam characteristics or for producing synchrotron radiation.

(a) Show that the rectangular magnet beam translation unit is achromatic to all orders, and show that the R_{56} element of the transport matrix, in small angle approximation, is

$$R_{56}=2\theta^2(\ell_1+\frac{2}{3}\rho\theta).$$

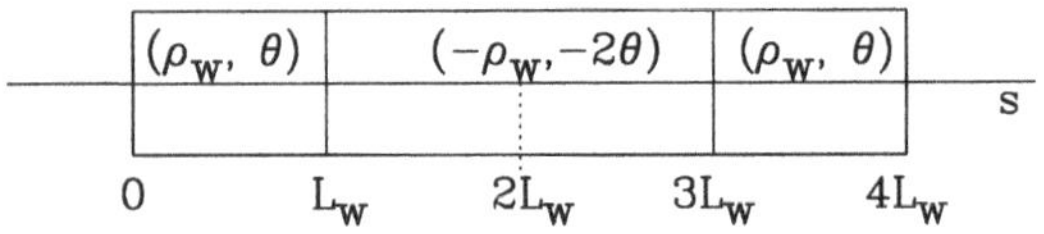

(b) A simplified compact geometry with $l_1 = l_2 = 0$ (shown in the figure above) is often used as a unit of the wiggler magnet in electron storage rings. Assuming that $D_0 = D_0' = 0$, show that the dispersion function created by the wiggler magnet, including he edge focusing effect, is

$$\frac{D(s)}{\rho_{\rm w}} = \begin{cases} -(1-\cos\phi), & 0 < s < L_{\rm w} \\ (1-\cos\bar{\phi}) - (1-\cos\theta)\cos\bar{\phi} & \\ \quad -[\sin\theta + 2\tan\theta(1-\cos\theta)]\sin\bar{\phi}, & L_{\rm w} < s < 2L_{\rm w} \end{cases}$$

where $\phi = s/\rho_{\rm w}$ and $\bar{\phi} = (s - L_{\rm w})/\rho_{\rm w}$, and

$$D'(s) = \begin{cases} -\sin\phi, & 0 < s < L \\ -\sin\theta - 2\tan\theta(1-\cos\theta), & s = L_{\rm w+} \\ \sin\bar{\phi} + (1-\cos\theta)\sin\bar{\phi} & \\ \quad -[\sin\theta + 2\tan\theta(1-\cos\theta)]\cos\bar{\phi}, & L_{\rm w} < s < 2L_{\rm w}. \end{cases}$$

Show that

$$D(s = 2L_{\rm w}) = 2\rho_{\rm w}\frac{1-\cos\theta}{\cos\theta}, \quad '(s = 2L_{\rm w}) = 0.$$

Since $D' = 0$ at the symmetry point, the wiggler is an achromat. In small bending-angle approximation, show that the dispersion function becomes

$$D(s) = \begin{cases} -s^2/2\rho_{\rm w}, & 0 < s < L_{\rm w} \\ -(2L_{\rm w}^2 - (2L_{\rm w} - s)^2)/2\rho_{\rm w}, & L_{\rm w} < s < 2L_{\rm w}, \end{cases}$$

$$D'(s) = \begin{cases} -s/\rho_{\rm w}, & 0 < s < L_{\rm w} \\ -(2L_{\rm w} - s)/\rho_{\rm w}, & L_{\rm w} < s < 2L_{\rm w}. \end{cases}$$

16. An accelerator with circumference 240 m is made of 24 FODO cells. The betatron tunes of the synchrotron are $\nu_x = 4.9$ and $\nu_z = 4.8$ respectively.

    (a) What are the maximum values of the betatron amplitude function and dispersion function? If one of the 48 dipoles has an error of 1estimate the maximum closed orbit deviation from the designed orbit in mm.

    (b) If one of the 24 focusing quadrupoles has 1answer the flowing questions using thin lens and small angle approximation. Estimate the maximum change in ΔD_x in meters, and estimate the maximum change of $\Delta\beta_x/\beta_x$ and $\Delta\beta_z/\beta_z$.

V Chromatic Aberration

A particle with momentum p executes betatron oscillations around an off-momentum closed orbit $x_{\rm co}(s)+D(s)\delta$, where $x_{\rm co}$ is the closed orbit for the on-momentum particle, D is the dispersion function, and $\delta = (p-p_0)/p_0$ is the fractional momentum deviation from the on-momentum p_0. Equation (2.148) is Hill's equation of the horizontal betatron motion. A higher energy particle with $\delta > 0$ has a larger momentum rigidity and thus a weaker effective focusing strength; a lower energy particle with $\delta < 0$ has a smaller momentum rigidity and a stronger effective focusing strength. This is reflected in the gradient error ΔK_x in Eq. (2.148). Similar gradient error exists in the vertical betatron motion. The resulting gradient errors ΔK_x and ΔK_z are[67]

$$\begin{cases} \Delta K_x = \left[-\dfrac{2}{\rho^2} + K(s)\right]\delta + O(\delta^2) \approx -K_x\delta, \\ \Delta K_z = -K(s)\delta + O(\delta^2) \approx -K_z\delta, \end{cases} \tag{2.198}$$

where $K = B_1/B\rho$ and $B_1 = \partial B_z/\partial x$. The chromatic gradient error is essentially equal to the product of the momentum deviation δ and the main focusing functions $-K_x$ and $-K_z$. The dependence of the focusing strength on the momentum of a circulating particle is called "chromatic aberration," which is proportional to the designed focusing functions K_x and K_z, and thus it is called "systematic" error. Systematic perturbations can alter the designed betatron amplitude functions and reduce the dynamical aperture for off-momentum particles. The effects of chromatic aberration include chromaticity, "beta-beat" associated with the half-integer stopbands, etc.

This section studies the effects of systematic chromatic aberration and its correction. In Sec. V.1 we define chromaticity and discuss its measurement and correction; in Sec. V.2 we examine the nonlinear perturbation due to chromatic sextupoles; in Sec. V.3 we study systematic half-integer stopbands and their effects on higher-order chromaticity; and in Sec. V.4 we outline basic machine design strategy.

---

[67]Including the effect of off-momentum orbits, the chromatic gradient error should include the effects of dispersion functions, fringe fields, etc. Some of these terms are included below:

$$\Delta K_x = \left[-\frac{2}{\rho^2} + K + 2\frac{D}{\rho}\left(\frac{1}{\rho^2} - K\right) - \left(\frac{1}{\rho}\right)' D' + \frac{\gamma_x}{\beta_x\rho}D\right](\delta - \delta^2 + \cdots) + \cdots,$$

$$\Delta K_z = \left[-K + \frac{K}{\rho}D + \left(\frac{1}{\rho}\right)' D' + \frac{\gamma_x}{\beta_x\rho}D\right](\delta - \delta^2 + \cdots) + \cdots,$$

where $K = B_1/B\rho$ is the gradient function of quadrupoles. Note that the higher-order gradient error depends on the betatron amplitude and dispersion functions. We neglect all chromatic effects arising from the dispersion function and fringe fields of magnets. For details see, e.g., K. Steffen, *High Energy Beam Optics* (Wiley, New York, 1965); S. Guiducci, *Proc. CERN Accelerator School*, CERN **91-04**, p. 53, 1991.

V.1 Chromaticity Measurement and Correction

The gradient error can induce betatron tune shift and betatron amplitude function perturbation. The chromatic gradient error of Eq. (2.198) gives rise to betatron tune shifts:

$$\begin{cases} \Delta\nu_x = \dfrac{1}{4\pi}\oint \beta_x \Delta K_x ds \approx \left(\dfrac{-1}{4\pi}\oint \beta_x K_x ds\right)\delta, \\ \Delta\nu_z = \dfrac{1}{4\pi}\oint \beta_z \Delta K_z ds \approx \left(\dfrac{-1}{4\pi}\oint \beta_z K_z ds\right)\delta. \end{cases} \tag{2.199}$$

The *chromaticity*, C_x or C_z, is defined as the derivative of the betatron tunes vs fractional momentum deviation, and the "natural chromaticity" arises solely from lattice quadrupoles:

$$C_y \equiv \frac{d(\Delta\nu_y)}{d\delta}, \qquad C_{y,\rm nat} \approx \frac{-1}{4\pi}\oint \beta_y K_y ds, \tag{2.200}$$

where the subscript y stands for either x or z. Because the focusing function is weaker for higher energy particles, the betatron tune decreases with particle momentum, and the natural chromaticity is negative.

The magnitude of the *natural chromaticity* $C_{y,\rm nat}$ depends on the lattice design. The natural chromaticity of a FODO lattice is (see Exercise 2.5.3)

$$C_{y,\rm nat}^{\rm FODO} \approx -\frac{1}{4\pi}N\left(\frac{\beta_{\rm max}}{f} - \frac{\beta_{\rm min}}{f}\right) = -\frac{\tan(\Phi_y/2)}{\Phi_y/2}\nu_y \approx -\nu_y, \tag{2.201}$$

where N is the number of cells, f is the focal length, Φ_y is the phase advance per cell, and $\nu_y = N\Phi_y/2\pi$ is the betatron tune of the machine. The "specific chromaticity," defined as $\xi_y = C_y/\nu_y$, is nearly equal to -1 for FODO lattices. The specific natural chromaticity of a high luminosity collider or a low-emittance electron storage ring can be as large as -4.

A beam is composed of particles with different momenta. The momentum spread of the beam is typically of the order of $\sigma_\delta \sim 10^{-5} - 10^{-2}$ depending on the applications and types of accelerator. Because of the chromaticity, the momentum spread gives rise to tune spread in the beam. If the chromaticity and the momentum spread of the beam become large enough that the betatron tunes overlap low-order nonlinear resonances, particle loss may imminently occur. Furthermore, the growth rate of transverse head-tail instabilities depends on the sign of the chromaticity (see Sec. VIII and Ref. [5]).

A. Chromaticity measurement

Machine chromaticities can be derived from measurements of betatron tunes vs beam momentum. Since beam momentum is related to rf frequency, the chromaticity can be obtained from measurements of betatron tune vs rf frequency, i.e.

$$C_y = \frac{d\nu_y}{d\delta} = -\eta\omega_{\rm rf}\frac{d\nu_y}{d\omega_{\rm rf}}, \tag{2.202}$$

where η is the phase-slip factor, and $\omega_{\rm rf}$ is the angular frequency of the rf system (see Exercise 2.5.8).

Figure 2.42 shows the "measured specific" chromaticities of the AGS.[68] Note that the vertical chromaticity becomes positive above about 22 GeV. Since the data obey $\frac{1}{2}(\xi_x + \xi_z) \approx -\frac{2}{\Phi}\tan\frac{\Phi}{2}$ (shown as the dashed line), where $\Phi \approx 53.8°$ is the phase advance of an AGS FODO cell, we have $\beta_x \approx \beta_z$ at these sextupole locations.

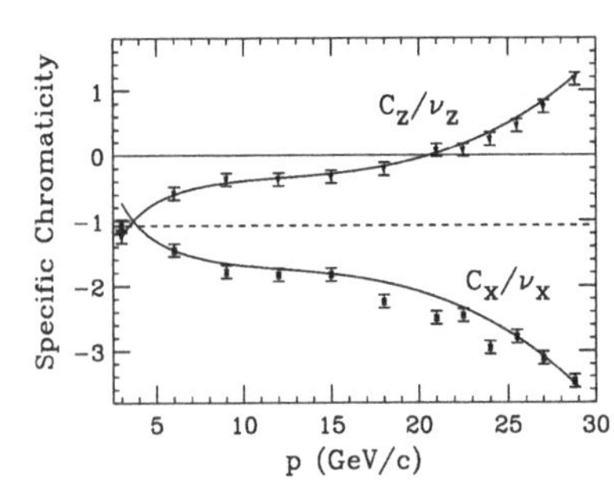

Figure 2.42: The measured chromaticities divided by the betatron tunes vs the beam momentum for the AGS. Deviation of the chromaticity from that of FODO lattice arises from sextupoles. At high energy, saturation of dipole magnets produces defocusing sextupole field. The solid curved line is obtained by modeling the sextupole field in the dipoles, as discussed in subsection C.

B. Chromatic correction

The natural chromaticity of a high-luminosity collider with low-β^* insertions is usually large. For example, the natural chromaticity for the Superconducting Super Collider (SSC) was expected to be about $C_{y,\rm nat} \approx -250$, which can lead to a natural tune spread of about $\Delta\nu \approx 0.1$ for a beam with an rms spread of $\delta = \pm 2 \times 10^{-4}$. Similarly, the natural chromaticity for the RHIC injection lattice is about $C_{y,\rm nat} \approx -50$, and the resulting tune spread will be $\Delta\nu \approx 0.5$ with a beam momentum spread of $\delta = \pm 5 \times 10^{-3}$. A circulating beam with such a large tune spread can encounter many nonlinear resonances, chromatic correction is needed to ensure good performance of a storage ring. This requires a magnet whose focusing function increases linearly with momentum in order to compensate the loss of focusing in quadrupoles.

First we examine the possibility of using sextupole magnets for chromaticity correction. The magnetic flux density of a sextupole magnet is

$$\frac{\Delta B_z}{B\rho} = \frac{B_2}{2B\rho}(x^2 - z^2), \quad \frac{\Delta B_x}{B\rho} = \frac{B_2}{B\rho}xz, \tag{2.203}$$

[68] E. Bleser, AGS Tech Note No. 288 (1987); E. Auerbach, E. Bleser, R. Thern, AGS Tech Note No. 276 (1987).

where $B_2 = \partial^2 B_z/\partial x^2|_{x=z=0}$. Substituting the transverse displacement of an off-momentum particle, $x = x_\beta(s) + D(s)\delta$, where x_β is the betatron displacement and $D(s)\delta$ is the off-momentum closed orbit, into Eq. (2.203), we obtain

$$\begin{cases} \dfrac{\Delta B_z}{B\rho} = -[S(s)D(s)\delta]x_\beta - \dfrac{S(s)}{2}(x_\beta^2 - z_\beta^2) - \dfrac{S(s)}{2}D^2(s)\delta^2, \\ \dfrac{\Delta B_x}{B\rho} = -[S(s)D(s)\delta]z_\beta - S(s)x_\beta z_\beta, \end{cases} \tag{2.204}$$

where $S(s) = -B_2/B\rho$ is the effective sextupole strength. Note that the first term of Eq. (2.204) depends linearly on the transverse betatron displacement. The effective quadrupole focusing functions $\Delta K_x = S(s)D(s)\delta$ and $\Delta K_z = -S(s)D(s)\delta$ depend linearly on the off-momentum deviation, sextupoles can be used for chromaticity correction. The second term of Eq. (2.204) can produce nonlinear perturbation in betatron motion, called geometric aberration, to be discussed in Sec. VII. Placement of sextupoles is important in minimizing nonlinear resonance strengths.

Including the contribution of sextupoles, the chromaticity becomes

$$C_x = \frac{-1}{4\pi}\oint \beta_x[K_x(s) - S(s)D(s)]ds,$$
$$C_z = \frac{-1}{4\pi}\oint \beta_z[K_z(s) + S(s)D(s)]ds.$$

Chromatic sextupoles located at nonzero dispersion function locations can be used to correct chromaticity. Generally, two families of sextupoles are needed to correct horizontal and vertical chromaticities.

For example, we consider a lattice of N repetitive FODO cells, where sextupoles are located near the focusing and defocusing quadrupoles. Let $S_{\rm F} = -B_2({\rm F})\ell_{\rm sf}/B\rho$ and $S_{\rm D} = -B_2({\rm D})\ell_{\rm sd}/B\rho$ be the integrated sextupole strengths at QF and QD respectively, where $\ell_{\rm sf}$, $\ell_{\rm sd}$, and $B_2({\rm F})$, $B_2({\rm D})$ are the length and the sextupole field strength at QF and QD. The sextupole strength needed to obtain zero chromaticity is (see Exercise 2.5.3)

$$S_{\rm F} = \frac{1}{2f^2\theta}\frac{\sin\frac{\Phi}{2}}{(1+\frac{1}{2}\sin\frac{\Phi}{2})}, \qquad S_{\rm D} = -\frac{1}{2f^2\theta}\frac{\sin\frac{\Phi}{2}}{(1-\frac{1}{2}\sin\frac{\Phi}{2})},$$

where f is the focal length, Φ the phase advance per cell, and θ the bending angle per half-cell.

For colliders or low-emittance storage rings, chromatic sextupoles are also arranged in families, located in the arcs, which consist mainly of FODO cells or DBA/TBA type cells. Since the low-β^* values in these lattices give rise to a large chromaticity, strong sextupoles are needed to correct it. If the intrinsic systematic half-integer stopband widths are large, the simple chromatic correction scheme using two families of sextupoles may not be sufficient to correct the higher-order chromatic effects.

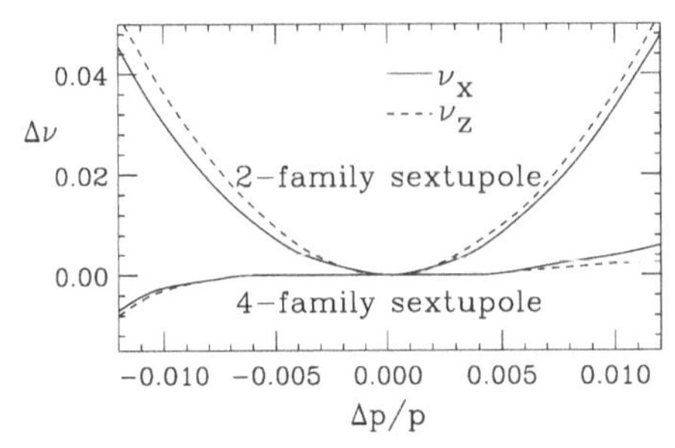

Figure 2.43: Variation of the betatron tune vs $\Delta p/p$ after chromatic correction with two and four families of sextupoles in RHIC. The chromatic gradient error can create a large betatron amplitude function modulation (betabeat), which in turn produces large second-order chromaticity.

Figure 2.43 shows an example of chromatic correction with two and four families of sextupoles in RHIC. Note that the second-order chromaticity $\Delta\nu_{x,z} \sim C^{(2)}_{x,z}\,\delta^2$ can cause substantial tune spread in a beam with a large momentum spread. In Sec. V.3, we will show that the chromatic gradient error can also create a large betatron amplitude function modulation (betabeat), which in turn induces a large second-order chromaticity. The second-order chromaticity and the betabeat can be simultaneously corrected by a proper chromatic stopband correction. Rules for their placement are as follows.

- In order to minimize sextupole strength, the chromatic sextupoles should be located near quadrupoles, where $\beta_x D_x$ and $\beta_z D_x$ are maximum.
- A large ratio of β_x/β_z for the focusing sextupole and a large ratio of β_z/β_x for the defocusing sextupole are needed for optimal independent chromaticity control.
- The families of sextupoles should be arranged to minimize the systematic half-integer stopbands and the strength of third and higher order betatron resonances.

C. Nonlinear modeling from chromaticity measurement

The measurements of chromaticities can be used to model nonlinear sextupole fields in an accelerator. For example, we discuss the nonlinear sextupole modeling of the AGS based on the measured chromaticities shown in Fig. 2.42. Since $C_{x,\rm data} < C_{x,\rm fodo}$, $C_{z,\rm data} > C_{z,\rm fodo}$, and $C_{x,\rm data} + C_{z,\rm data} = C_{x,\rm fodo} + C_{z,\rm fodo}$, the horizontally defocusing sextupoles must be located in dipoles, where $\beta_x \approx \beta_z$. To model the AGS, we assume that the sextupole fields arise from systematic error at both ends of each dipole, the eddy current sextupole due to the vacuum chamber wall, and the iron saturation sextupole at high field. The systematic error is independent of the beam momentum; the eddy current sextupole field depends inversely on beam momentum; and the saturation sextupole field that depends on a higher power of beam momentum. The solid lines in Fig. 2.42 represent theoretical calculations with the integrated sextupole strengths

$$\begin{aligned} S_{\rm b} &= -5.2\times10^{-4}+5.8\times10^{-2}/p \\ &\quad -(3.6\times10^{-4}p-7.0\times10^{-5}p^2+2.8\times10^{-6}p^3)\ ({\rm m}^{-2}), \\ S_{\rm e} &= -0.017\ ({\rm m}^{-2}), \end{aligned}$$

for the body and the ends of the short AGS bending magnets (2.0066 m) respectively. Here p is the beam momentum in unit of (GeV/c), $S_{\rm b}$ is the integrated sextupole field in each dipole distributed in the whole dipole, and $S_{\rm e}$ is the integrated sextupole field distributed only at the end of each dipole. $S_{\rm e}$ and the first term in $S_{\rm b}$ may be considered as the systematic error in dipoles, and they are momentum independent. The second term in $S_{\rm b}$ is due to the eddy current on the vacuum chamber wall, which is inversely proportional to the beam momentum, and proportional to $\dot{B}$, where $\dot{B}=2$ T/s in this experiment. The saturation term is nonlinear with respect to the momentum p. For the long magnets (2.3876 m) in the AGS, the integrated sextupole strength of the $S_{\rm b}$ term is assumed to be proportional to their length.

A chromaticity of about $-3\nu_x$ does not appear to cause difficulties in the AGS operation, which has recently attained an intensity of 6×10^{13} protons per pulse. Many low energy synchrotrons do not use chromatic correction sextupoles. However, chromaticity correction is absolutely essential in high energy synchrotrons and storage rings.

V.2 Nonlinear Effects of Chromatic Sextupoles

The Hamiltonian including sextupole nonlinearity is

$$H=\frac{1}{2}\left(x_\beta'^2+K_x x_\beta^2+z_\beta'^2+K_z z_\beta^2\right)+\frac{S(s)}{6}(x_\beta^3-3x_\beta z_\beta^2).$$

This Hamiltonian can drive third-order and higher-order nonlinear resonances at $3\nu_x=\ell,\nu_x\pm2\nu_z=\ell,\ldots$, where ℓ is an integer. However, the nonlinear resonance strength can be minimized by properly arranged sextupole families. In Sec. VII, we will show that if chromatic sextupoles are separated by an odd multiple of 180° in the betatron phase advance, their contributions to the third-order stopband width cancel each other in the first-order perturbation theory. Thus, four families of sextupoles can be arranged in a lattice with 90° phase advance per cell, and six families of sextupoles can be used in a lattice with 60° phase advance per cell. Such arrangements can also be used to correct the systematic half-integer stopband discussed in the next section.

V.3 Chromatic Aberration and Correction

The systematic chromatic gradient error can produce a large perturbation in the betatron amplitude functions for all off-momentum particles. Defining the betatron amplitude difference functions A and B (see Exercise 2.3.10), we find

$$A = \frac{\alpha_1\beta_0 - \alpha_0\beta_1}{\sqrt{\beta_0\beta_1}}, \quad B = \frac{\beta_1 - \beta_0}{\sqrt{\beta_0\beta_1}},$$
$$\frac{dB}{ds} = -A\left(\frac{1}{\beta_0} + \frac{1}{\beta_1}\right), \quad \frac{dA}{ds} = +B\left(\frac{1}{\beta_0} + \frac{1}{\beta_1}\right) + \sqrt{\beta_0\beta_1}\,\Delta K, \tag{2.205}$$

where $\Delta K = K_1 - K_0$ is the gradient error. The betatron amplitude functions β_0 and β_1 satisfy the Floquet equation

$$\beta_0' = -2\alpha_0, \quad \alpha_0' = K_0\beta_0 - \gamma_0, \quad d\psi_0/ds = 1/\beta_0,$$
$$\beta_1' = -2\alpha_1, \quad \alpha_1' = K_1\beta_1 - \gamma_1, \quad d\psi_1/ds = 1/\beta_1,$$

and ψ_0 and ψ_1 are the unperturbed and perturbed betatron phase functions.

From Eq. (2.205), we find that $A^2 + B^2 =$ constant in regions where $\Delta K = 0$. We consider the chromatic perturbation in quadrupoles and sextupoles with $\Delta K = -K\delta$ and $\Delta K = K_2(s)D\delta$ respectively. The changes of A are respectively

$$\Delta A = \int \sqrt{\beta_0\beta_1}\,\Delta K\, ds \approx \frac{\beta_0}{f}\frac{\Delta p}{p_0},$$
$$\Delta A = \int \sqrt{\beta_0\beta_1}\,K_2(s)D(s)ds\delta \approx -\beta_0\, g_{\rm eff}\,\frac{\Delta p}{p_0},$$

where f is the focal length of quadrupole. $g_{\rm eff} = (-B_2\Delta s/B\rho)D$ is the effective chromatic gradient error due to feed down of the off momentum orbit, $(B_2\Delta s/B\rho)$ is the integrated sextupole strength, and D is the dispersion function. Since the phase of A or B propagates at twice the betatron phase advance (see Exercise 2.3.10), two identical quadrupoles (sextupoles) separated by odd multiples of 90° in betatron phase advance cancel each other. Similarly, two identical quadrupoles (sextupoles) separated by an integer multiple of 180° in betatron phase advance will produce additive coherent kicks. By using sextupole families, the global chromatic perturbation function of the lattice can be minimized. The treatment is identical to the stopband integral to be discussed next.

A. Systematic chromatic half-integer stopband width

We have found that the perturbation of betatron function is most sensitive to stopband integrals near $p \approx [2\nu]$ harmonics (see Sec. III.4). The effect of systematic chromatic gradient error on betatron amplitude modulation can be analyzed by using the chromatic stopband integrals of Eq. (2.120):

$$J_{p,x} = \frac{1}{2\pi}\oint \beta_x \Delta K_x e^{-jp\phi_x} ds, \qquad J_{p,z} = \frac{1}{2\pi}\oint \beta_z \Delta K_z e^{-jp\phi_z} ds. \tag{2.206}$$

We consider a lattice made of P superperiods, where L is the length of a superperiod with $K(s+L) = K(s)$, $\beta(s+L) = \beta(s)$. Let $C = PL$ be the circumference of the accelerator. The integral of Eq. (2.206) becomes

$$\begin{aligned} J_{p,y} &= -\left\{\frac{\delta}{2\pi}\int_0^L \beta_y K_y e^{-jp\phi} ds\right\}\left[1 + e^{-jp\frac{2\pi}{P}} + e^{-j2p\frac{2\pi}{P}} + e^{-j3p\frac{2\pi}{P}} + \cdots\right] \\ &= -\left\{\frac{\delta}{2\pi}\int_0^L \beta_y K_y e^{-jp\phi} ds\right\}\ \zeta_P(\frac{p}{P})\ e^{-j\pi p\frac{P-1}{P}}, \qquad (2.207) \\ \zeta_P(u) &= \frac{\sin(Pu\pi)}{\sin(u\pi)}. \qquad (2.208) \end{aligned}$$

where y stands for either x or z. The diffraction-function $\zeta_P(u) \to P$ as $u \to$ integer. Thus the stopband integral $J_{p,y} = 0$ unless $p = 0$ (Mod P). At $p = 0$ (Mod P), the half-integer stopband integral increases by a factor of P, i.e. each superperiod contributes additive to the chromatic stopband integral.

Since the perturbation of betatron functions is most sensitive to the chromatic stopbands near $p \approx [2\nu_x]$ and $[2\nu_z]$, a basic design principle of strong-focusing synchrotrons is to avoid important systematic chromatic stopbands. This can be achieved by choosing the betatron tunes such that $[2\nu_x]$ and $[2\nu_z]$ are not divisible by the superperiod P. For example, the AGS lattice has $P = 12$, and the betatron tune should avoid a value of 6, 12, 18, etc. The actual betatron tunes at $\nu_{x/z} = 8.8$ are indeed far from systematic half-integer stopbands at $p = 6$ and 12, and the resulting chromatic perturbation is small. In fact, the AGS lattice can be approximated by a lattice made of 60 FODO cells. The important stopbands are located at $p = 30, 60, 90\cdots$, which are far from the betatron tunes. Similarly, the TEVATRON has a super-periodicity of $P = 6$, and the betatron tune should avoid 18, 24, 30, etc.

Generally, it is beneficial to design an accelerator with high super-periodicity so that the betatron tunes can be located far from the important chromatic stopbands. Some examples of high superperiod machines are $P = 12$ for the ALS, $P = 40$ for the APS, $P = 36$ for the ESRF, and $P = 44$ for the SPRING-8 at JSRF. However, a high energy accelerator or storage ring with large super-periodicity is costly. Thus the goal is to design an accelerator such that the chromatic stopband integral of each module is zero, or stopband integrals of two modules cancel each other.

B. Chromatic stopband integrals of FODO cells

Now we examine the chromatic stopband integral of the arc, which is composed of N FODO cells. The chromatic stopband integral in thin-lens approximation is

$$\begin{aligned} J_p &= -\frac{\delta}{2\pi}\left(\frac{\beta_{\max}}{f} - \frac{\beta_{\min}}{f}e^{-jp\frac{\Phi}{2\nu}}\right)\left[1 + e^{-jp\frac{\Phi}{\nu}} + e^{-j2p\frac{\Phi}{\nu}} + e^{-j3p\frac{\Phi}{\nu}} + \cdots\right] \\ &= -\frac{2\delta}{\pi\cos\frac{\Phi}{2}}\left(\sin\frac{\Phi}{2}\cos\frac{p\Phi}{4\nu} + j\sin\frac{p\Phi}{4\nu}\right)\ \zeta_N(\frac{p\Phi}{2\pi\nu})\ e^{-j\frac{(2N-1)p\pi}{2N}}, \end{aligned}$$

where Φ is the phase advance per cell, $\beta_{\max}$ and $\beta_{\min}$ are values of the betatron amplitude function at the focusing and defocusing quadrupoles respectively, f is the focal length of each quadrupole, and the diffracting function $\zeta_N(u)$ is given by Eq. (2.208). If $p\Phi/2\pi\nu = 0$ (Mod N), the diffracting function is equal to N. This means that each FODO cell contributes additive to the stopband integral. Fortunately, since $\Phi/2\pi$ is normally about 1/4 (90^0 phase advance) so that $p\Phi/2\pi\nu \approx p/4\nu \approx 1/2$, the chromatic stopband integral at $p \approx 2\nu$ due to N FODO cells is small. If $N\Phi =$ integer $\times\pi$, the transfer matrix of the arc is a unit matrix I or a half-unit matrix $-I$ and the chromatic stopband of the arc adds up to zero at harmonics $p \approx 2\nu$.

C. The chromatic stopband integral of insertions

Because of its small β^* value, the insertion may contribute a substantial amount to the chromatic stopband integral. The high-β triplets or doublets on both sides of the IP contribute additive to the systematic half-integer stopband near $p \approx 2\nu_{x/z}$. Since it is difficult to design an insertion with zero chromatic half-integer stopband width, cancellation of the chromatic stopband integrals between two adjacent insertions is desirable.

Let $\Phi^{\rm ins}$ and $J_p^{\rm ins}$ be respectively the phase advance and the chromatic stopband integral of an insertion. The total contribution of two adjacent insertions becomes

$$J_p = J_p^{\rm ins}\left[1 + \exp\left(j\frac{p\Phi^{\rm ins}}{\nu}\right)\right].$$

At the harmonic $p \approx [2\nu]$, we obtain $J_p = 0$ if $\Phi^{\rm ins} = (2n+1)\pi/2$. The chromatic stopband integrals of two adjacent insertions cancel each other for two adjacent quarter-wave modules. This cancellation principle remains valid when two insertions are separated by a unit transfer matrix. Such a procedure was extensively used in the design of RHIC and SSC lattice.[69]

D. Effect of the chromatic stopbands on chromaticity

The chromatic stopband integrals for large colliders, such as the SSC and RHIC, remain important even after careful manipulation of piecewise cancellation, particularly when the beam momentum spread is large. They give rise to a large betatron amplitude modulation, called betabeat, and second-order chromaticity for off-momentum particles. The following example illustrates the effect of betatron amplitude function modulation on chromaticity.

[69] S.Y. Lee, J. Claus, E.D. Courant, H. Hahn, G. Parzen, *IEEE Trans. Nucl. Sci.* **NS-32**, 1626 (1985); S.Y. Lee, G.F. Dell, H. Hahn, G. Parzen, *Proc 1987 Part. Accel. Conf.*, p. 1328, (1987); A. Garren, private communications; see also SSC reports.

We consider a lattice dominated by a single p harmonic half-integer chromatic stopband. The beta-beat of the lattice is

$$\frac{\Delta\beta}{\beta} \approx -\frac{|J_p|\cos(p\phi+\chi)}{2(\nu - p/2)},$$

where the chromatic stopband integral J_p given by Eq. (2.206), is proportional to δ. Substituting $\beta = \beta_0(1 + \Delta\beta/\beta_0)$ into Eq. (2.199), we obtain

$$\begin{aligned}
\Delta\nu_y &= C_y^{(1)}\delta + C_y^{(2)}\delta^2 + \cdots, \\
C_y^{(1)} &= -\frac{1}{4\pi}\oint \beta_y(K_y - S_y D)ds, \\
C_y^{(2)} &= -C_y^{(1)} - \frac{|J_{p,y}|^2/\delta^2}{4(\nu_y - p/2)},
\end{aligned}$$

where y stands for either x or z, If the first-order chromaticity is corrected, then $C_y^{(1)} = 0$. The remaining second-order tune shift $C_y^{(2)}\delta^2$ can arise from the chromatic stopband integral. Figure 2.43 shows an example of the second-order chromatic tune shift with δ. The stopband correction that minimizes the β-beat also minimizes the second-order chromaticity.

E. Effect of sextupoles on the chromatic stopband integrals

The chromatic sextupoles also contribute to the systematic chromatic stopbands. Here we present an example of chromatic correction for a collider lattice. First we evaluate the stopband integral due to the chromatic sextupoles. Let $S_{\rm F}$ and $S_{\rm D}$ be the integrated sextupole strength at QF and QD of FODO cells in the arc. The p-th harmonic stopband integral from these chromatic sextupoles is

$$J_{p,\rm sext} = \frac{\delta}{2\pi}\zeta_N\left(\frac{p\Phi}{2\pi\nu}\right)\left[\beta_{\rm F}S_{\rm F}D_{\rm F} + \beta_{\rm D}S_{\rm D}D_{\rm D}e^{-jp\Phi/2\nu}\right]e^{-j(N-1)p\Phi/2\nu}, \tag{2.209}$$

where N is the number of cells, and the diffraction function ζ_N is given by Eq. (2.208). As in Eq. (2.207), the stopband integral is zero or small if $N\Phi/\pi$ = integer, i.e. the chromatic sextupole does not contribute significantly to the chromatic stopband integral if the transfer matrix of the arc is I or $-I$.

To obtain a nonzero chromatic stopband integral, sextupoles are organized in families. We consider an example of a four-family scheme with

$$\{S_{\rm F1} = S_{\rm F} + \Delta_{\rm F},\ S_{\rm D1} = S_{\rm D} + \Delta_{\rm D},\ S_{\rm F2} = S_{\rm F} - \Delta_{\rm F},\ S_{\rm D2} = S_{\rm D} - \Delta_{\rm D}\},$$

that is commonly used in FODO cells with 90° phase advance. Here the parameters $S_{\rm F}$ and $S_{\rm D}$ are determined from the first-order chromaticity correction, Since $\beta(s)$ and $D(s)$ are periodic functions of s in the repetitive FODO cells, the parameters

Δ_F, Δ_D will not affect the first-order chromaticity, which is proportional to the zeroth harmonic of the stopband integral. However, the chromatic stopband integrals due to the parameters Δ_F and Δ_D are

$$\Delta J_{p,\rm sext} = \frac{\delta}{2\pi}\zeta_N(\frac{p\Phi}{2\nu\pi} - \frac{1}{2})\left[\beta_F\Delta_F D_F + \beta_D\Delta_D D_D e^{-jp\Phi/4\nu}\right] e^{-j(N-1)[(p\Phi/2\nu\pi)-(1/2)]\pi}.$$

At $p \approx [2\nu]$ and $\Phi/2\pi \approx 1/4$ (90° phase advance), we have $\zeta_N \to N$, i.e. every FODO cell contributes additive to the chromatic stopband. The resulting stopband width is proportional to Δ_F and Δ_D parameters. By adjusting Δ_F and Δ_D parameters, the betabeat and the second-order chromaticity can be minimized. The scheme works best for a nearly 90° phase advance per cell with $N\Phi$ = integer $\times\,\pi$, where the third-order resonance-driving term vanishes also for the four-family sextupole scheme. Fig. 2.43 shows an example of chromatic correction with four families of sextupoles in RHIC, where the second-order chromaticity and the betatron amplitude modulation can be simultaneously corrected.

Similarly, the six-family sextupole scheme works for 60° phase advance FODO cells, where the six-family scheme: $\{S_{F1}, S_{D1}, S_{F2}, S_{D2}, S_{F3}, S_{D3}\}$ has two additional parameters.

V.4 Lattice Design Strategy

Based on the linear betatron motion of previous Sections, the lattice design of accelerator can be summarized as follows. The lattice is generally classified into three categories: low energy booster, collider lattice, and low-emittance lattice storage rings.

- The betatron tunes should be chosen to avoid systematic integer and half-integer stopbands and systematic low-order nonlinear resonances.
- The chromatic sextupoles should be located at high dispersion function locations. The focusing and defocusing sextupole families should be located in regions where $\beta_x \gg \beta_z$, and $\beta_x \ll \beta_z$ respectively in order to gain independent control of the chromaticities.
- The chromatic aberration can be corrected by sextupole families via half-integer stopband integrals. In colliders, chromatic sextupoles located at high luminosity mini-β insertion region can provide effective chromatic-aberration correction.
- The betatron amplitude function and the betatron phase advance between the kicker and the septum should be optimized to minimize the kicker angle and maximize the injection or extraction efficiency. Local orbit bumps can be used to alleviate the demand for a large kicker angle. Furthermore, the injection line and the synchrotron optics should be properly "matched" or "mismatched" to optimize the emittance control. To improve the slow extraction efficiency,

the β value at the (wire) septum location should be optimized. The β_x and β_z values at the injection area, particularly in the strip injection scheme, should be adjusted to minimize emittance blow-up due to multiple Coulomb scattering. The local vacuum pressure at the high-β value locations should be minimized to minimize the effect of beam gas scattering.

- It is advisable to avoid the transition energy for low to medium energy synchrotrons in order to minimize the beam dynamics problems during acceleration.
- Experience with low energy synchrotrons indicates that the Laslett space-charge tune shift should be limited to about 0.3 (see Exercise 2.3.2). This criterion usually determines beam emittance and intensity.

Besides these design issues, problems regarding the dynamical aperture, nonlinear betatron detuning, collective beam instabilities, rf system, vacuum requirement, beam lifetime, etc., should be addressed. Some of these issues will be addressed in this introductory textbook. The design of minimum emittance electron storage rings will be discussed in Chap. 4, Sec. III.

Exercise 2.5

1. Show that the chromaticity of an accelerator consisting of N FODO cells in thin-lens approximation is

$$C_{\rm nat}^{\rm FODO} = -\frac{\tan(\Phi/2)}{\Phi/2}\nu,$$

where Φ is the phase advance per cell and $\nu = N\Phi/2\pi$ is the betatron tune.

2. A set of three quadrupoles ($\{{\rm Q_{F1}\ Q_{D2}\ Q_{F3}}\}$ or $\{{\rm Q_{D1}\ Q_{F2}\ Q_{D3}}\}$), called a low-$\beta$ triplet, is commonly used in insertion regions to provide horizontal and vertical low-β squeeze.

   (a) Show that the low-beta triplets contribute about

$$-\frac{2\Delta s}{4\pi\beta^*} \approx -\frac{1}{2\pi}\sqrt{\frac{\beta_{\rm max}}{\beta^*}}$$

   units of natural chromaticity, where Δs is the effective distance between the triplet and the interaction point (IP), β^* is the value of the betatron amplitude function at IP, and $\beta_{\rm max}$ is the maximum betatron amplitude function at the triplet.

   (b) If $\beta_{\rm max} \gg \beta^*$, show that the betatron phase advance between the triplet and IP is $\pi/2$.

   (c) Show that the triplets on both sides of IP contribute additive to the stopband integral at $p \approx 2\nu$, where ν is the betatron tune.

3. Show that the strengths of two sextupole families used to correct the chromaticities of FODO cells are

$$S_{\rm F} = \frac{1}{2f^2\theta}\frac{\sin(\Phi/2)}{(1+\frac{1}{2}\sin(\Phi/2))}; \quad S_{\rm D} = -\frac{1}{2f^2\theta}\frac{\sin(\Phi/2)}{(1-\frac{1}{2}\sin(\Phi/2))},$$

where f is the focal length of the quadrupole in the FODO cell, θ is the dipole bending angle of a half FODO cell, and Φ is the phase advance of the FODO cell. Note that the required sextupole strength is larger at the defocusing quadrupole.

4. Show that the chromatic stopband integrals for a lattice made of N FODO cells in thin-lens approximation are

$$J_{p,x} = -\frac{\delta}{2\pi}\left(\frac{\beta_x({\rm F})}{f_{\rm F}} - \frac{\beta_x({\rm D})}{f_{\rm D}}e^{-jp\Phi_x/2\nu_x}\right)\zeta_N(\frac{p}{N})e^{-jp\pi(N-1)/N},$$

$$J_{p,z} = -\frac{\delta}{2\pi}\left(-\frac{\beta_z({\rm F})}{f_{\rm F}} + \frac{\beta_z({\rm D})}{f_{\rm D}}e^{-jp\Phi_z/2\nu_z}\right)\zeta_N(\frac{p}{N})e^{-jp\pi(N-1)/N},$$

where Φ_x, Φ_z, and ν_x, ν_z are the phase advances per cell and the betatron tunes, β_x, β_z are betatron amplitude functions, $f_{\rm F}$ and $f_{\rm D}$ are focal lengths for focusing and defocusing quadrupoles, and the diffraction function $\zeta_N(u)$ is given by Eq. (2.208). Assuming $f_{\rm F} = f_{\rm D}$ with $\Phi_x = \Phi_z = \Phi = 2\pi\nu/N$, show that the chromatic stopband integral is

$$J_{p,x} = -\frac{2\delta}{\pi\cos\frac{\Phi}{2}}\left[\sin\frac{\Phi}{2}\cos\frac{p\pi}{2N} + j\sin\frac{p\pi}{2N}\right]\zeta_N(p/N)e^{-jp\pi(2N-1)/2N}.$$

5. Verify Eq. (2.209).

6. The AGS is composed of 12 superperiods with 5 nearly identical FODO cells per superperiod. The betatron tunes are $\nu_z = 8.8$ and $\nu_x = 8.7$. Calculate the systematic stopband widths for harmonics 17 and 18 respectively. What region of betatron tunes should be avoided to minimize the effect of systematic stopbands?

7. The Fermilab booster is a combined function synchrotron. The lattice is made of 24 cells, as shown below.

   FNALBSTCELL : LINE = (BF S120 BF S050 BD S600 BD S050)
   BF : SBEND L = 2.889612 K1 = 0.0542203 ANGLE = 0.070742407
   BD : SBEND L = 2.889612 K1 = −0.0577073 ANGLE = 0.060157561
   Sabc : DRIFT L = a.bc

   Find the systematic stopband width and discuss the choice of the betatron tunes.

8. Use the experimental data below to calculate the chromaticity of the IUCF cooler ring, where $\gamma_{\rm T} = 4.6$ and $C = 86.82$ m, at 45 MeV proton kinetic energy.

Betatron tunes vs revolution frequencies of the cooler

Frequency [MHz]	1.032680	1.031680	1.030680
Q_x	3.7156	3.7243	3.7364
Q_z	4.6790	4.6913	4.7080

VI Linear Coupling

We have discussed uncoupled linear betatron motion, but in reality betatron motions are coupled through solenoidal and skew-quadrupole fields. The solenoidal field exists in electron cooling storage rings, and in high-energy detectors at the interaction point (IP). The skew-quadrupole field arises from quadrupole roll, vertical closed-orbit error in sextupoles or horizontal closed-orbit error in skew sextupoles, fringe field of a Lambertson septum, and feed-downs from higher-order multipoles.

Linear betatron coupling is both a nuisance and a benefit in the operation of synchrotrons: the available dynamical aperture for particle motion may be reduced, but the vertical emittance of electron beams in storage rings can be adjusted, and the Touschek lifetime limitation can be alleviated by linear coupling.

This Section discusses beam dynamics associated with linear betatron coupling arising from skew quadrupoles and solenoids. The effective linear coupling Hamiltonian and resonance strength will be derived based on perturbation approximation. Here we find that the linear coupling can induce energy exchange between horizontal and vertical betatron motions. Furthermore, we show, in Sec. IV, that a skew quadrupole at a high horizontal dispersion location can produce vertical dispersion, which can generate vertical emittance for electron beams and result in lower luminosity for colliders (see Exercise 2.4.10). Thus measurement and correction of linear coupling are important.

VI.1 The Linear Coupling Hamiltonian

The vector potentials for skew quadrupoles and solenoids are given by

$$\begin{cases} A_x = A_z = 0, \quad A_s = \frac{1}{2}(\frac{\partial B_z}{\partial z} - \frac{\partial B_x}{\partial x})xz, & \text{for skew quadrupoles,} \\ A_x = \frac{1}{2}B_\parallel(s)z, \quad A_z = -\frac{1}{2}B_\parallel(s)x, \quad A_s = 0, & \text{for solenoids,} \end{cases} \tag{2.210}$$

where $B_\parallel(s)$ and $\frac{1}{2}(\frac{\partial B_z}{\partial z} - \frac{\partial B_x}{\partial x})$ are solenoid field strength and skew-quadrupole gradient. Substituting the components of the vector potential into the Hamiltonian in Eq. (2.15), we obtain the linearized equations of motion (see Exercise 2.6.3):

$$\begin{cases} x'' + K_x(s)x + 2gz' - (q - g')z = 0, \\ z'' + K_z(s)z + 2gx' - (q + g')x = 0, \end{cases} \tag{2.211}$$

where the primes are derivatives with respect to independent variable s, K_x and K_z are quadrupole-like focusing functions, and

$$g(s) = \frac{B_\parallel(s)}{2B\rho}, \quad q(s) = \frac{1}{2B\rho}\left[\frac{\partial B_z}{\partial z} - \frac{\partial B_x}{\partial x}\right] = \frac{a_1}{\rho} \tag{2.212}$$

are effective solenoid and skew quadrupole strengths. The skew quadrupole can also arise from "feed-down" of an off-centered vertical closed orbit in sextupoles. Let $z_{\rm co}$ be the closed orbit at a sextupole with sextupole strength $B_2 = \partial^2 B_z/\partial x^2$. The effective skew quadrupole strength becomes $q = B_2 z_{\rm co}/B\rho$.

Let $(x, p_x/p; z, p_z/p)$ be the conjugate phase space coordinates. The betatron motion of Eq. (2.211) can be derived from the linearized Hamiltonian: $\tilde{H} = \tilde{H}_0 + V_{\rm lc}$, where

$$\tilde{H}_0 = \frac{1}{2}\left[\left(\frac{p_x}{p}\right)^2 + \left(K_x(s) + g^2(s)\right)x^2 + \left(\frac{p_z}{p}\right)^2 + \left(K_z(s) + g^2(s)\right)z^2\right]$$

$$V_{\rm lc} = -q(s)xz - g(s)\left[\frac{p_x}{p}z - \frac{p_z}{p}x\right]. \tag{2.213}$$

When the linear coupling potential $V_{\rm lc}$ is small, we carry out perturbation expansion based on betatron motion of $\tilde{H}_0$. Applying Floquet transformation of Eq. (2.55) to the uncoupled Hamiltonian $\tilde{H}_0$, we obtain the coupling-potential:

$$\begin{aligned} V_{\rm lc} &= (\beta_x\beta_z J_x J_z)^{1/2}\left\{\left[-q + g\left(\frac{\alpha_x}{\beta_x} - \frac{\alpha_z}{\beta_z}\right)\right][\cos(\Phi_x + \Phi_z) + \cos(\Phi_x - \Phi_z)]\right. \\ &\quad \left. + g\left(\frac{1}{\beta_x} - \frac{1}{\beta_z}\right)\sin(\Phi_x + \Phi_z) + g\left(\frac{1}{\beta_x} + \frac{1}{\beta_z}\right)\sin(\Phi_x - \Phi_z)\right\}, \end{aligned} \tag{2.214}$$

where $\Phi_x = \phi_x + \chi_x(s) - \nu_x\theta$, $\chi_x = \int_0^s \frac{ds}{\beta_x}$, $\Phi_z = \phi_z + \chi_z(s) - \nu_z\theta$, $\chi_z = \int_0^s \frac{ds}{\beta_z}$.The conjugate-pair action-angle coordinates are (J_x, ϕ_x) and (J_z, ϕ_z) The Hamiltonian produces two resonances with driving terms listed in Table 2.2.

Table 2.2: Linear coupling resonances and their driving terms

Resonance	Driving phase	Amplitude-dependent factor	Classification
$\nu_x + \nu_z = \ell$	$(\Phi_x + \Phi_z)$	$J_x^{1/2} J_z^{1/2}$	sum resonance
$\nu_x - \nu_z = \ell$	$(\Phi_x - \Phi_z)$	$J_x^{1/2} J_z^{1/2}$	difference resonance

Since $V_{\rm lc}(s)$ is a periodic function of the "time coordinate" $\theta = s/R$, it can be expanded in Fourier harmonics as

$$V_{\rm lc}(\theta) = \frac{\sqrt{J_x J_z}}{2R}\sum_\ell \left\{G_{1,-1,\ell}e^{j(\phi_x - \phi_z - \ell\theta + \chi_{1,-1,\ell})} + c.c + G_{1,1,\ell}e^{j(\phi_x + \phi_z - \ell\theta + \chi_{1,1,\ell})} + c.c\right\},$$

where R is the average radius of accelerator, ℓ is the integer Fourier harmonic. The Fourier coefficients of the difference and sum resonances, $G_{1,\mp1,\ell}e^{j\chi_{1,\mp1,\ell}}$, are

$$\begin{aligned} G_{1,\pm1,\ell}\, e^{j\chi_{1,\pm1,\ell}} &= \frac{1}{2\pi}\oint \sqrt{\beta_x\beta_z}\, A_{\rm lc\pm}(s)\, e^{j[\chi_x \pm \chi_z - (\nu_x \pm \nu_z - \ell)\theta]} ds \\ A_{\rm lc\pm}(s) &= -\frac{a_1}{\rho} + g(s)\left(\frac{\alpha_x}{\beta_x} - \frac{\alpha_z}{\beta_z}\right) + jg(s)\left(\frac{1}{\beta_x} \mp \frac{1}{\beta_z}\right), \end{aligned} \tag{2.215}$$

where ν_x, ν_z are the unperturbed betatron tunes, and $A_{\rm lc\pm}(s)$ is the linear coupling kernel of the linear coupling potential Eq. (2.214). Both skew quadrupoles and solenoids can drive the sum and difference linear coupling resonances. We choose the resonance phase $\chi_\pm$ so that the Fourier amplitude (also called resonance strength) is $G_{1,\pm1,\ell} \geq 0$.

The linear coupling potential has been decomposed into terms of the difference and sum resonances located respectively at $\nu_x - \nu_z = \ell$ and $\nu_x + \nu_z = \ell'$, where ℓ and ℓ' are integers. In general, the coupling betatron sum-resonances are dangerous to the stable betatron motion. We will show, in Sec. VII, that the horizontal and vertical betatron amplitudes can grow without bound near a betatron sum resonance. Thus the betatron tunes are normally designed to avoid sum resonances.

If the linear coupling kernel $A_{\rm lc\pm}$ satisfies a periodic condition similar to that in a synchrotron with P superperiods, the resonance coupling coefficient $G_{1,\pm1,\ell}$ will be zero unless ℓ is an integer multiple of P. If ℓ is an integer multiple of P, each superperiod contributes additive to the linear coupling resonance strength. This is called the systematic linear coupling resonance. For example, since the superperiodicity of the LEP lattice is 8, the difference between the integer part of the horizontal and vertical betatron tunes should *not* be 0, 8, 16, $\cdots$, to minimize the effect of the systematic linear coupling resonance. The strength of the linear coupling resonance due to random errors such as quadrupole roll and vertical closed orbit in sextupoles is smaller. It occurs at all integer ℓ.

Near a difference linear coupling resonance, the horizontal and vertical betatron motions are coupled. The coupling resonance can cause beam size increase and decrease the beam lifetime. Thus the linear-coupling resonance-strength should be minimized, and the resonance strength is usually small. Thus the effective Hamiltonian for betatron tunes near an isolated coupling resonance will be discussed in the following sections.

VI.2 Effects of an Isolated Linear Coupling Resonance

Since the betatron tunes are normally near a linear coupling line and the resonance strength $G \equiv G_{1,-1,\ell}$ is also normally small, the effects of the linear coupling resonance on betatron motion can be studied in perturbation theory, by considering only an isolated coupling resonance. Near an isolated coupling resonance $\nu_x - \nu_z = \ell$, the Hamiltonian Eq. (2.213), in action-angle phase space coordinates, can be approximated by

$$H \approx \nu_x J_x + \nu_z J_z + G\sqrt{J_x J_z}\cos(\phi_x - \phi_z - \ell\theta + \chi), \tag{2.216}$$

where, for simplicity, we use the Fourier amplitude $G \equiv G_{1,-1,\ell} \geq 0$ and the phase factor $\chi \equiv \chi_{1,-1,\ell}$ of Eq. (2.215) hereafter in this chapter.

A. Normal modes at a single linear coupling resonance

The Hamiltonian Eq. (2.216) corresponds to two coupled linear oscillators, which can be expanded in terms of two normal modes with tunes (see Exercise 2.6.5)

$$\nu_{1,\pm} = \frac{1}{2}(\nu_x + \nu_z + \ell) \pm \frac{1}{2}\lambda, \quad \nu_{2,\pm} = \frac{1}{2}(\nu_x + \nu_z - \ell) \pm \frac{1}{2}\lambda; \tag{2.217}$$

$$\lambda = \sqrt{(\nu_x - \nu_z - \ell)^2 + G^2}. \tag{2.218}$$

The betatron tunes are separated by λ, and the minimum separation between the normal mode tunes is G. Figure 2.44 shows an example of measured betatron tunes vs quadrupole strength at IUCF cooler ring. By varying the strength of a focusing quadrupole, one changes the resonance proximity parameter $\delta = \nu_x - \nu_z + 1$. The normal mode tunes approach each other, reaching a minimum value of normal mode tune-separation, G as demonstrated in Fig. 2.44. This method is commonly used to measure the linear coupling strength and to correct linear coupling by minimizing tune-split with skew quadrupoles.

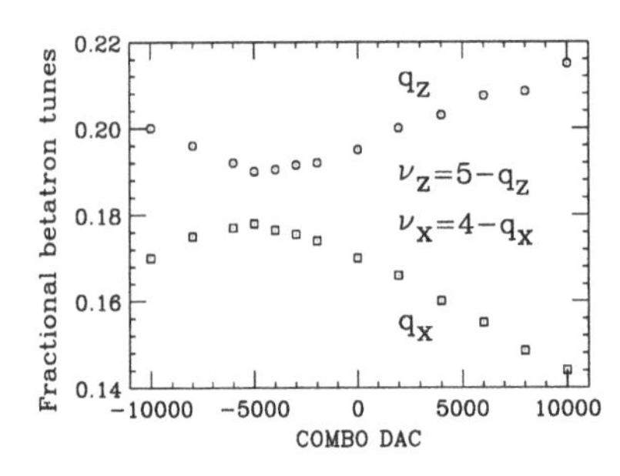

Figure 2.44: The measured betatron normal mode tunes vs the strength of an IUCF cooler quadrupole, showing that the horizontal and vertical motion are coupled. The minimum distance between two normal modes is equal to the coupling strength G. The vertical axis is the fractional part of the betatron tunes, and the horizontal axis is the digital to analog conversion (DAC) unit of a COMBO power supply for a set of horizontally focusing quadrupoles.

B. Resonance precessing frame and Poincaré surface of section

To study the linear coupling, we transform the Hamiltonian Eq. (2.216) into a "resonant precessing frame" by using the generating function:

$$F_2(\phi_x, \phi_z, J_1, J_2) = (\phi_x - \phi_z - \ell\theta + \chi)J_1 + \phi_z J_2,$$
$$\phi_1 = \phi_x - \phi_z - \ell\theta + \chi, \quad \phi_2 = \phi_z, \quad J_1 = J_x, \quad J_2 = J_x + J_z.$$

The new Hamiltonian is

$$H = H_1(J_1, \phi_1, J_2) + H_2(J_2), \tag{2.219}$$
$$H_1 = \delta_1 J_1 + G\sqrt{J_1(J_2 - J_1)}\cos\phi_1, \qquad H_2(J_2) = \nu_z J_2,$$

where $\delta_1 = \nu_x - \nu_z - \ell$ is the resonance proximity parameter.

Hamilton's equations of motion are

$$\begin{cases} \dot{J}_1 = -\dfrac{\partial H}{\partial \phi_1} = G\sqrt{J_1(J_2 - J_1)}\sin\phi_1, \\ \dot{\phi}_1 = \dfrac{\partial H}{\partial J_1} = \delta_1 + G\dfrac{J_2 - 2J_1}{2\sqrt{J_1(J_2 - J_1)}}\cos\phi_1, \end{cases} \qquad \begin{cases} \dot{J}_2 = -\dfrac{\partial H}{\partial \phi_2}, \\ \dot{\phi}_2 = \dfrac{\partial H}{\partial J_2}. \end{cases} \tag{2.220}$$

Since $\dot{J}_2 = -\partial H/\partial\phi_2 = 0$, we find

$$J_x + J_z = J_2 = \text{constant}. \tag{2.221}$$

The horizontal and vertical betatron motions exchange their actions while the sum of actions is conserved. The Hamiltonian $H_1(J_1\phi_1)$ is autonomous, i.e. independent of the time coordinate θ, the Hamiltonian is itself a constant of motion. For a given J_2, all tori can be described by a single parameter $H_1(J_1, \phi_1, J_2) = E_1$, which is determined by the initial condition. The system is integrable with two invariants J_2 and $H_1 = E_1$.

C. Initial horizontal orbit

We first consider a simple orbit with "energy" $E_1 = \delta_1 J_2$, which corresponds to a beam with an initial horizontal betatron kick. The particle trajectory obeys $H_1 = \delta_1 J_2$, or

$$\sqrt{J_2 - J_1}\left[\delta_1\sqrt{J_2 + J_1} - G\sqrt{J_1}\cos\phi_1\right] = 0 \quad \Longrightarrow \quad \begin{cases} P^2 + Q^2 = 2J_2, \\ Q^2 + \dfrac{\delta_1^2}{\lambda^2}P^2 = 2\dfrac{\delta_1^2}{\lambda^2}J_2, \end{cases} \tag{2.222}$$

where $\lambda = \sqrt{\delta_1^2 + G^2}$, and $(Q = \sqrt{2J_1}\cos\phi_1, P = -\sqrt{2J_1}\sin\phi_1)$ are phase space coordinates in the resonance rotating frame. The phase space portrait can be decomposed into two ellipses: a Courant Synder circle and a coupling ellipse. Figure 2.45 shows a schematic plot of the Courant-Snyder circle and the coupling ellipse of Eq (2.222). The coupling line is equivalent to the coupling line in the betatron tune space.

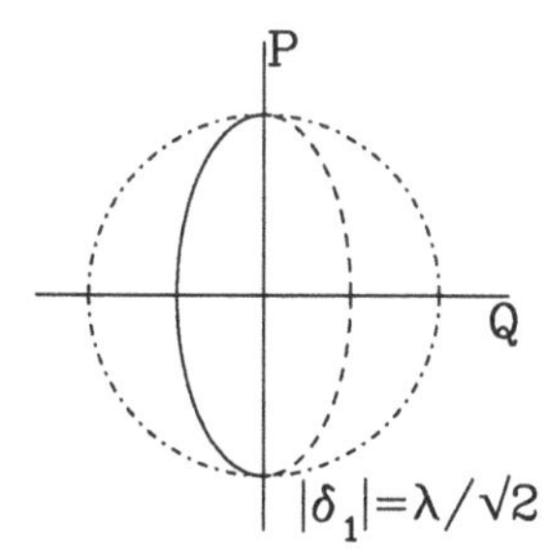

Figure 2.45: Schematic drawing of the Courant-Snyder circle (dashed-dots) of Eq. (2.222) and the coupling ellipse of (2.222) with $|\delta_1| = \lambda/\sqrt{2}$. Particle motion on the coupling ellipse follows the path of solid line for $\delta_1 < 0$ and the dashed line for $\delta_1 > 0$. At $\delta_1 = 0$, the coupling line is the vertical line at Q=0.

When the particle trajectory moves along the Courant-Snyder circle shown in Eq. (2.222) with $J_1 = J_2$, the phase ϕ_1 varies very rapidly. As the betatron oscillation reaches $(Q = 0,\ P = \sqrt{2J_2})$, particle trajectory follows the coupling ellipse of Eq. (2.222), which is inside the Courant-Snyder circle. The minimum horizontal amplitude is

$$|Q_{\min}| = \frac{|\delta_1|}{\lambda}\sqrt{2J_2}. \tag{2.223}$$

If $|\delta_1| \gg G$, then $|Q_{\min}| \approx \sqrt{2J_2}$ and the betatron coupling is negligible. If $\delta_1 = 0$, the coupling ellipse becomes a straight line cutting through the origin $Q = 0$ and $P = 0$. This means that the horizontal action can be fully converted to vertical action and vice versa.

D. General linear coupling solution

The (stable) fixed points of the Hamiltonian are determined by the conditions $\dot{J}_1 = 0$ and $\dot{\phi}_1 = 0$. They are located at $\phi_1 = 0$ or π with

$$\delta_1 \pm G\frac{J_2 - 2J_1}{2\sqrt{J_1(J_2 - J_1)}} = 0, \qquad \text{or} \qquad J_{1,\text{sfp}} = \begin{cases} (\frac{1}{2} + \delta_1/2\lambda)J_2, & (\phi_1 = 0) \\ (\frac{1}{2} - \delta_1/2\lambda)J_2, & (\phi_1 = \pi) \end{cases} \tag{2.224}$$

At SFPs, the horizontal and vertical betatron motions are correlated in phase without exchange in betatron amplitudes. Figure 2.46 shows 6 Poincaré surfaces of section in the resonance rotating frame with a given value of $J_2 = J_x + J_z$. The results are obtained from simple tracking calculations of particle motion in a synchrotron with perfect linear decoupled betatron motion everywhere except a localized skew quadrupole kick, where the betatron tunes are $\nu_x = 4.820$, $\nu_z = 4.825$, i.e. the resonance proximity parameter is $\delta_1 = -0.005$. The integrated skew quadrupole strength is $a_1\Delta s/\rho = 0.00628\ \text{m}^{-1}$. Using the values of betatron amplitude functions are $\beta_x = 10$ m and $\beta_z = 10$ m at the skew quadrupole location, we find the effective resonance strength is about $G = 0.010$.

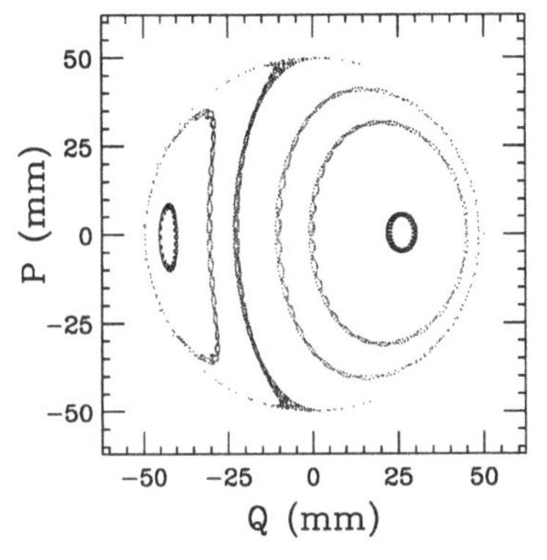

Figure 2.46: Normalized phase space ellipses of P vs Q in the resonance rotating frame obtained from numerical simulations of particle motion in a synchrotron with linear betatron motion and a localized skew quadrupole kick. The values of the betatron amplitude functions at the skew quadrupole location is $\beta_x = 10$ m, and $\beta_z = 10$ m; the betatron tunes of the machine are $\nu_x = 4.820$ and $\nu_z = 4.825$, i.e. $\delta_1 = -0.005$; the integrated skew quadrupole strength is $(a_1\Delta s)/\rho = 0.00628\ \text{m}^{-1}$. These ellipses correspond to various initial J_1 and ϕ_1 values with $J_2 = 90\pi$ mm-mrad. Note that the structure of the phase space ellipses remains the same if J_2 is varied.

With the coupling ellipse Eq. (2.222) rewritten as $G\sqrt{J_1}\cos\phi_1 = \delta_1\sqrt{J_2 - J_1}$, the Courant-Snyder ellipse Eq. (2.222) is divided into two halves (see Figs. 2.45 and 2.46).

Using Hamilton's equations [Eq. (2.220)], we obtain

$$\ddot{J}_1 + \lambda^2 J_1 = \lambda^2 \bar{J}, \tag{2.225}$$

where $\bar{J} = (2\delta_1 E + G^2 J_2)/2\lambda^2$ and $E = \delta_1 J_1 + G\sqrt{J_1(J_2 - J_1)}\cos\phi_1$. The solution of Eq. (2.225) is pure sinusoidal:

$$J_1 = \sqrt{\bar{J}^2 - (E/\lambda)^2}\cos[\lambda\theta + \varphi] + \bar{J}, \tag{2.226}$$

where $|E| \le \lambda\bar{J}$, and φ is an initial phase factor. The SFPs of Eq. (2.224) correspond to the orbit with $E = \pm\lambda\bar{J} = \pm\lambda J^{\pm}_{1,\mathrm{sfp}}$. The tune of the linear coupling motion is independent of betatron amplitude.[70]

If particles in a given bunch distribution have identical betatron tunes, linear coupling can cause bunch shape oscillations; the bunch will resume its original shape after λ^{-1} revolutions. But, if particles have different betatron tunes, they will orbit around different fixed points at different island tunes, and the motion will decohere after some oscillation periods.

VI.3 Experimental Measurement of Linear Coupling

To measure the effect of linear coupling, the horizontal and vertical betatron tunes are tuned to the linear coupling resonance line at $\nu_x - \nu_z = \ell$. In the following, we discuss an experimental study of linear coupling at the IUCF cooler ring. The cooler is a proton storage ring with electron cooling. The circumference is about 86.82 m, and the betatron tunes for this experiment were chosen to be $\nu_x = 3.826, \nu_z = 4.817$ with $\nu_x - \nu_z \approx -1$.

The experiment started with a single bunch of about 5×10^8 protons with kinetic energy of 45 MeV at the Indiana University Cyclotron Facility cooler ring. The cycle time was 10 s, and the injected beam was electron-cooled for about 3 s before the measurement, producing a full-width at half-maximum bunch length of about 9 m (or 100 ns) depending on the rf voltage. The rf system used in the experiment was operating at harmonic number $h = 1$ with frequency 1.0309 MHz. Since the emittance of the beam in the cooler is small (0.05 π-mm-mrad), the motion of the beam can be visualized as a macro-particle.

The coherent betatron oscillation of the beam was excited by a single-turn transverse dipole kicker. For the IUCF cooler ring, we used a kicker with rise and fall

[70] The motion about SFPs of a nonlinear Hamiltonian resembles islands in the phase space and is thus called island motion. Stable islands are separated by the separatrix orbit that passes through unstable fixed points (UFPs). However, there is no UFP for the linear coupling Hamiltonian Eq. (2.219). The number of complete island motions in one revolution is called the "island tune." Here we find that the island tune of coupling motion around SFPs is equal to λ, which is independent of the betatron amplitude.

times at 100 ns and a 600 ns flat top. This is sufficient for a single bunch with a bunch length less than 100 ns at 1.0309 MHz revolution frequency. The subsequent bunch transverse oscillations from a BPM are detected and recorded. Figure 2.47 shows a typical example of the beating oscillations due to the linear betatron coupling following a kick in the horizontal plane. The linear coupling in the IUCF cooler ring arose mainly from the solenoid at the electron cooling section, and possibly also from quadrupole roll and vertical closed-orbit deviations in sextupoles. The Lambertson septum magnet at the injection area also contributed a certain amount of skew quadrupole field, which was locally corrected.

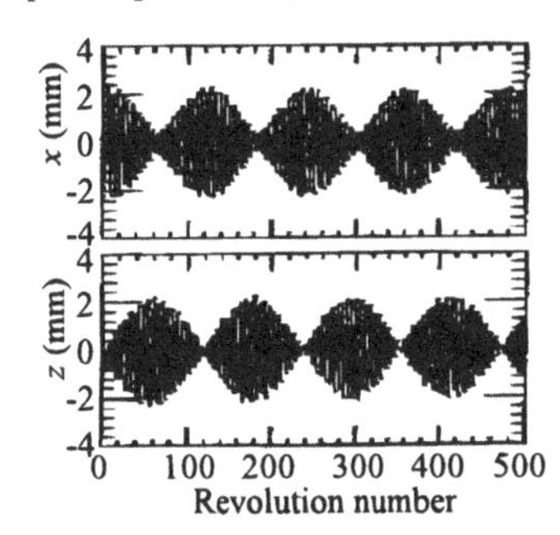

Figure 2.47: The measured coherent betatron oscillations excited by a horizontal kicker. The linear coupling gives rise to beating between the horizontal and vertical betatron oscillations. Note that the betatron beating between the x and z betatron motion gives rise to energy (action) exchange between the horizontal and vertical betatron oscillations.

In the presence of linear coupling, the measured betatron tunes correspond to normal modes of the betatron oscillations. The beat period were measured to be about 120 revolutions, which corresponds to $\lambda \approx 0.0083$, the tune separation between these two normal modes of Eq. (2.217).

Determination of the linear coupling phase

To measure the linear coupling phase χ, we can transform the horizontal and the vertical Poincaré maps into the resonant precessing frame discussed in Eq. (2.220). Figure 2.48 shows the normalized phase space $x, \mathcal{P}_x$ and $(z, \mathcal{P}_z)$ of the data shown in Fig. 2.47. The amplitude modulation of betatron motion is translated into breathing motion in the Poincaré map. The vertical map $(z, \mathcal{P}_z)$ is 180° out of phase with that of the horizontal map.

Transforming the phase space into a resonant precessing frame, we obtain a torus of 2D Hamiltonian shown in the middle-right plot of Fig. 2.48, where particle motion follows the Courant-Snyder invariant circle and a coupling ellipse (see also Fig. 2.45). The resonance phase was fitted to obtain an upright torus with a coupling phase of $\chi = 1.59$ rad. The coupling ellipse shown in the Poincaré surface section (middle-right plot of Fig. 2.48) has a small curvature, i.e. the proximity parameter δ_1 in this experiment is small.

The orientation of the resonant line was used to determine the coupling phase $\chi = 1.59$ rad, where the relative betatron phase advances at the locations of horizontal

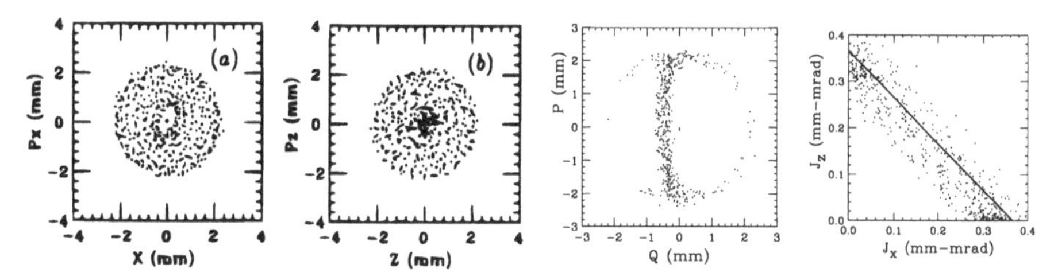

Figure 2.48: The betatron oscillations of Fig. 2.47 are transformed to the Poincaré map in the normalized coordinates $(x, \mathcal{P}_x)$ (left-most) and $(z, \mathcal{P}_z)$ (middle-left). Middle-right: The Poincaré surface of section in the resonant precessing frame derived from (x, x') and (z, z'), where $Q = \sqrt{2J_1\beta_x}\cos\phi_1$, $P = -\sqrt{2J_1\beta_x}\sin\phi_1$, and $\beta_x = 7.55$ m. Right: the actions J_z vs J_x showing the invariant of the linear coupling.

and vertical BPMs were included. Converting the phase space coordinates into action-angle variables, the right plot of Fig. 2.48 shows an invariant of the linear coupling, i.e. $J_x + J_z =$ constant.

The Poincaré map derived from experimental data at a 2D linear coupling resonance shows invariant tori of the Hamiltonian flow by comparing the middle-right plot in Fig. 2.48 to the solid line in Fig. 2.45. The particle appears to stay longer at the coupling circle because any position $J_x = J_1 < J_{1,\max}$ is on the coupling circle. The coupling circle is not the separatrix of the Hamiltonian! Using these invariant tori and Hamilton's equations of motion, we can determine the magnitude and the phase of the linear betatron coupling.

Determination of coupling strength $G_{1,-1,\ell}$

The measured action J_1 as a function of time and its time derivative $dJ_1/dN = 2\pi\dot{J}_1$ are plotted in Fig. 2.49, where a five-point moving average of J_1 is used to obtain a better behaved time derivative of the action J_1. The data of the time derivative dJ_1/dN are fitted with Eq. (2.220) to obtain $G_{1,-1,\ell} = 0.0078 \pm 0.0006$ and $\chi = 1.59$ rad, shown as a solid line in Fig. 2.49.

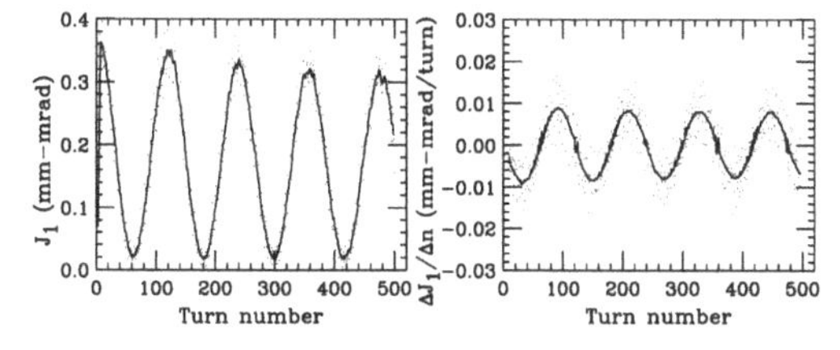

Figure 2.49: The action J_1 (left) in [π-mm-mrad] and its time derivative, dJ_1/dN (right) in [π-mm-mrad/turn]. The solid line in the left plot shows a five-point running average. The solid line in the right plot shows a fit by using Eq. (2.220) to obtain the coupling strength $G_{1,-1,\ell} = 0.0078$.

The magnitude of the linear coupling obtained from the invariant tori agrees well with that obtained by the traditional method of finding the minimum separation of the betatron tunes with combos of quadrupole strengths. The unperturbed betatron tune difference is $\delta_1 = -\sqrt{\lambda^2 - G_{1,-1,\ell}^2} \approx -0.0028$, where the $-$ sign arises from the shape of the coupling ellipse. Since $|\delta_1|$ is small, J_1 can reach nearly 0 in the coupled motion. In the above example, we show that a single digitized measurement can be used to obtain the magnitude and phase of the linear coupling. The exchange of the horizontal and vertical motion would be nearly 100% when $|\delta_1| \ll G$ *and* the initial condition is $J_{x0} = J_2$ and $J_{z0} = 0$. We also note that the linear coupling motion depends on the initial condition of the horizontal and the vertical motion as shown in Fig. 2.46. It is possible to have no action exchange under the same coupling condition as verified by experiments.[71]

Knowing the dynamics of the linear coupling of a single-particle motion may also help unravel questions concerning the dynamical evolution of the bunch distribution when the betatron tunes ramp through a coupling resonance. Such a problem is important for polarized proton acceleration in a low to medium energy synchrotron, where the vertical betatron tune jump method is used to overcome intrinsic depolarizing resonances. When the betatron tunes cross each other adiabatically after the tune jump, the increase in vertical emittance due to linear coupling may cause difficulty in later stages of polarized proton acceleration.

VI.4 Linear Coupling Correction with Skew Quadrupoles

The linear coupling resonance is usually corrected by maximizing the beat period of the transverse betatron oscillations using a pair, or at least two families, of skew quadrupoles. Figure 2.50 shows the output from a spectrum analyzer using the Δ-signal of a horizontal beam position monitor (BPM) as the input. The difference signal or a Δ-signal from BPMs carries the information of betatron oscillations around the closed orbit. To measures the power of betatron motion, a spectrum analyzer was tuned to a horizontal betatron sideband at zero span mode and was triggered 1.5 ms before the beam was coherently excited by a horizontal kicker. The beat period shown in Fig. 2.47 corresponds to the time interval between the dips of Fig. 2.50. The procedure for linear coupling correction is as follows

1. Maximize the peak to valley ratio in the spectrum by using quadrupole combos. This is equivalent to setting $\delta_1 = 0$ for attaining 100% coupling.
2. Maximize the time interval between dips (or peaks) of the spectrum by using families of skew quadrupoles. This reduces the coupling strength $G_{1,-1,\ell}$.

Repeated iteration of the above steps can efficiently correct linear coupling provided that skew quadrupole families have proper phase relations. This procedure is however hindered by betatron decoherence and by the 60 Hz power supply ripple, which is

[71]See J. Liu *et al.*, PRE **49**, 2347 (1993).

evident in Fig. 2.50. Other possible complications are closed-orbit changes due to off-center orbits in the quadrupoles and skew quadrupoles. However, the most important issue is that there is no guarantee *a priori* that the set of skew quadrupoles can properly correct the magnitude and phase of the linear coupling. Thus measurement of the coupling phase is also important.

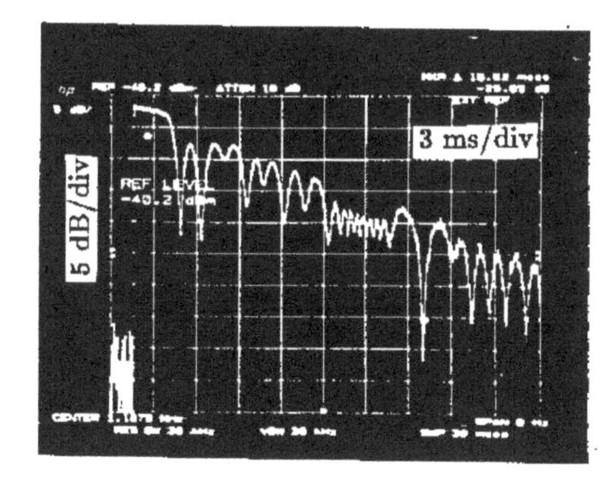

Figure 2.50: The spectrum of the Δ-signal from a horizontal BPM from a spectrum analyzer tuned to a betatron sideband frequency with resolution bandwidth 30 kHz and video bandwidth 30 kHz triggered 1.5 ms before a coherent horizontal kick. Note that (1) the time interval between these dips corresponded to the beat period of Fig. 2.47, (2) the decay of the power spectrum corresponded to betatron decoherence, and (3) the characteristic change in features at a 17 ms interval corresponded to a strong 60 Hz ripple, which altered betatron tunes.

VI.5 Linear Coupling Using Transfer Matrix Formalism

So far, our analysis of linear coupling has been based on single-resonance approximation in perturbation approach. The transfer matrix method of Sec. II can be expanded into 4×4 matrix by using transfer matrices for skew quadrupoles (Exercise 2.6.1) and solenoids (Exercise 2.6.2).

The 4×4 transfer matrix in one complete revolution can be diagonalized to obtain normal-mode betatron amplitude functions, and the coupling angle at each position in the ring.[72] This procedure has been implemented in MAD [23] and SYNCH [24] programs (see Exercise 2.6.6).

Exercise 2.6

1. This exercise derives the linear transfer matrix for a skew quadrupole, where the magnetic field is

$$B_z = -B_0 a_1 z, \quad B_x = B_0 a_1 x, \quad B_s = 0; \quad \text{with} \quad B_0 a_1 = \frac{1}{2}\left(\frac{\partial B_x}{\partial x} - \frac{\partial B_z}{\partial z}\right)_{x=z=0},$$

[72] D.A. Edwards and L.C. Teng, *IEEE Trans. Nucl. Sci.* **NS20**, 885 (1973); F. Willeke and G. Ripken, p. 758 in Ref. [15] (1988); J.P. Gourber *et al.*, *Proc. 1990 EPAC*, p. 1429 (1990); G. Guignard, *et al.*, *ibid.* p. 1432 (1990); L.C. Teng, PAC1997, p.1359 1361; D. Sagan and D. Rubin, PRSTAB Vol. 2, 074001 (1999).

where B_0 is the main dipole field strength, and a_1 is the skew quadrupole coefficient in multipole expansion of Eq. (2.19). Apparently, the skew quadrupole field satisfies Maxwell's equation $\partial B_z/\partial z + \partial B_x/\partial x = 0$. The vector potential is

$$A_s = -B_0 a_1 xz, \quad A_x = 0, \quad A_z = 0.$$

(a) Show that the equation of motion in a skew quadrupole is

$$x'' + qz = 0, \quad z'' + qx = 0, \quad \text{where} \qquad q = -\frac{1}{B\rho}\frac{\partial B_z}{\partial z} = \frac{a_1}{\rho}.$$

(b) Show that the transfer matrix of a skew quadrupole is

$$M = \begin{pmatrix} C_+ & S_+/\sqrt{q} & C_- & S_-/\sqrt{q} \\ -\sqrt{q}S_- & C_+ & -\sqrt{q}S_+ & C_- \\ C_- & S_-/\sqrt{q} & C_+ & S_+/\sqrt{q} \\ -\sqrt{q}S_+ & C_- & -\sqrt{q}S_- & C_+ \end{pmatrix}$$

where

$$C_+ = \frac{\cos\theta + \cosh\theta}{2}, \quad C_- = \frac{\cos\theta - \cosh\theta}{2},$$
$$S_+ = \frac{\sin\theta + \sinh\theta}{2}, \quad S_- = \frac{\sin\theta - \sinh\theta}{2},$$

$\theta = \sqrt{q}L$, and L is the length of the skew quadrupole.

(c) The coordinate rotation from (x, z) to $(\tilde{x}, \tilde{z})$ by an angle ϕ is

$$\begin{pmatrix} \tilde{x} \\ \tilde{x}' \\ \tilde{z} \\ \tilde{z}' \end{pmatrix} = R(\phi)\begin{pmatrix} x \\ x' \\ z \\ z' \end{pmatrix}, \quad R(\phi) = \begin{pmatrix} \cos\phi & 0 & \sin\phi & 0 \\ 0 & \cos\phi & 0 & \sin\phi \\ -\sin\phi & 0 & \cos\phi & 0 \\ 0 & -\sin\phi & 0 & \cos\phi \end{pmatrix}.$$

Show that the transfer matrix of a skew quadrupole is

$$M_{\text{skew quad}} = R(-45°)M_{\text{quad}}R(45°),$$

where M_{quad} is the transfer matrix of a quadrupole. This means that a skew quadrupole is equivalent to a quadrupole rotated by 45°.

(d) In the thin-lens limit, i.e. $L \to 0$ and $qL \to 1/f$, where f is the focal length, show that the 4×4 coupling transfer matrix reduces to

$$M = \mathbf{1} + \frac{-1}{f}\mathbf{U}, \quad \mathbf{U} = \begin{pmatrix} 0 & U \\ U & 0 \end{pmatrix}, \quad U = \begin{pmatrix} 0 & 0 \\ 1 & 0 \end{pmatrix}.$$

2. *Linear transfer Matrix of a Solenoid:* The particle equation of motion in an ideal solenoidal field is

$$x'' + 2gz' + g'z = 0, \qquad z'' - 2gx' - g'x = 0,$$

where the solenoidal field strength is $g = \frac{eB_\parallel(s)}{2p}$.

(a) Show that the coupled equation of motion becomes

$$y'' - j2gy' - jg'y = 0,$$

where $y = x + jz$, and j is the complex imaginary number.

(b) Transforming coordinates into rotating frame with

$$\bar{y} = ye^{-j\theta(s)}, \quad \text{where} \quad \theta = \int_0^s g ds,$$

show that the system is decoupled, and the equation of motion becomes

$$\bar{y}'' + g^2\bar{y} = 0.$$

Thus both horizontal and vertical planes are focused by the solenoid.

(c) Show that the transfer matrix in the rotating frame is

$$\tilde{M} = \begin{pmatrix} \cos\theta & \frac{1}{g}\sin\theta & 0 & 0 \\ -g\sin\theta & \cos\theta & 0 & 0 \\ 0 & 0 & \cos\theta & \frac{1}{g}\sin\theta \\ 0 & 0 & -g\sin\theta & \cos\theta \end{pmatrix},$$

where $\theta = gs$.[73]

(d) Transforming the coordinate system back to the original frame, i.e. $y = e^{j\theta}\bar{y}$, show that the transfer matrix for the solenoid becomes

$$M = \begin{pmatrix} \cos^2\theta & \frac{1}{g}\sin\theta\cos\theta & -\sin\theta\cos\theta & -\frac{1}{g}\sin^2\theta \\ -g\sin\theta\cos\theta & \cos^2\theta & g\sin\theta^2 & -\sin\theta\cos\theta \\ \sin\theta\cos\theta & \frac{1}{g}\sin^2\theta & \cos^2\theta & \frac{1}{g}\sin\theta\cos\theta \\ -g\sin^2\theta & \sin\theta\cos\theta & -g\sin\theta\cos\theta & \cos^2\theta \end{pmatrix}$$

3. Show that Hamilton's equations of motion for the Hamiltonian (2.213) in the presence of skew quadrupoles and solenoids are

$$x'' + K_x(s)x + 2gz' - (q - g')z = 0,$$
$$z'' + K_z(s)z + 2gx' - (q + g')x = 0.$$

where $g = B_{\parallel}(s)/2B\rho$ and $q = -(\partial B_z/\partial z)/B\rho = a_1/\rho$.

(a) Show that the perturbation potential due to skew quadrupoles and solenoids is

$$V_{\rm lc} = -\frac{a_1}{\rho}xz + g(s)\left(\frac{p_z}{p}x - \frac{p_x}{p}z\right).$$

(b) Expand the perturbation potential in Fourier series and show that the coupling coefficient $G_{1,-1,\ell}$ for the ℓ-th harmonic is given by Eq. (2.215).

[73]Note here that the solenoid, in the rotating frame, acts as a quadrupole in both planes. The focusing function is equal to g^2. In small rotating angle approximation, the corresponding focal length is $f^{-1} = g^2L = \Theta^2/L$, where L is the length of the solenoid, and $\Theta = gL$ is the rotating angle of the solenoid.

(c) If the accelerator lattice has P superperiods, show that $G_{1,-1,\ell} = 0$ unless $\ell = 0$ (Mod P).

4. Using the generating function

$$F_2(\phi_x, \phi_z, I_1, I_2) = (\phi_x - \phi_z - \ell\theta + \chi)I_1 + \phi_z I_2,$$

show that the linear coupling equation of motion for the Hamiltonian (2.216) can be transformed into the Hamiltonian (2.219) in *resonance rotating frame.*

(a) Show that the new conjugate phase-space variables are

$$I_1 = J_x, \quad I_2 = J_x + J_z, \quad \phi_1 = \phi_x - \phi_z - \ell\theta + \chi, \quad \phi_2 = \phi_z.$$

(b) Find the invariants of the Hamiltonian (2.216).

(c) Show that the equation of motion for I_1 is

$$\ddot{I}_1 + \lambda^2 I_1 = E_1\delta_1 + I_2 G_{1,-1,\ell}^2/2,$$

where the overdot corresponds to the derivative with respect to orbiting angle θ, $\delta_1 = \nu_x - \nu_z - \ell$ is the resonance proximity parameter, $\lambda = \sqrt{\delta_1^2 + G_{1,-1,\ell}^2}$, and $E_1 = \delta_1 I_1 + G_{1,-1,\ell}\sqrt{I_1(I_2 - I_1)}\cos\phi_1$ is a constant of motion.

(d) Discuss the solution in the resonance rotating frame.[74]

5. The Hamiltonian

$$H = \nu_x J_x + \nu_z J_z + G_{1,-1,\ell}\sqrt{J_x J_z}\cos(\phi_x - \phi_z + \chi)$$

for a single linear coupling resonance can be transformed to the normalized phase-space coordinates by

$$\begin{cases} X = \sqrt{2J_x}\cos(\phi_x + \chi_x), & P_x = -\sqrt{2J_x}\sin(\phi_x + \chi_x), \\ Z = \sqrt{2J_z}\cos(\phi_z + \chi_z), & P_z = -\sqrt{2J_z}\sin(\phi_z + \chi_z), \end{cases}$$

where $\chi_x - \chi_z = \chi$ is a constant linear coupling phase (Mod 2π) that depends on the location in the ring.

(a) Show that the Hamiltonian in the new phase-space coordinates is

$$H = \frac{1}{2}\nu_x(X^2 + P_x^2) + \frac{1}{2}\nu_z(Z^2 + P_z^2) + \frac{1}{2}G_{1,-1,\ell}(XZ + P_xP_z).$$

(b) Show that the eigen-frequency of the Hamiltonian is

$$\nu_\pm = \frac{1}{2}(\nu_x + \nu_z) \pm \frac{1}{2}\lambda, \quad \lambda = \sqrt{(\nu_x - \nu_z)^2 + |G_{1,-1,\ell}|^2}.$$

---

[74]For a general discussion on linear coupling with nonlinear detuning, see J.Y. Liu *et al.*, *Phys. Rev.* **E 49**, 2347 (1994).

(c) Solve X and Z in terms of the normal modes, and show that

$$\begin{cases} X = A_+ \cos(\nu_+\varphi + \xi_+) - \dfrac{G_{1,-1,\ell}}{\lambda + |\delta|} A_- \cos(\nu_-\varphi + \xi_-), \\ Z = \dfrac{G_{1,-1,\ell}}{\lambda + |\delta|} A_+ \cos(\nu_+\varphi + \xi_+) + A_- \cos(\nu_-\varphi + \xi_-), \end{cases}$$

where $A_\pm, \xi_\pm$ are obtained from the initial conditions. Particularly, we note that the "horizontal" and "vertical" betatron oscillations carry both normal-mode frequencies.

6. Analyze the linear stability of the simple tracking model shown in Fig. 2.46, i.e. the particle motion in a synchrotron with linear betatron motion and a localized skew quadrupole kick. Show that the condition of linear stability for betatron motion is

$$\frac{\sqrt{\beta_x\beta_z}}{f} \le \text{Min}\left\{ 2\sqrt{\frac{(1+\cos\Phi_x)(1+\cos\Phi_z)}{|\sin\Phi_x \sin\Phi_z|}},\ 2\sqrt{\frac{(1-\cos\Phi_x)(1-\cos\Phi_z)}{|\sin\Phi_x \sin\Phi_z|}} \right\},$$

where f is the focal length of the skew quadrupole, β_x and β_z are values of betatron amplitude functions at the skew quadrupole location, and Φ_x and Φ_z are betatron phase advances of the machine without the skew quadrupole. Based on your study of this problem, can you find the stability limit of a linearly coupled machine with superperiod P?

7. Consider the sum resonance driven by skew quadrupoles, where the coupling constant $G_{1,1,\ell}$ is given by Eq. (2.215). The Hamiltonian in the action-angle variables is given by

$$\mathcal{H} = \nu_x J_x + \nu_z J_z + |G_{1,1,\ell}|\sqrt{J_x, J_z}\cos(\phi_x + \phi_z - \ell\theta + \chi_+),$$

where χ_+ is the phase of the coupling constant, θ is the independent variable serving as the time coordinate, and ℓ is an integer near $\nu_x + \nu_z$, i.e. $|\nu_x + \nu_z - \ell| \ll 1$.

(a) Show that the difference of the actions, $J_x - J_z$, is a constant of motion.

(b) Let $g = |G_{1,1,\ell}|$. Show that

$$\ddot{J}_x = (g^2 - \delta^2)J_x + \text{constant}$$
$$\ddot{J}_z = (g^2 - \delta^2)J_x + \text{constant}$$

where overdots are derivative with respect to the independent variable θ, and the resonance proximity parameter is $\delta = \nu_x + \nu_x - \ell$. This means that if the tunes of a particle is within the sum resonance stopband width, i.e. $|\delta| < g$, the actions of the particle will grow exponentially at a growth rate of $\sqrt{g^2 - \delta^2}$.

VII Nonlinear Resonances

Chromaticity correction discussed in Sec. V requires sextupoles. Thus sextupole magnets are an integral part of accelerator lattice design. Furthermore, modern high energy storage rings frequently use high field (superconducting) magnets that inherently have systematic and random multipole fields. Normally, the nonlinearity is of the order of $10^{-3} - 10^{-4}$ compared with that of the linear component. However, when a resonance condition is encountered, the nonlinear magnetic field can give rise to geometric aberration on beam ellipses and particle loss. Careful analysis of the nonlinear beam dynamics is instrumental in determining the *dynamic aperture*, which is defined as the maximum amplitude of a stable particle motion. This section addresses nonlinear geometric aberration due to sextupoles and higher order multipoles, and provides an introduction to this important subject using the first-order perturbation treatment.

VII.1 Nonlinear Resonances Driven by Sextupoles

Since sextupoles are indispensable for chromatic correction, we begin with their effects on beam dynamics. The vector potential for a 2D sextupole magnet is $A_x = A_z = 0$, $A_s = \frac{B_2}{6}(x^3 - 3xz^2)$, where $B_2 = \partial^2 B_z/\partial x^2|_{x=z=0}$. The Hamiltonian in the presence of sextupole field is

$$H = \frac{1}{2}\left[x'^2 + K_x x^2 + z'^2 + K_z z^2\right] + V_3(x, z, s), \tag{2.227}$$

$$V_3(x, z, s) = \frac{1}{6}K_2(s)(x^3 - 3xz^2), \qquad K_2(s) = \mp\frac{B_2(s)}{B\rho},$$

where $\mp$ signs correspond to positive/negative charged particles. For a particle with fractional off-momentum deviation $\Delta p/p_0$, the sextupole strength is $K_2(s) = -\frac{B_2(s)}{B\rho[1+\delta]} \approx -\frac{B_2(s)}{B\rho}$ with $\delta = \Delta p/p_0$.

A. Tracking methods

In the presence of sextupole magnetic field, Hill's equation becomes

$$x'' + K_x(s)x = -\frac{1}{2}K_2(s)(x^2 - z^2), \qquad z'' + K_z(s)z = +K_2(s)xz. \tag{2.228}$$

The evolution of phase space coordinates of a particle can be obtained by tracking the equation of motion, where thin lens (kick-map) approximation has often been used because sextupole magnets used in accelerator are usually short. Let $S = \int K_2(s)ds = K_2\ell_{\rm sextupole}$ be the integrated sextupole strength. The changes of phase space coordinates at the sextupole magnet are[75]

$$\Delta x' = -\frac{1}{2}S(x^2 - z^2), \qquad \Delta z' = Sxz. \tag{2.229}$$

The propagation of phase space coordinates outside the sextupole magnet is transformed by the transfer mapping matrix (2.42). Figure 2.51 shows the Poincaré maps near a third order resonance at $\nu_x = 3.66$ and $\nu_x = 3.672$ respectively with one sextupole in an otherwise perfectly linear accelerator. The betatron amplitude functions at the thin lens sextupole location are $\beta_x = 20$ m and $\alpha_x = 0$. The topology of the phase space maps forms mirror reflection when the tune moves across the third integer resonance. The region of stability decreases as the betatron tune approaches the third order resonance.

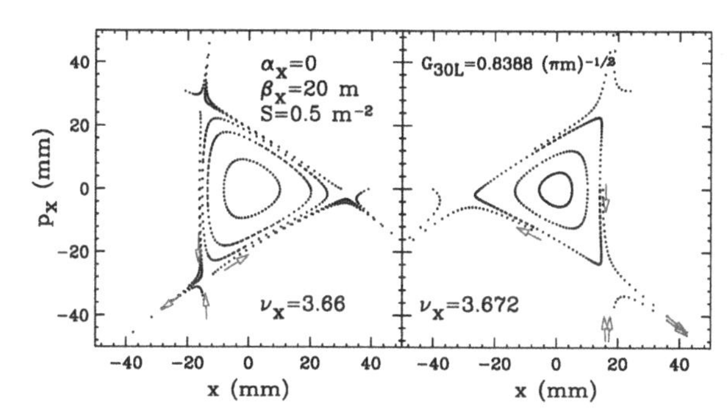

Figure 2.51: The Poincaré maps for betatron motion perturbed by a single sextupole magnet at a tune below (left) and above (right) a third order resonance. The integrated sextupole strength is $S = 0.5$ m^{-2} with lattice parameters $\beta_x = 20$ m, and $\alpha_x = 0$. Arrows indicate directions of motion near a separatrix.

B. The leading order resonances driven by sextupoles

In order to analyze the third order resonance analytically, we carry out Floquet transformation to the Hamiltonian (2.227). With Eq. (2.55) for coordinate transformation, the nonlinear perturbing potential $V_3(x, z, s)$ becomes

$$\begin{aligned} V_3 &= -\frac{\sqrt{2}}{4}J_x^{1/2}J_z\beta_x^{1/2}\beta_z K_2(s)[2\cos\Phi_x + \cos(\Phi_x + 2\Phi_z) + \cos(\Phi_x - 2\Phi_z)] \\ &\quad + \frac{\sqrt{2}}{12}J_x^{3/2}\beta_x^{3/2}K_2(s)[\cos 3\Phi_x + 3\cos\Phi_x], \end{aligned} \tag{2.230}$$

where

$$\Phi_x = \phi_x + \chi_x(s) - \nu_x\theta, \quad \chi_x = \int_0^s \frac{ds}{\beta_x}; \qquad \Phi_z = \phi_z + \chi_z(s) - \nu_z\theta, \quad \chi_z = \int_0^s \frac{ds}{\beta_z}.$$

[75]In this mapping equation for betatron motion, we disregard the effect of sextupoles on orbit length. Using Eq. (2.100), we find $\Delta C = (x\Delta x' + z\Delta z')$.

Here (J_x, ϕ_x) and (J_z, ϕ_z) are pairs of conjugate phase-space coordinates, $\theta = s/R$ is the orbiting angle, and R is the mean radius of the accelerator. Since V_3 is a periodic function of s, it can be expanded in Fourier harmonics.[76] The Hamiltonian (2.227) expressed in action-angle variables with θ as "time coordinate" becomes

$$\begin{aligned} H &= \nu_x J_x + \nu_z J_z + \sum_\ell \{G_{3,0,\ell} J_x^{3/2} \cos(3\phi_x - \ell\theta + \xi_{3,0,\ell}) \\ &+ G_{1,2,\ell} J_x^{1/2} J_z \cos(\phi_x + 2\phi_z - \ell\theta + \xi_{1,2,\ell}) \\ &+ G_{1,-2,\ell} J_x^{1/2} J_z \cos(\phi_x - 2\phi_z - \ell\theta + \xi_{1,-2,\ell}) \\ &+ g_{1,0,3,0,\ell} J_x^{3/2} \cos(\phi_x - \ell\theta + \eta_{3,0,3,0,\ell}) \\ &+ g_{1,0,1,2,\ell} J_x^{1/2} J_z \cos(\phi_x - \ell\theta + \eta_{3,0,1,2,\ell})\} \end{aligned} \tag{2.231}$$

where ℓ's are integers. Because the potential V_3 is an odd function of the betatron coordinates, sextupoles will not, in linear approximation, generate betatron detuning. Table 2.3 lists nonlinear resonances that can be excited by sextupoles in first-order perturbation theory.

Table 2.3: Resonances due to sextupoles and their driving terms

Resonance	Driving term	Lattice	Amplitude	Classification
$\nu_x + 2\nu_z = \ell$	$\cos(\Phi_x + 2\Phi_z)$	$\beta_x^{1/2}\beta_z$	$J_x^{1/2} J_z$	sum resonance
$\nu_x - 2\nu_z = \ell$	$\cos(\Phi_x - 2\Phi_z)$	$\beta_x^{1/2}\beta_z$	$J_x^{1/2} J_z$	difference resonance
$\nu_x = \ell$	$\cos\Phi_x$	$\beta_x^{1/2}\beta_z;\ \beta_x^{3/2}$	$J_x^{1/2} J_z,\ J_x^{3/2}$	parametric resonance
$3\nu_x = \ell$	$\cos 3\Phi_x$	$\beta_x^{3/2}$	$J_x^{3/2}$	parametric resonance

The third order resonances occur at $3\nu_x = \ell$, $\nu_x + 2\nu_z = \ell$ and $\nu_x - 2\nu_z = \ell$ and a peculiar integer resonance at $\nu_x = \ell$ (see Exercise 2.7.7) driven by the sextupole. The resonance strengths are respectively

$$G_{3,0,\ell}\, e^{j\xi_{3,0,\ell}} = \frac{\sqrt{2}}{24\pi} \oint \beta_x^{3/2}\, K_2(s)\, e^{j[3\chi_x(s) - (3\nu_x - \ell)\theta]} ds \tag{2.232}$$

$$G_{1,\pm 2,\ell}\, e^{j\xi_{1,\pm 2,\ell}} = \frac{\sqrt{2}}{8\pi} \oint \beta_x^{1/2}\beta_z\, K_2(s)\, e^{j[\chi_x(s) \pm 2\chi_z(s) - (\nu_x \pm 2\nu_z - \ell)\theta]} ds, \tag{2.233}$$

$$g_{1,0,3,0,\ell}\, e^{j\eta_{1,0,3,0,\ell}} = \frac{\sqrt{2}}{8\pi} \oint \beta_x^{3/2}\, K_2(s)\, e^{j[\chi_x(s) - (\nu_x - \ell)\theta]} ds$$

$$g_{1,0,1,2,\ell}\, e^{j\eta_{1,0,1,2,\ell}} = \frac{\sqrt{2}}{4\pi} \oint \beta_x^{1/2}\beta_z\, K_2(s)\, e^{j[\chi_x(s) - (\nu_x - \ell)\theta]} ds.$$

All sextupolar resonance strengths: $G_{3,0,\ell}$, $G_{1,\pm 2,\ell}$, $g_{1,0,3,0,\ell}$, and $g_{1,0,1,2,\ell}$, have the unit $(\pi \mathrm{m})^{-1/2}$, where the factor π arises from unit of the action J discussed in Eq. (2.50).

[76] G. Guignard, p. 822 in Ref. [14] (1988); G. Guignard, CERN 76-06, (1976).

The subscripts of $G_{3,0,3,0,\ell}$ and $G_{1,\pm 2,1,2,\ell}$ of dominant resonances have been shorten to $G_{3,0,\ell}$ and $G_{1,\pm 2,\ell}$ for simplicity. The resonance phases $\xi's$ in Eq. (2.232) depend on the reference position in accelerator that we choose to calculate these integrals. We can choose the amplitudes of resonance strengths $G's$ or $g's$ to be positive.

C. The third order resonance at $3\nu_x = \ell$

As demonstrated in Fig. 2.51, the Hamiltonian (2.231) near a third-order resonance at $3\nu_x = \ell$ can be well approximated by

$$H \approx \nu_x J_x + G_{3,0,\ell}\, J_x^{3/2} \cos{(3\phi_x - \ell\theta + \xi_{3,0,\ell})}, \tag{2.234}$$

where $G_{3,0,\ell}$ and $\xi_{3,0,\ell}$ are given by Eq. (2.232), (J_x, ϕ_x) are conjugate phase-space coordinates, θ is the orbiting angle serving as "time coordinate," ν_x is the horizontal betatron tune. The sextupole cause "geometric aberration" to the betatron motion. The magnitude of the geometric aberration depends on the resonance strength $G_{3,0,\ell}$ and the "resonance proximity" $\delta = \nu_x - \frac{\ell}{3}$.

If an accelerator has a superperiod P and the sextupole field satisfies the same periodic condition, the resonance strength $G_{3,0,\ell}$ is zero unless ℓ is an integer multiple of P (see Exercise 2.7.1). For example, if sextupoles are identical in each superperiod of the AGS, the systematic third-order resonance strength will be zero except at $\ell = 12, 24$, etc. Thus nonlinear resonances are classified into systematic and random resonances. Systematic nonlinear resonances are located at $\ell = P\times$integer. At a systematic resonance, each superperiod contributes coherently additive to the resonance strength. Since chromatic sextupoles are usually arranged according to accelerator-superperiod, one should pay great attention to systematic sextupolar nonlinear resonances by choosing the betatron tune to avoid systematic third order resonances. Random sextupole fields induce nonlinear resonances at all integer ℓ, and their resonance strengths are usually weak. Nevertheless, the betatron tunes should avoid low-order nonlinear resonances.

Transform the phase space coordinate to a *resonance rotating frame* with a generating function to obtain new phase-space coordinates:

$$F_2(\phi_x, J) = (\phi_x - \frac{\ell}{3}\theta + \frac{\xi_{3,0,\ell}}{3})J, \quad \Longrightarrow \quad \phi = \phi_x - \frac{\ell}{3}\theta + \frac{\xi_{3,0,\ell}}{3}, \quad J = J_x. \tag{2.235}$$

The Hamiltonian Eq. (2.234) and Hamilton's equations of motion are

$$H = \delta J + G_{3,0,\ell} J^{3/2} \cos 3\phi, \tag{2.236}$$

$$\dot\phi \equiv \frac{d\phi}{d\theta} = \delta + \frac{3}{2} G_{3,0,\ell} J^{1/2} \cos 3\phi, \qquad \dot J \equiv \frac{dJ}{d\theta} = 3G_{3,0,\ell} J^{3/2} \sin 3\phi.$$

where $\delta = \nu_x - \ell/3$ is the resonance proximity parameter. Since the Hamiltonian (2.236) is "autonomous" (meaning independent of the time (θ) coordinate), the "Hamiltonian" is invariant. Particle motion in the phase space follows the contour of a constant Hamiltonian.

Stable and unstable fixed points

The fixed points (FPs) of the Hamiltonian Eq. (2.236) are determined by $\dot{J} = 0$ and $\dot{\phi} = 0$ (see Appendix A Sec. I.2). Without nonlinear detuning, there is no stable fixed point for the third order resonance. The action at the UFP, equation of motion near the UFP, and Hamiltonian value at the UFP are

$$J_{\rm UFP}^{1/2} = \left|\frac{2\delta}{3G_{3,0,\ell}}\right| \quad \text{with} \quad \begin{cases} \phi_{\rm FP} = 0,\ \pm 2\pi/3 & \text{if } \delta/G_{3,0,\ell} < 0 \\ \phi_{\rm FP} = \pm\pi/3,\ \pi & \text{if } \delta/G_{3,0,\ell} > 0 \end{cases}, \tag{2.237}$$

$$\ddot{K} - 3\delta^2 K - 6\frac{\delta^2}{J_{\rm UFP}}K^2 = 0, \qquad E_{\rm UFP} = \frac{\delta}{3}\left(\frac{2\delta}{3G_{3,0,\ell}}\right)^2, \tag{2.238}$$

where $K = J - J_{\rm UFP}$. The motion near the unstable fixed point is hyperbolic. Because of the nonlinear term in Eq. (2.238), the amplitude grows faster than an exponential. Particle motion near a separatrix is marked with arrows in Fig. 2.51.

Separatrix

The separatrix is a Hamiltonian torus that passes through the UFP, i.e. $H = E_{\rm UFP}$. With Eq. (2.236), the separatrix orbit, for $\delta/G_{3,0,\ell} > 0$, is

$$[2X - 1]\left[P - \frac{1}{\sqrt{3}}(X+1)\right]\left[P + \frac{1}{\sqrt{3}}(X+1)\right] = 0, \quad \text{or}$$

$$X = 1/2,\ P = \frac{X+1}{\sqrt{3}},\ P = -\frac{X+1}{\sqrt{3}} \quad \Longrightarrow \quad (X,P)_{\rm UFP} = \begin{cases} (-1, 0), \\ (\frac{1}{2}, \frac{\sqrt{3}}{2}), \\ (\frac{1}{2}, -\frac{\sqrt{3}}{2}). \end{cases} \tag{2.239}$$

where X and P are $X = \sqrt{J/J_{\rm UFP}}\cos\phi$, $\quad P = -\sqrt{J/J_{\rm UFP}}\sin\phi$. Three straight lines divide the phase space into stable and unstable regions. The unstable fixed points $(X,P)_{\rm UFP}$ are the intersections of these three lines (see Fig. 2.51, where the right plot corresponds to the condition of Eq. (2.239). At $\delta > 0$, particles near $(x = 0,\ P_x = 0)$ moves clockwise in the resonance rotating frame and counter-clockwise at $\delta < 0$ (see Eq.(2.52) for the direction of betatron motion in phase space). At $\delta < 0$, particle motion lags behind the resonance rotating frame and the phase space motion appears counter-clockwise.

The dynamic aperture is defined as the maximum phase-space area for stable betatron motion. Near a third-order resonance, the stable phase-space area in (x, x') is the area of the triangle bounded by Eq. (2.239), i.e. $\frac{3\sqrt{3}}{2}J_{\rm UFP} = \frac{2}{\sqrt{3}}(\delta/G_{3,0,\ell})^2$. Beam loss may occur when particles wander beyond the separatrix. Without a nonlinear detuning term, the third-order resonance appears at all values of δ. The stable motion is bounded by the curve of $J_{\rm UFP}^{1/2}$ shown in Fig. 2.52. For a given aperture $J_{\rm max}$, we can define the third-order betatron resonance width as $|\delta|_{\rm width} = 2 \times [3^{1/4}G_{3,0,\ell}J_{\rm max}^{1/2}]$.

The third-order resonance can be applied to slow extraction of beam particles from a synchrotron. The resonance phase $\xi_{3,0,\ell}$ depends on the position of sextupoles relative to the extraction septum. The stable phase-space area is proportional to the resonance proximity parameter δ^2 (see Eq. (2.237)). If the betatron tune ν_x is ramped slowly towards a third-order resonance, beam particles can be slowly squeezed out of stable area and extracted to achieve high duty cycle in physics experiments and other applications.

Effect of nonlinear detuning

Nonlinear magnetic multipoles also generate nonlinear betatron detuning, i.e. the betatron tunes depend on the betatron amplitudes (actions):

$$Q_x = \nu_x + \alpha_{xx} J_x + \alpha_{xz} J_z + \cdots, \qquad Q_z = \nu_z + \alpha_{xz} J_x + \alpha_{zz} J_z + \cdots. \tag{2.240}$$

The unit of the first order nonlinear detuning parameters α's is $(\pi\text{m})^{-1}$. Typically magnitude of the detuning parameters is $10 \sim 10^4$ $(\pi\text{m})^{-1}$. With the betatron detuning, the third-order resonance Hamiltonian in the resonance rotating frame is

$$H = \delta J + \frac{1}{2}\alpha J^2 + G J^{3/2} \cos 3\phi, \tag{2.241}$$

where we use $\alpha = \alpha_{xx}$ and $G = G_{3,0,\ell}$ to simplify our notation. The fixed points of the Hamiltonian for $\alpha > 0$ and $G > 0$ are

$$\frac{\alpha J_{\rm FP}^{1/2}}{G} = \begin{cases} -\frac{3}{4} + \frac{3}{4}\sqrt{1 - \frac{16\alpha\delta}{9G^2}}, & \phi = 0, \pm 2\pi/3 \quad \delta < 0 \qquad \text{(UFP)} \\ +\frac{3}{4} - \frac{3}{4}\sqrt{1 - \frac{16\alpha\delta}{9G^2}}, & \phi = \pi, \pm\pi/3 \quad 0 \le \delta \le \frac{9G_{3,0,\ell}^2}{16\alpha} \qquad \text{(UFP)} \\ +\frac{3}{4} + \frac{3}{4}\sqrt{1 - \frac{16\alpha\delta}{9G^2}}, & \phi = \pi, \pm\pi/3 \quad \delta \le \frac{9G^2}{16\alpha} \qquad \text{(SFP)} \end{cases}$$

Figure 2.52 shows $|\alpha/G| \times J_{\rm UFP}^{1/2}$ vs $\alpha\delta/G^2$. Stable fixed points appear in the presence of nonlinear detuning. The bifurcation of third-order resonance occurs at $16\alpha\delta = 9G^2$. Note that $J_{\rm SFP} \ge J_{\rm UFP}$. A similar analysis can be carried out for $\alpha < 0$. Resonance exists only in tune space with $16\alpha\delta \le 9G^2$ for either $\alpha > 0$ or $\alpha < 0$. The dash-dot line in Fig. 2.52 shows the scaled $J_{\rm UFP}^{1/2}/|G|$ vs the scaled proximity parameter δ/G^2 for zero detuning case.

D. Experimental measurement of a $3\nu_x = \ell$ resonance

Because beam particles may be unstable at a nonlinear resonance, experimental measurements are generally difficult. It is easy to observe degradation of beam intensity

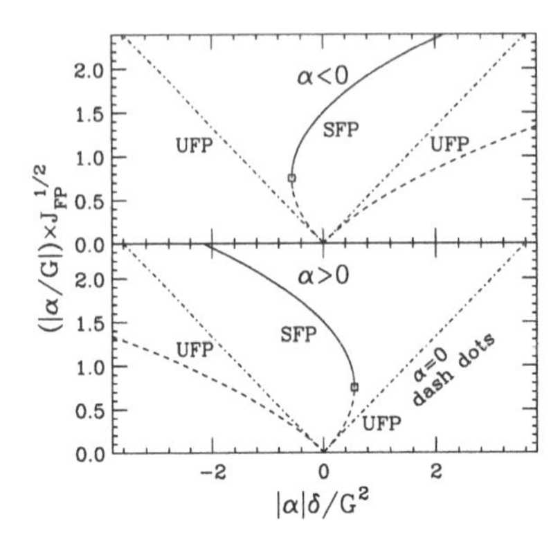

Figure 2.52: The scaled fixed point action of the third-order betatron resonance vs the scaled resonance proximity δ. The scaling parameters are resonance strength $G = G_{3,0\ell}$, and the nonlinear detuning $\alpha = \alpha_{xx}$. The dash-dots line is the UFP in the zero detuning limit, where $J_{\rm UFP}^{1/2} = |2\delta/3G|$. There is no SFP for $\alpha = 0$. The solid and dashed curves are $|\alpha/G|J_{\rm SFP}^{1/2}$ and $|\alpha/G|J_{\rm UFP}^{1/2}$ vs $|\alpha|\delta/G^2$ for $\alpha < 0$ (top plot) and $\alpha > 0$ (bottom plot). Note that $J_{\rm SFP} \geq J_{\rm UFP}$. Bifurcation of the third-order resonance occurs at $16\alpha\delta = 9G^2$ marked by a rectangle. Resonance islands appear only in tune space $16\alpha\delta \leq 9G^2$ for either $\alpha > 0$ or $\alpha < 0$.

and lifetime near a resonance. Measurements of Poincaré maps near a third-order resonance have been successful at SPEAR, TEVATRON, Aladdin, and the IUCF cooler ring. Figure 2.53 shows a Poincaré map obtained from a nonlinear beam dynamics experiment at the IUCF cooler ring. Converting into action-angle variables, we can fit these data by the Hamiltonian (2.241) to obtain parameters $G_{3,0,\ell}$ and ξ, and obtain the parameter δ by measuring the betatron tune at a small betatron amplitude. Using these measured nonlinear resonance parameters, one can model sextupole strengths of the storage ring.[77]

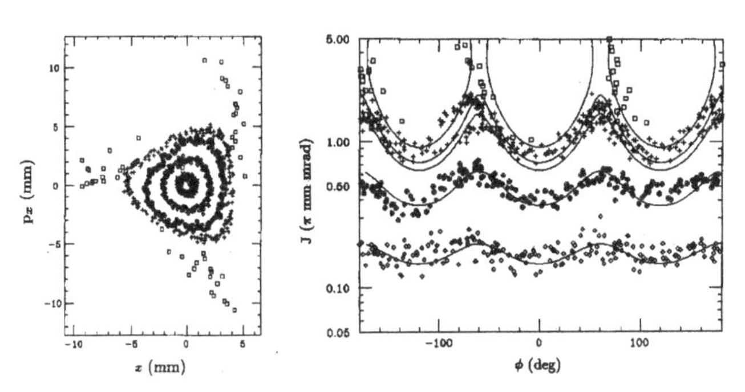

Figure 2.53: Left: The measured Poincaré map of the normalized phase-space coordinates (x, p_x) of betatron motion near a third-order resonance $3\nu_x = 11$ at the IUCF cooler ring. Right: the Poincaré map in action-angle variables (J, ϕ).

Particles outside the separatrix survive only about 100 turns. Tori for particles inside the separatrix are distorted by the third order resonance. The orientation of the Poincaré map, determined by sextupoles, rotates at a rate of betatron phase advance along the ring. The solid lines are Hamiltonian tori of Eq. (2.241) with $\delta = -0.0060$, $|G_{3,0,11}| = 2.2\ (\pi{\rm m})^{-1/2}$, and $\alpha_{xx} = 0$, which gives $J_{\rm UFP} = 3.3\ \pi\mu{\rm m}$. Equally good fix can be obtained by using the measured $\alpha_{xx} \approx 650\ (\pi{\rm m})^{-1}$ to obtain $J_{\rm UFP} = 3.1\ \pi\mu{\rm m}$, and $J_{\rm SFP} \approx 27.6\ (\pi\mu{\rm m})$.

[77]D.D. Caussyn, *et al.*, PRA **46**, 7942 (1992).

E. Other 3rd-order resonances driven by sextupoles

Besides the third-order resonance at $3\nu_x = \ell$, sextupoles contribute importantly to the nonlinear coupling resonances at $\nu_x \pm 2\nu_z = \ell$ with integer ℓ.[78] The third-order resonance strength can generally be obtained by taking the Fourier transform of Eq. (2.230). The *difference* resonance at $\nu_x - 2\nu_z = \ell$ produces betatron coupling (see Exercise 2.7.2). The invariant of this difference resonance is $2J_x + J_z =$ constant, i.e. There is an exchange of the horizontal and vertical motion, but each plane is limited by the invariant quantity. If the initial horizontal emittance is large, the difference resonance can cause a large increase in the vertical emittance and thus can also cause beam loss in the vertical plane.

Isochronous cyclotrons, discussed in Chapter 1, require the phase-slip-factor $\eta = 0$ of Eq. (2.165). This means that the transition energy $\gamma_T \approx \nu_x = \gamma$, where γ is the Lorentz energy factor of the beam. Thus all isochronous cyclotrons are designed to have $\nu_x \sim \gamma$ during the acceleration. Since the vertical betatron tune of the isochronous cyclotrons lies between 0.5 to 1. Passing the $\nu_x - 2\nu_z = 0$ resonance is almost un-avoidable. This resonance in the cyclotron community is called the Walkinshaw resonance (see Exercise 2.7.2).

The invariant of the *sum* resonance at $\nu_x + 2\nu_z = \ell$ is $2J_x - J_z =$ constant (see Exercise 2.7.3). This means that the actions of the horizontal and vertical planes can increase simultaneously to a large value, and the particle can get lost at large amplitudes. All sum resonance can cause beam emittance to blow-up in both horizontal and vertical planes and leads to beam loss

Furthermore, Eq. (2.231) shows that sextupoles also drive nonlinear resonances at $\nu_x = \ell$. Exercise 2.7.7 illustrates the difference between the nonlinear $\nu_x = \ell$ resonance and the linear betatron resonance discussed in Sec. III.1.

VII.2 Higher-Order Resonances

It appears, from the Hamiltonian in Eq. (2.231), that sextupoles will not produce resonances higher than the third order ones listed in Table 2.3. However, strong sextupoles are usually needed to correct chromatic aberration. Concatenation of strong sextupoles can generate high-order resonances such as $4\nu_x, 2\nu_x \pm 2\nu_z, 4\nu_z, 5\nu_x, \ldots$, etc. For example, carry out the second order canonical perturbation to the Hamiltonian in Eq. (2.231), the resonances of $3\nu_x = \ell_3$ and $\nu_x = \ell_1$ can be combined to produce a resonance at $4\nu_x = \ell_1 + \ell_3$, which is a 4th order resonance (see Exercise 2.7.8). Figure 2.54 shows the Poincaré maps of the single sextupole model of Fig. 2.51 at $\nu_x = 3.7496$ and $\nu_x = 3.795$, produced by a single strong sextupole in the accelerator. Exercise 2.7.8

[78]See M. Ellison *et al.*, PRE **50**, 4051 (1994) for the $\nu_x - 2\nu_z = \ell$ resonance, and J. Budnick *et al.*, *Nucl. Inst. Methods* **A368**, 572 (1996) for the $\nu_x + 2\nu_z = \ell$ resonance at the IUCF cooler ring. Experimental measurements of sum resonances are particularly difficult because of short lifetime and small beam current.

illustrates a canonical perturbation method to explain the tracking result. Similarly, sextupolar resonances can drive higher order resonances at $2\nu_x \pm 2\nu_z = \ell$, etc.

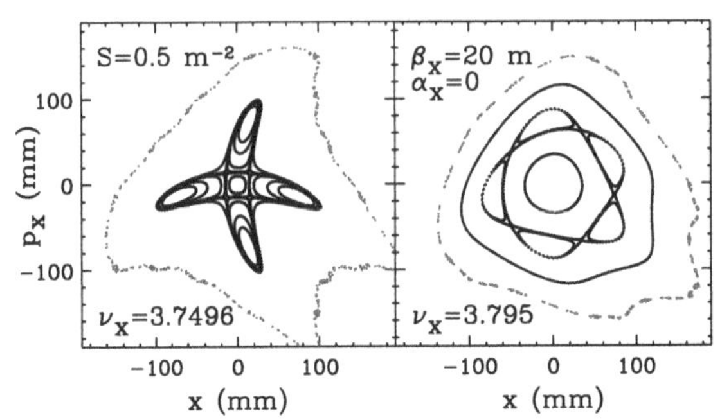

Figure 2.54: The normalized phase space maps for the same sextupole model at $\nu_x = 3.7496$ (left) and $\nu_x = 3.795$ (right) are shown in Fig. 2.51. Since resonance islands only exist with $\nu_x < 3.75$ or $\nu_x < 3.8$, the effective nonlinear detuning must be positive. The largest phase space map marks the boundary of stable motion. See Exercise 2.7.8.

Accelerator magnets may have many nonlinear magnetic multipoles, which can introduce nonlinear perturbation to betatron motion. Employing strong sextupoles, nonlinear beam dynamics experiments at Fermilab TEVATRON were used to study the concept of smear, nonlinear detuning, decoherence, and dynamic aperture.[79] Similarly, nonlinear beam dynamics studies at the IUCF cooler ring show the importance of nonlinear resonances. Nonlinear beam dynamics is beyond the scope of this book. Here we give an example of the fourth-order parametric resonance at $4\nu_x = 15$.[80] The nonlinear perturbation plays an important role on the stability of particle beams in accelerators. Experimental measurements of the effects on particle motion are important.

Near a *weak* fourth-order 1D resonance $4\nu_x \approx \ell$, the perturbation potential of octupoles and the effective Hamiltonian can normally be approximated by

$$V(s) = \cdots + \frac{1}{24}K_3(s)(x^4 - 6x^2z^2 + z^4) + \cdots$$

$$H = \nu_x J_x + \frac{1}{2}\alpha_{xx}J_x^2 + G_{4,0,\ell}J_x^2\cos(4\psi_x - \ell\theta + \xi_{4,0,\ell}), \tag{2.242}$$

$$G_{4,0,\ell}\, e^{j\xi_{4,0,\ell}} = \frac{1}{96\pi}\oint \beta_x^2\, K_3(s)\, e^{j[4\chi_x(s)-(4\nu_x-\ell)\theta]}ds, \tag{2.243}$$

where $K_3(s) = \mp B_3(s)/B\rho$ is the octupole field strength for positive/negative charged particles respectively, $G_{4,0,\ell}$ and $\xi_{4,0,\ell}$ are the resonance strength and phase, and α_{xx} is the nonlinear detuning. The phase space ellipses of Fig. 2.54 driven by a single sextupole can not be described by the Hamiltonian Eq. (2.242).

Figure 2.55 shows the measured Poincaré map near a fourth-order resonance $4\nu_x = 15$ at the IUCF cooler ring. The left plot shows the Poincaré map in the normalized

[79] A. Chao, *et al.*, PRL **61**, 2752 (1988); N. Merminga, *et al.*, EPAC1988, 791 (1988); T. Satogata, *et al.*, PRL **68**, 1838 (1992); T. Chen *et al.*, PRL **68**, 33 (1992).

[80] S.Y. Lee, *et al.*, PRL **67**, 3768 (1991); M. Ellison *et al.*, AIP Conf. Proc. No. 292, p. 170 (1992).

$(x, P_x \equiv \alpha_x x + \beta_x x'))$ phase space. The phase space map is transformed to the action-angle variables $(J = J_x, \phi = \psi_x)$ (see middle plot). The solid lines shows the fitted Hamiltonian tori of Eq. (2.242).

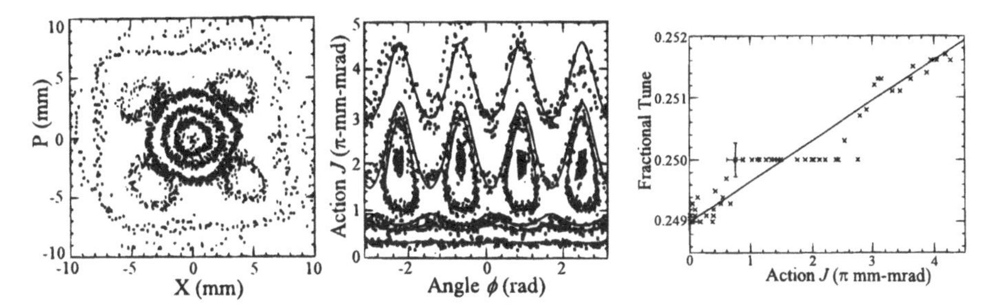

Figure 2.55: The measured betatron Poincaré map (surface of section) in the normalized phase space (left), and in action-angle variables (middle), and the fractional betatron tunes of particles (right) near a fourth-order resonance at the IUCF cooler ring. Be aware that action is not a constant motion near the resonance.

In this example, the fourth-order resonance islands are enclosed by stable invariant tori. The phase-space ellipse is distorted into four island when the betatron tune sits exactly on resonance. The solid lines are the Hamiltonian tori of Eq. (2.242) with parameters $\alpha_{xx} = 650\ (\pi\text{m})^{-1}$, $G_{4,0,15} = 80\ (\pi\text{m})^{-1}$, and $\nu_x - 3.75 = -7.8 \times 10^{-4}$. Because $|\alpha_{xx}| \gg 2|G_{4,0,15}|$, the resonance islands are bounded by stable tori, and because $\alpha > 0$, the resonance occurs at $\delta \equiv \nu_x - 3.75 < 0$, where $\delta = 0$ is called the bifurcation point (see Exercise 2.7.5 for the parametric dependence of the 4th order resonances). The right plot of Fig. 2.55 shows the fractional betatron tune. Note that particles are locked onto the resonance tune across the phase space within the resonance islands. In the nonlinear physics, the constant tune step is called the "devil's staircase," or resonance line shown in the "frequency map."[81]

The tunes of beam particles trapped inside the resonance islands are modulated by the *island tunes* (see Exercise 2.7.6). Particles trapped inside the resonance islands can be driven out by a transverse kicker at the island tune frequency.[82]

If we define the particle instantaneous tune as

$$q_x(n) = \frac{\phi_x(n) - \phi_x(n-1)}{2\pi}, \qquad q_z(n) = \frac{\phi_z(n) - \phi_z(n-1)}{2\pi}, \tag{2.244}$$

the "tune" will oscillate around the resonance. Figure 2.56 shows the normalized Poincaré maps (left) and particle instantaneous tune (right) for a beam with 50 particles in the presence of a fourth order resonance with $\nu_x = 0.2494$, $\alpha = 37.4$ (m^{-1}), $G_{4,0,\ell} = 10.6\ (\text{m}^{-1})$. The resulting tune map differs from that deduced from

[81] J. Laskar, *Physica* D **67**, 257 (1993); and J. Laskar and D. Robin, *Part. Acc.* 54, 183 (1996).
[82] See e.g. Wang, *et al.*, *PRE* **49**, 5697 (1994).

the frequency map, which extracts only the major component of the particle motion during the particle tracking. The solid line is the expected betatron tune vs amplitude. Since the action is not an invariant near a betatron resonance, the particle tunes are not given simply by Eq. (2.240). The dashed line is the tune of particles locked on the resonance. The frequency map method will put those particles at the resonance tune. However, their turn-by-turn particle tune of Eq. (2.244) varies as shown in Fig. 2.56.

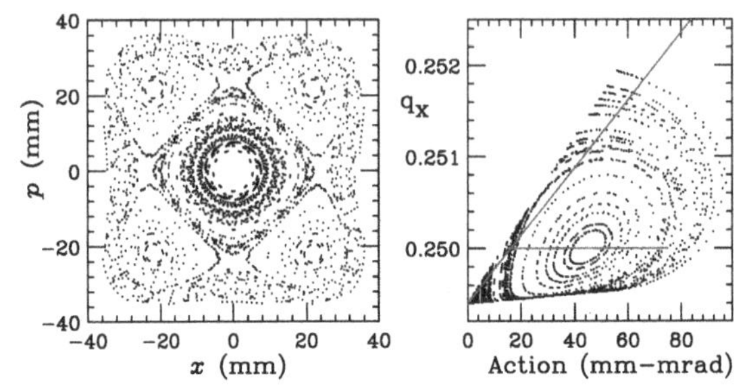

Figure 2.56: Left: The Poincaré maps normalized phase space map of a beam of 50 particles near a 4th order resonance. Right: The corresponding instantaneous particle tune map (see Eq. (2.244)). For particles locked on the resonance, their tune is 1/4 modulated by their associated island tune.

VII.3 Nonlinear Detuning from Sextupoles and Octupoles

Octupoles and multipoles with even order in phase space coordinates can produce nonlinear detuning. For example, the detuning parameters resulting from octupoles with $K_3 = -B_3(s)/(B\rho)$ are

$$\alpha_{xx} = \frac{1}{16\pi}\oint \beta_x^2 K_3 ds, \quad \alpha_{xz} = \frac{-1}{8\pi}\oint \beta_x\beta_z K_3 ds, \quad \alpha_{zz} = \frac{1}{16\pi}\oint \beta_z^2 K_3 ds. \quad (2.245)$$

Concatenation of strong sextupoles in high energy collider and storage rings can also induce substantial nonlinear betatron detuning. For sextupoles, the detuning coefficients α_{xx}, α_{xz}, and α_{zz} due to sextupoles are

$$\alpha_{xx} = \frac{-1}{16\pi}\sum_{i,j} S_i S_j \beta_{x,i}^{3/2}\beta_{x,j}^{3/2}\left[\frac{\cos 3(\pi\nu_x - |\psi_{x,ij}|)}{\sin 3\pi\nu_x} + 3\frac{\cos(\pi\nu_x - |\psi_{x,ij}|)}{\sin \pi\nu_x}\right],$$

$$\alpha_{xz} = \frac{-1}{8\pi}\left\{\sum_{i,j} S_i S_j \beta_{x,i}^{1/2}\beta_{x,j}^{1/2}\beta_{z,i}\beta_{z,j}\left[\frac{\cos[2(\pi\nu_z - |\psi_{z,ij}|) + \pi\nu_x - |\psi_{x,ij}|]}{\sin\pi(2\nu_z + \nu_x)}\right.\right.$$
$$\left.\left. + \frac{\cos[2(\pi\nu_z - |\psi_{z,ij}|) - \pi\nu_x + |\psi_{x,ij}|]}{\sin\pi(2\nu_z - \nu_x)}\right] - 2\sum_{i,j} S_i S_j \beta_{x,i}^{3/2}\beta_{x,j}^{1/2}\beta_{z,j}\frac{\cos(\pi\nu_x - |\psi_{x,ij}|)}{\sin\pi\nu_x}\right\},$$

$$\alpha_{zz} = \frac{-1}{16\pi}\sum_{i,j} S_i S_j \beta_{x,i}^{1/2}\beta_{x,j}^{1/2}\beta_{z,i}\beta_{z,j}\left[\frac{\cos[2(\pi\nu_z - |\psi_{z,ij}|) + \pi\nu_x - |\psi_{x,ij}|]}{\sin\pi(2\nu_z + \nu_x)}\right.$$
$$\left. - \frac{\cos[2(\pi\nu_z - |\psi_{z,ij}|) - (\pi\nu_x - |\psi_{x,ij}|)]}{\sin\pi(2\nu_z - \nu_x)} + 4\frac{\cos(\pi\nu_x - |\psi_{x,i} - \psi_{x,j}|)}{\sin\pi\nu_x}\right],$$

where $\psi_{x,ij} = \psi_{x,i} - \psi_{x,j}$ and $\psi_{z,ij} = \psi_{z,i} - \psi_{z,j}$ are betatron phase advances from s_j to s_i. Since the tune depends on the zeroth harmonic of a perturbed quadrupole field, the nonlinear detuning parameter is proportional to the superperiod of the accelerator. These coefficients can be evaluated from sextupole strengths distributed in one superperiod.

VII.4 Betatron Tunes and Nonlinear Resonances

A beam is composed of many particles. The betatron tunes of each particle depend on its off-momentum coordinate $\Delta p/p_0$ due to chromaticity, and on its betatron-actions due to magnetic multipoles, mean field of the space charge force, and beam-beam interaction for colliding beams. The incoherent Laslett space-charge tune spread of low energy high intensity accelerators can be as large as 0.5 (see Exercise 2.3.2). The beam-beam parameter ξ of Eq. (5.52) is about 0.1 for high luminosity colliders. The detuning parameters of Eq. (2.240) are usually minimized to optimize the dynamic aperture.

The betatron tunes should avoid the linear betatron resonances at $\nu_x = m$ or $\nu_z = n$, where m, n are integers, and half-integer integer betatron resonances at $2\nu_x = m$ or $2\nu_z = n$ due to the linear imperfections. Similarly, the betatron tunes should avoid linear coupling resonances at $\nu_x \pm \nu_z = \ell$ driven by skew quadrupoles and solenoids discussed in Sec. VI.

Sextupoles are important to chromatic correction discussed in Sec. V. We have discussed several low order resonances driven by sextupoles in previous sections. Besides the sextupole, magnetic multipoles do exist in accelerator magnets. The Coulomb (space charge) force of the beam is also highly nonlinear. These nonlinear forces produce higher order resonances. Figure 2.57 shows the betatron resonances up to the 4th (left) and the 8th order (right), where the solid lines correspond to resonances due to normal multipoles, while the dashed lines arise from skew multipoles. The available resonance-free tune space becomes small. When the betatron tune spread of the beam becomes large,[83] resonance (stopband) correction becomes important for attaining beam stability.

The lifetime of beams in many storage rings and colliders may suffer if the betatron tunes sit near a betatron resonance. The beam-beam interaction can drive higher order resonances observed at the SP$\bar{\text{P}}$S colliders, proton-proton colliders, and many e^+e^- colliders. For example, lifetime degradation has been observed near the 7th order resonance at the SP$\bar{\text{P}}$S driven by beam-beam interaction with linear beam-beam tune shift parameter of $\xi_{\rm bb} = 3.3 \times 10^{-3}$ per crossing (see Eq. (5.52) in Chap. 5).When the beam-beam tune spread is large, the tune space that is free from high order

[83]The betatron tune spread of a beam may arise from the incoherent space charge (Laslett) tune shift, chromaticity, beam-beam interaction, betatron amplitude detuning, etc.

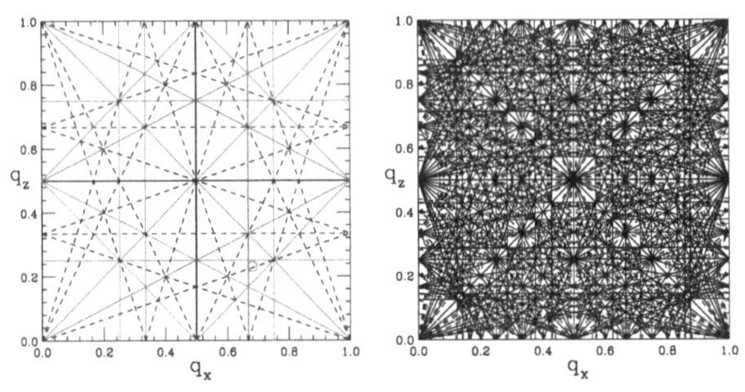

Figure 2.57: The resonance lines: $m_x\nu_x + m_z\nu_z =$ integer with $|m_x|+|m_z| \le 4$ (left) and $|m_x| + |m_z| \le 8$ (Right). The symbol q_x and q_z are the fractional parts of betatron tunes ν_x and ν_z. When higher order resonances are important, resonance-free tune space becomes small.

resonances becomes very small. Betatron tune stability has becomes an important issue for successful operation of storage rings.

A. Emittance growth, beam loss and dynamic aperture

As illustrated in previous Sections, resonances can form resonance islands with phase space area $16\sqrt{g/\alpha}$ in the asymptotic region (see Exercise 2.7.6e), where g is the effective resonance strength and α is the effective detuning parameter. A larger betatron detuning has a smaller resonance perturbation. If these resonance islands are bounded by invariant tori, the beam bunch is bounded, and the result is emittance growth, or "emittance dilution." However, when both the betatron detuning and beam emittance are large, the betatron tune spread of the beam may cover many resonances and result in particle loss and limited dynamic aperture.

When the betatron tunes of particles in a beam bunch sit on a resonance condition $m\nu_x + n\nu_z = \ell$, the betatron motion is strongly perturbed by a resonance, some particles may be trapped in resonance islands, some may drift beyond dynamic aperture and lost. We design and operate accelerators to avoid all low order betatron resonances up to $|m|+|n| \le 4$ (see Fig. 2.57). The left plot of Fig. 2.58 shows phase space distribution of a Gaussian beam with rms emittance 4.64 $\pi\mu$m sitting on a third order resonance at $\nu_x = 6.33$ with $G_{3,0,\ell} = 0.1483$ $(\pi\text{m})^{-1/2}$ and $\alpha = 391$ $(\pi\text{m})^{-1}$. Beam density is diluted by the existence of a nearby resonance. Accelerator magnets are designed to minimize higher order multipoles, and thus higher order resonances are normally weaker.

Similar experiments of beam loss and stability region in tune space have been carried out in electron analog of the AGS accelerator and the FFA accelerators.[84] These experiments confirm the Kolmogorov Arnold Moser (KAM) theorem: the particle motion in accelerator is stable and quasi-periodic if the betatron tunes can avoid low order resonances.

[84] E.D. Courant, Proc. of the CERN Symposium on High Energy Accelerators and Pion Physics Vol. 1, p. 257 (CERN, 1956); F. T. Cole, *et al.*, Review of Scientific Instruments **28**, 403 (1957).

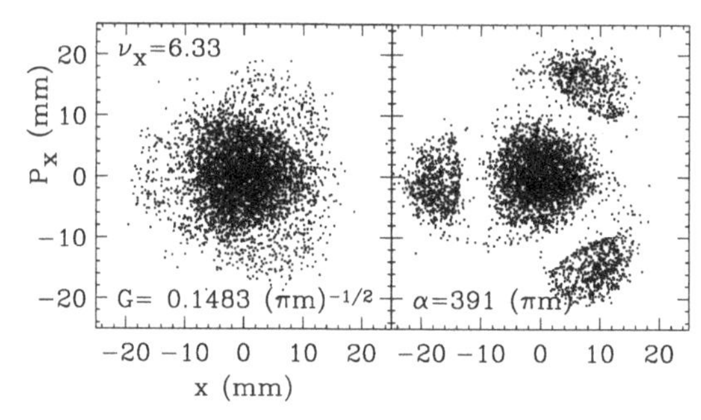

Figure 2.58: Left: The final phase space distribution of a beam with rms initial emittance 4.6 $\pi\mu$m sitting at $\nu_x = 6.33$ near a 3rd order resonance with $G = 0.1483\ (\pi\mathrm{m})^{-1/2}$ and $\alpha = 391\ (\pi\mathrm{m})^{-1}$. Right: phase space distribution of the same beam crossing through the same third order resonance with ramping rate $d\nu_x/dn = 3\times10^{-5}$.

The betatron tunes of a rapid cycling accelerator, including cyclotrons, may ramp through betatron resonances. Emittance of the beam can increase during the passage of resonances, and particles can also be trapped by resonance islands as the betatron tune moves away from the resonance. The right plot of Fig. 2.58 shows the phase space distribution when the betatron tune passes through a third order resonance.

Experimental measurement of emittance growth and beam loss are useful in the understanding of nonlinear beam dynamics in accelerators. The energy change rate of FFA accelerators can be very fast because the magnetic field does not change. Figure 2.59 shows the beam loss and betatron tunes vs time of the Kyoto University Research Reactor Institute (KURRI) scaling FFA. Beam loss may occur when the betatron tunes cross resonance lines. Since the orbit of the FFA accelerator moves across the magnet aperture, the magnetic field multipoles may also change. It requires detailed magnetic field map to analyze beam loss mechanisms. Note that in this example the acceleration rate is not particularly fast. Further measurements of the emittance growth and beam loss vs the beam acceleration rate would be very useful.

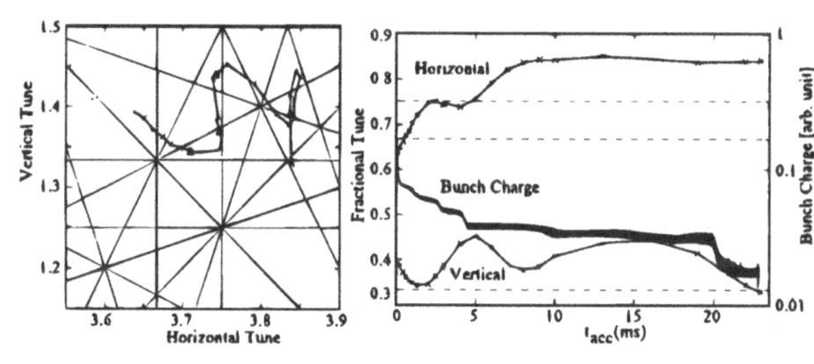

Figure 2.59: Movement of betatron tunes during ramp (Left) and fractional betatron tunes and bunch charge vs time (Right) during the ramp of the KURRI scaling FFA (see Figs. 22 and 23 in S.L. Sheehy *et. al.*, Prog. Theor. Exp. Phys. 2016, 073G01 Figs. 22, 23).

B. Tune diffusion rate and dynamic aperture

An off-momentum particle executes synchrotron motion at synchrotron frequency with $\Delta p/p_0 = (\widehat{\Delta p}/p_0)\cos\omega_s t$, where ω_s is the angular synchrotron frequency of Eq. (2.169). Chromaticity can induce modulation to the betatron phase at the synchrotron frequency, i.e. $\nu = \nu_0 + (\hat{\Delta\nu}) \cos \omega_s t$. If the chromaticity is large, the betatron tunes may cross resonances causing emittance growth or particles being carried away in resonance islands.

Tune modulation can also create overlapping resonances inducing chaos in regions of phase space where resonances overlap. Particularly, this can occur around the separatrix orbit. If the chaos is bounded by invariant tori, one may achieve emittance dilution without beam loss. When betatron tune is modulated, particles trapped in resonance islands can also be excited and driven out of resonance islands into separatrix and to an ever increasing amplitude. Such mechanism of particle diffusion causes emittance dilution and particle loss.[85] In most cases, when chaos occur, beam loss is inevitable and beam lifetime will be reduced.

The dynamic aperture (DA), defined as the maximum phase space boundary that beam particles can survive within the usable time in accelerators, is normally carried out by numerical simulations. There are a number of accelerator design and tracking programs [23, 24, 25]. Particles revolve about $10^3 - 10^4$ turns in rapid cycling synchrotrons, and up to 10^{10} turns in storage rings. Normally, the DA numerical tracking is carry out in 10^3 to 10^6 revolutions. Long term tracking is time consuming. Numerical tricks, used in testing the stability vs chaoticity with shorter term tracking, are Lyapunov exponent and tune diffusion rate.

The Lyapunov exponent measures the growth rate of the distance between two phase space points, initially infinitesimally small, in a dynamical system, i.e. $\lambda = \frac{1}{t}\ln(|\delta\mathbf{Y}(t)|/|\delta\mathbf{Y}(0)|)$, where $|\delta\mathbf{Y}(t)|$ and $|\delta\mathbf{Y}(0)|$ are the distances between two adjacent particle phase space coordinates at time t and at time $t = 0$, and λ is called the Lyapunov exponent. Chaotic phase space region exhibit positive Lyapunov exponent. In the stable phase space region, the Lyapunov exponent is 0, i.e. the distance is a linear or a sinusoidal function of time.

The tune diffusion rate is defined as the logarithm of betatron tune distance between the first half and the second half of a particle tracking simulation, i.e. $D_{\rm tune} = \log_{10}\sqrt{(\nu_x(2) - \nu_x(1))^2 + (\nu_z(2) - \nu_z(1))^2}$, where $(\nu_x(2), \nu_z(2))$ and $(\nu_x(1), \nu_z(1))$ are tunes obtained from the frequency map analysis from the second half and the first half of the particle tracking turns respectively. The tune diffusion rate would be small at the stable region, and become large near the chaotic or unstable region. Advance in accelerator physics understanding, numerical simulations becomes a reliable tool for the design of many accelerators.

[85]See e.g. A. Gerasimov and S.Y. Lee, *PRE* **49**, 3881 (1994); Y. Wang, *et al.*, *PRE* **49**, 5697 (1994). For controlled emittance dilution see e.g. D. Jeon, et al., *PRL* **80**, 2314 (1998); C.M. Chu *et al.*, *PRE* **60**, 6051 (1999); S.Y. Lee, K.Y. Ng, *Proc. of HB2010*, p. 639 (2010).

C. Space charge effects

The Coulomb force between charged particles in a beam can cause emittance growth and beam loss. The intrabeam Coulomb scattering (IBS) can produce emittance growth measured by the IBS parameter of the 6D phase space density. This is a non-resonance effect, to be discussed in Chapter 4.

The mean field of the Coulomb force can interact with the vacuum chamber and produce quadrupole field that can change the betatron tunes of the beam, called the coherent space charge tune shift.[86] The mean field of the Coulomb force can also produce a defocusing force on each particle within the beam. The defocusing force depends on the particle's position relative to the beam center; particles at the center of the beam bunch will experience the largest defocusing force; and particles farther away from the beam center receive a smaller defocusing force. The effect causes an "incoherent" space charge tune spread to the beam The strength is measured by the Laslett tune spread parameter [see Eq. (2.88) and Exercise 2.3.2]. If the tune spread is large, the betatron tunes will encounter resonances and emittance growth or beam loss will occur. The space charge spread parameter becomes a key parameter in the design of low energy synchrotrons. Some space charge resonance effects are listed below:

Effect of betatron resonances on beam emittances: As shown in Fig. 2.30, the vertical emittance grew rapidly in a rapid cycling Fermilab Booster synchrotron, where the space charge tune shift can be larger than 0.5. Based on the envelope modes of beam oscillation discussed in Sec. III.9, the resonance growth may arise from the envelope instability at a half-integer tune. However, careful analysis of emittance evolution data showed that the emittance growth of the Fermilab Booster was mainly caused by skew quadrupole resonances at $\nu_x + \nu_z = \ell$ and $\nu_x - \nu_z = 0$.

Systematic space charge resonances: Since the space charge potential for a beam with symmetric distribution has terms proportional to x^4, x^2z^2, z^4, $x^6 \ldots$ etc., higher order resonances can be important in causing emittance growth, particularly near the systematic resonances. As an example, we consider a perturbation of x^4 or z^4 term. the space charge kicks to particles from FODO cells with 90° phase advance will add up coherently and trap particles in *space charge induced resonance islands*. Experimental measurements at KEK PS observed the effect of the 4th order systematic resonance on beam distribution.[87] Systematic 4th order systematic resonances are located at $4\nu_x = \ell$, where ℓ is divisible by the accelerator superperiod P.

In recent years, non-scaling FFA accelerators have been considered for high power proton drivers in applications such as muon-collider, neutrino factory, nuclear waste transmutation, etc. The betatron tunes of a non-scaling FFA will cross many integer units, including systematic space charge resonances, such as $4\nu_x = P$, $4\nu_z = P$,

[86]See e.g. M.A. Plum *et al.*, *Proc. of PAC 1997*, p. 1611.
[87]S. Igarashi *et al.*, *Proc. of PAC2003*, p. 2610 (IEEE, NY, 2003)

$6\nu_x = P, \cdots$, where P is the superperiod of the accelerator. Simulations have been carried out to investigate their effects on beam distribution and emittance growth. Figure 2.60 shows an example of multiparticle simulation demonstrating the emittance growth and beam distribution in the phase space after passing through a systematic 4th order space charge resonance. More importantly, the emittance growth obeys a simple scaling law:[88] $(d\nu/dn)_c = 8.4(\Delta\nu_{\rm sc})^2 g \exp(31g)$, where $(d\nu/dn)_c$ is the critical tune ramp rate, $\Delta\nu_{\rm sc}$ is the space charge tune spread of the beam (Exercise 2.3.2), g is the systematic octupolar resonance strength due to the nonlinear space charge force.

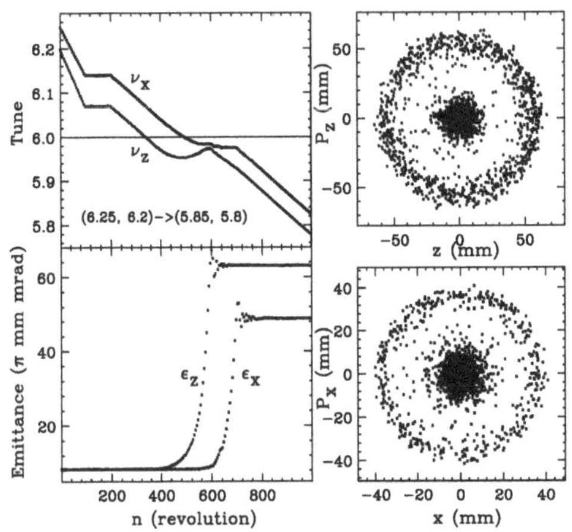

Figure 2.60: Left plots: The evolution of the small amplitude betatron tunes and the normalized emittances vs revolution numbers for a beam with $\Delta\nu_{\rm sc} = 0.109$ while ramping the bare betatron tunes through the fourth order systematic space-charge nonlinear resonance at $4\nu_x = P$. The emittance growth during the down ramp is much more than that of up ramp because the space charge detuning make the tunes stay longer at the resonance. The betatron tunes of most accelerators decrease during acceleration. Right plots: the normalized phase-space maps at the end of down ramp, where the phase space islands decohere into a ring in the phase space.

Simulations and experiments for Linac transport line showed that the $4\nu_x = 1$ systematic space charge resonance could produce sizable emittance growth while the envelope instability is not important.[89] These studies further confirmed that the half-integer stopband and the envelope instability were less important, as discussed in Sec. III.9.

Tune modulation induced by synchrotron motion: Because of synchrotron motion, a nonzero off-momentum particle oscillates between the front and the end of the beam. When the particle reaches the center of the beam, where the linear current density is highest, the betatron tunes of the particle are highly depressed due to space charge force. The space charge force can cause betatron tune modulation with a maximum modulation amplitude of $(\Delta\nu)_{\rm sc}$. This tune modulation still exist even when the chromaticity is corrected to zero. This tune modulation can also cause emittance blow-up or drive particles into resonance islands.

[88] S.Y. Lee, PRL **97**, 104801 (2006); S.Y. Lee, *et al.*, New J. Phys. **8**, 291 (2006); X. Pang *et al.*, Proceedings of HB2008 (http://accelconf.web.cern.ch/AccelConf/HB2008/papers/wega21.pdf).
[89] D. Jeon *et al.*, PRSTAB **12**, 054204.(2009). L. Groening *et al.*, PRL **102**, 234801 (2009).

Exercise 2.7

1. Show that the $3\nu_x = \ell$ resonance strength is given by Eq. (2.232) in the first-order perturbation approximation. Show that, for systematic resonances, $G_{3,0,\ell} = 0$ if $\ell \neq 0$ (Mod P), where P is the superperiodicity of the machine. Show that the resonance strength of the third-order resonance at $3\nu_x = \ell$ due to two sextupoles at s_1 and s_2 is proportional to

$$[\beta_x(s_1)]^{3/2}\,[K_2(s_1)\Delta s] + [\beta_x(s_2)]^{3/2}\,[K_2(s_2)\Delta s]\,e^{j[3\psi_{21}-(3\nu_x-\ell)\Delta\theta]},$$

where $\psi_{21} = \int_{s_1}^{s_2} ds/\beta_x$ is the betatron phase advance, $\Delta\theta = (s_2 - s_1)/R_0$, and R_0 is the average radius of the accelerator. Show that, at the $3\nu_x = \ell$ resonance, the "geometric aberrations" of these two sextupoles cancel each other if $\psi_{21} = \pi$ and $[\beta_x(s_1)]^{3/2}\,[K_2(s_1)\Delta s] = [\beta_x(s_2)]^{3/2}\,[K_2(s_2)\Delta s]$. The geometric aberration of two identical chromatic sextupoles located in the arc of FODO cells separated by 180° in phase advance cancel each other.

2. Near a third-order coupling *difference* resonance at $\nu_x - 2\nu_z = \ell$, where ℓ is an integer, the Hamiltonian can be approximated by

$$H = \nu_x J_x + \nu_z J_z + g J_x^{1/2} J_z \cos(\phi_x - 2\phi_z - \ell\theta + \xi),$$

i here ν_x, ν_z are the betatron tunes, $g = G_{1,-2,\ell} > 0$ and $\xi = \xi_{1,-2,\ell}$ of Eq. (2.233) are the effective resonance strength and phase, and $(J_x, \phi_x, J_z, \phi_z)$ are the horizontal and vertical action-angle coordinates.

   (a) Using the generating function

$$F_2(\phi_x, \phi_z, J_1, J_2) = (\phi_x - 2\phi_z - \ell\theta + \xi)J_1 + \phi_z J_2,$$

   transform the phase-space coordinates from $(J_x, \phi_x, J_z, \phi_z)$ to $(J_1, \phi_1, J_2, \phi_2)$ of the resonance rotating frame, and show that the new Hamiltonian becomes

$$\tilde{H} = H_1 + H_2,$$
$$H_1(J_1, \phi_1) = \delta_1 J_1 + g J_1^{1/2}(J_2 - 2J_1)\cos\phi_1, \quad H_2(J_2, \phi_2) = \nu_z J_2,$$

   where $\delta_1 = \nu_x - 2\nu_z - \ell$ is the resonance proximity parameter. Since the Hamiltonian $\tilde{H}$ is independent of "time" or θ, the Hamiltonian is a constant of motion. In the resonance rotating frame, the equation of motion is given by

$$\dot{\phi}_1 = \frac{\partial H}{\partial J_1}, \quad \dot{J}_1 = -\frac{\partial H}{\partial \phi_1}; \qquad \dot{\phi}_2 = \frac{\partial H}{\partial J_2}, \quad \dot{J}_2 = -\frac{\partial H}{\partial \phi_2}.$$

   Show that $2J_x + J_z$ is also a constant of motion.

   (b) In passing through the resonance for a particle with an initial J_{x0} and J_{z0}, the particle has a constant $J_2 = 2J_{x0} + J_{z0}$. The resonance can cause exchange of J_x and J_z, while maintaining a constant $J_2 = 2J_x + J_z$. Particle motion depends essentially only on the Hamiltonian H_1. Show that the unstable fixed points (UFP) of the Hamiltonian H_1 are located at

$$\left(J_{1,\text{ufp}} = \frac{J_2}{2},\ \phi_{1,\text{ufp}} = \arccos(\xi)\right), \quad \text{where } \xi \equiv \frac{\delta_1}{g\sqrt{2J_2}} \text{ with } |\xi| \leq 1.$$

Bifurcation occurs at $|\xi| = 1$. The UFPs do not exist at $|\xi| > 1$. The separatrix is the Hamiltonian flow passing through the UFP. Show that the UFP is $\sqrt{2J_1}\cos\phi_1 = \xi \cdot J_2$, which is a line cut through the phase space ellipse circle of $2J_1 = J_2$. At the UFP, $\dot{J}_1 = 0$ and $\dot{\phi}_1 = 0$, particles take long time to move away from it. The particle *frequency* map may appears to lock on the coupling line in the tune space. However, the instantaneous tune can vary widely around the resonance line.

(c) Show that the stable fixed points (SFP) of the Hamiltonian H_1 are

$$\sqrt{J_{1,\rm sfp}} = \frac{1}{6g}\left\{\pm\delta_1 + \sqrt{\delta_1^2 + 6g^2 J_2}\right\}, \qquad (\phi_1 = 0\text{and } \pi \text{ respectively}).$$

We can express the SFP as:

$$\sqrt{\frac{2J_{1,\rm sfp}}{J_2}} = \frac{1}{3}\left\{\pm\eta + \sqrt{\eta^2+3}\right\}, \qquad \text{with} \qquad \eta = \frac{\delta_1}{g\sqrt{2J_2}}.$$

Two SFPs are inside the the Courant-Snyder (CS) circle at $|\eta| \le 1$ on each side of the coupling line. At $\delta_1 = 0$, the SFP is located at a radius $\frac{1}{\sqrt{3}}$ from the the origin inside the Courant-Snyder circle. At $|\delta_1| > g\sqrt{2J_2}$, we note that its SFP is not zero, i.e. there is still J_x and J_z exchange such that J_2 is constant! An example of the Poincaré maps of 10 particles in a beam, plotted in every 5 revolutions for 1500 revolutions in the resonance rotating frame: $X = \sqrt{2\beta_x J_1}\cos\phi_1$ and $P = -\sqrt{2\beta_x J_1}\sin\phi_1$ is shown in the Figure below. The proximity parameters are $\delta = +0.001, 0,$ and -0.001 respectively. At $\delta =$, the resonance line cut through the middle of the beam phase space. The resonance strength is $g = 0.42$ (m$^{-1/2}$). When the resonance line cut through the beam, particle phase space map changes. Since these particles have different J_x and J_z, their Poincaré maps may appear to cross each other in 1-D projection, where the resonance line and the SFPs depend on J_2 is the particle.

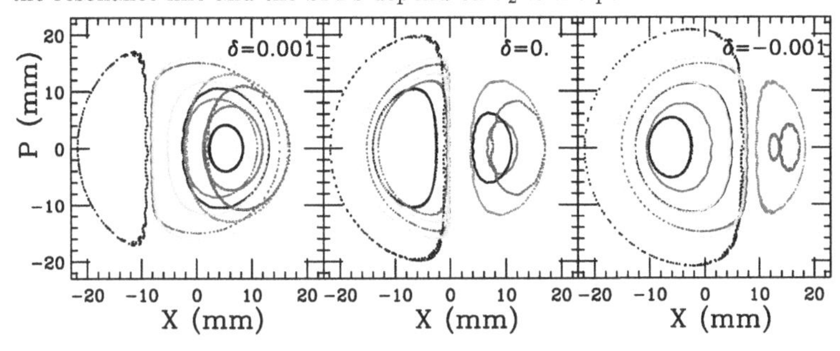

(d) Since the beam in isochronous cyclotrons may encounter or cross the sextupolar coupling resonance (called Walkinshaw resonance), we examine the effects of crossing this resonance. The beam emittance obeys the conservation law: $2\epsilon_x + \epsilon_z =$ constant. The beam distribution function of a 2D-Gaussian beam is $\rho(J_x, J_z) = \frac{1}{\epsilon_x\epsilon_z}\exp\{\frac{J_x}{\epsilon_x} + \frac{J_z}{\epsilon_z}\}$, which is a function of $\frac{J_x}{\epsilon_x} + \frac{J_z}{\epsilon_x}$. Expressing the distribution function in J_1 and J_2, show that $\rho(J_1, J_2)$ is independent of J_1 if $\epsilon_z = 2\epsilon_x$. This means that there is no emittance exchange in passing through the resonance when the emittances obey $\epsilon_z = 2\epsilon_x$.[90] Any beam distribution that depends on $\frac{J_x}{\epsilon_x} + \frac{J_z}{\epsilon_z}$, e.g. KV beam distribution of Eq. (2.71), will have a similar conclusion, i.e. when a beam with $\epsilon_z = 2\epsilon_x$ passes through a Walkinshaw resonance, there is no emittance exchange. A beam in the presence of quantum fluctuation and dissipation will also reach this equilibrium emittance condition.

3. Near a *sum* resonance at $\nu_x + 2\nu_z = \ell$, the Hamiltonian can be approximated by

$$H = \nu_x J_x + \nu_z J_z + g J_x^{1/2} J_z \cos(\phi_x + 2\phi_z - \ell\theta + \xi),$$

where ν_x, ν_z are betatron tunes, $g = G_{1,+2,\ell}$ and $\xi = \xi_{1,2,\ell}$ are the effective resonance strength and phase of Eq. (2.233), and $(J_x, \phi_x, J_z, \phi_z)$ are horizontal and vertical action-angle phase-space coordinates. Discuss the difference between the sum and difference resonances.

4. In general, betatron motion in storage rings can encounter many nonlinear resonances. Normally only low-order resonances are important. If the betatron tune of the machine is chosen such that $m\nu_x + n\nu_z \approx \ell$, where $m > 0, n$, and ℓ are integers, the Hamiltonian can be approximated by

$$H = H_0(J_x, J_z) + g J_x^{|m|/2} J_z^{|n|/2} \cos(m\phi_x + n\phi_z - \ell\theta + \xi) + \Delta H.$$

This is called a sum resonance if $mn > 0$, and a difference resonance if $mn < 0$. Here g and ξ are the resonance strength and phase. Neglecting the perturbing term ΔH that includes contributions from other resonances, derive the invariants of the approximated Hamiltonian.

Transform the phase-space coordinates from $(J_x, \phi_x, J_z, \phi_z)$ to $(J_1, \phi_1, J_2, \phi_2)$ by using the generating function

$$F_2(\phi_x, \phi_z, J_1, J_2) = (m\phi_x + n\phi_z - \ell\theta + \xi)J_1 + \phi_z J_2,$$

and find the new Hamiltonian. Show that the new Hamiltonian is a constant of motions, and show that the action J_2 is invariant, i.e. $nJ_x - mJ_z =$ constant. Discuss the difference between the sum and the difference resonances.

5. The Hamiltonian (2.242) can be transformed to a "time independent" or "resonance rotating" form:

$$H = \delta J + \frac{1}{2}\alpha J^2 + G J^2 \cos 4\psi,$$

where $J = J_x$, $\psi = \psi_x - \frac{\ell}{4}\theta + \frac{\xi}{4}$, $\alpha = \alpha_{xx}$, $G = G_{4,0,\ell} > 0$, $\xi = \xi_{4,0,\ell}$ and $\delta = \nu_x - \frac{\ell}{4}$ is the resonance proximity parameter.

(a) Show that the stable and unstable fixed points of the Hamiltonian are

$$\alpha > 0 \text{ and } \delta < 0: \begin{cases} J_{\rm SFP} = -\dfrac{\delta}{\alpha - 2G}, & \psi_{\rm SFP} = \dfrac{\pi}{4}, \dfrac{3\pi}{4}, \dfrac{5\pi}{4}, \dfrac{7\pi}{4} \quad \text{for } \alpha > 2G \\ J_{\rm UFP} = -\dfrac{\delta}{\alpha + 2G}, & \psi_{\rm UFP} = 0, \dfrac{\pi}{2}, \pi, \dfrac{3\pi}{2} \end{cases}$$

$$\alpha < 0 \text{ and } \delta > 0: \begin{cases} J_{\rm UFP} = -\dfrac{\delta}{\alpha - 2G}, & \psi_{\rm UFP} = \dfrac{\pi}{4}, \dfrac{3\pi}{4}, \dfrac{5\pi}{4}, \dfrac{7\pi}{4} \\ J_{\rm SFP} = -\dfrac{\delta}{\alpha + 2G}, & \psi_{\rm SFP} = 0, \dfrac{\pi}{2}, \pi, \dfrac{3\pi}{2} \quad \text{for } |\alpha| > 2G \end{cases}$$

[90]S.Y. Lee, K.Y. Ng, H. Liu, and H.C. Chao, PRL **110**, 094801 (2013).

Note that stable 4th order resonance islands exists only if $|\alpha| > 2G$. If the Stable resonance islands exist, we have $J_{\rm SFP} > J_{\rm UFP}$. The actions of both fixed points are proportional to the proximity parameter δ. To estimate the resonance width, we set $J_{x,\rm UFP} = J_{\rm UFP} = 6\epsilon_{\rm rms}$. The betatron tune width of unstable motion is resonance width: $\Delta\nu_{4,\rm width} = 6(|\alpha| + 2|G|)\epsilon_{\rm rms}$.

(b) The separatrix is obtained by setting $H = H(\psi_{\rm UFP}, J_{\rm UFP})$:

$$\delta J + \frac{1}{2}\alpha J^2 + GJ^2 \cos 4\psi = \begin{cases} -\dfrac{\delta^2}{2(\alpha - 2G)} & \text{if } \alpha > 0 \text{ and } \delta < 0 \\ -\dfrac{\delta^2}{2(\alpha + 2G)} & \text{if } \alpha < 0 \text{ and } \delta > 0 \end{cases}$$

for $|\alpha| > 2G > 0$. Show that the area of 4th order resonance islands is

$$4\oint_{-\pi/4}^{\pi/4} J(\psi)d\psi = \frac{8\,|-\delta|}{\sqrt{(\alpha - 2G)(\alpha + 2G)}} \tan^{-1}\sqrt{\frac{|4G|}{\alpha + 2G}}.$$

Furthermore, if we define the phase space coordinates: $(X = \sqrt{J}\cos\psi,\ P = \sqrt{J}\sin\psi)$, the separatrix can be decomposed into two ellipses:

$$A\left[(B + \sqrt{B^2 - 1})X^2 + P^2 + \frac{\delta}{A(B + 1 - \sqrt{B^2 - 1})}\right] \times \left[(B - \sqrt{B^2 - 1})X^2 + P^2 + \frac{\delta}{A(B + 1 + \sqrt{B^2 - 1})}\right] = 0,$$

where $A = \frac{1}{2}\alpha + G$, $B = (\alpha - 6G)/2A$. The UFPs are intersections of these two ellipses.

6. The Hamiltonian in the action-angle variables near a single resonance ($m\nu = \ell$) can be approximated by

$$H = \nu J + \frac{1}{2}\alpha J^2 + GJ^{m/2}\cos(m\phi - \ell\theta + \xi),$$

where J and ϕ are the conjugate action-angle variables for the betatron oscillations, θ is the orbital angle serving for the time coordinate, ν is the betatron tune (either horizontal or vertical), α is the nonlinear betatron detuning parameter arising from higher order multipoles, G in unit of $(\pi\text{m})^{-(m-2)/2}$ and ξ are the resonance strength and phase of the 1D resonance at $m\nu = \ell$ with integer m and ℓ.

(a) Using the generating function, $F_2 = (\phi - \frac{\ell}{m}\theta + \frac{\xi}{m})I$, show that the new Hamiltonian is

$$\tilde{H} = \delta I + \frac{1}{2}\alpha I^2 + GI^{\frac{m}{2}}\cos m\psi.$$

where the new action-angle coordinates are $I = J$ and $\psi = \phi - \frac{\ell}{m}\theta + \frac{\xi}{m}$, and $\delta = \nu - \frac{\ell}{m}$ is the proximity of the betatron tune to the resonance line.

(b) Show that the fixed points $(I_{\rm fp}, \psi_{\rm fp})$ of the Hamiltonian are

$$\sin m\psi_{\rm fp} = 0, \quad \text{and} \quad \delta + \alpha I_{\rm fp} \pm \frac{m}{2}GI_{\rm fp}^{\frac{m}{2}-1} = 0.$$

(c) In particle accelerators, the betatron tunes may be time dependent due to quadrupole current supply ripple. With a small tune modulation, the parameters α and G do not vary appreciably. Show that the equation for the phase oscillations becomes,

$$\ddot{\psi}_m + F\sin\psi_m = m\dot{\delta},$$

where $\psi_m = m\psi$ signifies the *island* phase angle and F is the spring constant for the phase oscillation given by

$$F = \frac{m^3}{2}EGI^{\frac{m}{2}-2} + \frac{m^2}{4}(m-4)\alpha GI^{\frac{m}{2}}.$$

(d) Using Hamilton's equation, show that the actual island tune $\tilde{\nu}_{island}$ of a given torus is

$$\tilde{\nu}_{island} = 2\pi m\left[\oint\frac{dI}{(G^2I^m - [E-\delta I - \frac{1}{2}\alpha I^2]^2)^{1/2}}\right]^{-1}.$$

The small amplitude island tune is $\sqrt{F}=\frac{m}{2}\left[2mE_{\rm sfp}GI_{\rm sfp}^{\frac{m}{2}-2}+(m-4)\alpha GI_{\rm sfp}^{\frac{m}{2}}\right]^{1/2}$.

(e) When the resonance strength G is small, the resonance island is small. The resonance Hamiltonian above can be approximated by $H\approx\frac{1}{2}\alpha(I-I_r)^2+g\cos(m\psi)$, where $I_r\approx|\delta/\alpha|$ is the resonance action, $g\approx GI_r^{m/2}\approx G(|\delta/\alpha|)^{m/2}$ is the effective resonance strength. The equations of motion of this Hamiltonian resemble Eqs. (2.167) and (2.168) of the synchrotron motion. It has m resonance islands. Show that the small amplitude island tune, the island width, and the total phase space area of m islands, in the unit of (πm), are

$$\nu_{\rm island} = m\sqrt{\alpha g}, \qquad \Delta I = 4\sqrt{g/\alpha},$$
$$\mathcal{A} = \frac{16}{\pi}\sqrt{|g/\alpha|} \approx \frac{16}{\pi}|G^{1/2}\delta^{m/4}\alpha^{-(m+2)/4}|.$$

7. Near an integer tune, the Hamiltonian Eq. (2.231) can be approximated by

$$H = \nu J + gJ^{3/2}\cos(\phi - \ell\theta + \xi),$$

where J and ϕ are the conjugate action-angle variables for the betatron oscillations, θ is the orbital angle serving for the time coordinate, ν is the horizontal betatron tune, $g = |g_{1,0,3,0,\ell}|$ and ξ are the resonance strength and phase given in Eq. (2.232). Using the generating function, $F_2 = (\phi - \ell\theta + \xi)I$, we find the new Hamiltonian in the "resonance rotating frame" as

$$\tilde{H} = \delta I + gI^{\frac{3}{2}}\cos\psi.$$

where the new action-angle coordinates are $I = J$ and $\psi = \phi - \ell\theta + \xi$, and $\delta = \nu - \ell$ is the proximity of the betatron tune to the resonance line.

(a) Show that the fixed point of the Hamiltonian is $J_{\rm UFP}^{1/2} = |\frac{2}{3}w|$ and $\psi_{\rm UFP} = 0$ for $\delta < 0$ and $\psi_{\rm UFP} = \pi$ for $\delta > 0$. where $w = \delta/g$.

(b) Show that the separatrix is given by the Hamiltonian Torus

$$\tilde{H} = \delta I + gI^{\frac{3}{2}} \cos\psi = -\frac{4}{27}|\delta|w^2$$

If one define the $X = \sqrt{J}\cos\psi$ and $P = -\sqrt{J}\sin\psi$, the separatrix orbit becomes

$$P^2 = \frac{1}{w \pm x}\left(X^2 \pm wX^2 + \frac{4}{27}w^3\right) = \frac{1}{w \pm x}\left(X \mp \frac{1}{3}w\right)\left(X - \frac{2}{3}w\right)^2$$

(c) Explain the difference between the nonlinear integer resonance above and the linear integer resonance of Sec. III.1, where the Hamiltonian can be approximated by

$$H = \nu J + gJ^{1/2}\cos(\phi - \ell\theta + \xi).$$

Transform the coordinates to the resonance rotating frame, the new Hamiltonian becomes $\tilde{H} = \delta I + gI^{1/2}\cos\psi$, where $I = J$, $\psi = \phi - \ell\theta + \xi$, and $\delta = \nu - \ell$. Define $X = \sqrt{I}\cos\psi$ and $P = -\sqrt{I}\sin\psi$, show that the betatron motion becomes

$$(X - X_{\rm co})^2 + P^2 = E/\delta + X_{\rm co}^2$$

where the "closed orbit" is $X_{\rm co} = -g/[2\delta]$ and E is the value of the Hamiltonian $\tilde{H}$. Let $g = \nu f_\ell/\sqrt{2}$, where f_ℓ is the stopband integral of Eq. (2.94), show that $X_{\rm co}$ is equal to $x_{\rm co}/\sqrt{2\beta_x}$ of Eq. (2.96) in the single stopband approximation.

8. **Canonical Perturbation method:** The 1D resonance Hamiltonian of (2.231) due to sextupoles can be approximated by

$$H = \nu_x J_x + G_3 J_x^{3/2}\cos(3\phi_x - \ell_3\theta) + g_1 J_x^{3/2}\cos(\phi_x - \ell_1\theta),$$

where $G_3 = G_{3,0,\ell_1}$ and $g_1 = g_{1,0,3,0,\ell_1}$ with phases $\xi_{3,0,\ell} = 0$ and $\xi_{1,0,3,0,\ell_1} = 0$. Examples of these 1D resonances are shown in Fig. 2.51 for $3\nu_x \sim \ell_3 = 11$ and $\nu_x \sim \ell_1 = 4$ for the Exercise 2.7.7. Using Eq. (2.232), we find $G_3 = \frac{\sqrt{2}}{24\pi}\beta_x^{3/2}(K_2\Delta\ell)$ and $g_1 = 3G_3$, where $K_2\Delta\ell$ is the integrated sextupole strength. The example shown in Fig. 2.51 gives $G = 0.8388$ $(\pi\text{m})^{-1/2}$, and $g_1 = 3G_3$ for $\beta_x = 20$ m and $K_2\Delta\ell = 0.5$ m^{-2}. We would like use this example to illustrate the combination of $3\nu_x = \ell_3$ and $\nu_x = \ell_1$ resonances can produce a combined resonance $3\nu_x + \nu_x = \ell_3 + \ell_1$ or $4\nu_x = 15$ resonance of shown in Fig. 2.54. Consider the canonical transformation:

$$F_2(\phi_x, J) = \phi_x J + B_3(J)\sin(3\phi_x - \ell_3\theta) + B_1(J)\sin(\phi_x - \ell_1\theta).$$

The coordinates are transformed according to:

$$\begin{aligned} \phi &= \phi_x + B_3'(J)\sin(3\phi_x - \ell_3\theta) + B_1'(J)\sin(\phi_x - \ell_1\theta) \\ J_x &= J + 3B_3(J)\cos(3\phi_x - \ell_3\theta) + B_1(J)\cos(\phi_x - \ell_1\theta). \end{aligned}$$

or

$$\begin{aligned} \Delta\phi = \phi_x - \phi \approx\ & -B_3'\sin(3\phi - \ell_3\theta) - B_1'\sin(\phi - \ell_1\theta) \\ & -3B_3'\cos(3\phi - \ell_3\theta)\left[B_3'\sin(3\phi - \ell_3\theta) + B_1'\sin(\phi - \ell_1\theta)\right] \\ & -B_1'\cos(\phi - \ell_1\theta)\left[B_3'\sin(3\phi - \ell_3\theta) + B_1'\sin(\phi - \ell_1\theta)\right] \end{aligned}$$

$$\begin{aligned}
\Delta J = J_x - J &\approx +3B_3\cos(3\phi-\ell_3\theta)+B_1\cos(\phi-\ell_1\theta)\\
&\quad +9B_3\sin(3\phi-\ell_3\theta)\left[B_3'\sin(3\phi-\ell_3\theta)+B_1'\sin(\phi-\ell_1\theta)\right]\\
&\quad +B_1\sin(\phi-\ell_1\theta)\left[B_3'\sin(3\phi-\ell_3\theta)+B_1'\sin(\phi-\ell_1\theta)\right]\\
\frac{\partial F_2}{\partial\theta} &\approx -\ell_3B_3\cos(3\phi-\ell_3\theta)-\ell_1B_1\cos(\phi-\ell_1\theta)\\
&\quad +3\ell_3B_3\sin(3\phi-\ell_3\theta)\left[-B_3'\sin(3\phi-\ell_3\theta)-B_1'\sin(\phi-\ell_1\theta)\right]\\
&\quad +\ell_1B_1\sin(\phi-\ell_1\theta)\left[-B_3'\sin(3\phi-\ell_3\theta)-B_1'\sin(\phi-\ell_1\theta)\right].
\end{aligned}$$

Let $C_3=\cos(3\phi-\ell_3\theta)$, $S_3=\sin(3\phi-\ell_3\theta)$, $C_1=\cos(\phi-\ell_1\theta)$, $S_1=\sin(\phi-\ell_1\theta)$, $3\delta_3=3\nu_x-\ell_3$, $\delta_1=\nu_x-\ell_1$, the transformed Hamiltonian becomes

$$\begin{aligned}
\tilde H &= \nu_xJ+3\delta_3B_3C_3+\delta_1B_1C_1+9\delta_3B_3S_3\left[B_3'S_3+B_1'S_1\right]+\delta_1B_1S_1\left[B_3'S_3+B_1'S_1\right]\\
&\quad +G_3J^{3/2}(1+\frac{3}{2J}\Delta J)(C_3-3S_3\Delta\phi)+g_1J^{3/2}(1+\frac{3}{2J}\Delta J)(C_1-S_1\Delta\phi)
\end{aligned}$$

Now, we choose B_3 and B_1 to cancel the first order terms:

$$B_3=-\frac{G_3J^{3/2}}{3\delta_3},\quad B_1=-\frac{g_1J^{3/2}}{\delta_1},\qquad B_3'=-\frac{3G_3J^{1/2}}{6\delta_3},\quad B_1'=-\frac{3g_1J^{1/2}}{2\delta_1}.$$

Take the "time" average to obtain $\langle C_3C_3\rangle=\frac{1}{2}$ $\langle S_3S_3\rangle=\frac{1}{2}$, $\langle C_1C_1\rangle=\frac{1}{2}$, $\langle S_1S_1\rangle=\frac{1}{2}$. Finally collect terms that drive the $4\nu_x=\ell_1+\ell_3$ resonance and obtain

$$\tilde H \approx \nu_xJ+\frac{1}{2}\alpha J^2+\left[G_4J^2+g_{4,0,6,0}J^3\right]\cos[4\nu_x-(\ell_1+\ell_3)\theta]+\cdots.$$

Using $\delta_1\approx-0.25$, $3\delta_3\approx0.25$, $g_1=3G_3$, we find

$$\alpha\approx+36G_3^2,\qquad G_4\approx18G_3^2,\qquad g_{4,0,6,0}=-972G_3^4.$$

For a single sextupole, the strengths of all resonances at different ℓ_3 or $\ell-1$ are identical. Including these ℓ_1 and ℓ_3 terms in the Hamiltonian, one can sum up all terms to give $\alpha\to\frac{\pi}{4}\alpha$ and $G_4\to\frac{\pi}{4}G_4$, where $\frac{\pi}{4}=1-\frac{1}{3}+\frac{1}{5}-\frac{1}{7}+\cdots$. Our results can be summarized as follows.

- The detuning parameter in the second order perturbation is positive, $\alpha>0$, the resonance islands appears in $\nu_x<3.75$ for the 4th order resonance and $\nu_x<3.8$ for the 5th order resonance in Fig. 2.54.
- Since $G_4=\frac{1}{2}\alpha$, the stable fixed points should not exist as discussed in Exercise 2.7.5. However, Fig. 2.54 shows clearly the existence of 4 resonance islands. The term $g_{4,0,6,0}$ reduces the effective resonance strength to guarantee the existence of resonance islands, and gives rise to non-symmetric islands. Because $\alpha\gtrsim 2G_4$, we find $J_{\rm SFP}\gg J_{\rm UFP}$.

The 5th order resonance in Fig. 2.54 arises from a higher order term in the above canonical perturbation expansion as a combination of $3\nu_x=\ell_3$, $\nu_x=\ell_1$, and $\nu_x=\ell_1$ resonances, i.e. a resonance at $5\nu_x=\ell_3+\ell_1+\ell_1$.

VIII Collective Instability and Landau Damping

So far we have discussed only single-particle motion in synchrotrons, where each particle can be described by a simple harmonic oscillator. In reality, a circulating charged particle beam resembles an electric circuit, where the impedance plays an important role in determining the induced voltage on circulating current. Likewise, the *impedance* of an accelerator is related to the voltage drop with respect to the motion of charged particle beams. The impedance is more generally defined as the Fourier transform of the electromagnetic waves induced by the passing charged particle beam, called *wakefield*. The induced electromagnetic field can, in turn, impart a force on the motion of each individual particle. Thus, particle motion is governed by the external focusing force and the wakefield generated by beams, and the beam distribution is determined by the motion of all particles. A self-consistent distribution function may be obtained by solving the Poisson-Vlasov equation.

Particle motion in an accelerator is classified into the transverse betatron motion and the longitudinal synchrotron motion. Wakefields are also classified into transverse or longitudinal modes. Likewise, the impedance is classified as longitudinal or transverse respectively. The effect of longitudinal impedance will be discussed in Chap. 3, Sec. VII. Here, we discuss some basic aspects of transverse collective beam instabilities and Landau damping. For a complete treatment of the subject, see Ref. [5, 6, 7]. In Sec. VIII.1, some properties of impedance are listed. In Sec VIII.2 we discuss transverse wave modes, where waves are classified as fast, backward, or slow waves. In Sec. VIII.3, we will show that a slow wave can become unstable in a simple impedance model. Landau damping and dispersion relation will be discussed in Sec. VIII.4.

VIII.1 Impedance

The impedance that a charged-particle beam experiences inside a vacuum chamber resembles impedance in a transmission wire. For beams, there are transverse and longitudinal impedance. The longitudinal impedance has the dimension ohm, and by definition, is equal to the voltage drop per revolution in a unit beam current. The corresponding energy loss per revolution is equal to the voltage drop times the charge of a particle. The transverse impedance is related to the transverse force on betatron motion, and has a dimension of ohm/meter.

The transverse impedance arises from accelerator components such as the resistive wall of vacuum chamber, space charge, image charge on vacuum chamber, broad-band impedance due to bellows, vacuum ports, and BPMs, and narrow band impedance due to high-Q resonance modes in rf cavities, septum and kicker tanks, etc. Without deriving them, we list these impedance as follows.

A. Resistive wall impedance

The transverse resistive wall impedance for a *cylindrical* vacuum chamber with radius b is related to the longitudinal resistive wall impedance via Panofsky-Wenzel theorem. With Eq. (3.290) for a vacuum chamber circumference $2\pi R$, the transverse resistive wall impedance is[91]

$$Z_{\perp,\mathrm{rw}}(\omega) = \frac{2c}{b^2}\frac{Z_{\parallel,\mathrm{rw}}}{\omega} = (1 + j\,\mathrm{sgn}(\omega))\frac{2Rc}{b^3\sigma_{\mathrm{c}}\delta_{\mathrm{skin}}}\frac{1}{\omega}, \tag{2.246}$$

where c is the speed of light, $\delta_{\mathrm{skin}} = \sqrt{2/(\sigma_{\mathrm{c}}\mu_{\mathrm{c}}|\omega|)}$ is the skin depth of electromagnetic wave in metal, σ_{c} is the conductivity of the vacuum chamber materials, μ_{c} is the permittivity in the vacuum chamber, ω is the wave frequency, and $\mathrm{sgn}(\omega)$ is the sign function of the frequency ω.

B. Space-charge impedance

Let a be the radius of a uniformly cylindrical distributed beam inside a cylinder vacuum chamber of radius b. Let x_0 be an infinitesimal displacement from the center of the cylindrical vacuum chamber. The resulting beam current density is

$$i(r,\phi) = \frac{I_0}{\pi a^2}\Theta(a-r) + \frac{I_0 x_0\cos\phi}{\pi a^2}\delta(r-a), \tag{2.247}$$

where $\Theta(x) = 1$ if $x \geq 0$ and 0 otherwise, $\delta(x)$ is the Dirac δ-function, and ϕ is the angle measured from the x-axis. Here, the first term is unperturbed beam current, and the second term arises from an infinitesimal horizontal beam displacement. The perturbing current is a circular current sheet with cosine-theta current distribution. Using the result of Exercise 1.9, we find the induced dipole field inside the beam cross section to be $\Delta B_{z,\mathrm{b}} = \mu_0 I_0 x_0/2\pi a^2$. Similarly the induced image current is (see Exercise 1.16) $I_{\mathrm{w}} = -(I_0 x_0\cos\phi_{\mathrm{w}}/\pi b^2)\delta(r-b)$, and the induced dipole field due to the image wall current is $\Delta B_{z,\mathrm{w}} = -\mu_0 I_0 x_0/2\pi b^2$. The total induced vertical dipole field due to the beam displacement is $\Delta B_z = \frac{\mu_0 I_0 x_0}{2\pi}\left(\frac{1}{a^2} - \frac{1}{b^2}\right)$, and, by definition, the resulting impedance per unit length become

$$Z_{\perp,\mathrm{mag}} = j\frac{\beta c\Delta B_z}{\beta I_0 x_0} = j\frac{Z_0}{2\pi}\left(\frac{1}{a^2} - \frac{1}{b^2}\right),$$

where βc is the velocity of the beam, and $Z_0 = \mu_0 c$ is the vacuum impedance. Similarly, the impedance due to the electric field is $Z_{\perp,\mathrm{elec}} = -j\frac{Z_0}{2\pi\beta^2}\left(\frac{1}{a^2} - \frac{1}{b^2}\right)$, and the resulting total impedance in one revolution becomes

$$Z_{\perp,\mathrm{sc}} = -j\frac{RZ_0}{\beta^2\gamma^2}\left(\frac{1}{a^2} - \frac{1}{b^2}\right), \tag{2.248}$$

[91]The imaginary number $j = -i$ in engineering convention is used throughout this textbook. Here, the resistive wall impedance consists of a resistive and an inductive component.

where a is the beam transverse radius, b is beam-pipe radius, R is the average radius of accelerator, and γ is the Lorentz relativistic factor. Because of the $\beta^2\gamma^2$ factor in the denominator, the space-charge impedance is important for low energy beams.

Note that the space charge impedance is capacitive because the beam radius a is less than the vacuum chamber radius b. However, when the oscillatory amplitude x_0 is large, the perturbation current of Eq. (2.247) is invalid, and the self space-charge force term may disappear. The remaining space charge impedance is the image current term, which is inductive.

C. Broad-band impedance

All vacuum chamber gaps and breaks, BPMs, bellows, etc., can be lumped into a term called broad-band impedance, which is usually assumed to take the form of a RLC circuit:

$$Z_{\perp,\rm bb} = \frac{2c}{b^2}\frac{Z_{\parallel,\rm bb}}{\omega} = \frac{2c}{b^2\omega}\frac{R_s}{1+j{\rm Q}\left(\omega/\omega_r - \omega_r/\omega\right)}, \tag{2.249}$$

where Q ≈ 1 is the quality factor, R_s is the shunt resistance, $\omega_r \approx (R/b)\omega_0$ is the cut-off frequency of vacuum chamber, R is the average radius of accelerator, and b is the beam pipe radius. The space-charge impedance can be considered as a broad-band impedance because it is independent of wave frequency.

D. Narrow-band impedance

Narrow-band impedance is usually represented by a sum of Eq. (2.249), where the corresponding Q-factor is usually large. Narrow-band impedance may arise from parasitic rf cavity modes, septum and kicker tanks, vacuum ports, etc.

E. Properties of the transverse impedance

When the beam centroid is displaced from the closed orbit, the motion can be expressed as a dipole current. The dipole current will set up a wakefield that acts on the beam. The transverse impedance of a ring is

$$Z_\perp(\omega) \equiv \frac{j}{e\beta I_0\langle y\rangle}\oint F_\perp ds = \frac{j}{I_1\beta}\oint (\vec{E}+\vec{v}\times\vec{B})_\perp ds = \frac{jC}{\beta}\int W_\perp(\tau)\, e^{-j\omega\tau}\, d\tau, \tag{2.250}$$

where $I_1 = I_0\langle y\rangle$ is the dipole current, $\langle y\rangle$ is the centroid of the beam in the betatron motion, and C is the circumference of the accelerator. The imaginary number j included in the definition of the impedance is needed to conform a real loss for a real positive resistance. This occurs because the driving force is leading the dipole current by a phase of $\pi/2$. The factor β in the denominator is included by convention. The wake function is related to the impedance with the causality condition by

$$W_\perp(t) = -\frac{j\beta}{2\pi}\int_{-\infty}^{\infty} Z_\perp(\omega)e^{j\omega t}d\omega \qquad \text{with} \qquad W_\perp(t) = 0 \quad (t<0). \tag{2.251}$$

Thus, the impedance can not have singularities in the lower half of the complex ω plane; however, it may have poles in the upper half plane. For example, the impedance of RLC resonator circuit in Eq. (2.249) has two poles located at

$$\omega = \omega_r \left[\pm\sqrt{1-(1/2\mathrm{Q})^2} + j(1/2\mathrm{Q})\right]. \tag{2.252}$$

The analytic properties of impedance provide us with the Kramer-Kronig relation, i.e. the real and imaginary parts are related by a Hilbert transform

$$\mathrm{Re}\, Z_\perp(\omega) = -\frac{1}{\pi}\int_{\mathrm{P.V.}} d\omega' \frac{\mathrm{Im}\, Z_\perp(\omega')}{\omega'-\omega}, \tag{2.253}$$

$$\mathrm{Im}\, Z_\perp(\omega) = \frac{1}{\pi}\int_{\mathrm{P.V.}} d\omega' \frac{\mathrm{Re}\, Z_\perp(\omega')}{\omega'-\omega}, \tag{2.254}$$

where P.V. means taking the principal value integral. Since the wake function is real, the impedance at a negative frequency is related to that at a positive frequency by $Z_\perp(-\omega) = -Z_\perp^*(\omega)$, or

$$\mathrm{Re}Z_\perp(-\omega) = -\mathrm{Re}Z_\perp(\omega), \quad \mathrm{Im}Z_\perp(-\omega) = +\mathrm{Im}Z_\perp(\omega). \tag{2.255}$$

Thus the real part of the transverse impedance is negative at negative frequency.

To summarize, various components of the transverse impedance $Z_\perp(\omega)$ are schematically shown in Fig. 2.61, where the real part of the impedance is an odd function of ω, and the imaginary part is an even function.

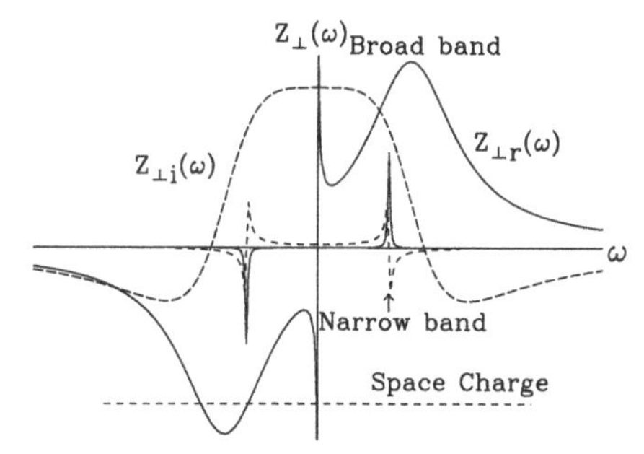

Figure 2.61: A schematic drawing of the transverse broad-band and narrow-band impedance. The real parts are shown as solid lines and the imaginary parts as dashed lines. A broad-band and a narrow-band impedance are represented by peaks in the real parts. The resistive wall impedance is important in the low-frequency region, and the space-charge impedance is independent of frequency.

VIII.2 Transverse Wave Modes

A coasting (DC) beam is defined as a beam made of particles continuously fill accelerator. The transverse coordinate at any instant of time is

$$y(t_0,\theta) = \sum_{n=-\infty}^{\infty} y_n e^{-jn\theta}, \tag{2.256}$$

where θ is the orbiting angle, and n is the mode number. At a fixed azimuth angle θ_0, the betatron oscillation of the transverse motion is

$$y(t, \theta_0) = \sum_{n=-\infty}^{\infty} y_n e^{j(Q\omega_0 t - n\theta_0)}, \tag{2.257}$$

where Q is the betatron tune, and ω_0 is the angular revolution frequency. Since $\theta = \theta_0 + \omega_0 t$, the nth mode of transverse motion and its angular phase velocity are

$$\begin{aligned} y(t, \theta) &= y_n e^{j[(n+Q)\omega_0 t - n\theta]}, \\ \dot{\theta}_{n,\mathrm{w}} &= \left(1 + \frac{Q}{n}\right)\omega_0. \end{aligned} \tag{2.258}$$

There are three possible transverse wave modes: the fast wave, the backward wave, and the slow wave. The corresponding angular phase velocity is

$$\dot{\theta}_{n,\mathrm{w}} = \begin{cases} \left(1 + \dfrac{Q}{n}\right)\omega_0, & \text{if } n > 0: \quad \text{fast wave} \\ -\left(\dfrac{Q}{|n|} - 1\right)\omega_0, & \text{if } n < 0 \text{ and } |n| < Q: \quad \text{backward wave} \\ \left(1 - \dfrac{Q}{|n|}\right)\omega_0, & \text{if } n < 0 \text{ and } |n| > Q: \quad \text{slow wave.} \end{cases} \tag{2.259}$$

The phase velocity of a fast wave is higher than the particle velocity, that of a slow wave is slower than the particle velocity, and a backward wave travels in the backward direction. The signal picked up at a transverse beam position monitor (BPM) is composed of frequencies located at $|n + Q|\omega_0$ (n integer), sidebands of rotation harmonics $n\omega_0$. Figure 2.62 shows a schematic drawing of a transverse beam spectra for a betatron tune of 3.2.

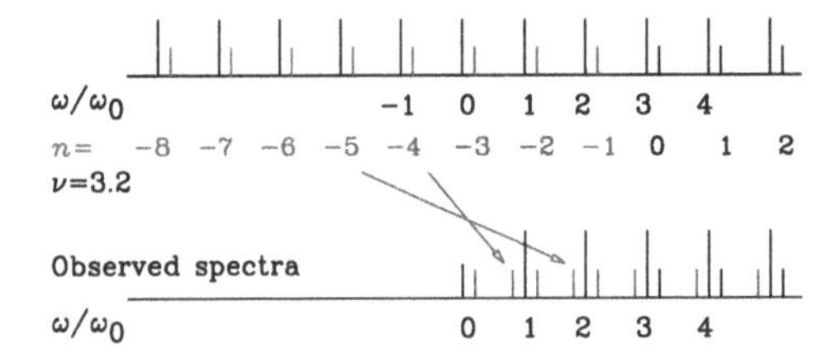

Figure 2.62: Top: a schematic drawing of transverse beam spectra for a beam with betatron tune 3.2, showing all sidebands $(n + \nu)\omega_0$. The observed spectra $|(n + \nu)|\omega$ is shown at the bottom plot.

VIII.3 Effect of Wakefield on Transverse Wave

In a global sense, the betatron motion is well approximated by simple harmonic motion. Let y_k be the horizontal or vertical transverse betatron displacement of the

kth particle, and let $Q_k\omega_0$ be the corresponding angular frequency of betatron motion. The equation of motion in the presence of a wakefield can be expressed as a force oscillator equation:

$$\ddot{y}_k + (Q_k\omega_0)^2 y_k = \frac{F_\perp(t)}{\gamma m}, \qquad F_\perp(t) = -j\frac{e\beta I Z_\perp}{2\pi R}\langle y\rangle, \tag{2.260}$$

where the overdot corresponds to the derivative with respect to time t, m is the mass, γ is the Lorentz relativistic factor, and R is the mean radius of the accelerator. The time-dependent transverse electromagnetic force $F_\perp$ comes from Eq. (2.250) due to a broad-band impedance, $Z_\perp$.

If beam particles encounter collective instability of mode n, and execute collective motion with a coherent frequency ω, we have

$$\begin{aligned} y_k &= Y_k\, e^{j(\omega t - n\theta)}, \\ \dot{y}_k &= \frac{\partial y_k}{\partial t} + \dot{\theta}\frac{\partial y_k}{\partial\theta} = j(\omega - n\omega_0)y_k, \qquad \ddot{y}_k = -(\omega - n\omega_0)^2 y_k, \end{aligned} \tag{2.261}$$

where n is the mode number, and Y_k is the amplitude of collective motion for the kth particle. Substituting into Eq. (2.260), we obtain

$$\begin{aligned} &\left[(Q\omega_0)^2 - (\omega - n\omega_0)^2\right] y_k = -j\frac{e\beta I Z_\perp}{2\pi R\gamma m}\langle y\rangle. \qquad \text{or} \\ &(\omega - \omega_{n,\mathrm{w}})\, y_k = j\frac{e\beta I Z_\perp}{4\pi R\gamma m Q\omega_0}\langle y\rangle, \\ &\omega_{n,\mathrm{w}} = (n+Q)\omega_0, \qquad \text{with} \quad \begin{cases} n > 0 & \text{fast wave,} \\ n < 0, \quad |n| < Q & \text{backward wave,} \\ n < 0, \quad |n| > Q & \text{slow wave.} \end{cases} \end{aligned} \tag{2.262}$$

where $\langle y\rangle = \int \rho(\xi)\, y_k\, d\xi$ is the centroid of the beam, $\rho(\xi)$ is the normalized beam distribution function with $\int \rho(\xi) d\xi = 1$, ξ represents a set of parameters that describe the dependence of the betatron tune on its amplitude, $\omega_{n,\mathrm{w}}$ is the mode wave frequency, and we have used the relation $\omega - n\omega_0 + Q\omega_0 \approx \omega_{n,\mathrm{w}} - n\omega_0 + Q\omega_0 = 2Q\omega_0$. The real part of the slow wave frequency $\omega_{n,\mathrm{w}}$ is *negative*. Averaging over the beam distribution, we obtain a *dispersion relation* for the collective frequency ω

$$1 = j(V + jU)\int \frac{\rho(\xi)}{\omega - \omega_{n,\mathrm{w}}(\xi)}\, d\xi, \qquad V + jU = \frac{e\beta I Z_\perp}{4\pi R\gamma m Q\omega_0}, \tag{2.263}$$

where U and V parameters are the scaled impedance with V is related to growth rate, and U is related to collective frequency shift.

The set of parameters ξ represents any variables that $\omega_{n,\mathrm{w}}$ and the beam distribution function depend on. Since betatron tunes depend on betatron amplitudes due to space-charge force, sextupoles, and other higher-order magnetic multipoles, the betatron amplitude can serve as a ξ parameter. Since Q and ω_0 depend on the off-momentum parameter $\delta = \Delta p/p$, δ can also be chosen as a possible ξ parameter.

We assign the fractional off-momentum coordinate as a ξ parameter. The dependence of the coherent frequency on the off-momentum variable δ is

$$\omega_{n,\mathrm{w}} = \omega_{n,\mathrm{w}0} + [C_y - n\eta]\,\omega_0\,\delta. \tag{2.264}$$

where η is the phase-slip factor and C_y is the chromaticity, A beam is made of particles with different off-momentum δ and thus a spread in the coherent frequencies. However, the wave frequency spread vanishes at mode number

$$n_0 = \frac{C_y}{\eta}. \tag{2.265}$$

At this mode, the revolution frequency shift is canceled by the betatron frequency shift for all particles with different off-momentum δ, i.e. no mode frequency spread. Since $\omega_{n,w} = \omega_{n_0,w0}$, the solution of the dispersion relation in Eq. (2.263) is $\omega = \omega_{n_0,w0} - U + jV$, where U, proportional to the imaginary part of the impedance, produces frequency shift, and V, related to the real part of the impedance, can generate collective beam instability if $V < 0$. For a slow wave with $n_0 = C_y/\eta < 0$ and $|n_0| > Q$, we find $\omega_{n_0,w0} < 0$, where $Z_\perp^{\mathrm{real}} < 0$. The beam may become unstable against transverse collective instability. Here we discuss the modes of coasting beams. The frequency $n_0\omega_0$ in a bunched-beam is the betatron frequency shift from the bunch head to tail (see Exercise 3.2.14).

A. Beam with zero frequency spread

For a beam with zero frequency spread, i.e. $\rho(\xi) = \delta(\xi - \xi_0)$, we obtain

$$\omega = \omega_{n,\mathrm{w}0} - U + jV. \tag{2.266}$$

The imaginary part of impedance gives rise to a frequency shift, and the resistive part generates an imaginary coherent frequency ω. If the imaginary part of the coherent frequency is negative, the betatron amplitude grows exponentially with time, and the beam encounters collective instability.

For fast and backward waves, $\omega_{n,\mathrm{w}0}$ is positive. The real part of the impedance $Z_\perp(\omega)$ is positive (see Fig. 2.61), thus the imaginary part of the coherent frequency is positive, and there is no growth of collective instability. On the other hand, the collective frequency $\omega_{n,\mathrm{w}0}$ of a slow wave is negative, where the real part of the transverse impedance is negative. *Since the imaginary part of collective frequency is negative, a beam with zero frequency spread can suffer slow wave collective instability.*

$$\mathrm{Re}\,[\omega_{\mathrm{coll}}] = \omega_{n,\mathrm{w}0} - U, \quad \mathrm{Im}\,[\omega_{\mathrm{coll}}] = V.$$

B. Beam with finite frequency spread

With parameters U and V, the dispersion relation for coherent dipole mode frequency ω becomes

$$(-U + jV)^{-1} = \int \frac{\rho(\xi)}{\omega - \omega_{n,\mathrm{w}}(\xi)} d\xi. \tag{2.267}$$

The solution of the dispersion relation corresponds to a coherent eigenmode of collective motion. If the imaginary part of the coherent frequency is negative, the amplitude of the coherent motion grows with time. On the other hand, if the imaginary part of each eigenmode is positive, coherent oscillation is damped. The threshold of collective instability can be obtained by finding the solution with $\omega = \omega - j|0^+|$, where 0^+ is an infinitesimal positive number. The remarkable thing is that there are solutions of real ω even when $-U + jV$ is complex.

In general, the growth rate can be solved from the dispersion integral with known impedance and distribution function. Similarly, for a given growth rate, the dispersion relation provides a relation between $U(\omega)$ vs $V(\omega)$. If the distribution function is symmetric in betatron frequency, the U vs V contour plot will have reflection symmetry with respect to the V axis. For any beam distribution $\rho(\xi)$ that does not have infinite tails, the threshold curve contains two straight vertical lines lying on the U axis. This means that if the coherent frequency shift is beyond the distribution tails, the beam can be, but may not necessarily be, unstable against collective instability, and the growth rate is proportional to the real part of the impedance.

C. A model of collective motion

We consider a macro-particle model of a beam with $\langle Y \rangle = \sum \rho_k Y_k$, where ρ_k is the distribution function with $\sum \rho_k = 1$. In matrix form, Eq. (2.262) becomes

$$[\omega - \omega_{n,\mathrm{w}}(k)]\, Y_k = W \sum_i \rho_i Y_i, \tag{2.268}$$

where $W = -U + jV$ for a broad-band impedance. If $\omega_{n,\mathrm{w}}(k) = \omega_{n,\mathrm{w}0}$ is independent of k, i.e. no frequency spread, the collective frequency is trivially given by[92]

$$\omega_{\mathrm{coll}} = \omega_{n,\mathrm{w}0} + W, \tag{2.269}$$

which is identical to the solution of Eq. (2.266). The corresponding eigenvector for collective mode is $Y_{k,\mathrm{coll}} = \rho_k Y_k$. Thus any amount of a negative real part of the impedance can produce a negative imaginary collective frequency and lead to collective instability. The external force is coupled only to the collective mode. All other incoherent solutions have random phase with eigenvalue $\omega_{n,\mathrm{w}0}$, no frequency shift!

[92]The collective mode occurs frequently in almost all many-body systems. In nuclear physics, the giant dipole resonance where protons oscillate coherently against neutrons presents a similar physical picture.

Now, if there is a frequency spread between different particles, we have to diagonalize the matrix of Eq. (2.268). This is equivalent to solve the collective mode frequency from the dispersion relation of Eq. (2.267). In general, if the frequency spread $\Delta\omega_{n,\mathrm{w}}$ among beam particles is larger than the coherent frequency shift parameter W, the collective mode disappears, and there is no coherent motion. The disappearance of the collective mode due to tune spread is called *Landau damping.*

The requirement of a large frequency spread for Landau damping is a necessary condition but not a sufficient one. We consider a frequency spread model

$$\omega_{n,\mathrm{w}}(k)Y_k = \omega_{n,\mathrm{w}0}Y_k + \Delta\Omega(Y_k - \langle Y\rangle),$$

where $\Delta\Omega$ is a constant frequency spread of the beam. In this model, the frequency shift of a particle is proportional to local beam density. This model of tune spread resembles space-charge tune shift. The resulting collective mode frequency is

$$\omega_{\mathrm{colla}} = \omega_{n,\mathrm{w}0} + W. \tag{2.270}$$

This means that the frequency spread that is proportional to the distribution function can not damp the collective motion. This is equivalent to the argument that the space-charge tune shift can not damp the transverse collective instability. Since the space-charge tune shift is a tune shift relative to the center of a bunch, and the coherent motion is relative to the closed orbit of the machine, the space-charge tune shift alone can not provide Landau damping against transverse collective instability.

VIII.4 Frequency Spread and Landau Damping

From Eq. (2.266), we see that a slow wave can suffer transverse collective instability for a beam with zero frequency spread. What is the effect of frequency spread on collective instabilities? The key is the Landau damping mechanism discussed below. The examples illustrate the essential physics of Landau damping.

A. Landau damping

The equation of collective motion (2.260) can be represented by a forced oscillator:

$$\ddot{y} + \omega_\beta^2 y = \hat{F}\sin\omega t, \tag{2.271}$$

where ω is the collective frequency, and $\omega_\beta = Q\omega_0$ is the betatron frequency. The solution is

$$\begin{aligned} y(t) &= +\frac{\hat{F}}{\omega_\beta^2 - \omega^2}\left(\sin\omega t - \frac{\omega}{\omega_\beta}\sin\omega_\beta t\right) + \left\{y_0\cos\omega_\beta t + \frac{\dot{y}_0}{\omega_\beta}\sin\omega_\beta t\right\} \\ &\to +\frac{\hat{F}}{\omega_\beta^2 - \omega^2}\left(\sin\omega t - \frac{\omega}{\omega_\beta}\sin\omega_\beta t\right). \end{aligned} \tag{2.272}$$

We remove the solution of the homogeneous equation inside $\{...\}$ by setting the initial values $y_0 = 0$ and $\dot{y}_0 = 0$ at $t = 0$. We are interested in the response of the particle under external force. The lower plot of Fig. 2.63 shows $y(t)$ for three particles with $\omega_\beta = 0.85$ (dash-dots), $\omega_\beta = 0.8$ (dashes), and $\omega_\beta = 0.76$ (line) under the action of an external force with $\omega = 0.75, \hat{F} = 0.01$.

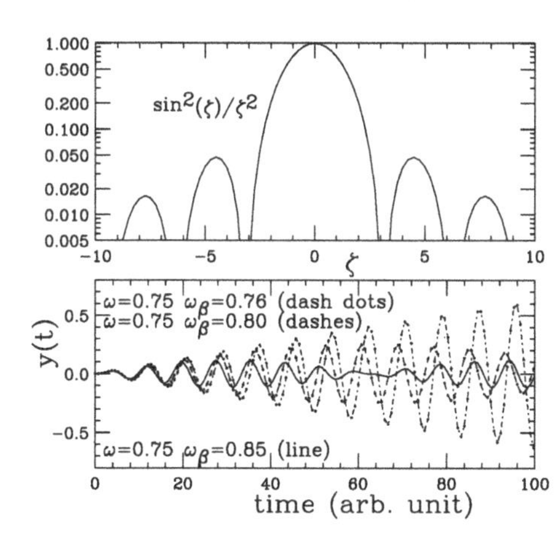

Figure 2.63: Top plot: the coherent function $(\sin^2\zeta)/\zeta^2$. Note that the function becomes smaller as the ζ variable increases. This means that the external force can not coherently act on a particle if $(\omega_\beta - \omega)T$ becomes large. Bottom plot: $y(t)$ vs time t due to an external sinusoidal driving force $F(t) = \hat{F}\sin\omega t$. Here the units of ω and t are related: if ω is in rad/s, t is in s, and if ω is in 10^6 rad/s, then t is in μs. The frequency differences of these three particles are respectively $\Delta\omega = 0.01, 0.05$, and 0.1. The particle with a larger frequency difference will fall out of coherence with the external force earlier.

If ω_β differs substantially from the driving frequency ω, the external force can not deliver energy to the system forever. The particle motion will be out of phase with the external force sooner or later, and the energy will be transferred back to the external force. A smaller $\omega - \omega_\beta$ results in a longer in-phase time, as shown in the lower plot of Fig. 2.63. As the time increases, the number of particles (oscillators) that remain in phase with the external force becomes smaller and smaller. This is the essence of Landau damping.

When an external force $F(t)$ is applied to a Hamiltonian system, the equation of motion is $\ddot{y} + dV/dy = F(t)$, where V is the potential energy. The total energy imparted by the external force for a time T is $\Delta H = \int_0^T \dot{y}F(t)dt$, where H is the Hamiltonian of the system. The average power delivered to our dynamic system (2.271) is

$$\langle P(T)\rangle = \frac{1}{T}\int_0^T \dot{y}\,(\hat{F}\sin\omega t)\,dt \approx \left[\frac{\hat{F}^2\omega}{4(\omega+\omega_\beta)}\frac{\sin^2\zeta}{\zeta^2}\right]T + \cdots, \qquad (2.273)$$

where $\zeta = \frac{1}{2}(\omega - \omega_\beta)T$. Here we retain only the leading term, proportional to time T. The upper plot of Fig. 2.63 shows the coherent functions $(\sin^2\zeta)\,/\,\zeta^2$. As time T increases, the coherent frequency window decreases, i.e.

$$|\omega - \omega_\beta| \sim 1/T, \qquad (2.274)$$

or equivalently, fewer and fewer oscillators will be affected by the external force. When the external force can not pump energy into a beam with a finite frequency spread, collective instability disappears, i.e. the system is Landau damped.

The power dissipation to the oscillator in Eq. (2.273) comes entirely from the second term in parentheses in Eq. (2.272). The term $y(t)$ that is in phase with the force is a reactive term, which does not dissipate energy. The second term in Eq. (2.272) that absorbs energy from the external force is a resistive term, which can induce collective instability.

Now we consider the case that the external force arisen from a wakefield due to the collective motion of the beam. Even if there were an initial collective motion to produce an external field at the beginning, the collective motion of the beam could not be sustained for a long time if there were a substantial frequency spread. The collective instability is thus suppressed by Landau damping.

B. Solutions of dispersion integral with Gaussian distribution

We consider a beam with Gaussian distribution given by

$$\rho(\delta) = \frac{1}{\sqrt{2\pi}\sigma_\delta} e^{-\delta^2/2\sigma_\delta^2}, \tag{2.275}$$

where $\delta = \Delta p/p_0$ is the fractional off-momentum coordinate, and σ_δ is the rms momentum width of the beam. With $\omega_{n,\rm w}$ of Eq. (2.264), the dispersion relation of Eq. (2.267) becomes

$$-u + jv = j\left[w\left(\frac{\omega - \omega_{n,\rm w0}}{\sigma_\omega}\right)\right]^{-1}, \qquad \sigma_\omega = \sqrt{2}\,|C_y - n\eta|\,\omega_0\sigma_\delta \tag{2.276}$$
$$w(z) = e^{-z^2}\mathrm{erfc}(-jz) = \frac{j}{\pi}\int_{-\infty}^{\infty}\frac{e^{-t^2}}{z-t}dt, \qquad u = \frac{\sqrt{\pi}U}{\sigma_\omega}, \quad v = \frac{\sqrt{\pi}V}{\sigma_\omega}.$$

where σ_w is the rms frequency spread of the beam for mode n, $w(z)$ is the complex error function, u and v are the reduced imaginary and real parts of the impedance of Eq. (2.263). The curves u vs v for Gaussian distribution for $\mathrm{Im}(\omega/\sigma_\omega) = 0$ and -0.5 are shown in Fig. 2.64, where the rectangular symbols in each curve represent coherent frequency shifts at $\mathrm{Re}(\omega - \omega_{\rm n,w0}) = \pm\sigma_\omega$ (inner ones) and $\pm 2\sigma_\omega$ (outer ones).

From Fig. 2.64, we observe that if a coherent mode frequency falls within the width of the spectrum, the threshold of the collective beam instability requires a finite resistive impedance. This is because the coherent mode excites only a small fraction of the particles in the beam, and most of the beam particles are off resonance. Thus the collective beam motion is damped. Landau damping differs in essence from phase-space damping due to beam cooling, or phase space decoherence due to tune spread (see Exercise 2.8.5).

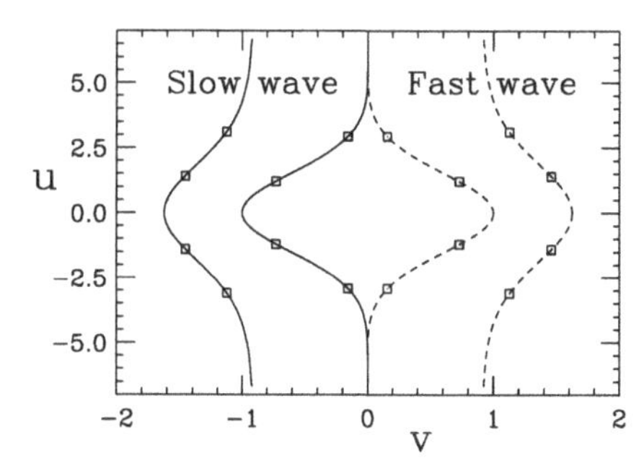

Figure 2.64: The normalized u vs v for Gaussian distribution is plotted with two different growth rates with $\mathrm{Im}(\omega) = 0$ and $-0.5\sigma_\omega$. The rectangular symbols represent the coherent frequency shifts at $\mathrm{Re}(\omega - \omega_{\mathrm{n,w0}}) = \pm\sigma_\omega$ and $\pm 2\sigma_\omega$. The solid line with $\mathrm{Im}(\omega)$=0 shows that a beam with frequency spread can tolerate a finite amount of the real part of impedance at the threshold of collective instability.

We note particularly that the frequency spread can vanish for mode number n_0 of Eq. (2.265). Because such modes have vanishing frequency spread, collective instabilities may not be Landau damped. However, if chromaticity C_y is negative below transition energy or if C_y is positive above transition energy, then mode n_0 with vanishing frequency spread is a fast wave. Since the real part of a fast wave is positive, the imaginary part of collective frequency is positive, and thus there is no collective instability. This has commonly been employed to overcome transverse collective instabilities.[93]

Exercise 2.8

1. Verify the wave angular velocity of Eq. (2.258), and show that fast, slow, and backward waves travel faster, slower, and backward relative to particle angular velocity, respectively.

2. Show that the solution given by Eq.(2.266) for the dispersion relation at zero frequency spread is identical to the collective frequency solution by matrix diagonalization of Eq. (2.268), and show that the eigenvector of collective motion is $Y_{k,\mathrm{coll}} = \rho_k Y_k$.

3. Using the Gaussian distribution of Eq. (2.275) to show that the dispersion relation becomes an algebraic equation, Eq. (2.276).

4. The solution of Eq. (2.271) with initial condition $y_0 = \dot{y}_0 = 0$ is

$$y = \frac{\hat{F}}{\omega_\beta^2 - \omega^2}\left(\sin\omega t - \frac{\omega}{\omega_\beta}\sin\omega_\beta t\right).$$

   (a) Plot $y(t)$ as a function of t for $\omega = 1$, $\hat{F} = 0.01$ with three particles at $\omega_\beta = 0.8$, 0.9, and 0.99.

   (b) Let $\omega_\beta = \omega + \epsilon$, where $|\epsilon|$ is a small frequency deviation. Show that

$$y(t) \approx \frac{\hat{F}}{2\omega}\left[\frac{1-\cos(\epsilon t)}{\epsilon}\sin\omega t - \frac{\sin(\epsilon t)}{\epsilon}\cos\omega t\right].$$

[93]For *bunched beams*, the head-tail instability has been observed in SPS and Fermilab Main Ring above transition energy if the chromaticity is negative (see J. Gareyte, p. 134 in Ref. [18]). Treatment of head-tail instability is beyond the scope of this book.

Show that the first term in square brackets does not absorb energy from the external force but the second term can. The first term corresponds to a reactive coupling and the second term is related to a resistive coupling.

(c) If a beam has a distribution function given by $\rho(\xi)$ with $\int \rho(\xi)d\xi = 1$, discuss the centroid of beam motion, i.e. $\langle y(t)\rangle = \int y(t)\rho(\xi)d\xi$. For example, we choose $\xi = \epsilon$ and $\rho(\epsilon) = 1/2\Delta$, if $|\epsilon| \le \Delta$; and 0, otherwise.

5. Consider a beam with uniform momentum distribution

$$\rho(\delta) = \begin{cases} 1/(2\Delta) & \text{if } |\delta| \le \Delta, \\ 0 & \text{otherwise.} \end{cases}$$

(a) Show that the dispersion relation Eq. (2.267) becomes

$$(-u + jv)^{-1} = \ln \frac{\omega - \omega_{n,\mathrm{w}0} + \sigma_w}{\omega - \omega_{n,\mathrm{w}0} - \sigma_w},$$

where $u = U/2\sigma_w$, $v = V/2\sigma_w$, and $\sigma_w = |C_y - n\eta|\omega_0\Delta$.

(b) Show that the imaginary part of the coherent frequency is

$$\mathrm{Im}\,(\omega) = \sigma_w \frac{e^{-u}\sin[v/(u^2+v^2)]}{1 + e^{-2u/(u^2+v^2)} - 2e^{-u/(u^2+v^2)}\cos[v/(u^2+v^20]}.$$

Show that the condition that $\mathrm{Im}\omega = 0^-$ is

$$u^2 + (v + \frac{1}{2\pi})^2 = \frac{1}{4\pi^2}.$$

Plot u vs v, and compare your result with that shown in Fig. 2.64.

6. A beam is usually composed of particles with different frequencies.[94] Let $\rho(\omega_\beta)$ be the frequency distribution of the beam with $\int \rho(\omega_\beta)d\omega_\beta = 1$. If initially all particles are located at $y = \dot{y} = 0$, and at time $t = 0$ all particles are kicked to an amplitude A, i.e.

$$y(t) = A\cos\omega_\beta t,$$

and begin coherent betatron motion, find the centroid of beam motion as a function of time with the following frequency distribution functions.

(a) If the frequency distribution of the beam is

$$\rho(\omega_\beta) = \frac{1}{\sqrt{2\pi}\sigma} e^{-(\omega_\beta - \omega_0)^2/2\sigma^2},$$

where σ is the rms frequency spread, show that $\langle y\rangle = Ae^{-\sigma^2 t^2/2}\cos\omega_0 t$.

[94] This exercise illustrates the difference between Landau damping and beam decoherence (or filamentation). Note that coherent beam motion will decohere within a time range $\Delta t \sim 1/\sigma$, where σ is the rms frequency spread of the beam. As the coherent motion is damped, particles are not damped to the center of phase space. See also Fig. 2.17 for coherent betatron oscillation induced by an rf dipole kicker.

(b) If the frequency distribution is a one-sided exponential

$$\rho(\omega_\beta) = \begin{cases} (1/\sigma)e^{-(\omega_\beta-\omega_0)/\sigma}, & \text{if } \omega_\beta > \omega_0, \\ 0 & \text{otherwise,} \end{cases}$$

where σ is the rms frequency spread, and $\langle \omega_\beta \rangle = \omega_0 + \sigma$, show that

$$\langle y \rangle = \frac{A}{1+\sigma^2 t^2}\left[\cos \omega t - \sigma t \sin \omega t\right].$$

(c) If the frequency distribution is a Lorentzian

$$\rho(\omega_\beta) = \frac{\Gamma}{\pi[(\omega_\beta - \omega_0)^2 + \Gamma^2]},$$

where Γ is the width, show that $\langle y \rangle = Ae^{-\Gamma t} \cos \omega t$.

(d) If the frequency distribution is uniform with

$$\rho(\omega_\beta) = \begin{cases} 1/(2\Gamma), & \text{if } \omega_0 - \Gamma < \omega_\beta < \omega_0 + \Gamma, \\ 0, & \text{otherwise,} \end{cases}$$

where $\Gamma = \sqrt{3}\sigma$ and σ is the rms width, show that

$$\langle y \rangle = A \frac{\sin \Gamma t}{\Gamma t} \cos \omega_0 t.$$

(e) If the frequency distribution is parabolic with

$$\rho(\omega_\beta) = \begin{cases} (3/(4\Gamma))\left[1 - ((\omega_\beta - \omega_0)/\Gamma)^2\right], & \text{if } \omega_0 - \Gamma < \omega_\beta < \omega_0 + \Gamma, \\ 0, & \text{otherwise,} \end{cases}$$

where $\Gamma = \sqrt{5}\sigma$ and σ is the rms width, show that

$$\langle y \rangle = 3A\left(\frac{\sin \Gamma t}{(\Gamma t)^3} - \frac{\cos \Gamma t}{(\Gamma t)^2}\right) \cos \omega_0 t.$$

(f) If the frequency distribution is quadratic with

$$\rho(\omega_\beta) = \begin{cases} (2/\pi\Gamma)\sqrt{1 - ((\omega_\beta - \omega_0)/\Gamma)^2}, & \text{if } \omega_0 - \Gamma < \omega_\beta < \omega_0 + \Gamma, \\ 0, & \text{otherwise,} \end{cases}$$

where $\Gamma = 2\sigma$ and σ is the rms width, show that

$$\langle y \rangle = A\left[J_0(\Gamma t) + J_2(\Gamma t)\right] \cos \omega_0 t,$$

where J_0 and J_2 are Bessel functions.

IX Synchro-Betatron Hamiltonian

So far, we have discussed particle motion only in (x, p_x, z, p_z) phase-space coordinates. The remaining phase-space coordinates $(t, -H)$ or $(t, -E)$ have not been mentioned. Here we will study the "synchrotron" equation of motion for phase-space coordinates $(t, -E)$. The terminology of synchrotron motion is derived from the synchronization of particle motion with rf electric field. The name "synchrotron" has been broadly used for all circular accelerators that employ rf electric field for beam acceleration.

This unified description has the advantage of treating synchrotron motion and betatron motion on an equal footing. It is particularly useful in the study of synchro-betatron coupling resonances.[95]

To simplify algebra, we disregard vertical betatron coordinates (z, p_z) and consider only a planar synchrotron. Neglecting vertical betatron phase-space coordinates, the Hamiltonian is [see Eq. (2.14)]

$$\begin{aligned} H_0 &= -(1+\frac{x}{\rho})\left[\left(\frac{E-e\Phi}{c}\right)^2 - m^2c^2 - p_x^2 - p_z^2\right]^{1/2} - eA_s \\ &\approx -(1+\frac{x}{\rho})\left(p - \frac{e\Phi}{\beta c}\right) + (1+\frac{x}{\rho})\left(\frac{p_x^2+p_z^2}{2p}\right) - eA_s, \end{aligned} \tag{2.277}$$

where the orbital length s is used as an independent variable, $p = \sqrt{(E/c)^2 - (mc)^2}$ is the momentum of a particle, ρ is the bending radius of the Frenet-Serret coordinate system, Φ is the scaler potential, A_s is the longitudinal vector potential, and $(x, p_x, z, p_z, t, -E)$ are canonical phase-space coordinates.

The static transverse magnetic field is

$$B_z = \frac{1}{1+(x/\rho)}\frac{\partial A_s}{\partial x}, \quad B_x = -\frac{1}{1+(x/\rho)}\frac{\partial A_s}{\partial z},$$

and the longitudinal varying electric field can be obtained from

$$E_s = -\frac{\partial A_s}{\partial t} = \sum_k V_k \delta_{\rm p}(s - s_k)\sin(\omega_{\rm rf}t + \phi_{0k}), \tag{2.278}$$

where $\delta_{\rm p}(s - s_k) = \sum_n \delta(s - s_k - 2\pi nR)$ is a periodic delta function with period $2\pi R$, V_k is the rf voltage, $\omega_{\rm rf}$ is the angular frequency of the rf field, and ϕ_{0k} is the initial phase of the kth cavity. Thus the rf accelerating field can be represented by

$$A_{s,\rm rf} = \frac{1}{\omega_{\rm rf}}\sum_k V_k \delta_{\rm p}(s - s_k)\cos(\omega_{\rm rf}t + \phi_{0k}). \tag{2.279}$$

[95]See T. Suzuki, *Part. Accel.* **18**, 115 (1985); see also S.Y. Lee and H. Okamoto, *Phys. Rev. Lett.* **80**, 5133 (1998) for the effects of space charge dominated beams.

The Hamiltonian is an implicit function of energy E. Let $E = E_0 + \Delta E$ and $p = p_0 + \Delta p$, where E_0 and p_0 are the energy and momentum of the reference particle. We obtain then

$$\frac{\Delta p}{p_0} \approx \frac{\Delta E}{\beta^2 E_0} - \frac{1}{2\gamma^2}(\frac{\Delta E}{\beta^2 E_0})^2, \qquad \frac{\Delta E}{\beta^2 E_0} \approx \frac{\Delta p}{p_0} + \frac{1}{2\gamma^2}(\frac{\Delta p}{p_0})^2. \tag{2.280}$$

Expanding the dipole field B_z in power series with $B_z = B_0 + B_1 x + \cdots$, where $B_1 = \partial B_z/\partial x$, we obtain

$$A_s = B_0 x + \frac{B_0}{2\rho}x^2 + \frac{1}{2}B_1(x^2 - z^2) + \cdots + A_{s,\mathrm{rf}} + A_{s,\mathrm{sc}}, \tag{2.281}$$

where $A_{s,\mathrm{rf}}$ given by Eq. (2.279) stands for the vector potential of rf cavities. The space charge force of the beam particles gives rise to a mean field, that can be represented by a scaler and vector potentials $\Phi = V_{\mathrm{sc}}$ and $A_{s,\mathrm{sc}}$ with $A_{s,\mathrm{sc}} = \beta^2 V_{\mathrm{sc}}/\beta c$. Substituting the scaler and vector potentials into the Hamiltonian, we obtain

$$\begin{aligned} H_0 &= -p_0 - p_0\frac{\Delta E}{\beta^2 E_0} + p_0\frac{1}{2\gamma^2}(\frac{\Delta E}{\beta^2 E_0})^2 - p_0\frac{\Delta E}{\beta^2 E_0}\frac{x}{\rho} \\ &\quad + \frac{p_x^2 + p_z^2}{2p_0} + \frac{p_0}{2}(K_x x^2 + K_z z^2) - eA_{s,\mathrm{rf}} \end{aligned} \tag{2.282}$$

up to second order in phase-space coordinates, where $K_x = 1/\rho^2 - B_1/B\rho$ is the focusing function for the horizontal plane, and we used the identity condition $B_0 = -p_0/e\rho$, which signifies the expansion of x around the closed orbit at the reference energy.

The next step is to transform the coordinate system onto the closed orbit for a particle with off-energy ΔE. This procedure cancels the cross-term proportional to $(\Delta E/\beta^2 E_0)\cdot x$ in the Hamiltonian. Using the generating function

$$\begin{aligned} F_2(x, \bar{p}_x, t, -\bar{\Delta E}) &= (x - D\frac{\bar{\Delta E}}{\beta^2 E_0})\bar{p}_x - (E + \bar{\Delta E})t \\ &\quad + x\frac{D'}{\beta c}\bar{\Delta E} - \frac{1}{2}DD'p_0(\frac{\bar{\Delta E}}{\beta^2 E_0})^2, \end{aligned}$$

where the new phase-space coordinates are

$$\begin{aligned} \bar{p}_x &= p_x - D'\tfrac{\bar{\Delta E}}{\beta c}, & \bar{x} &= x - D\frac{\bar{\Delta E}}{\beta^2 E_0} \\ \bar{\Delta E} &= E - E_0, & \bar{t} &= t + (\frac{D}{\beta^2 E_0}\bar{p}_x - \frac{D'}{\beta c}\bar{x}), \end{aligned}$$

and the dispersion function D satisfies $D'' + K_x D = \frac{1}{\rho}$, we obtain a new Hamiltonian,

$$\begin{aligned} H_1 &= -p_0\frac{\bar{\Delta E}}{\beta^2 E_0} - \frac{1}{2}p_0\left(\frac{D}{\rho} - \frac{1}{\gamma^2}\right)\left(\frac{\bar{\Delta E}}{\beta^2 E_0}\right)^2 + \frac{\bar{p}_x^{\,2}}{2p_0} + \frac{p_0}{2}K_x\bar{x}^2 \\ &\quad - \frac{dE_0}{ds}\left(\bar{t} - \frac{D}{\beta^2 E_0}\bar{p}_x + \frac{D'}{\beta c}\bar{x}\right) - eA_{s,\mathrm{rf}}. \end{aligned} \tag{2.283}$$

Note that $\bar{x}$ is the betatron phase-space coordinate around the off-momentum closed orbit, and the rf vector potential is

$$eA_{s,\rm rf} = \frac{1}{\omega_{\rm rf}} \sum_k eV_k\, \delta_{\rm p}(s-s_k)\, \cos\left[\omega_{\rm rf}\left(\bar{t} - \frac{D}{\beta^2 E_0}\bar{p_x} + \frac{D'}{\beta c}\bar{x}\right) + \phi_{0k}\right]. \quad (2.284)$$

Now we expand the standing wave of the rf field into a traveling wave, i.e.

$$\delta_{\rm p}(s-s_k)\cos(\omega_{\rm rf}\bar{t}+\phi_{0k}) = \frac{1}{4\pi R}\sum_{n=-\infty}^{\infty}[e^{j(n\theta+\omega_{\rm rf}\bar{t}+\phi_{0k}-n\theta_k)} + e^{j(n\theta-\omega_{\rm rf}\bar{t}-\phi_{0k}-n\theta_k)}], \quad (2.285)$$

Keeping only terms that synchronize the beam arrival time with $n = \pm h$, we obtain

$$\delta_{\rm p}(s-s_k)\cos(\omega_{\rm rf}\bar{t}+\phi_{0k}) = \frac{1}{2\pi R}\cos(\omega_{\rm rf}\bar{t} - \frac{hs}{R} + \phi_{0k} + h\theta_k), \quad (2.286)$$

where $\phi_{0k} + h\theta_k$ should be an integer multiple of 2π. Using the generating function and coordinate transformation:

$$F_2 = x\bar{p_x} + (\omega_{\rm rf}\bar{t} - \frac{hs}{R})W,$$
$$p_x = \bar{p_x}, \quad x = \bar{x}, \quad W = -\frac{\Delta \bar{E}}{\omega_{\rm rf}}, \quad \phi = \omega_{\rm rf}\bar{t} - \frac{hs}{R},$$

we obtain the Hamiltonian

$$\begin{aligned} H_2 \;=\; & -\frac{\omega_{\rm rf}^2}{2\beta^3 cE_0}(\frac{D}{\rho} - \frac{1}{\gamma^2})W^2 + \frac{p_x^2}{2p_0} + \frac{p_0}{2}K_x x^2 \\ & -\frac{1}{2\pi R\omega_{\rm rf}}\sum_k eV_k\, \cos(\phi - D\frac{\omega_{\rm rf}p_x}{\beta c p_0} + D'\frac{\omega_{\rm rf}x}{\beta c}) \\ & -\frac{\sin\phi_s}{2\pi R\omega_{\rm rf}}\sum_k eV_k\, (\phi - D\frac{\omega_{\rm rf}p_x}{\beta c p_0} + D'\frac{\omega_{\rm rf}x}{\beta c}). \end{aligned}$$

Making a scale change to canonical phase-space coordinates with

$$(x, p_x,\, \phi, W) \quad \to \quad (x, x' = \frac{p_x}{p_0},\, \phi, \frac{W}{p_0}),$$

we obtain the Hamiltonian

$$\begin{aligned} H_3 = \frac{H_2}{p_0} \;=\; & \frac{1}{2}(x'^2 + K_x x^2) - \frac{1}{2}\left(\frac{D}{\rho} - \frac{1}{\gamma^2}\right)\left(\frac{\omega_{\rm rf}W}{\beta^2 E_0}\right)^2 \\ & -\frac{1}{2\pi h\beta^2 E_0}\sum_k eV_k\, \cos\left(\phi - \frac{h}{R}Dx' + \frac{h}{R}D'x\right) \\ & -\frac{\sin\phi_s}{2\pi h\beta^2 E_0}\sum_k eV_k\, \left(\phi - \frac{h}{R}Dx' + \frac{h}{R}D'x\right). \end{aligned}$$

Since ϕ and (x, x') are coupled through dispersion function (D, D') in rf cavities, synchrotron and betatron motions are coupled. This is called synchro-betatron coupling (SBC). If a resonance condition is encountered, it is called synchro-betatron resonance (SBR).

In general, the SBC potential must satisfy the Panofsky-Wenzel theorem, which relates the transverse kicks to the longitudinal energy gain. Consider a particle of charge e and velocity $\vec{v} = d\vec{s}/dt$ experiencing a kick from a component in an accelerator. The total transverse momentum and energy changes are

$$\Delta\vec{p}_\perp = e\int_{t_a}^{t_b}(\vec{E} + \vec{v}\times\vec{B})_\perp dt, \qquad \Delta E = e\int_{s_a}^{s_b}\vec{E}\cdot d\vec{s},$$

where $\vec{E}$ and $\vec{B}$ are electromagnetic fields, and $t_b - t_a$ is the transit time of the kicker component, s_a, s_b are the entrance and exit azimuthal coordinates of the kicker. Then the Panofsky-Wenzel theorem yields a relation between the transverse kick and the energy gain[96]

$$\frac{h}{R}\frac{\partial}{\partial\phi}\left(\frac{\Delta p_\perp}{p_0}\right) = \nabla_\perp\left(-\frac{\Delta E}{\beta^2 E_0}\right), \tag{2.287}$$

where $\Delta p_\perp/p_0$ is the transverse kick, $R\phi/h$ is the longitudinal phase-space coordinate of the particle, and $\nabla_\perp$ is the transverse gradient. Thus if the transverse kick depends on the longitudinal coordinates, then the energy gain depends on the transverse coordinates.

This synchro-betatron coupling potential, which satisfies the Panofsky-Wenzel theorem Eq. (2.287), can generally be expressed as a function of 6D phase-space coordinates. The synchrotron phase-space coordinates are chosen naturally to be $(R\phi/h, -\Delta p/p_0)$, and the Hamiltonian in 6D phase-space coordinates becomes

$$\begin{aligned} H_4 &= \frac{1}{2}(x'^2 + K_x x^2) + \frac{1}{2}(z'^2 + K_z z^2) - \frac{1}{2}(\frac{D}{\rho} - \frac{1}{\gamma^2})(\frac{\Delta p}{p_0})^2 \\ &\quad -\frac{1}{2\pi h\beta^2 E_0}\sum_k eV_k\cos(\phi - \frac{h}{R}Dx' + \frac{h}{R}D'x) \\ &\quad -\frac{\sin\phi_s}{2\pi\beta^2 E_0}\sum_k eV_k(\phi - \frac{h}{R}Dx' + \frac{h}{R}D'x). \end{aligned} \tag{2.288}$$

It is worth noting that if RF cavities are located in a straight section, the phase factor $-Dx' + D'x$ will be the same for all cavities (see Exercise 2.9.1). The driving terms for the synchro-betatron coupling in Eq. (2.288) coherently add up in all cavities arranged in one straight section. Thus it is beneficial to put rf cavities in dispersion free regions.

[96]D.A. Goldberg and G.L. Lambertson, *AIP Conf. Proc.* No. 249, p. 537 (1992).

Neglecting synchro-betatron coupling, the Hamiltonian for canonical phase-space variables $(x, x', z, z', R\phi/h, -\Delta p/p_0)$ is

$$\begin{aligned} H &= H_\perp(x, x', z, z') + H_s(\frac{R}{h}\phi, \frac{\Delta p}{p_0}) \\ H_\perp &= \frac{1}{2}(x'^2 + K_x x^2 + z'^2 + K_z z^2) + \cdots \\ H_s &= \frac{1}{2}(\frac{1}{\gamma^2} - \frac{D}{\rho})(\frac{\Delta p}{p_0})^2 - \sum_k \frac{eV_k}{h\beta^2 E}[\cos\phi + (\phi - \phi_s)\sin\phi_s]\delta(\theta - \theta_k), \end{aligned} \tag{2.289}$$

where rf cavities are assumed to be at dispersion free locations. Averaging over one revolution around the ring, the Hamiltonian for synchrotron motion becomes

$$\begin{aligned} \langle H_s \rangle &= -\frac{1}{2}\eta(\frac{\Delta p}{p_0})^2 - \frac{eV}{2\pi h\beta^2 E_0}[\cos\phi - \cos\phi_s + (\phi - \phi_s)\sin\phi_s] \\ &= -\frac{1}{2}\eta(\frac{\Delta p}{p_0})^2 - \frac{\nu_s^2}{h^2|\eta|}[\cos\phi - \cos\phi_s + (\phi - \phi_s)\sin\phi_s], \end{aligned} \tag{2.290}$$

where η is the phase slip factor, and $\nu_s = \sqrt{\frac{h|\eta|eV}{2\pi\beta^2 E_0}}$ is the synchrotron tune of the stationary bucket with $\phi_s = 0$.

The action of the synchrotron oscillations and the linearized betatron oscillations can be defined on an equal footing as

$$I_s = \frac{R}{2\pi h}\oint \frac{\Delta p}{p_0} d\phi, \quad I_x = \frac{1}{2\pi}\oint x' dx, \quad I_z = \frac{1}{2\pi}\oint z' dz. \tag{2.291}$$

The synchrotron action I_s (π-mm-mrad) is related to the commonly used phase area A (eV-s) of the phase-space coordinates $(\phi, \Delta E/h\omega_0)$ by

$$I_s = \frac{R\omega_0}{\beta^2 E_0} A = 3.2 \times 10^5 \frac{A[\text{eVs}]}{\beta\gamma} \quad [\pi\mu\text{m}]. \tag{2.292}$$

Since the typical longitudinal phase-space area is about 0.1 – 1.0 eV-s, the corresponding longitudinal action is 100–1000 times as large as the transverse action. This result has important implications for the synchro-betatron coupling (SBC) resonances.[97]

Exercise 2.9

1. Show that the function $-Dx' + D'x$ in the Hamiltonian H_4 is invariant in a straight section, where D is the dispersion function, x is the horizontal betatron function, and the primes are derivative with respect to the longitudinal coordinate s. Show that rf cavities located in a straight section contribute coherently to SBC if the dispersion function is not zero.
2. Show that if the SBC potential is an analytic function of 6D phase space coordinates $(x, x', z, z', R\phi/h, -\Delta p/p_0)$, it satisfies the Panofsky-Wenzel theorem.

[97]See e.g. S.Y. Lee, *Phys. Rev.* E **49**, 5706 (1994).

Chapter 3

Synchrotron Motion

In general, particles gain energy from electric field in longitudinal direction.[1] Since the electric field strength of an electrostatic accelerator is limited by field breakdown and by the length of the acceleration column, electrostatic accelerators have mainly been used for low energy accelerators. Alternatively, a low-loss radio-frequency (rf) cavity operating at a resonance condition can be used to provide accelerating voltage with $V\sin(\phi_{\rm s}+\omega_{\rm rf}t)$, where V is the amplitude of the rf voltage, $\phi_{\rm s}$ is a phase factor, and $\omega_{\rm rf}$ is the angular frequency synchronized with the arrival time of beam particles. In this chapter we study particle dynamics in the presence of rf accelerating voltage waves.

Although we can derive a 6D Hamiltonian for both synchrotron and betatron oscillations (see Chap. 2, Sec. IX), here, for simplicity, we will derive the synchrotron Hamiltonian based only on the revolution frequency and energy gain relations. This formalism lacks the essential connection between synchrotron and betatron motions, but it simplifies the choice of synchrotron phase-space coordinates.

A particle synchronized with rf phase $\phi=\phi_{\rm s}$ at revolution period T_0 and momentum p_0 is called a *synchronous particle*. A synchronous particle will gain or lose energy, $eV\sin\phi_{\rm s}$, per passage through an rf cavity. Normally the magnetic field is ideally arranged in such a way that the synchronous particle moves on a closed orbit that passes through the center of all magnets. Particles with different betatron amplitudes execute betatron motion around this ideal closed orbit.

A beam bunch consists of particles with slightly different momenta. A particle with momentum p has its own off-momentum closed orbit, $D\delta$, where D is the dispersion function and $\delta=(p-p_0)/p_0$ is the fractional momentum deviation. Since energy gain depends sensitively on the synchronization of rf field and particle arrival time, what happens to a particle with a slightly different momentum when the synchronous particle is accelerated?

[1]This statement also applies to charged particle acceleration in the betatron and the induction linac, in which the induced electromotive force is given by the time derivative of the magnetic flux.

The phase focusing principle of synchrotron motion was discovered by McMillan and Veksler [21]: If the revolution frequency f is higher for a higher momentum particle, i.e. $df/d\delta > 0$, the higher energy particle will arrive at the rf gap earlier, i.e. $\phi < \phi_s$. Therefore if the rf wave synchronous phase is chosen such that $0 < \phi_s < \pi/2$, higher energy particles will receive less energy gain from the rf gap. Similarly, lower energy particles will arrive at the same rf gap later and gain more energy than the synchronous particle. This process provides the phase stability of synchrotron motion. In the case of $df/d\delta < 0$, phase stability requires $\pi/2 < \phi_s < \pi$.

The discovery of phase stability paved the way for all modern high energy accelerators, called "synchrotrons," and after half a century of research and development, it remains the cornerstone of modern accelerators. Particle acceleration without phase stability is limited to low energy accelerators, e.g. Cockcroft-Walton, Van de Graaff, betatron, etc. Furthermore, bunched beams can be shortened, elongated, combined, or stacked to achieve many advanced applications by using rf manipulation schemes. Phase-space gymnastics have become essential tools in the operation of high energy storage rings.

In this chapter we study the dynamics of synchrotron motion. In Sec. I, we derive the synchrotron equation of motion in various phase-space coordinates. Section II deals with adiabatic synchrotron motion, where an invariant *torus* corresponds to a constant Hamiltonian value. In Sec. III, we study the perturbation of synchrotron motion resulting from rf phase and amplitude modulation, synchro-betatron coupling through dipole field error, ground vibration, etc. In Sec. IV, we treat non-adiabatic synchrotron motion near transition energy, where the Hamiltonian is not invariant. In Sec. V, we study beam injection, extraction, stacking, bunch rotation, phase displacement acceleration, beam manipulations with double rf systems and barrier rf systems, etc. Section VI treats fundamental aspects of rf cavity design. In Sec. VII, we introduce collective longitudinal instabilities. In Sec. VIII, we provide an introduction to the linac.

I Longitudinal Equation of Motion

Let the longitudinal electric field at an rf gap be

$$\mathcal{E} = \mathcal{E}_0 \sin(\phi_{\rm rf}(t) + \phi_s), \quad \phi_{\rm rf} = h\omega_0 t, \tag{3.1}$$

where $\omega_0 = \beta_0 c/R_0$ is the angular revolution frequency of a reference (synchronous) particle, $\mathcal{E}_0$ is the amplitude of the electric field, $\beta_0 c$ and R_0 are respectively the speed and the average radius of the reference orbiting particle, h is an integer called the harmonic number, and ϕ_s is the phase angle for a synchronous particle with respect to the rf wave. We assume that the reference particle passes through the cavity gap in time $t \in nT_0 + (-g/2\beta c, g/2\beta c)$ (n = integer), where g is the rf cavity gap width.

The energy gain for the reference particle per passage is

$$\Delta E = e\mathcal{E}_0\beta c \int_{-g/2\beta_0 c}^{g/2\beta_0 c} \sin(h\omega_0 t + \phi_s)dt = e\mathcal{E}_0 gT \sin\phi_s, \tag{3.2}$$

$$T = \frac{\sin(hg/2R_0)}{(hg/2R_0)}. \tag{3.3}$$

where e is the charge of the circulating particles, and T is the transit time factor. The effective voltage seen by the orbiting particle is $V = \mathcal{E}_0 gT$. The transit time factor arises from the fact that a particle passes through the rf gap within a finite time interval so that the energy gain is the time average of the electric field in the gap during the transit time (see also Exercise 3.1). If the gap length is small, the transit time factor is approximately equal to 1. However, a high electric field associated with a small gap may cause sparking and electric field breakdown.

Since a synchronous particle synchronizes with the rf wave with a frequency of $\omega_{\rm rf} = h\omega_0$, where $\omega_0 = \beta_0 c/R_0$ is the revolution frequency and h is an integer, it encounters the rf voltage at the same phase angle ϕ_s every revolution. The acceleration rate for this synchronous particle is $\dot{E}_0 = \frac{\omega_0}{2\pi}eV\sin\phi_s$, where the dot indicates the derivative with respect to time t.

Now we consider a non-synchronous particle with small deviations of rf parameters from the synchronous particle, i.e.

$$\begin{cases} \omega = \omega_0 + \Delta\omega, & \phi = \phi_s + \Delta\phi, \quad \theta = \theta_s + \Delta\theta, \\ p = p_0 + \Delta p, & E = E_0 + \Delta E. \end{cases}$$

Here $\phi_s, \theta_s, \omega_0, p_0, E_0$ are respectively the rf phase angle, azimuthal orbital angle, angular revolution frequency, momentum, and energy of a synchronous particle, and $\phi, \theta, \omega, p, E$ are the corresponding parameters for an off-momentum particle.

The phase coordinate is related to the orbital angle by $\Delta\phi = \phi - \phi_s = -h\Delta\theta$, or

$$\Delta\omega = \frac{d}{dt}\Delta\theta = -\frac{1}{h}\frac{d}{dt}\Delta\phi = -\frac{1}{h}\frac{d\phi}{dt}. \tag{3.4}$$

The energy gain per revolution for this non-synchronous particle is $eV\sin\phi$, where ϕ is its rf phase angle, and the acceleration rate is $\dot{E} = \frac{\omega}{2\pi}eV\sin\phi$. The equation of motion for the energy-difference becomes[2]

$$\frac{d}{dt}\left(\frac{\Delta E}{\omega_0}\right) = \frac{1}{2\pi}eV(\sin\phi - \sin\phi_s). \tag{3.5}$$

[2]We use the relation

$$\frac{1}{\omega}\dot{E} - \frac{1}{\omega_0}\dot{E}_0 = \frac{1}{\omega_0}\dot{\Delta E} - \dot{E}\frac{\Delta\omega}{\omega_0^2} \approx \frac{1}{\omega_0}\dot{\Delta E} + \left[\dot{E}\frac{\Delta(1/\omega_0)}{\Delta E}\right]\Delta E + \cdots = \frac{d}{dt}\left(\frac{\Delta E}{\omega_0}\right).$$

Using the fractional off-momentum variable, we obtain

$$\delta = \frac{\Delta p}{p_0} = \frac{\omega_0}{\beta^2 E}\frac{\Delta E}{\omega_0}, \qquad \dot{\delta} = \frac{\omega_0}{2\pi\beta^2 E} eV(\sin\phi - \sin\phi_{\rm s}). \tag{3.6}$$

The next task is to find the time evolution of the phase angle variable ϕ. Using Eq. (3.4), we find

$$\dot{\phi} = -h(\omega - \omega_0) = -h\Delta\omega. \tag{3.7}$$

Using the relation $\omega R/\omega_0 R_0 = \beta/\beta_0$ and the result in Chap. 2, Sec. IV, we obtain

$$\frac{\Delta\omega}{\omega_0} = \frac{\beta R_0}{\beta_0 R} - 1. \tag{3.8}$$

$$R = R_0(1 + \alpha_0\delta + \alpha_1\delta^2 + \alpha_2\delta^3 + \cdots),$$

$$\alpha_{\rm c} = \frac{1}{R_0}\frac{dR}{d\delta} = \alpha_0 + 2\alpha_1\delta + 3\alpha_2\delta^2 + \cdots \equiv \frac{1}{\gamma_{\rm T}^2}, \tag{3.9}$$

where R is the mean radius of a circular accelerator and $\alpha_{\rm c}$ is the momentum compaction factor,[3] $\gamma_{\rm T}mc^2$, or simply $\gamma_{\rm T}$, is called the transition energy. Most accelerator lattices have $\alpha_0 > 0$ and the closed-orbit length for a higher energy particle is longer than the reference orbit length. Some specially designed synchrotrons can achieve the condition $\alpha_0 = 0$, where the circumference, up to first order, is independent of particle momentum. Recently, medium energy proton synchrotrons have been designed to have an imaginary $\gamma_{\rm T}$ or a negative momentum compaction (see Chap. 2, Sec. IV.8). The orbit length in a negative compaction lattice is shorter for a higher energy particle.

Let $p = mc\beta\gamma = p_0 + \Delta p$ be the momentum of a non-synchronous particle. The fractional off-momentum coordinate δ is

$$\delta = \frac{\Delta p}{p_0} = \frac{\beta\gamma}{\beta_0\gamma_0} - 1. \tag{3.10}$$

Expressing β and γ in terms of the off-momentum coordinate δ, we find

$$\frac{\gamma}{\gamma_0} = \sqrt{1 + 2\beta_0^2\delta + \beta_0^2\delta^2},$$

$$\frac{\beta}{\beta_0} = \frac{1+\delta}{\sqrt{1 + 2\beta_0^2\delta + \beta_0^2\delta^2}} = 1 + \frac{1}{\gamma_0^2}\delta - \frac{3\beta_0^2}{2\gamma_0^2}\delta^2 + \frac{\beta_0^2(5\beta_0^2 - 1)}{2\gamma_0^2}\delta^3 + \cdots.$$

Combining Eqs. (3.8) and (3.9), we obtain

$$\frac{\Delta\omega}{\omega_0} = -\eta(\delta)\delta, \tag{3.11}$$

[3]Typically, we have $\alpha_1\gamma_{\rm T}^2 \approx \frac{1}{\nu_x}\frac{d\nu_x}{d\delta} \approx 1$ for accelerators without chromatic corrections. The α_1 term depends on the sextupole field in the accelerator.

where the phase slip factor is $\eta(\delta) = \eta_0 + \eta_1\delta + \eta_2\delta^2 + \cdots$ with

$$\begin{cases} \eta_0 = (\alpha_0 - \frac{1}{\gamma_0^2}), \\ \eta_1 = \frac{3\beta_0^2}{2\gamma_0^2} + \alpha_1 - \alpha_0\eta_0, \\ \eta_2 = -\frac{\beta_0^2(5\beta_0^2-1)}{2\gamma_0^2} + \alpha_2 - 2\alpha_0\alpha_1 + \frac{\alpha_1}{\gamma_0^2} + \alpha_0^2\eta_0 - \frac{3\beta_0^2\alpha_0}{2\gamma_0^2}. \end{cases} \tag{3.12}$$

In linear approximation. we find $\Delta\omega = -\eta_0\omega_0\delta = (\frac{1}{\gamma^2} - \frac{1}{\gamma_{\rm T}^2})\omega_0\delta$ Below the transition energy ($\gamma < \gamma_{\rm T}$) a higher energy particle with $\delta > 0$ has a higher revolution frequency. The speed of the higher energy particle more than compensates the difference in path length. At transition energy, the revolution frequency is independent of particle momentum. The AVF cyclotron operates in this isochronous condition. The nonlinear term in Eq. (3.11) becomes important near transition energy, to be addressed in Sec. IV. Above transition energy ($\gamma > \gamma_{\rm T}$) a higher energy particle with $\delta > 0$ has a smaller revolution frequency, i.e. the particle appears to have a "negative mass." Combining Eqs. (3.7) and (3.11), we obtain the phase equation of motion:

$$\dot{\phi} = h\omega_0\eta\delta = \frac{h\omega_0^2\eta}{\beta^2 E}\left(\frac{\Delta E}{\omega_0}\right), \tag{3.13}$$

where (ϕ, $\Delta E/\omega_0$) or equivalently (ϕ, δ) are pairs of conjugate phase-space coordinates. Equations (3.5) and (3.13) form the "synchrotron equation of motion."

I.1 The Synchrotron Hamiltonian

The synchrotron equations of motion (3.5) and (3.13) can be derived from a "Hamiltonian"

$$\begin{aligned} H &= \frac{1}{2}\frac{h\eta\omega_0^2}{\beta^2 E}\left(\frac{\Delta E}{\omega_0}\right)^2 + \frac{eV}{2\pi}[\cos\phi - \cos\phi_{\rm s} + (\phi - \phi_{\rm s})\sin\phi_{\rm s}] \\ &= \frac{1}{2}h\omega_0\eta_0\delta^2 + \frac{\omega_0 eV}{2\pi\beta^2 E}[\cos\phi - \cos\phi_{\rm s} + (\phi - \phi_{\rm s})\sin\phi_{\rm s}] \end{aligned} \tag{3.14}$$

with time t as an independent variable and for phase-space coordinates $(\phi, \Delta E/\omega_0)$ and (ϕ, δ) respectively. This Hamiltonian, although legitimate, is inconsistent with the "Hamiltonian" for transverse betatron oscillations, where s is the independent coordinate. To simplify our discussion, we will disregard the inconsistency and study synchrotron motion of Eq. (3.14). A fully consistent treatment is needed when we study synchro-betatron coupling resonances.[4]

With this simplified Hamiltonian, we now discuss the stability condition for small amplitude oscillations, where the linearized equation of motion is

$$\frac{d^2}{dt^2}(\phi - \phi_{\rm s}) = \frac{h\omega_0^2 eV\eta_0\cos\phi_{\rm s}}{2\pi\beta^2 E}(\phi - \phi_{\rm s}). \tag{3.15}$$

---

[4]S.Y. Lee, *Phys. Rev.* **E49**, 5706 (1994).

The stability condition for synchrotron oscillation is $\eta_0 \cos\phi_s < 0$, discovered by McMillan and Veksler [21]. Below the transition energy, with $\gamma < \gamma_T$ or $\eta_0 < 0$, the synchronous phase angle should be $0 < \phi_s < \pi/2$. Similarly the synchronous phase angle should be shifted to $\pi - \phi_s$ above the transition energy. This can be accomplished by a phase shift of $\pi - 2\phi_s$ to the rf wave. The *angular synchrotron frequency* and *synchrotron tune* (the number of synchrotron oscillations per revolution) are

$$\omega_s = \omega_0\sqrt{\frac{heV|\eta_0\cos\phi_s|}{2\pi\beta^2 E}} = \frac{c}{R}\sqrt{\frac{heV|\eta\cos\phi_s|}{2\pi E}}, \tag{3.16}$$

$$Q_s = \frac{\omega_s}{\omega_0} = \sqrt{\frac{heV|\eta_0\cos\phi_s|}{2\pi\beta^2 E}} \equiv \nu_s\sqrt{|\cos\phi_s|}, \quad \text{with} \quad \nu_s = \sqrt{\frac{heV|\eta_0|}{2\pi\beta^2 E}}. \tag{3.17}$$

where c is the speed of light and R is the average radius of synchrotron. The rf bucket at the synchronous rf phase $\phi_s = 0$ or π is called the stationary bucket because there is no net change of beam energy. The synchrotron tune at the stationary bucket is usually denoted by ν_s. The rf buckets with $\phi_s \neq 0$ or π are called running buckets.

Typically the synchrotron tune is of the order of $\sim 10^{-3}$ for proton synchrotrons and 10^{-1} for electron storage rings. Figure 3.1 shows the measured synchrotron tune of the Fermilab Booster in a ramping cycle from 400 MeV to 8 GeV. The inset shows the rf voltage and the corresponding rf synchronous phase during the ramping cycle.

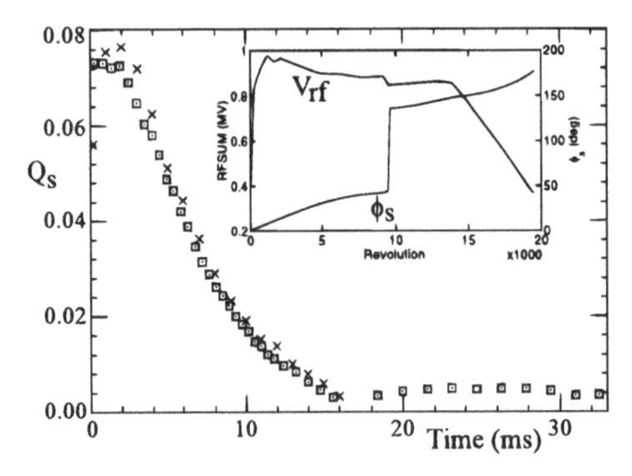

Figure 3.1: The synchrotron tune of a ramping cycle for the Fermilab Booster, which is a rapid cycling accelerator at 15 Hz. At $t = 17$ ms, the beam crosses transition energy. Rectangular symbols are results obtained by the ICA measurement method (see Appendix Sec. III) and $\times$'s are obtained from spectrum analyzer to a difference signal. The synchrotron tune for a rapid cycling synchrotron at low energy can be large (Courtesy of X. Huang).

The stability of particle motion in rf force potential can be understood from the left plot of Fig. 3.2, where the rf potentials are shown in the left plots for $\phi_s = 0$ and $\phi_s = \pi/6$. The potential well near the synchronous phase angle provides restoring force for quasi-harmonic oscillations. The horizontal dashed line shows the maximum Hamiltonian value for a stable synchrotron orbit. The corresponding stable phase-space (bucket) area is shown in the middle plots in the normalized phase space coordinates $(h|\eta|/\nu_s)\delta_{sx}$ vs ϕ. Particles inside the rf bucket execute stable synchrotron motion, while particles fall outside the bucket will be lost.

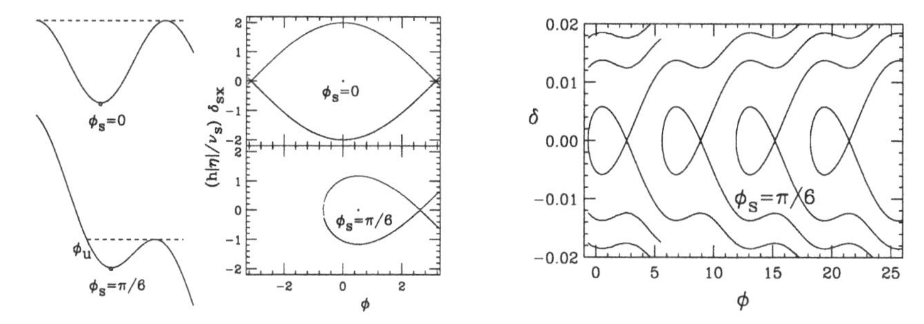

Figure 3.2: Left: schematic drawing of the rf potentials for $\phi_s = 0$ and $\pi/6$. The dashed line shows the maximum "energy" for stable synchrotron motion. Middle: the corresponding separatrix orbits in $(h|\eta|/\nu_s) \times \delta_{sx}$ vs ϕ. The phase ϕ_u is the turning point of the separatrix orbit. Right: an example of stable rf buckets, called fish diagram, in δ vs ϕ at $\phi_s = \pi/6$.

For an accelerator with a harmonic number h, there are h buckets. Particles can fill some of these stable buckets. The filling pattern can be arranged by injection schemes. The right plot of Fig. 3.2 shows a fish diagram of rf buckets at $\phi_s = \pi/6$.

I.2 The Synchrotron Mapping Equation

In Hamiltonian formalism, the rf electric field is considered to be uniformly distributed in an accelerator. In reality, rf cavities are localized in a short section of a synchrotron, the synchrotron motion is more realistically described by the symplectic mapping equation:

$$\begin{cases} \delta_{n+1} = \delta_n + \dfrac{eV}{\beta^2 E}(\sin\phi_n - \sin\phi_s), \\ \phi_{n+1} = \phi_n + 2\pi h\eta(\delta_{n+1})\delta_{n+1}. \end{cases} \tag{3.18}$$

The physics of the mapping equation can be visualized as follows. First, the particle gains or loses energy at its nth passage through the rf cavity, then the rf phase ϕ_{n+1} depends on the new off-momentum coordinate δ_{n+1}. Since the Jacobian of the mapping from (ϕ_n, δ_n) to $(\phi_{n+1}, \delta_{n+1})$ is equal to 1, the mapping preserves phase-space area.

Note that Eq. (3.18) treats the rf cavity as a single lumped element in an accelerator. In reality, the rf cavities may be distributed non-uniformly. The rf phase change between different cavities may not be uniform. Because synchrotron motion is usually slow, Hamiltonian formalism and mapping equations are equivalent. Because of the simplicity of the mapping equations, they are usually used in particle tracking calculations.

I.3 Evolution of Synchrotron Phase-Space Ellipses

The phase-space area enclosed by a trajectory (ϕ, δ) obtained from Eq. (3.18) is independent of energy. Therefore, Eq. (3.18) can not be used in tracking simulations of particle beam in acceleration. During acceleration, the phase-space area in $(\phi, \Delta E/\omega_0)$ is invariant. The phase-space mapping equation for phase-space coordinates $(\phi, \Delta E/\omega_0)$ should be used. The adiabatic damping of phase-space area can be obtained by transforming phase-space coordinates $(\phi, \Delta E/\omega_0)$ to (ϕ, δ).

The separatrix for the rf bucket shown in Fig. 3.2 is a closed curve. In a rapid cycling synchrotron or electron linac where the acceleration gradient is high, the separatrix is not a closed curve. The mapping equations for synchrotron phase-space coordinates $(\phi, \Delta E)$ are

$$\Delta E_{n+1} = \Delta E_n + eV(\sin\phi_n - \sin\phi_s), \tag{3.19}$$

$$\phi_{n+1} = \phi_n + \frac{2\pi h\eta}{\beta^2 E}\Delta E_{n+1}. \tag{3.20}$$

The quantity $\eta/\beta^2 E$ in Eq. (3.20) depends on energy, which is obtained from Eq. (3.19), i.e. $E = E_{0,n+1} = E_{0,n} + eV\sin\phi_s$, $\gamma = E/mc^2$, $\beta = \sqrt{1 - 1/\gamma^2}$, and $\eta = \alpha_c - 1/\gamma^2$. If the acceleration rate is low, the factor $h\eta/\beta^2 E$ is nearly constant, and the separatrix orbit shown in Fig. 3.2 can be considered as a closed curve.

When the acceleration rate is high, tori of the synchrotron mapping equations are not closed curves. Figure 3.3 shows two tori in phase-space coordinates $(\phi, \Delta E/\beta^2 E)$ with parameters $V = 100$ kV, $h = 1$, $\alpha_c = 0.04340$, $\phi_s = 30°$ at 45 MeV proton kinetic energy. Note that the actual attainable rf voltage V is about 200-1000 V in a low energy proton synchrotron. When the acceleration rate is high, the separatrix is not a closed curve. The phase-space tori change from a fish-like to a golf-club-like shape. This is equivalent to the *adiabatic damping* of phase-space area discussed in Chapter 2, Sec. II. Since the acceleration rate for proton (ion) beams is normally low, the separatrix torus shown in Fig. 3.2 is a good approximation. When the acceleration rate is high, e.g. in many electron accelerators, the tori near the separatrix may resemble those in Fig. 3.3.

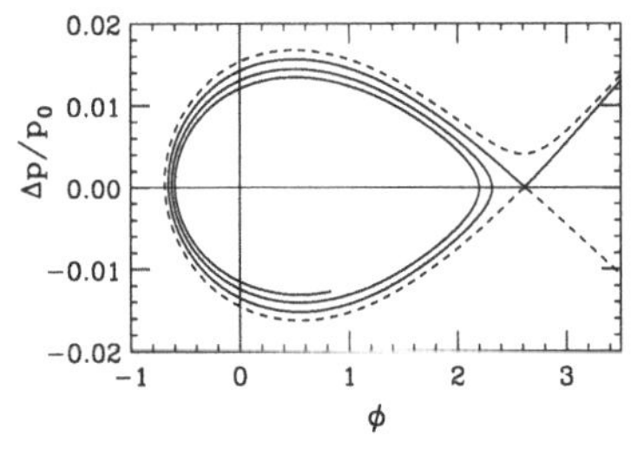

Figure 3.3: Two tori in phase-space coordinates $(\phi, \Delta E/\beta^2 E)$ obtained from mapping equations (3.19) and (3.20) with parameters $V = 100$ kV, $h = 1$, $\alpha_c = 0.04340$, and $\phi_s = 30°$ at 45 MeV proton kinetic energy. IUCF Cooler Ring has typical rf voltage at about 1–2 kV. Note that the dashed line become the separatrix orbit, while the solid line is trapped into the bucket. This phenomenon is adiabatic damping of synchrotron motion.

I.4 Some Practical Examples

A basic rf cavity requirement for beam acceleration rate is

$$B\rho = \frac{p}{e}, \quad \dot{p} = \frac{1}{\beta c}\dot{E}, \quad f = \frac{\beta c}{2\pi R}, \quad \Longrightarrow \quad V \sin\phi_s = 2\pi R\rho\dot{B}. \tag{3.21}$$

For example, proton acceleration in the IUCF cooler ring from 45 MeV to 500 MeV in one second requires $\dot{B} = \frac{\Delta[B\rho]}{\rho\Delta t} \approx 1.1$ Tesla/sec, and $V \sin\phi_s \approx 240$ Volts, where we use $\rho \approx 2.4$ m, $R \approx 14$ m. The result is independent of the harmonic number used.

Similarly, acceleration of protons from 9 GeV to 120 GeV in 1 s at the Fermilab Main Injector would require $\dot{B} \approx 1.6$ Tesla/s. The circumference is 3319.4 m with $\rho = 235$ m. The voltage requirement becomes $V \sin\phi_s = 1.2$ MV.

I.5 Summary of Synchrotron Equations of Motion

A. Using t as independent variable

Using time t as an independent variable, the equations of motion and the Hamiltonian are listed as follows.

- Using $(\phi, \Delta E/\omega_0)$ as phase-space coordinates:

$$\frac{d\phi}{dt} = \frac{h\omega_0^2\eta}{\beta^2 E}\left(\frac{\Delta E}{\omega_0}\right), \quad \frac{d\,(\Delta E/\omega_0)}{dt} = \frac{1}{2\pi}eV(\sin\phi - \sin\phi_s), \tag{3.22}$$

$$H = \frac{1}{2}\frac{h\eta\omega_0^2}{\beta^2 E}\left(\frac{\Delta E}{\omega_0}\right)^2 + \frac{eV}{2\pi}[\cos\phi - \cos\phi_s + (\phi - \phi_s)\sin\phi_s]. \tag{3.23}$$

- Using (ϕ, δ) as phase-space coordinates:

$$\frac{d\phi}{dt} = h\omega_0\eta\delta, \quad \frac{d\delta}{dt} = \frac{\omega_0 eV}{2\pi\beta^2 E}(\sin\phi - \sin\phi_s), \tag{3.24}$$

$$H = \frac{1}{2}h\omega_0\eta\delta^2 + \frac{\omega_0 eV}{2\pi\beta^2 E}[\cos\phi - \cos\phi_s + (\phi - \phi_s)\sin\phi_s]. \tag{3.25}$$

- Using $(\phi, \mathcal{P} = -(h|\eta|/\nu_s)\delta)$ as the normalized phase-space coordinates:

$$\frac{d\phi}{dt} = \omega_0\nu_s\mathcal{P}, \quad \frac{d\mathcal{P}}{dt} = \frac{\eta}{|\eta|}\omega_0\nu_s(\sin\phi - \sin\phi_s), \tag{3.26}$$

$$H = \frac{1}{2}\omega_0\nu_s\mathcal{P}^2 + \frac{\eta}{|\eta|}\omega_0\nu_s[\cos\phi - \cos\phi_s + (\phi - \phi_s)\sin\phi_s]. \tag{3.27}$$

- Using $(\tau = (\phi - \phi_s)/h\omega_0, \dot{\tau})$ as phase-space coordinates:

$$\frac{d\tau}{dt} = \dot{\tau}, \quad \frac{d\dot{\tau}}{dt} = -\frac{\eta\omega_0 eV}{2\pi\beta^2 E}[\sin(\phi_s - h\omega_0\tau) - \sin\phi_s], \tag{3.28}$$

$$H = \frac{1}{2}\dot{\tau}^2 + \frac{\eta eV}{2\pi h\beta^2 E}[\cos(\phi_s - h\omega_0\tau) - \cos\phi_s - h\omega_0\tau\sin\phi_s]. \tag{3.29}$$

The corresponding normalized phase space is $(\tau, \dot{\tau}/\omega_s)$.

B. Using longitudinal distance s as independent variable

- Using $(R\phi/h, -\Delta p/p_0)$ as phase-space coordinates, the Hamiltonian is

$$H = -\frac{1}{2}\eta\left(\frac{\Delta p}{p_0}\right)^2 - \frac{\nu_s^2}{h^2|\eta|}\left[\cos\phi - \cos\phi_s + (\phi - \phi_s)\sin\phi_s\right], \tag{3.30}$$

where $\nu_s = \sqrt{h|\eta|eV/2\pi\beta^2 E_0}$ is the synchrotron tune at $\phi_s = 0$. This synchrotron Hamiltonian is on an equal footing with the transverse betatron motion. In particular, the negative sign in the first term corresponds to negative mass above the transition energy, where $\eta > 0$.

Exercise 3.1

1. An rf cavity consists of an insulating gap g across which the rf voltage is applied. The gap length is finite and the rf field changes with time during transit time Δt. The total energy gain of a particle passing through the gap is the time average of the rf voltage during the transit time, i.e.

$$\Delta E = \frac{e}{\Delta t}\int_{-\Delta t/2}^{\Delta t/2} V(t)dt, \quad V(t) = V_g\sin(\phi + h\omega_0 t)$$

where V_g is the peak gap voltage, and ϕ the rf phase of the particle. Show that the effective voltage is

$$V = V_g T, \quad T = \frac{\sin(hg/2R)}{(hg/2R)},$$

where R is the mean radius of the accelerator. Thus the transit time factor T is the same for all particles.

2. Show that the relation between the rf frequency of an accelerator and the magnetic flux density $B(t)$ during particle acceleration at a constant radius is given by

$$\omega_{\rm rf} = \frac{hc}{R_0}\left[\frac{B^2(t)}{B^2(t) + (mc^2/ec\rho)^2}\right]^{1/2},$$

where h is the harmonic number, ρ is the bending radius of the dipoles, R_0 is the mean radius of the accelerator, e and m are the charge and mass of the particle, and c is the speed of light.

3. Calculate synchrotron tunes for the proton synchrotrons listed in the following table with $\phi_s = 0$.

RF parameters of some proton synchrotrons

P-synchrotron	AGS	RHIC	FNAL-MI	FNAL-BST	SSC	Cooler
K.E. [GeV/u]	0.2	28	8	0.4	2000	0.045
V_{rf} [MV]	0.3	0.3	2	0.95	10	0.0001
h	12	342	588	84	17424	1
γ_T	8.5	24.5	21.8	5.446	140	4.6
C [m]	807.12	3833.84	3319.4	474.2	87120	86.8
ν_s						

4. Electrons in storage rings emit synchrotron radiation. The energy loss per turn is given by

$$U_0 = C_\gamma \beta^3 E_0^4/\rho,$$

where E_0 is the beam energy, βc is the beam velocity, ρ is the bending radius of dipoles, and

$$C_\gamma = 8.85 \times 10^{-5} \text{ m/(GeV)}^3.$$

The energy loss due to synchrotron radiation is compensated by the rf accelerating field, i.e. $U_0 = eV_{rf} \sin\phi_s$. Calculate synchrotron tunes for the electron storage rings listed in the following table.

RF parameters of some electron synchrotrons

	LEP	ALS	APS	NLC DR	BEPC	TRISTAN
C [m]	26658.9	196.8	1060	223	240.4	3018
Energy [GeV]	50	1.2	7.0	1.98	2.2	30.
ρ [m]	3096.2	4.01	38.96	4.35	10.35	246.5
V_{rf} [MV]	400	1.5	10	1.0	0.8	400
h	31320	328	1248	531	160	5120
γ_T	50.86	26.44	64.91	46.1	5.0	25.5
ν_x	76.2	14.28	35.22	23.81	6.18	36.8
ν_z	70.2	8.18	14.3	8.62	7.12	38.7
ϕ_s [deg]						
Q_s						

5. The synchrotron tune of the Fermilab Booster during the ramping cycle is shown in Fig. 3.1. We note that the rf voltage is ramped from a low value to a high value at injection for adiabatic capture (see Sec. V.2 in Chap. 3).

   (a) Compare the measured synchrotron tune at the injection energy with that obtained from Exercise 5.1.3. Explain the difference.

   (b) At time $t = 17$ ms, what happens to the the measured synchrotron tune? What is the beam energy and what happens to the synchronous phase angle?

   (c) Why the synchronous phase angle is moved to 180° at extraction?

6. Compare and discuss the synchrotron mapping equations of Eq. (3.18) vs Eqs. (3.19) and (3.20).

7. Verify Eq. (3.21) of rf voltage requirement during the beam energy ramping process.

8. Find the dimension of phase space area in various phase space coordinates discussed in Sec. I.5

9. Write a computer program to track synchrotron motion near the separatrix, explore the dependence of the separatrix on the acceleration rate, and verify the golf-club-like tori in Fig. 3.3, where the torus of the solid line is captured into the bucket, while the torus of dashed line is outside the bucket.

10. Redefine $y \equiv h|\eta|\delta/\nu_s$.

    (a) Show that Eq. (3.18) of the symplectic mapping equation for a stationary bucket synchrotron motion can be transformed into the standard map:

$$y_{n+1} = y_n + 2\pi\nu_s(\sin\phi_n - \sin\phi_s),$$
$$\phi_{n+1} = \phi_n + \frac{\eta}{|\eta|}2\pi\nu_s y_{n+1},$$

    where $\nu_s = \sqrt{h|\eta|eV/2\pi\beta^2 E}$ is the synchrotron tune.

    (b) Write a program to track the phase-space points (ϕ, y) such that $\phi \in [-\pi, \pi]$, $y \in [-2, 2]$. Examine the symmetry of the tracking equation in (ϕ, y) space.

    (c) Explore the phase-space evolution at $\nu_s = \nu_{s,c} = 0.39324366$, where $\nu_{s,c}$ is the critical synchrotron tune for global chaos in the synchrotron phase space.

II Adiabatic Synchrotron Motion

With time t as an independent variable, Hamilton's equations of motion for the synchrotron Hamiltonian for phase-space coordinates (ϕ, δ) shown in Eq. (3.14) are

$$\dot{\phi} = h\eta\omega_0\delta, \qquad \dot{\delta} = \frac{\omega_0 eV}{2\pi\beta^2 E}(\sin\phi - \sin\phi_{\rm s}), \tag{3.31}$$

where the over-dots indicate derivatives with respect to time t. For simplicity in notation, hereafter, the subscript of the energy E_0 of the beam has been neglected. If $|\eta| \neq 0$, the small amplitude synchrotron tune is given by Eq. (3.17). The synchrotron period is $T_{\rm s} = T_0/Q_{\rm s}$, where T_0 is the revolution period.

The typical synchrotron tune in proton synchrotrons is of the order of 10^{-3}, i.e. it takes about 1000 revolutions to complete one synchrotron oscillation. The typical synchrotron tune in electron storage rings is of the order of 10^{-1}. If the rf parameters V and $\phi_{\rm s}$ vary only slowly with time so that the gain in beam energy in each revolution is small, and η differs substantially from 0, the Hamiltonian is time independent or nearly time independent.

During beam acceleration, the Hamiltonian (3.14) generally depends on time. However, if the acceleration rate is low, the Hamiltonian can be considered as quasi-static. This corresponds to adiabatic synchrotron motion, where parameters in the synchrotron Hamiltonian change slowly so that the particle orbit is a torus of constant Hamiltonian value. The condition for adiabatic synchrotron motion is

$$\alpha_{\rm ad} = \left|\frac{1}{\omega_{\rm s}^2}\frac{d\omega_{\rm s}}{dt}\right| = \frac{1}{2\pi}\left|\frac{dT_{\rm s}}{dt}\right| \ll 1, \tag{3.32}$$

where $\omega_{\rm s}$ is the angular synchrotron frequency and $\alpha_{\rm ad}$ is called the adiabaticity coefficient. Typically, when $\alpha_{\rm ad} \leq 0.05$, the time variation of synchrotron period is small and the trajectories of particle motion can be approximately described by tori of constant Hamiltonian values.

II.1 Fixed Points

The Hamiltonian for adiabatic synchrotron motion has two fixed points $(\phi_{\rm s}, 0)$ and $(\pi - \phi_{\rm s}, 0)$, where $\dot{\phi} = 0$ and $\dot{\delta} = 0$. The phase-space point $(\phi_{\rm s}, 0)$ is the stable fixed point (SFP). Small amplitude phase-space trajectories around the stable fixed point are ellipses. Therefore the SFP is also called an elliptical fixed point.

The phase-space trajectories near the unstable fixed point (UFP) $(\pi - \phi_{\rm s}, 0)$ are hyperbola. Thus the UFP is also called a hyperbolic fixed point. The torus that passes through the UFP is called the *separatrix*; it separates phase space into regions of bound and unbound oscillations. Figure 3.4 shows the separatrix orbit in the normalized coordinates $(h|\eta|/\nu_s)\delta_{\rm sx}$ vs ϕ for $\eta < 0$ with $\phi_{\rm s} = 0, \pi/6, \pi/3$ and for $\eta > 0$

with $\phi_s = 2\pi/3, 5\pi/6, \pi$. The synchrotron phase space is divided into stable and unstable regions, and only particles in the stable region can be accelerated to high energy. Particles in synchrotrons are naturally bunched. A beam in which particles are grouped together forming bunches is called a *bunched beam.*

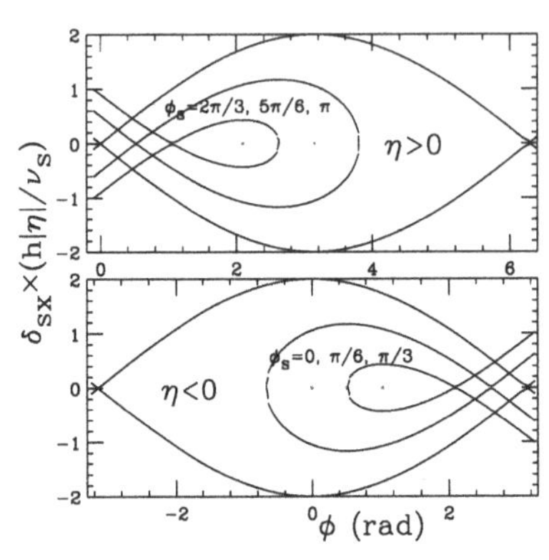

Figure 3.4: Separatrix orbits of the normalized phase space coordinates $(h|\eta|/\nu_s)\delta_{\rm sx}$ vs ϕ for $\eta > 0$ (above transition energy) with $\phi_s = 2\pi/3, 5\pi/6, \pi$, (top) and for $\eta < 0$ (below transition energy) with $\phi_s = 0, \pi/6, \pi/3$ (bottom). The UFP of each separatrix is $\pi - \phi_s$. The phase space area enclosed by the separatrix is called the *bucket area.* The maximum off-momentum deviation of the rf bucket is called *bucket height.* The acceleration rate is zero for a stationary bucket at $\phi_s = 0$ or π. When $\sin\phi_s \neq 0$ or π, the bucket is called a *running bucket.* When an rf system has a harmonic number h applied to an accelerator, there are h identical buckets and a maximum of h bunches can be stored in the accelerator.

For a slowly time-varying Hamiltonian, particle motion adiabatically follows a phase-space ellipse, called a "Hamiltonian torus" at a "constant" Hamiltonian value. The phase-space area enclosed by a Hamiltonian torus is $\tilde{\mathcal{A}} = \oint \delta(\phi)d\phi$. The phase-space area enclosed by the separatrix orbit is called the *bucket area.* The maximum momentum deviation of the separatrix orbit is called the *bucket height.* Particles outside the rf bucket drift along the longitudinal direction, and particles inside the rf bucket execute quasi-harmonic motion within the bucket.

II.2 Bucket Area

The separatrix passes through the unstable fixed point $(\pi - \phi_s, 0)$. Therefore the Hamiltonian value of the separatrix and the phase-space trajectory of the separatrix, where $H = H_{\rm sx}$, are

$$H_{\rm sx} = \frac{\omega_0 eV}{2\pi\beta^2 E}\,[-2\cos\phi_s + (\pi - 2\phi_s)\sin\phi_s]. \tag{3.33}$$

$$\delta_{\rm sx}^2 + \frac{eV}{\pi\beta^2 Eh\eta}[\cos\phi + \cos\phi_s - (\pi - \phi - \phi_s)\sin\phi_s] = 0.$$

The separatrix has two turning points, ϕ_u and $\pi - \phi_s$ with $\cos\phi_u + \phi_u\sin\phi_s = -\cos\phi_s + (\pi - \phi_s)\sin\phi_s$. For $\phi_s = 0$, the turning points are $-\pi$ and π.

The phase-space area enclosed by the separatrix is called the bucket area, i.e.

$$\tilde{\mathcal{A}}_{\rm B} = \oint \delta_{\rm sx}(\phi)d\phi = 16\sqrt{\frac{eV}{2\pi\beta^2 Eh|\eta|}}\;\alpha_{\rm b}(\phi_s) = \frac{16Q_s}{h|\eta|\sqrt{|\cos\phi_s|}}\,\alpha_{\rm b}(\phi_s), \tag{3.34}$$

$$\alpha_{\rm b}(\phi_{\rm s}) = \frac{1}{4\sqrt{2}}\int_{\phi_{\rm u}}^{\pi-\phi_{\rm s}}\left[-\frac{|\eta|}{\eta}\left[\cos\phi + \cos\phi_{\rm s} - (\pi - \phi - \phi_{\rm s})\sin\phi_{\rm s}\right]\right]^{1/2} d\phi, \quad (3.35)$$

where the factor $\alpha_{\rm b}(\phi_{\rm s})$ is the ratio of bucket areas of a running bucket ($\phi_{\rm s} \neq 0$) to a stationary bucket ($\phi_{\rm s} = 0$). Table 3.1 lists $\alpha_{\rm b}(\phi_{\rm s})$ as a function of the synchronous phase angle $\phi_{\rm s}$. Naturally $\alpha_{\rm b}(0) = 1$, and $\alpha_{\rm b}(\pi/2) = 0$, i.e. the bucket area vanishes at 90° synchronous phase angle. We note that $\alpha_{\rm b}(\phi_{\rm s}) \approx (1 - \sin\phi_{\rm s})/(1 + \sin\phi_{\rm s})$.

Table 3.1: Turning point ($\phi_{\rm u}$ in degree), UFP ($\pi - \phi_{\rm s}$ in degree), Bucket length ($\pi - \phi_{\rm s} - \phi_{\rm u}$), bucket height factor $Y(\phi_{\rm s})$ and bucket area factor $\alpha(\phi_{\rm s})$.

$\sin\phi_{\rm s}$	$\phi_{\rm u}$	$\pi - \phi_{\rm s}$	$Y(\phi_{\rm s})$	$\alpha_{\rm b}(\phi_{\rm s})$	$\frac{1-\sin\phi_{\rm s}}{1+\sin\phi_{\rm s}}$
0.00	−180.00	180.00	1.0000	1.0000	1.0000
0.10	−118.90	174.26	0.9208	0.8041	0.8182
0.20	−93.71	168.46	0.8402	0.6611	0.6667
0.30	−73.59	162.54	0.7577	0.5388	0.5385
0.40	−55.66	156.42	0.6729	0.4305	0.4286
0.50	−38.69	150.00	0.5852	0.3333	0.3333
0.60	−21.88	143.13	0.4936	0.2460	0.2500
0.70	−4.48	135.57	0.3967	0.1679	0.1765
0.80	14.59	126.87	0.2919	0.0991	0.1111
0.90	37.77	115.84	0.1731	0.0408	0.0526
1.00	90.00	90.00	0.	0.	0.

The corresponding invariant bucket area in $(\phi, \Delta E/\omega_0)$ phase-space variables,

$$\mathcal{A}_{\rm B,rms} = \frac{\beta^2 E}{\omega_0}\tilde{\mathcal{A}}_{\rm B,rms} = h\pi\sigma_{\Delta t}\sigma_{\Delta E}, \quad (3.36)$$

is the phase-space area of h buckets in accelerator, where Δt is the bucket width in time (s), ΔE is the bucket energy height (eV), and the resulting bucket phase-space area is in eV-s. Table 3.2 lists relevant formulas for rf bucket properties.

Table 3.2: Formula for bucket area in conjugate phase space coordinates

	$(\phi, \frac{\Delta E}{\omega_0})$	(ϕ, δ)	$(\phi, \frac{h\lvert\eta\rvert}{\nu_{\rm s}}\delta)$
Bucket Area	$16\left(\frac{\beta^2 EeV}{2\pi\omega_0^2 h\lvert\eta\rvert}\right)^{1/2}\alpha_{\rm b}(\phi_{\rm s})$	$16\left(\frac{eV}{2\pi\beta^2 Eh\lvert\eta\rvert}\right)^{1/2}\alpha_{\rm b}(\phi_{\rm s})$	$16\,\alpha_{\rm b}(\phi_{\rm s})$
Bucket Height	$2\left(\frac{\beta^2 EeV}{2\pi\omega_0^2 h\lvert\eta\rvert}\right)^{1/2} Y(\phi_{\rm s})$	$2\left(\frac{eV}{2\pi\beta^2 Eh\lvert\eta\rvert}\right)^{1/2} Y(\phi_{\rm s})$	$2\,Y(\phi_{\rm s})$

The bucket length is $|(\pi - \phi_{\rm s}) - \phi_{\rm u}|$, and the bucket height or the maximum momentum width is

$$\delta_{\rm B} = \left(\frac{2eV}{\pi\beta^2 Eh|\eta|}\right)^{1/2} Y(\phi_{\rm s}) = \frac{2Q_{\rm s}}{h|\eta|}\,\tilde{Y}(\phi_{\rm s}). \quad (3.37)$$

Here bucket height factors $Y(\phi_s)$ and $\tilde{Y}(\phi_s)$ are ratios of the maximum momentum height to that of a stationary bucket, i.e.

$$Y(\phi_s) = \left|\cos\phi_s - \frac{\pi - 2\phi_s}{2}\sin\phi_s\right|^{1/2}, \quad \tilde{Y}(\phi_s) = \left|1 - \frac{\pi - 2\phi_s}{2}\tan\phi_s\right|^{1/2}. \qquad (3.38)$$

Table 3.1 also lists turning point and bucket height factor, $Y(\phi_s)$, of rf bucket.

II.3 Small-Amplitude Oscillations and Bunch Area

The linearized synchrotron Hamiltonian around the SFP is simple harmonic with

$$\begin{aligned}
H &= \frac{1}{2}h\omega_0\eta\delta^2 - \frac{\omega_0 eV\cos\phi_s}{4\pi\beta^2 E}\varphi^2, \\
\varphi &= \hat{\phi}\cos(\omega_s t + \chi), \quad \delta = -\frac{Q_s}{h\eta}\hat{\phi}\sin(\omega_s t + \chi), \\
&\left(\frac{\delta}{\hat{\delta}}\right)^2 + \left(\frac{\varphi}{\hat{\phi}}\right)^2 = 1, \quad \frac{\hat{\delta}}{\hat{\phi}} = \left(\frac{eV|\cos\phi_s|}{2\pi\beta^2 Eh|\eta|}\right)^{1/2} = \frac{Q_s}{h|\eta|},
\end{aligned} \qquad (3.39)$$

where $\varphi = \phi - \phi_s$, and the synchrotron frequency is given by Eq. (3.17) with $\omega_s = Q_s\omega_0$, and $\hat{\delta}$ and $\hat{\phi}$ are maximum amplitudes of the phase-space ellipse. The phase-space area of the ellipse is $\pi\hat{\delta}\hat{\phi}$.

A. Gaussian beam distribution

The equilibrium beam distribution is a function of the invariant ellipse of Eq. (3.39). In many beam applications, we use the normalized Gaussian distribution given by

$$\rho(\delta, \phi) = \frac{1}{2\pi\sigma_\delta\sigma_\phi}\exp\left\{-\frac{1}{2}\left[\frac{\phi^2}{\sigma_\phi{}^2} + \frac{\delta^2}{\sigma_\delta{}^2}\right]\right\}, \qquad (3.40)$$

where σ_δ and σ_ϕ are rms momentum spread and bunch length respectively. The corresponding rms phase-space area is $\tilde{\mathcal{A}}_{\rm rms} = \pi\sigma_\delta\sigma_\phi$. The phase-space area that contains 95% of the particles in a Gaussian beam distribution is $\tilde{\mathcal{A}}_{95\%} = 6\tilde{\mathcal{A}}_{\rm rms}$, where the factor 6 depends on the distribution function.

The synchrotron phase-space area (or emittance) $\mathcal{A}$, usually measured in eV-s, is defined as the area in the phase space $(\phi/h, \Delta E/\omega_0)$ for **one** bunch,[5] it is related to

[5]The energy of a heavy ion beam is usually expressed as [MeV/u] or [GeV/u], the E in the denominator of Eq. (3.41) can be expressed as $A \times (E/A)$, where A is the atomic mass number or the number of nucleons in a nucleus, $E/A = \gamma uc^2$, and $u = 0.931494$ GeV/c^2 is the atomic mass unit. Thus the phase-space area is commonly defined as phase-space area per amu expressed as [eV-s/u] for heavy ion beams. The factors eV and E in this chapter should be modified by ZeV and $A \times (E/A)$ for heavy ion beams, where Z is the ion's charge number, A is the atomic mass number, and E/A is the energy per nucleon.

$\tilde{\mathcal{A}}$ by a factor $h\omega_0/\beta^2 E$. Using Eq. (3.39), we find the maximum momentum width and bunch length of a bunch as

$$\tilde{\mathcal{A}} = \pi\hat{\delta}\hat{\phi} = h\mathcal{A}\left(\frac{\omega_0}{\beta^2 E}\right). \tag{3.41}$$

$$\hat{\delta} = \mathcal{A}^{1/2}\left(\frac{\omega_0}{\pi\beta^2 E}\right)^{1/2}\left(\frac{heV|\cos\phi_s|}{2\pi\beta^2 E|\eta|}\right)^{1/4} = \left(\mathcal{A}\frac{\omega_0 Q_s}{\pi\beta^2 E|\eta|}\right)^{1/2},$$

$$\hat{\theta} = \frac{1}{h}\hat{\phi} = \mathcal{A}^{1/2}\left(\frac{\omega_0}{\pi\beta^2 E}\right)^{1/2}\left(\frac{2\pi\beta^2 E|\eta|}{heV|\cos\phi_s|}\right)^{1/4} = \left(\mathcal{A}\frac{\omega_0|\eta|}{\pi\beta^2 E Q_s}\right)^{1/2},$$

$$\frac{\hat{\delta}}{\hat{\theta}} = \left(\frac{heV|\cos\phi_s|}{2\pi\beta^2 E|\eta|}\right)^{1/2} = \frac{Q_s}{|\eta|}. \tag{3.42}$$

The invariant rms phase-space area for one bunch in eV-s is $\mathcal{A}_{\rm rms} = \pi\sigma_{\Delta t}\sigma_{\Delta E}$. The scaling properties of bunch length and bunch height of Eq. (3.42) become

$$\hat{\delta} \sim \mathcal{A}^{1/2}V^{1/4}h^{1/4}|\eta|^{-1/4}\gamma^{-3/4}, \qquad \hat{\theta} \sim \mathcal{A}^{1/2}V^{-1/4}h^{-1/4}|\eta|^{1/4}\gamma^{-1/4}, \tag{3.43}$$

where the adiabatic damping is also explicitly shown. As the energy approaches the transition energy with $\eta \to 0$, we expect that $\hat{\delta} \to \infty$, and $\hat{\theta} \to 0$. This is not true because the synchrotron motion around the transition energy is non-adiabatic. It will be discussed in Sec. IV.

Similarly, the invariant phase-space ellipse in (θ, δ) phase space is

$$\left(\frac{\delta^2}{\hat{\delta}}\right)^2 + \left(\frac{\theta}{\hat{\theta}}\right)^2 = 1, \qquad \frac{\hat{\delta}}{\hat{\theta}} = \frac{Q_s}{|\eta|}, \tag{3.44}$$

where $\hat{\delta}$ and $\hat{\theta}$ are the maximum amplitudes of phase-space ellipse. The normalized Gaussian distribution in (θ, δ) space becomes

$$\rho(\delta,\theta) = \frac{1}{2\pi\sigma_\delta\sigma_\theta}\exp\left\{-\frac{1}{2}\left[\frac{\theta^2}{\sigma_\theta{}^2} + \frac{\delta^2}{\sigma_\delta{}^2}\right]\right\}. \tag{3.45}$$

Here σ_θ and σ_δ are respectively the rms bunch angular width and rms fractional momentum spread. The bunch length is $\sigma_s = R\sigma_\theta$ in meters, where R is the average radius of the accelerator, or $\sigma_t = \sigma_\theta/\omega_0$ in s.

Now we consider $N_{\rm B}$ particles distributed in a bunch, where $N_{\rm B}$ may vary from 10^8 to 10^{14} particles. The line distribution and the peak current (in Amperes) of the bunch are

$$\rho(\phi) = \frac{N_{\rm B}}{\sqrt{2\pi}\sigma_\phi}e^{-\phi^2/2\sigma_\phi^2} \qquad \text{or} \qquad \rho(\theta) = \frac{N_{\rm B}}{\sqrt{2\pi}\sigma_\theta}e^{-\theta^2/2\sigma_\theta^2}, \tag{3.46}$$

$$\hat{I} = \frac{N_{\rm B}\,e}{\sqrt{2\pi}\sigma_t} = \frac{N_{\rm B}\,e\omega_0}{\sqrt{2\pi}\sigma_\theta} = \left(\frac{2\pi}{\sqrt{2\pi}\sigma_\theta}\right)\frac{N_{\rm B}\,e}{T_0}, \tag{3.47}$$

where $N_B e/T_0$ is the average current, and $2\pi/(\sqrt{2\pi}\sigma_\theta)$ is the *bunching factor.*

B. Synchrotron motion in reference time coordinates

In collective beam instabilities, we use the particle arrival time τ and its time derivative $\dot{\tau}$ for synchrotron phase-space coordinates, i.e.

$$\tau = -\frac{\theta - \theta_0}{\omega_0}, \quad \text{and} \quad \dot{\tau} = \frac{\Delta\omega}{\omega_0} = +\eta\delta. \tag{3.48}$$

The linearized synchrotron Hamiltonian becomes $H = \frac{1}{2}\left[\dot{\tau}^2 + \omega_s^2\tau^2\right]$, where ω_s is the angular synchrotron frequency shown in Eq. (3.16). The phase-space ellipse that corresponds to a constant Hamiltonian and the solutions are

$$\tau^2 + \frac{\dot{\tau}^2}{\omega_s^2} = \hat{\tau}^2; \qquad \tau = \hat{\tau}\cos\psi, \quad \frac{\dot{\tau}}{\omega_s} = -\hat{\tau}\sin\psi, \qquad \psi = \psi_0 + \omega_s t, \tag{3.49}$$

where $\hat{\tau}$ and ψ are respectively the synchrotron amplitude and phase. See Eq. (3.273) for its application.

C. Approximate action-angle variables

Expanding the phase coordinate around SFP with $\phi = \phi_s + \varphi$, the synchrotron Hamiltonian becomes (see Exercise 3.2.11)

$$H = \frac{1}{2}h\omega_0\eta\delta^2 + \frac{1}{2h\eta}\omega_0 Q_s^2\left[\varphi^2 - \frac{1}{3}\tan\phi_s\,\varphi^3 - \frac{1}{12}\,\varphi^4 + \cdots\right], \tag{3.50}$$

where $Q_s = \sqrt{heV|\eta\cos\phi_s|/2\pi\beta^2 E} = \nu_s\sqrt{|\cos\phi_s|}$ is the small amplitude synchrotron tune. For simplicity, we assume $\eta > 0$ in this section.

We would like to transform the phase space coordinates (φ, δ) to action-angle coordinates (ψ, J) by a generating function as:

$$F_1(\varphi, \psi) = -\frac{Q_s}{2h|\eta|}\varphi^2\tan\psi \implies \varphi = \sqrt{\frac{2h\eta J}{Q_s}}\cos\psi, \; \delta = -\sqrt{\frac{2Q_s J}{h\eta}}\sin\psi. \tag{3.51}$$

With the approximate action-angle variables, the Hamiltonian for synchrotron motion becomes

$$H = \omega_0 Q_s J + \omega_0\frac{\sqrt{2h\eta Q_s}}{12}\tan\phi_s\; J^{3/2}\left[\cos 3\psi + 3\cos\psi\right] - \omega_0\frac{h\eta}{6}J^2\cos^4\psi. \tag{3.52}$$

If we apply the canonical perturbation method (see Exercise 3.2.11), the averaged synchrotron Hamiltonian and the amplitude dependent synchrotron tune become

$$\begin{aligned} \langle H\rangle &= \omega_0 Q_s J - \frac{\omega_0 h\eta}{16}\left(1 + \frac{5}{3}\tan^2\phi_s\right)J^2 + \cdots \\ \tilde{Q}_s(J) &\equiv \frac{1}{\omega_0}\frac{\partial\langle H\rangle}{\partial J} \approx Q_s\left[1 - \frac{h\eta}{8Q_s}\left(1 + \frac{5}{3}\tan^2\phi_s\right)J\right]. \end{aligned} \tag{3.53}$$

II.4 Small-Amplitude Synchrotron Motion at the UFP

Small amplitude synchrotron motion around an unstable fixed point (UFP) is also of interest in accelerator physics. Expanding the Hamiltonian around the UFP, i.e. $\varphi = \phi - (\pi - \phi_s)$, we obtain

$$\dot{\delta} = -\frac{\omega_0 eV\cos\phi_s}{2\pi\beta^2 E}\varphi, \qquad \dot{\varphi} = h\eta\omega_0\delta,$$
$$\ddot{\delta} = \omega_s^2\delta, \qquad \ddot{\varphi} = \omega_s^2\varphi. \tag{3.54}$$

The particle motion is hyperbolic around the UFP.

Now, we study the evolution of an elliptical torus of Eq. (3.39) at the UFP. We would like to find the evolution of bunch shape when the center of the beam bunch is instantaneously kicked[6] onto the UFP at time $t = 0$. With normalized coordinates, the solutions of Eq. (3.54) are

$$\tilde{\varphi} = \frac{\varphi}{\hat{\phi}}, \quad \tilde{\delta} = \frac{\delta}{\hat{\delta}}, \qquad \tilde{\varphi} = ae^{\omega_s t} + be^{-\omega_s t}, \quad \tilde{\delta} = ae^{\omega_s t} - be^{-\omega_s t}, \tag{3.55}$$

where a and b are determined from the initial condition. With the constants a and b eliminated, the evolution of the bunch shape ellipse is

$$\tilde{\varphi}^2 - 2\left(\frac{\eta}{|\eta|}\tanh 2\omega_s t\right)\tilde{\varphi}\tilde{\delta} + \tilde{\delta}^2 = (\cosh 2\omega_s t)^{-1}. \tag{3.56}$$

Thus the upright phase-space ellipse will become a tilted phase-space ellipse encompassing the same phase-space area. The width and height of phase-space ellipse increase or decrease at a rate $e^{\pm\omega_s t}$, where t is the length of time that the bunch stays at UFP (see Exercise 3.2.5). This scheme of bunch deformation can be used for bunch rotation or bunch compression. At $\omega_s t \gg 1$, the ellipse becomes a line $\tilde{\varphi} \pm \tilde{\delta} = 0$. However, the nonlinear part of synchrotron Hamiltonian will distort the ellipse.

II.5 Synchrotron Motion for Large-Amplitude Particles

The phase space trajectory of adiabatic synchrotron motion follows a Hamiltonian torus $H(\phi, \delta) = H_0$, where $H(\phi, \delta)$ is the synchrotron Hamiltonian in Eq. (3.14) and the constant Hamiltonian value H_0 is

$$H_0 = \frac{1}{2}h\omega_0\eta\hat{\delta}^2 = \frac{\omega_0 eV}{2\pi\beta^2 E}[\cos\hat{\phi} - \cos\phi_s + (\hat{\phi} - \phi_s)\sin\phi_s].$$

[6]In reality, the beam does not jump in the phase space; instead, the phase of the rf wave is being shifted so that the UFP is located at the center of a bunch.

Here $\hat{\phi}$ and $\hat{\delta}$ are respectively the maximum phase coordinate and fractional momentum deviation of synchrotron motion. Using Hamilton's equation $\dot{\phi} = h\omega_0\eta\delta$, we find the synchrotron oscillation period as

$$T = \oint \left(2h\omega_0\eta\left[H_0 - \frac{\omega_0 eV}{2\pi\beta^2 E}[\cos\phi - \cos\phi_s + (\phi - \phi_s)\sin\phi_s]\right]\right)^{-1/2} d\phi, \qquad (3.57)$$

where H_0 is the Hamiltonian value of a torus. The angular synchrotron frequency is $2\pi/T$. The action of the torus is

$$J = \frac{1}{2\pi}\oint \left(\frac{2}{h\omega_0\eta}[H_0 - \frac{\omega_0 eV}{2\pi\beta^2 E}(\cos\phi - \cos\phi_s + (\phi - \phi_s)\sin\phi_s)]\right)^{1/2} d\phi. \qquad (3.58)$$

The synchrotron period of Eq. (3.57) can also be derived by differentiating Eq. (3.58) with respect to J, and using $dH_0/dJ = \omega(J)$ to find the synchrotron frequency.

A. Stationary synchrotron motion

For simplicity, we consider the stationary synchrotron motion above the transition energy with $\eta > 0$, or $\phi_s = \pi$. The Hamiltonian value for a torus with a maximum phase coordinate $\hat{\phi}$ (or maximum off-momentum coordinate $\hat{\delta}$) is

$$H_0 = \frac{1}{2}h\omega_0\eta\hat{\delta}^2 = \frac{\omega_0}{h\eta}\nu_s^2(1 - \cos\hat{\phi}) \quad \Longrightarrow \quad \hat{\delta} = \frac{2\nu_s}{h\eta}\sin\frac{\hat{\phi}}{2}. \qquad (3.59)$$

The maximum off-momentum coordinate $\hat{\delta}$ is related to the maximum phase coordinate $\hat{\phi}$ for a Hamiltonian torus, where the phase space trajectory is given by $(\phi, \pm\delta(\phi))$: $\delta(\phi) = \frac{\nu_s}{h\eta}\sqrt{2(\cos\phi - \cos\hat{\phi})}$. The phase space area $\mathcal{A}$, or the action J enclosed by the Hamiltonian torus is

$$\mathcal{A} = 2\pi J = 2\int_{-\hat{\phi}}^{\hat{\phi}} \delta(\phi)d\phi = 16\frac{\nu_s}{h\eta}\left[E(k) - (1 - k^2)\,K(k)\right],$$

$$K(k) = \int_0^{\pi/2} \frac{dw}{\sqrt{1 - k^2\sin^2 w}}, \qquad E(k) = \int_0^{\pi/2} \sqrt{1 - k^2\sin^2 w}\, dw, \qquad (3.60)$$

where $K(k)$ and $E(k)$ are the complete elliptic integrals of the first and second kinds, and $k = \sin(\hat{\phi}/2)$ is the modulus of these integrals.

B. Synchrotron tune

The synchrotron tune of the Hamiltonian torus with maximum phase amplitude $\hat{\phi}$ becomes (see Exercise 3.2.8)

$$\tilde{Q}_s(\hat{\phi}) = \frac{\pi\nu_s}{2K(\sin(\frac{\hat{\phi}}{2}))} \qquad (3.61)$$

Figure 3.5 compares the theoretical curve of Eq. (3.61) with a measured synchrotron tune at the IUCF Cooler. When the value of the Hamiltonian H_0 approaches that of the separatrix $H_{\rm sx}$ of Eq. (3.33), the synchrotron tune becomes zero and the synchrotron period becomes infinite. In the small angle approximation, we find $\tilde{Q}_{\rm s}(\hat{\phi}) \approx (1 - \frac{1}{16}\hat{\phi}^2)\nu_{\rm s}$, which is identical to Eq. (3.53) at $\phi_{\rm s} = 0$.

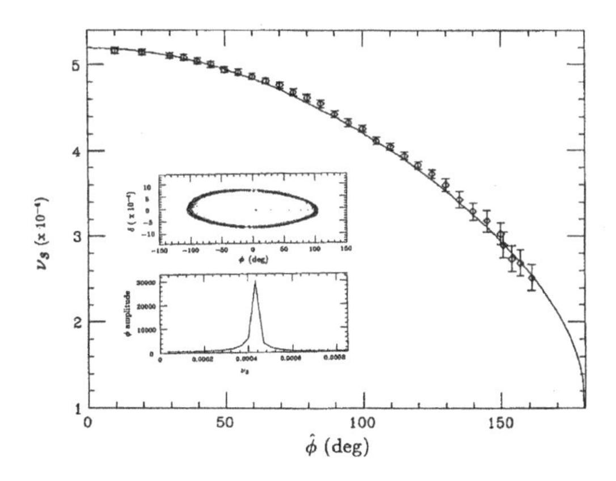

Figure 3.5: The measured synchrotron tune obtained by taking the FFT of the synchrotron phase coordinate is plotted as a function of the maximum phase amplitude of the synchrotron oscillations. The solid line shows the theoretical prediction of Eq. (3.61). The inset shows an example of the synchrotron phase-space map measured at the IUCF Cooler, and the corresponding FFT spectrum. The zero amplitude synchrotron tune was $\nu_{\rm s} = 5.2 \times 10^{-4}$.

Since the synchrotron tune is nonlinear, particles having different synchrotron amplitudes in a beam bunch can have different synchrotron tunes. If the bunch area is a substantial fraction of the bucket area, the synchrotron tune spread may be large. For a mismatched beam bunch, synchrotron tune spread can cause beam *decoherence*, a *filamentation* process, where beam particles spread out in the synchrotron phase space. Beam filamentation causes a mismatched beam bunch to evolve into spirals bounded by a Hamiltonian torus. The final bunch area is determined by the initial beam distribution and parameters of the rf system (see Fig. 3.20 in Sec. V). Filamentation can dilute the phase-space density of the beam. When the beam encounters longitudinal collective beam instability, or mis-injection in the rf bucket, or rf voltage and phase modulations, etc., the mismatched phase-space distribution will decohere and result in beam dilution. This process is important to rf capture in low energy synchrotrons during injection. On the other hand, synchrotron tune spread is useful in providing Landau damping for collective beam instabilities.

II.6 Experimental Tracking of Synchrotron Motion

Experimental measurements of synchrotron phase-space coordinates are important in improving the performance of synchrotrons. For example, a phase detector is needed in implementing a phase feedback loop to damp dipole or higher-order synchrotron modes. In this section we discuss the methods of measuring the off-momentum and rf phase coordinates of a beam.

The fractional off-momentum coordinate of a beam can be derived by measuring the closed orbit of transverse displacement $\Delta x_{\rm co}$ at a high dispersion function location. The off-momentum coordinate is

$$\frac{\Delta p}{p} = \frac{\Delta x_{\rm co}}{D}, \tag{3.62}$$

where D is the horizontal dispersion function. Since synchrotron oscillation is relatively slow in proton synchrotrons, the signal-to-noise ratio can be enhanced by using a low-pass filter at a frequency slightly higher than the synchrotron frequency.

The inset in Fig. 3.5 shows a synchrotron phase space ellipse measured at the IUCF Cooler Ring. The phase coordinate is obtained by a phase detector, and the fractional off-momentum coordinate is obtained from the displacement of the beam centroid measured with a beam position monitor (BPM).

The BPM system had an rms position resolution of about 0.1 mm. By averaging the position measurements the stability of the horizontal closed orbit was measured to be within 0.02 mm. The momentum deviation is related to the off-momentum closed orbit, $\Delta x_{\rm co}$, by $\Delta x_{\rm co} = D\delta$, where $\delta = \Delta p/p_0$ is the fractional momentum deviation, and the horizontal dispersion function D is about 4.0 m at a high-dispersion location. The position signals from the BPM were passed through a 3 kHz low-pass filter before digitization to remove effects due to coherent betatron oscillations and high frequency noise. Since the synchrotron frequency at the IUCF Cooler in this experiment was less than 1 kHz for an rf system with $h = 1$, a 3 kHz low-pass filter could be used to average out betatron oscillations of a few hundred kHz.

The synchrotron phase coordinate can be measured by comparing the bunch arrival time with the rf cavity wave. First, we examine the characteristics of beam current signal from a beam position monitor. We assume that the bunch length is much shorter than the circumference of an accelerator. With the beam bunch approximated by an ideal δ-function pulse, the signal from a beam position monitor (BPM) or a wall gap monitor (WGM)[7] is

$$I(t,\tau) = N_{\rm B}e\sum_{\ell=-\infty}^{\infty}\delta(t+\tau-\ell T_0) = \frac{N_{\rm B}e}{T_0}\sum_{n=-\infty}^{\infty}e^{in\omega_0(t+\tau)}, \tag{3.63}$$

where $N_{\rm B}$ is the number of particles in a bunch, T_0 is the revolution period, $\omega_0 = 2\pi/T_0$ is the angular revolution frequency, and $\tau = (\theta-\theta_{\rm s})/\omega_0$ is the arrival time relative to the synchronous particle. Equation (3.63) shows that the periodic delta-function pulse, in time domain, is equivalent to sinusoidal waves at all integer harmonics of the revolution frequency.

[7]A wall gap monitor consists of a break in the vacuum chamber. The wall current that flows through a resistor, typically about 50 Ohms with a stray capacitance of about 30 pF, can then be measured. The bandwidth is about 100 MHz.

To measure the phase coordinate or equivalently the relative arrival time τ, we first select a sinusoidal wave by using the band-pass filter, or we select the fundamental harmonic with a low-pass filter including only the fundamental harmonic. The sinusoidal signal is compared with the rf wave; and the phase between the beam and the reference rf wave can be obtained by using phase detectors.[8] Normally, the BPM sum signal or the WGM signal can be used to measure the relative phase of the beam.[9] Since the rf frequency was 1.03 MHz for the 45 MeV protons in this measurement at the IUCF Cooler, the BPM signal was passed through a 1.4 MHz low-pass filter to eliminate high harmonics noise before it was compared with an rf signal in a phase detector.

The phase-space map of synchrotron oscillations can be obtained by plotting $\Delta p/p_0$ vs ϕ in each revolution. Since the synchrotron tune of a proton synchrotron is small, the synchrotron motion can be tracked at N revolution intervals, where $N \ll 1/\nu_s$. The top inset in Fig. 3.5 shows the Poincaré map of the longitudinal phase space at 10 turn intervals; the bottom inset shows the FFT of the phase data. The resulting synchrotron tune as a function of peak phase amplitude is compared with the theoretical prediction in Fig. 3.5.

Exercise 3.2

1. Write a simple program to calculate $\alpha_b(\phi_0)$.

2. This exercise concerns the acceleration of protons in the AGS booster. The injection kinetic energy is 200 MeV from the linac. The circumference of the booster ring is 201.78 m, the transition energy is $\gamma_T = 4.5$, the extraction energy is 1.5 GeV kinetic energy, the acceleration time is 160 ms, and the harmonic number is $h = 3$.

   (a) Find the rf voltage needed for acceleration of a proton bunch in the booster.

   (b) The bunch area is determined by several factors, such as line charge density, microwave instabilities, transition crossing in the AGS, etc. If we need a bunch area of about 1 eV-s per bunch in $(\Delta E/\omega_0, \phi/h)$ phase space, and the bucket area is about 1.2 times as large as the bunch area, what is the minimum rf voltage needed?

   (c) What is the rf frequency swing needed to accelerate protons from 200 MeV to 1.5 GeV?

   (d) How does the rf bucket area change during the acceleration process?

[8]See Roland E. Best, *Phase Locked Loops, Theory, Design, and Applications*, pp. 7-9 (McGraw-Hill, New York, 1984). The type II phase detector utilizes XOR logic, and has a range of $\pm 90°$; the type III utilizes the edge triggered JK-master-slave flip-flop circuit, and has a range of $\pm 180°$. The type III has a phase error of about $\pm 10°$ near $0°$, but can adequately measure the synchrotron tune. For more accurate measurement of phase amplitude response, the type II can be used. To extend the range of beam phase detection, a type IV phase detector with a range of $\pm 360°$ can be used.

[9]M. Ellison *et al.*, *Phys. Rev.* **E 50**, 4051 (1994).

3. For a constant rf voltage and synchronous phase angle, show that the rf bucket area in the $(\Delta E/\omega_0, \phi/h)$ phase space has a minimum at $\gamma = \sqrt{3}\gamma_T$.

4. Particle acceleration at a constant bucket is a possible "rf program" in synchrotrons. Find the relation between rf voltage and beam energy.

5. Verify Eq. (3.54). Rotate the phase-space ellipse of Eq. (3.56) into the upright position, show that the width and height of the bunch change by a factor $e^{\pm\omega_s t}$, where t is the time the bunch stays at the UFP, and estimate the time needed to double the bunch height.

6. The anti-protons produced from the Main Injector (Main Ring) pulses have the following characteristics: $p_0 = 8.9$ GeV/c, $\sigma_t = 0.15$ ns, $\sigma_E = 180$ MeV, or $\Delta p/p_0 = \pm 2\%$. The antiprotons are captured in the Debuncher into the 53.1 MHz ($h = 90$) rf bucket with $V = 5$ MV, $\phi_s = 180°$, $\gamma_T = 7.7$, and circumference $C = 505$ m.[10]

   (a) Find the bucket height, synchrotron tune, and synchrotron period of the Debuncher ring with the rf system.

   (b) At 1/4 of the synchrotron period after antiproton injection, the rf voltage is lowered suddenly to match the bunch shape. Show that the final rf voltage V_2 related to the initial voltage V_1 and the final energy spread in this debunching process are respectively

$$\frac{V_2}{V_1} = \left(\frac{\nu_{s1}}{|\eta|}(\omega_0\sigma_{t1})\right)^2 \left(\frac{\sigma_{E1}}{\beta^2 E}\right)^{-2}, \qquad \sigma_{E2} = \left[\frac{\nu_{s1}}{|\eta|}(\omega_0\sigma_{t1})\right]\beta^2 E.$$

   Find the final matched rf voltage for the Debuncher and the final energy spread of the antiproton beams.

7. Assuming stationary bucket, fill out the beam properties of the proton synchrotrons in the table below.

P-synchrotron	AGS	RHIC	FNAL-MI	FNAL-BST	Cooler
K.E. [GeV]	25	250	120	8	0.045
$V_{\rm rf}$ [MV]	0.3	0.3	2	0.95	0.0001
h	12	342	588	84	1
γ_T	8.5	24.5	21.8	5.446	4.6
C [m]	807.12	3833.84	3319.4	474.2	86.8
$\mathcal{A}_{\rm rms}$ [eV-s]	1.5	0.5	0.15	0.15	0.0001
σ_E [MeV]					
σ_t [ns]					

8. Show that the synchrotron tune of a particle with phase amplitude $\hat{\phi}$ in a stationary bucket is

$$Q_s(\hat{\phi}) = \pi\nu_s / 2K(\sin\frac{\hat{\phi}}{2}),$$

   where $K(x)$ is the complete elliptical integral of the first kind given in Eq. (3.60).

[10] See A.V. Tollestrup and G. Dugan, p. 954 in Ref. [17] (1983). Note that 5 MV is the maximum voltage that the Debuncher rf system can deliver.

9. Define $p_\phi = h\omega\eta\delta$, and show that the synchrotron equations of motion become

$$\dot{\phi} = p_\phi, \quad \dot{p}_\phi = \frac{\eta}{|\eta|}\omega_s^2(\sin\phi - \sin\phi_s),$$

where $\omega_s = \omega_0\sqrt{h|\eta|eV/2\pi\beta^2 E}$, and the overdot indicates the derivative with respect to time t. The Hamiltonian for a stationary rf system with ϕ_s becomes

$$H = \frac{1}{2}p_\phi^2 + \frac{\eta}{|\eta|}\omega_s^2(\cos\phi - 1).$$

(a) Using the generating function show that phase-space coordinates are

$$F_1 = -\frac{\omega_s}{2}\phi^2\tan\psi \quad \Longrightarrow \quad \phi = \sqrt{2J/\omega_s}\cos\psi, \quad p_\phi = -\sqrt{2J\omega_s}\sin\psi,$$

where J and ψ are action-angle coordinates. Show that the Hamiltonian below the transition energy becomes

$$H = \omega_s J + \omega_s^2\left[1 - \frac{J}{\omega_s}\cos^2\psi - \cos\left(\sqrt{\frac{2J}{\omega_s}}\cos\psi\right)\right].$$

(b) Using the phase averaging method, show that the synchrotron tune is approximately given by

$$Q_s(J) = \nu_s\left[\frac{1}{2} + \frac{J_1(w)}{w}\right], \quad w = \sqrt{\frac{2J}{\omega_s}},$$

where $J_1(w)$ is the Bessel function.

(c) Compare the accuracy of the above approximated synchrotron tune to that of the exact formula given by Exercise 3.2.8.

10. Let $\hat{\phi}$ be the maximum synchrotron phase amplitude. Show that the maximum off-momentum deviation is

$$\hat{\delta} = \frac{\sqrt{2}Q_s}{h|\eta|}\left[-1 + \frac{\cos\hat{\phi}}{\cos\phi_s} + (\hat{\phi} - \phi_s)\tan\phi_s\right].$$

11. Expanding the phase coordinate around the SFP with $\phi = \phi_s + \varphi$, the synchrotron Hamiltonian becomes

$$H = \frac{1}{2}h\omega_0\eta\delta^2 + \frac{1}{2h\eta}\omega_0 Q_s^2\left[\varphi^2 - \frac{1}{3}\tan\phi_s\varphi^3 - \frac{1}{12}\varphi^4 + \cdots\right],$$

$$Q_s = \sqrt{\frac{heV|\eta\cos\phi_s|}{2\pi\beta^2 E}} = \nu_s\sqrt{|\cos\phi_s|}.$$

For simplicity, we assume $\eta > 0$ in this exercise.

(a) Using the generating function to transform (φ, δ) to angle-action (ψ, J) coordinates, show that the coordinate transformation and the resulting Hamiltonian in action-angle are

$$F_1(\phi, \psi) = -\frac{Q_s}{2h\eta}\varphi^2 \tan\psi$$
$$\varphi = \sqrt{2h\eta J/Q_s}\cos\psi, \ \delta = -\sqrt{2Q_s J/h\eta}\sin\psi,$$
$$H = \omega_0 Q_s J - \frac{\omega_0\sqrt{2h\eta Q_s}}{12}\tan\phi_s\, J^{3/2}\left[\cos 3\psi + 3\cos\psi\right] - \frac{\omega_0 h\eta}{6}J^2\cos^4\psi.$$

(b) Using the generating function

$$F_2(\psi, I) = \psi I + G_3(I)\sin 3\psi + G_1(I)\sin\psi,$$

show that terms proportional to $J^{3/2}$ in the Hamiltonian can be canceled if G_3 and G_1 are chosen to be

$$G_3 = \frac{\sqrt{2h\eta}}{36\sqrt{Q_s}}\tan\phi_s\, I^{3/2}, \quad G_1 = \frac{\sqrt{2h\eta}}{4\sqrt{Q_s}}\tan\phi_s\, I^{3/2}.$$

Finding new canonical variables to cancel low-order perturbation terms is called the *canonical perturbation technique.*

(c) Show that the new Hamiltonian is

$$\begin{aligned} H &= \omega_0 Q_s I - \frac{\omega_0 h\eta}{6} I^2\cos^4\psi \\ &-\frac{\omega_0\sqrt{2h\eta Q_s}}{8}\tan\phi_s I^{1/2}[\cos 3\psi + 3\cos\psi][3G_3\cos 3\psi + G_1\cos\psi]. \end{aligned}$$

Now the perturbation in the new action variable is proportional to I^2.

(d) Show that the average Hamiltonian and thus the synchrotron tune for a particle with a synchrotron amplitude are

$$\langle H\rangle = \omega_0 Q_s I - \frac{\omega_0 h\eta}{16}\left(1 + \frac{5}{3}\tan^2\phi_s\right) I^2 + \cdots.$$
$$\tilde{Q}(\hat{\varphi}) = Q_s\left[1 - \frac{1}{16}\left(1 + \frac{5}{3}\tan^2\phi_s\right)\hat{\varphi}^2\right],$$

where $\hat{\varphi}$ is the maximum synchrotron amplitude in the quasi-harmonic approximation. Compare your result with that of Eq. (3.61) for $\phi_s = 0$.

12. The natural rms fractional momentum spread of electron beams in a storage ring is $\sigma_E/E = \sqrt{C_q\gamma^2/\mathcal{J}_E\rho}$, where $C_q = 3.83\times 10^{-13}$ m, ρ is the bending radius, and $\mathcal{J}_E \approx 2$ is the damping partition. In the NLC damping ring (DR) parameter list shown in Exercise 3.1.3, the rms fractional momentum spread of the electron beam is $\sigma_\delta = 0.000813$. Find the bunch length and rms phase-space area in eV-s.

13. The equilibrium distribution in linearized synchrotron phase space is a function of the invariant ellipse given by Eq. (3.44), where $\hat{\theta} = |\eta|\hat{\delta}/\nu_s$. When a mismatched Gaussian beam

$$\rho(\delta,\theta) = \frac{N_B e}{2\pi\sigma_\delta\sigma_\theta}\exp\left\{-\frac{1}{2}\left[\frac{\theta^2}{\sigma_\theta{}^2}+\frac{\delta^2}{\sigma_\delta{}^2}\right]\right\}$$

is injected into the synchrotron at time $t = 0$, what is the time evolution of the beam? Here σ_θ and σ_δ are respectively the initial rms bunch angular width and fractional momentum spread, and the mismatch condition for the linearized synchrotron motion is given by $\sigma_\theta \neq |\eta|\sigma_\delta/\nu_s$.

(a) Show that the projection of the beam distribution function onto the θ axis is[11]

$$\rho(\theta,t) = \frac{N_B e}{\sqrt{2\pi}\tilde{\sigma}}e^{-\theta^2/2\tilde{\sigma}^2} \qquad \tilde{\sigma}^2 = \sigma_\theta^2\cos^2\omega_s t + (|\eta|\sigma_\delta/\nu_s)^2\sin^2\omega_s t.$$

Show that the peak current is $\hat{I}(t) = N_B e\omega_0/\sqrt{2\pi}\tilde{\sigma}$.

(b) For a weakly mismatched beam, show that

$$\sigma_\theta^2 \approx \sigma_0^2(1-\Delta V/2V), \quad (\eta\sigma_\delta/\nu_s)^2 \approx \sigma_0^2(1+\Delta V/2V),$$

where $\sqrt{2}\sigma_0 = \sqrt{\sigma_\theta^2 + (\eta\sigma_\delta/\nu_s)^2}$ is the matched rms beam width, ΔV is the mismatched voltage, and V is the voltage for the matched beam profile. Show that the peak current for the weakly mismatched beam is

$$\hat{I}(t) \approx \frac{N_B e\omega_0}{\sqrt{2\pi\sigma_0^2}}\left(1 - \frac{\Delta V}{4V}\cos 2\omega_s t\right).$$

Discuss your result. Because the bunch tumbles at twice the synchrotron frequency, the resulting coherent beam motion is called the quadrupole synchrotron mode. The nonlinear synchrotron tune will cause the mismatched injection to filament and the resulting phase-space area will be larger.

14. An off momentum particle executes synchrotron motion with $\delta = \hat{\delta}\sin\omega_s t$, where ω_s is the synchrotron tune, $\hat{\delta}$ is the amplitude of synchrotron motion. Show that the accumulated betatron phase advance in a half-synchrotron period

$$\Delta\Phi_\beta = 4\pi\frac{C\hat{\delta}}{\omega_s T_0} = \frac{C\omega_0}{\eta}2\hat{\tau},$$

where C is the chromaticity, $2\hat{\tau} = 2\eta\hat{\delta}/\omega_s$ is the time width of synchrotron motion, and η is the phase slip factor. Note that the accumulated betatron phase shift is proportional to the synchrotron time-width. The quantity $C\omega_0/\eta$ is the betatron frequency shift from the head to tail for all particle. If $C/\eta < 0$, the frequency shift is negative, and beam may sample impedance in the negative frequency and cause transverse head-tail instability.

[11] Transform the (θ,δ) coordinate system into the normalized coordinate system ($x = \theta$ and $p = |\eta|\delta/\nu_s$), where the matched beam profile is a circle. Make coordinate transformation into the synchrotron rotating frame. The beam profile in the x plane is equal to $\rho(x) = \int \rho(x,p)dp$.

III RF Phase and Voltage Modulations

Particle motion in accelerators experiences perturbations from rf phase and amplitude noise, power supply ripple, wakefields, etc. These perturbation sources cause rf phase or voltage modulations. In general, the frequency spectrum of rf noise may contain high frequency arising from random thermal (white) noise, low frequency from power supply ripple and ground motion, and medium frequency from mechanical vibration etc. In this section, we study the effects of a single frequency sinusoidal rf phase and voltage modulation on particle motion and beam distribution. Physics of beam response to a single frequency modulation can be applied to more complicated multi-frequency perturbations. In Sec. III.1, the longitudinal phase space coordinates $(\phi, \Delta p/p)$ will be expanded in action-angle variables (J, ψ). With these results, the perturbed Hamiltonian of phase and voltage modulation can easily be expressed in action-angle variables (see Secs. III.2 and III.5). Once this is accomplished, effects of rf phase and voltage modulation can be readily obtained.

III.1 Normalized Phase-Space Coordinates

Using normalized momentum deviation coordinate $\mathcal{P} = -(h|\eta|/\nu_{\rm s})(\Delta p/p)$, the Hamiltonian for a stationary synchrotron motion is

$$H_0 = \frac{1}{2}\nu_{\rm s}\mathcal{P}^2 + 2\nu_{\rm s}\sin^2\frac{\phi}{2}, \tag{3.64}$$

where $\nu_{\rm s} = \sqrt{h|\eta|eV/2\pi\beta^2 E}$ is the synchrotron tune at $|\cos\phi_{\rm s}| = 1$, the orbital angle θ is the independent variable, and $(\phi, \mathcal{P})$ are normalized conjugate phase-space coordinates. The Hamiltonian has fixed points at

$$(\phi, \mathcal{P})_{\rm SFP} = (0, 0) \quad \text{and} \quad (\phi, \mathcal{P})_{\rm UFP} = (\pi, 0).$$

The synchrotron Hamiltonian is autonomous (time independent), and thus the Hamiltonian value is a constant of motion.

Expressing the synchrotron coordinates in parameters k and w as

$$\sin\frac{\phi}{2} = k\sin w, \quad \frac{\mathcal{P}}{2} = k\cos w, \tag{3.65}$$

we obtain $H_0 = 2\nu_{\rm s}k^2$, where $k = 0$ corresponds to the SFP and $k = 1$ corresponds to the separatrix orbit that passes through the UFP. The action is

$$J = \frac{1}{2\pi}\oint \mathcal{P}d\phi = \frac{8}{\pi}\,[E(k) - (1-k^2)K(k)], \tag{3.66}$$

where the complete elliptical function integrals are [30]

$$E(k) = \int_0^{\pi/2}\sqrt{1-k^2\sin^2 w}\;dw, \quad K(k) = \int_0^{\pi/2}\frac{1}{\sqrt{1-k^2\sin^2 w}}\,dw.$$

In the normalized phase-space coordinates, the maximum action ($k = 1$) is $J_{\rm max} = 8/\pi$, and the maximum bucket area is $\mathcal{A} = 2\pi J_{\rm max} = 16$ (see Table 3.2).

For synchrotron motion with a small action, the power series expansions of elliptical integrals are

$$K(k) = \frac{\pi}{2}\left[1 + (\frac{1}{2})^2k^2 + (\frac{1\cdot 3}{2\cdot 4})^2k^4 + (\frac{1\cdot 3\cdot 5}{2\cdot 4\cdot 6})^2k^6 + \cdots\right],$$
$$E(k) = \frac{\pi}{2}\left[1 - (\frac{1}{2})^2\frac{k^2}{1} - (\frac{1\cdot 3}{2\cdot 4})^2\frac{k^4}{3} - (\frac{1\cdot 3\cdot 5}{2\cdot 4\cdot 6})^2\frac{k^6}{5} - \cdots\right].$$

The action is related to the parameter k by

$$J = 2k^2(1 + \frac{1}{8}k^2 + \frac{3}{64}k^4 + \cdots), \tag{3.67}$$
$$2k^2 = J(1 - \frac{1}{16}J - \frac{1}{256}J^2 - \cdots). \tag{3.68}$$

In terms of the action, the Hamiltonian is $H_0(J)$. The *synchrotron tune* becomes

$$\tilde{Q}_{\rm s}(J) = \frac{\partial H_0}{\partial J} = \frac{\pi\nu_{\rm s}}{2K(k)} = \nu_{\rm s}(1 - \frac{J}{8} - \frac{3J^2}{256} - \cdots), \tag{3.69}$$

where we have used the identities

$$2k^2\frac{dE(k)}{dk^2} = E(k) - K(k), \quad 2k^2\frac{dK(k)}{dk^2} = \frac{1}{1-k^2}E(k) - K(k).$$

Using the generating function to transform from (ϕ, δ) to (ψ, J), i.e.

$$F_2(\phi, J) = \int_0^\phi \mathcal{P}(\tilde{\phi})\, d\tilde{\phi}, \quad \Longrightarrow \quad \psi = \frac{\partial F_2}{\partial J} = \frac{\tilde{Q}_{\rm s}(J)}{\nu_{\rm s}}\int_0^\phi \frac{d\tilde{\phi}}{\mathcal{P}(\tilde{\phi})}. \tag{3.70}$$

The angle variable ψ, conjugate to the action J, can also be obtained by integrating Hamilton's equation Eq. (3.69):

$$\psi = \frac{\pi\nu_{\rm s}}{2K}\theta + \psi_0. \tag{3.71}$$

The next task is to express the normalized off-momentum coordinate $\mathcal{P}$, and the synchrotron phase coordinate ϕ, in Fourier harmonics of the conjugate angle parameter ψ. First, using Hamilton's equation $\dot{\phi} = \nu_{\rm s}\mathcal{P}$, we can relate the orbital angle θ to the w parameter of Eq. (3.65) as

$$\nu_{\rm s}(\theta - \theta_0) = \int_{\phi_0}^{\phi}\frac{d\phi}{\mathcal{P}} = u - u_0,$$

where

$$u = \int_0^w \frac{1}{\sqrt{1-k^2\sin^2 w}}dw, \quad u_0 = \int_0^{w_0} \frac{1}{\sqrt{1-k^2\sin^2 w}}dw.$$

The Jacobian elliptical functions, cn and sn, are then defined as

$$\sin w = \mathrm{sn}(u|k), \quad \cos w = \mathrm{cn}(u|k), \tag{3.72}$$

and the synchrotron phase-space coordinates are related to the Jacobian elliptical function by

$$\mathcal{P} = 2k\ \mathrm{cn}(u|k), \quad \sin\frac{\phi}{2} = k\ \mathrm{sn}(u|k). \tag{3.73}$$

Thus the expansion of $\mathcal{P}$ and $\sin(\phi/2)$ in Fourier harmonics of ψ is equivalent to the expansion of $\mathrm{cn}(u|k)$ and $\mathrm{sn}(u|k)$ in $\psi = \pi u/2K$. This can be achieved by using Eq. (16.23.2) in Ref. [30], i.e.

$$\begin{aligned} \mathcal{P} &= 2k\mathrm{cn}(u|k) = \frac{4\pi\sqrt{k}}{K(k)}\sum_0^\infty \frac{q^{n+1/2}}{1+q^{2n+1}}\cos(2n+1)\psi \\ &\approx (2J)^{1/2}\cos\psi + \frac{(2J)^{3/2}}{64}\cos 3\psi + \frac{(2J)^{5/2}}{4096}\cos 5\psi + \cdots, \end{aligned} \tag{3.74}$$

where ψ is the synchrotron phase with the q parameter given by

$$q = e^{-\pi K'/K} = \frac{k^2}{16} + 8(\frac{k^2}{16})^2 + 84(\frac{k^2}{16})^3 + 992(\frac{k^2}{16})^4 + \cdots,$$

with $K' = K(\sqrt{1-k^2})$. Similarly, using the identity $k^2\mathrm{sn}^2(u|k) = 1 - \mathrm{dn}^2(u|k)$, we obtain

$$2\sin^2\frac{\phi}{2} = \sum_{n=-\infty}^{\infty} G_n(J)e^{jn\psi} \approx -\frac{J}{2}\cos 2\psi - \frac{J^2}{32}\cos 4\psi + \cdots, \tag{3.75}$$

$$G_n(J) = \frac{1}{2\pi}\int_0^{2\pi}(1-\cos\phi)e^{-jn\psi}d\psi, \tag{3.76}$$

where $G_{-n} = G_n^*$. Because $1 - \cos\phi$ is an even function, $G_n = 0$ for odd n. The expansion of normalized coordinates in action-angle variables is useful for evaluating the effect of perturbation on synchrotron motion, discussed below.

Sum rule theorem

The solutions of many dynamical systems can be obtained by expanding the perturbation potential in action angle variables. For the case of rf phase modulation, the expansion of the normalized off-momentum coordinate is

$$\mathcal{P} = \sum_{n=-\infty}^{\infty} f_n(J)e^{in\psi}, \tag{3.77}$$

where $f_{-n} = f_n^*$ and, from Eq. (3.74), the strength functions f_n are

$$f_{2m+1} = \frac{2\pi\sqrt{k}q^{m+1/2}}{K(k)(1+q^{2m+1})}, \qquad f_{2m} = 0.$$

Because $\mathcal{P}$ is an odd function, only odd harmonics exist. Furthermore, the sum of all strength functions is (see Exercise 3.3.2)

$$\sum_{n=-\infty}^{\infty} |f_n|^2 = \frac{\tilde{Q}_s(J)}{\nu_s} J. \tag{3.78}$$

We observe that the strength functions are zero at the center of the rf bucket where $J = 0$ and at the separatrix where $\tilde{Q}_s(J_{sx}) = 0$.

III.2 RF Phase Modulation and Parametric Resonances

If the phase of the rf wave changes by an amount $\varphi(\theta)$, where $\theta = \omega_0 t$ is the orbiting angle serving as time coordinate, the synchrotron mapping equation is

$$\phi_{n+1} = \phi_n + 2\pi h\eta\delta_n + \Delta\varphi(\theta), \tag{3.79}$$
$$\delta_{n+1} = \delta_n + \frac{eV}{\beta^2 E}(\sin\phi_{n+1} - \sin\phi_s), \tag{3.80}$$

where $\Delta\varphi(\theta) = \varphi(\theta_n + 2\pi) - \varphi(\theta_n)$ is the difference in rf phase error between successive turns in the accelerator. In this section, we consider only a sinusoidal rf phase modulation with[12]

$$\varphi = a\sin(\nu_m\theta + \chi_0),$$

where ν_m is the modulation tune, a is the modulation amplitude, and χ_0 is an arbitrary phase factor. The resulting rf phase difference in every revolution is $\Delta\varphi = 2\pi\nu_m a\cos(\nu_m\theta + \chi_0)$.

For simplicity, we consider the case of a stationary bucket with $\phi_s = 0$ for $\eta < 0$. Using the normalized off-momentum coordinate $\mathcal{P} = -(h|\eta|/\nu_s)\delta$, we obtain the perturbed Hamiltonian

$$H = H_0 + H_1 = \frac{1}{2}\nu_s\mathcal{P}^2 + 2\nu_s\sin^2\frac{\phi}{2} + \nu_m a\mathcal{P}\cos(\nu_m\theta + \chi_0), \tag{3.81}$$

where the perturbation potential of rf phase modulation is

$$H_1 = \nu_m a\mathcal{P}\cos(\nu_m\theta + \chi_0). \tag{3.82}$$

[12] M. Ellison *et al.*, *Phys. Rev. Lett.* **70**, 591 (1993); M. Syphers *et al.*, *Phys. Rev. Lett.* **71**, 719 (1993); H. Huang *et al.*, *Phys. Rev.* E **48**, 4678 (1993); Y. Wang *et al.*, *Phys. Rev.* E **49**, 1610 (1994).

Expressing the phase-space coordinate $\mathcal{P}$ in action-angle coordinates with Eq. (3.74), we can expand the perturbation in action-angle variables

$$\begin{aligned} H_1 &= \nu_{\rm m} a\sqrt{J/2}\,[\cos(\psi+\nu_{\rm m}\theta+\chi_0)+\cos(\psi-\nu_{\rm m}\theta-\chi_0)] \\ &\quad +\nu_{\rm m} a\frac{(2J)^{3/2}}{128}[\cos(3\psi+\nu_{\rm m}\theta+\chi_0)+\cos(3\psi-\nu_{\rm m}\theta-\chi_0)]+\dots, \end{aligned} \tag{3.83}$$

where J and ψ are conjugate action-angle variables. The rf phase error generates only odd order parametric resonances because $\mathcal{P}$ is an odd function. However, two nearby strong parametric resonances can drive secondary and tertiary resonances. For example, the 1:1 and 3:1 parametric resonances driving by a strong phase modulation can produce a secondary 4:2 resonance at $\nu_{\rm m} \approx 2\nu_{\rm s}$. In the following, we discuss only the primary parametric resonances, particularly the 1:1 dipole mode.

A. Effective Hamiltonian near a parametric resonance

When the modulation tune is near an odd multiple of synchrotron sideband, i.e. $\nu_{\rm m} = (2m+1)\nu_{\rm s}$, stationary phase condition exists for a parametric resonance term. We neglect all non-resonance terms in H_1 to obtain an approximate synchrotron Hamiltonian

$$H \approx \nu_{\rm s} J - \frac{1}{16}\nu_{\rm s} J^2 + \nu_{\rm m} f_{2m+1} J^{m+1/2}\cos\left((2m+1)\psi - \nu_{\rm m}\theta - \chi_0\right), \tag{3.84}$$

where $f_1 = a/\sqrt{2}$, $f_3 = a/32\sqrt{2}$, etc. The effect of rf phase modulation on phase-space distortion can be solved by using the effective parametric resonance Hamiltonian, that resembles the Hamiltonian for 1-D betatron resonances discussed in Sec. VII, Chap. 2. This primary parametric resonance is called $(2m+1)$:1 resonance. In this section, we consider only the dominant dipole mode below.

B. Dipole mode

If the phase modulation amplitude is small, the dominant contribution arises from the $m = 0$ sideband. Near the first-order synchrotron sideband with $\nu_{\rm m} \approx \nu_{\rm s}$, the Hamiltonian for the dipole mode is

$$H \approx \nu_{\rm s} J - \frac{1}{16}\nu_{\rm s} J^2 + \frac{\nu_{\rm s} a}{\sqrt{2}} J^{1/2}\cos(\psi - \nu_{\rm m}\theta - \chi_0). \tag{3.85}$$

The Hamiltonian can be transformed into the resonance rotating frame:

$$\begin{aligned} &F_2(\psi, I) = (\psi - \nu_{\rm m}\theta - \chi_0 - \pi)\, I, \\ &\chi = \psi - \nu_{\rm m}\theta - \chi_0 - \pi, \quad I = J; \\ &\tilde{H} = (\nu_{\rm s} - \nu_{\rm m}) I - \frac{1}{16}\nu_{\rm s} I^2 - \nu_{\rm s}\frac{a}{\sqrt{2}} I^{1/2}\cos\chi. \end{aligned} \tag{3.86}$$

where (ψ, J) are transformed to the new phase-space coordinates (χ, I). Since the new Hamiltonian $\tilde{H}$ is "time" independent in the resonance rotating frame, a torus of particle motion will follow a constant Hamiltonian contour, where Hamilton's equations of motion are

$$\dot{\chi} = \nu_{\rm s} - \nu_{\rm m} - \frac{1}{8}\nu_{\rm s} I - \nu_{\rm s}\frac{a}{2\sqrt{2I}}\cos\chi, \qquad \dot{I} = -\nu_{\rm s}\frac{a}{2}\sqrt{2I}\sin\chi. \tag{3.87}$$

The fixed points of the Hamiltonian, which characterize the structure of resonant islands, are given by the solution of $\dot{I} = 0$, $\dot{\chi} = 0$. Using $g = \sqrt{2J}\cos\chi$, with $\chi = 0$ or π, to represent the phase coordinate of a fixed point, we obtain the equation for g as

$$g^3 - 16\left(1 - \frac{\nu_{\rm m}}{\nu_{\rm s}}\right) g + 8a = 0. \tag{3.88}$$

When the modulation tune is below the *bifurcation tune* $\nu_{\rm bif}$ given by[13]

$$\nu_{\rm m} \le \nu_{\rm bif} = \nu_{\rm s}\left[1 - \frac{3}{16}(4a)^{2/3}\right], \tag{3.89}$$

Eq. (3.88) has three solutions:

$$\begin{cases} g_a(x) = -\frac{8}{\sqrt{3}}x^{1/2}\cos\frac{\xi}{3}, & (\psi = \pi) \\ g_b(x) = \frac{8}{\sqrt{3}}x^{1/2}\sin(\frac{\pi}{6} - \frac{\xi}{3}), & (\psi = 0) \\ g_c(x) = \frac{8}{\sqrt{3}}x^{1/2}\sin(\frac{\pi}{6} + \frac{\xi}{3}), & (\psi = 0) \end{cases} \tag{3.90}$$

where

$$x = 1 - \nu_{\rm m}/\nu_{\rm s}, \quad x_{\rm bif} = 1 - \nu_{\rm bif}/\nu_{\rm s}, \quad \xi = \arctan\sqrt{\left(\frac{x}{x_{\rm bif}}\right)^3 - 1}, \quad x_{\rm bif} = \frac{3}{16}(4a)^{2/3}.$$

Here g_a and g_b are respectively the outer and the inner stable fixed points (SFPs) and g_c is the unstable fixed point (UFP). The reason that g_a and g_b are SFPs and g_c is the UFP will be discussed shortly. Particle motion in the phase space can be described by tori of constant Hamiltonian around SFPs. The lambda-shaped phase amplitudes of the SFPs ($|g_a|$ and $|g_b|$, solid lines) and UFP ($|g_c|$, dashed line) shown in the left plot of Fig. 3.6 vs the modulation frequency is a characteristic property of the dipole mode excitation with nonlinear detuning. In the limit $\nu_{\rm m} \ll \nu_{\rm bif}$, we have $\xi \to \pi/2$, thus $g_a \to -4x^{1/2}$, $g_c \to 4x^{1/2}$, and $g_b \to 0$.

The Hamiltonian tori in phase space coordinates $\mathcal{P} = -\sqrt{2I}\sin\chi$ vs $X = \sqrt{2I}\cos\chi$ are shown in the right plot of Fig. 3.6. The actual Hamiltonian tori rotate about the

[13]Find the root of the discriminant of the cubic equation (3.88).

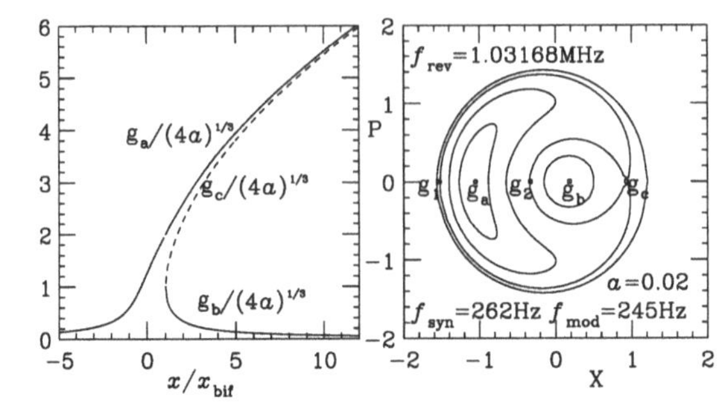

Figure 3.6: Left: fixed point amplitudes $|g_a|$, $|g_b|$, and $|g_c|$ (in unit of $(4a)^{1/3}$). Right: Poincaré surfaces of section for $f_{\rm m}=245$ Hz and $f_{\rm s}=262$ Hz at $a=0.02$. The SFPs are g_a and g_b, and the UFP is g_c.

center of the phase space at the modulation tune $\nu_{\rm m}$, i.e. the phase space ellipses return to this structure in $1/\nu_{\rm m}$ revolutions. The torus passing through the UFP is the separatrix, which separates the phase space into two stable islands. The intercept of the the separatrix with the phase axis is denoted by g_1 and g_2.

When the modulation frequency approaches the bifurcation frequency from below ($x/x_{\rm bif} > 1$), the UFP and the outer SFP move in and the inner SFP moves out. At the bifurcation frequency, where $x = x_{\rm bif}$ and $\xi = 0$, the UFP collides with the inner SFP with $g_b = g_c = (4a)^{1/3}$; and they disappear together. Beyond the bifurcation frequency, $\nu_{\rm m} > \nu_{\rm bif}$ ($x < x_{\rm bif}$), there is only one real solution to Eq. (3.88):

$$g_a(x) = -(4a)^{1/3}\left[\left(\sqrt{1-\left(\frac{x}{x_{\rm bif}}\right)^3}+1\right)^{1/3} - \left(\sqrt{1-\left(\frac{x}{x_{\rm bif}}\right)^3}-1\right)^{1/3}\right]. \quad (3.91)$$

In particular, $g_a = -(8a)^{1/3}$ at $x = 0$ ($\nu_{\rm m} = \nu_{\rm s}$), and $g_a = -2(4a)^{1/3}$ at $x = x_{\rm bif}$. The characteristics of bifurcation appear in all orders of resonances with nonlinear detuning. As the modulation tune approaches the bifurcation tune, resonance islands can be created or annihilated.

C. Island tune

Let y, p_y be the local coordinates about a fixed point of the Hamiltonian, i.e.

$$y = \sqrt{2I}\cos\chi - g, \qquad p_y = -\sqrt{2I}\sin\chi, \quad (3.92)$$

where g is a fixed point of the Hamiltonian. With a local coordinate expansion, the Hamiltonian (3.86) becomes

$$H_{\rm island} = \frac{\nu_{\rm s}a}{4g}\left(1-\frac{g^3}{4a}\right)y^2 + \frac{\nu_{\rm s}a}{4g}p_y^2 + \cdots. \quad (3.93)$$

Therefore the fixed point g is a stable fixed point if $(1 - g^3/4a) \geq 0$. Because $g_a^3/4a \leq 0$ and $0 \leq g_b^3/4a \leq 1$, g_a and g_b are SFPs. Since $g_c^3/4a \geq 1$, g_c is the

UFP. The equilibrium beam distribution (see Appendix A, Sec. II.3), which satisfies the Fokker-Planck-Vlasov equation, is generally a function of the local Hamiltonian, Eq. (3.93) can also provide information on the local distortion of the bunch profile.

The island tune for the small-amplitude oscillations is

$$\nu_{\rm island} = \left|\nu_{\rm s}\left(1 - \frac{g^2}{16}\right) - \nu_{\rm m}\right|\left(1 - \frac{g^3}{4a}\right)^{1/2}. \tag{3.94}$$

The island tune around the inner SFP given by g_b at $\nu_{\rm m} \ll \nu_{\rm bif}$ is approximately given by $\nu_{\rm island} \approx |\nu_{\rm s}(1-\frac{1}{16}g^2)-\nu_{\rm m}|$. This means that the solution of the equations of motion can be approximated by a linear combination of the solution of the homogeneous equation with tune $\nu_{\rm s}(1-\frac{1}{16}g^2)$ and the particular solution with tune $\nu_{\rm m}$.[14] Thus *the island tune is the beat frequency between these two solutions.* When the modulation tune $\nu_{\rm m}$ approaches $\nu_{\rm bif}$, with $(1-g_b^3/4a)^{1/2} \to 0$, the island tune for small-amplitude oscillation about the inner SFP approaches 0 and the small-amplitude island tune for the outer SFP at $\nu_{\rm m} = \nu_{\rm bif}$ is $\nu_{\rm island} = 3|\nu_{\rm s}(1-\frac{1}{16}g^2) - \nu_{\rm m}|$. In this region of the modulation frequency, the linear superposition principle fails. When the modulation frequency becomes larger than the bifurcation frequency so that $[1-(g^3/4a)]^{1/2} \to 1$, we obtain again $\nu_{\rm island} = |\nu_{\rm s}(1-\frac{1}{16}g^2) - \nu_{\rm m}|$, and the linear superposition principle is again applicable. The island tune for large-amplitude motion about a SFP can be obtained by integrating the equation of motion along the corresponding torus of the Hamiltonian in Eq. (3.86).

D. Separatrix of resonant islands

The Hamiltonian torus that passes through the UFP is the separatrix. With the UFP g_c substitutes into the Hamiltonian (3.86), the separatrix torus is

$$H(J,\psi) = \nu_{\rm s}\left[\frac{1}{2}xg_c^2 - \frac{1}{64}g_c^4 - \frac{1}{2}ag_c\right], \tag{3.95}$$

where $x = 1 - \nu_{\rm m}/\nu_{\rm s}$. The separatrix orbit intersects the phase axis at g_1 and g_2. These intercepts, shown in Figs. 3.6 and 3.7, are useful in determining the maximum phase amplitude of synchrotron motion with external phase modulation. With the notation $h_i = g_i/(4a)^{1/3}$, the intercepts of the separatrix are

$$h_1 = -h_c - \frac{2}{\sqrt{h_c}}, \quad h_2 = -h_c + \frac{2}{\sqrt{h_c}}.$$

The intercepts of the separatrix with the phase axis, h_1 and h_2, and the fixed points, h_a, h_b and h_c are shown in 3.7.

[14]M. Ellison *et al.*, *Phys. Rev. Lett.* **70**, 591 (1993).

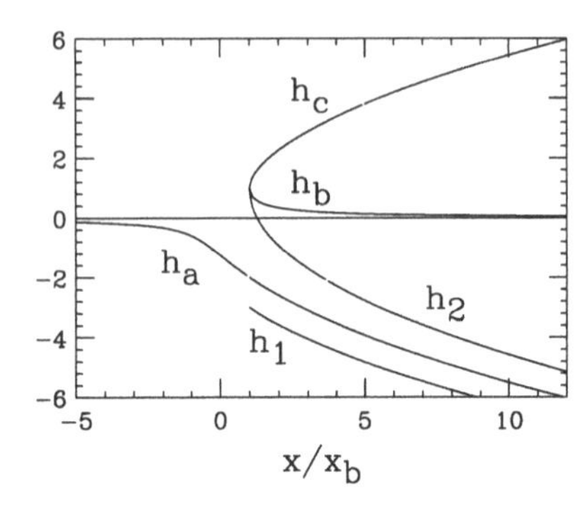

Figure 3.7: The fixed points in units of $(4a)^{1/3}$ are plotted as a function of the modulation frequency in $x/x_{\rm bif}$, where $x = 1 - \nu_{\rm m}/\nu_{\rm s}$ and $x_{\rm bif} = \frac{3}{16}(4a)^{2/3}$ with a as the amplitude of the phase modulation. The SFPs are $h_a = g_a/(4a)^{1/3}$ and $h_b = g_b/(4a)^{1/3}$ and the UFP is $h_c = g_c/(4a)^{1/3}$. The intercepts of the separatrix with the phase axis are shown as $h_1 = g_1/(4a)^{1/3}$ and $h_2 = g_2/(4a)^{1/3}$.

III.3 Measurements of Synchrotron Phase Modulation

Here we discuss an example of experimental measurements of rf phase modulation at the IUCF Cooler. The experimental procedure started with a single bunch of about 3×10^8 protons with kinetic energy 45 MeV. The corresponding revolution period was 969 ns with an rf frequency of 1.03148 MHz. The cycle time was 10 s. The injected beam was electron-cooled for about 3 s. The full width at half maximum bunch length was about 5.4 m (or 60 ns). The low-frequency rf system of the IUCF Cooler at $h = 1$ was used in this experiment.

For the longitudinal rf phase shift, the beam was kicked longitudinally by a phase shifter and the data acquisition system was started 2000 turns before the phase kick. The principle of the phase shifter used is as follows. The rf signal from an rf source is split into a 90° phase shifted channel and a non-phase shifted channel. A separate function generator produces two modulating voltages, each proportional to the sine and cosine of the intended phase shift $\varphi_{\rm mod}$. As a result of the amplitude modulation, the two rf channels are multiplied by $\sin\varphi_{\rm mod}$ and $\cos\varphi_{\rm mod}$ respectively. These two modulated signals were added, using an rf power combiner, resulting in an rf phase shift $\varphi_{\rm mod}$ in the rf wave. The control voltage versus actual phase shift linearity was experimentally calibrated. Both the phase error due to control nonlinearity and the parasitic amplitude modulation of the IUCF Cooler rf systems were controlled to less than 10%.

The phase lock feedback loop was switched off in our experiment. The response time of the step phase shift was limited primarily by the inertia of the resonant cavity. At 1 MHz, the quality factor Q of the rf cavity was about 40, resulting in a half-power bandwidth of about 25 kHz. The corresponding response time for a step rf phase shift was about 40∼50 revolutions. In this experiment, the synchrotron oscillation frequency was chosen to be about 540 Hz, or about 1910 revolutions (turns) in the accelerator. Measurements of subsequent beam-centroid displacements have been discussed in Chap. 3, Sec. II.6.

A. Sinusoidal rf phase modulation

When the bunch, initially at $\phi_i = 0$, $\delta_i = 0$, experiences the rf phase sinusoidal modulation with $\varphi_{\rm mod} = a \sin \nu_{\rm m}\theta$, where $\nu_{\rm m}$ is the modulation tune and a the modulation amplitude with $a \ll 1$. The synchrotron motion, in terms of a differential equation, is

$$\dot{\phi} = h\eta\delta + \nu_{\rm m} a \cos \nu_{\rm m}\theta, \quad \dot{\delta} = \frac{eV}{2\pi\beta^2 E} \sin\phi - \lambda\delta, \tag{3.96}$$

where ϕ is the particle phase angle relative to the modulated rf phase, the overdot indicates the derivative with respect to the variable θ, and λ is the damping decrement due to electron cooling. Thus the synchrotron equation of motion becomes

$$\ddot{\phi} + \frac{2\alpha}{\omega_0}\dot{\phi} + \nu_{\rm s}^2 \sin\phi = -a\nu_{\rm m}^2 \sin\nu_{\rm m}\theta + \frac{2\alpha}{\omega_0}\nu_{\rm m} a \cos\nu_{\rm m}\theta. \tag{3.97}$$

The measured damping coefficient α at the IUCF Cooler was $\alpha = \omega_0\lambda/4\pi \approx 3 \pm 1$ s^{-1}. Since the measurement time was typically within 150 ms after the phase kick or the start of rf phase modulation, the effect of electron cooling was not important in these measurements.

The subsequent beam centroid phase-space coordinates are tracked at 10 revolution intervals. The left plots of Fig. 3.8 show examples of measured ϕ and $\mathcal{P} = \frac{h\eta}{\nu_{\rm s}}\frac{\Delta p}{p}$ vs turn number at 10-turn intervals for an rf phase modulation amplitude of 1.45° after an initial phase kick of 42° at modulation frequencies of 490 Hz (upper) and 520 Hz (lower). The resulting response can be characterized by the beating amplitude and period. The beating period is equal to $T_0/\nu_{\rm island}$, where T_0 is the revolution period and $\nu_{\rm island}$ is the island tune, and the beating amplitude is equal to the maximum intercept of Poincaré surface of section with the phase axis.

B. Action angle derived from measurements

For small-amplitude synchrotron motion, Eq. (3.51) can be used to deduce the action and angle variables, i.e.[15]

$$J = \frac{1}{2}(\phi^2 + \mathcal{P}^2), \qquad \tan\psi = -\frac{\mathcal{P}}{\phi} \tag{3.98}$$

in the $(\phi, \mathcal{P})$ phase space.

For large-amplitude synchrotron motion, we need to use the following procedure to deduce the action-angle variables from the measured synchrotron phase-space coordinates. This procedure can improve the accuracy of data analysis.

[15]Note that the action in the (ϕ, δ) phase space is related to the action in the $(\phi, \mathcal{P})$ space by a constant factor $h|\eta|/(\nu_{\rm s}\sqrt{|\cos\phi_{\rm s}|})$.

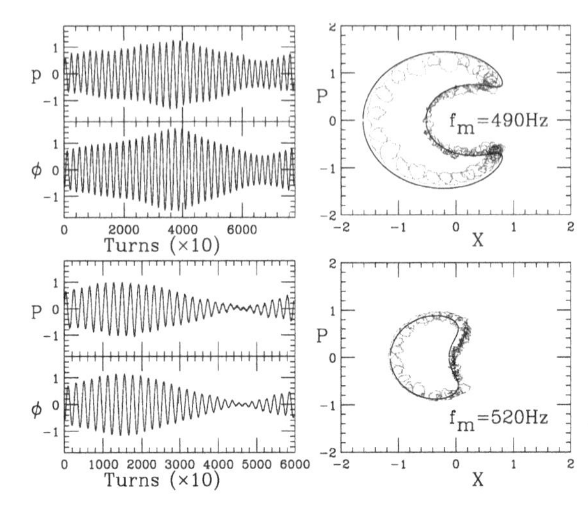

Figure 3.8: Left plots: normalized off-momentum coordinate $\mathcal{P}$ and the phase ϕ vs revolutions at 10-turn intervals. Right plots: the corresponding Poincaré surfaces of section. The upper and lower plots correspond to $f_{\rm m} = 490$ and 520 Hz respectively. The modulation amplitude was $a = 1.45°$, and the initial phase kick amplitude was 45°. The solid line shows the Hamiltonian torus of Eq. (3.85).

1. The k value at the phase-space coordinates $(\phi, \mathcal{P})$ is

$$k^2 = \frac{\mathcal{P}^2}{4} + \sin^2\frac{\phi}{2}. \tag{3.99}$$

   The action can be obtained from Eq. (3.66) or Eq. (3.67).

2. The synchrotron phase, ψ, can be obtained from the expansion

$$\frac{\mathcal{P}}{2\sin\frac{\phi}{2}} = \frac{\pi}{2K}\tan(\frac{\pi}{2} - \psi) - \frac{2\pi}{K}\sum_{n=1}^{\infty}\frac{q^{2n}}{1+q^{2n}}\sin 2n\psi. \tag{3.100}$$

For synchrotron motion with relatively large k, a better approximation for data analysis can be obtained through polynomial approximation of Eqs. (17.3.34) and (17.3.36) of Ref. [30] to evaluate $K(k), E(k)$ and q functions and obtain J and ψ. For each data point $(\phi, \mathcal{P})$, Eq. (3.99) is used to calculate k, and finally, the action J is obtained from Eq. (3.66). The corresponding angle variable ψ is obtained from Eq. (3.100).

C. Poincaré surface of section

The Poincaré map in the resonance frame is formed by phase-space points in

$$(\sqrt{2J}\cos(\psi - \nu_{\rm m}\theta), -\sqrt{2J}\sin(\psi - \nu_{\rm m}\theta)).$$

The resulting invariant tori are shown in the right plots in Fig. 3.8. It becomes clear that the measured response period corresponds to the period of island motion around a SFP, and the response amplitude is the intercept of the invariant torus with the phase axis. The trajectory of a beam bunch in the presence of external rf phase modulation traces out a torus determined by the initial phase-space coordinates of the bunch. Since the torus, which passes through fixed initial phase-space coordinates, depends on the rf phase modulation frequency, the measured tori depend on the driven frequency. Figure 3.8 shows invariant tori deduced from experimental data. The solid lines are invariant Hamiltonian tori of Eq. (3.86), where the synchrotron frequency was fitted to be about 535±3 Hz.

III.4 Effects of Dipole Field Modulation

Ground motion of quadrupoles and power supply ripple in dipoles can cause dipole field modulation. Equation (2.102) in Chap. 2, Sec. III, shows that the change of path length of a reference orbit is $\Delta C = D_x\theta$, where θ is angular kick resulting from dipole field errors and D_x is the dispersion function. If the dipole field is modulated, the path length and thus the arrival time at rf cavities of particles are also modulated. This effect is equivalent to rf phase modulation, which gives rise to parametric resonances in synchrotron motion. The effect is a special type of "synchro-betatron coupling" that may limit the performance of high energy colliders.

Here we discuss experimental measurements of dipole field modulation at the IUCF Cooler. For this experiment, the harmonic number was $h = 1$, the phase slip factor was $\eta \approx -0.86$, the stable phase angle was $\phi_0 = 0$, and the revolution frequency was $f_0 = 1.03168$ MHz at 45 MeV proton kinetic energy. The rf voltage was chosen to be 41 V to obtain a synchrotron frequency of $f_s = \omega_s/2\pi = 262$ Hz in order to avoid harmonics of the 60 Hz ripple. The synchrotron tune was $\nu_s = \omega_s/\omega_0 = 2.54 \times 10^{-4}$. We chose $\nu_x = 3.828$, $\nu_z = 4.858$ to avoid nonlinear betatron resonances. The corresponding smallest horizontal and vertical betatron sideband frequencies were 177 and 146 kHz respectively.

With horizontal dipole (vertical field) modulation at location s_0, the horizontal closed-orbit deviation is $x_{co}(s, s_0, t) = G(s, s_0)\theta(t)$ (see Chap. 2, Sec. III), where $G(s, s_0)$ is the Green's function, $\theta(t) = \hat{\theta}\sin(\omega_m t + \chi_0)$, $\hat{\theta} = \hat{B}_m\ell/B\rho$, and $\hat{B}_m$ is the peak modulation dipole field. Furthermore, if the dispersion function at the modulating dipole location is not zero, the path length is also modulated. The change in the circumference is

$$\Delta C = D_x\theta(t) = D_x\hat{\theta}\ \sin(\omega_m t + \chi_0), \tag{3.101}$$

where D_x is the dispersion function at the modulation dipole location. The corresponding rf phase difference becomes $\Delta\phi = 2\pi h(\Delta C/C)$, where $C = 86.82$ m is the circumference of the IUCF Cooler. In our experiment, the maximum rf phase shift per turn $\hat{\Delta\phi}$ was $0.78 \times 10^{-5}\hat{B}_m$ radians, where the magnetic field $\hat{B}_m$ is in Gauss.

The longitudinal phase-space coordinates $(\phi, \Delta p/p_0)$ at the nth and $(n+1)$th revolutions are transformed according to mapping equations:

$$\phi_{n+1} = \phi_n + 2\pi h\eta \left(\frac{\Delta p}{p}\right)_n + \Delta\phi, \tag{3.102}$$

$$\left(\frac{\Delta p}{p}\right)_{n+1} = \left(\frac{\Delta p}{p}\right)_n + \frac{eV}{\beta^2 E}\sin\phi_{n+1} - \lambda\left(\frac{\Delta p}{p}\right)_n, \tag{3.103}$$

where the fractional momentum deviation of particles $(\Delta p/p_0)$ is the conjugate coordinate to synchrotron phase angle ϕ, and λ is the phase-space damping parameter related to electron cooling. Thus the synchrotron equation of motion, in the presence of transverse dipole field modulation, becomes

$$\frac{d^2\phi}{dt^2} + 2\alpha\frac{d\phi}{dt} + \omega_{\rm s}^2\sin\phi = \omega_{\rm m}^2 a\cos\omega_{\rm m}t + 2\alpha\omega_{\rm s}a\sin\omega_{\rm m}t, \tag{3.104}$$

where the damping coefficient is $\alpha = \lambda\omega_0/4\pi$. With an electron current of 0.75 A, the damping time for 45 MeV protons was measured to be about 0.33 ± 0.1 s or $\alpha = 3 \pm 1\ {\rm s}^{-1}$, which was indeed small compared with $\omega_{\rm s} = 1646\ {\rm s}^{-1}$.

Because the synchrotron frequency is much smaller than the revolution frequency in proton storage rings, the phase errors of each turn accumulate. The equivalent phase modulation amplitude is enhanced by a factor $\omega_0/2\pi\omega_{\rm m}$, i.e. the effective phase modulation amplitude parameter a is

$$a = \frac{h\omega_0 D_x\hat{\theta}}{\omega_{\rm m}C} = \frac{\omega_0}{2\pi\omega_{\rm m}}\hat{\Delta\phi}. \tag{3.105}$$

Although the cooling was weak, the transient solution of Eq. (3.104) was damped out by the time of measurement. We therefore measured the steady state solution, in contrast to the experiment discussed in the previous section, where we measured the transient solutions. Let the steady state solution of the nonlinear parametric dissipative resonant system, Eq. (3.104), be

$$\phi \approx g\sin(\omega_{\rm m}t - \chi), \tag{3.106}$$

where we used the approximation of a single harmonic. Expanding the term $\sin\phi$ in Eq. (3.104) up to the first harmonic, we obtain the equation for the modulation amplitude g and the phase χ as

$$\left[-\omega_{\rm m}^2 g + 2\omega_{\rm s}^2 J_1(g)\right]^2 + \left[2\alpha\omega_{\rm m}g\right]^2 = \left[\omega_{\rm m}\omega_{\rm s}a\right]^2 + \left[2\alpha\omega_{\rm s}a\right]^2 \tag{3.107}$$

$$\chi = \arctan\left[\frac{g\omega_{\rm m}(\omega_{\rm m}^2 + 4\alpha^2) - 2\omega_{\rm s}^2\omega_{\rm m}J_1(g)}{4\alpha\omega_{\rm s}^2 J_1(g)}\right], \tag{3.108}$$

where J_1 is the Bessel function [30] of order 1. Steady state solutions of Eq. (3.107) are called *attractors* for the dissipative system. The existence of a unique phase factor χ for solutions of the dissipative parametric resonant equation implies that the attractor is a single phase-space point rotating at modulation frequency $\omega_{\rm m}$.

When the modulation frequency is below the bifurcation frequency, $\omega_{\rm bif}$, which is given by

$$\left.\frac{\partial \omega_{\rm m}}{\partial g}\right|_{\omega_{\rm m}=\omega_{\rm bif}} = 0,$$

Eq. (3.107) has three solutions. A stable solution with a large phase amplitude g_a and phase factor $\chi_a \approx \pi/2$ is the outer attractor. The stable solution at a smaller phase amplitude g_b with $\chi_b \approx -\pi/2$ is the inner attractor. The third solution g_c with $\chi_c \approx -\pi/2$ corresponds to the unstable (hyperbolic) solution, which is associated with the UFP of the effective non-dissipative Hamiltonian. When the damping parameter α is small, these two stable solutions are nearly equal to the SFPs of the effective Hamiltonian, and are almost opposite to each other in the synchrotron phase space, as shown in Fig. 3.6. They rotate about the origin at the modulation frequency [see Eq. (3.106)]. When the damping parameter α is increased, the stable solution (g_a, χ_a) and the unstable solution (g_c, χ_c) approach each other. At a large damping parameter, they collide and disappear, i.e. the outer attractor solution disappears. When the modulation frequency is larger than the bifurcation frequency, only the outer attractor solution exists.

When the modulation frequency is far from the bifurcation frequency, the response amplitude for the inner attractor at $\omega_{\rm m} \ll \omega_{\rm bif}$, or for the outer attractor at $\omega_{\rm m} \gg \omega_{\rm bif}$, can be approximated by solving the linearized equation (3.107), i.e.

$$g = \left(\frac{(\omega_{\rm m}\omega_{\rm s})^2 + (2\alpha\omega_{\rm s})^2}{(\omega_{\rm s}^2 - \omega_{\rm m}^2)^2 + (2\alpha\omega_{\rm m})^2}\right)^{1/2} a. \tag{3.109}$$

A. Chaotic nature of parametric resonances

In the presence of a weak damping force, *fixed points of the time-averaged Hamiltonian become attractors.* A weak damping force does not destroy the resonance island created by external rf phase modulation. Because of phase-space damping, these fixed points of the Hamiltonian become attractors. Particles in the phase space are damped incoherently toward these attractors, while the attractors rotate about the center of the bucket at the modulation frequency. As the damping force becomes larger, the outer SFP and the UFP may collide and disappear.

Numerical simulations based on Eq. (3.103) were done to demonstrate the coherent and incoherent nature of the single particle dynamics of the parametric resonance system. One of the results is shown in Fig. 3.9, where each black dot corresponds to initial phase-space coordinates that converge toward the outer attractor. Complementary phase-space coordinates converge mostly to the inner attractor except for a small patch of phase-space coordinates located on the boundary of the separatrix, which will converge toward two attractors located near the separatrix.

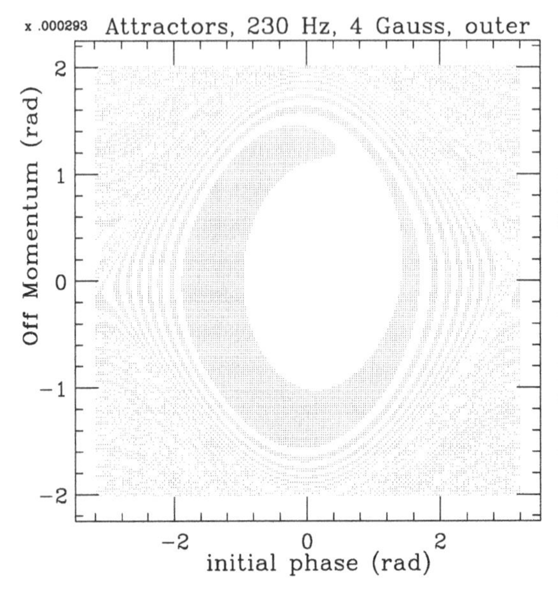

Figure 3.9: Initial normalized phase-space coordinates, obtained from a numerical simulation of Eq. (3.103), which converge to the outer attractor are shown for $\hat{B}_{\rm m} = 4$ Gauss and $f_{\rm m} = 230$ Hz. The synchrotron frequency is 262 Hz. The number of phase-space points that converge to the inner or the outer attractors can be used to determine the beamlet intensity.

The basin of attraction for the inner and the outer attractors forms non-intersecting intervolving spiral rings. To which attractor a particle will converge depends sensitively on the initial phase-space coordinates, especially for particles outside the bucket. The orientation of initial phase-space coordinates converging toward the inner or the outer attractor depends on the initial driving phase χ_0 of the dipole field in Eq. (3.101). Numerical simulations indicate that all particles located initially inside the rf bucket will converge either to the inner or to the outer attractor. However, initial phase-space coordinates in a small patch located at the separatrix of the rf bucket converge toward two attractors moving along the separatrix.

B. Observation of attractors

Since the injected beam from the IUCF K200 AVF cyclotron is uniformly distributed in the synchrotron phase space within a momentum spread of about $(\Delta p/p) \approx \pm 3 \times 10^{-4}$, all attractors can be populated. The phase coordinates of these attractors could be measured by observing the longitudinal beam profile from BPM sum signals on an oscilloscope. Figure 3.10 shows the longitudinal beam profile accumulated through many synchrotron periods with modulation field $\hat{B}_{\rm m} = 4$ G for modulation frequencies of 210, 220, 230, 240, 250, and 260 Hz; it also shows the rf waveform for reference.

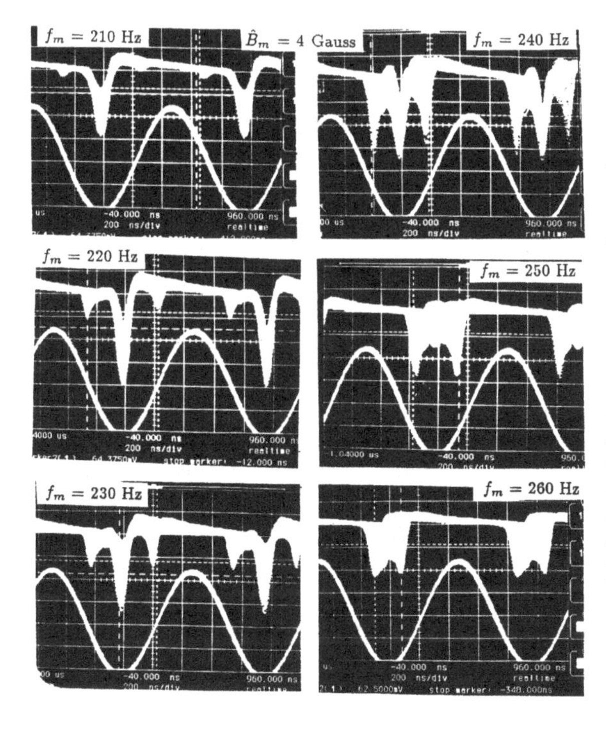

Figure 3.10: Modulation frequencies of left panel: 210, 220, 230 Hz; of right panel: 240, 250, 260 Hz. Synchrotron frequency is 262 Hz. Oscilloscope traces of accumulated BPM sum signals showing the splitting of a beam bunch into beamlets below the bifurcation frequency. The modulation amplitude was $\hat{B}_{\rm m} = 4$ G. The sine waves are the rf waveform. The relative populations of the inner and outer attractors can be understood qualitatively from numerical simulations of the attractor basin.

It was puzzling at first why the longitudinal profile exhibited gaps in time domain, as if there were no synchrotron motion for the beam bunch located at a relatively large phase amplitude. However, using a fast sampling digital oscilloscope (HP54510A) for a single trace, we found that the beam profile was not made of particles distributed in a ring of large synchrotron amplitude, but was composed of two beamlets. Both beamlets rotated in the synchrotron phase space at the modulating frequency, as measured from the fast Fourier transform (FFT) of the phase signal. If the equilibrium distribution of the beamlet was elongated, then the sum signal, which measured the peak current of the beam, would show a large signal at both extremes of its phase coordinate, where the peak current was large. When the beamlet rotated to the central position in the phase coordinate, the beam profile became flat with a smaller peak current. Therefore the profile observed with the oscilloscope offered an opportunity to study the equilibrium distribution of charges in these attractors.

If we assume an equilibrium elliptical beamlet profile with Gaussian distribution, the current density distribution function becomes

$$\rho(\phi,t) = \frac{\rho_1}{\sqrt{2\pi}\sigma_1} e^{-[\phi-\phi_1(t)]^2/2\sigma_1^2} + \frac{\rho_2}{\sqrt{2\pi}\sigma_2} e^{-[\phi-\phi_2(t)]^2/2\sigma_2^2}, \tag{3.110}$$

where ρ_1 and ρ_2 represent the populations of the two beamlets with $\rho_1+\rho_2 = 1$. Since each particle in the two beamlets rotates in the phase space at modulating frequency $\omega_{\rm m}$, the parameters $\phi_{1,2}$ and $\sigma_{1,2}$ are

$$\phi_1(t) = g_a \sin(\omega_{\rm m} t - \chi_a), \quad \phi_2(t) = g_b \sin(\omega_{\rm m} t - \chi_b),$$

and

$$\sigma_1^2 = \sigma_{10}^2(1 + r_1 \sin^2 \omega_{\rm m} t), \quad \sigma_2^2 = \sigma_{20}^2(1 + r_2 \sin^2 \omega_{\rm m} t).$$

Here $g_{a,b}$ and $\chi_{a,b}$ are the amplitudes and phases of the two beamlets, obtained by solving Eqs. (3.107) and (3.108). Since the profile observed on the oscilloscope was obtained by accumulation through many synchrotron periods, it did not depend on the parameters $\chi_{a,b}$, i.e. these profiles were not sensitive to the relative positions of the two beamlets. The eccentricity parameters r_1 and r_2 signify the aspect ratio of the two beamlets, and σ_{10} and σ_{20} represent the average rms bunch length. For example, the aspect ratio, given by $1 : 1 + r_1$ of the outer beamlet at modulation frequency 220 Hz was found to be about 1:3 from the profile in Fig. 3.10. This means that the peak current for the outer beamlet was reduced by a factor of 3 when this beamlet rotated to the center of the phase coordinate. The relative populations of the two beamlets was about 75% for the inner and 25% for the outer, obtained by fitting the data. As the modulating frequency increased toward the synchrotron frequency, the phase amplitude of the outer beamlet became smaller and its population increased. When the modulating frequency was higher than the bifurcation frequency $\omega_{\rm bif}$, the center peak disappeared (see 260 Hz data of Fig. 3.10).

C. The hysteretic phenomena of attractors

The phase amplitudes of attractors shown in Fig. 3.11 also exhibited hysteresis phenomena. When the modulation frequency, which was initially above the bifurcation frequency, was ramped downward, the phase amplitude of the synchrotron oscillations increased along the outer attractor solution. When it reached a frequency far below the bifurcation frequency, the phase amplitude jumped from the outer attractor to the inner attractor solution. On the other hand, if the modulation frequency, originally far below the bifurcation frequency, was ramped up toward the bifurcation frequency, the amplitude of the phase oscillations followed the inner attractor solution. At a modulation frequency near the bifurcation frequency, the amplitude of the synchrotron oscillations jumped from the inner to the outer attractor solution.

The hysteresis depended on beam current and modulation amplitude a. Since a large damping parameter could destroy the outer attractor, the hysteresis depended also on the dissipative force. The observed phase amplitudes were found to agree well with the solutions of Eq. (3.107). Similar hysteretic phenomena have been observed in electron-positron colliders, related to beam-beam interactions, where the amplitudes of the coherent π-mode oscillations showed hysteretic phenomena.[16] At a large beam-beam tune shift, the vertical beam size exhibited a flip-flop effect with respect to the relative horizontal displacement of two colliding beams.[17]

D. Systematic property of parametric resonances

The formalism discussed so far seems complicated by the transformation of phase-space coordinates into action-angle variables. However, the essential physics is rather simple. In this section, we will show that the global property of parametric resonances can be understood simply from Hamiltonian dynamics.

The circles in Fig. 3.11 show a compilation of beamlet phase amplitude vs modulation frequency for four different experimental phase modulation amplitudes. The solid lines show the synchrotron tune and its third harmonic. We note that the bifurcation of the 1:1 resonance islands follows the tune of the *unperturbed* Hamiltonian system, and the measured third order 3:1 resonance islands fall on the curve of the third harmonic of the synchrotron tune. The sideband around the first order synchrotron tune corresponds to the 60 Hz power supply ripple. Because the rf phase modulation does not excite 2:1 resonance, we did not find parametric resonances at the second synchrotron harmonic.

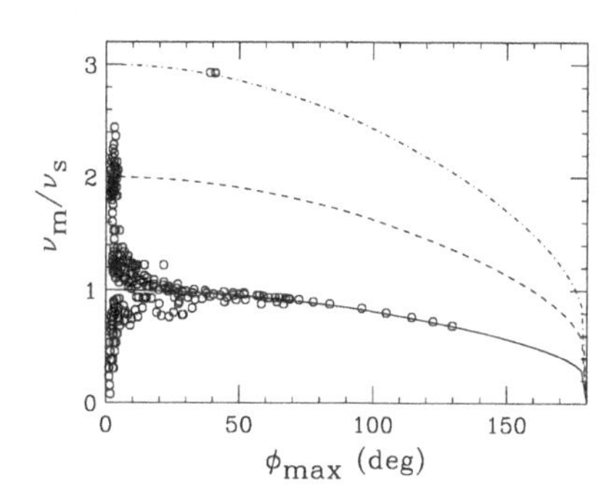

Figure 3.11: The phase amplitudes of beamlets excited by rf phase modulation, measured from the oscilloscope trace, are compared with the theoretical synchrotron tune. The bifurcation of the resonance islands follows the unperturbed tune of the synchrotron Hamiltonian, shown as the lower solid line (see also Fig. 3.6 on the bifurcation of 1:1 parametric resonance). The third order resonance island falls also on the third harmonic of the synchrotron tune.

When an external time dependence perturbation is applied to a Hamiltonian system, the perturbed Hamiltonian contains a perturbing term similar to that in Eq. (3.83). On the basis of the Kolmogorov-Arnold-Moser (KAM) theorem, many Hamiltonian tori are mildly perturbed and survived, while some tori encountering resonance condition are destroyed. Thus the external perturbation excites only particles locally in the phase space where the amplitude dependent synchrotron tune falls exactly at the modulation tune, where the particle motion can be described by the

[16]See T. Ieiri and K. Hirata, *Proc. 1989 Part. Accel. Conf.* p. 926 (IEEE, New York, 1989).

[17]See M.H.R. Donald and J.M. Paterson, *IEEE Trans. Nucl. Sci.* **NS-26**, 3580 (1979); G.P. Jackson and R.H. Siemann, *Proc. 1987 Part. Accel. Conf.* p. 1011 (IEEE, New York, 1987).

effective parametric resonance Hamiltonian (3.84).[18] The size of a resonance island depends on the slope of tune vs amplitude, strength function $g_n(J)$, and amplitude of perturbation.

In fact, the external perturbation creates a local minimum in the potential energy at the SFP locations. When a weak damping force is applied to the dynamical system, the SFP becomes an attractor, and the local potential well becomes the basin for stable particle motion. Thus a beam inside an rf bucket can split into beamlets.

When the modulation frequency is varied, SFPs (attractors) are formed along the tune of the unperturbed Hamiltonian, i.e.

$$\nu_{\rm m} = n\tilde{Q}_{\rm s}(J_{\rm SFP}). \tag{3.111}$$

The measurement of attractor amplitude vs modulation tune is equivalent to the measurement of synchrotron tune vs synchrotron amplitude, as clearly seen in Fig. 3.11. Since the rf phase modulation does not excite even synchrotron harmonics, we do not observe a 2:1 attractor in Fig. 3.11. If, however, a stronger phase modulation is applied to the dynamical system, a 2:1-like (4:2) parametric resonance can be formed by 1:1 and 3:1 resonances through second order perturbation.

An important implication of the above parametric excitation theorem is that chaos at the separatrix orbit is induced by overlapping parametric resonances. This can be understood as follows. Let $Q(J)$ be the tune of a dynamical system, where the tune is zero at the separatrix, i.e. $Q(J_{\rm sx}) = 0$. Now a time dependent perturbation can induce a series of parametric resonances in the perturbed Hamiltonian. These parametric resonances, located at $nQ(J)$ with integer n, can be excited by time dependent perturbation. Since $nQ(J_{\rm sx}) \approx 0$ for all n near the separatrix, a perturbation with low frequency modulation can produce many overlapping parametric resonances near the separatrix and lead to local chaos. This result can be applied to synchrotron motion as well as to betatron motion, where higher order nonlinear resonances serve as the source of time dependent modulation.

Now, we apply this result to evaluate the effect of low frequency modulations on particle motion. If the amplitude of low frequency modulation is not large, it will induce overlapping parametric resonances only near the separatrix. If the beam size is relatively small, the stochasticity at the separatrix will do little harm to the beam motion. However, when the modulation frequency approaches the tune of particles at the center of the bucket, particle orbits near the center of the bucket will be strongly perturbed, forming islands within the bucket.

In reality, the perturbation arising from wakefields, rf phase error, dipole field error, ground vibration, etc., consists of a spectrum of frequency distributions. The

[18]The remaining terms play the role of time dependent perturbations to the effective Hamiltonian of Eq. (3.84). Based on KAM theorem, many higher order resonance islands exist within each parametric resonance island.

mean field of the perturbation gives rise to the effect called potential well distortion, which, solved self-consistently in the Vlasov equation, modifies the unperturbed tune of the system. The remaining time dependent perturbation can generate further bunch deformation, bunch splitting, hysteresis, etc., depending on its frequency spectrum. The complicated collective instability phenomenon is in fact closely related to nonlinear beam dynamics. A series of *beam transfer function* measurements were made at electron storage rings. Sweeping the rf phase modulation frequency and measuring the response by measuring either the centroid of the beam, or the beam profile from a synchrotron light monitor using a streak camera, the response of the beam to external rf phase modulation can be obtained.[19]

III.5 RF Voltage Modulation

The beam lifetime limitation due to rf noise has been observed in many synchrotrons, e.g., the super proton synchrotron (SPS) in CERN.[20] There has been some interest in employing rf voltage modulation to induce super slow extraction through a bent crystal for very high energy beams,[21] rf voltage modulation to stabilize collective beam instabilities, rf voltage modulation for extracting beam with a short bunch length, etc. Since the rf voltage modulation may be used for enhancing a desired beam quality, we will study the physics of synchrotron motion with rf voltage modulation, that may arise from rf noise, power supply ripple, wakefields, etc. Beam response to externally applied rf voltage modulation has been measured at the IUCF Cooler.[22]

A. The equation of motion with rf voltage modulation

In the presence of rf voltage modulation, the synchrotron equations of motion are

$$\phi_{n+1} = \phi_n - 2\pi\nu_{\rm s}\frac{\eta}{|\eta|}\mathcal{P}_n, \tag{3.112}$$

$$\mathcal{P}_{n+1} = \mathcal{P}_n - 2\pi\nu_{\rm s}[1 + b\sin(\nu_{\rm m}\theta_{n+1} + \chi)]\sin\phi_{n+1} - \frac{4\pi\alpha}{\omega_0}\mathcal{P}_n, \tag{3.113}$$

where $\mathcal{P} = -h|\eta|\delta/\nu_{\rm s}$ is the normalized off-momentum coordinate conjugate to ϕ; $\delta = \Delta p/p_0$ is the fractional momentum deviation from the synchronous particle; η is the phase slip factor; $\nu_{\rm s} = \sqrt{h|\eta|eV/2\pi\beta^2E_0}$ is the synchrotron tune at zero

[19]See e.g. M.H. Wang, *et al.*, *Proc. 1997 Part. Accel. Conf.* (1997); J. Byrd, *ibid.* (1997); M.G. Minty *et al.*, *ibid.* (1997); D. Rice, private communications.

[20]D. Boussard, *et al.*, *IEEE Trans. Nucl. Sci.* **NS-26**, 3484 (1979); D. Boussard, *et al.*, *Proc. 11th Int. Conf. on High Energy Accelerators*, p. 620 (Birkhauser, Basel, 1980); G. Dôme, CERN **87-03**, p. 370 (1987); S. Krinsky and J.M. Wang, *Part. Accel.* **12**, 107 (1982).

[21]H.J. Shih and A.M. Taratin, SSCL-389 (1991); W. Gabella, J. Rosenzweig, R. Kick, and S. Peggs, *Part. Accel.* **42**, 235 (1993).

[22]D. Li *et al.*, *Phys. Rev.* E **48**, R1638 (1993); D.D. Caussyn *et al.*, *Proc. Part. Acc. Conf.* p. 29 (IEEE, Piscataway, NJ, 1993); D. Li *et al.*, *Nucl. Instrum. Methods* A **364**, 205 (1995).

amplitude; E_0 is the beam energy; $b = \Delta V/V$ is the fractional rf voltage modulation strength ($b > 0$); $\nu_{\rm m}$ is the rf voltage modulation tune; χ is a phase factor; θ is the orbital angle used as time variable; $\omega_0 = 2\pi f_0$ is the angular revolution frequency; and α is the phase-space damping factor resulting from phase-space cooling.

At the IUCF Cooler, the phase-space damping rate was measured to be about $\alpha \approx 3.0 \pm 1.0$ s^{-1}, which is much smaller than $\omega_0\nu_{\rm s}$, typically about 1500 s^{-1} for the $h = 1$ harmonic system. Without loss of generality, we discuss the case for a particle energy below the transition energy, i.e. $\eta < 0$.

Neglecting the damping term, i.e. $\alpha = 0$, the equation of motion for phase variable ϕ is

$$\ddot{\phi} + \nu_{\rm s}^2[1 + b\sin(\nu_{\rm m}\theta + \chi)]\sin\phi = 0, \tag{3.114}$$

where the overdot indicates the time derivative with respect to θ. In linear approximation with $\sin\phi \approx \phi$, Eq. (3.114) reduces to Mathieu equation. By choosing $\chi = -\pi/2$ and $z = \frac{1}{2}\nu_{\rm m}\theta$, $p = 4\nu_{\rm s}^2/\nu_{\rm m}^2$, and $q = 2b\nu_{\rm s}^2/\nu_{\rm m}^2$, we can linearize Eq. (3.114) into Mathieu's equation [30]

$$\frac{d^2\phi}{dz^2} + (p - 2q\cos 2z)\phi = 0. \tag{3.115}$$

In accelerator physics applications, p and q are real with $q \ll 1$. The stable solutions of Mathieu's equation are obtained with the condition that the parameter p is bounded by the characteristic roots $a_r(q)$ and $b_{r+1}(q)$, where $r = 0, 1, 2, \cdots$. In other words, unstable solutions are in the region $b_r(q) \le p \le a_r(q)$, where $r = 1, 2, \cdots$. The first order unstable region and the second order unstable region respectively

$$2\nu_{\rm s}(1 - \frac{1}{4}b) \le \nu_{\rm m} \le 2\nu_{\rm s}(1 + \frac{1}{4}b). \tag{3.116}$$
$$\nu_{\rm s}(1 - \frac{5}{24}b^2) \le \nu_{\rm m} \le \nu_{\rm s}(1 + \frac{1}{24}b^2),$$

which can be obtained from the second order perturbation theory.[23] The width of the instability decreases rapidly with increasing order for small b. In our application, we need to consider only the lowest order Mathieu instability. Since synchrotron motion is nonlinear, the linear Mathieu instability analysis can be extended to nonlinear synchrotron motion as follows.

B. The perturbed Hamiltonian

The synchrotron equation of motion with rf voltage modulation can be derived from the Hamiltonian $H = H_0 + H_1$ with

$$H_0 = \frac{1}{2}\nu_{\rm s}\mathcal{P}^2 + \nu_{\rm s}(1 - \cos\phi), \tag{3.117}$$
$$H_1 = \nu_{\rm s}b\,\sin(\nu_{\rm m}\theta + \chi)\,[1 - \cos\phi], \tag{3.118}$$

[23] L.D. Landau and E.M. Lifschitz, *Mechanics*, 3rd. ed. (Pergamon Press, Oxford, 1976).

where H_0 is the unperturbed Hamiltonian and H_1 the perturbation. For a weakly perturbed Hamiltonian system, we expand H_1 in action-angle coordinates of the unperturbed Hamiltonian

$$H_1 = \nu_s b \sum_{n=-\infty}^{\infty} |G_n(J)| \ \sin(\nu_m\theta - n\psi - \gamma_n), \tag{3.119}$$

where we choose $\chi = 0$ for simplicity, and $|G_n(J)|$ is the Fourier amplitude of the factor $(1 - \cos\phi)$ with γ_n its phase, defined in Eq. (3.76).

Since $(1 - \cos\phi)$ is an even function of ψ in $[-\pi, \pi]$, the Fourier integral for G_n from Eq. (3.76) is zero except for n even with $G_{-n} = G_n^*$. Thus rf voltage modulation generates only even-order synchrotron harmonics in H_1. Expanding $G_n(J)$ in power series, we obtain

$$G_0 \approx \frac{1}{2}J + \frac{1}{2048}J^3 + \cdots \quad \Longrightarrow \quad \Delta\tilde{Q}_s \approx \frac{1}{2}\nu_s b \sin\nu_m\theta.$$
$$G_2 \approx -\frac{1}{4}J + \frac{1}{128}J^2 + \cdots, \quad G_4 \approx -\frac{1}{64}J^2 + \frac{1}{2048}J^3 + \cdots, \quad G_6 \approx \frac{3}{4096}J^3 + \cdots.$$

Note that the $G_0(J)$ term in the perturbation contributes to synchrotron tune modulation ΔQ_s.

C. Parametric resonances

When the modulation frequency is near an even harmonic of the synchrotron frequency, i.e. $\nu_m \approx nQ_s$ ($n =$ even integers), particle motion can be coherently perturbed by the rf voltage modulation resulting from a resonance driving term (stationary phase condition). The resonances, induced by the external harmonic modulation of the rf voltage, are called parametric resonances. Using the generating function

$$F_2 = (\psi - \frac{\nu_m}{n}\theta + \frac{\gamma_n}{n} + \frac{\pi}{2n})\tilde{J},$$

we obtain the Hamiltonian in a resonance rotating frame as

$$\tilde{H} = E(\tilde{J}) - \frac{\nu_m}{n}\tilde{J} + \nu_s b|G_n(\tilde{J})|\cos n\tilde{\psi} + \Delta\tilde{H}(\tilde{J}, \tilde{\psi}, \theta), \tag{3.120}$$

where the remaining small time dependent perturbation term ΔH oscillates at frequencies $\nu_m, 2\nu_m, \cdots$. In the time average, we have $\langle\Delta\tilde{H}\rangle \approx 0$. Thus the time averaged Hamiltonian $\langle\tilde{H}\rangle$ for the nth order parametric resonance becomes

$$\langle\tilde{H}\rangle = E(\tilde{J}) - \frac{\nu_m}{n}\tilde{J} + \nu_s b|G_n(\tilde{J})|\cos n\tilde{\psi}. \tag{3.121}$$

The phase-space contour may be strongly perturbed by a parametric resonance. Since $|G_{n+2}/G_n| \sim J$ for $n > 0$, the resonance strength is greatest at the lowest harmonic for particles with small phase amplitude. The system is most sensitive to the rf voltage modulation at the second synchrotron harmonic.

D. Quadrupole mode

When the rf voltage modulation frequency is near the second harmonic of synchrotron frequency, particle motion is governed by the $n = 2$ parametric resonance Hamiltonian

$$\langle \tilde{H} \rangle = (\nu_{\rm s} - \frac{\nu_{\rm m}}{2})\tilde{J} - \frac{\nu_{\rm s}}{16}\tilde{J}^2 + \frac{\nu_{\rm s}}{4}b\tilde{J}\cos 2\tilde{\psi} \tag{3.122}$$

in the resonance rotating frame. Since the Hamiltonian (3.122) is autonomous, Hamiltonian is a constant of motion. For simplicity, we drop the tilde notations. Hamilton's equations are

$$\dot{J} = \frac{\nu_{\rm s}}{2}bJ\sin 2\psi, \tag{3.123}$$

$$\dot{\psi} = \nu_{\rm s} - \frac{\nu_{\rm m}}{2} - \frac{\nu_{\rm s}}{8}J + \frac{\nu_{\rm s}}{4}b\cos 2\psi. \tag{3.124}$$

The fixed points that determine the locations of islands and separatrix of the Hamiltonian are obtained from $\dot{J} = 0$, $\dot{\psi} = 0$. The stable fixed points (SFPs) ($\psi = 0$ and π) and the unstable fixed points (UFPs) ($\psi = \pi/2$ and $3\pi/2$) are

$$J_{\rm SFP} = \begin{cases} 8(1 - \frac{\nu_{\rm m}}{2\nu_{\rm s}}) + 2b, & \text{if } \nu_{\rm m} \le 2\nu_{\rm s} + \frac{1}{2}b\nu_{\rm s} \\ 0, & \text{if} \nu_{\rm m} > 2\nu_{\rm s} + \frac{1}{2}b\nu_{\rm s} \end{cases} \tag{3.125}$$

$$J_{\rm UFP} = \begin{cases} 8(1 - \frac{\nu_{\rm m}}{2\nu_{\rm s}}) - 2b, & \text{if } \nu_{\rm m} \le 2\nu_{\rm s} - \frac{1}{2}b\nu_{\rm s} \\ 0, & \text{if } 2\nu_{\rm s} - \frac{1}{2}b\nu_{\rm s} \le \nu_{\rm m} \le 2\nu_{\rm s} + \frac{1}{2}b\nu_{\rm s} \end{cases} \tag{3.126}$$

Examples of Hamiltonian Tori around SFPs are shown in Fig. 3.12.

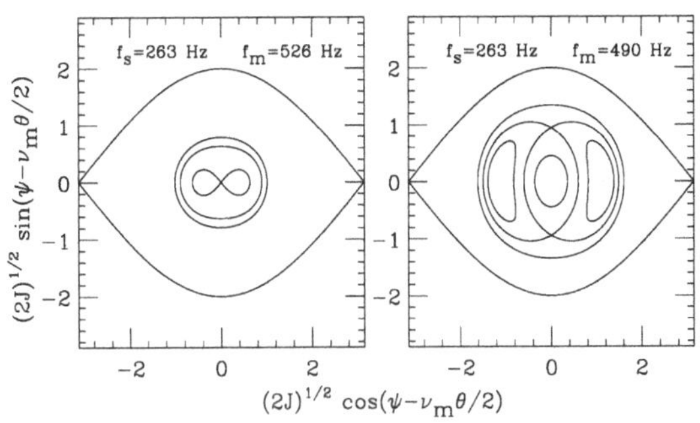

Figure 3.12: The separatrix and tori of the Hamiltonian (3.122) in the resonance rotating frame. The synchrotron frequency is $f_{\rm s} = 263$ Hz, the voltage modulation amplitude is $b = 0.05$, and the modulation frequencies are $f_{\rm m} = 526$ Hz (left plot) and $f_{\rm m} = 490$ Hz (right plot).

We note that the second harmonic rf voltage modulation can induce an instability at $J_{\rm UFP} = 0$ in the frequency domain $2\nu_{\rm s} - \frac{1}{2}b\nu_{\rm s} \le \nu_{\rm m} \le 2\nu_{\rm s} + \frac{1}{2}b\nu_{\rm s}$. This is the first order Mathieu resonance of Eq. (3.116). Nonlinear synchrotron motion extends the instability to lower modulation frequency at larger synchrotron amplitude, according to

$$\nu_{\rm m} = 2\nu_{\rm s}(1 - \frac{J_{\rm UFP}}{8}) - \frac{1}{2}\nu_{\rm s}b, \tag{3.127}$$

which is a nonlinear extension of Mathieu instability.

Modulation of rf voltage at the second harmonic of the synchrotron tune had been found useful in damping the multi-bunch instabilities for the damping ring at the Stanford linear collider (SLC), and stabilizing coupled bunch instabilities induced by parasitic rf cavity modes with high brightness beams at the Taiwan Light Source.[24] By adjusting the amplitude and phase of the rf voltage modulation, the collective instability of high brightness electron beams in the SLC damping ring can be controlled. The damping mechanism may be understood as follows. When the voltage modulation at $\nu_{\rm m} = 2\nu_{\rm s}$ is applied, the Mathieu resonance gives rise to an UFP at the origin of the phase space and the SFP is displaced to $J_{\rm SFP} = 2b$. Since electrons are damped incoherently into the SFP by the synchrotron radiation damping, the beam distribution becomes dumbbell-shaped in phase space, rotating in the longitudinal phase space at half the modulation frequency, i.e. the synchrotron frequency. The size and orientation of the dumbbell can be controlled by parameter b and phase χ.

E. The separatrix

The separatrix torus, which passes through the UFPs, is given by

$$H(J,\psi) = H(J_{\rm UFP},\psi_{\rm UFP}). \tag{3.128}$$

The separatrix intersects the phase axis at the actions J_1 and J_2 given by

$$J_1 = \begin{cases} J_{\rm SFP} + \sqrt{J_{\rm SFP}^2 - J_{\rm UFP}^2} & \text{if } \nu_{\rm m} \le 2\nu_{\rm s} - \frac{1}{2}b\nu_{\rm s} \\ 2J_{\rm SFP} & \text{if } 2\nu_{\rm s} - \frac{1}{2}b\nu_{\rm s} \le \nu_{\rm m} \le 2\nu_{\rm s} + \frac{1}{2}b\nu_{\rm s} \end{cases} \tag{3.129}$$

and

$$J_2 = \begin{cases} J_{\rm SFP} - \sqrt{J_{\rm SFP}^2 - J_{\rm UFP}^2} & \text{if } \nu_{\rm m} \le 2\nu_{\rm s} - \frac{1}{2}b\nu_{\rm s} \\ 0 & \text{if } 2\nu_{\rm s} - \frac{1}{2}b\nu_{\rm s} \le \nu_{\rm m} \le 2\nu_{\rm s} + \frac{1}{2}b\nu_{\rm s}. \end{cases} \tag{3.130}$$

The intercepts can be used to determine the maximum synchrotron phase oscillation due to rf voltage modulation. Figure 3.12 shows also the intercepts of separatrix with phase axis. The island size $\Delta\phi_{\rm island}$ is $\sqrt{2J_1} - \sqrt{2J_2}$.

F. The amplitude dependent island tune of 2:1 parametric resonance

For an autonomous dynamical system governed by the Hamiltonian (3.122), the Hamiltonian is a constant of motion. The Hamiltonian value is $E_{\rm s} = \frac{1}{16}\nu_{\rm s}J_{\rm SFP}^2$ at SFP, and $E_{\rm u} = \frac{1}{16}\nu_{\rm s}J_{\rm UFP}^2$ at UFP. Using Hamilton's equations of motion, we obtain $\dot{J} = f(J,E)$, where

$$f(J,E) = 2\nu_{\rm s}\sqrt{\left[\frac{J^2}{16} + \frac{E}{\nu_{\rm s}} - (1 - \frac{\nu_{\rm m}}{2\nu_{\rm s}} - \frac{b}{4})J\right]\left[(1 - \frac{\nu_{\rm m}}{2\nu_{\rm s}} + \frac{b}{4})J - \frac{E}{\nu_{\rm s}} - \frac{J^2}{16}\right]}. \tag{3.131}$$

[24] M.H. Wang, and S.Y. Lee, *Journal of Applied Physics*, **92**, 555 (2002); J.D. Fox and P. Corredoura, *Proc. European Part. Accel. Conf.* p. 1079 (Springer-Verlag, Heidelberg, 1992).

For a given Hamiltonian value E, the action J is limited by $J_{\rm min}$ and $J_{\rm max}$ given by

$$J_{\rm min} = (1-\sqrt{1-x})J_{\rm SFP}, \quad J_{\rm max} = (1+\sqrt{1-x})J_{\rm SFP},$$

where $x = E/E_{\rm s}$, with $x \in [J_{\rm UFP}^2/J_{\rm SFP}^2\,,\,1]$. Note that $J_{\rm SFP} = \frac{1}{2}(J_{\rm min}+J_{\rm max})$. The island tune becomes

$$Q_{\rm island} = \frac{2\pi}{\oint d\theta} = 2\pi\left[\oint\frac{dJ}{f(J,E)}\right]^{-1} = \frac{\pi\nu_{\rm s}\sqrt{2bJ_{\rm SFP}}}{8K(k)}x^{1/4}, \tag{3.132}$$

$$k = \frac{1}{\sqrt{2}}\sqrt{1-\frac{xJ_{\rm SFP}-J_{\rm UFP}}{\sqrt{x}(J_{\rm SFP}-J_{\rm UFP})}},$$

where $K(k)$ is the complete elliptical integral of the first kind [30]. At $x = 1$, the island tune becomes $\nu_s\sqrt{bJ_{\rm SFP}/8}$. At the separatrix with $x = J_{\rm UFP}^2/J_{\rm SFP}^2$, the island tune is zero.

III.6 Measurement of RF Voltage Modulation

We describe here an rf voltage modulation measurement at the IUCF Cooler. The experiment started with a single bunch of about 5×10^8 protons with kinetic energy 45 MeV. The cycle time was 10 s, with the injected beam electron-cooled for about 3 s, producing a full width at half maximum bunch length of about 9 m (or 100 ns) depending on rf voltage. The low frequency rf system used in the experiment was operating at harmonic number $h = 1$ with frequency 1.03168 MHz.

A. Voltage modulation control loop

The voltage control feedback of the IUCF Cooler rf system works as follows. The cavity rf voltage is picked up and rectified into DC via synchronous detection. The rectified DC signal is compared to a preset voltage. The error found goes through a nearly ideal integrator that has very high DC gain. The integrated signal is then used to control an attenuator regulating the level of rf signal being fed to rf amplifiers. Because of the relatively low Q of the cavity at the IUCF Cooler, the effect of its inertia can be ignored if the loop gain is rolled off to unity well before $f_0/2$Q, where f_0 is the resonant frequency of the rf cavity and Q ≈ 50 is the cavity Q value. Thus, no proportional error feedback is needed to stabilize the loop. The overall loop response exhibits the exponential behavior prescribed by a first order differential equation, i.e. $dV/dt = -V/\tau$, where V is the rf voltage and the characteristic relaxation time τ is about $10-200$ μ s.

The amplitude modulation is summed with the reference and compared to the cavity sample signal. The modulation causes a change in the error voltage sensed by the control loop and results in modulation of the attenuator around a preset cavity voltage. The maximum modulation rate is limited by the loop response time of about 10 kHz. The modulation rates in our experiments are well within this limit. The modulation amplitude was measured and calibrated.

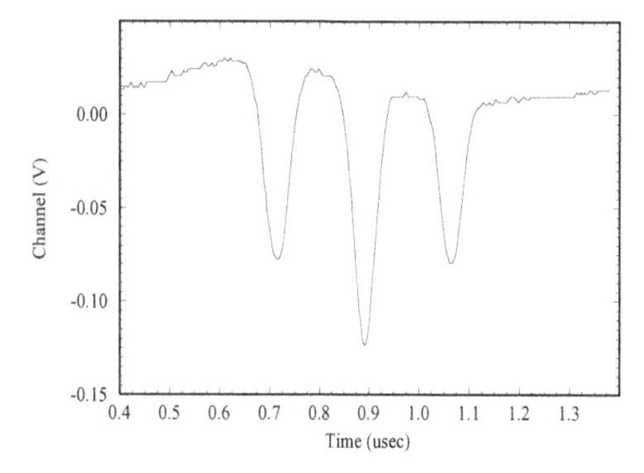

Figure 3.13: The beam bunch was observed to split into three beamlets in a single rf bucket measured from a fast sampling scope in (μs). The voltage modulation amplitude is $b = 0.05$ at modulation frequency $f_{\rm m} = 480$ Hz with synchrotron tune $f_{\rm s} = 263$ Hz. Note that the outer two beamlets rotated around the center beamlet at a frequency equal to half the modulation frequency.

B. Observations of the island structure

Knowing that the beam bunch will be split into beamlets, as shown in Sec. III.2, we first measured the phase oscillation amplitude of the steady state solution by using the oscilloscope. The beam was injected, the rf voltage was modulated, and the beam was cooled with electron current 0.75 A. Then the steady state bunch distribution was measured. Figure 3.13 shows that the sum signals from a beam position monitor (BPM) on a fast oscilloscope triggered at the rf frequency exhibited two peaks around a central peak. A fast 1×10^9 sample per second oscilloscope was used to measure the profile of the beam in a single pass. The profile shown in Fig. 3.13 indicated that there were three beamlets in the $h = 1$ rf bucket. The beam particles were damped to attractors of the dissipative parametric resonant system. Thus the phase amplitude of the outer peaks measured from the oscilloscope can be identified as the phase amplitude of the SFP.

Since the attractors (or islands) rotate around the origin of the rf bucket with half the modulation frequency, the observed beam profile in an oscilloscope is a time average of the BPM sum signal. Because the equilibrium beamlet distribution in a resonance island has a large aspect ratio in the local phase-space coordinates, the resulting beam profile will exhibit two peaks at the maximum phase amplitude, resembling that in Fig. 3.13. This implies that when a beamlet rotates to the upright position in the phase coordinate, a larger peak current can be observed. On the other hand, when a beamlet rotates to the flat position, where the SFPs are located on the $\mathcal{P}$ axis, the aspect ratio becomes small and the line density is also small.

The measured action J of the outer beamlets as a function of modulation frequency is shown in Fig. 3.14, where $J_{\rm SFP}$ of the Hamiltonian (3.122) is also shown

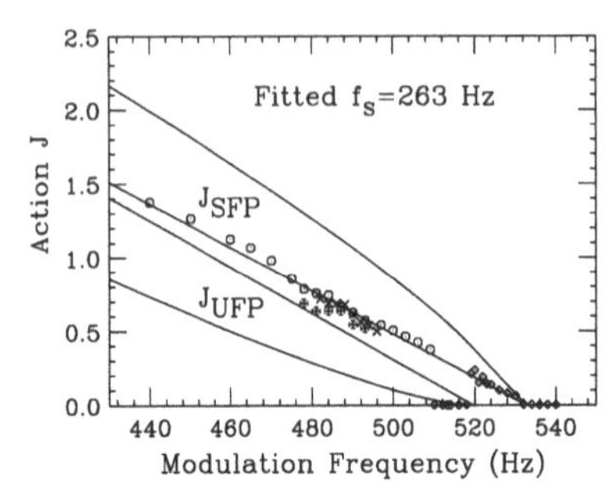

Figure 3.14: The measured action J of outer beamlets as a function of modulation frequency. Here $J \approx \frac{1}{2}\hat{\phi}^2$ with $\hat{\phi}$ as the peak phase amplitude of attractors. Different symbols correspond to measurements at different times for an almost identical rf voltage. The solid line for $J_{\rm SFP}$ obtained from Eq. (3.125) fits data with $f_s = 263$ Hz. The actions of UFP $J_{\rm UFP}$ and intercepts J_1 and J_2 of the separatrix with the phase axis are also shown.

for comparison. Experimentally, we found that the action of the outer attractor varied linearly with modulation frequency. Similarly, $J_{\rm SFP}$ is also a linear function of modulation frequency, where the slope depends sensitively on synchrotron frequency. Using this sensitivity, the synchrotron frequency was determined more accurately to be about 263±1 Hz for this run. Our experimental results agreed well with the theoretical prediction except in the region $f_{\rm m} \in [510, 520]$ Hz, where we did not observe beam splitting. A possible explanation is that the actual beam size was larger than the separation of islands. In this case, the SFPs were about 100 ns from the center of the bucket. Once $f_{\rm m}$ reached $2f_s - \frac{1}{2}bf_s \approx 520$ Hz, where $J_{\rm UFP} = 0$, the beam was observed to split into only two beamlets. It was also clearly observed that all parametric resonance islands ceased to exist at $f_{\rm m} = 2f_s + \frac{1}{2}bf_s \approx 532$ Hz.

Exercise 3.3

1. Prove the identity of the action integral in Eq. (3.66).
2. We consider a general Hamiltonian

$$H = \frac{1}{2}\nu_s \mathcal{P}^2 + V(\phi),$$

where $(\phi, \mathcal{P})$ are conjugate phase-space variables with orbiting angle θ as time variable, ν_s is the small amplitude synchrotron tune, and $V(\phi)$ is the potential.[25] The action is $J = (1/2\pi)\oint \mathcal{P}d\phi$. Using the generating function

$$F_2 = \int_0^\phi \mathcal{P}d\phi,$$

show that the coordinate transformation between phase variable ψ and coordinate ϕ is

$$d\psi = \frac{Q(J)}{\nu_s}d\phi,$$

[25] In linear approximation, the potential can be expressed as $V(\phi) = \frac{1}{2}\nu_s\phi^2 + \cdots$. However, small amplitude behavior of the potential is not a necessary condition for the sum rule theorem stated in this exercise.

where ψ is the conjugate phase variable to the action J. Expanding $\mathcal{P}$ in action-angle variables with

$$\mathcal{P} = \sum_{n=-\infty}^{\infty} f_n e^{in\psi},$$

prove the sum rule theorem

$$\sum_{n=-\infty}^{\infty} |f_n|^2 = \frac{Q(J)}{\nu_s} J.$$

3. From Exercise 2.4.8, we find that the change of orbit length due to a modulating dipole kicker is given by

$$\Delta C = D(s_0)\,\theta(t) = D(s_0)\,\hat{\theta}\,\sin(\omega_{\rm m} t + \chi_0),$$

where $D(s_0)$ is the dispersion function at the dipole location, $\hat{\theta}$ is the maximum dipole kick angle, $\omega_{\rm m}$ is the modulating angular frequency, and χ_0 is an arbitrary initial phase. The modulating tune is $\nu_{\rm m} = \omega_{\rm m}/\omega_0$, where ω_0 is the angular revolution frequency.

(a) Show that the modulating dipole field produces an equivalent rf phase error

$$\Delta\phi = \frac{2\pi h D(s_0)\hat{\theta}}{C} \sin(\omega_{\rm m} t + \chi_0) = \hat{\Delta\phi} \sin(\omega_{\rm m} t + \chi_0),$$

where C is the circumference of the synchrotron, and h is the harmonic number.

(b) Show that the amplitude of the equivalent rf wave phase error is

$$a = \hat{\Delta\phi}/2\pi\nu_{\rm m}.$$

Give a physical argument that the amplitude of the equivalent rf wave phase error a is amplified as the modulation tune $\nu_{\rm m}$ becomes smaller.

(c) Evaluate the effective rf modulation amplitude a for the accelerators listed in the table below, where C is the circumference, $\Delta B\ell$ is the integrated dipole field error, $f_{\rm mod}$ is the modulation frequency, D is the dispersion function at the dipole, γ is the Lorentz relativistic factor, and h is the harmonic number.

	IUCF Cooler	RHIC	MI	Recycler
C (m)	86.8	3833.8	3319.4	3319.4
$\Delta B\ell$ (Gm)	1	1	1	1
$f_{\rm mod}$ (Hz)	262	60	60	4
D (m)	4	1	1	1
γ	1.04796	24	21.8	9.5
h	1	342	588	1
a				

4. Using the conjugate phase space coordinates

$$Q = \sqrt{2J}\cos(\psi - \frac{1}{2}\nu_{\rm m}\theta), \qquad P = -\sqrt{2J}\sin(\psi - \frac{1}{2}\nu_{\rm m}\theta),$$

show that the Hamiltonian (3.122) for the quadrupole mode is

$$H = \frac{1}{2}(\delta + \frac{\nu_{\rm s}b}{4})Q^2 + \frac{1}{2}(\delta - \frac{\nu_{\rm s}b}{4})P^2 - \frac{\nu_{\rm s}}{64}(Q^2 + P^2)^2,$$

where $\delta = \nu_{\rm s} - (\nu_{\rm m}/2)$ and, without loss of generality, we assume $b > 0$. Show that the fixed points of the Hamiltonian are located at

$$\begin{array}{ll} P_{\rm SFP} = 0, \quad Q_{\rm SFP} = 0 & (\nu_{\rm m} > 2\nu_{\rm s} + \nu_{\rm s}b/2) \\ P_{\rm SFP} = 0, \quad Q_{\rm SFP} = \sqrt{16(1 - \nu_{\rm m}/2\nu_{\rm s}) + 4b} & (\nu_{\rm m} > 2\nu_{\rm s} + \nu_{\rm s}b/2) \\ Q_{\rm UFP} = 0, \quad P_{\rm UFP} = 0 & (2\nu_{\rm s} - \nu_{\rm s}b/2 \le \nu_{\rm m} \le 2\nu_{\rm s} + \nu_{\rm s}b/2) \\ Q_{\rm UFP} = 0, \quad P_{\rm UFP} = \sqrt{16(1 - \nu_{\rm m}/2\nu_{\rm s}) - 4b} & (\nu_{\rm m} \le 2\nu_{\rm s} - \frac{1}{2}\nu_{\rm s}b). \end{array}$$

Compare this result with Eqs. (3.125) and (3.126). Show that the separatrix for $\nu_{\rm m} \le 2\nu_s - \nu_s b/2$ is given by two circles

$$(Q - Q_{\rm c})^2 + P^2 = r^2, \qquad (Q + Q_{\rm c})^2 + P^2 = r^2$$

with

$$Q_{\rm c} = \sqrt{4b}, \qquad r = \sqrt{16\delta/\nu_s}.$$

The separatrix in the betatron phase space for slow beam extraction that employs a half integer stopband is identical to that given in this exercise. Quadrupoles are used to provide resonance driving term b, and octupoles are used to provide nonlinear detuning α_{xx}. The resulting effective Hamiltonian is

$$H_{\rm eff} = \nu_x J_x + \frac{1}{2}\alpha_{xx}J_x^2 + bJ_x\cos(\psi_x - \frac{\ell}{2}\theta),$$

where $\alpha_{xx} = (-1/16\pi B\rho)\oint \beta_x^2 B_3 ds$ is the detuning parameter, B_3 is the octupole strength, and b is the half integer stopband width.

5. Show that the equation of motion for rf dipole on betatron motion in Eq. (2.107) near a betatron sideband can be cast into an effective Hamiltonian

$$H_{\rm eff} = \nu J + \frac{1}{2}\alpha J^2 + gJ^{1/2}\cos(\psi - \nu_{\rm m}\theta + \chi),$$

where ν, (J, ϕ), α are the tune, the action-angle coordinates, and the detuning parameter of the betatron motion, g is proportional to the rf dipole field strength, and $\nu_{\rm m}$ is the rf dipole modulation tune. Find the fixed points of the Hamiltonian and discuss the dependence of the fixed point on parameters $\nu_{\rm m} - \nu$, and α.

IV Nonadiabatic and Nonlinear Synchrotron Motion

Transition energy has been both a nuisance in machine operation and a possible blessing for attaining beam bunches with some desired properties, such as enhanced beam separation for filtering ion beams having nearly equal charge to mass ratios, and beam bunches with ultra-small beam width.[26] However, the synchrotron frequency spread vanishes at transition energy, and the circulating beams can suffer microwave instabilities and other collective instabilities for lack of Landau damping, to be discussed in Sec. VII.

Near the transition energy region, the adiabaticity condition (3.32) is not satisfied, i.e. the Hamiltonian is time dependent and is not a constant of motion. This results in non-adiabatic synchrotron motion, where the bucket area increases dramatically, and the phase-space area occupied by the beam bunch is a small fraction of the bucket area. The linearized rf potential is a good approximation. If the phase slip factor is independent of the off-momentum variable, we will obtain analytic solutions for the linearized synchrotron motion near transition energy in Sec. IV.1. The integral of the linearized Hamiltonian is also an ellipse, and the action is a constant of motion. We will discuss the scaling properties of the beam at the transition energy crossing.

However, when the phase slip factor η_0 of Eq. (3.11) becomes small, the nonlinear phase slip factor term η_1 can be important. This again raises another nonlinear problem in synchrotron motion, i.e. parts of a beam bunch can encounter a defocusing force during transition energy crossing. In Sec. IV.2 we study nonlinear synchrotron motion due to nonlinearity in phase slip factor. Although the action of a Hamiltonian flow is invariant, the torus is highly distorted and particles in a beam may be driven out of the rf bucket after crossing the transition energy. In Sec. IV.3 we examine beam manipulation techniques for particle acceleration through transition energy. In Sec. IV.4 we study the effects of nonlinear phase slip factor and examine the properties of the so-called α-bucket, and in Sec. IV.5 we study problems associated with quasi-isochronous (QI) storage rings, which may provide beam bunches with ultra-short bunch length.

[26]Since the bunch width becomes very short and the momentum spread becomes large at transition energy, transition energy may be used to generate short bunches. See e.g., R. Cappi, J.P. Delahye, and K.H. Reich, *IEEE Trans. Nucl. Sci.* **NS-28**, 2389 (1981). Using the sensitivity of the closed orbit to beam momentum at transition energy, one can filter beam momentum from nearly identical Z/A (charge to mass ratio) ion beams. Oxygen and sulfur ions have been filtered at transition energy in the CERN PS.

IV.1 Linear Synchrotron Motion Near Transition Energy

Since the energy gain per revolution in rf cavities is small, we assume $\gamma = \gamma_{\rm T} + \dot\gamma t$, where $\dot\gamma = d\gamma/dt$ is the acceleration rate, and t is the time coordinate. The phase slip factor becomes

$$\eta_0 = \alpha_0 - \gamma^{-2} \approx \frac{2\dot\gamma t}{\gamma_{\rm T}^3}. \tag{3.133}$$

Here we have neglected the dependence of the phase slip factor on the off-momentum coordinate δ, and assume that all particles in a bunch pass through transition energy at the same time. Substituting Eq. (3.133) into Eq. (3.17), we obtain

$$\omega_{\rm s}^2 = \frac{|t|}{\tau_{\rm ad}^3}, \qquad \tau_{\rm ad} = \left(\frac{\pi\beta^2 mc^2\gamma_{\rm T}^4}{\dot\gamma\omega_0^2 heV|\cos\phi_{\rm s}|}\right)^{1/3}. \tag{3.134}$$

Here $\tau_{\rm ad}$ is the *adiabatic time*. At $|t| \gg \tau_{\rm ad}$, the adiabaticity condition (3.32) is satisfied. At $|t| \geq 4\tau_{\rm ad}$ the adiabatic condition is approximately fulfilled because $\alpha_{\rm ad} = |d(\omega_{\rm s}^{-1})/dt| = \frac{1}{2}(\tau_{\rm ad}/|t|)^{3/2} \approx 0.06$. Table 3.3 lists the adiabatic time for some proton synchrotrons. Typically $\tau_{\rm ad}$ is about 1–10 ms. Note that the beam parameters for RHIC correspond to those of a typical gold beam injected from the AGS with charge number $Z = 79$, and atomic mass number $A = 197$. The injection energy for proton beams in RHIC is above transition energy.

Table 3.3: The adiabatic and nonlinear times of some proton synchrotrons.

	FNAL Booster	FNAL MI	AGS	RHIC	KEKPS	CPS
C (m)	474.2	3319.4	807.12	3833.8	339.29	628.32
V (kV)	950	4000	300	300	90	200
h	84	588	12	360	9	6-20
$\gamma_{\rm T}$	5.4	20.4	8.5	22.5	6.76	6.5
$\dot\gamma$ (s^{-1})	200	190	70	1.6	40	60
$\mathcal{A}$ (eVs/u)	0.04	0.04	1.	0.3	0.3	0.5
$\hat\delta$ ($\times 10^{-3}$)	6.4	2.5	6.7	4.5	5.4	6.6
$\tau_{\rm ad}$ (ms)	0.2	2.0	2.5	36	1.8	1.5
$\tau_{\rm nl}$ (ms)	0.13	0.19	0.61	63	0.7	0.5

In linear approximation, the synchrotron equations of motion near the transition energy region become

$$\dot\delta = \frac{\omega_0 eV}{2\pi\beta^2 E}\cos\phi_{\rm s}(\Delta\phi), \qquad (\dot{\Delta\phi}) = \frac{2h\omega_0\dot\gamma}{\gamma_{\rm T}^3}\, t\,\delta, \tag{3.135}$$

where the overdot indicates the derivative with respect to time t, and $\delta = \Delta p/p_0$ and $\Delta\phi = \phi - \phi_s$ are the fractional off-momentum and phase coordinates of a particle. Taking into account the synchronous phase change from ϕ_s to $\pi - \phi_s$ across transition energy, we obtain

$$\frac{d}{dt}\left(\frac{\tau_{\rm ad}^3}{|t|}\frac{d}{dt}\Delta\phi\right) + \Delta\phi = 0. \tag{3.136}$$

Defining a new time variable y as

$$y = \int_0^x x^{1/2}dx = \frac{2}{3}x^{3/2} \qquad \text{with} \qquad x = \frac{|t|}{\tau_{\rm ad}}, \tag{3.137}$$

Eq. (3.136) can be transformed into Bessel's equation of order 2/3,

$$\varphi'' + \frac{1}{y}\varphi' + (1 - \frac{(2/3)^2}{y^2})\varphi = 0, \tag{3.138}$$

where $\varphi = y^{-2/3}\Delta\phi$, and the primes indicate derivatives with respect to time variable y. The solution of Eq. (3.138) can be written readily as

$$\Delta\phi = bx\left[\cos\chi\, J_{2/3}(y) + \sin\chi\, N_{2/3}(y)\right], \tag{3.139}$$

where χ and b are constants to be determined from the initial condition. Here the Neumann function is $N_\nu(z) = [J_\nu(z)\cos\pi\nu - J_{-\nu}(z)]/\sin\pi\nu$. It is also called the Bessel function of the second kind. In Ref. [30], the notation is $Y_\nu(z)$. The off-momentum coordinate δ can be obtained from Eq. (3.135), i.e.

$$\dot{\Delta\phi} = \frac{2h\omega_0\dot\gamma t}{\gamma_{\rm T}^3}\delta = \frac{\Delta\phi}{\tau_{\rm ad}x} + \frac{bx^{2/3}}{\tau_{\rm ad}}\left(\cos\chi\left[\frac{2J_{2/3}}{3y} - J_{5/3}\right] + \sin\chi\left[\frac{2N_{2/3}}{3y} - N_{5/3}\right]\right).$$

Combining this with Eq. (3.139), we obtain the constant of motion

$$\alpha_{\phi\phi}(\Delta\phi)^2 + 2\alpha_{\phi\delta}\Delta\phi\delta + \alpha_{\delta\delta}\delta^2 = 1, \tag{3.140}$$

where

$$\alpha_{\phi\phi} = \frac{\pi^2}{9b^2x^2}\left[\left(\frac{3}{2}yN_{5/3} - 2N_{2/3}\right)^2 + \left(2J_{2/3} - \frac{3}{2}yJ_{5/3}\right)^2\right],$$
$$\alpha_{\phi\delta} = \frac{\pi^2}{9b^2}\left(\frac{2h\dot\gamma\omega_0\tau_{\rm ad}^2}{\gamma_{\rm T}^3}\right)\left[N_{2/3}\left(\frac{3}{2}yN_{5/3} - 2N_{2/3}\right) - J_{2/3}\left(2J_{2/3} - \frac{3}{2}yJ_{5/3}\right)\right],$$
$$\alpha_{\delta\delta} = \frac{\pi^2}{9b^2}x^2\left(\frac{2h\dot\gamma\omega_0\tau_{\rm ad}^2}{\gamma_{\rm T}^3}\right)^2[J_{2/3}^2 + N_{2/3}^2].$$

There is no surprise that the constant of motion for a time dependent *linear* Hamiltonian is an ellipse. In (ϕ, δ) phase-space coordinates, the shape of the ellipse changes with time. The phase-space area enclosed in the ellipse of Eq. (3.140) is a *constant of motion* given by

$$\tilde{\mathcal{A}} = \frac{\pi}{\sqrt{\alpha_{\phi\phi}\alpha_{\delta\delta} - \alpha_{\phi\delta}^2}} = \frac{3b^2\gamma_{\rm T}^3}{2h\dot{\gamma}\omega_0\tau_{\rm ad}^2} = h\mathcal{A}\frac{\omega_0}{\beta^2\gamma_{\rm T}mc^2}, \tag{3.141}$$

where $\mathcal{A}$ is the phase-space area of the bunch in eV-s. Thus the parameter b is

$$b = \left(\frac{2\mathcal{A}h^2\omega_0^2\dot{\gamma}\tau_{\rm ad}^2}{3mc^2\beta^2\gamma_{\rm T}^4}\right)^{1/2}. \tag{3.142}$$

A. The asymptotic properties of the phase space ellipse

The phase-space ellipse is tilted in the transition energy region. Using a Taylor series expansion around $y = 0$, we obtain

$$\alpha_{\phi\phi} = \frac{\pi^2}{9b^2}\frac{4}{3^{1/3}[\Gamma(\frac{2}{3})]^2}, \tag{3.143}$$

$$\alpha_{\phi\delta} = -\frac{\pi^2}{9b^2}\left(\frac{2h\dot{\gamma}\omega_0\tau_{\rm ad}^2}{\gamma_{\rm T}^3}\right)\frac{\sqrt{3}}{\pi}, \tag{3.144}$$

$$\alpha_{\delta\delta} = \frac{\pi^2}{9b^2}\left(\frac{2h\dot{\gamma}\omega_0\tau_{\rm ad}^2}{\gamma_{\rm T}^3}\right)^2\frac{3^{4/3}[\Gamma(\frac{2}{3})]^2}{\pi^2}. \tag{3.145}$$

The tilt angle, the maximum momentum spread, and the maximum bunch width of the ellipse are

$$\psi = \frac{1}{2}\tan^{-1}\frac{2\alpha_{\phi\delta}}{\alpha_{\phi\phi} - \alpha_{\delta\delta}}, \tag{3.146}$$

$$\hat{\delta}\Big|_{\gamma=\gamma_{\rm T}} = \frac{\gamma_{\rm T}}{3^{1/6}\beta\tau_{\rm ad}\Gamma(\frac{2}{3})}\left(\frac{2\mathcal{A}}{3mc^2\dot{\gamma}}\right)^{1/2} \approx 0.502\frac{\gamma_{\rm T}}{\beta\tau_{\rm ad}}\left(\frac{\mathcal{A}}{mc^2\dot{\gamma}}\right)^{1/2}, \tag{3.147}$$

$$\hat{\phi}\Big|_{\gamma=\gamma_{\rm T}} = \sqrt{\frac{\alpha_{\delta\delta}}{\alpha_{\phi\phi}\alpha_{\delta\delta} - \alpha_{\phi\delta}^2}} = \frac{3^{2/3}\Gamma(\frac{2}{3})}{\pi}\left(\frac{2\mathcal{A}h^2\omega_0^2\dot{\gamma}\tau_{\rm ad}^2}{3mc^2\beta^2\gamma_{\rm T}^4}\right)^{1/2}. \tag{3.148}$$

Note that $\hat{\delta}$ is finite at $\gamma = \gamma_{\rm T}$ for a nonzero acceleration rate. At a higher acceleration rate, the maximum momentum width of the beam will be smaller. Substituting the adiabatic time $\tau_{\rm ad}$ of Eq. (3.134) into Eq. (3.147), we obtain the following scaling property:

$$\hat{\delta}\Big|_{\gamma=\gamma_{\rm T}} \sim h^{1/3}V^{1/3}\mathcal{A}^{1/2}\dot{\gamma}^{-1/6}\gamma_{\rm T}^{-1/3}. \tag{3.149}$$

The scaling property is important in the choice of operational conditions.

In the adiabatic region where $x \gg 1$, we can use asymptotic expansion of Bessel functions to obtain

$$\alpha_{\phi\phi} \to \frac{\pi^2}{3b^2} x^{-1/2}, \quad \alpha_{\phi\delta} \to 0, \quad \alpha_{\delta\delta} \to \frac{\pi^2}{3b^2} \left(\frac{2h\dot{\gamma}\omega_0 \tau_{\rm ad}^2}{\gamma_{\rm T}^3} \right)^2 x^{1/2}.$$

The phase-space ellipse is restored to the upright position.

B. The Gaussian distribution function at transition energy

The distribution function that satisfies the Vlasov equation is a function of the invariant ellipse (3.140). Using the Gaussian distribution function model, we obtain

$$\begin{aligned} \Psi_0(\Delta\phi, \delta) &= \frac{3N_{\rm B}(\alpha_{\phi\phi}\alpha_{\delta\delta} - \alpha_{\phi\delta}^2)^{1/2}}{\pi} e^{-3[\alpha_{\phi\phi}(\Delta\phi)^2 + 2\alpha_{\phi\delta}\delta(\Delta\phi) + \alpha_{\delta\delta}\delta^2]} \\ &= N_{\rm B} G_1(\Delta\phi) G_2(\delta), \end{aligned} \tag{3.150}$$

where $N_{\rm B}$ is the number of particles in the bunch, the factor 3 is chosen to ensure that the phase-space area $\mathcal{A}$ of Eq. (3.141) corresponds to 95% of the beam particles, and the normalized distribution functions $G_1(\Delta\phi)$ and $G_2(\delta)$ are

$$G_1(\Delta\phi) = \sqrt{\frac{3(\alpha_{\phi\phi}\alpha_{\delta\delta} - \alpha_{\phi\delta}^2)}{\pi\alpha_{\delta\delta}}} \exp\{-\frac{3(\alpha_{\phi\phi}\alpha_{\delta\delta} - \alpha_{\phi\delta}^2)}{\alpha_{\delta\delta}} (\Delta\phi)^2\}$$

$$G_2(\delta) = \sqrt{\frac{3\alpha_{\delta\delta}}{\pi}} \exp\{-3\alpha_{\delta\delta}(\delta + \frac{\alpha_{\phi\delta}}{\alpha_{\delta\delta}}\Delta\phi)^2\}.$$

Note here that $G_1(\Delta\phi)$ is the line charge density, and the peak current is still located at $\Delta\phi = 0$. Using the ellipse of Eq. (3.140), we can evaluate the evolution of the peak current at the transition energy crossing.

IV.2 Nonlinear Synchrotron Motion at $\gamma \approx \gamma_{\rm T}$

In Sec. IV.1, all particles were assumed to cross transition energy at the same time. This is not true, because the phase slip factor depends on the off-momentum coordinate δ. Near the transition energy region, the nonlinear phase slip factor of Eq. (3.11) becomes quite important. Expanding the phase slip factor up to first order in δ, the synchrotron equations of motion become

$$\dot{\Delta\phi} = h\omega_0 \left(\frac{2\dot{\gamma}t}{\gamma_{\rm T}^3} + \eta_1\delta \right) \delta, \qquad \dot{\delta} = \frac{\omega_0 eV\cos\phi_{\rm s}}{2\pi\beta^2 E}(\Delta\phi), \tag{3.151}$$

where the synchronous particle crosses transition energy at time $t = 0$, and, to a good approximation, the phase slip factor has been truncated to second order in δ. At time

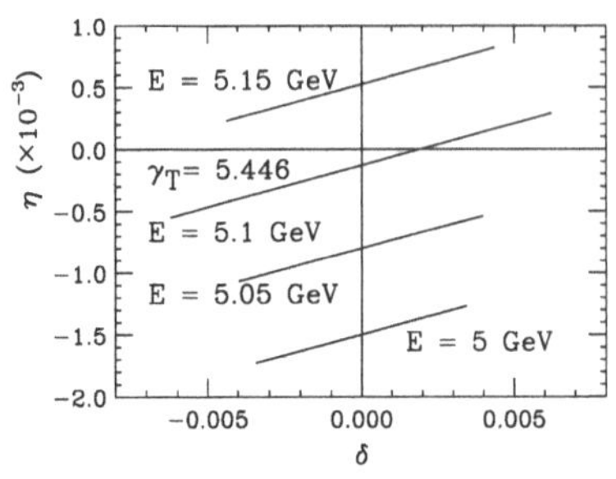

Figure 3.15: Schematic plot of η vs δ near the transition energy region for the Fermilab Booster, where $\gamma_{\rm T} = 5.446$, $\gamma_{\rm T}^2\alpha_1 = 0.5$, and a phase-space area of 0.05 eV-s are used to calculate $\eta(\delta)$ for the beam. A beam bunch is represented by a line of $\eta(\delta)$ vs δ. The synchrotron motion corresponds to particle motion along this line. At the beam synchronous energy of $E = 5.1$ GeV, which is below the transition energy of 5.11 GeV, particles at $\delta > 0.0018$ will experience unstable synchrotron motion due to the nonlinear phase slip factor.

$t = 0$, the synchronous phase is also shifted from $\phi_{\rm s}$ to $\pi - \phi_{\rm s}$ in order to achieve stable synchrotron motion.

Figure 3.15 shows the phase slip factor η vs the fractional off-momentum coordinate δ near transition energy for a beam in the Fermilab Booster. A beam bunch with momentum width $\pm\hat{\delta}$ is represented by a short tilted line. At a given time (or beam energy), particles are projected onto the off-momentum axis represented by this line. Since the phase slip factor is nonlinear, the line is tilted. When the beam is accelerated (or decelerated) toward transition energy, a portion of the beam particles can cross transition energy and this leads to unstable synchrotron motion, as shown in the example at 5.1 GeV beam energy in Fig. 3.15. Since the synchrotron motion is slow, we hope that the unstable motion does not give rise to too much bunch distortion before particles are recaptured into a stable bucket.

To characterize nonlinear synchrotron motion, we define the *nonlinear time* $\tau_{\rm nl}$ as the time when the phase slip factor changes sign for the particle at the maximum momentum width $\hat{\delta}$ of the beam, i.e. $\eta_0 + \eta_1\hat{\delta} = -(2\dot{\gamma}\tau_{\rm nl}/\gamma_{\rm T}^3) + \eta_1\hat{\delta} = 0$, or

$$\tau_{\rm nl} = \gamma_{\rm T}^3\frac{\eta_1}{2\dot{\gamma}}\hat{\delta} = \gamma_{\rm T}\frac{\frac{3}{2}\beta_0^2 + \gamma_{\rm T}^2\alpha_1}{2\dot{\gamma}}\hat{\delta}, \tag{3.152}$$

where $\hat{\delta}$ is the maximum fractional momentum spread of the beam, η_1 is obtained from Eq. (3.11), and the α_1 term can be adjusted by sextupoles. For a lattice without sextupole correction, we typically have $\gamma_{\rm T}^2\alpha_1 \approx 1$. Within the nonlinear time $\pm\tau_{\rm nl}$, some portions of the beam could experience unstable synchrotron motion. Note that the nonlinear time depends on the off-momentum width of the beam. Table 3.3 lists the nonlinear time of some accelerators, where $\alpha_1 = 0$ is assumed. Note that the nonlinear time for RHIC is particularly long because superconducting magnets can tolerate only a slow acceleration rate.

When the beam is accelerated toward transition energy to within the range

$$\gamma_{\rm T} - \dot{\gamma}\tau_{\rm nl} \le \gamma \le \gamma_{\rm T} + \dot{\gamma}\tau_{\rm nl},$$

the phase equation begins to change sign for particles at higher momenta while the phase angle ϕ_s has not yet been shifted. Therefore these particles experience defocusing synchrotron motion. After the synchronous energy of the bunch reaches transition energy and the synchronous phase has also been shifted from ϕ_s to $\pi - \phi_s$, lower momentum portions of the bunch will experience defocusing synchrotron motion. The problem is most severe for accelerators with a slow acceleration rate.

The relative importance of non-adiabatic and nonlinear synchrotron motions depends on the adiabatic time of Eq. (3.134) that governs the adiabaticity of the synchrotron motion, and the nonlinear time $\tau_{\rm nl}$, within which some portion of the beam particles experiences unstable synchrotron motion. Using Eq. (3.151), we obtain

$$\delta'' = -x\delta + \frac{\tau_{\rm nl}}{\tau_{\rm ad}}\frac{\delta^2}{\hat{\delta}}, \tag{3.153}$$

where the primes indicate derivatives with respect to $x = |t|/\tau_{\rm ad}$. Note that when the nonlinear time $\tau_{\rm nl}$ vanishes, the solution of Eq. (3.153) is an Airy function, discussed in Sec. IV.1. Since the solution of the nonlinear equation is not available, we estimate the growth of momentum width by integrating the unstable exponent. The growth factor is

$$G = \exp\left\{\int_0^{\tau_{\rm nl}/\tau_{\rm ad}} \left(-x + \frac{\tau_{\rm nl}}{\tau_{\rm ad}}\right)^{1/2} dx\right\} = \exp\left\{\frac{2}{3}\left(\frac{\tau_{\rm nl}}{\tau_{\rm ad}}\right)^{3/2}\right\} \tag{3.154}$$

for a particle with $\delta = \hat{\delta}$. The maximum momentum height is increased by the growth factor G, which depends exponentially on $\tau_{\rm nl}/\tau_{\rm ad}$. Depending on the adiabatic and nonlinear times, important beam dynamics problems are nonlinear synchrotron motion and of microwave instability to be discussed in Sec. VII.

We have seen that the momentum width will increase due to the nonlinear phase slip factor. However, we should bear in mind that the synchrotron motion can be derived from a Hamiltonian

$$H = \frac{1}{2}h\omega_0\left[\eta_0 + \frac{2}{3}\eta_1\delta\right]\delta^2 - \frac{\omega_0 eV}{4\pi^2\beta^2 E}\cos\phi_s(\Delta\phi)^2, \tag{3.155}$$

where $\eta_0 = 2\dot{\gamma}t/\gamma_{\rm T}^3$. Expressing Hamilton's equation as a difference mapping equation, we can easily prove that the Jacobian is 1. Therefore the area of the phase-space ellipse of each particle is conserved, and the 1D dynamical system is integrable. The action integral is a distorted curve in phase-space coordinates. When the bunch is accelerated through transition energy, some portions of the phase-space torus may lie outside the stable ellipse of the synchrotron Hamiltonian. They may be captured by other empty buckets of the rf system, or may be lost because of the aperture limitation. For a modern high intensity hadron facility, the loss would cause radiation problems; therefore efforts to eliminate transition energy loss are important.

IV.3 Beam Manipulation Near Transition Energy

Near the transition energy, the revolution frequencies of all particles are nearly identical, i.e. the beam is isochronous or quasi-isochronous. Since there is no frequency spread for Landau damping, the beam can suffer microwave instability. The tolerance of microwave instability near transition energy will be discussed in Sec. VII.

The nonlinear phase slip factor can cause defocusing synchrotron motion for a portion of the bunch. The growth of the bunch area is approximately $G^2 = \exp\{\frac{4}{3}(\tau_{\rm nl}/\tau_{\rm ad})^{3/2}\}$ shown in Eq. (3.154). The 5% beam loss at transition energy found for proton synchrotrons built in the 60's and 70's may arise mainly from this nonlinear effect. Bunched beam manipulation are usually needed to minimize beam loss and uncontrollable emittance growth. Minimizing both $\tau_{\rm ad}$ and $\tau_{\rm nl}$ provides cleaner beam acceleration through the transition energy.

A. Transition energy jump

By applying a set of quadrupoles, transition energy can be changed suddenly in order to attain fast transition energy crossing (see Chap. 2, Sec. IV.8). The effective $\gamma_{\rm T}$ crossing rate is $\dot\gamma_{\rm eff} = \dot\gamma - \dot\gamma_{\rm T}$. For example, if $\gamma_{\rm T}$ is changed by one unit in 1 ms,[27] the effective transition energy crossing rate is 1000 s^{-1}, which is much larger than the beam acceleration rates listed in Table 3.3.

Transition $\gamma_{\rm T}$ jump has been employed routinely in the CERN PS. The scheme has also been studied in the Fermilab Booster and Main Injector, the KEK PS, and the AGS. The minimum $\gamma_{\rm T}$ jump width is

$$\Delta\gamma_{\rm T} = 2\dot\gamma \times {\rm Max}(\tau_{\rm ad}, \tau_{\rm nl}). \tag{3.156}$$

B. Momentum aperture for faster beam acceleration

The synchronization of dipole field with synchronous energy is usually accomplished by a "radial loop," which provides a feedback loop for rf voltage and synchronous phase angle. In most accelerators, the maximum $\dot B$ is usually limited, but the rf voltage and synchronous phase angle can be adjusted to move the beam across the momentum aperture. The radial loop can be programmed to keep the beam closed orbit inside the nominal closed orbit below transition energy, and to attain faster acceleration across transition energy so that the beam closed orbit is outside the nominal closed orbit above transition energy. For an experienced machine operator to minimize the beam loss with a radial loop, the essential trick is to attain a faster transition energy crossing rate.

[27]The $\gamma_{\rm T}$ jump time scale is non-adiabatic with respect to synchrotron motion. However, the time scale can be considered as adiabatic in betatron motion so that particles adiabatically follow the new betatron orbit.

C. Flatten the rf wave near transition energy

Near transition energy, partial loss of focusing force in synchrotron motion can be alleviated by flattening the rf wave. This can be done by choosing $\phi_s = \pi/2$ or employing a second or third harmonic cavity.[28] In the flattened rf wave, all particles gain an equal amount of energy each turn, and thus δ of each particle is approximately constant in a small energy range. The solution of Eq. (3.151) with $\dot{\delta} = 0$ is

$$\Delta\phi = \Delta\phi_1 + \frac{h\omega_0\dot{\gamma}\delta}{\gamma_T^3}(t^2 - t_1^2) + \eta_1 h\omega_0\delta^2\,(t - t_1), \tag{3.157}$$

where t_1 is the rf flattening period, $(\Delta\phi_1, \delta_1)$ are the initial phase-space coordinates of the particle, and $\delta = \delta_1$.

Figure 3.16 shows the evolution of the phase-space torus when the rf wave is flattened across the transition energy region; the parameters used in this calculations are $\gamma_T = 22.5$, $\omega_0 = 4.917 \times 10^5$ rad/s, $\dot{\gamma} = 1.6$ s^{-1}, $h = 360$, $t_1 = -63$ ms, and $\gamma_T^2\eta_1 \approx 2$. Note that the ellipse evolves into a boomerang shaped distribution function with an equal phase-space area. The rf flattening scheme is commonly employed in isochronous cyclotrons.

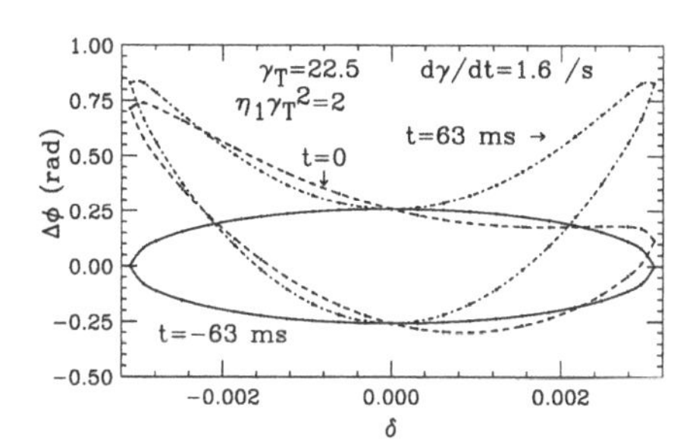

Figure 3.16: The evolution of a phase-space ellipse in the flattened rf wave near the transition energy region. Note that the off-momentum coordinates of each particle are unchanged, while the bunch length elongates along the ϕ axis.

IV.4 Synchrotron Motion with Nonlinear Phase Slip Factor

In the production of secondary beams, very short proton bunches are needed for attaining small emittance. Very short electron bunches, e.g. sub-millimeter in bunch length, have many applications such as time resolved experiments with synchrotron light sources, coherent synchrotron radiation, and damping rings for the next linear colliders. Since the ratio of bunch length to bunch height is proportional to $\sqrt{|\eta|}$, a possible method of producing short bunches is to operate the accelerator in an

[28] See e.g., C.M. Bhat *et al.*, *Phys. Rev.* E **55**, 1028 (1997). The AVF cyclotron has routinely employed this method for beam acceleration. This concept was patented by G.B. Rossi, U.S. Patent 2778937 (1954).

isochronous condition for proton synchrotrons, or to reduce the momentum compaction factor for electron storage rings. Because of its potential benefit of the low η condition, we carefully study the physics of the QI dynamical system.[29]

Table 3.2 and Eq. (3.43) show that the synchrotron bucket height and momentum spread become very large when $|\eta|$ is small. This requires careful examination because when the phase slip factor η is small, its dependence on the fractional momentum deviation δ becomes important. The synchrotron Hamiltonian needs to take into account the effects of nonlinear phase slip factor.

Expanding the phase slip factor as $\eta = \eta_0 + \eta_1\delta + \cdots$ and using the orbiting angle θ as the independent variable, we obtain the Hamiltonian for synchrotron motion as

$$H = \frac{1}{2}h\left(\eta_0 + \frac{2}{3}\eta_1\delta\right)\delta^2 + \frac{eV}{2\pi\beta^2 E}[\cos\phi - \cos\phi_s + (\phi - \phi_s)\sin\phi_s], \tag{3.158}$$

where we have truncated the phase slip factor to the second order in δ. The fixed points of the nonlinear synchrotron Hamiltonian are

$$(\phi, \delta)_{\rm SFP} = (\phi_s,\ 0), \qquad (\pi - \phi_s,\ -\eta_0/\eta_1), \tag{3.159}$$

$$(\phi, \delta)_{\rm UFP} = (\pi - \phi_s,\ 0), \quad (\phi_s,\ -\eta_0/\eta_1). \tag{3.160}$$

Note that the nonlinear phase slip factor introduces another set of fixed points in the phase space. The fixed points with $\delta_{\rm FP} = 0$ are the nominal fixed points. The fixed points with $\delta_{\rm FP} = -\eta_0/\eta_1$ arising from the nonlinear-phase-slip factor are called nonlinear-phase-slip-factor (NPSF) fixed points. These fixed points play important role in determining the dynamics of synchrotron motion.

We define $\nu_s = \sqrt{h|\eta_0|eV/2\pi\beta^2 E}$ for small amplitude synchrotron tune, and use the normalized phase space coordinates ϕ and $\mathcal{P} = (h\eta_0/\nu_s)\delta$. The Hamiltonian of synchrotron motion becomes

$$H = \frac{1}{2}\nu_s\mathcal{P}^2 + \frac{1}{2y}\nu_s\mathcal{P}^3 + \nu_s[\cos\phi - \cos\phi_s + (\phi - \phi_s)\sin\phi_s]. \tag{3.161}$$

The parameter

$$y = 3h\eta_0^2/2\eta_1\nu_s \tag{3.162}$$

signifies the relative importance of the linear and nonlinear parts of the phase slip factor. If $|y| \gg 1$, the nonlinear phase slip factor is not important, and if $|y|$ is small, the phase space tori will be deformed.

[29] A. Riabko *et al.*, *Phys. Rev.* **E54**, 815 (1996); D. Jeon *et al.*, *Phys. Rev.* **E54**, 4192 (1996); M. Bai *et al.*, *Phys. Rev.* **E55**, 3493 (1997); C. Pellegrini and D. Robin, *Nucl. Inst. Methods*, A **301**, 27 (1991); D. Robin, *et. al.*, *Phys. Rev.* **E48**, 2149 (1993); H. Bruck *et al.*, *IEEE Trans. Nucl. Sci.* **NS20**, 822 (1973); L. Liu *et al.*, *Nucl. Instru. Methods*, **A329**, 9 (1993); H. Hama, S. Takano and B. Isoyama, *Nucl. Instru. Methods*, **A329**, 29 (1993); S. Takano, H. Hama and G. Isoyama, *Japan J. Appl. Phys.* **32**, 1285 (1993); A. Nadji *et al.*, *Proc. EPAC94* p. 128 (1994); D. Robin, H. Hama, and A. Nadji, LBL-37758 (1995).

Figure 3.17 shows the separatrix of the nonlinear Hamiltonian in normalized phase space coordinates for $\phi_s = 150°$ and $180°$ respectively, where, without loss of generality, we have assumed $\eta_0 > 0$ and $\eta_1 > 0$. The separatrix that passes through the nominal fixed points are nominal separatrix. When the nominal separatrix crosses the unstable NPSF fixed point, the separatrix of two branches will become one (see the middle plots of Fig. 3.17 and Exercise 3.4.6). This condition occurs at $y = y_{cr}$, given by

$$y_{cr} = \sqrt{27[(\pi/2 - \phi_s)\sin\phi_s - \cos\phi_s]}. \tag{3.163}$$

For $y \gg y_{cr}$, the stable buckets of the upper and lower branches are separated by a distance of $\Delta\mathcal{P} = 2y/3$. Particle motion can be well described by neglecting the $\mathcal{P}^3$ term in the Hamiltonian.

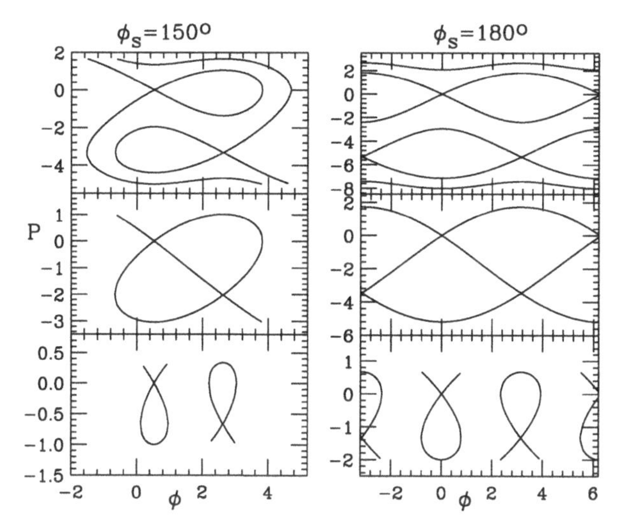

Figure 3.17: Left: Separatrix in the normalized phase space (ϕ, $P = \mathcal{P}$) for the synchrotron Hamiltonian with parameters $\phi_s = 150°$ and $y = 5$ (top), $y = y_{cr} = 3.0406$ (middle) and 1 (bottom); Right: Separatrix with parameters $\phi_s = 180°$ and $y = 8$ (top), $y = y_{cr} = 5.1962$ (middle), and 3 (bottom). In this example, we assume $\eta_0 > 0$ and $\eta_1 > 0$. Note the dependence of the Hamiltonian tori on the parameter y.

For $y < y_{cr}$, the separatrix ("fish") is deformed into up-down shape (see lower plots). They are called "α-bucket." Since the α-bucket is limited in a small region of the phase coordinate ϕ, small angle expansion is valid. The particle motion inside such a quasi-isochronous (QI) dynamical system can be analytically solved as follows.

IV.5 The QI Dynamical Systems

The synchrotron equation of motion for the rf phase coordinate ϕ of a particle is

$$\dot{\phi} = h\eta\delta, \qquad \eta = \eta_0 + \eta_1\delta + \cdots, \tag{3.164}$$

where h is the harmonic number, $\delta = \Delta p/p_0$ is the fractional momentum deviation from a synchronous particle, the overdot indicates the derivative with respect to the orbiting angle $\theta = s/R_0$, and η is the phase slip factor, η_0 and η_1 are the first order and second order phase slip factors. In many storage rings, truncation of the phase slip factor at the η_1 term is a good approximation. Similarly, the equation of motion for the fractional off-momentum deviation is

$$\dot{\delta} = \frac{eV_0}{2\pi\beta^2 E_0}[\sin(\phi+\phi_s) - \sin\phi_s] \approx \frac{eV_0\cos\phi_s}{2\pi\beta^2 E}\phi, \tag{3.165}$$

where V_0 and ϕ_s are the rf voltage and synchronous phase angle, βc is the speed, and E_0 is the energy of the beam. Here, the linearized phase coordinate in Eq. (3.165) is a good approximation because the (up-down) synchrotron bucket is limited in a small range of the phase coordinate (see Fig. 3.17).

With $t = \nu_s\theta$ as the time variable, where $\nu_s = \sqrt{heV_0|\eta_0\cos\phi_s|/2\pi\beta^2 E_0}$ is the small amplitude synchrotron tune, and with (x,p) as conjugate phase-space coordinates, where

$$x = -\frac{\eta_1}{\eta_0}\frac{\Delta p}{p_0}, \quad p = \frac{\nu_s\eta_1}{h\eta_0^2}\phi, \tag{3.166}$$

the synchrotron Hamiltonian for particle motion in QI storage rings becomes

$$H_0 = \frac{1}{2}p^2 + V(x), \qquad V(x) = \frac{1}{2}x^2 - \frac{1}{3}x^3. \tag{3.167}$$

This universal Hamiltonian is autonomous and the Hamiltonian value E is a constant of motion with $E \in [0, \frac{1}{6}]$ for particles inside the bucket.

The equation of motion for the QI Hamiltonian with $H_0 = E$ is the standard Weierstrass equation,

$$\left(\frac{d\wp(u)}{du}\right)^2 = 4(\wp - e_1)(\wp - e_2)(\wp - e_3), \tag{3.168}$$

where $u = t/\sqrt{6}$, $\wp = x$, and the turning points are

$$e_1 = \frac{1}{2} + \cos(\xi), \quad e_2 = \frac{1}{2} + \cos(\xi - 120^\circ), \quad e_3 = \frac{1}{2} + \cos(\xi + 120^\circ)$$

with $\xi = \frac{1}{3}\arccos(1 - 12E)$. The ξ parameter for particles inside the bucket varies from 0 to $\pi/3$. Figure 3.18 shows the separatrix of the QI bucket QI potential, and the turning points, where e_2 and e_3 are turning points for stable particle motion.

The Weierstrass elliptic $\wp$-function is a single valued doubly periodic function of a single complex variable. For particle motion inside the separatrix, the discriminant $\Delta = 648E(1-6E)$ is positive, and the Weierstrass $\wp$ function can be expressed in terms of the Jacobian elliptic function [30]

$$x(t) = e_3 + (e_2 - e_3)\,\mathrm{sn}^2\left(\sqrt{\frac{e_1 - e_3}{6}}t|m\right), \tag{3.169}$$

$$k = \frac{e_2 - e_3}{e_1 - e_3} = \frac{\sin\xi}{\sin(\xi + 60^\circ)}. \tag{3.170}$$

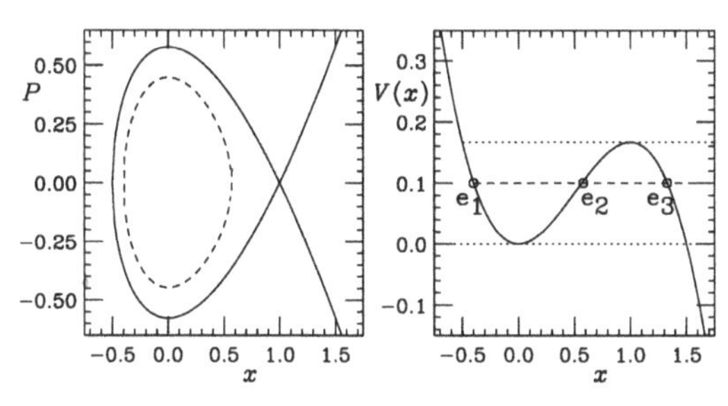

Figure 3.18: Schematic plots of the QI bucket (left) and the QI potential (right). The dotted lines are the lower and upper limits of the "energy" 0 and 1/6 respectively. The turning points e_1, e_2, and e_3 are also shown for energy $E = 0.1$, associated with the dashed-line beam bunch at the left plot. The separatrix of the QI bucket is one of the separatrix, plotted side-way, shown in Fig. 3.17.

The separatrix orbit, which corresponds to $k = 1$, is

$$x_{\rm sx}(t) = 1 - \frac{3}{\cosh t + 1}, \quad p_{\rm sx}(t) = \frac{3\sinh t}{(\cosh t + 1)^2}. \tag{3.171}$$

The tune of the QI Hamiltonian is

$$Q(E) = \frac{\pi[\sqrt{3}\sin(\xi + 60^\circ)]^{1/2}}{\sqrt{6}K(k)}, \tag{3.172}$$

where $K(k)$ is the Jacobian Elliptical function. The normalized tune of the QI Hamiltonian is compared with that of the normal synchrotron Hamiltonian in Fig. 3.19.

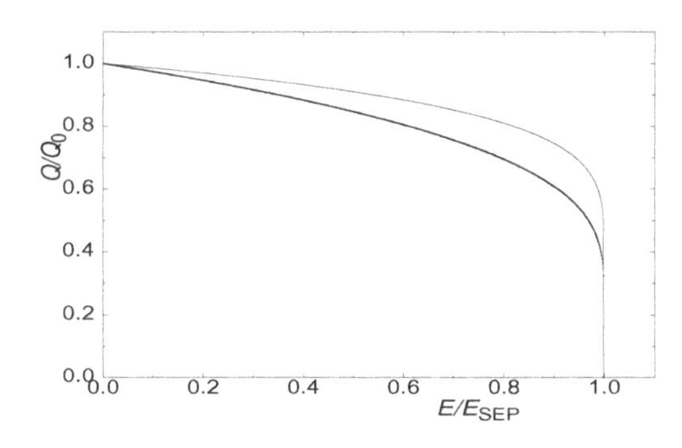

Figure 3.19: The synchrotron tune of the QI dynamical system (upper curve) is compared with that of the nominal rf potential (lower curve); plotted in relative Hamiltonian value, or "energy" $E/E_{\rm sep}$. Note that the sharp drop of the QI synchrotron tune at the separatrix can cause chaotic motion for particles with large synchrotron amplitudes under the influence of the low-frequency time-dependent perturbation.

Here, we note that the synchrotron tune decreases to zero very sharply near the separatrix. Because of the sharp decrease in synchrotron tune, time dependent perturbation will cause overlapping parametric resonances and chaos near the separatrix.[30]

[30]H. Huang, *et al.*, *Phys. Rev.* E**48**, 4678 (1993); M. Ellison, *et al.*, *Phys. Rev. Lett.* **70**, 591 (1993); M. Syphers, *et al.*, *Phys. Rev. Lett.* **71**, 719 (1993); Y. Wang, *et al.*, *Phys. Rev.* **E49**, 1610 (1994). D. Li, *et al.*, *Phys. Rev.* **E48**, R1638 (1993); D. Li, *et al.*, *Nucl. Inst. Methods*, **A364**, 205 (1995).

The action of a torus is

$$J = \frac{1}{2\pi}\oint p dx = \frac{1}{8}\sqrt{\frac{2}{3}}(e_2 - e_3)^2(e_1 - e_3)^{1/2} F\left(\frac{3}{2}, -\frac{1}{2}; 3; k\right), \tag{3.173}$$

where F is the hypergeometric function [30]. The action of the separatrix orbit is $J_{\rm sx} = 3/5\pi$, or equivalently the bucket area is 6/5. Using the generating function

$$F_2(x, J) = \int_{e_3}^{x} p\, dx, \tag{3.174}$$

the angle variable is $\psi = \partial F_2/\partial J = Qt$. The resulting Hamiltonian is

$$H_0(J) \approx J - \frac{5}{12}J^2 + \cdots.$$

Because of the synchrotron radiation damping, the equation of motion for QI electron storage rings is $x'' + Ax' + x - x^2 = 0$,where the effective damping coefficient is

$$A = \frac{\lambda}{\nu_{\rm s}} = \frac{U_0 J_E}{2\pi E_0 \nu_{\rm s}}. \tag{3.175}$$

Here λ is the damping decrement, U_0 is the energy loss per revolution, and J_E is the damping partition number. In QI storage rings, the effective damping coefficient is enhanced by a corresponding decrease in synchrotron tune, i.e. $A \sim |\eta_0|^{-1/2}$, where the value of A can vary from 0 to 0.5.

Including the rf phase noise, the Hamiltonian in normalized phase-space coordinates is

$$H = \frac{p^2}{2} + \frac{1}{2}x^2 - \frac{1}{3}x^3 + \omega_{\rm m} B x \cos\omega_{\rm m} t, \tag{3.176}$$

where

$$B = \frac{\eta_1 a}{\eta_0 \nu_{\rm s}} \tag{3.177}$$

is the effective modulation amplitude, a is the rf phase modulation amplitude, $\omega_{\rm m} = \nu_{\rm m}/\nu_{\rm s}$ is the normalized modulation tune, and $\nu_{\rm m}$ is the modulation tune of the original accelerator coordinate system. Note that the effective modulation amplitude B is greatly enhanced for QI storage rings by the smallness of η_0, i.e. $B \sim |\eta_1|/|\eta_0|^{3/2}$. Including the damping force, the equation of motion becomes

$$x'' + A\,x' + x - x^2 = -\omega_{\rm m}\, B \cos\omega_{\rm m} t. \tag{3.178}$$

The stochasticity of such a dynamical system has been extensively studied.[31] Experimental verification of the QI dynamical system has not been fully explored. Detailed discussions of this topic is beyond this introductory textbook.

---

[31] A. Riabko *et al.*, *Phys. Rev.* **E54**, 815 (1996); D. Jeon *et al.*, *Phys. Rev.* **E54**, 4192 (1996); M. Bai *et al.*, *Phys. Rev.* **E55**, 3493 (1997).

Exercise 3.4

1. Verify the adiabatic time, the nonlinear time, and the momentum spread of the beam $\hat{\delta}$ at $\gamma = \gamma_T$ for the accelerators listed in Table 3.3.

2. Show that Eq. (3.135) can be reduced to

$$\delta'' + x\delta = 0$$

where the primes indicate derivatives with respect to the variable $x = |t|/\tau_{\rm ad}$, where $\tau_{\rm ad}$ is the adiabatic time of Eq. (3.134).

   (a) Express the solution in terms of Airy functions and find the equation for the invariant torus.

   (b) Verify Eq. (3.140).

3. Show that $\tau_{\rm nl}/\tau_{\rm ad} \propto \dot{\gamma}^{-5/6}\gamma_{\rm T}^{-2/3}$. Discuss the effects of high vs low $\gamma_{\rm T}$ lattices on the dynamics of synchrotron motion near the transition energy.

4. The Fermilab Main Injector accelerates protons from 8.9 GeV to 120 GeV in 1 s. Assuming $\gamma_T = 20.4$, calculate the characteristic time and the maximum momentum spread for a phase space area of 0.04 eV-s.

5. Show that the phase space area enclosed by $(\Delta\phi, \delta)$ of Eq. (3.157) is equal to the phase space area enclosed by $(\Delta\phi_1, \delta_1)$ of the initial ellipse.

6. Using the normalized phase space coordinates ϕ and $\mathcal{P}$, show that the Hamiltonian (3.158) with nonlinear phase slip factor depends only on a single parameter $y = 3h\eta_0^2/2\nu_{\rm s}\eta_1$. Show that the separatrices of the Hamiltonian are

$$\nu_s\mathcal{P}^3 + y\mathcal{P}^2 + 2y[\cos\phi + \cos\phi_{\rm s} + (\phi + \phi_{\rm s} - \pi)\sin\phi_{\rm s}] = 0,$$
$$\nu_s\mathcal{P}^3 + y\mathcal{P}^2 + 2y[\cos\phi - \cos\phi_{\rm s} + (\phi - \phi_{\rm s})\sin\phi_{\rm s}] - \frac{4}{27}y^3 = 0.$$

Show that when $y = y_{\rm cr}$ of Eq. (3.163) the separatrix of the upper branch passes through the UFP of the lower branch.

7. Show that the QI Hamiltonian can be reduced to Eq. (3.167) and that the solution is given by the Weierstrass elliptical function.

V Beam Manipulation in Synchrotron Phase Space

A charged particle beam is usually produced by an intense ion source, pre-accelerated by an electrostatic Cockcroft-Walton or an RFQ, prebunched and injected into a linac to reach an injection energy for low energy synchrotrons, called booster synchrotrons or boosters. The beam is accumulated, phase-space painted, stacked in a low energy booster, and accelerated toward higher energies by a chain of synchrotrons of various sizes. The reasons for this complicated scheme are economics and beam dynamics issues. Since dipole and quadrupole magnets have low and high field operational limits, the range of beam energy for a synchrotron is limited. The mean-field Coulomb force can also have a large effect on the stability of low energy beams in boosters, where the space-charge tune shift, proportional to circumference of the synchrotron, is limited to about 0.3–0.4.

For the acceleration of ion beams, the fractional change of beam velocity in low energy boosters can be large. The rf frequency for a low energy booster has to be tuned in a wide range. The rf voltage requirement is determined by technical issues such as rf cavity design, rf power source, and the requirements in the momentum aperture and phase-space area. During beam acceleration, phase-space area is normally conserved. The beam distribution function can thus be manipulated to some desirable properties for experiments. Careful consideration is thus needed to optimize the operation and construction costs of accelerators.

On the other hand, electrons are almost relativistic at energies above 10 MeV, and the required range of rf frequency change is small. However, electrons emit synchrotron radiation, which must be compensated by the longitudinal rf electric field in a storage ring. Since synchrotron radiation power depends on particle energy, and the mean energy loss of a beam is compensated by the rf field, particle motion in synchrotron phase-space is damped. The synchrotron radiation emitted by a relativistic electron is essentially concentrated in a cone with an angular divergence of $1/\gamma$ along its path, and the energy compensation of the rf field is along the longitudinal direction; the betatron motion is also damped. Equilibrium is reached when the quantum fluctuation due to the emission of photons and the synchrotron radiation damping are balanced. The resulting momentum spread is independent of the rf voltage, and the transverse emittance depends essentially on the lattice arrangement.

In this section we examine applications of the rf systems in the bunched beam manipulations, including phase displacement acceleration, phase-space stacking, adiabatic capture, bucket to bucket transfer, bunch rotation, and debunching. We carefully study the double rf systems, that have often been applied in the space charge dominated beams and high brilliance electron storage rings for providing a larger tune spread for Landau damping. We also study the barrier rf systems that have been proposed for low energy proton synchrotrons. In general, innovative bunched beam manipulation schemes can enhance beam quality for experiments.

V.1 RF Frequency Requirements

Particle acceleration in synchrotrons requires synchronism between rf frequency and particle revolution frequency. Thus the rf frequency is an integer multiple of the revolution frequency $\omega_{\rm rf} = h\omega_0(B, R_0)$, where h is the harmonic number, and the angular revolution frequency ω_0 is a function of the magnetic field B and the average radius of the synchrotron R_0. The momentum p_0 of a particle is related to the magnetic field by $p_0 = e\rho B$, where ρ is the bending radius, and e is the particle's charge. Thus the rf frequency is

$$\begin{aligned} \omega_{\rm rf} &= h\frac{\beta c}{R_0} = \frac{he\rho B}{R_0\gamma m} = \frac{hc}{R_0}\left[\frac{B^2(t)}{B^2(t) + (mc^2/ec\rho)^2}\right]^{1/2}, \qquad (3.179) \\ \frac{mc^2}{ec\rho} &= \begin{cases} 3.1273/\rho\,[\mathrm{m}]\ \text{Tesla} & \text{for protons,} \\ 0.001703/\rho\,[\mathrm{m}]\ \text{Tesla} & \text{for electrons,} \end{cases} \end{aligned}$$

where m is the particle's mass. The rf frequency is a function of the dipole magnetic field, particularly particularly important for low energy proton or ion accelerators. In low to medium energy synchrotrons, the rf system is usually limited by the range of required frequency swing. Table 3.4 lists parameters of some proton synchrotrons.

Table 3.4: RF parameters of some proton synchrotrons

	AGS BST	AGS	RHIC	FNALBST	FNALMI
Inj. K.E. [GeV/u]	0.001/0.2	0.2(1.5)	12	0.2(0.4)	8.0
Acc. Rate [GeV/s]	100	60	3.7	200	100
Max. K.E. [GeV]	1.5	30	250	8	500
$f_{\rm rf}$ [MHz]	0.18–4.1	2.4–4.6	26.68–26.74	30.0–52.8	52.8–53.1
Av. Radius [m]	$(1/4)R_{\rm ags}$	128.457	$(19/4)R_{\rm ags}$	75.47	528.30
h	1–3 (2)	12 (8)	6×60	84	7×84
$V_{\rm rf}$ [kV]	90	300	300	950	4000

In some applications, the magnetic field can be ramped linearly as $B = a + bt$, or resonantly as $B = (\hat{B}/2)(1 - \cos\omega t) = \hat{B}\sin^2(\omega t/2)$, with ramping frequency $\omega/2\pi$ varying from 1 Hz to 50 Hz; the rf frequency should follow the magnetic field ramp according to Eq. (3.179), for which cavities with ferrite tuners are usually used. On the other hand, electrons are nearly relativistic at all energies, and the rf frequency swing is small. High frequency pill-box-like cavities are usually used. Normally the frequency range can be in the 200, 350, 500, and 700 MHz regions, where rf power sources are readily available. In recent years, wideband solid state rf power sources and narrowband klystron power sources have been steadily improved. New methods of beam manipulation can be employed.

Requirements of rf systems depend on their applications. To achieve high beam power in meson factories and proton drivers for spallation neutron sources, a fast acceleration rate is important. For example, the ISIS at the Rutherford Appleton Laboratory has a 50 Hz ramp rate, whereas the rf systems in the Spallation Neutron Source (SNS) provide only beam capture. On the other hand, acceleration rate is less important in storage rings used for internal target experiments.

A. The choice of harmonic number

The harmonic number determines the bunch spacing and the maximum number of particles per bunch obtainable from a given source, which can be important for colliding beam facilities. The harmonic numbers are related by the mean radii of the chain of accelerators needed to reach an efficient box-car injection scheme, which is equivalent to bucket to bucket transfer from one accelerator to another. For example, the average radius of the AGS Booster is 1/4 that of the AGS, and the ratio of harmonic numbers is 4. Similar reasoning applies to the chain of accelerators.

Since the damping time of electron beams in electron storage rings (see Table 4.2, Chap. 4, Sec. I.4) is short, the injection scheme of damping accumulation at full energy is usually employed in high performance electron storage rings. The choice of harmonic number for high energy electron storage rings is determined mainly by the availability of the rf power source, efficient high quality cavity design, and the size of the machine. Since rf power sources are available at 200, 350, 500, 700 MHz regions, most of the rf cavities of electron storage rings are operating at these frequencies. The harmonic number is then determined by the rf frequency and circumference of the storage ring.

B. The choice of rf voltage

High intensity beams usually require a larger bunch area to control beam instabilities. Since the rf bucket area and height are proportional to $\sqrt{V_{\rm rf}}$, a minimum voltage is needed to capture and accelerate charged particles efficiently.

In electron storage rings, the choice of rf voltage is important in determining the beam lifetime because of quantum fluctuation and Touschek scattering, a large angle Coulomb scattering process converting the horizontal momenta of two electrons into longitudinal momenta.

In general, the rf voltage is limited by the rf power source and the Kilpatrick limit of sparking at the rf gap. The total rf voltage of synchrotrons and storage rings is usually limited by the available space for the installation of rf cavities.

V.2 Capture and Acceleration of Proton and Ion Beams

At low energy, the intensity and brightness of an injected beam are usually limited by space-charge forces, intrabeam scattering, microwave instability, etc.; phase-space painting for beam distribution manipulation can be used to alleviate some of these problems (see Chap. 2, Sec. III.8, for transverse phase-space painting).

Since the injected beam from a linac normally has a large energy spread, the rf voltage requirement in booster synchrotrons needs enough bucket height for beam injection. The peak voltage is usually limited by the power supply and electric field breakdown at the rf cavity gap. A debuncher or a bunch rotator in the transfer line can be used to lower the momentum width of injected beams. The resulting captured beam brightness depends on the rf voltage manipulation. The following example illustrates the difference between adiabatic capture and non-adiabatic capture processes.

A. Adiabatic capture

During multi-turn injection (transverse or longitudinal phase-space painting or charge exchange strip injection), very little beam loss in the synchrotron phase-space can theoretically be achieved by adiabatically ramping the rf voltage with $\phi_s = 0$. The right plots, (e) to (h), of Fig. 3.20 show an example of adiabatic capture in the IUCF cooler injector synchrotron (CIS). The proton beam was accelerated from 7 to 200 MeV at 1 Hz repetition rate, and the rf voltage $V_{\rm rf}(t)$ was increased from a small value to 240 V adiabatically, plots (e) and (f), while the synchronous phase was kept at zero. The adiabaticity coefficient of Eq. (3.21) becomes

$$\alpha_{\rm ad} = \frac{T_s}{4\pi V_{\rm rf}}\frac{dV_{\rm rf}}{dt} = \frac{T_s}{2\pi \mathcal{A}_{\rm B}}\frac{d\mathcal{A}_{\rm B}}{dt}, \tag{3.180}$$

where T_s is the synchrotron period and $\mathcal{A}_{\rm B}$ is the bucket area. In order to satisfy the adiabatic condition, the initial rf voltage should have a small finite initial voltage V_0, and the rf voltage is ramped to a final voltage smoothly (see also Exercise 3.5.1).

After beam capture, the synchronous phase was ramped adiabatically to attain a desired acceleration rate. Good acceleration efficiency requires adiabatic ramping of V and ϕ_s while providing enough bucket area during beam acceleration. In this numerical example, we find that the capture efficiency is about 99.6%. In reality, the momentum spread of the injected beam is about 0.5% instead of 0.1% shown in this example. The maximum voltage is only barely able to hold the momentum spread of the injected beam from linac. The actual capture efficiency is much lower. A possible solution is to install a debuncher in the injection transfer line for lowering the momentum spread of the injected beam.

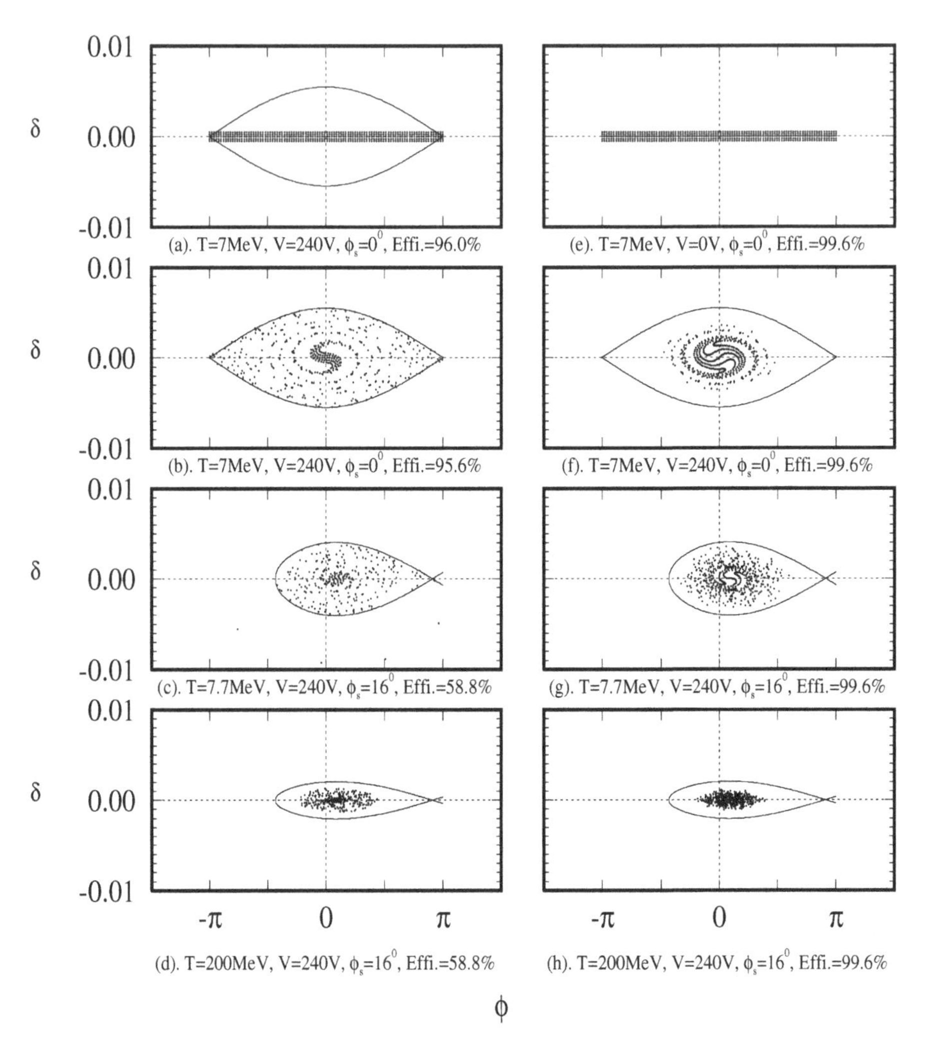

Figure 3.20: The left plots, (a) to (d), show non-adiabatic beam capture during injection and acceleration. The right plots, (e) to (h), show adiabatic capture of the injected beam: the rf voltage is ramped from 0 to 240 V adiabatically to capture the injected beam with a momentum spread of 0.1%. The rf synchronous phase is then ramped adiabatically to achieve the required acceleration rate. The actual momentum spread of the injected beam is about ±0.5%, and thus the actual adiabatic capture efficiency is substantially lower. Space-charge force and microwave instability are not included in the calculation. This calculation was done by X. Kang (Ph.D. Thesis, Indiana University, 1998).

B. Non-adiabatic capture

The left plots, (a) to (d), of Fig. 3.20 show an example of non-adiabatic capture with nonzero initial rf voltage. When the rf voltage is set to 240 V to capture the injected beam, the beam fills up the entire phase-space, as shown in plot (b). Beam loss occurs during acceleration, the final phase-space area is larger, and the capture efficiency is low. With microwave instability and space-charge effects included, the capture efficiency may be even lower.

As seen in plots (b) and (f), the injected beam particles decohere and fill up the entire bucket area because of synchrotron tune spread. The decoherence results in emittance growth.

C. Chopped beam at the source

Many fast cycling synchrotrons require nonzero rf voltage and nonzero rf synchronous phase $\phi_s > 0$ to achieve the desired acceleration rate. In this case, capture efficiency is reduced by the nonadiabatic capture process. To circumvent low efficiency, a beam chopper consisting of mechanical or electromagnetic deflecting devices, located at the source, can be used to paint the phase-space of the injected beam and eliminate beam loss at high energy.

V.3 Bunch Compression and Rotation

When a bunch is accelerated to its final energy, it may be transferred to another accelerator or used for research. When the beam is transferred from one accelerator to another, the beam profile matching condition is

$$\left[\frac{\hat{\delta}}{R\hat{\theta}}\right]_{\text{acc.1}} = \left[\frac{\hat{\delta}}{R\hat{\theta}}\right]_{\text{acc.2}}, \quad \text{or} \quad \left[\frac{1}{R}\sqrt{\frac{hV}{|\eta|}}\right]_{\text{acc.1}} = \left[\frac{1}{R}\sqrt{\frac{hV}{|\eta|}}\right]_{\text{acc.2}}. \tag{3.181}$$

This matching condition may be higher than the limit of a low frequency rf system. Similarly, the bunch length of a beam may need to be shortened in many applications. A simple approach is to raise the voltage of the accelerator rf system. However, the peak voltage of an rf system is limited by the breakdown of electric field at the acceleration gap. According to the empirical Kilpatrick criterion, the rf frequency f [MHz] is related to the peak electric field gradient $E_{\rm K}$ [MV/m] by

$$f = 1.64\, E_{\rm K}^2\, e^{-8.5/E_{\rm K}}. \tag{3.182}$$

Because of this limitation, we have to use different beam manipulation techniques such as bunch compression by rf gymnastics, etc.

Bunched beam gymnastics are particularly important for shortening the proton bunch before the protons hit their target in antiproton or secondary beam production.

Generally, the emittance of secondary beams is equal to the product of the momentum aperture of the secondary-beam capture channel and the bunch length of the primary beam. When the bunch length of a primary proton beam is shortened, the longitudinal emittance of the secondary antiproton beam becomes smaller. The antiproton beam can be further debunched through phase-space rotation in a debuncher by converting momentum spread to phase spread, and the final antiproton beam is transported to an accumulator for cooling accumulation (see Exercise 3.5.3).

Beam bunch compression is also important in shortening the electron bunch in order to minimize the beam breakup head-tail instabilities in a linac (see Sec. VIII). A few techniques of bunch compression are described below.

A. Bunch compression by rf voltage manipulation

The first step it to lower the rf voltage adiabatically, e.g. $V_0 \to V_1$, so that the bucket area is about the same as the bunch area. Then the rf voltage is increased non-adiabatically from V_1 to V_2. The unmatched beam bunch rotates in synchrotron phase-space. At 1/4 or 3/4 of the synchrotron period, a second rf system at a higher harmonic number is excited to capture the bunch, or a kicker is fired to extract beams out of the synchrotron. Figure 3.21 shows schematic phase-space ellipses during the bunch compression process. The lower-left plot shows the final phase-space ellipse in an idealized linear synchrotron motion. In reality, the maximum attainable rf voltage is limited, and the final phase-space ellipse is distorted by the nonlinear synchrotron motion that causes emittance dilution.

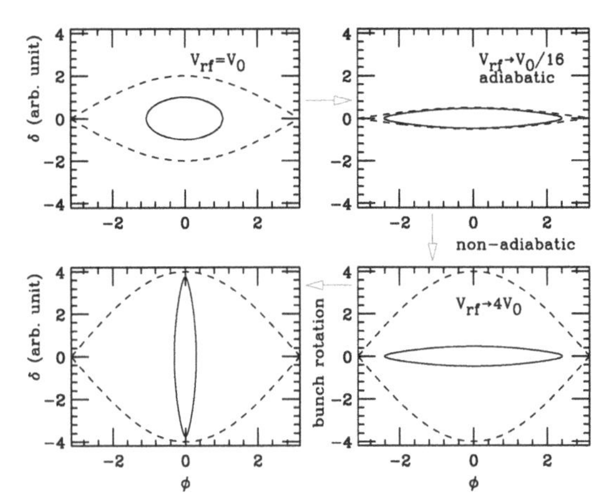

Figure 3.21: Schematic drawings (clockwise) of bunch compression scheme using rf voltage manipulation. The bunch area is initially assumed to be about 1/5 of the bucket area (top-left). The voltage is adiabatically reduced by 16 times so that the bunch is almost fill the bucket area (top-right). As the voltage is non-adiabatically raised to four times the original rf voltage, the mismatched bunch begins to rotate. When the bunch length is shortened (lower-left) at 1/4 of the synchrotron period, a kicker can be fired to extract the beam.

For a given bunch area, the rms bunch width and height are obtained from Eq. (3.42) during the adiabatic rf voltage compression from V_0 to V_1. After the rf voltage is jumped to V_2, the bunch height will become bunch width according to

Eq. (3.42). The maximum bunch compression ratio, defined as the ratio of the bunch lengths at $(V_0 = V_2) \to V_1 \to V_2$, becomes

$$r_c = \frac{\sigma_{\phi,\mathrm{i}}}{\sigma_{\phi,\mathrm{f}}} = \left(\frac{V_2}{V_1}\right)^{1/4} = \left(\frac{\mathcal{A}_{\mathrm{bucket,max}}}{\mathcal{A}_{\mathrm{bunch}}}\right)^{1/2} = \frac{2\sqrt{2}}{\sqrt{3\pi}\sigma_\phi}, \tag{3.183}$$

where we have used the properties that the bunch area supposedly fills up the bucket area at $V_{\mathrm{rf}} = V_1$, and the fact that the bucket area is 16 (see Table 3.2), and the bunch area containing 95% of the beam is $6\pi\sigma_\phi^2$ in the normalized synchrotron phase space coordinates.

B. Bunch compression using unstable fixed point

If the rf phase is shifted so that the unstable fixed point (UFP) is located at the center of the bunch, the bunch will begin compressing in one direction and stretching in the other direction along the separatrix orbit (see Sec. II.4 and Exercise 3.2.5). In linear approximation, the bunch length and bunch height change according to $\exp\{\pm\omega_s t_{\mathrm{ufp}}\} = \exp\{\pm 2\pi t_{\mathrm{ufp}}/T_s\}$, where ω_s is the small amplitude synchrotron angular frequency, T_s is the synchrotron period, and t_{ufp} is the time-duration that the bunch stays at the UFP. The length of stay at the UFP can be adjusted to attain a required aspect ratio of the beam ellipse.

When the SFP of the rf potential is shifted back to the center of the bunch. The mis-matched bunch profile will begin to execute synchrotron motion. At 3/8 of the synchrotron period, the bunch can be captured by a matched high frequency rf system or kicked out of the accelerator by fast extraction.

We now derive the ultimate bunch compression ratio for the rf phase shift method as follows. In the normalized phase-space coordinates, ϕ and $\mathcal{P} = -(h|\eta|/\nu_s)(\Delta p/p)$, the Hamiltonian for stationary synchrotron motion is given by Eq. (3.64).

Near the UFP, the separatrix of the Hamiltonian in Eq. (3.64) can be approximated by two straight lines crossing at 45° angles with the horizontal axis ϕ. When the rf phase is shifted so that the beam sits on the UFP, the bunch width and height will stretch and compress along the separatrix. The rate of growth is equal to $\exp(\omega_s t_{\mathrm{ufp}})$. The maximum rf phase coordinate $\phi_{\max}$ that a bunch width can increase and still stay within the bucket after the rf phase is shifted back to SFP is given approximately by

$$\frac{1}{2}\phi_{\max}^2 + 2\sin^2\left(\frac{\phi_{\max}}{2}\right) \approx 2, \tag{3.184}$$

where we assume linear approximation for particle motion near SFP. Thus we obtain $\phi_{\max} \approx \sqrt{2}$. Using Liouville's theorem, conservation of phase-space area, we find

$$\pi\sigma_{\phi,\mathrm{i}}^2 = \pi\sigma_{\mathcal{P},\mathrm{f}}\sigma_{\phi,\mathrm{f}}. \tag{3.185}$$

Assuming that 95% of the beam particles reach $\phi_{\rm max} = \sqrt{2}$ so that $\sigma_{\mathcal{P},{\rm f}} = \frac{\sqrt{2}}{\sqrt{6}}\phi_{\rm max} = \sqrt{2/3}$, we find the compression ratio as

$$r^{\rm p}_{\rm c,max} = \frac{\sigma_{\phi,{\rm i}}}{\sigma_{\phi,{\rm f}}} = \frac{\sigma_{\mathcal{P},{\rm f}}}{\sigma_{\phi,{\rm i}}} \approx \frac{\sqrt{2}}{\sqrt{3}\sigma_{\phi,{\rm i}}}, \tag{3.186}$$

The time needed to reach this maximum compression ratio is

$$\omega_{\rm s}\hat{t}_{\rm ufp} = \ln\frac{1}{\sigma_{\phi,{\rm i}}} - 0.203. \tag{3.187}$$

A difficulty associated with bunch compression using rf phase-shift is that the rf voltage may remain at a relatively low value during the bunch rotation stage. The effect of non-linear synchrotron motion will be more important because the ratio of bucket-area to the bunch-area is small.

The difficulty of nonlinear synchrotron motion in the final stage of bunch rotation can be solved by using the buncher in the transport line. After proper bunch compression, the beam is kicked out of the synchrotron and the R_{56} transport matrix element will compress bunch, i.e. lower energy particles travel shorter path, and the higher energy particle travel a longer path. However, the resulting compression ratio is reduced by a factor of $1/\sqrt{2}$. Since there is no constraint that the final bunch size should fit into the bucket, one can regain the factor of $\sqrt{2}$ in staying longer at the UFP.

C. Bunch rotation using buncher/debuncher cavity

The principle of bunch rotation by using a buncher/debuncher cavity is based on the correlation of the time and off-momentum coordinates (the transport element R_{56}). By employing a cavity to accelerate and decelerate parts of the beam bunch, the bunch length and the momentum spread can be adjusted. This method is commonly used in the beam transfer line. For example, a simple debuncher used to decrease the energy spread of a non-relativistic beam out of a linac can function as follows. First, let the beam drift a distance L so that higher energy particles are ahead of lower energy particles. A cavity that decelerates leading particles and accelerates trailing particles can effectively decrease the energy spread of the beam.

For relativistic particles, a drift space can not provide the correlation for the transport element R_{56} because all particles travel at almost the same speed. It requires bending magnets for generating local dispersion functions so that the path length is correlated with the off-momentum coordinate.[32] A buncher/debuncher cavity can then be used to shorten or lengthen the bunch.

[32]See e.g., T. Raubenheimer, P. Emma, and S. Kheifets, *Proc. 1993 PAC*, p. 635 (1993).

V.4 Debunching

When rf systems are non-adiabatically turned off, particles drift and fill up the entire ring because the rotation frequency depends on the off-momentum variable. The debunching rate is $\dot{\phi} = h\omega_0\eta\delta$. Neglecting synchrotron radiation loss, the momentum spread will not change. The bunch shape will be distorted because particles of higher and lower momenta drift in different directions. The debunching time can be expressed as

$$T_{\rm db} = 2\pi/h\omega_0\eta\hat{\delta}, \tag{3.188}$$

where $\hat{\delta}$ is the maximum momentum spread of a beam. Note that the momentum spread of the entire beam remains the same in this non-adiabatic debunching process.

To reduce the momentum spread in the debunching process, we can adiabatically lower the rf voltage. In this case, the resulting debunched beam has a smaller momentum spread. The phase-space area remains the same if we can avoid collective beam instability.

V.5 Beam Stacking and Phase Displacement Acceleration

The concept of beam stacking is that groups of particles are accelerated to a desired energy and left to circulate in a fixed magnetic field; and subsequent groups are accelerated and deposited adjacent to each other. The accumulated beam will overlap in physical space at special locations, e.g. small β and zero $D(s)$ locations, which increases the density and the collision rate. In a successful example of beam stacking in the ISR *pp* collider, a single beam current of 57.5 A was attained. To accomplish phase-space stacking, phase displacement acceleration is usually employed.[33]

In a Hamiltonian system, particles can not cross the separatrix, therefore particles outside the bucket can not be captured during acceleration. Since the magnetic field depends on rf frequency, only particles inside the stable rf bucket are accelerated toward high energy. Particles outside the rf bucket are lost in the vacuum chamber because of the finite magnet aperture.

What happens to the unbunched coasting beam outside the separatrix when an empty moving bucket is accelerated through the beam? Since the beam is outside the separatrix, it may not be captured into the bucket if the rf bucket acceleration is adiabatic. Particles flow along lines of constant action, and their energies are lowered. The change in energy is $\Delta E = \omega_0 \mathcal{A}/2\pi$. Similarly, when a bucket is decelerated toward lower energy, the beam energy will be displaced upward in phase-space, i.e. accelerated. Phase displacement acceleration has been used to accelerate coasting beams

[33] K.R. Symon and A.M. Sessler, *Methods of radio-frequency acceleration in fixed field accelerators with applications to high current and intersecting beam accelerators*, p. 44, CERN Symp. 1956; L.W. Jones, C.H. Pruett, K.R. Symon and K.M. Terwilliger, in *Proc. of Int. Conf. on High-Energy Accelerators and Instrumentation*, p. 58 (CERN, 1959).

in the Intersecting Storage Ring (ISR) at CERN[34] and to compensate synchrotron radiation loss in electron storage rings.

In a storage rings with electron cooling or stochastic cooling, a newly injected beam accelerated by phase displacement can be moved toward the cooling stack to achieve a high cooling rate. This method has been successfully used to accumulate polarized protons at low energy cooling storage rings, and to accumulate antiprotons at antiproton accumulators. For example, the cooling stacking method can enhance polarized proton intensity by a factor of 1000 in the IUCF Cooler.[35] Similarly, with phase displacement acceleration, antiprotons can be moved to the cooling stack for cooling accumulation.

V.6 Double rf Systems

Space charge has been an important limitation to beam intensity in many low energy proton synchrotrons. Space charge induces potential well distortion and generates coherent and incoherent betatron tune shifts, which may lower the thresholds for transverse and longitudinal collective instabilities. Fast beam loss may occur during accumulation and storage when the peak beam current exceeds a threshold value.

To increase the threshold beam intensity, a double rf system has often been used to increase the synchrotron frequency spread, which enhances Landau damping in collective beam instabilities. As early as 1971, an attempt was made to increase Landau damping by installing a cavity operating at the third harmonic of the accelerating frequency in the Cambridge Electron Accelerator (CEA).[36] This technique was also successfully applied to cure coupled bunch mode instabilities at ISR, where an additional cavity was operated at the sixth harmonic of the primary rf frequency.[37]

Adding a higher harmonic rf voltage to the main rf voltage can flatten the potential well. Since the equilibrium beam profile follows the shape of the potential well, a double rf system can provide a smaller *bunching factor*, defined as the fraction of the circumference occupied by a beam or the ratio of peak current to average current, than that of a single rf system. Therefore, for a given DC beam current in a synchrotron, the peak current and consequently the incoherent space-charge tune shift are reduced. For example, a double rf system with harmonics 5 and 10 was successfully used in

[34] A high current stack at the ISR has a momentum spread of about 3%, that can be handled by a low power rf system in the ISR. By employing the phase displacement acceleration, the circulating beams in ISR were accelerated from 26 GeV to 31.4 GeV without loss of luminosity. The installation of low-β superconducting quadrupoles in 1981 brought a record luminosity of 1.4×10^{32} cm^{2}s^{-1}. The machine stopped operation in December 1983, giving its way to a fully operational SP$\bar{\text{P}}$S, that observed its first $p\bar{p}$ collision at the center of mass energy of 540 GeV on July 10, 1981.

[35] A. Pei, Ph.D. Thesis, Indiana University (1993).

[36] R. Averill *et al., Proc. 8th Int. Conf. on High Energy Accelerators*, p. 301 (CERN 1971).

[37] P. Bramham *et al., Proc. 9th Int. Conf. on High Energy Accel.* (CERN, 1974); P. Bramham *et al., IEEE Trans. Nucl. Sci.* **NS-24**, 1490 (1977).

the Proton Synchrotron Booster (PSB) at CERN to increase the beam intensity by 25 – 30% when the coherent longitudinal sextupole and decapole mode instabilities were suppressed by beam feedback systems.[38] At the Indiana University Cyclotron Facility (IUCF), a recent beam dynamics experiment showed that with optimized electron cooling the beam intensity in the cooler ring was quadrupled when two rf cavities were used.[39]

A. Synchrotron equation of motion in a double rf system

For a given particle at angular position θ relative to the synchronous angle θ_s, the phase angle of the primary rf system can be expressed as

$$\phi = \phi_{1s} - h_1(\theta - \theta_s), \tag{3.189}$$

where ϕ is the phase coordinate relative to the primary rf cavity, ϕ_{1s} is the phase angle for the synchronous particle, and h_1 is the harmonic number for the primary rf system. Similarly, the rf phase angle for the second rf system is

$$\phi_2 = \phi_{2s} - h_2(\theta - \theta_s) = \phi_{2s} + \frac{h_2}{h_1}(\phi - \phi_{1s}), \tag{3.190}$$

where h_2 is the harmonic number for the second rf system and ϕ_{2s} is the corresponding synchronous phase angle. The equation of motion becomes

$$\dot{\delta} = \frac{\omega_0 e V_1}{2\pi\beta^2 E}\left\{\sin\phi - \sin\phi_{1s} + \frac{V_2}{V_1}\left(\sin\left[\phi_{2s} + \frac{h_2}{h_1}(\phi - \phi_{1s})\right] - \sin\phi_{2s}\right)\right\}, \tag{3.191}$$

where the overdot is the derivative with respect to orbiting angle θ, and V_1 and V_2 are the voltages of the rf cavities. Equations (3.13) and (3.191) are Hamilton's equations of motion for a double rf system.

Using the normalized momentum coordinate $\mathcal{P} = -(h_1|\eta|/\nu_s)(\Delta p/p_0)$, the Hamiltonian is

$$\begin{aligned} H &= \frac{1}{2}\nu_s\mathcal{P}^2 + V(\phi), \\ V(\phi) &= \nu_s\{(\cos\phi_{1s} - \cos\phi) + (\phi_{1s} - \phi)\sin\phi_{1s} \\ &\quad -\frac{r}{h}\left[\cos\phi_{2s} - \cos\left(\phi_{2s} + h(\phi - \phi_{1s})\right) - h(\phi - \phi_{1s})\sin\phi_{2s}\right]\}. \end{aligned} \tag{3.192}$$

Here $\nu_s = \sqrt{h_1 e V_1|\eta|/2\pi\beta^2 E_0}$ is the synchrotron tune at zero amplitude for the primary rf system, $h = h_2/h_1$, $r = -V_2/V_1$, and ϕ_{1s} and ϕ_{2s} are the corresponding rf

[38]See J.M. Baillod *et al.*, *IEEE Trans. Nucl. Sci.* **NS-30**, 3499 (1983); G. Galato *et al*, *Proc. PAC*, p. 1298 (1987).

[39]See S.Y. Lee *et al.*, *Phys. Rev.* E**49**, 5717 (1994); J.Y. Liu *et al.*, *Phys. Rev.* E**50**, R3349 (1994); J.Y. Liu *et al.*, *Part. Accel.* **49**, 221-251 (1995).

phase angles of a synchronous particle. Here, the conditions $r = 1/h$ and $h \sin\phi_{2s} = \sin\phi_{1s}$ are needed to obtain a flattened potential well. For $r > 1/h$, there are two inner buckets on the ϕ axis. The effective acceleration rate for the beam is $\Delta E = eV_1(\sin\phi_{1s} - r\sin\phi_{2s})$ per revolution.

Because the rf bucket is largest at the lowest harmonic ratio, we study the double rf system with $h = 2$. To simplify our discussion, we study a stationary bucket with $\phi_{1s} = \phi_{2s} = 0°$. However, the method presented in this section can be extended to more general cases with $\phi_{1s} \neq 0$ and $\phi_{2s} \neq 0$.

B. Action and synchrotron tune

When the synchrotron is operating at $\phi_{1s} = \phi_{2s} = 0$, the net acceleration is zero and the Hamiltonian becomes

$$H = \frac{\nu_s}{2}\mathcal{P}^2 + \nu_s\left[(1-\cos\phi) - \frac{r}{2}(1-\cos 2\phi)\right]. \tag{3.193}$$

The fixed points $(\phi_{FP}, \mathcal{P}_{FP})$ are listed in Table 3.5.

Table 3.5: SFP and UFP of a double rf system.

	SFP	UFP
$0 \le r \le \frac{1}{2}$	(0,0)	$(\pi, 0)$
$\frac{1}{2} < r$	$(\pm\arccos(\frac{1}{2}), 0)$	$(0,0)$, $(\pm\pi, 0)$

Since the Hamiltonian is autonomous, the Hamiltonian value E is a constant of motion with $E/\nu_s \in [0, 2]$. The action is

$$J(E) = \frac{1}{\pi}\int_{-\hat{\phi}}^{\hat{\phi}} \mathcal{P}\, d\phi, \tag{3.194}$$

where $\hat{\phi}$ is the maximum phase angle for a given Hamiltonian torus. The value E is related to $\hat{\phi}$ by $E = 2\nu_s(1 - 2r\cos^2(\hat{\phi}/2))\sin^2(\hat{\phi}/2)$; the phase-space area is $2\pi J$; and the synchrotron tune is $Q_s = (\partial J/\partial E)^{-1}$. The bucket area $\mathcal{A}_b$ is

$$\mathcal{A}_b = 2\pi\hat{J} = 8\left[\sqrt{1+2r} + \frac{1}{\sqrt{2r}}\ln(\sqrt{1+2r} + \sqrt{2r})\right], \tag{3.195}$$

which is a monotonic increasing function of the ratio r. The corresponding bucket area for the single rf system is $\mathcal{A}_b(r \to 0) = 16$ (see Table 3.2).

C. The $r \le 0.5$ case

Changing the variables with

$$t = \tan\frac{\phi}{2}, \quad d\phi = \frac{2}{1+t^2}dt, \quad t_0 = \tan\frac{\hat{\phi}}{2}, \quad \tau = \frac{t}{t_0},$$

we obtain

$$\frac{\partial J}{\partial E} = \frac{2(1+t_0^2)}{\pi\nu_s t_0[1+(1+2r)t_0^2]^{1/2}} \int_0^1 \left[(1-\tau^2)\left(\frac{1-2r+t_0^2}{t_0^2[1+(1+2r)t_0^2]} + \tau^2\right)\right]^{-1/2} d\tau. \quad (3.196)$$

Thus the synchrotron tune becomes [31]

$$\frac{Q_s}{\nu_s} = \frac{\pi\sqrt{(1-2r)+2t_0^2+(1+2r)t_0^4}}{2(1+t_0^2)K(k_1)}, \quad (3.197)$$

where $K(k_1)$ is the complete elliptic integral of the first kind with modulus

$$k_1 = \frac{t_0\sqrt{1+(1+2r)t_0^2}}{\sqrt{(1-2r)+2t_0^2+(1+2r)t_0^4}}. \quad (3.198)$$

In fact, this formula is also valid for $r > 0.5$ and $\hat{\phi} > \phi_b$, where ϕ_b is the intercept of the inner separatrix with the phase axis.

D. The $r > 0.5$ case

For $r > 0.5$, the origin of phase-space $\mathcal{P} = \phi = 0$ becomes a UFP of the unperturbed Hamiltonian. Two SFPs are located at $\mathcal{P} = 0$ and $\phi = \pm\phi_f$, where $\cos(\phi_f/2) = 1/2r$. The inner separatrix, which passes through the origin, intersects the phase axis at $\pm\phi_b$ with $\cos(\phi_b/2) = 1/\sqrt{2r}$. Figure 3.22 shows ϕ_b and ϕ_f and some phase space ellipses for $r = 0.6$.

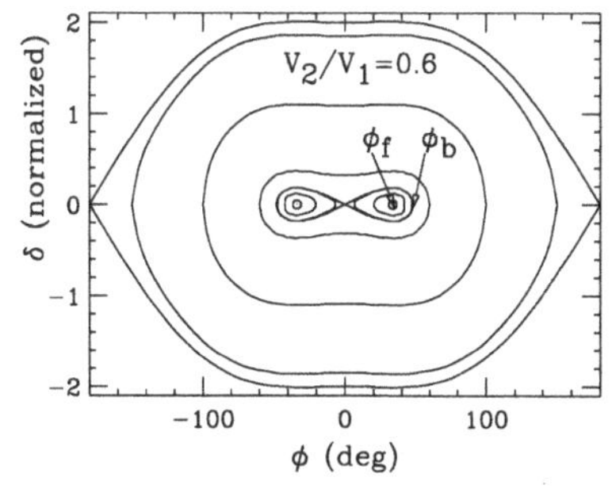

Figure 3.22: RF bucket and phase space ellipses for a double rf system with $h_2/h_1 = 2$ and $r = V_2/V_1 = 0.6$. Two stable fixed points are located at $(\pm\phi_f, 0)$, where $\phi_f = \cos^{-1}(1/2r)$. The maximum phase amplitude of the inner separatrix is $\phi_b = \cos^{-1}(1/\sqrt{2r})$. For $r \gg 1$, the SFP $\phi_f \to \pm\frac{\pi}{2}$, and the bucket is split into 2.

A given torus inside the inner bucket corresponds to a Hamiltonian flow of constant Hamiltonian value. Let ϕ_l and ϕ_u be the lower and upper intercepts of a torus with the phase axis, where $\phi_{\rm u} = \hat{\phi}$ and $\sin(\phi_{\rm l}/2) = \sqrt{\sin^2(\phi_{\rm b}/2) - \sin^2(\phi_{\rm u}/2)}$. The derivative of the action with respect to the energy for the torus becomes

$$\frac{\partial J}{\partial E} = \frac{\sqrt{(1+t_{\rm u}^2)(1+t_{\rm l}^2)}}{\pi\nu_{\rm s}\sqrt{2r}} \int_{t_{\rm l}}^{t_{\rm u}} \frac{dt}{\sqrt{(t_{\rm u}^2 - t^2)(t^2 - t_{\rm l}^2)}}, \tag{3.199}$$

where $t_{\rm u} = \tan(\phi_{\rm u}/2)$, $t_{\rm l} = \tan(\phi_{\rm l}/2)$, $t = \tan(\phi/2)$, and $d\phi = 2dt/(1+t^2)$. Thus the synchrotron tune is

$$\frac{Q_{\rm s}}{\nu_{\rm s}} = \frac{\sqrt{2r}\pi t_{\rm u}}{\sqrt{(1+t_{\rm u}^2)(1+t_{\rm l}^2)}} \frac{1}{K(k_2)}, \tag{3.200}$$

where modulus $k_2 = \sqrt{t_{\rm u}^2 - t_{\rm l}^2}/t_{\rm u}$.

Figure 3.23 shows the synchrotron tune as a function of the amplitude of synchrotron oscillation for various voltage ratios. At $r = 0$, the system reduces to a single primary rf cavity, where the synchrotron tune is $Q_{\rm s}/\nu_{\rm s} = 1$ at zero amplitude. As r increases, the derivative of synchrotron tune vs action becomes large near the origin. Since large tune spread of the beam is essential for Landau damping of collective beam instabilities, an optimal rf voltage ratio is $r = 0.5$, where the synchrotron tune spread of the beam is maximized for a given bunch area.

At $r = 0.5$, the synchrotron tune becomes

$$\frac{Q_{\rm s}}{\nu_{\rm s}} = \frac{\pi t_0}{\sqrt{2(1+t_0^2)}K(k)} = \frac{\pi(E/2\nu_{\rm s})^{1/4}}{\sqrt{2}K(k)} \tag{3.201}$$

with modulus

$$k = \sqrt{\frac{1}{2}\left(1 + \frac{t_0^2}{1+t_0^2}\right)} = \sqrt{\frac{1}{2}\left(1 + \sqrt{\frac{E}{2\nu_{\rm s}}}\right)},$$

where $t_0 = \tan(\hat{\phi}/2)$. For small amplitude synchrotron motion, $t_0 = 0$ and $k_0 = 1/\sqrt{2}$. In this case, the maximum synchrotron tune is $\hat{Q}_{\rm s} = 0.7786\nu_{\rm s}$, located at $\hat{\phi} = 117°$ (or $E = 1.057\nu_{\rm s}$). Near this region, $\partial Q_{\rm s}/\partial\hat{\phi}$ is very small or zero. When the voltage ratio is $r > 0.5$, a dip in $Q_{\rm s}(J)$ appears at the inner separatrix of inner buckets, and two small potential wells are formed inside the inner separatrix.

E. Action-angle coordinates

Although analytic solutions for action-angle variables, presented in this section, are valid only for the case with $r = 0.5$, the method can be extended to obtain similar solutions for other voltage ratios. With the generating function, the angle coordinate becomes

$$F_2(\phi, J) = \int_{\hat{\phi}}^{\phi} \mathcal{P}(\phi')d\phi', \tag{3.202}$$

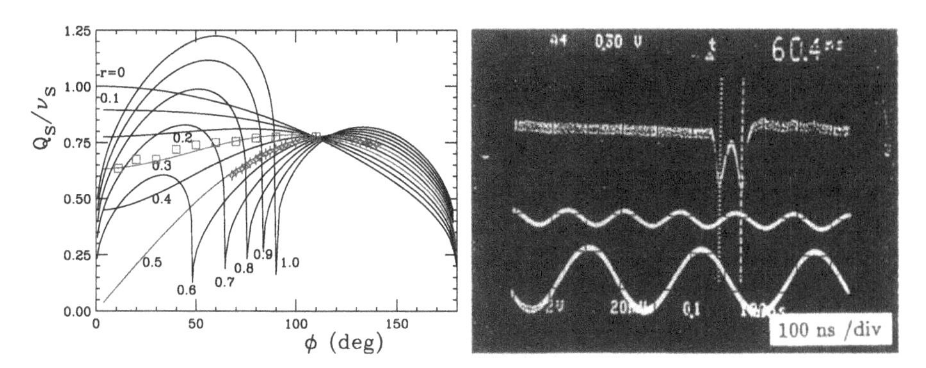

Figure 3.23: Left: The normalized synchrotron tune as a function of the peak phase $\phi = \hat{\phi}$ for various values of voltage ratio r. The rectangular and star symbols are data measured at the IUCF Cooler Ring. Note that when $r > 0.5$, the center of the bucket becomes an UFP. Two SFPs are located at the phase amplitude $\phi = \arccos(1/2r)$, where the synchrotron tune is maximum. Right: Beam profile of a proton beam bunch that was cooled by electron-cooling system and damped to 2 beamlets at stable fixed points. The separation between beamlets can be used to calibrate voltage ratio of 2 rf systems resulting $r = 0.6$ in this experiment.

$$\begin{aligned}\psi &= \frac{\partial F_2}{\partial J} = \frac{\partial E}{\partial J}\int_{\hat{\phi}}^{\phi}\frac{\partial \mathcal{P}}{\partial E}d\phi' = \frac{Q_s}{\nu_s}\int_{\hat{\phi}}^{\phi}\frac{d\phi}{\mathcal{P}} \\ &= \frac{Q_s}{\nu_s}\frac{(1+t_0^2)}{t_0\sqrt{1+2t_0^2}}\int_1^{\tau}\left((1-\tau^2)(\frac{1}{1+2t_0^2}+\tau^2)\right)^{-1/2}d\tau = \frac{Q_s}{\sqrt{2}\nu_s}\frac{\sqrt{1+t_0^2}}{t_0}u,\end{aligned}$$

where

$$u = \int_1^{\mathrm{cn}u}\frac{dx}{\sqrt{(1-x^2)(k'^2+k^2x^2)}}, \quad k = \sqrt{(1+2t_0^2)/[2(1+t_0^2)]}, \quad k' = \sqrt{1-k^2},$$

and the Jacobian elliptical function cn$u$ is

$$\mathrm{cn}u = \frac{\tan(\phi/2)}{\tan(\hat{\phi}/2)} = \frac{2\pi}{kK(k)}\sum_{n=0}^{\infty}\frac{q^{n+1/2}}{1+q^{2n+1}}\cos[(2n+1)\psi], \tag{3.203}$$

with $q = e^{-\pi K'/K}$, $K' = K(\sqrt{1-k^2})$, and

$$\psi = \frac{Q_s}{\nu_s}\frac{\sqrt{1+t_0^2}}{\sqrt{2}t_0}u = \frac{\pi u}{2K(k)}, \tag{3.204}$$

From Eq. (3.203), we obtain

$$\phi = 2\arctan\left(\tan\frac{\hat{\phi}}{2}\,\mathrm{cn}u\right), \quad \text{or} \quad \tan\frac{\phi}{2} = \tan\frac{\hat{\phi}}{2}\mathrm{cn}u, \tag{3.205}$$

and from Hamilton's equation of motion, we get

$$\mathcal{P} = -2\sqrt{2}\sin\left(\frac{\hat{\phi}}{2}\right)\tan\left(\frac{\hat{\phi}}{2}\right)\frac{\mathrm{sn}u\,\mathrm{dn}u}{1+[\tan(\hat{\phi}/2)\,\mathrm{cn}u]^2}. \tag{3.206}$$

When the voltage ratio is not 0.5, Eqs. (3.204) to (3.206) remain valid provided that the modulus is replaced by k_1 of Eq. (3.198) or k_2 of Eq. (3.200).

Thus the transformation of the phase-space coordinates $(\phi, \mathcal{P})$ to the action-angle variables (J, ψ) can be accomplished by using Eqs. (3.205) and (3.206) or equivalently

$$\sin\frac{\phi}{2} = \frac{\xi^{1/4}\,\mathrm{cn}u}{\sqrt{1-\xi^{1/2}\,\mathrm{sn}^2 u}}, \qquad \frac{\mathcal{P}}{2} = \pm\xi^{1/2}\sqrt{1-\frac{\mathrm{cn}^4 u}{(1-\xi^{1/2}\,\mathrm{sn}^2 u)^2}}, \tag{3.207}$$

where $\xi = E/2\nu_s$.

F. Small amplitude approximation

A tightly bunched beam occupies a small phase-space area. The formulas for small amplitude approximation are summarized as follows:

$$J \approx \frac{8\sqrt{2}K}{3\pi}\left(\frac{E}{2\nu_s}\right)^{3/4} = \frac{8\sqrt{2}K}{3\pi}\sin^3\frac{\hat{\phi}}{2}, \tag{3.208}$$

$$\frac{Q_s}{\nu_s} \approx \frac{\pi}{\sqrt{2}K}\sin\frac{\hat{\phi}}{2},$$

$$\phi \approx \hat{\phi}\frac{2\pi}{kK}\sum_{n=0}^{\infty}\frac{q^{n+1/2}}{1+q^{2n+1}}\cos(2n+1)\psi,$$

$$\mathcal{P} = -\hat{\phi}^2\frac{\pi^2}{\sqrt{2}kK^2}\sum_{n=0}^{\infty}\frac{(2n+1)q^{n+1/2}}{1+q^{2n+1}}\sin(2n+1)\psi, \tag{3.209}$$

where $k \approx 1/\sqrt{2}$, $K = K(k) \approx 1.8541$, and $q \approx e^{-\pi}$.

Let $\mathcal{A}$ be the rms phase-space area of the bunch, and $\sigma_{\mathcal{P}}$ and σ_ϕ the rms conjugate phase-space coordinates. We then obtain

$$\sigma_\phi = \left(\frac{3\mathcal{A}}{2\sqrt{2}K}\right)^{1/3}, \quad \sigma_{\mathcal{P}} = \left(\frac{3\mathcal{A}}{8K}\right)^{2/3}. \tag{3.210}$$

The rms tune spread of the beam is then

$$\Delta Q = \frac{\pi}{\sqrt{2}K}\left(\frac{3\mathcal{A}}{16\sqrt{2}K}\right)^{1/3}\nu_s. \tag{3.211}$$

G. Sum rule theorem and collective instabilities

The perturbing potential due to rf phase modulation is linearly proportional to $\mathcal{P}$ of Eq. (3.206). Expanding $\mathcal{P}$ in action-angle coordinates as $\mathcal{P} = \sum_n f_n(J)e^{jn\psi}$, we find that the strength functions $f_n(J)$ satisfy the sum rule shown in Eq. (3.78). The sum rule can be used to identify the region of phase-space that is sensitive to rf phase modulation (see Exercise 3.3.1).

Since $dQ/dJ = 0$ occurs inside the bucket, it may be of concern that large amplitude particles can become unstable against collective instabilities. When an rf phase or voltage noise is applied to beams in a double rf system, particle motion near the center of the bucket may become chaotic because of overlapping resonances. However, the chaotic region is bounded by invariant tori, and the effect on beam dilution may not be important. A most critical situation arises when the synchrotron amplitude of the beam reaches the region where Q_s is maximum or near the rf bucket boundary, where the tune spread is small. The beam may be susceptible to collective instabilities, and feedback systems may be needed for a high intensity beam that occupies a sizable phase-space area.

V.7 The Barrier RF Bucket

Bunch beam gymnastics have been important in antiproton production, beam coalescence for attaining high bunch intensity, multi-turn injection, accumulation, phase-space painting, etc. The demand for higher beam brightness in storage rings and higher luminosity in high energy colliders requires intricate beam manipulations at various stages of beam acceleration. In particular, a flattened rf wave form can be employed to shape the bunch distribution in order to alleviate space-charge problems in low energy proton synchrotrons and to increase the tune spread in electron storage rings. The extreme of the flattened rf wave form is the barrier bucket.[40]

For achieving high luminosity in the Fermilab TeV collider Tevatron, a Recycler has been built, which would recycle unused antiprotons from the Tevatron. The recycled antiprotons can be cooled by stochastic cooling or electron cooling to attain high phase-space density. At the same time, the Recycler would also accumulate newly produced, cooled antiprotons from the antiproton Accumulator. To maintain the antiproton bunch structure, a barrier rf wave form can be used to confine the beam bunch and shape the bunch distribution waiting for the next collider refill. The required bunch length and the momentum spread of the beam can be adjusted more easily by gymnastics with barrier rf waves than with the usual rf cavities.

[40]See J. Griffin, C. Ankenbrandt, J.A. MacLachlan, and A. Moretti, *IEEE Trans. Nucl Sci.* **NS-30**, 3502 (1983); V.K. Bharadwaj, J.E. Griffin, D.J. Harding, and J.A. MacLachlan, *IEEE Trans. Nucl. Sci.* **NS-34**, 1025 (1987); S.Y. Lee and K.Y. Ng, *Phys. Rev.* **E55**, 5992 (1997); M. Fujieda *et al.*, PRSTAB **2**, 122001 (1999).

The barrier rf wave is normally generated by a solid state power amplifier, which has intrinsic wide bandwidth characteristics. An arbitrary voltage wave form can be generated across a wideband cavity gap. Figure 3.24 shows some possible barrier rf waves with half sine, triangular, and square function forms. These wave forms are characterized by voltage amplitude $V(\tau)$, pulse duration T_1, pulse gap T_2 between positive and negative voltage pulses, and integrated pulse strength $\int V(\tau)d\tau$. For example, the integrated pulse strength for a square wave form is V_0T_1. The rf wave form is applied to a wideband cavity with frequency hf_0, where h is an integer, and f_0 is the revolution frequency of synchronous particles. The effect on the beam depends mainly on the integrated voltage of the rf pulse. Acceleration or deceleration of the beam can be achieved by employing a biased voltage wave in addition to the bunch-confining positive and negative voltage pulses.

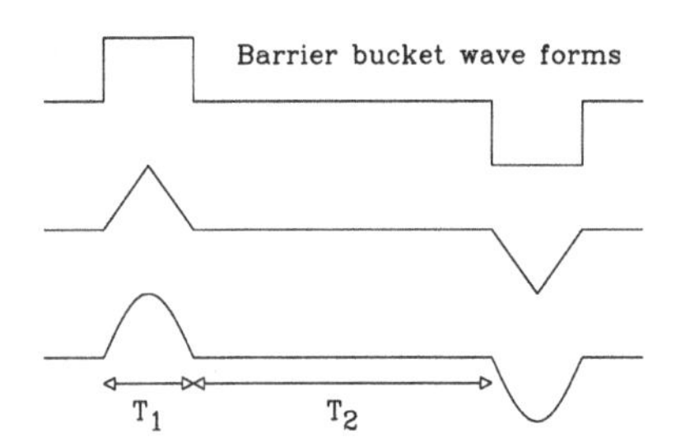

Figure 3.24: Possible wave forms for the barrier bucket. The barrier rf wave is characterized by a voltage height $V(\tau)$, a pulse width T_1, and a pulse gap T_2.

Most of the time, orbiting particles see no cavity field in passing through the cavity gap. When a particle travels in the time range where the rf voltage is not zero, its energy can increase or decrease depending on the sign of the voltage it encounters. In this way, the accelerator is divided into stable and unstable regions. Thus the wide bandwidth rf wave can create a barrier bucket to confine orbiting particles.

A. Equation of motion in a barrier bucket

For a particle with energy deviation ΔE, the fractional change of the orbiting time $\Delta T/T_0$ is

$$\frac{\Delta T}{T_0} = \eta\frac{\Delta E}{\beta^2 E_0}, \tag{3.212}$$

where η is the phase slip factor, βc and E_0 are the speed and the energy of a synchronous particle, and T_0 is its revolution period. Without loss of generality, we consider here synchrotron motion with $\eta < 0$. For $\eta > 0$, the wave form of the barrier bucket is reversed.

The time coordinate for an off-momentum particle $-\tau$ is given by the difference between the arrival time of this particle and that of a synchronous particle at the center of the bucket. The equations of motion for the phase-space coordinate τ and particle energy deviation are

$$\frac{d\tau}{dt} = -\eta\frac{\Delta E}{\beta^2 E_0}, \qquad \frac{d(\Delta E)}{dt} = \frac{eV(\tau)}{T_0}. \tag{3.213}$$

The equations of particle motion in a barrier rf wave are governed by Eq. (3.213).

B. Synchrotron Hamiltonian for general rf wave form

From the equations of motion (3.213), we obtain the general synchrotron Hamiltonian for an arbitrary barrier rf wave form:

$$H = -\frac{\eta}{2\beta^2 E_0}(\Delta E)^2 - \frac{1}{T_0}\int_0^{\tau} eV(\tau)d\tau. \tag{3.214}$$

Thus the maximum off-energy bucket height can be easily derived:

$$\Delta E_{\rm b} = \left(\frac{2\beta^2 E_0}{|\eta| T_0}\left|\int_{T_2/2}^{T_2/2+T_1} eV(\tau)d\tau\right|\right)^{1/2}, \tag{3.215}$$

where T_1 is the width of the barrier rf wave form. Since the barrier rf Hamiltonian is time independent, an invariant torus has a constant Hamiltonian value. We define the W parameter for a torus from the equation below:

$$\frac{|\eta|}{2\beta^2 E_0}(\hat{\Delta E})^2 = \frac{1}{T_0}\left|\int_{T_2/2}^{T_2/2+W} eV(\tau)d\tau\right|. \tag{3.216}$$

The synchrotron period of a Hamiltonian torus becomes

$$T_{\rm s} = 2\frac{T_2}{|\eta|}\left(\frac{\beta^2 E_0}{|\hat{\Delta E}|}\right) + 4T_{\rm c}, \tag{3.217}$$

$$T_{\rm c} = \frac{\beta^2 E_0}{|\eta|}\int_0^W \left[(\hat{\Delta E})^2 - \frac{2\beta^2 E_0}{|\eta| T_0}\int_{T_2/2}^{T_2/2+\tau} eV(\tau')d\tau'\right]^{-1/2} d\tau, \tag{3.218}$$

where $4T_{\rm c}$ is the time for a particle to be reflected in the potential well. Clearly, all physical quantities depend essentially on the integral $\int V(\tau)d\tau$. Thus, the essential physics is independent of the exact shape of the barrier rf wave.

C. Square wave barrier bucket

Since the effect of the barrier rf wave on particle motion depends essentially on the integrated rf voltage wave, we consider only the square wave forms with voltage heights $\pm V_0$ and pulse width T_1 in time, separated by a gap of T_2. When the particle passes through the cavity gap at voltage $\pm V_0$, it gains (loses) an equal amount of energy eV_0, i.e. $d(\Delta E)/dt = \pm eV_0/T_0$ every turn. The number of cavity passages before the particle loses all its off-energy value $\hat{\Delta E}$ is

$$N = \frac{|\hat{\Delta E}|}{eV_0}. \tag{3.219}$$

Thus the phase-space trajectory for a particle with maximum off-energy $\hat{\Delta E}$ is

$$(\Delta E)^2 = \begin{cases} (\hat{\Delta E})^2 & \text{if } |\tau| \le T_2/2 \\ (\hat{\Delta E})^2 - \left(|\tau| - \frac{T_2}{2}\right)\frac{\omega_0\beta^2 E_0 eV_0}{\pi|\eta|} & \text{if } T_2/2 \le |\tau| \le (T_2/2) + T_1, \end{cases} \tag{3.220}$$

where $\omega_0 = 2\pi f_0$ is the angular revolution frequency of the beam. The phase-space ellipse is composed of a straight line in the rf gap region and a parabola in the square rf wave region. The phase-space area of the invariant phase-space ellipse is

$$\mathcal{A} = 2T_2\hat{\Delta E} + \frac{8\pi|\eta|}{3\omega_0\beta^2 E_0 eV_0}(\hat{\Delta E})^3. \tag{3.221}$$

The maximum energy deviation or the barrier height that a barrier rf wave can provide is

$$\Delta E_{\rm b} = \left(\frac{eV_0 T_1}{T_0}\frac{2\beta^2 E_0}{|\eta|}\right)^{1/2}, \tag{3.222}$$

where T_1 is the pulse width of the rf voltage wave, and T_0 is the revolution period of the beam. The bucket height depends on $V_0 T_1$, which is the integrated rf voltage strength $\int V(\tau)d\tau$. The synchrotron period is

$$T_{\rm s} = 2\frac{T_2}{|\eta|}\left(\frac{\beta^2 E_0}{|\hat{\Delta E}|}\right) + 4\frac{|\hat{\Delta E}|}{eV_0}T_0 \tag{3.223}$$

for particles inside the bucket. The mathematical minimum synchrotron period of Eq. (3.223) and the corresponding maximum synchrotron tune are

$$T_{\rm s,min} = \left(\frac{32T_0T_2\beta^2 E_0}{|\eta|eV_0}\right)^{1/2}, \qquad \nu_{\rm s,max} = \left(\frac{T_0}{T_2}\frac{|\eta|eV_0}{32\beta^2 E_0}\right)^{1/2}. \tag{3.224}$$

Note here that $\pi T_0/(16T_2)$ plays the role of harmonic number h of a regular rf system. The synchrotron tune is a function of the off-energy parameter $\hat{\Delta E}$ given by

$$\nu_{\rm s} = 4\nu_{\rm s,max}\sqrt{\frac{T_1}{T_2}\frac{\hat{\Delta E}}{\Delta E_b}}\left(1 + 4\left[\frac{\hat{\Delta E}}{\Delta E_b}\right]^2\frac{T_1}{T_2}\right)^{-1}. \tag{3.225}$$

Note that when the rf pulse gap width decreases to $T_2/T_1 < 4$, the synchrotron tune becomes peaked at an amplitude within the bucket height. This feature is similar to that of a double rf system.Figure 3.25 shows ν_s vs ΔE with Fermilab Recycler parameters $E_0 = 8.9$ GeV, $\gamma_T = 20.7$, $f_0 = 89.8$ kHz, $T_1 = 0.5$ μs, $V_0 = 2$ kV, and $T_2/T_1 = 1, 2, 4$, and 8. For example, $\nu_{s,\max} = 3.7 \times 10^{-5}$ for $T_2 = T_1$, i.e. the synchrotron frequency is 3.3 Hz.

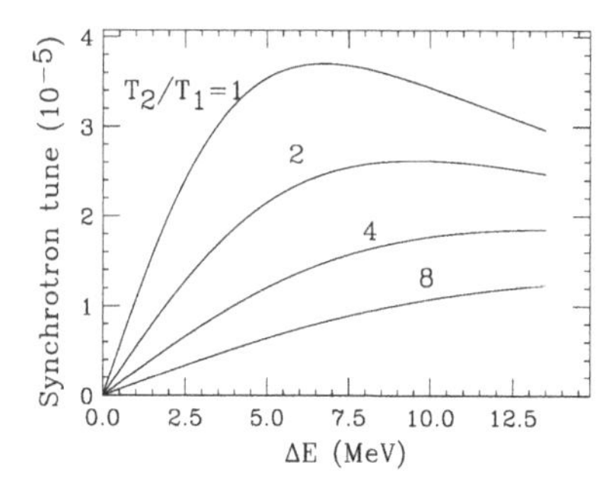

Figure 3.25: Synchrotron tune vs off-energy parameter ΔE. Parameters used are $E_0 = 8.9$ GeV, $f_0 = 89.8$ kHz, $V_0 = 2$ kV, $\gamma_T = 20.7$, and $T_1 = 0.5$ μs. Note that if $T_2 > 4T_1$, the synchrotron tune is a monotonic function of ΔE. On the other hand, if $T_2 < 4T_1$, the synchrotron tune is peaked at an off-energy ΔE smaller than the bucket height $\Delta E_{\rm b}$.

D. Hamiltonian formalism

The Hamiltonian for the phase-space coordinates $(\tau, \Delta E)$ is

$$H_0 = \frac{\eta}{2\beta^2 E_0}(\Delta E)^2 + \frac{\omega_0 e V_0 T_1}{2\pi} f_0(\tau, T_1, T_2), \tag{3.226}$$

where

$$\begin{aligned} f_0(\tau, T_1, T_2) &= -1 + \frac{1}{T_1}\left[(\tau + T_1 + \frac{T_2}{2})\theta(\tau + T_1 + \frac{T_2}{2}) - (\tau + \frac{T_2}{2})\theta(\tau + \frac{T_2}{2})\right. \\ &\left. -(\tau - \frac{T_2}{2})\theta(\tau - \frac{T_2}{2}) + (\tau - T_1 - \frac{T_2}{2})\theta(\tau - T_1 - \frac{T_2}{2})\right]. \end{aligned} \tag{3.227}$$

Here $\theta(x)$ is the standard step function with $\theta(x) = 1$ for $x > 0$ and $\theta(x) = 0$ for $x < 0$.

For a constant T_1, T_2 and V_0, the Hamiltonian H_0 is a constant of motion. The action of a Hamiltonian torus is

$$J = \frac{1}{2\pi}\oint \Delta E d\tau = \frac{1}{2\pi}\sqrt{\frac{\omega_0\beta^2 E_0 e V_0}{\pi|\eta|}}\oint \sqrt{W + f_0(\tau, T_1, T_2)}d\tau. \tag{3.228}$$

The parameter W with a dimension of time is related to the Hamiltonian value by

$$H_0 = -\frac{\omega_0 e V_0}{2\pi}W = \frac{\eta}{2\beta^2 E_0}\left(\hat{\Delta E}\right)^2. \tag{3.229}$$

For a given Hamiltonian torus, W has the physical meaning that it is equal to the maximum phase excursion $|\tau|$ in the rf wave region. Therefore $W = 0$ corresponds to an on-momentum particle, and $W = T_1$ is associated with particles on the bucket boundary.

The action for a particle torus inside the bucket and the bucket area of the maximum action with $W = T_1$ are

$$
\begin{aligned}
J &= \frac{1}{2\pi}\sqrt{\frac{\omega_0\beta^2 E_0 eV_0}{\pi|\eta|}}\left[2T_2\sqrt{W} + \frac{8}{3}W^{3/2}\right] = \frac{1}{2\pi}\left[2T_2 + \frac{8}{3}W\right]\hat{\Delta E}, \\
\mathcal{B} &= 2\pi\hat{J} = \left(2T_2 + \frac{8}{3}T_1\right)\Delta E_{\rm b}.
\end{aligned}
\tag{3.230}
$$

The bucket area depends only on the integrated rf voltage strength $\int V(\tau)d\tau = V_0T_1$.

E. Action-angle coordinates

Canonical transformation from the phase-space coordinates $(\tau, \Delta E)$ to the action-angle variable can be achieved by using the generating function:

$$
F_2(J,\tau) = \int_{-\hat{\tau}}^{\tau}\Delta E d\tau, \qquad \psi = \frac{\partial F_2}{\partial J} = \frac{\pi\sqrt{W}}{T_2 + 4W}\int_{-\hat{\tau}}^{\tau}\frac{d\tau}{\sqrt{W + f_0}}. \tag{3.231}
$$

where $\hat{\tau} = W + (T_2/2)$. The integral can be evaluated easily to obtain

$$
\psi = \begin{cases}
\frac{2\pi\sqrt{W}}{T_2+4W}\sqrt{W + \frac{1}{2}T_2 + \tau} & \text{if } -W - \frac{1}{2}T_2 \le \tau \le -\frac{1}{2}T_2,\ \Delta E > 0 \\
\psi_{\rm c} + \frac{\pi}{T_2+4W}\left(\tau + \frac{1}{2}T_2\right) & \text{if } -\frac{1}{2}T_2 \le \tau \le \frac{1}{2}T_2,\ \Delta E > 0 \\
2\psi_{\rm c} + \psi_s - \frac{2\pi\sqrt{W}}{T_2+4W}\sqrt{W + \frac{1}{2}T_2 - \tau} & \text{if } \frac{1}{2}T_2 \le \tau \le W + \frac{1}{2}T_2,\ \Delta E > 0 \\
2\psi_{\rm c} + \psi_s + \frac{2\pi\sqrt{W}}{T_2+4W}\sqrt{W + \frac{1}{2}T_2 - \tau} & \text{if } \frac{1}{2}T_2 \le \tau \le W + \frac{1}{2}T_2,\ \Delta E < 0 \\
3\psi_{\rm c} + \psi_s + \frac{\pi}{T_2+4W}\left(\frac{1}{2}T_2 - \tau\right) & \text{if } -\frac{1}{2}T_2 \le \tau \le \frac{1}{2}T_2,\ \Delta E < 0 \\
4\psi_{\rm c} + 2\psi_s - \frac{2\pi\sqrt{W}}{T_2+4W}\sqrt{W + \frac{1}{2}T_2 + \tau} & \text{if } -W - \frac{1}{2}T_2 \le \tau \le -\frac{1}{2}T_2,\ \Delta E < 0,
\end{cases}
$$

where

$$
\psi_{\rm c} = \frac{2\pi W}{T_2 + 4W}, \qquad \psi_s = \frac{\pi T_2}{T_2 + 4W} \tag{3.232}
$$

are respectively the synchrotron phase advances for a half orbit in the rf wave region and in the region between two rf pulses. Note that $2\psi_{\rm c} + \psi_s = \pi$ for one half of the synchrotron orbit, and that the motion of a stable particle orbit in the barrier bucket with $\eta < 0$ is clockwise. We choose the convention of $\dot{\psi} > 0$ corresponding to a clockwise motion in synchrotron phase-space.

When a perturbation, such as rf noise, is applied to the barrier rf system, stable bucket area may be reduced. The resonance strength functions and their associated sum rules can be derived analytically. The resonance strength function decreases slowly with mode number. The rf phase and voltage modulation can severely dilute bunch area if the modulation frequency is near the top of the synchrotron tune and its harmonics. The rf phase modulation due to orbit length modulation resulting from ground vibration can be important. Because the solid state amplifier is a low power device, it is important to avoid a large reduction of stable phase-space area. Active compensation may be used to compensate the effect of rf phase modulation.[41]

V.8 Beam-stacking in Longitudinal Phase space

Beam intensity is limited by space-charge effects at low energies. Rapid cycling synchrotrons (RCS) can be used to increase beam power. However, RCS is usually limited by its achievable energy and a second-stage accelerator is required to increase both energy and beam power. Slip-stacking injection may be used to double the beam power. The idea of slip-stacking was first proposed by F.E. Mills, where he studied the stability of particle motion under the influence of two rf systems at a nearby frequency.[42] During the slip stacking process, both systems are at the stationary phase condition. The Hamiltonian of the two-rf system in the normalized phase space coordinates is (see Eq. (3.192))

$$H = \frac{\nu_s \mathcal{P}^2}{2} + \nu_s\{[1 - \cos\phi] + [1 - \cos(\phi - \nu_{\rm slip}\theta)]\}, \qquad (3.233)$$

where $\mathcal{P}$ and ϕ are the normalized phase-space coordinates, $\nu_s = \sqrt{h|\eta|eV_{\rm rf}/2\pi\beta^2 E}$ is the small-amplitude synchrotron tune, $V_{\rm rf}$ is the rf voltage, and βc and E are the nominal speed and energy of the beam particles, $\nu_{\rm slip} = f_{\rm slip}/f_0$ is the *slip-tune*, $f_{\rm slip}$ is the slip frequency, and f_0 is the revolution frequency of the stacking-Ring. All physical quantities represent parameters of the slip-stacking ring. The rf phase ϕ and the normalized off-momentum $\mathcal{P}$ are conjugate canonical coordinates, while θ represents the independent 'time' variable, which increases by 2π in each revolution around the stacking-ring.

In order for slip-stacking to work, these two rf systems must generate buckets which slip by exactly one train or one batch of rf buckets in consecutive injections from the rapid-cycling booster synchrotron. This condition fixes the rf frequency difference to $f_{\rm slip} = h_{\rm B} f_{\rm B}$, where $f_{\rm B}$ is the repetition frequency of the RCS and $h_{\rm B}$ is

[41] See S.Y. Lee and K.Y. Ng, *Phys. Rev.* **E55**, 5992 (1997).

[42] F.E. Mills, BNL 15936 (1971).

its rf harmonics. For Fermilab, the RCS is the Booster with $h_{\rm B} = 84$ and $f_{\rm B} = 15$ Hz. The Recycler ring serves as the stacking ring with the harmonic number $h_{\rm R} = 588$.

We assume that these two rf systems have the same total rf cavity voltage $V_{\rm rf}$. One of the beam bunches that synchronize the rf system $(1 - \cos\phi)$, while the train of buckets that synchronize with the rf system $(1 - cos(\phi - \nu_{\rm slip}\theta))$ is moving at a different momentum. These to beams bunches slip against each other at the slip tune of $\nu_{\rm slip}$. If the phase-slip factor is $\eta < 0$, the rf buckets generated by the rf system corresponding to $\cos(\phi - \nu_{\rm slip}\theta)$ are at a slightly lower energy than the buckets generated by the rf system of $\cos\phi$ in the Hamiltonian, i.e. the lower-energy bucket series slips forward at the rate of $\Delta\phi = \nu_{\rm slip}\theta$.

The fractional momentum that separates the upper and lower bucket series is

$$\Delta\delta_{\rm sep} \equiv \frac{\Delta P_{\rm sep}}{P_0} = \frac{\nu_{\rm slip}}{h|\eta|} = \frac{h_B f_B}{h_R|\eta| f_0}, \tag{3.234}$$

where $\Delta P_{\rm sep}$ is the momentum difference of the two slip-stacking beams, P_0 is the nominal momentum of the beams, and η is the phase slip-factor of the slip-stacking ring. Once the repetition rate of the RCS, the phase-slip factors, and the revolution frequency of the slip-stacking ring are designed, the momentum separation of the two bucket series, $\Delta\delta_{\rm sep}$, is fixed. In terms of the normalized off-momentum coordinate $\mathcal{P}$, the separation of the centers of the upper and lower buckets is

$$\Delta\mathcal{P}_{\rm sep} = \frac{h|\eta|\Delta\delta_{\rm sep}}{\nu_s} = \frac{\nu_{\rm slip}}{\nu_s} \equiv \alpha_s, \tag{3.235}$$

where α_s is called the *slip-stacking parameter*. The unperturbed bucket height is $|\mathcal{P}| \leq 2$ for stationary bucket (see Table 3.2 and the bucket of Fig. 3.12). The unperturbed rf buckets of the two rf systems just touch each other at $\alpha_s = 4$. The two unperturbed rf buckets are separated from each other when $\alpha_s > 4$, and they overlap when $\alpha_s < 4$.

Slip-stacking had been tried in the CERN SPS to accumulate beams from the CERN PS, successfully applied in the Fermilab Tevatron Run IIB to increase antiproton production, and employed in the Fermilab Recycler Ring to increase the proton beam power for neutrino production.[43] Because of the presence of the two rf systems, the two series of rf buckets, upper and lower, are mutually perturbing each other. The stable bucket areas become smaller than those of the unperturbed rf buckets. When the upper and lower series of buckets overlap, resonance islands can be generated in-between the two series of rf buckets usually around $p = 0$ and $\phi = 0$ and $\pm\pi$. In addition, chaotic regions may be created, which can reduce the stable region of the rf buckets significantly.

[43] D. Boussard and Y. Mizumachi, IEEE Trans. Nucl. Sci. **NS-26**, 3623 (1979); K. Seiya, *et al.* PAC2005 347, (2005); I. Kourbanis, IPAC2014, 904, (2014).

Overlapping resonances can be avoided if the upper and lower buckets are widely separated or if $\alpha_s \gg 4$. Bigger α_s, however, implies smaller rf voltage and therefore smaller unperturbed bucket areas (in δ-ϕ coordinates), which may not be large enough to accommodate the beam injected from the RCS. On the other hand, smaller α_s implies larger rf voltage. One may think that there would be bigger unperturbed bucket areas to accept the beam injected from the RCS. When $\alpha_s < 4$, these two bucket series can produce strong overlapping resonances and chaos so that the stable parts of the buckets become smaller than the unperturbed buckets. Careful choice of the rf parameters provide successful doubling of beam intensity.

Numerical simulations can be carried out to analyze the interaction of these two rf buckets. Transforming the phase space into the Poincaré map, where the particles in the slipping bucket are shifted backward in the time coordinate, we will observe stationary resonance islands in the Poincaré map. Figure 3.26 shows the Poincaré maps with parameter $\alpha_s = 4.1$ (left) and 6.0 (right) respectively.

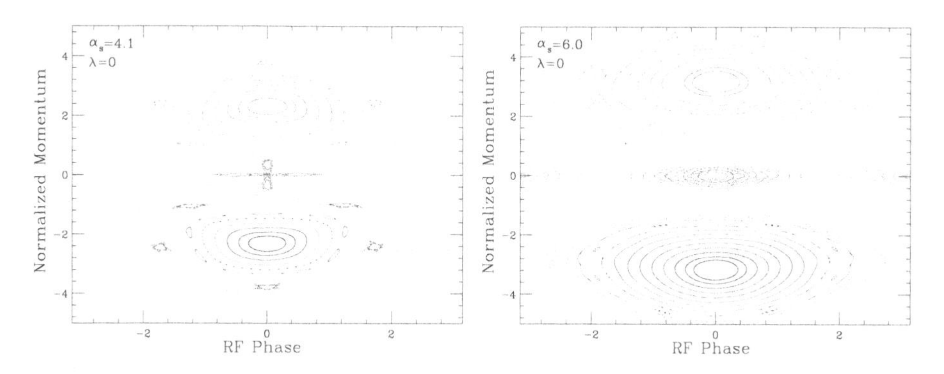

Figure 3.26: Left: Phase-space structure with slip-stacking parameter $\alpha_s = 4.1$; Right: phase-space structure with slip-stacking parameter $\alpha_s = 6.0$. Parametric resonances are excited by the mutual interaction between these two rf systems.

Note that there is little perturbation at $\alpha_s = 6$ in comparison with that of $\alpha_s = 4.1$. Furthermore, We note that there is a prominent 5th order resonance at $\alpha_s = 4.1$ and the 7th order resonance at $\alpha_s = 6.0$. These parametric resonances are produced by mutual interaction between these two rf systems discussed in Sec. III in Chapter 3.

For beam particles in one of the buckets, the other slip-stacking rf system produces a time-dependent modulation at the tune of $\nu_{\rm slip} = \alpha_s \nu_s$, which is a combination of phase and voltage modulations. If the modulation tune is equal to an integer multiple of the particle tune, the parametric resonance occurs. The synchrotron tune of a particle in the bucket depends on its synchrotron amplitude, i.e., the synchrotron tune is ν_s at small amplitude and decreases to zero at the separatrix of the synchrotron phase space, as shown in Fig. 3.27. Resonances will occur at different phase-space

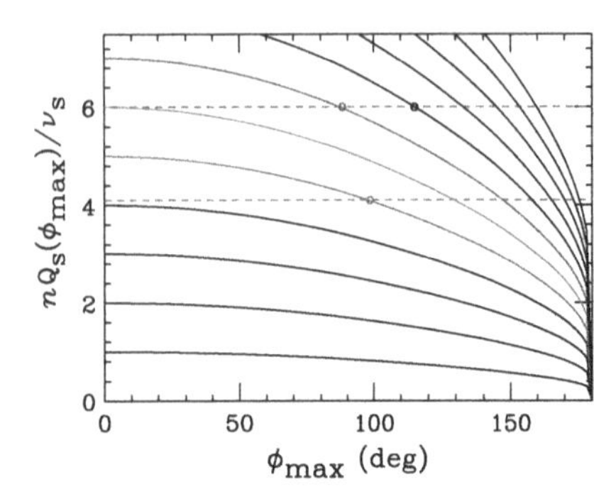

Figure 3.27: The harmonics of the synchrotron tune of a stationary rf system system versus the rf phase amplitude $\phi_{\rm max}$. The horizontal dashed lines correspond to the modulation tunes at slip-stacking parameters of $\alpha_s = 4.1$ and 6 of Fig. 3.26. When the slip-stacking modulation tune cut through an nth harmonic of the synchrotron tune, the nth-order resonance will occur at that phase-space amplitude, evidently seen in Fig. 3.26.

locations as the slip-tune $\alpha_s \nu_s$ changes.

If the modulation (slip-stacking) tune cuts through the nth harmonic of the synchrotron tune, the nth-order resonance, called the n:1 parametric resonance, will appear at the corresponding phase-space location. For example, the horizontal red dashed line in Fig. 3.27 corresponding to $\alpha_s = 4.1$ cuts through the 5th harmonic of the synchrotron tune to produce a 5th-order resonance at the maximum rf phase amplitude $\phi_{\rm max} \sim 95^\circ$. Similarly, the $\alpha_s = 6$ line cut through the 7 times the synchrotron tune line will produce the 7th order parametric resonance at its phase amplitude.

Two strong resonances can also concatenate into a second-order resonance, for example, the 4:1 and 5:1 resonances can interact to produce a 9:2 resonance at the phase space in between these two first-order resonances, evidently shown at the Left plot of Fig. 3.26. The size of the resonance islands depend on the resonance strength and the slope of the n-harmonic synchrotron tune versus amplitude.

We also note that the phase space region between the upper and lower buckets in Fig. 3.26 can cause overlapping resonances phase space region near the separatrix. Using the second canonical perturbation method, one can find a cavity to compensate the interaction between these two bucket at the overlapping region.[44]

Exercise 3.5

1. The Cooler Injector Synchrotron (CIS) accelerates protons from 7 MeV to 200 MeV in 1.0 Hz. The circumference is 17.364 m. The rf system operates at $h = 1$ with a maximum voltage 240 V. The momentum compaction factor is $\alpha_{\rm c} = 0.6191$. The momentum spread of the injection linac is about $\pm 5.0 \times 10^{-3}$.

   (a) Assuming that the rf voltage is ramped according to

$$V_{\rm rf}(t) = V_0 + (V_1 - V_0)\left(3\frac{t^2}{T_1^2} - 2\frac{t^3}{T_1^3}\right), \quad t \in [0, T_1],$$

[44]J. Eldred, Ph.D. Thesis, Department of Physics, Indiana University, Bloomington, IN, December 2015; FERMILAB-THESIS-2015-31; J. Eldred and R. Zwaska, Phys. Rev. ST Accel. Beams **19**, 104001 (2016); S.Y. Lee and K.Y. Ng, PRAB, **20**, 064202 (2017).

where V_0 and $V_1 = 240$ V are the initial and the maximum final rf voltages, T_1 is the voltage ramp time. Calculate the adiabaticity coefficient of Eq. (3.21), and the rf bucket height during the rf voltage ramping as functions of time t with $V_0 = 10$ V and $T_1 = 10$ ms. Change these parameters to see the variation of the adiabaticity coefficient.

(b) If the magnetic field of a proton synchrotron is ramped according to

$$B(t) = B_0 + \left(3\frac{t^2}{t_1^2} - 2\frac{t^3}{t_1^3}\right)(B_1 - B_0), \quad t \in [0, t_1]$$

where B_0 and B_1 are magnetic field at the injection and at the flat top, and $t = 0$ and $t = t_1$ are the time at the beginning of ramp and at the flat top, find the frequency ramping relation of the rf cavity, and find the maximum $\dot{B}$.

2. In proton accelerators, the rf gymnastics for bunch rotation is performed by adiabatically lowering the voltage from V_1 to V_2 and suddenly raising the voltage from V_2 to V_1 (see also Exercise 3.2.6). Using Eq. (3.42) and conservation of phase-space area, show that the bunch length in the final step is

$$\hat{\theta}_{\text{final}} = \left(\frac{\nu_{s2}}{\nu_{s1}}\right)^{1/2} \hat{\theta}_{\text{initial}},$$

where $\hat{\theta}_{\text{initial}}$ is the initial bunch length in orbital angle variable, and ν_{s1} and ν_{s2} are the synchrotron tune at voltages V_1 and V_2. Apply the bunch rotation scheme to proton beams at $E = 120$ GeV in the Fermilab Main Injector, where the circumference is 3319.4 m, the harmonic number is $h = 588$, the transition energy is $\gamma_T = 21.8$, and the phase-space area is $\mathcal{A} = 0.05$ eV-s for 6×10^{10} protons. Find the voltage V_2 such that the final bunch length is 0.15 ns with an initial voltage $V_1 = 4$ MV. The energy of the secondary antiprotons is 8.9 GeV. If the acceptance of the antiproton beam is $\pm 3\%$, what is the phase-space area of the antiproton beams? If the antiproton production efficiency is 10^{-5}, what is the phase-space density of the antiproton beams?

3. Neglecting wakefield and other diffusion mechanisms, the momentum spread of an electron beam in a storage ring is determined mainly by the equilibrium between the quantum fluctuation of photon emission and the radiation damping. For an isomagnetic ring, it is given by

$$(\frac{\sigma_E}{E})^2 = C_q \frac{\gamma^2}{\mathcal{J}_s \rho}, \qquad C_q = \frac{3C_u \hbar}{4mc} = \frac{55}{32\sqrt{3}} \frac{\hbar}{mc} = 3.83 \times 10^{-13} \text{ m},$$

where $\mathcal{J}_s$ is the damping partition number with $\mathcal{J}_s \approx 2$ for separate function machines. Using the electron storage ring parameters listed in Exercise 3.1.6, calculate the phase-space area in eV-s.

4. Verify Eq. (3.191) and derive the Hamiltonian for the double rf system. For a flattened potential well in the double rf system with $\phi_{1s} = \phi_{2s} = 0$, show that the Hamiltonian for small amplitude synchrotron motion is

$$H \approx \frac{1}{2}\mathcal{P}^2 + \nu_s b \phi^4,$$

where $b = (h^2 - 1)/24$, h is the ratio of the harmonic numbers, and the independent "time" variable is the orbital angle θ. We solve the synchrotron motion for the quartic potential below.

(a) Since the Hamiltonian is time independent, the Hamiltonian value E is a constant of motion. Show that the action variable is related to the Hamiltonian value by

$$J = \frac{4K}{3\pi b^{1/4}}\left(\frac{E}{\nu_{\rm s}}\right)^{3/4} = \frac{4K}{3\pi}b^{1/2}\hat{\phi}^3,$$

where $K = K(\sqrt{\frac{1}{2}}) = 1.85407468$ is the complete elliptical integral with modulus $k = 1/\sqrt{2}$, $\hat{\phi}$ is the amplitude of the phase oscillation.

(b) Show that the synchrotron tune is

$$\frac{Q_{\rm s}(J)}{\nu_{\rm s}} = \frac{\pi}{K}b^{1/4}\left(\frac{E}{\nu_{\rm s}}\right)^{1/4} = \frac{3^{1/3}\pi^{4/3}b^{1/3}}{4^{1/3}K^{4/3}}J^{1/3} = \frac{\pi}{K}b^{1/2}\hat{\phi}.$$

(c) Define the generating function

$$F_2(\phi, J) = \int_0^{\phi} \mathcal{P} d\phi,$$

and show that the solution of the synchrotron motion is given by

$$\phi = \hat{\phi}\ \mathrm{cn}\left(\frac{2K}{\pi}\psi|\frac{1}{2}\right),$$
$$\mathcal{P} = -\sqrt{2}\hat{\mathcal{P}}\ \mathrm{sn}\left(\frac{2K}{\pi}\psi|\frac{1}{2}\right)\ \mathrm{dn}\left(\frac{2K}{\pi}\psi|\frac{1}{2}\right),$$

where cn, sn, and dn are elliptical functions with modulus $k = 1/\sqrt{2}$. Compare your results with that of Eq. (3.209) for the $h = 2$ case.

5. Two strong resonances can interact to create a secondary resonance located in the phase space between these two primary resonance as shown in Exercise 2.7.8, where 3:1 and 1:1 resonances produce a 4:2 resonance. The slip-stacking rf buckets of Eq. (3.233 can also produce resonances at the phase space in the middle of the two buckets, besides the parametric resonances shown in Fig. 3.26. These secondary resonances can overlap with the primary parametric resonances so that the bucket overlapping region becomes chaotic. One can use the canonical perturbation method to understand these secondary resonance. This exercise explore the canonical perturbation technique.

   (a) To simplify the derivation, symmetrize the Hamiltonian to a frame with the upper and lower buckets centered at $p = +\frac{1}{2}\alpha_s$, or with frames moving at $+\frac{1}{2}\nu_{\rm slip}\theta$. Using the generating function

$$F_2(\phi, \tilde{p}) = \left(\phi - \frac{\nu_{\rm slip}\theta}{2}\right)\left(\tilde{p} + \frac{\alpha_s}{2}\right),$$

   show that the new Hamiltonian is

$$H = \frac{\nu_s\tilde{p}^2}{2} + \nu_s\left[2 - \cos\left(\tilde{\phi} + \frac{\nu_{\rm slip}\theta}{2}\right) - \cos\left(\tilde{\phi} - \frac{\nu_{\rm slip}\theta}{2}\right)\right]$$

where $\tilde{p}$ and $\tilde{\phi}$ are the conjugate canonical coordinates of the symmetrized slip-stacking rf systems, and we ignore a constant term $+\frac{\nu_s\alpha_s^2}{8}$ from the Hamiltonian. This Hamiltonian represents the upper and lower buckets moving at $\Delta\phi = \mp\nu_{\rm slip}\theta/2$ respectively, while the structures in-between the two buckets centered at $\tilde{p}=0$ is stationary.

(b) Resonances due to the interaction of these bucket occurs at the phase space near $\tilde{p}=0$, Because we wish to study the phase space structure in-between the upper and lower rf buckets, we perform a canonical transformation to cancel the potential-energy part of the Hamiltonian using the generating function

$$F_2(\tilde{\phi},\bar{p}) = \phi\bar{p} + a(\bar{p})\sin(\tilde{\phi}+\frac{\nu_{\rm slip}}{2}\theta) + b(\bar{p})\sin(\tilde{\phi}-\frac{\nu_{\rm slip}}{2}\theta),$$

where $a(\bar{p})$ and $b(\bar{p})$ are two functions of $\bar{p}$ to be determined, show that the new Hamiltonian is

$$H = \frac{1}{2}\nu_s\bar{p}^2 + \frac{2}{\alpha_s^2-4\bar{p}^2}\nu_s\left(1-\cos(2\bar{\phi})\right) + ...$$

where the ... represents either constants or time dependent terms that are average to zero in the new Hamiltonian, and we have chosen

$$a(\bar{p}) = \frac{2}{\alpha_s+2\bar{p}} \qquad \text{and} \qquad b(\bar{p}) = -\frac{2}{\alpha_s-2\bar{p}},$$

to cancel the primary rf bucket potentials. The combined effect of two rf systems is a 2nd order bucket located at the phase space of these two rf buckets. The strength of this bucket is $2/\alpha_s^2$ of that of the primary bucket. A cavity at the frequency $2hf_0 + f_s lip$ at the voltage of $-\frac{2}{\alpha_s^2}V_0$ can be used to cancel the 2nd order bucket induced by two primary cavities, and help to eliminate the chaos near the separatrices of these two primary buckets.

VI Fundamentals of RF Systems

The basic function of rf cavities is to provide a source of electric field for beam manipulations, including acceleration, deceleration, bunching and debunching, and deflection. The longitudinal electric field must be synchronized with the particle arrival time. Resonance cavities, where only electromagnetic fields at resonance frequencies can propagate, are a natural choice in rf cavity design.

Cavities are classified according to their operational frequencies. For cavities operating at a few hundred MHz or higher, pillbox cavities with nose-cone or disk loaded geometry can be used. At lower frequencies, coaxial geometry is commonly employed. Some fundamental parameters of cavities are transit time factor, shunt impedance, and quality factor.

The transit time factor of Eq. (3.3) reflects the finite passage time for a particle to traverse the rf cavity, while the accelerating field varies with time. The transit time factor reduces the effective voltage seen by passing particles. We may reduce the accelerating voltage gap to increase the transit time factor, but a smaller gap can cause electric field breakdown due to the Kilpatrick limit (see Sec. V.3).

The quality factor (Q-factor) depends on the resistance of the cavity wall and the characteristic impedance of the rf cavity structure. It is defined as the ratio of the rf power stored in the cavity to the power dissipated on the cavity wall. The shunt impedance, defined as the ratio of the square of the rf voltage seen by the beam to the dissipated power, is an important figure of merit in cavity design. Generally, the ratio of shunt impedance to Q-factor depends only on the geometry of the cavity and the characteristic impedance, i.e. a higher Q-factor cavity has a higher shunt impedance.

In this section we examine some basic principles in cavity design. Properties of pillbox and coaxial-geometry cavities will be discussed. Some fundamental characteristic parameters, the shunt impedance, the Q-factor, and the filling time, of a resonance cavity will be defined and discussed. At a given resonance frequency, we will show that a resonance cavity can be well approximated by an equivalent RLC circuit. Beam loading and Robinson dipole-mode instability will be addressed. High frequency cavities of linacs will be discussed in Sec. VIII.

VI.1 Pillbox Cavity

We first consider a cylindrically symmetric pillbox cavity [22] of radius b and length ℓ (left plot of Fig. 3.28). Maxwell's equations (see Appendix B Sec. IV) for electromagnetic fields inside the cavity are

$$\nabla\cdot\vec{B}=0,\quad \nabla\times\vec{B}=\mu\epsilon\frac{\partial\vec{E}}{\partial t},\quad \nabla\cdot\vec{E}=0,\quad \nabla\times\vec{E}=-\frac{\partial\vec{B}}{\partial t},\tag{3.236}$$

where ϵ and μ are dielectric permittivity and permeability of the medium. The EM waves in the cavity can conveniently be classified into transverse magnetic (TM) mode, for which the longitudinal magnetic field is zero, and transverse electric (TE) mode, for which the longitudinal electric field is zero. The TM modes are of interest for beam acceleration in the rf cavity, while the TE modes can be used for beam deflection.

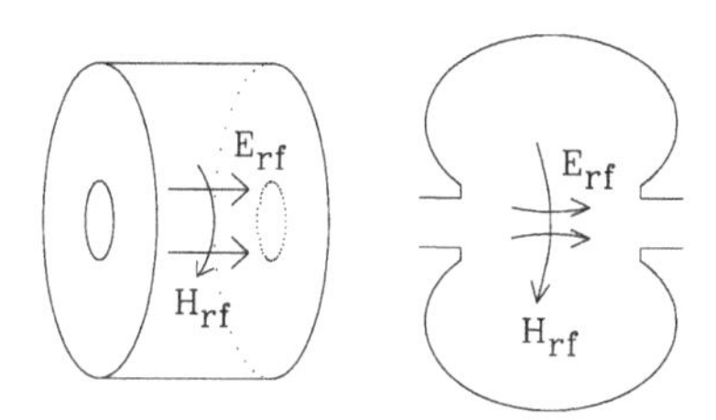

Figure 3.28: Schematic drawings of high frequency cavities. Left: pill-box cavity with disk load; right: nose-cone cavity. Although their names and shapes are different, these high frequency cavities have similar basic features.

An an ideal conducting surface with infinite conductivity, the electromagnetic fields satisfy $\hat{n}\times\vec{E}=0$ and $\hat{n}\cdot\vec{H}=0$, where $\hat{n}$ is the vector normal to the conducting surface. There is *no tangential component of electric field*, and *no normal component of magnetic field.* Assuming a time dependence factor $e^{j\omega t}$ for electric and magnetic fields, the TM standing wave modes in cylindrical coordinates (r,ϕ,s) are (see Appendix B Sec. IV)

$$\begin{cases} E_s = A\,k_r^2\; J_m(k_r r)\cos m\phi\cos ks \\ E_r = -A\,kk_r\; J'_m(k_r r)\cos m\phi\sin ks \\ E_\phi = A\,(mk/r)J_m(k_r r)\sin m\phi\sin ks \\ B_s = 0 \\ B_r = -jA\,(m\omega/c^2 r)\; J_m(k_r r)\sin m\phi\cos ks \\ B_\phi = -jA\,(\omega k_r/c^2)\; J'_m(k_r r)\cos m\phi\cos ks \end{cases} \tag{3.237}$$

where A is a constant, $s=0$ and ℓ correspond to the beginning and end of the pillbox cavity, m is the azimuthal mode number, k, k_r are wave numbers in the longitudinal and radial modes, and $\omega/c=\sqrt{k^2+k_r^2}$. The standing wave can be decomposed into traveling waves in the $+\hat{s}$ and $-\hat{s}$ directions. The solution is chosen so that $E_r=0$ and $E_\phi=0$ at $s=0$.

The longitudinal wave number k is determined by the boundary condition that $E_r=0$ and $E_\phi=0$ at $s=0$ and ℓ, and the radial wave number is determined by the boundary condition with $E_s=0$ and $E_\phi=0$ at $r=b$, i.e.

$$k_{s,p}=\frac{p\pi}{\ell},\quad p=0,1,2,\cdots,\qquad\qquad k_{r,mn}=\frac{j_{mn}}{b}, \tag{3.238}$$

where j_{mn}, listed in Table B.1, are zeros of Bessel functions $J_m(j_{mn})=0$. The resonance wave number k for mode number (m,n,p) is

$$k_{mnp}=\sqrt{k_{r,mn}^2+k_{s,p}^2}=\sqrt{\frac{j_{mn}^2}{b^2}+\frac{p^2\pi^2}{\ell^2}}=\frac{\omega_{mnp}}{c}=\frac{2\pi}{\lambda_{mnp}}. \tag{3.239}$$

The lowest frequency mode is usually called the fundamental mode. Other resonance frequencies are called high order modes (HOM). A good cavity design is to damp HOMs without affecting the fundamental mode. The EM field of the lowest mode TM_{010} ($k_{s,p} = 0$) is

$$E_s = E_0 J_0(kr), \quad B_\phi = j\frac{E_0}{c} J_1(kr), \quad k_{010} = \frac{2.405}{b}, \; \lambda = \frac{2\pi b}{2.405}. \tag{3.240}$$

For example, a 3 GHz structure corresponds to $\lambda = 10$ cm and $b = 3.8$ cm. Such a structure is usually used for high frequency cavities. Since $k_{s,p} = 0$ for the TM_{010} mode, the phase velocity $\omega/k_{s,p} = \infty$. Thus beam particles traveling at speed $v \le c$ can not synchronize with the electromagnetic wave and receive net acceleration.

To slow down the phase velocity, the cavity is loaded with one beam hole with an array of cavity geometries and shapes. Figure 3.28 shows high frequency cavities with disk and nose-cone loaded geometries. Many different geometric shapes are used in the design of high frequency cavities, but their function and analysis are quite similar. All cavities convert TEM wave energy into TM mode to attain a longitudinal electric field. We will return to this subject in Sec. VIII.

VI.2 Low Frequency Coaxial Cavities

Lower frequency rf systems usually resemble coaxial wave guides, where the length is much larger than the width. Figure 3.29 shows an example of a coaxial cavity. The TEM wave in the coaxial wave guide section is converted to the TM mode at the cavity gap through the capacitive load. When the cavity is operating in 50 to 200 MHz range, it requires a very small amount of ferrite for tuning.[45] When the cavity is operating at a few MHz range, ferrite rings in the cavity are needed to slow down EM waves. The ferrite is biased with magnetic field bias frequency tuning.[46]

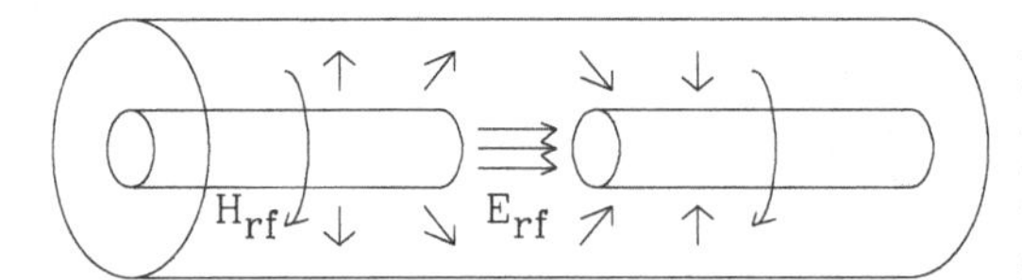

Figure 3.29: Schematic drawing of a low frequency coaxial cavity. Note that the TEM wave is matched to a TM wave at the capacitive loaded gap for the acceleration electric field.

Using the wave guide transmission line theory, characteristic properties of rf systems can be analyzed. Let r_1 and r_2 be the inner and outer radii of a wave guide.

[45]Ferrite is magnetic ceramic material that combines the property of high magnetic permeability and high electric resistivity. The material is made of double oxide spinel $\mathrm{Fe_2O_3}MO$, where M can be Mn, Zn, Cr, Ni, etc. Ferrites are commonly used in frequency synthesis devices, Touch-Tone telephone, low loss microwave devices, etc. Application in accelerator can be found in induction linac, frequency tuning for rf cavities, kickers, etc.

[46]W.R. Smythe, *IEEE Trans. Nucl. Sci.*, **NS-30**, 2173 (1983).

The inductance $\tilde{L}$ and the capacitance $\tilde{C}$ per unit length of the concentric coaxial wave guides are

$$\tilde{L} = \frac{\mu}{2\pi}\ln\frac{r_2}{r_1} + \frac{\mu_c\delta_{\rm skin}}{4\pi}(\frac{1}{r_1}+\frac{1}{r_2}), \qquad \tilde{C} = \frac{2\pi\epsilon}{\ln(r_2/r_1)}, \tag{3.241}$$

where μ_c is the permeability of the conductor, $\delta_{\rm skin} = \sqrt{2/\omega\mu_c\sigma}$ is the skin depth of flux penetration. The inductance and capacitance of the coaxial cavity structure are respectively $L = \tilde{L}\ell$ and $C = \tilde{C}\ell$, where ℓ the length of the structure. Neglecting the flux penetration in the conductor, the resonating frequency and the cavity length for the quarter-wave mode are

$$v = \frac{\omega}{k} = \frac{1}{\sqrt{\tilde{L}\tilde{C}}} = \frac{1}{\sqrt{\epsilon\mu}}, \qquad \ell = \frac{\lambda}{4} = \frac{v}{4f} = \frac{75}{f[\rm MHz]}\frac{v}{c}\ [\rm m]. \tag{3.242}$$

To shorten the length of the cavity ℓ, we need to slow down the wave speed by the ferrite materials. For a cavity operating beyond 20 MHz, ferrite can be used only for tuning purposes. At frequencies below tens of MHz, the rf cavities must be ferrite loaded in order to fit into the available free space in an accelerator. Typically the permittivity and magnetic permeability of ferrite are about 10 ϵ_0 and 10 – 500 μ_0. When a biased field is applied to the ferrite core, the magnetic permeability can be tuned to match the change of the particle revolution frequency.

To understand the capacitive loading that converts the TEM wave into the TM wave at the cavity gap, we study the rf electromagnetic wave in the wave guide. The characteristic impedance of a wave guide is

$$Z_c = R_c = \sqrt{\frac{L}{C}} \approx \frac{1}{2\pi}\sqrt{\frac{\mu}{\epsilon}}\ \ln\frac{r_2}{r_1}. \tag{3.243}$$

Now, we consider an ideal lossless transmission line, where the electromagnetic field has no longitudinal component. Assuming a time dependent factor $e^{j\omega t}$, the current and voltage across the rf structure are (see Exercise 3.6.3)

$$I(s,t) = I_0\cos ks - j(V_0/R_c)\sin ks, \quad V(s,t) = V_0\cos ks - jI_0R_c\sin ks, \tag{3.244}$$

where $k = \frac{2\pi}{\lambda} = \frac{\omega}{v}$ is the wave number of the line, $v = 1/\sqrt{\epsilon\mu}$ is wave speed in the medium, s is the distance from one end of the transmission line, V_0 and I_0 are the voltage and current at the cavity gap.

For a standing wave, where the end of the transmission line is shorted, the boundary condition at the shorted side is $V = 0$. we find $V_0\cos k\ell - jI_0R_c\sin k\ell = 0$, and the current at the shorted side is $I_\ell = I_0/\cos(k\ell)$. The resulting rf current and voltage become $I(s,t) = I_\ell\cos[k(s-\ell)]$ and $V(s,t) = -jI_\ell R_c\sin[k(s-\ell)]$. The line input impedance at the gap becomes

$$Z_{\rm in} = \frac{V(0,t)}{I(0,t)} = +jR_c\tan k\ell. \tag{3.245}$$

The line impedance is inductive if $k\ell < \pi/2$. The length of the line is chosen to match the gap capacitance at a required resonance frequency, i.e.

$$Z_{\rm in} + Z_{\rm gap} = 0, \quad \text{or} \quad \cot k\ell_{\rm r} = \omega C_{\rm gap} R_{\rm c} \equiv g, \tag{3.246}$$

where $Z_{\rm gap} = -j/(\omega C_{\rm gap})$ is the gap impedance, $C_{\rm gap}$ is the capacitance of a half gap, and g is the capacitive coupling factor of the cavity, For example, a total capacitance of 10 pF implies that $C_{\rm gap} = 20$ pF. The length $\ell_{\rm r}$ of one-half cavity, the gap capacitance, the biased current, and the external loading capacitance can be designed to attain a resonance condition for a given frequency range. A load capacitor may be shunted to decrease the resonance frequency or minimize the cavity length. The effective capacitance is $C_{\rm gap} + C_{\rm load}$.

In principle, for a given $\ell_{\rm r}, R_{\rm c}$, and $C_{\rm gap}$, there are many resonance frequencies that satisfy Eq. (3.246). The lowest frequency is called the fundamental TEM mode. If the loading capacitance is small, the resonance condition of Eq. (3.246) becomes $k\ell_{\rm r} = \pi/2$, i.e. $\ell_{\rm r} = \lambda/4$: the length of the coaxial cavity is equal to 1/4 of the wavelength of the TEM wave in the coaxial wave guide. Such a structure is also called a quarter-wave cavity. The gap voltage of the coaxial cavity is

$$V_{\rm rf} = +jI_\ell R_{\rm c} \sin k\ell_{\rm r} = +j\frac{I_\ell R_{\rm c}}{\sqrt{1+g^2}}. \tag{3.247}$$

A. Shunt impedance and Q-factor

The surface resistivity $R_{\rm s}$ of the conductor and the resistance R of a transmission line are

$$R_{\rm s} = \sqrt{\frac{\mu_c \omega}{2\sigma}}, \quad R = \frac{R_{\rm s}\ell}{2\pi}(\frac{1}{r_1} + \frac{1}{r_2}), \tag{3.248}$$

where σ is the conductivity of the material, ω is the rf frequency, r_1 and r_2 are the inner and outer radii of the transmission line, and ℓ is the length. Thus the quality factor becomes

$$\mathrm{Q} = \frac{R_{\rm c}}{R} = \frac{\omega L}{R} \approx \frac{2r_1 r_2}{(r_1+r_2)\delta_{\rm skin}}\frac{\mu}{\mu_c}\ln\frac{r_2}{r_1}. \tag{3.249}$$

Table 3.6 lists typical Q-factors for a copper cavity as a function of cavity frequency, where we have used $\ln(r_2/r_1) \approx 1$ and $r_1 \approx 0.05$ m, and $\sigma_{\rm Cu} \approx 5.8 \times 10^7$ $[\Omega\mathrm{m}]^{-1}$ at room temperature.

An important quantity in the design and operation of rf cavities is the shunt impedance. This is the resistance presented by the structure to the beam current at the resonance condition. The total power of dissipation $P_{\rm d}$ of the transmission line cavity and the shunt impedance become

$$P_{\rm d} = \frac{I_\ell^2 R}{2}\int_0^{k\ell_{\rm r}} \cos^2 x\, dx = \frac{I_\ell^2 R}{4(1+g^2)}[(1+g^2)\cot^{-1} g + g], \tag{3.250}$$

Table 3.6: Some characteristic properties of coaxial RF cavities made of copper.

f [MHz]	1	10	100
$\delta_{\rm skin}$ [μm]	66.	21.	6.6
r_1	0.05	0.05	0.05
Q	1100	3500	11000

$$R_{\rm sh} = \frac{|V_{\rm rf}|^2}{2P_{\rm d}} \qquad \text{or} \qquad \frac{R_{\rm sh}}{\rm Q} = \left\{\frac{4}{\pi}\frac{\pi/2}{(1+g^2)\cot^{-1}g + g}\right\} R_{\rm c} \tag{3.251}$$

The expression in brackets is a shunt-impedance geometry factor due to the equivalent gap capacitance loading, i.e. $C_{\rm eq} = C_{\rm gap} + C_{\rm load}$. Figure 3.30 shows the geometry factor (solid) and the phase advance $k\ell_{\rm r}$ vs the the capacitive coupling factor g.

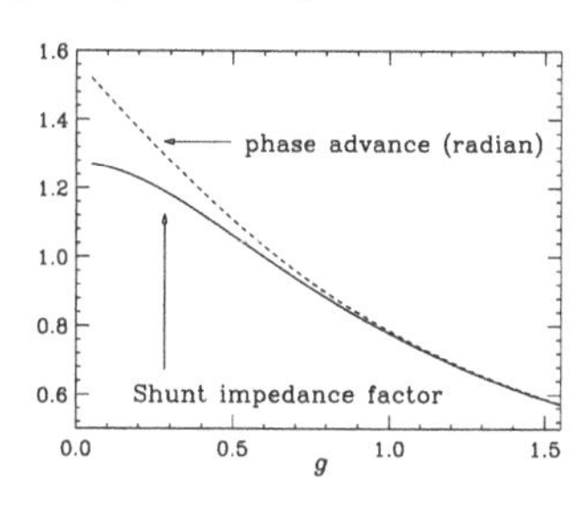

Figure 3.30: The shunt-impedance geometric factor of the bracket in Eq. (3.251) (solid) and the phase advance $k\ell_{\rm r} = \arctan(1/g)$ (dashed) vs the capacitive coupling factor. As the gap capacitance increases, the shunt impedance decreases.

From the transmission-line point of view, the cavity gap presents a capacitance and resistive load shown in Fig. 3.31, where $Z_{\rm in} = j\omega L_{\rm eq}$, and $C_{\rm eq} = C_{\rm gap} + C_{\rm load}$. The matching condition of Eq. (3.246) implies that the reactance of the cavity is zero on resonance, and the effective impedance is $R_{\rm sh}$. The resonance frequency and the Q-factor of the equivalent RLC-circuit are $\omega_{\rm r} = 1/\sqrt{L_{\rm eq}C_{\rm eq}}$ and $\text{Q} = R_{\rm sh}\sqrt{C_{\rm eq}/L_{\rm eq}}$.

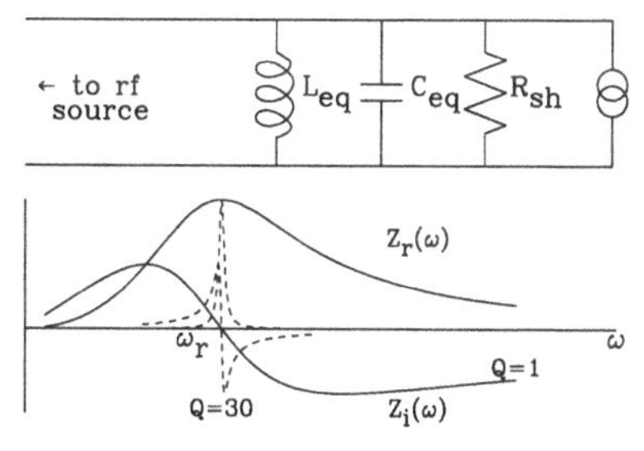

Figure 3.31: Top: Schematic drawing of an equivalent circuit of a cavity. The input impedance of the wave guide is represented by an equivalent inductance. The wave guide is loaded with capacitive cavity-gap and real shunt impedance. Bottom: Plot of the impedance of Eq. (3.252). The solid lines are the real and the imaginary parts of a resonance impedance with Q=1, and the dashed lines are the corresponding parts at Q=30.

The impedance of the rf system, represented by a parallel RLC circuit, is

$$Z = \left(\frac{1}{R_{\rm sh}} + j\omega C_{\rm eq} + \frac{1}{j\omega L_{\rm eq}}\right)^{-1} = \frac{R_{\rm sh}}{1 + j\text{Q}(\frac{\omega}{\omega_{\rm r}} - \frac{\omega_{\rm r}}{\omega})} \approx R_{\rm sh}\cos\psi e^{-j\psi}, \tag{3.252}$$

$$\psi = \tan^{-1}\frac{2\text{Q}(\omega-\omega_{\rm r})}{\omega_{\rm r}}, \qquad \omega_{\rm r} = \frac{1}{\sqrt{L_{\rm eq}C_{\rm eq}}}, \qquad \text{Q} = R_{\rm sh}\sqrt{\frac{C_{\rm eq}}{L_{\rm eq}}} = \frac{R_{\rm sh}}{R_{\rm c}},$$

where $L_{\rm eq}$ and $C_{\rm eq}$ are the equivalent inductance and capacitance, ψ is the cavity detuning angle, and $R_{\rm c}$ is the characteristic impedance. At the resonance frequency $\omega_{\rm r}$ particles see a pure resistive load with an effective resistance $R_{\rm sh}$. The rf system becomes capacitive at $\omega > \omega_{\rm r}$, and inductive at $\omega < \omega_{\rm r}$. The bottom plot of Fig. 3.31 shows the real and imaginary parts of Eq. (3.252) for Q=1 and Q=30.

Accelerator cavities usually contain also many parasitic HOMs. Each HOM has its shunt impedance and Q-value. If the frequency of one of the HOMs falls on a synchrotron or betatron sideband, the beam can be strongly affected by the parasitic rf driven resonance. Correction, detuning, and lowering the Q-factor of these sidebands are very important in rf cavity design and operation.

B. Filling time

The quality factor defined in Eq. (3.249) is equal to the ratio of the stored power $P_{\rm st}$ to the dissipated power $P_{\rm d}$, Using energy conservation, we find

$$\mathrm{Q} = \frac{P_{\rm st}}{P_{\rm d}} = \frac{\omega W_{\rm s}}{P_{\rm d}}, \tag{3.253}$$

$$\frac{dW_{\rm s}}{dt} = -P_{\rm d} = -\frac{\omega}{\mathrm{Q}} W_{\rm s},$$

$$W_{\rm s} = W_{\rm s0} e^{-\omega t/\mathrm{Q}} = W_{\rm s0} e^{-2t/T_{\rm f0}}, \qquad T_{\rm f0} = \frac{2\mathrm{Q}}{\omega}. \tag{3.254}$$

where $W_{\rm s}$ is the stored energy and the unloaded *filling time* $T_{\rm f0}$ is equal to the time for the electric field or voltage to decay to $1/e$ of its original value.

C. Qualitative feature of rf cavities

Qualitatively, the rf voltage is the time derivative of the total magnetic flux linking orbit (Faraday's law of induction). We assume that a sinusoidal time dependent magnetic flux density with $1/r$ dependence in a coaxial cavity structure. The induced voltage is

$$V_{\rm rf} = \frac{\Delta\Phi}{\Delta t} = f_{\rm rf}\ell \int_{r_1}^{r_2} B(r)dr = f_{\rm rf} B_1 [\ell r_1 \ln\frac{r_2}{r_1} \approx f_{\rm rf} B_1 A, \tag{3.255}$$

where $A = \ell r_1 \ln(r_2/r_1) \approx \ell(r_2 - r_1)$ is the effective area of the ferrite core and B_1 is the peak magnetic flux at $r = r_1$. Because of the logarithmic dependence on r_2, it is inefficient to increase the outer radius of the ferrite core to increase the rf voltage. The peak magnetic flux in Eq. (3.255) depends on the ferrite material.

The quality factor Q is the ratio of stored power to the dissipated power. When there are many dissipative power sources, the loaded Q-factor is

$$\frac{1}{\mathrm{Q_L}} = \frac{P_{\rm d}}{P_{\rm st}} = \frac{1}{\mathrm{Q}_1} + \frac{1}{\mathrm{Q}_2} + \frac{1}{\mathrm{Q}_3} + \cdots,$$

where $P_{\rm st}$ is the power stored in the cavity and $P_{\rm d}$ is the total dissipated power. The shunt impedance of an rf structure is the resistance presented to the beam current at the resonance condition, i.e.

$$R_{\rm sh} = \frac{|V_{\rm rf}|^2}{2P_{\rm d}} = \frac{R_{\rm c}P_{\rm st}}{P_{\rm d}} \approx R_{\rm c}{\rm Q_L}. \tag{3.256}$$

The quality factor Q of the ferrite loaded cavity is dominated by the Q value of the ferrite material itself, i.e. ${\rm Q_{ferrite}} \approx 10 - 300$, which alone is not adequate for the required frequency tuning range. Frequency tuning can be achieved by inducing a shunt capacitor and a DC magnetic field in the ferrite core. With an external magnet or bias current that encircles the ferrite without contributing a net rf flux, the effective permeability for rf field can be changed.

Since the Q-value of ferrite is relatively low, power dissipation in ferrite is important. The dissipation power is

$$P_{\rm d} = \frac{f_{\rm rf}}{Q}\left[\frac{\ell}{2\mu}\int_{r_1}^{r_2} B^2(r)2\pi r dr\right] = \frac{\pi r_1 f_{\rm rf} B_1^2 A}{\mu {\rm Q}} \approx \frac{\pi r_1 V_{\rm rf}^2}{A\,[\mu f_{\rm rf}{\rm Q}]} \tag{3.257}$$

where the bracket is the average magnetic energy dissipated in each cycle. The power dissipation in a ferrite cavity is inversely proportional to $(\mu f_{\rm rf}{\rm Q})$, which characterizes ferrite materials. Since the power is inversely proportional to the effective area A, we need a large volume of ferrite to decrease the flux density in order to minimize energy loss for achieving high rf voltage at low frequencies,

At rf frequencies above tens of MHz, the cavity size (normally 1/2 or 1/4 wavelength) becomes small enough that a resonant structure containing little or no ferrite may be built with significantly lower power loss at ${\rm Q} \approx 10^4$ with a narrower bandwidth. At frequencies of a few hundred MHz, where adequate and efficient rf power sources are commercially available, the main portion of the rf cavity can be made of copper or aluminum with a small amount of ferrite used for tuning. The cavity can still be considered as a coaxial wave guide, and Eqs. (3.241) to (3.255) remain valid. The characteristic impedance $R_{\rm c}$ of Eq. (3.243) is about 60 Ω. The stored power is $I^2R_{\rm c}$ and the power dissipation is I^2R, where I is the surface current and R of Eq. (3.248) is the surface resistance of the structure. At frequencies above a few hundred MHz, resonance frequency can be tuned only by a slotted tuner or by physically changing the size of the cavity.

D. The rf cavity of the IUCF cooler injector synchrotron

The IUCF cooler injector synchrotron (CIS) is a low energy booster for the IUCF cooler ring. It accelerates protons (or light ions) from 7 MeV to 225 MeV. The cavity is a quarter-wave coaxial cavity with heavy capacitance loading.To make the cavity length reasonably short and to achieve rapid tuning, required for synchrotron acceleration, ten Phillips 4C12 type ferrite rings are used. The μ of the ferrite material is changed by a superimposed DC magnetic field provided by an external quadrupole magnet. The ferrite rings return the magnet flux between the two adjacent quadrupole tips (Fig. 3.32).

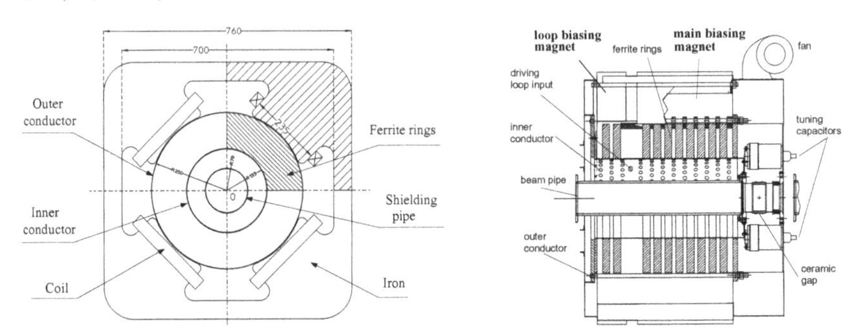

Figure 3.32: The cross section (left) and the longitudinal view of the CIS rf cavity. The external quadrupole magnet provides biased field in ferrite rings to change the effective permeability (courtesy of Alex Pei).

Analysis of such a field shows that the field direction is mostly parallel to the rf field, i.e. along the azimuthal direction, except near the tips of the quadrupole, where the biased fields in the ferrite rings are perpendicular to the rf field. In the working region of the ferrite biasing strength, the effect of the perpendicular component on ferrite rf-μ is small. The effective rf-μ, to first order in wave propagation, is determined by dB/dH, as in parallel biasing analysis, rather than B/H, as in perpendicular analysis. The phenomenon of gyromagnetic resonance associated with perpendicular biasing, however, needs to be considered and avoided in the design of the cavity.

The advantages of using an external biasing magnet include making it possible to separate the rf field from the biasing elements, and the rf field in the cavity will not be affected by the biasing structure. As many windings of the bias coils as practical can be used — resulting in a smaller amperage requirement for the bias supply. In CIS and the IUCF cooler ring, the bias supplies for these external quadrupole biasing magnet type cavities are rated at only 20 A. If the biasing field is to be produced only by a bias winding threaded through the rf cavity, the number of windings is usually limited to no more than a few turns because of possible resonance and arcing. It usually takes 1000 A or more to bias such a cavity.

As the frequency changes, the power loss in ferrite material varies (usually increasing as frequency increases). As a result, it has been difficult to feed the rf generator power to the cavity efficiently because of the high voltage standing wave ratio (VSWR) caused by impedance mismatch (see Appendix B.IV.3). In the CIS cavity this problem was solved by dividing the ferrite rings into two sections; the strength of the biasing magnet in each section can be adjusted by the coupling loop.

The coupling coefficient can be used to compensate the change in the gap impedance, and the input impedance can be maintained constant to match the transmission line impedance of the rf amplifier. The CIS cavity is thus able to operate with a 10:1 frequency ratio with high efficiency, due to the higher impedance of a resonant structure and optimized amplifier coupling.

The loading capacitor can further reduce the length requirement of coaxial cavities and can also be used conveniently to switch frequency bands. For example, the CIS cavity can be operated at 0.5 – 5 MHz or 1 – 10 MHz by varying its loading capacitor. Including the load capacitance, $C_{\rm load}$, the circuit matching (resonance) condition of Eq. (3.246) becomes $\cot(k\ell_{\rm r}) = R_{\rm c}\omega(C_{\rm gap} + C_{load})$.

E. Wake-function and impedance of an RLC resonator model

If we represent a charged particle of charge q by $I(t) = q\delta(t) = (1/2\pi)\int qe^{j\omega t}d\omega$, the energy loss due to the passage of an rf gap, represented by an RLC resonator model, is

$$\Delta U = \int |I(\omega)|^2 Z(\omega)d\omega = 2k_{\rm r}q^2, \qquad k_{\rm r} = \frac{\omega_{\rm r}R_{\rm sh}}{4\pi Q}, \tag{3.258}$$

where $k_{\rm r}$ is the *loss factor* of the impedance at frequency $\omega_{\rm r}$. This means that the passing particle loses energy and induces a wakefield in the cavity.

The longitudinal impedance is the Fourier transform of the wake function, and thus the wake function is the inverse Fourier transform of the impedance, For the RLC resonator model, the wake function becomes (see Exercise 3.6.5)

$$\begin{aligned} Z(\omega) &= \int_0^\infty W(t)e^{-j\omega t}dt = \int_{-\infty}^\infty W(t)e^{-j\omega t}dt, \\ W(t) &= \frac{1}{2\pi}\int Z(\omega)e^{j\omega t}d\omega = 4\pi k_{\rm r}\left[\cos\tilde{\omega}_{\rm r}t - \frac{1}{\tilde{\omega}_{\rm r}T_{\rm f0}}\sin\tilde{\omega}_{\rm r}t\right]e^{-t/T_{\rm f0}}\,\Theta(t), \end{aligned} \tag{3.259}$$

where $\Theta(t) = 1$ if $t > 0$, and 0 if $t < 0$; $T_{\rm f0} = 2Q/\omega_{\rm r}$ is unloaded filling time defined in Eq. (3.254); and $\tilde{\omega}_{\rm r} = \omega_{\rm r}\ (1 - (1/4Q^2))^{1/2}$. If the filling time is long, then the wake potential is a sinusoidal function with angular frequency $\tilde{\omega}_{\rm r}$, i.e.

$$W(t) \approx 4\pi k_{\rm r}e^{-t/T_{\rm f0}}\cos\omega_{\rm r}t.$$

Thus the filling time corresponds also to the wakefield decay time. When beams pass repetitively through the cavity, the effective voltage is the sum of the voltage supplied by the generator current and the wakefields of all beams. Beam loading is important in the design and operation of rf cavities.

VI.3 Beam Loading

A passing beam charge can induce wakefield in an rf cavity. The beam induced rf voltage can alter the effective voltage at the rf gap. Without proper compensation, the resulting rf voltage acting on the passing beam may cause beam deceleration in an uncontrollable manner. Thus beam loading needs to be considered in the operation of rf cavities.

A. Phasor

The sinusoidal electromagnetic fields and voltages in a standing wave rf structure can be expressed as complex quantities, i.e. $V = V_0 e^{j(\omega t+\theta)}$, where ω is the frequency, θ is a phase angle, and V_0 is the amplitude of the rf voltage. The rf voltage seen by the beam is the projection of the rotating vector on the real axis. Now, we choose a coordinate system that rotates with the rf frequency, and thus the rf voltage is stationary in this rotating coordinate system. In the rotating coordinate system, the voltage vector is called a phasor: $\tilde{V} = Ve^{j\theta}$ with $V_0 \cos\theta = V_0 \sin\phi_s$, where ϕ_s is the synchronous phase angle. Phasors are manipulated by using usual rules of complex vector algebra. The properties of rf fields can be studied by using graphic reconstruction in phasor diagrams.

B. Fundamental theorem of beam loading

The cavity provides a longitudinal electric field for particle acceleration. However, when a charged particle passes through the cavity, the image current on the cavity wall creates an electric field that opposes the particle motion. The question arises: what fraction of the electric field or voltage created by the beam affects the beam motion? The question can be addressed by the fundamental theorem of beam loading due to Wilson [see P. B. Wilson, *AIP Conf. Proc.* **87**, 452 (1981).]: *A charged particle sees exactly $\frac{1}{2}$ of its own induced voltage.*

To prove this fundamental theorem, we assume that the stored energy in a cavity in any given mode is $W = \alpha V^2$. We assume that a fraction f of the induced voltage is seen by the inducing particle, and the effective voltage is $V_{\rm e} = fV_{\rm b}$, where $V_{\rm b}$ is the induced voltage in each passage. We assume further that the induced voltage lies at phase angle χ with respect to the inducing current or charge.

Now, we consider two identical charged particles of charge q, separated by phase angle θ, passing through the cavity. The total energy deposited in the cavity and the energy loss by these two particles are respectively

$$W_{\rm c} = \alpha|\tilde{V}_{\rm b}(1) + \tilde{V}_{\rm b}(2)|^2 = \alpha\left(2V_{\rm b}\cos\frac{\theta}{2}\right)^2 = 2\alpha V_{\rm b}^2(1+\cos\theta),$$
$$\Delta U = [qV_{\rm e}] + [qV_{\rm e} + qV_{\rm b}\cos(\chi+\theta)],$$

where the first and second brackets are the energy losses due to the first and second particles respectively. From the conservation of energy, $\Delta U = W_{\rm c}$, we obtain

$$\chi = 0, \quad V_{\rm b} = \frac{q}{2\alpha}, \quad V_{\rm e} = \frac{1}{2}V_{\rm b}, \quad f = \frac{1}{2}. \tag{3.260}$$

The result can be summarized as follows:

1. The induced voltage of a beam must have a phase maximally opposite the motion of the charge, i.e. the phase angle $\chi = 0$.
2. $V_{\rm e} = V_{\rm b}/2$. The particle sees exactly 1/2 of its own induced voltage.
3. $W_{\rm c} = \alpha V_{\rm b}^2 = q^2/4\alpha = kq^2$, where k is the loss factor, $k = V_{\rm b}^2/(4W_{\rm c})$.
4. $V_{\rm b} = 2kq$ or $V_{\rm e} = kq$.

C. Steady state solution of multiple bunch passage

Consider an infinite train of bunches, separated by time $T_{\rm b}$, passing through an rf cavity gap. When the cavity is on resonance, the induced voltage seen by the particle is

$$V_{\rm b} = \frac{1}{2}V_{\rm b0} + V_{\rm b0}(e^{-(\lambda+j\phi)} + e^{-2(\lambda+j\phi)} + \cdots) = V_{\rm b0}(-\frac{1}{2} + \frac{1}{1-e^{-(\lambda+j\phi)}}), \tag{3.261}$$

where $\phi = (\omega - \omega_{\rm r})T_{\rm b}$ is the relative bunch arrival phase with respect to the cavity phase at the rf gap, $\omega_{\rm r}$ is the resonance frequency of the rf cavity, and $\lambda = T_{\rm b}/T_{\rm f}$ is the decay factor of the induced voltage between successive bunch passages, and $T_{\rm f} = 2{\rm Q_L}/\omega_{\rm r}$ is the cavity time constant or the cavity filling time. Here $\rm Q_L$ is the loaded cavity quality factor, taking into account the generator resistance $R_{\rm g}$ in parallel with the RLC circuit of the cavity, i.e.

$${\rm Q_L} = \frac{R_{\rm sh}R_{\rm g}}{(R_{\rm sh}+R_{\rm g})R_{\rm c}} = \frac{\rm Q_0}{1+d}, \quad d = \frac{R_{\rm sh}}{R_{\rm g}}. \tag{3.262}$$

The filling time of the loaded cavity is reduced by a factor $1/(1+d)$.

The cavity detuning angle ψ and the rf phase shift are

$$\psi = \tan^{-1}\left[\frac{2{\rm Q_L}(\omega-\omega_{\rm r})}{\omega_{\rm r}}\right] = \tan^{-1}\left[(\omega-\omega_{\rm r})T_{\rm f}\right], \tag{3.263}$$

$$\phi = (\omega-\omega_{\rm r})T_{\rm b} = +(T_{\rm b}/T_{\rm f})\tan\psi = +\lambda\tan\psi, \tag{3.264}$$

where ω is the cavity operation frequency. For rf cavities used in accelerators, we have $\lambda = T_{\rm b}/T_{\rm f} = \omega_{\rm r}T_{\rm b}/2{\rm Q_L} \ll 1$, and the induced voltage seen by the beam is

$$V_{\rm b} = I_{\rm i}R_{\rm sh}\lambda(-\frac{1}{2} + \frac{1}{1-e^{-(\lambda+j\phi)}}) \approx I_{\rm i}\frac{R_{\rm sh}}{(1+d)}\cos\psi e^{-j\psi} \quad (\lambda\to 0), \tag{3.265}$$

where $I_{\rm i}$ is the rf image current, $V_{\rm b0} = I_{\rm i}R_{\rm sh}T_{\rm b}/T_{\rm f}$, and the term $-1/2$ is neglected. *The beam induced voltage across the rf gap at the steady state is exactly the rf image current times the impedance of the rf cavity* (see Eq. (3.252)).

VI.4 Beam Loading Compensation and Robinson Instability

To provide particle acceleration in a cavity, we need a generator rf current $\tilde{I}_0 = I_0 e^{j\theta}$ with phase angle θ so that the voltage acting on the beam is $V_{\rm acc} = V_{\rm g}\cos\theta = V_{\rm g}\sin\phi_{\rm s}$, where $\phi_{\rm s}$ is the synchronous phase angle. It appears that the rf system would be optimally tuned if it were tuned to on-resonance so that it had a resistive load with $V_{\rm g} = I_0 R_{\rm sh}$. However, we will find shortly that the effect of beam loading would render such a scheme unusable.

When a *short* beam bunch passes through the rf system, the image rf current $I_{\rm i}$ generated by the beam is twice the DC current, as shown in Eq. (2.125). The beam will induce $I_{\rm i}R_{\rm sh}$ across the voltage gap (see dashed line in Fig. 3.33). The voltage seen by the beam is the sum of the voltage produced by the generator current and the beam induced current. Thus the stable phase angle $\phi_{\rm s}$ of the synchrotron motion will be changed by the induced voltage. This is shown schematically in Fig. 3.33 (left), where the required gap voltage $I_0 R_{\rm sh}$ and the synchronous phase angle $\phi_{\rm s}$ are altered by the voltage induced by the image current. The projection of the resultant vector V_0 on the real axis is negative, and results in deceleration of the beam.

One way to compensate the image current is to superimpose, on the generator current, current directly opposite to the image current. Such a large rf generator current at a phase angle other than that of the rf acceleration voltage is costly and unnecessary.

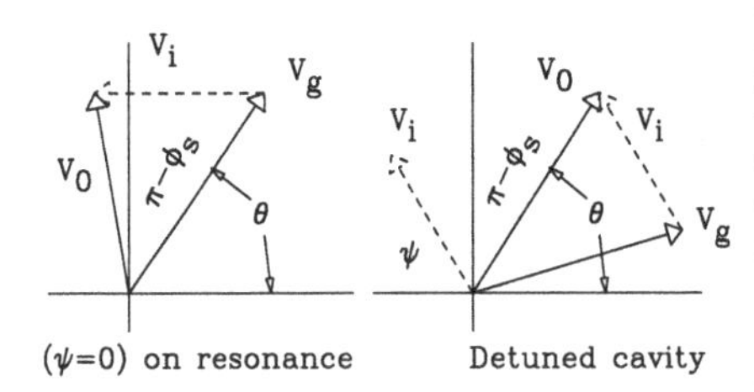

Figure 3.33: Phasor diagrams for beam loading compensation. Left: The beam loading voltage for a cavity tuned on resonance. The combination of generator voltage $V_{\rm g}$ and induced voltage $V_{\rm i}$ gives rise to a decelerating field V_0. Right: When the cavity is detuned to a detuning angle ψ, the superposition of the generator voltage $V_{\rm g}$ and the beam loading voltage $V_{\rm i}$ gives a proper cavity voltage V_0 for beam acceleration.

An alternative solution is to detune the accelerating structure.[47] The detuning angle and the generator current are adjusted so that the resultant voltage has a correct magnitude and phase for beam acceleration. This scheme will minimize the generator current. We define the following phasor currents and voltages for the analysis of this

[47] J.E. Griffin, *AIP Conf. Proc.* **87**, 564 (1981); F. Pedersen, *IEEE Tran. Nucl. Sci.* **NS-32**, 2138 (1985); D. Boussard, CERN 91-04, p. 294 (1991).

problem.

$$\begin{cases} \tilde{I}_0 = I_0 e^{j\theta} & \text{generator current necessary for accelerating voltage in the absence of beam} \\ \tilde{I}_g = I_g e^{j(\theta+\theta_g)} & \text{required generator current with beam} \\ \tilde{I}_i = -I_i = -I_b & \text{rf beam image current, } I_i \text{ is a positive quantity} \\ \tilde{V}_g = \tilde{I}_g R_{sh} \cos\psi e^{-j\psi} & \text{voltage induced by generator current} \\ \tilde{V}_0 = V_0 e^{j\theta} & \text{required rf accelerating voltage} \\ \psi = \tan^{-1}\left[\frac{2Q(\omega-\omega_r)}{\omega_r}\right] & \text{detuning angle} \\ Y = I_i/I_0 & \text{ratio of image current to unloaded generator current} \end{cases}$$

The equation for a proper accelerating voltage is

$$\begin{aligned} \tilde{V}_0 &= \tilde{V}_g + \tilde{V}_i \\ I_0 R_{sh} e^{j\theta} &= [I_g e^{j(\theta+\theta_g)} - I_i] R_{sh} \cos\psi e^{-j\psi}. \end{aligned} \tag{3.266}$$

Here the induced voltage is derived from the steady state beam loading. By equating the real and imaginary parts, we obtain

$$\tan\theta_g = \frac{\tan\psi - Y\sin\theta}{1 + Y\cos\theta}, \quad I_g = I_0 \frac{1 + Y\cos\theta}{\cos\theta_g}, \tag{3.267}$$

where θ_g is the phase angle of the generator current relative to the ideal $\tilde{I}_0$. The optimal operating condition normally corresponds to $\theta_g = 0$, which minimizes I_g, i.e. the generator current is optimally chosen to be parallel to $\tilde{I}_0$, and Eq. (3.267) reduces to

$$I_g = I_0(1 + Y\sin\phi_s), \quad \tan\psi = Y\cos\phi_s. \tag{3.268}$$

Figure 3.33 (right) shows the beam loading phasor diagram with a detuned cavity angle ψ. The resultant vector of the generator voltage and the image current voltage is the effective accelerating voltage for the beam.

A. Robinson dipole mode instability

In accelerators, beams experience many sources of perturbation such as power supply ripple, mis-injection, mismatched beam profile, rf noise, voltage error, etc. Beam stability may sometimes need sophisticated active feedback systems. The topic of control and feedback is beyond the scope of this textbook. Here, we discuss only the dipole mode stability condition related to beam loading, studied by Robinson in 1964 [K.W. Robinson, CEA report CEAL-1010 (1964)].

We consider a small perturbation by shifting the arrival time of all bunches by a phase factor ξ. The accelerating rf voltage will be perturbed by the same phase factor,

$$V_{acc} = V_0\cos(\theta - \xi) = V_0\cos\theta + \xi V_0\sin\theta = V_0\sin\phi_s + \xi V_0\cos\phi_s. \tag{3.269}$$

where the first term is the intended accelerating voltage and the second term is the effect of phase perturbation due to an error in arrival time.

The wrong arrival time shifts the image beam current by a phase angle ξ. The perturbation to the image rf current its induced rf voltage are

$$\Delta\tilde{I}_{\rm i} = j\xi\tilde{I}_{\rm i} = -j\xi I_{\rm i},$$
$$\Delta\tilde{V}_{\rm i} = -j\xi I_{\rm i}R_{\rm sh}\cos\psi e^{-j\psi},$$
$$\Delta V_{\rm ir} = \Re\{\Delta\tilde{V}_{\rm i}\} = -\xi Y V_0 \cos\psi\sin\psi,$$

where the beam sees real part $\Delta V_{\rm ir}$ of the induced voltage or the projection of the phasor voltage onto the real axis. The net change in accelerating voltage seen by the *bunch* becomes

$$\Delta V_{\rm acc} = \xi V_0\cos\phi_{\rm s}\left[1 - Y\frac{\sin\psi\cos\psi}{\cos\phi_{\rm s}}\right]. \tag{3.270}$$

A small perturbation in arrival time causes a perturbation in acceleration voltage proportional to the phase shift. If the voltage induced by the image charge is not significant, the bunches in the accelerator will execute synchrotron motion. Thus the equation of motion for the phasor error ξ is (see Exercise 3.6.7)

$$\ddot{\xi} = -\nu_{\rm s}^2\left(1 - Y\frac{\sin\psi\cos\psi}{\cos\phi_{\rm s}}\right)\xi. \tag{3.271}$$

Using Eq. (3.268), we find that the Robinson stability condition becomes

$$1 - Y\frac{\sin\psi\cos\psi}{\cos\phi_{\rm s}} \geq 0 \quad \text{or} \quad 1 - \frac{\sin^2\psi}{\cos^2\phi_{\rm s}} \geq 0. \tag{3.272}$$

This means that Robinson stability requires $\psi < \theta = |\frac{1}{2}\pi - \phi_{\rm s}|$. In general, Eq. (3.272) is applicable to all high order modes. For those modes, Robinson stability can be described as follows.

Below transition energy, with $\cos\phi_{\rm s} > 0$, Robinson stability can be attained by choosing $\sin\psi < 0$, i.e. the cavity frequency is detuned with $\omega < \omega_{\rm r}$. Above transition energy, with $\cos\phi_{\rm s} < 0$, the cavity should be detuned so that $\sin\psi > 0$ or $\omega > \omega_{\rm r}$ in order to gain Robinson stability. Since the stability condition is a function of bunch intensity, instability is a self-adjusting process. Beam loss will appear until the Robinson stability condition can be achieved. Active feedback systems have been used to enhance the stability of bunched beam acceleration [See e.g. D. Boussard, CERN 87-03, p. 626 (1987); CERN 91-04, p. 294 (1991).]

B. Qualitative feature of Robinson instability

Robinson instability can be qualitatively understood as follows. The wakefield produced in a cavity by a circulating bunch is represented qualitatively in Fig. 3.34, where the impedance of the cavity is assumed to be real.

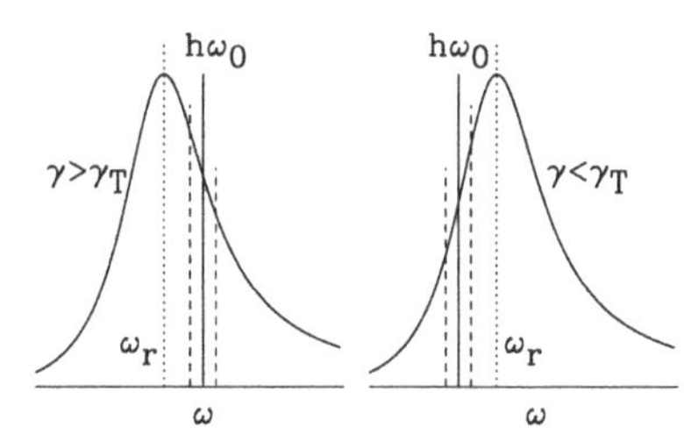

Figure 3.34: A schematic drawing of the real part of impedance arising from a wakefield induced by the circulating beam. To avoid Robinson instability, the cavity should be detuned to $h\omega_0 > \omega_r$ above transition energy and $h\omega_0 < \omega_r$ below transition energy. Above transition energy, higher energy particles have a smaller revolution frequency and thus lose more energy if the cavity detuning is $h\omega_0 > \omega_r$. A similar argument applies to rf cavities operating below transition energy.

Since the revolution frequency is related to the fractional momentum spread by

$$\frac{\Delta\omega}{\omega_0} = -\eta\frac{\Delta E}{\beta^2 E_0},$$

a higher beam energy has a smaller revolution frequency above the transition energy. If the cavity is detuned so that $h\omega_0 > \omega_r$, where ω_r is the resonance frequency of the cavity (Fig. 3.34, left), the beam bunch at higher energy sees a higher shunt impedance and loses more energy, and the beam bunch at lower beam energy sees a lower shunt impedance and loses less energy. Thus the centroid of the beam bunch will damp in the presence of beam loading, and the dipole mode of beam motion is Robinson damped. Similarly, if the cavity is detuned such that $h\omega_0 < \omega_r$, Robinson stability will be attained below transition energy.

Exercise 3.6

1. The skin depth $\delta_{\rm skin}$ of an AC current with angular frequency ω traveling on a conductor of bulk conductivity σ is $\delta_{\rm skin} = \sqrt{2/\mu\sigma\omega}$, where μ is the permeability.

   (a) Show that the surface resistivity defined as $R_s = 1/\sigma\delta_{\rm skin}$ [in Ohm] is given by

   $$R_s = \sqrt{\frac{\mu\omega}{2\sigma}}.$$

   The surface resistivity does not depend on the geometry of the conductor.

   (b) Show that the resistance of a coaxial structure is given by Eq. (3.248) with

   $$R = \frac{R_s\ell}{2\pi}(\frac{1}{r_1} + \frac{1}{r_2}),$$

   where ℓ is the length of the structure and r_1, r_2 are the inner and outer radii of the coaxial wave guide.

2. Show that the solution of Maxwell's equation in the cylindrical coordinate is given by Eq. (3.237).

3. In a lossless transverse electromagnetic (TEM) wave transmission line, the equation for the current and voltage is

$$\frac{\partial V}{\partial s} = -\tilde{L}\frac{\partial I}{\partial t}, \quad \frac{\partial I}{\partial s} = -\tilde{C}\frac{\partial V}{\partial t},$$

where $\tilde{L}$ and $\tilde{C}$ are respectively the inductance and capacitance per unit length.

   (a) Show that the general solution of the right/left traveling TEM wave is given by

$$V = f(t \mp \frac{s}{v}), \quad I = \pm\frac{1}{R_c} f(t \mp \frac{s}{v}),$$

   where f is an arbitrary wave form, $v = \frac{1}{\sqrt{\tilde{L}\tilde{C}}}$ is the wave speed, $R_c = \sqrt{\tilde{L}/\tilde{C}}$ is the characteristic impedance of the line.

   (b) For TEM sinusoidal waves in a transmission line, show that the current and voltage are related by

$$\begin{cases} I(s,t) = [I_0 \cos ks - j(V_0/R_c)\sin ks]e^{j\omega t} \\ V(s,t) = [V_0 \cos ks - jI_0 R_c \sin ks]e^{j\omega t} \end{cases}$$

   where ω is the wave angular frequency, k is the wave number with wave speed $v = \omega/k$, and I_0 and V_0 are the amplitudes of current and voltage at $s = 0$.

4. Verify Eqs.(3.252) and plot Z vs ω for $\omega_r = 200$ MHz, $Q = 10^4$, and $R_{sh} = 25$ MΩ.

5. Consider the excitation of an RLC circuit with a current impulse $I(t) = q\delta(t)$, where $\delta(t)$ is the Dirac delta-function.

   (a) Using Eq. (2.252) for pole decomposition of the RLC resonantor, Show that the induced voltage is

$$V(t) = \frac{1}{2\pi}\int \frac{q\, e^{j\omega t}}{j\omega C + \frac{1}{R} + \frac{1}{j\omega L}} d\omega = \frac{q\omega_r R_{sh} e^{-t/T_f}}{Q}\left[\cos\tilde{\omega}_r t - \frac{1}{\tilde{\omega}_r T_f}\sin\tilde{\omega}_r t\right]\Theta(t),$$

   where $\omega_r = 1/\sqrt{LC}$, $R_{sh} = R$, $R_{sh}/Q = \sqrt{L/C}$, $\tilde{\omega}_r = \omega_r\sqrt{1-(1/2Q)^2}$, $T_f = 2Q/\omega_r$, and $\Theta(t) = 1$ if $t \geq 0$ and 0 elsewhere.

   (b) Evaluate the integral of Eq. (3.258) and show that the loss factor of a parallel RLC resonator is

$$k_r = \frac{\omega_r R_{sh}}{4\pi Q},$$

   where R_{sh} is the shunt resistance, Q is the Q-factor, and ω_r is the resonance frequency.

6. Using the result of Exercise 3.6.5, show that $V_{b0} = I_i R_{sh}\lambda$ for a train of bunches separated by a time interval T_b, where $\lambda = T_b/T_f$, $I_i = 2q/T_b$, and $T_f = 2Q/\omega_r$. Verify Eq. (3.265)

7. Use the following steps to derive Eq. (3.271).

(a) Let ξ be the rf phase associated with the error in beam arrival time. Show that

$$\dot{\xi} = h\eta\delta_b,$$

where the overdot indicates the derivative with respect to orbiting angle θ, and δ_b is the momentum error of the beam centroid.

(b) Show that

$$\dot{\delta}_b = \frac{eV_0 \cos\phi_s}{2\pi\beta^2 E}\left[1 - Y\frac{\sin 2\psi}{2\cos\phi_s}\right]\xi.$$

Thus you have arrived at Eq. (3.271).

(c) Draw the Robinson stability region, i.e. $1 \geq (\sin 2\psi/2\cos\phi_s)Y$, in (Y, ψ) for $\phi_s = 0°, 30°, 60°, 120°, 150°, 180°$.

8. The current for charged-particle beams in an accelerator is

$$I(t) = \sum_{\ell}\sum_{m=1}^{h} q_m \delta(t - m\frac{T_0}{h} - \ell T_0),$$

where T_0 is the revolution period, h is the harmonic number, and q_m is the charge in the mth bucket. Show that the amplitude of the Fourier harmonic is

$$I(\omega) = \sum_m I_m e^{-jm\omega T_0/h}$$

where $I_m = q_m/T_0$ is the current of the mth bucket. (1) If all buckets are filled with equal charge, what happens to the spectrum? (2) If there is only one bunch in the ring with harmonic number h, what is the beam spectrum? (3) Verify the symmetric properties: $I(-\omega) = I(\omega)^*$; $I(n\omega_0) = I((h-n)\omega_0)^*$, and $I(n\omega_0) = I((h+n)\omega_0)$.

VII Longitudinal Collective Instabilities

As the demand for beam brightness increases, the physics of collective instabilities becomes more important. Indeed, almost all accelerators and storage rings have suffered some type of collective instability that limits beam intensity or beam brightness. This section provides an introduction to the collective instability in synchrotron motion induced by wakefield, similar to the transverse collective dipole mode instability discussed in Chap. 2, Sec. VIII. A beam bunch can produce wakefield that affects the particle motion and change the beam distribution, the beam distribution can further enhance the wakefield to cause a run-away collective instability.

In the frequency domain, the collective motion is governed by the impedance, which is the Fourier transform of the wakefield. The impedance responsible for collective instabilities can be experimentally measured by the beam transfer function measurements,[48] or from passive measurements of beam loss, coherent and incoherent tune shift, and equilibrium momentum spread and emittance. Collective instabilities can cause bunch lengthening, beam brightness dilution, luminosity degradation, beam loss in machine operation.

Longitudinal collective instabilities have many modes. The collective synchrotron motion can be classified according to synchrotron modes, as discussed in Sec. III, where the phase space are split into resonance islands. On the other hand, since the growth rate of the microwave instability is very large, it can be classified according to the longitudinal mode with density fluctuation. This causes a beam bunch to form microbunches. Decoherence due to nonlinear synchrotron motion generates emittance dilution.

In this introduction text, we discuss only single bunch effects without mode coupling. In Sec. VII.1, we discuss the coherent frequency spectra of beams in a synchrotron. Knowledge of coherent synchrotron modes provides useful information about possible sources, and about the signature at the onset of collective instabilities. An experimental measurement of coherent synchrotron mode will be discussed. Detecting the onset of instabilities and measuring coherent synchrotron modes can help us understand the mechanism of collective instabilities. In Sec. VII.2, we study the linearized Vlasov equation with a coasting beam, and derive a dispersion relation for the collective frequency in single mode approximation. In Sec. VII.3, we list possible sources of the longitudinal impedance. In Sec. VII.4, we examine the microwave instability for a beam with zero momentum spread and for a beam with Gaussian momentum spread, and discuss the Keil-Schnell criterion and the turbulent bunch lengthening. Mode coupling and coupled bunch collective instabilities and other advanced topics can be found in a specialized advanced textbooks [5, 6, 7].

[48] A. Hofmann, *Proc. 1st EPAC*, p. 181 (World Scientific, Singapore, 1988).

VII.1 Beam Spectra of Synchrotron Motion

The current observed at a wall gap monitor or a BPM from a circulating charged particle is represented by a periodic δ-function in Eq. (2.125). The corresponding frequency spectra occur at all harmonics of the revolution frequency f_0. Similarly, the current of N equally spaced circulating particles is described by Eq. (2.130), where the Fourier spectra are separated by Nf_0. Since $N \sim 10^8 - 10^{14}$, Nf_0 is well above the bandwidths of BPMs and detection instruments, the coherent rf signal is invisible. Such a beam is called a coasting or DC beam because only the DC signal is visible. Nevertheless, the Schottky signal of each individual charged particle can produce high frequency resistive power loss proportional to the number of particles.[49]

The analysis above is applicable to a single short bunch or equally spaced short bunches. The frequency spectra of a single short bunch occur at all harmonics of the revolution frequency f_0. For B equally spaced short bunches, the coherent frequency spectra are located at harmonics of Bf_0. For bunches separated by T_0/h, the dominant harmonics are located at harmonics of hf_0.

A. Coherent synchrotron modes

The synchrotron motion of beam particles introduces a modulation in the periodic arrival time. Modifying Eq. (3.63) with a periodic linear synchrotron motion and expanding it in Fourier series, we obtain

$$\begin{aligned} I_e(t) &= e \sum_{\ell=-\infty}^{\infty} \delta(t - \tau \cos(\omega_s t + \psi) - \ell T_0) \\ &= \frac{e}{T_0} \sum_{n=-\infty}^{\infty} \sum_{m=-\infty}^{\infty} j^{-m} J_m(n\omega_0 \tau) e^{j[(n\omega_0 + m\omega_s)t + m\psi]}, \end{aligned} \quad (3.273)$$

where e is the charge, τ and ψ are the amplitude and phase of the synchrotron motion, $\omega_s = \omega_0\sqrt{heV|\eta\cos\phi_s|/2\pi\beta^2 E}$ is the synchrotron angular frequency with ϕ_s as the rf phase of the synchronous particle, T_0 is the revolution period, and J_m is the Bessel function of order m. The resulting spectra of particle motion are classified into synchrotron modes, i.e. there are synchrotron sidebands around each orbital harmonic n. The amplitude of the mth synchrotron sideband is proportional to the Bessel function J_m.

For a beam with bunch length σ_τ, the coherent synchrotron mode frequency extends typically up to $\omega_{\text{roll off}} \sim 1/\sigma_\tau$ (see Eq. (2.129)), and thus $n\omega_0\tau \le \tau/\sigma_\tau$, where τ is the synchrotron oscillation amplitude, and σ_τ is the bunch length. For a stable beam with $\tau/\sigma_\tau \le 0.1$, the power of the first order synchrotron sideband is about -26 dB below that of the revolution harmonic. However, the coherent mode frequency

[49]See A. Hofmann and T. Risselada, *Proc. of PAC 1983*, p. 2400 (1983).

may extend beyond the bunch-length roll-off $1/\sigma_\tau$ due to micro-bunching, residual coherent synchrotron motion, etc. The measurement of the synchrotron sideband power can be used to infer the residual coherent synchrotron motion of a beam and other coherent synchrotron modes. Figure 3.35 shows the coherent spectrum of a production beam in the Taiwan Light Source in Taiwan. There were 154 bunches in 200 buckets separated by 2.0 ns. The coherent mode frequencies are mainly located at revolution harmonics multiples of 499.6438 MHz. Because of the resolution bandwidth of spectrum analyzer (SA), the observed peak power of each harmonic appeared to roll-off faster than the prediction of a Gaussian distribution with a bunch length of 24 ps. The $m = 1$ synchrotron sideband around the 499.6438 MHz was about 82 dB below the revolution harmonic, indicating that the amplitude of coherent synchrotron oscillation was about $\tau \sim 0.05$ ps, a very stable beam.

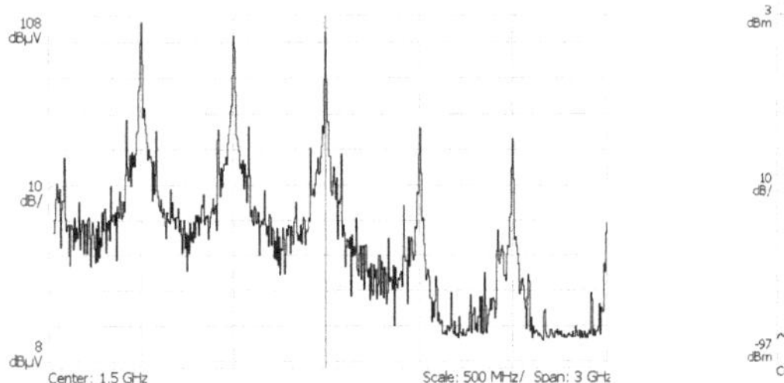

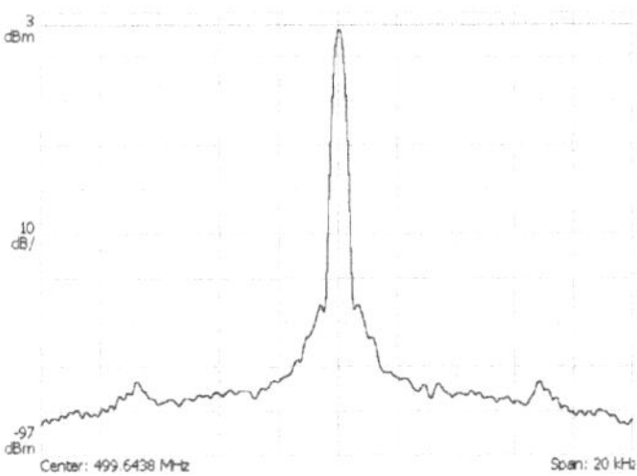

Figure 3.35: Beam power (10 dB/division) vs frequency. Left: The spectra (0-3 GHz) of a TLS production beam. There are 154 bunches in 200 buckets with $f_{\rm rf} = 499.6438$ MHz. The rms bunch length is about 24 ns. Right: The spectrum around the frequency $f_{\rm rf}$ with span 20 kHz. The power of the synchrotron mode is about 82 dB below that of the revolution harmonic (Graph courtesy of Yi-Chih Liu).

A bunch is made of particles with different synchrotron amplitudes and phases, the coherent synchrotron modes of the bunch can be obtained by averaging the synchrotron mode over the bunch distribution. For a ψ-independent beam distribution function $\rho(\tau,\psi) = \rho_0(\tau)$, the beam current becomes

$$I_0(t) = \int I_e(t)\rho_0(\tau)\tau d\tau d\psi = I_{\rm av}\sum_{n=-\infty}^{\infty} A_{n,0}e^{jn\omega_0 t}, \tag{3.274}$$

$$A_{n,0} = 2\pi\int_0^\infty J_0(n\omega_0\tau)\rho_0(\tau)\tau d\tau,$$

where $I_{\rm av} = N_{\rm B}ef_0$ is the average current, and $A_{n,0}$ is the Hankel transformation of ρ_0. Equation (3.274) contains only orbital harmonics $n\omega_0$, i.e. all synchrotron sidebands of individual particles are averaged to zero. The inverse Hankel transformation can be used to determine the unperturbed distribution function $\rho_0(\tau)$.

Now, consider a coherent synchrotron mode in the bunch distribution, e.g. $\rho(\tau,\psi) = \rho_0(\tau) + \Delta\rho(\tau,\psi)$ with the mth synchrotron mode at the coherent frequency Ω_c, the coherent density becomes $\Delta\rho(\tau,\psi) = \rho_m(\tau)e^{j(\Omega_c t - m\psi)}$, and the current signal is

$$I(t) = I_0(t) + \int I_e(t)\Delta\rho(\tau,\psi)\tau d\tau d\psi = I_0(t) + I_{\rm av}\sum_{n=-\infty}^{\infty} A_{n,m}e^{j(n\omega_0 + m\omega_s + \Omega_c)t},$$
$$A_{n,m} = 2\pi\int_0^\infty J_m(n\omega_0\tau)\rho_m(\tau)\tau d\tau, \tag{3.275}$$

where $A_{n,m}$ is the mth order Hankel transformation. The mth *coherent* synchrotron sideband appears around all coherent revolution harmonics. Using the inverse Hankel transformation, we can deduce the beam distribution function from the amplitudes of coherent modes integrals $A_{n,m}$, that form the kernel of the Sacherer integral equation to solve the coherent mode frequency of the longitudinal collective instability. The coherent synchrotron mode intensity can be obtain by taking the spectrum of a beam during the onset of coherent mode instability.[50] As an illustrative example, we measure the power of a synchrotron mode of a longitudinally kicked beam.

B. Coherent synchrotron modes of a kicked beam

We consider an initial Gaussian beam distribution (see Eq. (3.29) for the phase space coordinates) and a phase kick of time $\tau_{\rm k}$:

$$\begin{aligned}
\rho_0\left(\tau, \frac{\dot\tau}{\omega_s}\right) &= \frac{1}{2\pi\sigma_\tau^2}\exp\left(-\frac{\tau^2 + (\dot\tau/\omega_s)^2}{2\sigma_\tau^2}\right).\\
&\rightarrow \frac{1}{2\pi\sigma_\tau^2}\exp(-\frac{\tau_{\rm k}^2}{2\sigma_\tau^2} - \frac{\hat\tau^2}{2\sigma_\tau^2} - \frac{\hat\tau\tau_{\rm k}}{\sigma_\tau^2}\cos\psi),\\
&= \frac{1}{2\pi\sigma_\tau^2}\exp(-\frac{\tau_{\rm k}^2}{2\sigma_\tau^2} - \frac{\hat\tau^2}{2\sigma_\tau^2})\sum_{m=-\infty}^{\infty}(-1)^m I_m(\frac{\hat\tau\tau_{\rm k}}{\sigma_\tau^2})e^{jm\psi},
\end{aligned} \tag{3.276}$$

where $\tau_{\rm k}$ is the amplitude of an initial phase kicked and the coherent mode amplitudes obey $I_{-m} = I_m$. Using formula 6.633.4 in Ref. [31], we obtain the coherent distribution and the coherent mode integral of Eq. (3.275) as

$$\begin{aligned}
\rho_m(\hat\tau) &= \frac{(-1)^m}{2\pi\sigma_\tau^2}\exp(-\frac{\tau_{\rm k}^2}{2\sigma_\tau^2} - \frac{\hat\tau^2}{2\sigma_\tau^2})I_m(\frac{\hat\tau\tau_{\rm k}}{\sigma_\tau^2}),\\
A_{n,m} &= e^{-\frac{1}{2}(n\omega_0\tau_{\rm k})^2(\frac{\sigma_\tau}{\tau_{\rm k}})^2}J_m(n\omega_0\tau_{\rm k}).
\end{aligned} \tag{3.277}$$

[50] F. Sacherer, *IEEE Trans. Nucl. Sci.* **NS-20**, (1973), ibid **NS-24**, (1977); J.L. Laclare, CERN 87-03, p. 264 (1987). Because the measurement of $A_{n,m}$ peak power depends on the resolution bandwidth and the beam intensity, it is difficult to obtain these amplitudes in a single scan. Furthermore, for spectrum power in a very large frequency span, bandwidth limitation of amplifiers, attenuators, and other components should be carefully evaluated.

The power of the mth sideband of a kicked beam is proportional to the square of the mth order Bessel function. For non-Gaussian beams, the power spectrum is a weighted average of Bessel functions in Eq. (3.275). We describe below an experiment, measuring the coherent mode power at the IUCF Cooler.

C. Measurements of coherent synchrotron modes

The experiment started with a single bunched beam of about 5×10^8 protons at a kinetic energy of 45 MeV and harmonic number $h = 1$, revolution frequency $f_0 = 1.03168$ MHz, and phase slip factor $\eta = -0.86$. The cycle time was 5 s, while the injected beam was electron-cooled for about 3 s. The bunch length, could be adjusted by varying the rf voltage, was about 4.5 m (50 ns) FWHM, or $\sigma_\tau \approx 20$ ns. The bunched beam was kicked longitudinally by phase-shifting the rf cavity wave form (see Sec. III.3). A function generator was used to generate a 0 to 10 V square wave to control the phase kick. The rf phase lock feedback loop, which normally locks the rf cavity to the beam, was switched off. The resulting phase oscillations of the bunch relative to the rf wave form were measured by a phase detector, which was used to calibrate the control voltage for the phase shifters versus the actual phase shift. Both the phase error due to control nonlinearity and the parasitic amplitude modulation of the IUCF Cooler rf systems were kept to less than 10%. The response time of the step phase shifts was limited primarily by the inertia of the rf cavities, which had a quality factor Q of about 40. The magnitude of the phase shift was varied by the size of the applied step voltage.

The spectrum analyzer (SA), set at frequency span 0 Hz, video bandwidth 100 Hz, resolution bandwidth 100 Hz, was triggered about 5 ms before the phase shift. The power observed at a synchrotron sideband from the SA is shown in Fig. 3.36, where the top and bottom traces respectively show the SA responses at $f_0 - f_s$ and $6f_0 - f_s$ vs time. The kicked amplitude was 90 ns, or equivalently $\omega_0\tau_{\rm k} = 0.58$ rad. The resolution bandwidth of SA was 100 Hz, thus the measurement of the sideband power was taken at 10 ms after the phase kick. The sideband power shown in Fig. 3.37 was proportional to $|A_{1,1}|^2$ for the upper trace and $|A_{6,1}|^2$ for the lower trace.

Since $\omega_0\tau_{\rm k} \approx 0.58$ and $\sigma_\tau = 20$ ns for the case shown in Fig. 3.36, we find $A_{1,1} \sim e^{-0.0083}J_1(0.58)$, and $A_{6,1} \sim e^{-0.299}J_1(3.48)$. The initial power at the fundamental harmonic sideband, which is proportional to $|A_{1,1}|^2$, after the phase kick will be a factor of 6 larger than that of the 6th orbital harmonic. As the synchrotron phase amplitude decreases because of electron cooling, the power $A_{1,1}$ decreases because $J_1(\omega_0\tau_a)$ decreases with decreasing $\omega_0\tau_a$, where τ_a is the synchrotron amplitude. On the other hand, as $6\omega_0\tau_a$ decreases, $J_1(6\omega_0\tau_a)$ increases. Therefore the power spectrum shown in the lower plot of Fig. 3.36 increases with time. Figure 3.37 shows the power of the $m = 1$ sideband as a function of $\omega\tau = n\omega_0\tau_{\rm k}$, where n is the revolution

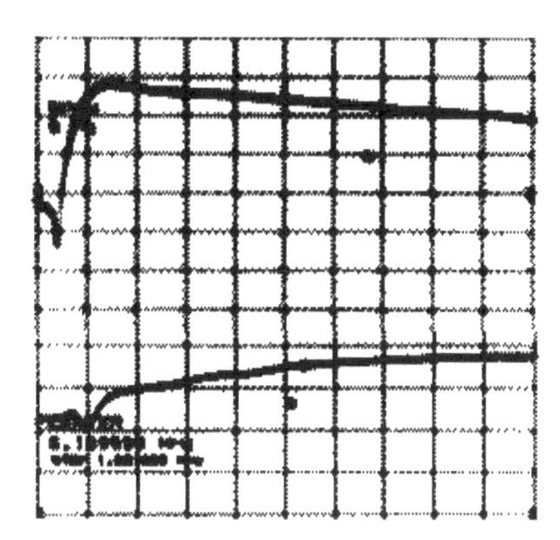

Figure 3.36: The synchrotron sideband power of a kicked beam observed from an SA tuned to the first revolution sideband (upper trace) and the 6th revolution sideband (lower trace) vs time. The revolution frequency was 1.03168 MHz. The setting of the SA was resolution bandwidth 100 Hz, video bandwidth 100 Hz, and frequency span 0 Hz. The sideband power decreased with time for the first harmonic and increased for the 6th harmonic, probably due to electron cooling in the IUCF Cooler. The vertical axis is coherent synchrotron power in dB per division, and the horizontal axis is time at 10 ms per division.

harmonic. For a kicked Gaussian beam, the power $P_{n,1}$ is proportional to $|A_{n,1}|^2$:

$$P_{n,1} \sim |A_{n,1}|^2 = e^{-(n\omega_0\tau_\mathrm{k})^2(\sigma_\tau/\tau_\mathrm{k})^2}|J_1(n\omega_0\tau_\mathrm{k})|^2. \tag{3.278}$$

Because the actual power depends on the beam intensity, all data are normalized at the first peak around $n\omega_0\tau_\mathrm{k} \approx 1.8$. Solid curves are obtained from Eq. (3.278) normalized to the peak of experimental data. Finite bunch length suppresses the power of higher order harmonics is clearly seen in the experimental data of Fig. 3.37.

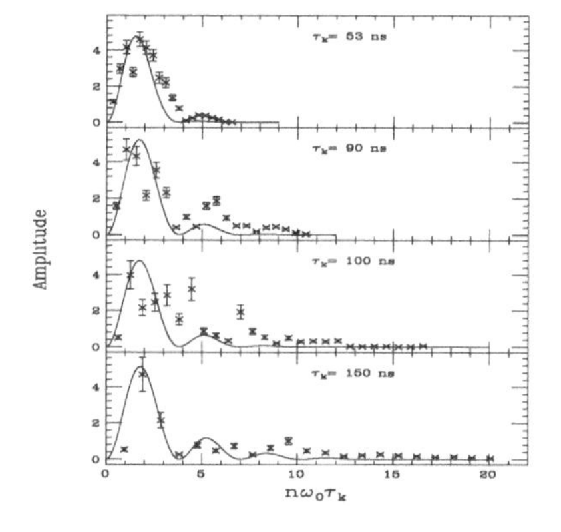

Figure 3.37: Measured $m = 1$ synchrotron sideband power vs frequency for different phase kicked amplitudes is compared with theory based on a Gaussian beam distribution. Plots from top to bottom correspond to a kicking amplitude (time) of 53, 90, 100 and 150 ns. These data were normalized to the peak of the theoretical predictions of Eq. (3.278) without other adjustable parameter.

When a bunched beam encounters collective instability, the observed sideband power $|A_{n,1}|^2$ is proportional to the weighted average of the coherent mode density $\rho(\hat{\tau})$ shown in Eq. (3.277). Measurement of $A_{n,1}$ for all orbital harmonics can be used to obtain the coherent mode distribution function. Similarly, setting up the central frequency at the second synchrotron harmonic, we can measure the $m = 2$ synchrotron modes for the kicked beam.

Difficulties of all spectra power measurements are (1) the measurement of power depends on the resolution bandwidth so that it can not be measured in one single sweep; (2) BPMs or wall-gap monitors, amplifiers and attenuators are bandwidth limited; and (3) the coherent signal is proportional to $N_{\rm B}^2$, and thus sensitive to beam intensity during the measurement. However, the experiments can be parasitic without interfering regular machine operation.

VII.2 Collective Microwave Instability in Coasting Beams

For coasting beams, there is no rf cavity and the unperturbed distribution function is a function only of the off-momentum coordinate $\delta = \Delta p/p_0$. Let $\Psi_0(\delta)$ be the normalized distribution function with $\int \Psi_0 d\delta = 1$. Because of the impedance of the ring, the beam generates wakefields, which in turn perturb particle motion.

A self-consistent distribution function obeys the Vlasov equation

$$\frac{d\Psi}{dt} = \frac{\partial \Psi}{\partial t} + \dot{\theta}\,\frac{\partial \Psi}{\partial \theta} + \dot{\delta}\,\frac{\partial \Psi}{\partial \delta} = 0, \tag{3.279}$$

where the overdot is the derivative with respect to time t. In the presence of a wakefield, we assume a single longitudinal mode with the distribution function:

$$\Psi = \Psi_0(\delta) + \Delta\Psi_n e^{j(\Omega t - n\theta)}, \tag{3.280}$$

where Ψ_0 is the unperturbed distribution, Ω is the coherent frequency, θ is the orbiting angle, and $\Delta\Psi_n(\delta)$ is the perturbation amplitude for the longitudinal mode n. The perturbation causes density fluctuation along the machine, i.e. the collective instability of mode number n can cause a coasting beam into n microbunches. In general, the perturbing distribution function should be written as a linear superposition of all possible modes. The frequencies of the collective motion are eigenfrequencies of the coupled system.

By definition, the energy gain/loss per revolution due to the wakefield is equal to the current times the longitudinal broadband impedance, and the time derivative of the fractional off-momentum coordinate δ of a coasting beam become

$$\begin{aligned} \Delta E\Big|_{\rm per\ turn} &= Z_\parallel \left(eI_0 \int \Delta\Psi_n d\delta\right) e^{j(\Omega t - n\theta)}, \\ \dot{\delta} &= \frac{\omega_0}{2\pi\beta^2 E}\left(eI_0 Z_\parallel \int \Delta\Psi_n d\delta\right) e^{j(\Omega t - n\theta)}. \end{aligned} \tag{3.281}$$

where the impedance is evaluated at the collective frequency Ω. Since $|\Delta\Psi_n| \ll \Psi_0$ at the onset threshold of collective instability, we linearize the Vlasov equation to obtain

$$j(\Omega - n\dot{\theta})\Delta\Psi_n = -\frac{\omega_0 e I_0 Z_\parallel}{2\pi\beta^2 E}\frac{\partial \Psi_0}{\partial \delta}\left(\int \Delta\Psi_n d\delta\right). \tag{3.282}$$

Using $\dot{\theta} = \omega$ and integrating Eq. (3.282), we obtain the dispersion relation

$$1 = j\frac{eI_0 n\omega_0 (Z_\parallel/n)}{2\pi\beta^2 E}\int \frac{\partial\Psi_0/\partial\delta}{\Omega - n\omega}d\delta = j\frac{eI_0 n^2\omega_0 (Z_\parallel/n)}{2\pi\beta^2 E}\int \frac{\Psi_0}{(\Omega - n\omega)^2}\frac{\partial\omega}{\partial\delta}d\delta, \quad (3.283)$$

where partial integration has been carried out in the second equality.

The eigenfrequency Ω of the collective motion is the solution of the dispersion relation. If the imaginary part of the coherent mode frequency is negative, i.e. $\mathrm{Im}\,\Omega < 0$, the perturbation amplitude grows exponentially, and the beam encounters the *collective microwave instability*, where the terminology is derived from the fact that the coherent frequency observed is in the microwave frequency range. With the relation $\omega = \omega_0 - \omega_0\eta\delta$, the dispersion integral can be analytically obtained for some distribution functions of the beam. First we examine possible sources of longitudinal impedance.

VII.3 Longitudinal Impedance

The impedance and the wake function are related by

$$Z_\parallel(\omega) = \int_{-\infty}^{\infty} W_\parallel(t)e^{-j\omega t}dt, \qquad W_\parallel(t) = \frac{1}{2\pi}\int_{-\infty}^{\infty} Z_\parallel(\omega)e^{j\omega t}d\omega. \quad (3.284)$$

Because the wake function is real and obeys the causality $W_\parallel(t) = 0$ for $t < 0$, the impedance has the property: $Z_\parallel(-\omega) = Z_\parallel^*(\omega)$, i.e. the real part of the longitudinal impedance is positive and is a symmetric function of the frequency. In fact, the property of $Z_\parallel(\omega)/\omega$ is similar to that of $Z_\perp(\omega)$. Without making the effort to derive them, we list below some sources of commonly used impedance models. Since the wakefield obeys the causality principle, the impedance does not have singularities in the lower complex plane. The real and imaginary parts of the impedance are related by the Hilbert transform

$$\mathrm{Re}\,Z_\parallel(\omega) = -\frac{1}{\pi}\int_{\mathrm{P.V.}} d\omega'\frac{\mathrm{Im}\,Z_\parallel(\omega')}{\omega' - \omega}, \qquad \mathrm{Im}\,Z_\parallel(\omega) = +\frac{1}{\pi}\int_{\mathrm{P.V.}} d\omega'\frac{\mathrm{Re}\,Z_\parallel(\omega')}{\omega' - \omega},$$

where P.V. stands for the principal value integral.

A. Space-charge impedance

Let a be the radius of a uniformly distributed coasting beam, and let b be the radius of a beam pipe (Fig. 3.38). The electromagnetic fields of the coasting beam are

$$E_r = \begin{cases} \dfrac{e\lambda r}{2\pi\epsilon a^2} \\[2ex] \dfrac{e\lambda}{2\pi\epsilon r} \end{cases} \qquad B_\phi = \begin{cases} \dfrac{\mu_0 e\lambda\beta c r}{2\pi a^2} & r \le a \\[2ex] \dfrac{\mu_0 e\lambda\beta c}{2\pi r} & r > a \end{cases} \quad (3.285)$$

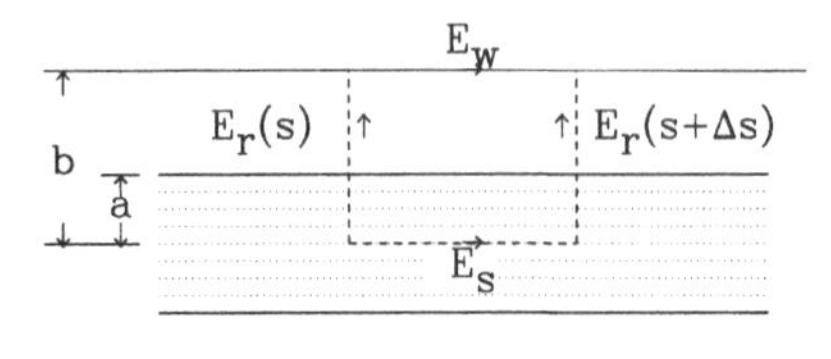

Figure 3.38: Geometry of a uniformly distributed beam with radius a in a beam pipe of radius b. The induced electric fields that arise from impedance are shown schematically. The rectangular loop is used for the path integral of Faraday's law.

where λ is the particle's line density, e is the charge, βc is the speed, and ϵ_0 and μ_0 are the permittivity and permeability of the vacuum.

Consider a small fluctuation in the line density $\lambda = \lambda_0 + \lambda_1 e^{j(\Omega t - n\theta)}$ and current $I = I_0 + I_1 e^{j(\Omega t - n\theta)}$, where $I_0 = e\beta c\lambda_0$ and $I_1 = e\beta c\lambda_1$. The perturbation generates an electric field on the beam. Using Faraday's law

$$\oint \vec{E}d\vec{\ell} = -\frac{\partial}{\partial t}\int \vec{B}\cdot d\vec{\sigma}$$

along the loop shown in Fig. 3.38, where $d\vec{\sigma}$ is the surface integral, we obtain

$$(E_s - E_{\rm w})\Delta s + \frac{eg_0}{4\pi\epsilon_0}[\lambda(s+\Delta s) - \lambda(s)] = -\Delta s\frac{\mu_0 e\beta c g_0}{4\pi}\frac{\partial\lambda}{\partial t},$$

where E_s and $E_{\rm w}$ are the electric fields at the center of the beam pipe and at the vacuum chamber wall, and the geometry factor $g_0 = 1 + 2\ln(b/a)$ is obtained from the integral along the radial paths from the beam center to the vacuum chamber wall. If the impedance is averaged over the beam cross section, the geometric factor becomes $g_0 = \frac{1}{2} + 2\ln(b/a)$. On the other hand, if the perturbation is on the surface of the beam, the geometry factor becomes $g_0 = 2\ln(b/a)$. Assuming that the disturbance is propagating at the same speed as the orbiting beam particles, i.e. $\partial\lambda/\partial t = -\beta c(\partial\lambda/\partial s)$, the electric field acting on the circulating beam becomes

$$E_s = E_{\rm w} - \frac{eg_0}{4\pi\epsilon_0\gamma^2}\frac{\partial\lambda}{\partial s}, \tag{3.286}$$

where the factor $1/\gamma^2$ arises from the cancellation of forces due to the electric and magnetic fields.

For most accelerators, the vacuum chamber wall is inductive at low and medium frequency range. Let $L/2\pi R$ be the inductance per unit length, then the induced wall electric field is

$$E_{\rm w} = \frac{L}{2\pi R}\frac{dI_{\rm w}}{dt} = \frac{e\beta^2c^2L}{2\pi R}\frac{\partial\lambda}{\partial s}. \quad \text{or} \quad E_s = -e\left[\frac{g_0}{4\pi\epsilon_0\gamma^2} - \frac{\beta^2c^2L}{2\pi R}\right]\frac{\partial\lambda}{\partial s}.$$

The total voltage drop in one revolution on the beam and the impedance are

$$\Delta U = -e\beta cR\frac{\partial\lambda}{\partial s}\left[\frac{g_0Z_0}{2\beta\gamma^2} - \omega_0 L\right]; \qquad \frac{Z_\parallel}{n} = -\frac{\Delta U}{nI_1} = -j\left[\frac{g_0Z_0}{2\beta\gamma^2} - \omega_0 L\right]. \tag{3.287}$$

where $\beta c = \omega_0 R$ is the speed of the orbiting particles, $Z_0 = 1/\epsilon_0 c = 377$ ohms is the vacuum impedance, and we use $R(\partial\lambda/\partial s) = (\partial\lambda/\partial\theta) = -jn\lambda_1$, and $e\beta c\lambda_1 = I_1$, The first term in Eq. (3.287) is the space-charge impedance and the second term is the inductance of the vacuum chamber wall. Typical values of the space-charge impedance at transition energy are listed in Table 3.7.

Table 3.7: Typical space-charge impedance at $\gamma = \gamma_{\rm T}$.

	AGS	RHIC	Fermilab BST	Fermilab MI	KEKPS
$\gamma_{\rm T}$	8.7	22.5	5.4	20.4	6.8
$\lvert Z_{\parallel,{\rm sc}}\rvert/n$ [Ω]	13	1.5	30	2.3	20

B. Resistive wall impedance

The vacuum chamber wall is normally not perfectly conducting, and $E_{\rm w}$ can also induce a resistive impedance part that depends on the conductivity, microwave frequency, and skin depth. Because the resistivity of the vacuum chamber wall is finite, part of the wakefield can penetrate the vacuum chamber and cause energy loss to the beam. Penetration of electromagnetic wave into the vacuum chamber can be described by Maxwell's equations

$$\nabla \times \vec{E} = -\mu\frac{\partial \vec{H}}{\partial t}, \qquad \nabla \times \vec{H} = \vec{J} = \sigma_{\rm c}\vec{E}, \quad \Longrightarrow \quad \nabla^2\vec{E} = \mu\sigma_{\rm c}\frac{\partial \vec{E}}{\partial t}, \tag{3.288}$$

where $\sigma_{\rm c}$ is the conductivity and μ is the permeability. Here we use Ohm's law, and neglect the contribution from the displacement current provided that the frequency of the electromagnetic wave is not very high.[51] The electric field inside the conductor becomes

$$\vec{E} = \hat{s}\ E_0 \exp\{j(\omega t - kx)\}, \qquad k = (1-j)\sqrt{|\omega|\sigma_{\rm c}\mu/2}.$$

where x is the depth into the vacuum chamber wall and k is the wave number. The imaginary part of the wave number is the inverse of the penetration depth, or equivalently, the skin depth is $\delta_{\rm skin} = \sqrt{2/\mu\sigma_{\rm c}\omega}$. The electromagnetic fields penetrate a skin depth inside the vacuum chamber wall. The resistance due to the electric field becomes

$$Z_{\parallel}^{\rm real} \approx \frac{2\pi R}{2\pi b\sigma_{\rm c}\delta_{\rm skin}} = \frac{Z_0\beta}{2b}\left(\frac{|\omega|}{\omega_0}\right)^{1/2}\delta_{\rm skin,0}, \tag{3.289}$$

[51]For frequencies $\omega \ll \sigma_{\rm c}/\epsilon \approx \sigma_{\rm c}Z_0 c \approx 10^{19}$ Hz, where ϵ is the permittivity, the displacement current contribution to Maxwell's equation is small.

where Z_0 is the vacuum impedance, β is Lorentz's relativistic velocity factor, b is the vacuum chamber radius, $\delta_{\rm skin,0} = \sqrt{2/\mu\sigma_{\rm c}\omega_0}$ is the skin depth at the revolution frequency ω_0. Since the magnetic energy is equal to the electric energy, the magnitude of the reactance is equal to the resistance. The resistive wall impedance becomes

$$Z_\|(\omega) = (1 + j\,{\rm sgn}(\omega))\,\frac{Z_0\beta}{2b}\left(\frac{|\omega|}{\omega_0}\right)^{1/2}\delta_{\rm skin,0}, \tag{3.290}$$

where the sign function, ${\rm sgn}(\omega) = +1$ if $\omega \geq 0$ and -1 if $\omega < 0$, is added so that the impedance satisfies the causality condition.

C. Narrowband and broadband impedance

Narrowband impedance arise from parasitic modes in rf cavities and cavity-like structures in accelerators. Broadband impedance arise from vacuum chamber breaks, bellows, and other discontinuities in accelerator components. The longitudinal narrowband and broadband impedance can conveniently be represented by an equivalent RLC circuit

$$Z(\omega) = \frac{R_{\rm sh}}{1 + jQ(\omega/\omega_{\rm r} - \omega_{\rm r}/\omega)}, \tag{3.291}$$

where $\omega_{\rm r}$ is the resonance frequency, $R_{\rm sh}$ is the shunt impedance, and Q is the quality factor. The high order mode (HOM) of rf cavities is a major source of narrowband impedance. Parameters for narrowband impedance depend on the geometry and material of cavity-like structures.

For a broadband impedance, the Q-factor is usually taken to be 1, and the resonance frequency to be the cut-off frequency $\omega_{\rm r,bb} = \omega_0 R/b = \beta c/b$,where ω_0 is the revolution frequency, R is the average radius of the accelerator, and b is the vacuum chamber size. The magnitude of the broadband shunt impedance can range from 50 ohms for machines constructed in the 60's and 70's to less than 1 ohm for recently constructed machines, where the vacuum chamber is carefully smoothed.

To summarize, the longitudinal impedance $Z_\|(\omega)/\omega$ or $Z_\|/n$ are schematically shown in Fig. 3.39, where the solid and dashed lines correspond to the real and imaginary parts respectively. The symmetry of the impedance as a function of ω is also shown.

VII.4 Single Bunch Microwave Instability

The negative mass instability was predicted in 1960's. Experimental observations were obtained in the intersecting storage rings (ISR), where microwave signal was detected in the beam debunching process. Subsequently, it was observed in almost all existing high intensity accelerators.

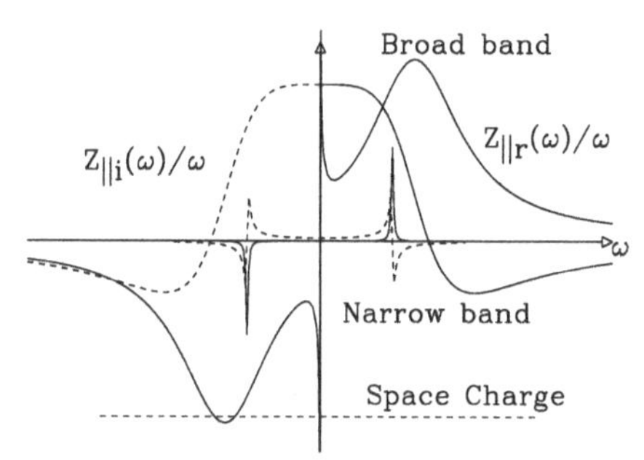

Figure 3.39: Schematic of a longitudinal impedance that includes broadband, narrow-band, and space-charge impedance. Including the resistive wall impedance in the longitudinal impedance, we find that $|\mathrm{Re}(Z_{\|}/\omega)|$ becomes large at $\omega \approx 0$.

A. Negative mass instability without momentum spread

First, we consider negative mass instability. In the absence of momentum spread with $\Psi_0(\delta) = \delta_{\rm d}(\delta)$, where $\delta = \Delta p/p_0$ and $\delta_{\rm d}(x)$ is the Dirac δ-function, the solution of Eq. (3.283) is

$$\left(\frac{\Omega}{n\omega_0}\right)^2 = -j\frac{eI_0\, Z_{\|}/n}{2\pi\beta^2 E}\,\eta\,. \tag{3.292}$$

The condition for having a real Ω is $-j(Z_{\|}/n)\eta > 0$. This condition is only satisfied for a space-charge (capacitive) impedance below the transition energy, or an inductive impedance above the transition energy. If $Z_{\|}/n$ is capacitive, e.g. space-charge impedance, the collective frequency is a real number below the transition energy with $\eta < 0$. This results in a collective frequency shift without producing collective instabilities. On the other hand, if the impedance is inductive, the collective frequency becomes a complex number below the transition energy, and the solution with a negative imaginary part gives rise to collective instability. For resistive impedance, the beam with a zero momentum spread is unstable. Table 3.8 shows the characteristic behavior of microwave collective instability.

Table 3.8: Characteristic behavior of collective instability without Landau damping.

	$Z_{\|}/n$	capacitive	inductive	resistive
Below transition	$\eta < 0$	stable	unstable	unstable
Above transition	$\eta > 0$	unstable	stable	unstable

The terminology of "negative mass instability" is derived from a pure space charge effect. Above the transition energy with $\eta > 0$, a higher energy particle takes longer time to complete one revolution, or it appears to have a negative mass. Since the "microwave instability" resulting from the space-charge impedance occurs when $\eta > 0$, it is also called *negative mass instability.* However, a beam with a small frequency spread can also encounter microwave instability at $\gamma < \gamma_{\rm T}$ if the impedance is inductive, or resistive.

B. Landau damping with finite frequency spread

For a beam with a finite momentum spread with $\eta \neq 0$, the coherent mode frequency can be obtained by solving the dispersion relation. In this case, there is a finite region of impedance value where the growth rate of collective instability is zero, and collective motion is Landau damped.

If the distribution function is a symmetric function of momentum deviation δ, the threshold impedance for microwave instability is reflectively symmetric with respect to the real part of the impedance. Depending on the actual distribution function, the threshold of collective instability can be estimated from the dispersion relation.

For example, we consider a Gaussian beam model of a coasting beam given by

$$\Psi_0 = \frac{1}{\sqrt{2\pi}\sigma_\delta} \exp\left\{-\frac{\delta^2}{2\sigma_\delta^2}\right\},$$

where $\delta = \Delta p/p_0$ and σ_δ is the rms momentum spread. In the limit of small frequency spread, the distribution becomes the Dirac δ-function. The rms frequency spread of the beam becomes $\sigma_\omega = \omega_0\eta\sigma_\delta$. The dispersion relation can be integrated to obtain

$$\frac{Z_\parallel}{n} = j\frac{4\pi\beta^2 E\sigma_\delta^2\eta}{eI_0} J_{\rm G}^{-1}, \tag{3.293}$$

$$J_{\rm G} = \sqrt{\frac{2}{\pi}} \int_{-\infty}^{\infty} \frac{xe^{-x^2/2}}{x + \tilde{\Omega}/(n\omega_0\eta\sigma_\delta)} dx = 2[1 + j\sqrt{\pi}yw(y)], \tag{3.294}$$

where $\tilde{\Omega} = \Omega - n\omega_0$, and $w(y)$ is the complex error function with $y = -\tilde{\Omega}/(\sqrt{2}n\omega_0\eta\sigma_\delta)$. Asymptotically, we have $J_{\rm G} \to y^{-2}$ as $y \to \infty$. Thus in the limit of zero detuning (or zero frequency spread), Eq. (3.293) reduces to Eq. (3.292).

We usually define the effective U and V parameters, or U' and V' parameters as

$$U + jV = \frac{eI_0\,(Z_\parallel/n)}{2\pi\beta^2 E\sigma_\delta^2\eta}, \qquad U' + jV' = \frac{eI_0\,(Z_\parallel/n)}{\beta^2 E\delta_{\rm FWHM}^2\eta}. \tag{3.295}$$

For the Gaussian beam, we find $\delta_{\rm FWHM} = \sqrt{8\ln 2}\,\sigma_\delta$. In terms of U and V parameters, Eq. (3.293) becomes $-j(U + jV)J_{\rm G}/2 = 1$.

The solid line in the left plot of Fig. 3.40 shows the threshold V' vs U' parameters of collective microwave instability with $\text{Im}(\Omega) = 0$. Dashed lines inside the threshold curve correspond to stable motion, and the dashed lines outside the threshold curve are unstable with growth rates $-(\text{Im}\,\Omega)/\sqrt{2\ln 2}\,\omega_0\eta\sigma_\delta = 0.1, 0.2, 0.3, 0.4$, and 0.5 respectively. The right plot of Fig. 3.40 shows the threshold V' vs U' parameters, from inside outward, for the normalized distribution functions $\Psi_0(x) = 3(1-x^2)/4$, $8(1-x^2)^{3/2}/3\pi$, $15(1-x^2)^2/16$, $315(1-x^2)^4/32$, and $(1/\sqrt{2\pi})\exp(-x^2/2)$. All distribution functions, except the Gaussian distribution, are limited to $x \le 1$. Note that a distribution function with a softer tail, i.e. a less sudden cutoff, gives a larger stability region in the parametric space.

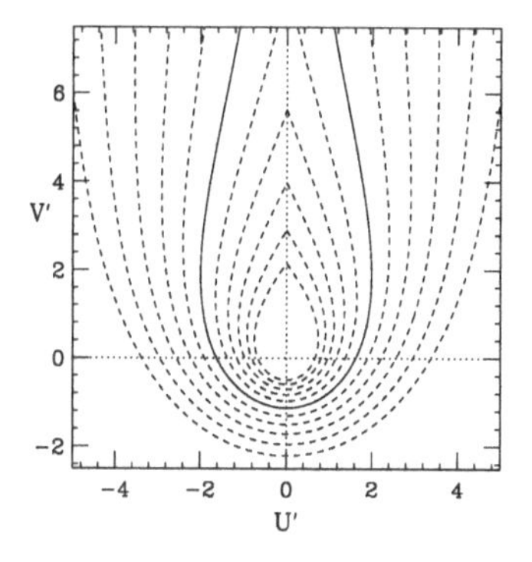

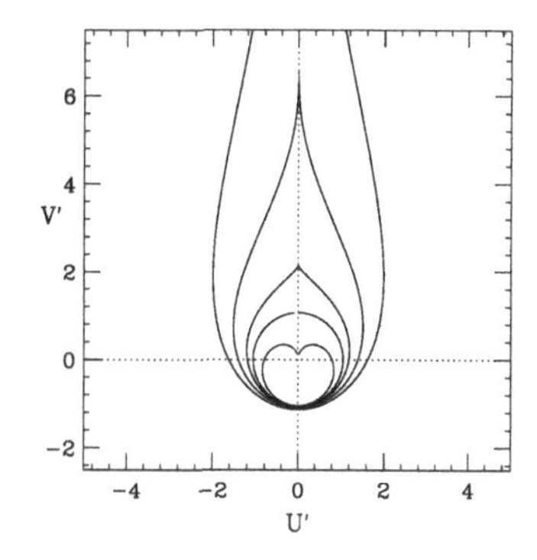

Figure 3.40: Left: The solid line shows the parameters V' vs U' for a Gaussian beam distribution at a zero growth rate. Dashed lines inside the threshold curve are stable. They correspond to $-\mathrm{Im}\,\Omega/(\sqrt{2\ln 2}\,\omega_0\eta\sigma_\delta) = -0.1, -0.2, -0.3, -0.4$, and -0.5. Dashed lines outside the threshold curve have growth rates $-\mathrm{Im}\,\Omega/(\sqrt{2\ln 2}\,\omega_0\eta\sigma_\delta) = 0.1, 0.2, 0.3, 0.4$, and 0.5 respectively. Right: The threshold V' vs U' parameters for various beam distributions.

C. Keil-Schnell criterion

Figure 3.40 show that the stability region depends on beam distribution. Based on experimental observations and numerical calculations of the dispersion relation, a simplified estimation of the stability condition is to draw a circle around the origin in the impedance plane, called the Keil-Schnell criterion:

$$\left|\frac{Z_\parallel}{n}\right| \le \frac{2\pi\beta^2 E\sigma_\delta^2|\eta|F}{eI_0}, \tag{3.296}$$

where F is a form factor that depends on the distribution function. For a Gaussian beam, $F = 1$; and for a tri-elliptical distribution with $\Psi_0(x) = 8(1-x^2)^{3/2}/3\pi$, $F \approx 0.94$ [5, 6, 7]. The total longitudinal energy drop from impedance, $eI_0|Z_\parallel|$, per unit frequency spread $n|\eta|\sqrt{2\pi}\sigma_\delta$ for mode number n should be less than the total energy spread $\sqrt{2\pi}\beta^2 E\sigma_\delta$ of the beam. Since the microwave growth rate is usually fast, and the the wavelength of the coherent wave is usually small compared with the bunch length, the Keil-Schnell criterion at threshold of instability can be applied to the bunched beam by replacing the average current I_0 by the peak current $\hat{I}$:[52]

$$\left|\frac{Z_\parallel}{n}\right| \le \frac{2\pi\beta^2 E\sigma_\delta^2|\eta|F}{e\hat{I}}, \tag{3.297}$$

[52]E. Keil and W. Schnell, CERN-ISR-TH-RF/69-48 (July 1969); A.G. Ruggiero and V.G. Vaccaro, CERN ISR TH/68-33 (1968); Since the growth rate of the microwave instability is normally very fast, the threshold condition can be obtained from the local peak current of the beam, called Boussard conjecture, which has been well tested in the Intersecting Storage Ring (ISR). See e.g. J.M. Wang and C. Pellegrini, *Proc. 11th HEACC*, p. 554 (1980).

where $I_0 = N_B ef$ is the average bunch current, $\hat{I} = F_B I_0$, and $F_B = 2\pi/\sqrt{2\pi}\sigma_\theta$ is the bunching factor, where $\sigma_\theta = \omega_0 \sigma_{\Delta t}$ is the bunch length in orbiting angle.

D. Microwave instability near transition energy

Near the transition energy, Landau damping for microwave single bunch instability vanishes because of a small synchrotron frequency spread. The Keil-Schnell criterion is not applicable in this region. For a pure capacitive impedance, e.g. space-charge impedance, instability occurs when $\gamma \geq \gamma_T$. For a pure inductance impedance, instability exists only below transition. Since the beam distribution function is non-adiabatic in the transition energy region, determination of microwave instability needs careful evaluation of the dispersion integral.

We assume a model of collective microwave instability such that the longitudinal modes are nearly decoupled and thus the coherent growth rate can be obtained by solving the dispersion relation Eq. (3.283). Furthermore, we assume a Gaussian beam model with the threshold impedance determined by the peak current. The peak current is located at the center of the bunch $\Delta\phi = 0$. The distribution function and the peak current become (see Sec. IV.1)

$$\Psi_0(\delta) = \sqrt{\frac{3\alpha_{\delta\delta}}{\pi}} e^{-3\alpha_{\delta\delta}\delta^2}, \qquad \hat{I} = I_0 \sqrt{\frac{3(\alpha_{\phi\phi}\alpha_{\delta\delta} - \alpha_{\phi\delta}^2)}{\pi\alpha_{\delta\delta}}} = I_0 \frac{\sqrt{3\pi}}{\tilde{A}\sqrt{\alpha_{\delta\delta}}}, \tag{3.298}$$

where $\alpha_{\delta\delta}$ is given by Eq. (3.140), I_0 is the average current and $\tilde{A}$ is the rms phase-space area of the beam. The dispersion integral can be integrated to obtain the coherent mode frequency given by

$$1 = j\frac{3eI_0\,(Z_\parallel/n)}{2\pi^{3/2}\beta^2 E\eta} \frac{\pi\sqrt{3\alpha_{\delta\delta}}}{\tilde{A}} J_G^{-1}, \tag{3.299}$$
$$J_G = 2[1 + j\sqrt{\pi} y w(y)], \quad y = -\frac{\Omega}{n\omega_0\eta}\sqrt{6\alpha_{\delta\delta}}.$$

For a given broadband impedance model with constant $Z_\parallel/n$, we can find the eigenvalue of the growth rate Im $(\Omega(\mathrm{t}))$ by solving Eq. (3.299).[53]

The solution of Eq. (3.299) shows that the growth rate near the transition energy is nearly equal to the growth rate without Landau damping. This is easy to understand: at $\gamma = \gamma_T$, the frequency spread of the beam becomes zero, and Landau damping vanishes. Fortunately, the growth rate is also small at $\gamma \approx \gamma_T$.

---

[53]See e.g. S.Y. Lee and J.M. Wang, *IEEE Trans. Nucl. Sci.* **NS-32**, 2323 (1985). The impedance model $Z_\parallel/n = 5 - j(Z_{\parallel,\mathrm{sc}}/n)$ ohms was used to study the growth rate around the transition energy for RHIC. Microwave instability below transition may arise from the real impedance. Because of a large space-charge impedance, the growth rate appears to be larger above the transition energy.

The total growth factor across the transition energy region can be estimated by

$$G = \exp\left\{\int (-\mathrm{Im}\Omega)_{\mathrm{unstable}} dt\right\}. \tag{3.300}$$

The total growth factor is a function of the scaling variable $|Z_\parallel/n|N_b/\tilde{A}$. Note that the growth factor is much smaller if the initial phase-space area is increased. Phase-space dilution below transition energy has become a useful strategy in accelerating high intensity proton beams through transition energy. The CERN PS and the AGS employ this method for high intensity beam acceleration. Bunched beam dilution can be achieved either by using a high frequency cavity as noise source or by mismatched injection at the beginning of the cycle.

The distribution function model Eq. (3.298) does not take into account nonlinear synchrotron motion near the transition energy. For a complete account of microwave instability, numerical simulation is an important tool near transition energy.[54] A possible cure for microwave instability is to pass through transition energy fast with a transition energy jump. Furthermore, blow-up of phase-space area before transition energy crossing can also alleviate the microwave growth rate.

We have discussed microwave instabilities induced by a broadband impedance. In fact, it can also be generated by a narrowband impedance. Longitudinal bunch shapes in the KEK proton synchrotron (PS) were measured by a fast bunch-monitor system, which showed the rapid growth of the microwave instability at the frequency of 1 GHz and significant beam loss just after transition energy.[55] Temporal evolution of the microwave instability is explained with a proton-klystron model. The narrowband impedance of the BPM system causes micro-bunching in the beam that further induces wakefield. The beam-cavity interaction produces the rapid growth of the microwave instability. This effect is particularly important near the transition energy, where the frequency spread of the beam vanishes, and the Landau damping mechanism disappears.

E. Microwave instability and bunch lengthening

When the current is above the microwave instability threshold, the instability can cause micro-bunching. The energy spread of the beam will increase until the stability condition is satisfied. For proton or hadron accelerators, the final momentum spread of the beam may be larger than that threshold value caused by decoherence of the synchrotron motion.

[54]W.W. Lee and L.C. Teng, *Proc. 8th Int. Conf. on High Energy Accelerators*, CERN, p. 327 (1971); J. Wei and S.Y. Lee, *Part. Accel.* **28**, 77-82 (1990); S.Y. Lee and J. Wei, *Proc. EPAC*, p. 764 (1989); J. McLachlan, private communications on ESME Program.

[55]See e.g. K. Takayama *et al.*, *Phys. Rev. Lett.*, **78**, 871 (1997).

Due to synchrotron radiation damping in electron storage rings, the final momentum spread and the bunch length are determined by the microwave instability threshold of Eq. (3.297):

$$\sigma_\theta = \omega_0\sigma_t = \frac{|\eta|}{\nu_s}\sigma_\delta = \left(\frac{eI_0|\eta Z_\parallel/n|}{(2\pi)^{3/2}F\nu_s^2\beta^2 E}\right)^{1/3}, \tag{3.301}$$

where ν_s is the synchrotron tune. Note that the bunch length depends only on the parameter $\xi = (I_0|\eta|/\nu_s^2\beta^2 E)$ provided that the impedance does not depend on the bunch length. Chao and Gareyte showed that the bunch lengths of many electron storage rings scaled as $\sigma_\theta \sim \xi^{1/(2+a)}$. This is called Chao-Gareyte scaling law. For a broadband impedance, we have $a = 1$. The scaling law is not applicable if the impedance depends on the beam current and bunch length.

F. Microwave instability induced by narrowband resonances

At low energy, the longitudinal space charge potential, shown as the first term in Eq. (3.287), can be large for high intensity beam bunch. It requires a costly large rf cavity potential to keep beam particles bunched inside the rf bucket. In particular, if it requires a beam gap for a clean extraction, and for minimizing the effect of the electron-cloud instability.

The longitudinal space charge potential can be compensated by the inductive impedance shown in the second term of Eq. (3.287). We consider a cavity with ferrite ring filling a pillbox. The inductance is

$$L = \frac{2\mu'\mu_0\ell}{4\pi}\ln\frac{R_2}{R_1}, \tag{3.302}$$

where μ' is the real part of the ferrite permittivity, R_1 and R_2 are the inner and outer radii of the ferrite rings, and ℓ is the length of the pillbox cavity. The inductive inserts carried out at PSR experiment employs coaxial pillbox cavity with 30 ferrite rings each with width 2.54 cm, 12.7 cm inner diameter (id), and 20.3 cm outer diameter (od). The Proton Storage Ring (PSR) at Los Alamos National Laboratory compresses high intensity proton beam from the 800 MeV linac into a bunch of the order of 250 ns. The parameters for PSR are $C = 90.2$ m, $\gamma_T = 3.1$, $\nu_x = 3.2$, $\nu_z = 2.2$, $\nu_s = 0.00042$, and $f_0 = 2.8$ MHz.

To cancel the space charge impedance at 800 MeV for PSR at the harmonic $h = 1$, one requires about 3 pillbox cavities. The experimental test for this experiment was indeed successful. Unfortunately, the beam also encounters collective microwave beam instability at high intensity. Figure 3.41 shows the microbunching of the beam under the action of three ferrite inserts when an initial bunched coasting beam injected into the ring. The instability is landau damped by heating up the ferrite core to change its permeability and lower the Q-value of the TM_010 mode.[56]

[56]M.A. Plum, *et al.*, Phys. Rev. Special Topics, Accelerators and Beams, **2**, 064201 (1999); C.

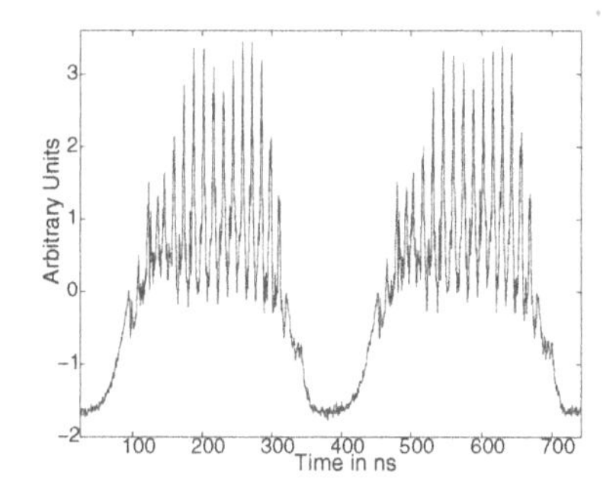

Figure 3.41: The longitudinal beam profile at PSR encountered microwave instability caused by inductive inserts, where three 1-m long ferrite ring cavities were installed in the PSR ring. The injected beam was a uniformly distributed bunched-coasting beam with cavity off. After threshold intensity is encountered, the impedance causes microbunching in the beam. [Courtesy of R. Macek, LANL]

The microwave instability is induced by a narrowband impedance with Q $\approx$ 1 at the center frequency of $f_{\rm res} \approx 27f_0$. Although the inductive inserts can be used to cancel the space charge impedance, the pillbox cavity can generate a narrowband impedance to cause microwave instability of the beam at higher harmonics. In order to alleviate this problem, it is necessary to broaden the narrowband impedance by either choosing different design geometries for different ferrite inserts, or by heating the ferrite so that the imaginary part (μ'') of the permittivity is larger at the cavity resonance frequency. At PSR, the cavities was heated to 125-150° C, so that the beam is below the microwave instability threshold.

Exercise 3.7

1. In synchrotrons, beam bunches are filled with a gap for ion-clearing, abort, extraction kicker rise time, etc. Show that the frequency spectra observed from a BPM for short bunches filled with a gap have a diffraction-pattern-like structure: $\sin(nM\pi/N)/\sin(n\pi/N)$ for M identical consecutive bunches in N buckets. Specifically, find the frequency spectra for 10 buckets filled with 9 equal intensity short bunches. The revolution frequency is assumed to be 1 MHz.

2. Show that the impedance of Eq. (3.291) has two poles in the upper half of the ω plane, and find their loci. Use the inverse Fourier transformation to show that the wake function of the RLC resonator circuit is ($\tilde\omega_{\rm r} = \omega_{\rm r}\sqrt{1-1/4{\rm Q}^2}$)

$$W_\parallel = \frac{R_{\rm sh}\omega_{\rm r}}{\rm Q}e^{-\omega_{\rm r}t/2{\rm Q}}\left[\cos\tilde\omega_{\rm r}t - \frac{1}{\sqrt{4{\rm Q}^2-1}}\sin\tilde\omega_{\rm r}t\right].$$

3. The parameters of the SLC damping ring are E = 1.15 GeV, ν_x = 8.2, ν_z = 3.2, $\alpha_{\rm c}$ = 0.0147, $\gamma\epsilon_{x,z}$ = 15 π mm-mrad, $\sigma_{\Delta p/p} = 7.1\times10^{-4}$, $V_{\rm rf}$ = 800 kV, C = 35.270 m, h = 84, $f_{\rm rf}$ = 714 MHz, ρ = 2.0372 m, and the energy loss per revolution is U_0 = 93.1 keV. If the threshold of bunch lengthening is $N_{\rm B} = 1.5\times10^{10}$, use the Keil-Schnell formula to estimate the impedance of the SLC damping ring.[57]

Beltran, Ph.D. thesis, Indiana University (2003).

[57]G.E. Fisher *et al., Proc. 12th HEACC*, p. 37 (1983); L. Rivkin, *et al., Proc. 1988 EPAC*, p. 634 (1988); see also P. Krejcik, *et al., Proc. 1993 PAC*, p. 3240 (1993). The authors of the last paper observed sawtooth instability at the threshold current $N_{\rm B} = 3\times10^{10}$.

4. Assuming that the microwave instability growth rate is equal to the damping rate at equilibrium, find the tolerable impedance as a function of the machine parameters. Use Eq. (3.292) for The growth rate of microwave instability in a quasi-isochronous electron storage ring and the damping rate $\tau_s = 2ET_0/\mathcal{J}_sU_0$, where E is the energy of the particle, T_0 is the revolution period, the damping partition $\mathcal{J}_s \approx 2$, $U_0 = C_\gamma E^4/\rho$, $C_\gamma = 8.85 \times 10^{-5}$ m/(GeV^3), and ρ is the bending radius.

5. Consider a pillbox-like cavity with length ℓ (see Sec. VII.4). The cavity is filled with ferrite rings with inner and outer radii a and b respectively. Show that the longitudinal impedance for TM_{010} mode is[58]

$$\frac{Z_\parallel}{\ell} = j\frac{Z_0}{2\pi a}\sqrt{\frac{\mu' - j\mu''}{\epsilon_r}}\,\frac{H_0^{(1)}(k_c a)H_0^{(2)}(k_c b) - H_0^{(1)}(k_c b)H_0^{(2)}(k_c a)}{H_1^{(1)}(k_c a)H_0^{(2)}(k_c b) - H_0^{(1)}(k_c b)H_1^{(2)}(k_c a)}.$$

where $H_m^{(n)}$ are Hankel functions which represent incoming and outgoing waves, $Z_0 = 377\Omega$ is the impedance of free space. $k_c = \omega\sqrt{\mu\epsilon} = k\sqrt{\epsilon_r(\mu' - j\mu'')}$, $k = \frac{\omega}{c} = \omega\sqrt{\mu_0\epsilon_0}$ in vacuum, ϵ_r is the relative permittivity and μ' and μ'' are the real and complex parts of the relative complex permeability.

6. The equation of motion for the fractional off-momentum deviation of a particle is

$$\frac{d\delta}{dt} = \frac{\omega_0}{2\pi\beta^2 E}\left[eV(\sin\phi - \sin\phi_s) + \Delta U\right],$$

where the second term in the bracket is the effect of voltage drop due to impedance of the accelerator. Using the voltage drop of the space charge in Eq. (3.287), show that the synchrotron Hamiltonian is

$$H(\phi,\delta) = \frac{h\eta\omega_0}{2}\delta^2 + \frac{\omega_0 eV}{2\pi\beta^2 E}[\cos\phi - \cos\phi_s + (\phi - \phi_s)\sin\phi_s] + \frac{h\omega_0 e^2 c g_0 Z_0 N_B}{4\pi\beta^2\gamma^2 RE}\rho(\phi),$$

where $\rho(\phi)$ is the normalized beam distribution in the synchrotron phase-angle, N_B is the number of particles in a bunch, $g_0 = 1 + 2\ln(b/a)$ is the geometric factor, R is the mean radius of the accelerator. For a Gaussian normalized distribution function, $\rho(\phi) = \frac{1}{\sqrt{2\pi}\sigma_\phi}\exp\{-\frac{(\phi-\phi_s)^2}{2\sigma_\phi^2}\}$. Show that the Hamiltonian becomes

$$H(\phi,\delta) \approx \frac{h\eta\omega_0}{2}\delta^2 - \frac{\omega_0 e}{4\pi\beta^2 E}\left[V\cos\phi_s - \frac{hecg_0 Z_0 N_B}{2\gamma^2 R\sigma_\phi^3}\right](\phi - \phi_s)^2.$$

Note that the space charge produce potential well distortion, and synchrotron detuning. At energies below the transition energy, where $\cos\phi_s \geq 0$, the space charge force tends to push particles away from the center. At energies above the transition energy, the space charge force tends to focus the beam. As the beam bunch is accelerated through the transition energy, the mis-match in the matched bunch length will set off quadrupole mode oscillations. This phenomenon is called Sorenssen effect.[59]

[58]The general formula to calculate the shunt impedance is $\Delta V = -IZ_\parallel = -E_s\ell$, with E_s the longitudinal electric field, ℓ the total length, and I obtained by Ampere's law: $I = \oint H dl = 2\pi a H_\phi$.

[59]A. Sorrenssen, Particle Accelerators **6**, 141 (1975).

VIII Introduction to Linear Accelerators

By definition, any accelerator that accelerates charged particles in a straight line is a linear accelerator (linac).[60] Linacs includes induction linacs; electrostatic accelerators such as the Cockcroft-Walton, Van de Graaff and Tandem; radio-frequency quadrupole (RFQ) linacs; drift-tube linacs (DTL); coupled cavity linacs (CCL); coupled cavity drift-tube linacs (CCDTL); high-energy electron linacs, etc. Modern linacs, almost exclusively, use rf cavities for particle acceleration in a straight line. For linacs, important research topics include the design of high gradient acceleration cavities, control of wakefields, rf power sources, rf superconductivity, and the beam dynamics of high brightness beams.

Linacs evolved through the development of high power rf sources, rf engineering, superconductivity, ingenious designs for various accelerating structures, high brightness electron sources, and a better understanding of high intensity beam dynamics. Since electrons emit synchrotron radiation in synchrotron storage rings, high energy e^+e^- colliders with energies larger than 200 GeV per beam can be effectively attained only by high energy linacs. Current work on high energy linear colliders is divided into two camps, one using superconducting cavities and the other using conventional copper cavities. In conventional cavity design, the choice of rf frequency varies from S band to millimeter wavelength at 30 GHz in the two beam acceleration scheme. Research activity in this line is active, as indicated by bi-annual linac, and annual linear collider conferences.

Since the beam in a linac is adiabatically damped, an intense electron beam bunch from a high brightness source will provide a small emittance at high energy. The linac has also been used to generate coherent synchrotron light. Many interesting applications will be available using high brilliance coherent photon sources.

This section provides an introduction to a highly technical and evolving branch of accelerator physics. In Sec. VIII.1 we review some historical milestones. In Sec. VIII.2 we discuss fundamental properties of rf cavities. In Sec. VIII.3 we present the general properties of electromagnetic fields in accelerating cavity structures. In Sec. VIII.4 we address longitudinal particle dynamics and in Sec. VIII.5, transverse particle dynamics. Since the field is evolving, many advanced school lectures are available.

VIII.1 Historical Milestones

In 1924 G. Ising published a first theoretical paper on the acceleration of ions by applying a time varying electric field to an array of drift tubes via transmission lines; subsequently, in 1928 R. Wideröe used a 1 MHz, 25 kV rf source to accelerate potassium ions up to 50 keV. The optimal choice of the distance between acceleration

[60]See Ref. [4]; G.A. Loew and R. Talman, *AIP Conf. Proc.* **105**, 1 (1982); J. Le Duff, CERN 85-19, p. 144 (1985).

gaps is $d = \beta\lambda/2 = \beta c/2f$, where d is the distance between drift tube gaps, βc is the velocity of the particle, and λ and f are the wavelength and frequency of the rf wave. A Wideröe structure is shown in the top plot of Fig. 1.4. In 1931–34 E.O. Lawrence, D. Sloan *et al.*, at U.C. Berkeley, built a Wiederöe type linac to accelerate Hg ions to 1.26 MeV using an rf frequency of about 7 MHz.[61] At the same time (1931-1935) K. Kingdon at the General Electric Company and L. Snoddy at the University of Virginia, and others, accelerated electrons from 28 keV to 2.5 MeV.

The drift tube distance could be minimized by using a high frequency rf source. For example, the velocity of a 1 MeV proton is $v = \beta c = 4.6 \times 10^{-2}c$, and the length of drift space in a half cycle at rf frequency $f_{\rm rf} = 7$ MHz is $\frac{1}{2}vf_{\rm rf}^{-1} \approx 1$ m. As the energy increases, the drift length becomes too long. The solution is to use a higher frequency system, which became available from radar research during WWII. In 1937 the Varian brothers invented the klystron at Stanford. Similarly, high power magnetrons were developed in Great Britain.[62]

However, the accelerator is almost capacitive at high frequency, and it radiates a large amount of power $P = IV$, where V is the accelerating voltage, $I = \omega CV$ is the displacement current, C is the capacitance between drift tubes, and ω is the angular frequency. The solution is to enclose the gap between the drift tubes in a cavity that holds the the electromagnetic energy in the form of a magnetic field by introducing an inductive load to the system. To attain a high electric field, the cavity is designed to have a resonant frequency that synchronizes with the particle motion. An acceleration cavity is a structure in which the longitudinal electric field can be stored at the gap for particle acceleration, as shown in Fig. 3.42.

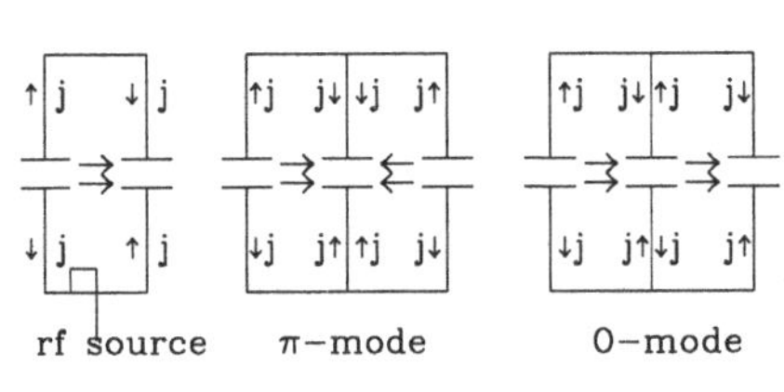

Figure 3.42: Left: Schematic drawing of a single gap cavity fed by an rf source. The rf currents are indicated by j on the cavity wall. Middle: A two-gap cavity operating at π-mode, where the electric fields at two gaps have opposite polarity. Right: A two-gap cavity operating at 0-mode, where the electric fields at all gaps have the same polarity. In 0-mode (or 2π-mode) operation, the rf currents on the common wall cancel, and the wall becomes unnecessary.

When two or more cavity gaps are adjacent to each other, the cavity can be operated at π-mode or 0-mode, as shown in Fig. 3.42. In 0-mode, the resulting current is zero at the common wall so that the common wall is useless. Thus a group of drift tubes can be placed in a single resonant tank, where the field has the same

[61] G. Ising, *Arkiv für Matematik o. Fisik* **18**, 1 (1924); R. Wideröe, *Archiv für Electrotechnik* **21**, 387 (1928); D.H. Sloan and E.O. Lawrence, *Phys. Rev.* **32**, 2021 (1931); D.H. Sloan and W.M. Coate, *Phys. Rev.* **46**, 539 (1934).

[62] The power source of present day household microwave ovens is the magnetron.

phase in all gaps. Such a structure (see Fig. 1.4) was invented by L. Alvarez in 1945.[63] In 1945–47 L. Alvarez, W.K.H. Panofsky, *et al.*, built a 32 MeV, 200 MHz proton drift tube linac (DTL). Drift tubes in the Alvarez structure are in one large cylindrical tank and powered at the same phase. The distances between the drift tubes, $d = \beta\lambda$,[64] are arranged so that the particles, when they are in the decelerating phase, are shielded from the fields.

In 1945 E.M. McMillan and V.I. Veksler discovered the phase focusing principle, and in 1952 J. Blewett invented electric quadrupoles for transverse focusing based on the alternating gradient focusing principle. These discoveries solved the 3D beam stability problem, at least for low intensity beams. Since then, Alvarez linacs has commonly been used to accelerate protons and ions up to 50–200 MeV kinetic energy.

In the ultra relativistic regime with $\beta \to 1$, cavities designed for high frequency operation are usually used to achieve a high accelerating field. At high frequencies, the klystron, invented in 1937, becomes a powerful rf power source. In 1947-48 W. Hansen *et al.*, at Stanford, built the MARK-I *disk loaded linac* yielding 4.5 MeV electrons in a 9 ft structure powered by a 0.75 MW, 2.856 GHz magnetron.[65] On September 9, 1967, the linac at Stanford Linear Accelerator Center (SLAC) accelerated electrons to energies of 20 GeV. In 1973 P. Wilson, D. Farkas, and H. Hogg, at SLAC, invented the rf energy compression scheme SLED (SLAC Energy Development) that provided the rf source for the SLAC linac to reach 30 GeV. In 1990's, SLAC has achieved 50 GeV in the 3 km linac.

Another important idea in high energy particle acceleration is acceleration by traveling waves.[66] The standing wave cavity in a resonant structure can be decomposed into two traveling waves: one that travels in synchronism with the particle, and the backward wave that has no net effect on the particle. Thus the shunt impedance of a traveling wave structure is twice that of a standing wave structure except at the phase advances 0 or π. To regain the factor of two in the shunt impedance for standing wave operation, E. Knapp and D. Nagle invented the side coupled cavity in 1964.[67] In 1972 E. Knapp *et al.* successfully operated the 800 MHz side coupled cavity linac (CCL) to produce 800 MeV energy at Los Alamos. In 1994 the last three tanks of

---

[63]L. Alvarez, *Phys. Rev.* **70**, 799 (1946).

[64]It appears that the distance between drift tubes for an Alvarez linac is twice that of a Wideröe linac, and thus less efficient. However, the use of a high frequency rf system in a resonance-cavity more than compensates the requirement of a longer distance between drift tubes.

[65]E.L. Ginzton, W.W. Hanson and W.R. Kennedy, *Rev. Sci. Instrum.* **19**, 89 (1948); W.W. Hansen *et al.*, *Rev. Sci. Instrum.* **26**, 134 (1955).

[66]J.W. Beams at the University of Virginia in 1934 experimented with a traveling-wave accelerator for electrons using transmission lines of different lengths attached to a linear array of tubular electrodes and fed with potential surges generated by a capacitor-spark gap circuit, similar to the system proposed by Ising. Burst of electrons were occasionally accelerated to 1.3 MeV. See J.W. Beams *et. al.*, *Phys. Rev.* **44**, 784 (1933); *Phys. Rev.*, **45**,849 (1934).

[67]E. Knapp *et al.*, *Proc. 1966 linac Conf.*, p. 83 (1966).

the DTL linac at Fermilab were replaced by CCL to upgrade its proton energy to 400 MeV. Above $\beta \geq 0.3$, CCL has been widely used for proton beam acceleration. A combination of CCL with DTL produces the CCDTL structure suitable for high gradient proton acceleration.

For the acceleration of ions, the Alvarez linac is efficient for $\beta > 0.04$. The acceleration of low energy protons and ions relies on DC accelerators such as the Cockcroft-Walton or Van de Graaff. In 1970 I. Kapchinskij and V. Teplyakov at ITEP Moscow invented the radio-frequency quadrupole (RFQ) accelerator. In 1980 R. Stokes *et al.* at Los Alamos succeeded in building an RFQ to accelerate protons to 3 MeV. Today RFQ is commonly used to accelerate protons and ions for injection into linacs or synchrotrons.

Since the first experiment on a superconducting linear accelerator at SLAC in 1965, the superconducting (SC) cavity has become a major branch of accelerator physics research. In the 1970's, many SC post linear accelerators were constructed for the study of heavy ion collisions in nuclear physics.[68] Recently, more than 180 m of superconducting cavities have been installed in CEBAF for the 4 GeV continuous electron beams used in nuclear physics research. More than 400 m of SC cavities at about 7 MV/m were installed in LEP energy upgrade, and reached 3.6 GV rf voltage for the operation of 104.5 GeV per beam in 2000.[69] The TESLA project had also successfully achieved an acceleration gradient of 35 MV/m.

VIII.2 Fundamental Properties of Accelerating Structures

Fundamental properties of all accelerating structures are the transit time factor, shunt impedance, and Q-value. These quantities are discussed below.

A. Transit time factor

We consider a standing wave accelerating gap, e.g. the Alvarez structure, and assume that the electric field in the gap is independent of the longitudinal coordinate s. If $\mathcal{E}$ is the maximum electric field at the acceleration gap, the accelerating field and the energy gain in traversing the accelerating gap are

$$\mathcal{E}_s = \mathcal{E} \cos \omega t, \tag{3.303}$$
$$\Delta E = e \int_{-\frac{g}{2}}^{\frac{g}{2}} \mathcal{E} \cos \frac{\omega s}{v}\, ds = e\mathcal{E} g T_{\rm tr} = eV_0, \quad T_{\rm tr} = \frac{\sin(\pi g/\beta\lambda)}{\pi g/\beta\lambda},$$

where $V_0 = \mathcal{E} g T_{\rm tr}$ is the effective voltage of the gap, $T_{\rm tr}$ is the transit time factor, $\lambda = 2\pi c/\omega$ is the rf wavelength, and $\pi g/\beta\lambda$ is the rf phase shift across the gap. If the

[68]See e.g., H. Piel, CERN **87-03**, p. 376 (1987); CERN **89-04**, p. 149 (1994), and references therein. These low energy SC cavities are essentially drift-tube type operating at $\lambda/4$ or $\lambda/2$ modes.

[69]P. Brown *et al.*, Proceedings of PAC2001, p. 1059 (IEEE, 2001).

gap length of a standing wave structure is equal to the drift tube length, i.e. $g = \beta\lambda/2$, the transit time factor is $T_{\rm tr} = \sin(\pi/2)/(\pi/2) = 0.637$. This means that only 63% of the rf voltage is used for particle acceleration. To improve the efficiency, the gap length g should be reduced. However, a small g can lead to sparking at the gap. Since there is relatively little gain for $g < \beta\lambda/4$, the gap g is designed to optimize linac performance. The overall transit time factor for standing wave structures in DTL is about 0.8. The transit time factor of Eq. (3.303) is valid for the standing wave structure. The transit time factor for particle acceleration by a guided wave differs from that of Eq. (3.303). An example is illustrated in Exercise 3.8.7.

B. Shunt impedance

Neglecting power loss to the transmission line and reflections between the source and the cavity, electromagnetic energy is consumed in the cavity wall and beam acceleration. The shunt impedance for an rf cavity is defined as

$$R_{\rm sh} = V_0^2/P_{\rm d}, \tag{3.304}$$

where V_0 is the effective acceleration voltage, and $P_{\rm d}$ is the dissipated power. For a multi-cell cavity structure, it is also convenient to define the shunt impedance per unit length $r_{\rm sh}$ as

$$r_{\rm sh} = \frac{R_{\rm sh}}{L_{\rm cav}} = \frac{\mathcal{E}^2}{P_{\rm d}/L_{\rm cav}} \qquad \text{or} \qquad \frac{dP_{\rm d}}{ds} = -\frac{\mathcal{E}^2}{r_{\rm sh}}, \tag{3.305}$$

where $\mathcal{E}$ is the effective longitudinal electric field that includes the transit time factor, and $dP_{\rm d}/ds$ is the fraction of input power loss per unit length in the wall. The power per unit length needed to maintain an accelerating field $\mathcal{E}$ is $P_{\rm d}/L = \mathcal{E}^2/r_{\rm sh}$ and the accelerating gradient for low beam intensity is $\mathcal{E} = \sqrt{r_{\rm sh}P_{\rm d}/L_{\rm cav}}$.

For a 200 MHz proton linac, we normally have $r_{\rm sh} \approx 15-50$ MΩ/m, depending on the transit time factors. For an electron linac at 3 GHz, $r_{\rm sh} \approx 100$ MΩ/m. For high frequency cavities, the shunt impedance is generally proportional to $\omega^{1/2}$ (see Exercise 3.8.4). A high shunt impedance with low surface fields is an important guideline in rf cavity design. For example, using a 50 MW high peak power pulsed klystron, the accelerating gradient of a 3 GHz cavity can be as high as 70 MV/m. The working SLC S-band accelerating structure delivers about 20 MV/m.[70]

C. The quality factor Q

The quality factor is defined by $\mathrm{Q} = \omega W_{\rm st}/P_{\rm d}$, and thus we obtain

$$dW_{\rm st}/dt = -P_{\rm d} = -\omega W_{\rm st}/\mathrm{Q}; \qquad W_{\rm st} = W_{\rm st,0}e^{-2t/t_{\rm F,sw}}, \quad t_{\rm F,sw} = 2\mathrm{Q_L}/\omega, \tag{3.306}$$

[70] P. Raimondi, *et al.*, Proceedings of the EPAC2000, (EPAC, 2000).

where $W_{\rm st}$ is the maximum stored energy, Q_L is the loaded Q-factor that includes the resistance of the power source, and $t_{\rm F,sw}$ is the filling time for the standing wave operation, which is the time for the field to decay to $1/e$ of its initial value. The Q-factor of an accelerating structure is independent of whether it operates in standing wave or traveling wave modes.

For a traveling wave structure, the stored energy per unit length, the power loss per unit length, and the filling time for a traveling wave structure are respectively[71]

$$w_{\rm st} = W_{\rm st}/L_{\rm cav}; \qquad \frac{dP_{\rm d}}{ds} = -\frac{\omega w_{\rm st}}{Q}, \quad \text{or} \quad Q = -\frac{\omega w_{\rm st}}{dP_{\rm d}/ds}; \tag{3.307}$$

$$t_{\rm F,tw} = L_{\rm cav}/v_{\rm g}, \tag{3.308}$$

where $L_{\rm cav}$ is the length of the cavity structure and $v_{\rm g}$ is the velocity of the energy flow. A useful quantity is the ratio $R_{\rm sh}/Q$:

$$\frac{R_{\rm sh}}{Q} = \frac{V_0^2}{\omega W_{\rm st}}, \quad \text{or} \quad \frac{r_{\rm sh}}{Q} = \frac{(V_0/L_{\rm cav})^2}{\omega(W_{\rm st}/L_{\rm cav})} = \frac{\mathcal{E}^2}{\omega w_{\rm st}}, \tag{3.309}$$

which depends only on the cavity geometry and is independent of the wall material, welds, etc.

VIII.3 Particle Acceleration by EM Waves

Charged particles gain or lose energy when the velocity is parallel to the electric field. A particle traveling in the same direction as the plane electromagnetic (EM) wave will not gain energy because the electric field is perpendicular to the particle velocity. On the other hand, if a particle moves along a path that is not parallel to the direction of an EM wave, it can gain energy. However, it will quickly pass through the wave propagation region unless a wiggler field is employed to bend back the particle velocity vector.[72] Alternatively, a wave guide designed to provide electric field along the particle trajectory at a phase velocity equal to the particle velocity is the basic design principle of rf cavities.

The rf cavities for particle acceleration can be operated in standing wave or traveling wave modes.[73] Standing wave cavities operating at steady state are usually used in synchrotrons and storage rings for beam acceleration or energy compensation of synchrotron radiation energy loss. The standing wave can also accelerate oppositely charged beams traveling in opposite directions. Its high duty factor can be used to

[71]We will show that the velocity of the energy flow is equal to the group velocity, $v_{\rm g} = P_{\rm d}/w_{\rm st}$. The conventional definition of standing wave filling time in Eq. (3.306) is twice that of the traveling wave in Eq. (3.308).

[72]This scheme includes inverse free electron laser acceleration and inverse Cerenkov acceleration.

[73]See G.A. Loew, R.H. Miller, R.A. Early, and K.L. Bane, *Proc. 1979 Part. Acc. Conf.*, p. 3701 (IEEE, 1979); R.H. Miller, SLAC-PUB-3935 (1988); see also Exercise 3.8.7.

accelerate long pulsed beams such as protons, and continuous wave (CW) electron beams in the Continuous Electron Beam Accelerator Facility (CEBAF). On the other hand, employing high power pulsed rf sources, a traveling wave structure can attain a very high gradient for the acceleration of an intense electron beam pulse.

In this section we study the properties of electromagnetic waves in cavities. These waves are classified into transverse magnetic (TM) or transverse electric (TE) modes. The phase velocity of the EM waves can be slowed down by capacitive or inductive loading. We will discuss the choice of standing wave vs traveling wave operation, the effect of shunt impedance, and the coupled cavity linac.

A. EM waves in a cylindrical wave guide

First we consider the propagation of EM waves in a cylindrical wave guide. Since there is no ends for the cylindrical wave guide, the EM fields can be described by the traveling wave component in Eq. (3.237) in Sec. VI.1 (see Appendix B Sec. IV). The EM fields of the lowest frequency TM_{01} mode, traveling in the $+\hat{s}$ direction, are

$$\begin{aligned} E_s &= E_0 J_0(k_r r) e^{-j[ks-\omega t]}, \\ E_r &= j\frac{k}{k_r} E_0 J_1(k_r r) e^{-j[ks-\omega t]}, \\ H_\phi &= j\frac{\omega}{cZ_0 k_r} E_0 J_1(k_r r) e^{-j[ks-\omega t]}, \\ E_\phi &= 0, \quad H_s = 0, \quad H_r = 0, \end{aligned} \tag{3.310}$$

where $Z_0 = \sqrt{\mu_0/\epsilon_0}$ is the vacuum impedance, (r, ϕ) is the cylindrical coordinate, s is the longitudinal coordinate, k is the propagation wave number in the $+\hat{s}$ direction, and k_r is the radial wave number:

$$k^2 = (\omega/c)^2 - k_r^2 \qquad\qquad k_{r,mn} = j_{mn}/b, \tag{3.311}$$

where the propagation modes are determined by the boundary condition for $E_s = E_\phi = 0$ at the pipe radius $r = b$ with j_{mn} are zeros of the Bessel functions $J_m(j_{mn}) = 0$ listed in Table B.1 in Appendix B Sec. IV.

The frequency of the TM_{01} mode is $\omega/c = \sqrt{k^2 + (2.405/b)^2}$, shown in Fig. 3.43. The subscript 01 stands for $m = 0$ in ϕ-variation, 1 radial-node at the boundary of the cylinder [see Eq. (3.237)]. This mode is a free propagation mode along the longitudinal $\hat{s}$ direction. We define $\omega_c = k_r c = 2.405c/b$. The wave number of the TM_{01} wave and the corresponding phase velocity $v_{\rm p}$ become

$$k = \frac{\omega}{c}\left[1 - \left(\frac{\omega_c}{\omega}\right)^2\right]^{1/2}, \qquad v_{\rm p} = \frac{\omega}{k} = \frac{c}{[1-(\omega_c/\omega)^2]^{1/2}} > c. \tag{3.312}$$

Unattenuated wave propagation at $\omega < \omega_c$ is not possible. Since the phase velocity propagates faster than the speed of light, the particle can not be synchronized with the EM wave during acceleration. At low frequency, the wave travels forward and backward with a very large phase velocity; it is not useful for particle acceleration. At high frequency, the phase velocity approaches c. However, the electromagnetic field is transverse; it becomes the transverse TEM wave, i.e.

$$\frac{E_s}{E_r} = j\frac{k_r}{k} \to 0, \qquad \frac{H_\phi}{E_r} = \frac{\omega}{ckZ_0} \to \frac{1}{Z_0}.$$

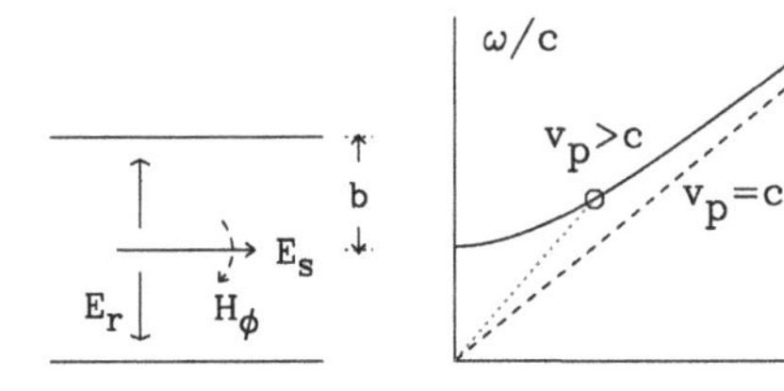

Figure 3.43: Left: Schematic drawing of a cylindrical cavity. Right: Dispersion curve $(\omega/c)^2 = k^2+(2.405/b)^2$ for the TM_{01} wave. The phase velocity ω/k for a wave without cavity load is always greater than the velocity of light. At high frequencies, where $k_r \to 0$, the phase velocity approaches the speed of light. However, the longitudinal component of the EM wave vanishes.

B. Phase velocity and group velocity

Equation (3.310) represents an infinitely long pulse of EM waves in the cylindrical wave guide. The phase of the plane wave, $ks - \omega t$, travels at a phase velocity of $v_{\rm p} = ds/dt = \omega/k$. In reality, we have to discuss a short pulse formed by a group of EM waves. Since the Maxwell equation is linear, the pulse can be decomposed in linear superposition of Fourier series.

For a quasi-monochromatic pulse at frequency ω_0 in free space, the electric field can be represented by

$$\begin{aligned} E(t,s) &= A(t)e^{j(\omega_0 t - ks)} = \frac{1}{2\pi}\int\int A(\xi)e^{j[\omega t - ks - \omega\xi + \omega_0\xi]}d\xi d\omega \\ &= \frac{1}{2\pi}\int\int A(\xi)e^{j[\omega t - k(\omega)s - \omega\xi + \omega_0\xi]}d\xi d\omega, \end{aligned} \tag{3.313}$$

where $A(t)$ is the amplitude with a short time duration and we include the dispersion of the wave number in Eq. (3.311). For a quasi-monochromatic wave at the angular frequency ω_0, we expand the dispersion wave number around ω_0:

$$k(\omega) = k(\omega_0) + \left.\frac{dk}{d\omega}\right|_{\omega_0}(\omega - \omega_0) = k_0 + k'(\omega - \omega_0). \tag{3.314}$$

Substituting Eq. (3.314) into Eq. (3.313), we obtain $E(t,s) = A(t - k's)\ e^{j(\omega_0 t - k_0 s)}$. Note that the phase of the pulse propagates at a "phase velocity" of $v_{\rm p} = \omega_0/k_0$, and the amplitude function of the EM pulse propagates at the "group velocity"

$$v_{\rm g} = \frac{1}{k'} = \left.\frac{d\omega}{dk}\right|_{\omega_0}. \tag{3.315}$$

Using Eq. (3.312) for single-mode wave propagation, we obtain $v_{\rm g} = kc^2/\omega$, or $v_{\rm p}v_{\rm g} = c^2$. From Fig. 3.43 we see that the group velocity is zero at $k = 0$.

In fact, the group velocity is equal to the velocity of energy flow in the wave guide to be shown as follows: The power of the TM wave, the total energy per unit length stored, and the velocity of the energy flow are

$$P = \frac{1}{2}Re\int_S E_r H_\phi^* dS = \frac{1}{2}E_0^2\frac{k\omega}{cZ_0k_r^2}\int_0^b J_1^2(k_r r)2\pi r dr,$$
$$W = 2W_{\rm m} = \frac{1}{2}E_0^2\frac{\mu\omega^2}{c^2Z_0^2k_r^2}\int_0^b J_1^2(k_r r)2\pi r dr,$$
$$v_{\rm e} = \frac{P}{W} = \frac{k}{\omega}c^2 = v_{\rm g},$$

where H_ϕ^* is the complex conjugate of H_ϕ, $W_{\rm m}$ is the magnetic energy. The *velocity of energy flow* is equal to the *group velocity*.

C. TM modes in a cylindrical pillbox cavity

Now we consider a cylindrical pillbox cavity, where both ends of the cylinder are nearly closed. The cylinder has a beam hole for the passage of particle beams (Fig. 3.44). Here we discuss the standing wave solution of Maxwell's equation for a "closed pillbox cavity," and the effect of beam holes. The effect of a chain of cylindrical cells on the propagation of EM waves is discussed in the next section.

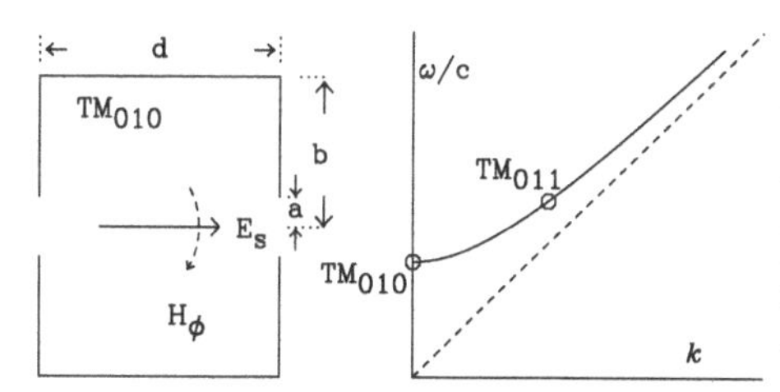

Figure 3.44: Left: Schematic of a cylindrical cavity. Right: Dispersion curve $(\omega/c)^2 = (p\pi/d)^2 + (2.405/b)^2$ for TM_{01p} resonance waves (marked as circles) for a closed cylindrical pillbox without beam holes. With proper design of pillbox geometry, the phase velocity of the TM_{010} mode can be slowed to the particle speed for beam acceleration.

We first discuss the standing wave solution of a closed pillbox cavity without beam holes. With a time dependent factor $e^{j\omega t}$, the TM mode solution of Eq. (3.237) in the closed cylindrical pillbox cavity is reproduced as follows:

$$\begin{cases} E_s = Ck_r^2\, J_m(k_r r)\cos m\phi\cos ks, \\ E_r = -Ckk_r\, J_m'(k_r r)\cos m\phi\sin ks, \\ E_\phi = Cnk\,\dfrac{1}{r}J_m(k_r r)\sin m\phi\sin ks, \end{cases}\qquad \begin{cases} H_s = 0, \\ H_r = -jC\dfrac{m\omega\epsilon_0}{r}\,J_m(k_r r)\sin m\phi\cos ks, \\ H_\phi = -jC\omega\epsilon_0 k_r\, J_m'(k_r r)\cos m\phi\cos ks, \end{cases}$$

where the longitudinal magnetic field is zero for TM modes, ω is the angular frequency, and k_r and k are wave numbers of the radial and longitudinal modes. The dispersion relation is $\omega/c = \sqrt{k_r^2 + k^2}$. Similarly, there are also TE modes where the longitudinal electric field is zero.

Using the boundary conditions that $E_r = 0$ and $E_\phi = 0$ at $s = 0$ and d, we obtain $kd = p\pi$ $(p = 0, 1, 2, \cdots)$, where d is the length of the pillbox, kd is the phase advance of the EM wave in the cavity cell. We also use the boundary conditions $E_s = 0$ and $E_\phi = 0$ at the pipe radius $r = b$ to obtain the radial modes $k_{r,mn}b = j_{mn}$, where b is the inner radius of the cylinder, and j_{mn} are zeros of the Bessel functions $J_m(j_{mn}) = 0$ listed in Table B.1 (Appendix B Sec. IV).

In summary, the resonance frequency ω for the TM_{mnp} mode is

$$\frac{\omega_{mnp}}{c} = \sqrt{\frac{j_{mn}^2}{b^2} + \frac{p^2\pi^2}{d^2}}. \tag{3.316}$$

For the TM_{010} mode, we have $k = 0$ and $\omega/c = k_r = 2.405/b$ shown as a circle on Fig. 3.44. The electromagnetic fields for this mode are

$$E_s = E_0 J_0(k_r r), \quad B_\phi = j\frac{E_0}{c} J_1(k_r r). \tag{3.317}$$

Figure 3.44 (right) show also the mode frequency of TM_{011} on the dispersion curve. Both these modes have phase velocities greater than c.

To lower the phase velocity, beam hole radius a and cylinder radius b are tailored to provide matched phase advance kd and phase velocity ω/k for the structure. Analytic solution of Maxwell's equations for an actual cavity geometry is difficult. The EM wave modes can be calculated by finite element or finite difference EM codes with a periodic boundary (resonance) condition and a prescribed phase advance kd across the cavity gap.

The solid lines in Fig. 3.45 are the dispersion curves of frequency f vs phase shift kd for TM_{0np} modes of a SLAC-like pillbox cavity with $a = 18$ mm, $b = 43$ mm, and $d = 34.99$ mm.[74] Because of the coupling between adjacent pillbox-cavities, the discrete mode frequencies become a continuous function of the phase advance kd, and the phase-velocity is effectively lowered. The dashed lines show the world line $v_{\rm p} = c$. The details of the TM_{010} mode are shown in the right plot. At $f = 2.856$ GHz, the phase shift per cell is about 120°, and the phase velocity $v_{\rm p}$ is equal to c.

The frequencies of the TM modes 010, 011, 020, 021, 030 for a closed cylindrical pillbox are shown as circles in the left plot of Fig. 3.45. Increasing the size of the beam

[74]The calculation was done by Dr. D. Li using MAFIA in 2D monopole mode. The wall thickness chosen was 6.027 mm. The wall thickness slightly influences the mode frequencies of TM_{0n1} modes, where the effective d parameter is reduced for a single cell structure. The actual SLAC structure is a constant gradient structure with frequency of $f = 2.856$ GHz, phase advance of $2\pi/3$, length of the structure of $L = 3.05$ m, inner diameter of $2b = 83.461 - 81.793$ mm, disk diameter of $2a = 26.22 - 19.24$ mm, and disk thickness of 5.842 mm. See also C.J. Karzmark, Xraig S. Nunan, and Eiji Tanabe, *Medical Electron Accelerators*, (McGraw-Hill, New York, 1993).

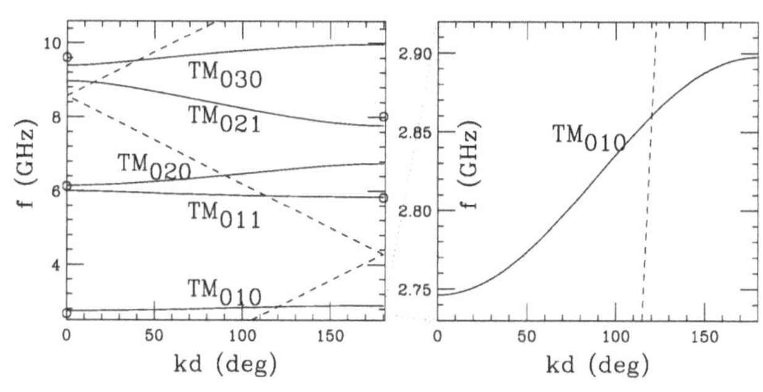

Figure 3.45: Left: Dispersion curves, f vs kd, for TM_{01p} modes for a pillbox cavity with a = 18 mm, b = 43 mm, and d = 34.99 mm. Circles show the TM_{0np} mode frequencies for a closed pillbox cavity. The dashed lines show the world line $v_p = c$. Right: Dispersion curve of TM_{010} mode.

hole decreases the coupling capacitance and increases the TM_{010} mode frequency. More importantly, it provides a continuous TM mode frequency as a function of wave number k. When the beam hole radius decreases, all mode frequencies become horizontal lines. When the beam hole is completely closed, the mode frequencies become discrete points, the circles in the left plot of Fig. 3.45.

Table 3.9: Parametric dependence of the SLAC cavity geometry

b (mm)	d (mm)	kd (deg)	f (GHz)	R_{sh} (MΩ)	Q	r_{sh} (MΩ/m)
42.475	17.495	60	2.8579	0.5107	7713	29.2
42.000	26.24	90	2.853	1.2	10947	45.73
41.805	30.616	105	2.857	1.559	12413	50.92
41.685	34.99	120	2.854	1.874	13700	53.56
41.580	39.36	135	2.857	2.14	14848	54.37
41.415	46.653	160	2.857	2.416	16507	51.79
41.290	52.485	180	2.857	2.466	17646	46.98

Table 3.9 shows parametric dependence of a SLAC-like pillbox cavity at $f = 2.856$ GHz. Note that the shunt impedance per unit length is maximum at a phase advance of about 135°. The phase advance per cell at a given frequency is mainly determined by the cell length.

D. Alvarez structure

The Alvarez linac cavity resembles the TM_{010} standing wave mode (see Table 3.10). The tank radius and other coupling structures, such as rods and slugs inside the cavity, are designed to obtain a proper resonance frequency for the TM_{010} mode, and thus we have $b \approx 2.405c/\omega$. The resulting electric field of Eq. (3.317) is independent of s. The total length is designed to have a distance $\beta\lambda$ between two adjacent drift tubes (cells), where βc is the speed of the accelerating particles. Since β increases along the line, the distance between drift tubes increases as well. Table 3.10 shows some properties of an Alvarez linac, the SLAC cavity, and the CEBAF cavity.

Table 3.10: Some parameters of basic cylindrical cavity cells

Machine	f (MHz)	b (cm)	d (cm)	$N_{\rm cell}$	$\mathcal{E}$ (MV/m)
Alvarez linac	201.25	57.0	$\sum_i \beta_i \lambda$		
Fermilab (cavity1)		47	744	55	1.60
Fermilab (cavity2)		45	1902	59	2.0
CEBAF SC cavity	1497	7.66	10.	5	$5-10$
SLAC linac	2856	4.2	3.5	≈ 100	20

E. Loaded wave guide chain and the space harmonics

The phase velocity must be brought to the level of the particle velocity, i.e. $v_{\rm p} \approx c$. A simple method of reducing the phase velocity is to load the structure with disks, or washers. Figure 3.45 shows, as an example, frequency f vs phase advance kd of the loaded SLAC-like pillbox cavity. Loaded cavity cells can be joined together to form a cavity module. Opening a beam hole at the center of the cavity is equivalent to a capacitive loading for attaining continuous bands of resonance frequencies. The question is, what happens to the EM wave in a chain of cavity cells?

If the wave guide is loaded with wave reflecting structures such as iris, nose-cone, etc., shown in Fig. 3.46 (top), the propagating EM waves can be reflected by obstruction disks. The size of the beam hole determines the degree of coupling and the phase shift from one cavity to the next. When the a, b parameters of the disk radii are tailored correctly, the phase change from cavity to cavity along the accelerator gives an overall phase velocity that is equal to the particle velocity. The reflected waves for a band of frequencies interfere destructively so that there is no radial field at the irises. Since the irises play no role in wave propagation, this gives rise to a minor perturbation in the propagating wave. The dispersion relation in this case resembles that in Fig. 3.43. At some frequencies the reflected waves from successive irises are exactly in phase so that the irises force a standing wave pattern. At these frequencies, unattenuated propagation is impossible, so that the EM wave becomes a standing wave and the group velocity again becomes zero, i.e. the phase advance $kd = \pi$. Such a chain of loaded wave guides can be used to slow the phase velocity of EM waves.

With the Floquet theorem for the periodic wave guide, the EM wave of an infinitely long disk loaded wave guide is

$$\tilde{E}_s(r,\phi,s,t) = e^{-j[k_0 s - \omega t]} E_s(r,\phi,s), \quad \tilde{H}_\phi(r,\phi,s,t) = e^{-j[k_0 s - \omega t]} H_\phi(r,\phi,s),$$
$$E_s(r,\phi,s+d) = E_s(r,\phi,s), \quad H_\phi(r,\phi,s+d) = H_\phi(r,\phi,s),$$

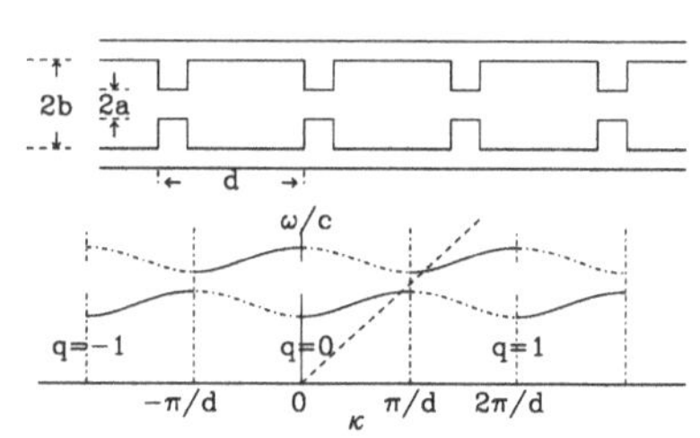

Figure 3.46: Top: Schematic of a chain of cylindrical cavities. Bottom: Dispersion curve (ω/c) vs k. The phase velocity ω/k with a cavity load is equal to the speed of light at a specific point of the dispersion curve, shown as the intersection of the dashed diagonal line and the solid dispersion curve. The solid line branches correspond to forward traveling waves and the dashed line branches are associated with backward traveling waves. The $q=0$ space harmonic corresponds to $kd \in (-\pi,\pi)$, and the $q=1$ space harmonic to $kd \in (\pi, 3\pi)$, etc.

$$\tilde{E}_s(r,\phi,s,t) = e^{-j[k_0 s-\omega t]}\sum_{q=-\infty}^{\infty} E_{s,q}(r,\phi)e^{-j2q\pi s/d} = e^{j\omega t}\sum_{q=-\infty}^{\infty} E_{s,q}(r,\phi)e^{-jk_q s},$$

where d is the period of the wave guide, the propagation wave number is $k_q = k_0 + \frac{2\pi q}{d}$ (q = integer) for the qth "space harmonic", and k_0 is the propagation wave number of the "fundamental space harmonic." These space harmonics are shown in Fig. 3.46. We note further that as $k_0 d \to 0$ or π, forward and backward traveling branches coincide and they will contribute to enhance the electric field.

The field components of the lowest TM_{0n} mode with cylindrical symmetry become

$$\tilde{E}_s = \sum_q E_{0q} J_0(k_{r,q} r)e^{-j[k_q s-\omega t]}, \tag{3.318}$$

$$\tilde{E}_r = j\sum_q \frac{k_q}{k_{r,q}} E_{0q} J_1(k_{r,q} r)e^{-j[k_q s-\omega t]}, \tag{3.319}$$

$$\tilde{H}_\phi = j\frac{1}{Z_0}\sum_q \frac{k_0}{k_{r,q}} E_{0q} J_1(k_{r,q} r)e^{-j[k_q s-\omega t]}, \tag{3.320}$$

where the wave number and the phase velocity at a given frequency ω are

$$k_q^2 = (\omega/c)^2 - k_{r,q}^2, \qquad v_{\mathrm{p},q} = \frac{\omega}{k_q} = \frac{\omega}{k_0 + 2\pi q/d}. \tag{3.321}$$

Note that $k_{r,q}=0$ and $J_0(k_{r,q}r)=1$ for $v_{\mathrm{p},q}=c$. This indicates that the electric field of the qth space harmonic is independent of the transverse position.[75]

The dispersion curve of a periodic loaded wave-guide structure (or slow wave structure) is a typical Brillouin-like diagram shown in Fig. 3.46, where the branches with solid lines correspond to forward traveling wave, and the branches with dashed

[75]One may wonder how to reconcile the fact that the tangential electric field component E_s must be zero at $r=b$. The statement that the electric field is independent of transverse position is valid only near the center axis of loaded wave-guide structures.

dots are backward traveling wave. Because the dispersion curve is a simple translation of $2\pi/d$, and these curves must join, they must have zero slope at the lower frequency ω_0/c, where $k_0 d = 0$, and at the upper frequency ω_π/c, where $k_0 d = \pi$ (see also Fig. 3.45). The range of frequencies $[\omega_0, \omega_\pi]$ is called the *pass band*, or the propagation band. The extreme of the pass band is $k_0 d = \pi$, where the group velocity is zero. At $k_0 d = \pi$, the cavity has lowest rf loss,[76] making this a favorable mode of operation for accelerator modules.

The electric field at a snapshot is shown schematically in Fig. 3.47. At an instant of time, it represents a traveling wave or the maximum of a standing wave. The upper plot shows the snapshot of an electromagnetic wave. The lengths of $kd = \pi, 2\pi/3$, and $\pi/2$ cavities are also shown. The arrows indicate the maximum electric field directions. The lower plot shows a similar snapshot for $kd = 0, \pi/2, 2\pi/3$ and π cavities.

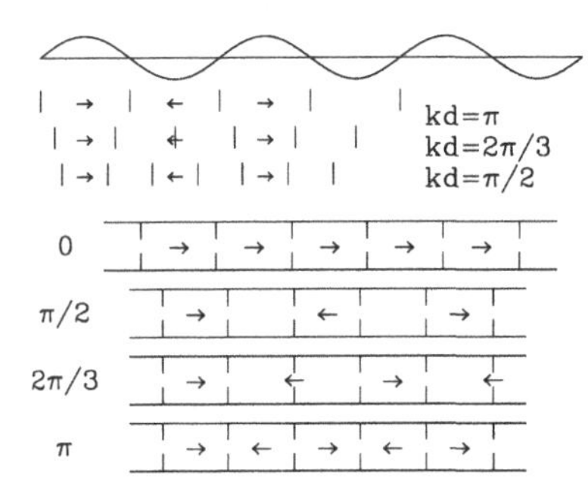

Figure 3.47: Top: Snapshot of a sinusoidal wave for phase advances $kd = \pi/2, 2\pi/3$, and π. Bottom: Snapshot at the maximum electric field configuration across each cell for $kd = 0, \pi/2, 2\pi/3$, and π phase shift structures. The actual electromagnetic fields must satisfy the periodic boundary conditions. The snapshot represents the field pattern of a traveling wave guide or the maximum field pattern of a standing wave. Note that only half of the $kd = \pi/2$ mode has longitudinal electric field in the standing wave mode. The resulting shunt impedance is half of that in traveling wave operation.

The condition for wave propagation is $-1 \le \cos k_0 d \le 1$. If we draw a horizontal line in the dispersion curve within the pass band of the frequency, there are infinite numbers of crossings between the horizontal line and the dispersion curve. These crossings are separated into space harmonics. Higher order space harmonics have no effect on a beam because they have very different phase velocity. Each point corresponds to the propagation factor k_q, which has an identical slope in the ω/c vs k curve, i.e. an identical group velocity:

$$v_{g,q} = \frac{d\omega}{dk_q} = \frac{d\omega}{dk_0} = v_g. \tag{3.322}$$

A module made of N cells resembles a chain of N weakly coupled oscillators. There are $N+1$ resonances located at

$$k_{0m} d = m\pi/N \quad (m = 0, 1, 2, \cdots, N). \tag{3.323}$$

[76]The rf loss is proportional to $|H_\phi|^2$ on the cavity wall.

In the coupled RLC circuit model, the resonance frequency of the electric coupled cavity is[77]

$$\omega_m = \omega_0 \left[1 + \kappa(1 - \cos k_{0m} d)\right]^{1/2}, \tag{3.324}$$

where ω_0 is the resonance frequency without beam hole coupling, and κ is the coupling coefficient. The resonance frequency can be more accurately calculated from powerful finite difference, or finite element, programs such as 2D URMEL, SUPERFISH, LALA, and 3D MAFIA. The size and the length of cavity cells are also tailored to actual rf sources for optimization.

The operating condition $v_{\rm p} = c$ is equivalent to

$$k_0 = \omega/c, \quad \text{or} \quad k_{r,0} = 0$$

for the fundamental space harmonic [see Eq. (3.321)]. Since $J_0(k_{r,0}r) = 1$, *the energy gain of a charged particle is independent of its transverse position*, i.e. the longitudinal electric field of the fundamental space harmonic is independent of the radial position within the radius of the iris. This implies that the transverse force on the particle vanishes as well (see Sec. VIII.5).

F. Standing wave, traveling wave, and coupled cavity linacs

We have shown that the Alvarez linac operates at the standing wave TM_{010} mode, with drift tubes used to shield the electric field at the decelerating phase. The effective acceleration gradient is reduced by the transit time factor and the time the particle spends inside the drift tube. On the other hand, a wave guide accelerator, where the phase velocity is equal to the particle velocity, can effectively accelerate particles in its entire length. A wave guide accelerator is usually more effective if the particle velocity is high. There are two ways to operate high-β cavities: standing wave or traveling wave.

The filling time of a standing wave structure is a few times the cavity filling time $2\text{Q}_\text{L}/\omega$, where Q_L is the loaded Q-factor, to allow time to build up its electric field strength for beam acceleration. Standing wave cavities are usually used to accelerate CW beams, e.g. the CEBAF rf cavity at the Jefferson Laboratory (see Table 3.10), and long pulse beams, e.g. in the proton linacs and storage rings. In a storage ring, a standing wave can be used to accelerate beams of oppositely charged particles moving in opposite directions.

Standing wave operation of a module made of many cells may have a serious problem of many nearby resonances. For example, if a cavity has 50 cells, it can have standing waves at

$$kd = \pi, \ \frac{49}{50}\pi, \ \frac{48}{50}\pi, \ \cdots.$$

[77]See Exercise 3.8.6. For magnetically coupled cavity, the resonance frequency is given by $\omega_0 = \omega\left[1 + \kappa(1 - \cos k_0 d)\right]^{1/2}$.

Since $d\omega/dk = 0$ for a standing wave at $kd = 0$ or π, these resonances are located in a very narrow range of frequency. A small shift of rf frequency will lead to a different standing wave mode. This problem can be minimized if the standing wave operates at the $kd = \pi/2$ condition, where $d\omega/dk$ has its highest value. However, the shunt impedance in $kd = \pi/2$ mode operation is reduced by a factor of 2, because only half of the cavity cells are used for particle acceleration. Similarly, the forward traveling wave component of a standing wave can accelerate particles, the resulting shunt impedance is 1/2 of that of a traveling wave structure except for the phase advance $kd = 0$ or π (see Fig. 3.48).

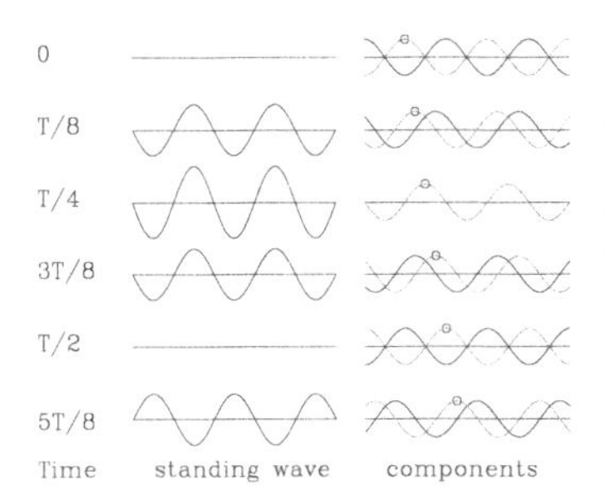

Figure 3.48: A standing wave (left) can be decomposed into forward and backward traveling waves (right). Since only the forward traveling wave can accelerate the beam, the shunt impedance is 1/2 of that of the traveling wave structure except for $kd = 0$ and π standing wave modes, where two neighboring space harmonics contribute to regain the factor of two in the shunt impedance. Note that the particle riding on top of the right-going wave that has the phase velocity equal to the particle velocity will receive energy gain

Since every other cavity cell has no electric fields in $kd = \pi/2$ standing wave operation, these empty cells can be shortened or moved outside. This led to the invention of the coupled cavity linac (CCL) by E. Knapp and D. Nagle in 1964. The idea is schematically shown in Fig. 3.49.

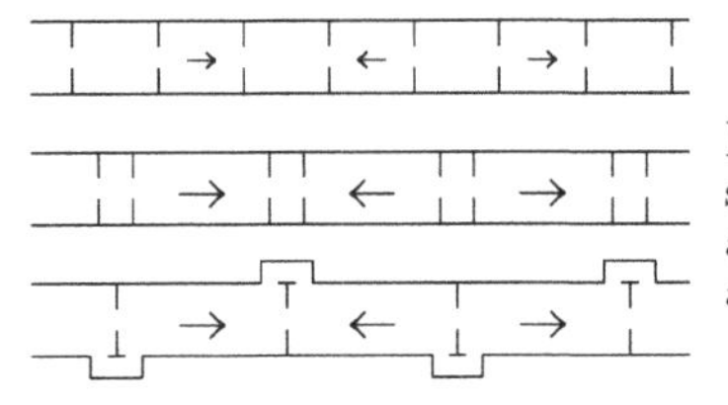

Figure 3.49: A schematic drawing of the $\pi/2$ phase shift cavity structure (top), where the field free regions are shortened (middle), and moved outside to become a coupled cavity structure (bottom).

The CCL cavities operate at $\pi/2$ mode, where field free cells are located outside the main cavity cells. These field free cells are coupled to the main accelerating cavity in the high magnetic field region. The electric field pattern of the main accelerating cavity cells looks like that of a π-mode cavity. Such a design regains the other half of the shunt impedance and provides very efficient proton beam acceleration for $\beta > 0.3$.

The high-β linac can also be operated as a traveling wave guide. There are divided into "constant gradient" and "constant impedance" structures (see Exercise 3.8.8). The accelerating cavities of a constant impedance structure are identical and the power attenuation along the linac is held constant. On the other hand, the geometry of accelerating cavities of a constant gradient structure are tapered to maintain a constant accelerating field along the linac. The filling time for a traveling wave guide is $L_{\rm cav}/v_{\rm g}$, where $L_{\rm cav}$ is the length of a cavity and $v_{\rm g}$ is the group velocity. Typical group velocity is about $0.05c$. Table 3.10 lists the properties of SLAC linac cavity, that is a constant gradient structure operating at a phase advance of $2\pi/3$. With a high peak power rf source, a traveling wave cavity can provide a high acceleration gradient for intense electron beams.

G. High Order Modes (HOMs)

So far we have discussed only the fundamental mode of a cavity. In reality, high order modes (HOMs) can be equally important in cavity design. Efforts are being made to design or invent new cavity geometries with damped HOMs or detuned and damped HOMs. Such efforts are instrumental for future linear colliders operating at high frequencies.

These HOMs, particularly TM_{11p}-like modes, can affect the threshold current of a linac. When a beam is accelerated in cavities, it also generates long range and short range wakefields. A long range wake can affect trailing bunches, and a short range wake can cause a bunch tail to break up. These instabilities are called BBU (beam break up, or beam blow up) instabilities, observed first in 1957.[78] The BBU is a transverse instability. Its threshold current can be increased by a quadrupole focusing system. It also depends strongly on the misalignment of accelerating structure and rf noise. Operation of the SLAC linac provides valuable information on transverse instability of intense linac beams.[79]

VIII.4 Longitudinal Particle Dynamics in a Linac

Phase focusing of charged particles by a sinusoidal rf wave provides longitudinal stability in a linac. Let $t_{\rm s}$, $\psi_{\rm s}$ and $W_{\rm s}$ be the time, rf phase, and energy of a synchronous particle, and let t, ψ, and W be the corresponding physical quantities for a non-synchronous particle. We define the synchrotron phase space coordinates as

$$\Delta t = t - t_{\rm s}, \quad \Delta\psi = \psi - \psi_{\rm s} = \omega(t - t_{\rm s}), \quad \Delta W = W - W_{\rm s}. \tag{3.325}$$

The accelerating electric field is

$$\mathcal{E} = \mathcal{E}_0 \sin \omega t = \mathcal{E}_0 \sin(\psi_s + \Delta\psi), \tag{3.326}$$

where the coordinate s is chosen to coincide with the proper rf phase coordinate. The change of the phase coordinate is

$$\frac{d\Delta\psi}{ds} = \omega\left(\frac{dt}{ds} - \frac{dt_{\rm s}}{ds}\right) = \omega\left(\frac{1}{v} - \frac{1}{v_{\rm s}}\right) \approx -\frac{\omega}{mc^3\beta_{\rm s}^3\gamma_{\rm s}^3}\Delta W, \tag{3.327}$$

[78] T.R. Jarvis, G. Saxon, and M.C. Crowley-Milling, *IEEE Trans. Nucl. Sci.* **NS-112**, 9 (1965).
[79] See J.T. Seeman, p. 255 in Ref. [19].

where $v = ds/dt$ and $v_s = ds/dt_s$ are the velocities of a particle and a synchronous particle, and the subscript s is used for physical quantities associated with a synchronous particle. This equation is in fact identical to Eq. (3.13), where $\omega/\beta_s c$ is equivalent to the harmonic number per unit length, $\Delta W/\beta_s^2 E$ is the fractional momentum spread, and $-1/\gamma_s^2$ is the equivalent phase slip factor. Since the momentum compaction in a linac is zero, the beam in a linac is always below transition energy.

The energy gain from rf accelerating electric fields is[80]

$$\frac{d\Delta W}{ds} = e\mathcal{E}_0 \left[\sin(\psi_s + \Delta\psi) - \sin\psi_s\right] \approx e\mathcal{E}_0 \, \cos\psi_s \, \Delta\psi. \tag{3.328}$$

The Hamiltonian for the synchrotron motion becomes

$$H = -\frac{\omega}{2mc^3\beta^3\gamma^3}(\Delta W)^2 + e\mathcal{E}_0 \left[\cos(\psi_s + \Delta\psi) + \Delta\psi \sin\psi_s\right]. \tag{3.329}$$

Hereafter, β_s and γ_s are replaced by β and γ for simplicity. The linearized synchrotron equation of motion is simple harmonic:

$$\frac{d^2\Delta W}{ds^2} = -k_{\rm syn}^2 \, \Delta W, \qquad k_{\rm syn} = \sqrt{\frac{e\mathcal{E}_0\omega\cos\psi_s}{mc^3\beta^3\gamma^3}}, \tag{3.330}$$

where $k_{\rm syn}$ is the wave number of the synchrotron motion. For medium energy proton linacs, $k_{\rm syn}$ is about 0.1 to 0.01 $\rm m^{-1}$, which is equal to the wave number of transverse motion. Synchro-betatron coupling can be an important beam dynamics issue. For high energy electrons, $k_{\rm syn} \sim 1/\sqrt{\gamma^3}$ is small. The beam particles move rigidly in synchrotron phase space, and thus the synchronous phase angle is normally chosen as $\phi_s = \frac{\pi}{2}$, i.e. electron bunches are riding on top of the crest of the rf wave. The beam will get the maximum acceleration and a minimum energy spread.

In contrast to synchrotrons, the linac usually do not have repetitive periodic structures, the concept of synchrotron tune is not necessary. However, if there is a quasi-periodic external focusing structures such as periodic solenoidal focusing systems, FODO focusing systems, or periodic doublet focusing systems, etc., the synchrotron tune can be defined as the $\nu_{\rm syn} = k_{\rm syn}L/(2\pi)$, where L is the length of the periodic focusing system. Parametric synchrotron resonances can occur if $m\nu_{\rm syn} = \ell$ is satisfied, where m and ℓ are integers. Near a parametric synchrotron resonance, the longitudinal phase space will form islands as discussed in Sec. III.

A. The capture condition in an electron linac with $v_p = c$

Since $\beta_s\gamma_s$ changes rapidly in the first few sections of electron linac, the Hamiltonian contour is not a constant of motion. Tori of phase space ellipses form a golf-club-like

[80]Note that the convention of the rf phase used in the linac community differs from that of the storage ring community by a phase of $\pi/2$. In this textbook, we use the rf phase convention of the storage ring community.

shape, shown in Fig. 3.3. This section will show that all captured particles ride on top of the rf wave.

In an electron linac operating at a phase velocity equal to c, what happens to the injected electrons with velocities less than c? Let ψ be the phase angle between the wave and the particle. Assuming constant gradient acceleration, the electric field seen by the electron is $\mathcal{E}_0 \sin\psi$. Since the phase velocity and the particle velocity are different, the path length difference between the EM wave and the particle in time interval dt is

$$d\ell = (c - v)dt = \frac{\lambda}{2\pi}d\psi, \qquad \frac{d\psi}{dt} = \frac{2\pi c}{\lambda}(1-\beta), \tag{3.331}$$

where $\lambda = 2\pi c/\omega$ is the rf wavelength, $d\ell/\lambda = d\psi/2\pi$, and $\beta = v/c$. The particle gains energy through the electric field, i.e.

$$\frac{d(\gamma m v)}{dt} = mc\frac{d}{dt}\left[\frac{\beta}{(1-\beta^2)^{1/2}}\right] = e\mathcal{E}_0 \sin\psi, \qquad \text{or} \qquad \frac{d\zeta}{dt} = -\frac{e\mathcal{E}_0}{mc}\sin\psi\sin^2\zeta,$$

where $\beta = \cos\zeta$. Using the chain rule $d\psi/dt = (d\psi/d\zeta)(d\zeta/dt)$, we can integrate the equation of motion to obtain

$$\cos\psi_2 - \cos\psi_1 = \frac{2\pi mc^2}{e\mathcal{E}_0\lambda}\tan\frac{\zeta}{2}\bigg|_1^2 = -\frac{2\pi mc^2}{e\mathcal{E}_0\lambda}\left(\frac{1-\beta_1}{1+\beta_1}\right)^{1/2} = -Y_{\rm inj}, \tag{3.332}$$

where the indices 1 and 2 specify the injection and the captured condition respectively, and we have used $\beta_2 = 1$ and the relation

$$\tan(\zeta/2) = ((1-\cos\zeta)/(1+\cos\zeta))^{1/2} = [(1-\beta)/(1+\beta)]^{1/2} = \gamma - \sqrt{\gamma^2-1}.$$

The capture condition, Eq. (3.332), favors a linac with a higher acceleration gradient $\mathcal{E}_0$. If $Y_{\rm inj} = 1.5$, particles within an initial phase $-\pi/3 < \psi_1 < \pi/3$ will be captured inside the phase region $\pi > \psi_2 > 2\pi/3$. If the factor $Y_{\rm inj} = 1$, all particles within $-\pi/2 < \psi_1 < \pi/2$ will be captured into the region $\pi > \psi_2 > \pi/2$. In particular, particles distributed within the range $\Delta > \psi_1 > -\Delta$ will be captured into the range $\pi/2 \le \psi_2 \le \pi/2 + \Delta^2/2$ ($\Delta \ll 1$). For example, all injected beam with phase length 20° will be compressed to a beam with a phase length 3.5° in the capture process.

The capture efficiency and energy spread of the electron beam can be optimized by a prebuncher. A prebuncher is usually used to prebunch the electrons from a source, which can be thermionic or rf gun. We assume a thermionic gun with a DC gun voltage V_0, which is usually about 80–150 kV. Let the electric field and the gap width of the prebuncher be $\mathcal{E}\sin(\omega t)$ and g. Electrons that arrive earlier are slowed and that arrive late are sped up. At a drift distance away from the prebuncher, the faster electrons catch up the slower ones. Thus electrons are prebunched into a smaller phase extension to be captured by the buncher and the main linac (see Exercise 3.8.9). All captured high energy electrons can ride on top of the crest of the rf wave in order to gain maximum energy from the rf electric field.

B. Energy spread of the beam

In a multi-section linac, individual adjustment of each klystron phase can be used to make a bunch with phase length Δ ride on top of the rf crest, i.e. $\psi_s = \frac{\pi}{2}$. The final energy spread of the beam becomes

$$\frac{\Delta W}{W_s} = 1 - \cos\frac{\Delta}{2} = \frac{\Delta^2}{8}. \tag{3.333}$$

This means that a beam with a phase spread of 0.1 rad will have an energy spread of about 0.13 %. Thus the injection match is important in minimizing the final energy spread of the beam. Other effects that can affect the beam energy are beam loading, wakefields, etc. A train of beam bunches extracts energy from the linac structure and, at the same time, the wakefield induced by the beam travels along at the group velocity. Until an equilibrium state is reached, the energies of individual beam bunches may vary.

C. Synchrotron motion in proton linacs

Since the speed of protons in linacs is not highly relativistic, the synchronous phase angle ψ_s can not be chosen as $\frac{\pi}{2}$. The synchrotron motion in ion linac is adiabatic. The longitudinal particle motion follows a torus of the Hamiltonian flow of Eq. (3.329). Table 3.11 lists bucket area and bucket height for longitudinal motion in proton linacs (see also Table 3.2 for comparison), where $\alpha_b(\psi_s)$ and $Y(\psi_s)$ are running bucket factors shown in Eqs. (3.35) and (3.38). The rf phase region for stable particle motion can be obtained from ψ_u and $\pi - \psi_s$ identical to those in the second the third columns of Table 3.1.

Table 3.11: Properties of rf bucket in conjugate phase space variables

	$(\psi, \frac{\Delta E}{\omega})$	(ψ, δ)
Bucket Area	$16\left(\frac{m\beta^3\gamma^3c^3e\mathcal{E}_0}{\omega^3}\right)^{1/2}\alpha_b(\psi_s)$	$16\left(\frac{\gamma e\mathcal{E}_0}{\omega\beta mc}\right)^{1/2}\alpha_b(\psi_s)$
Bucket Height	$2\left(\frac{mc^3\beta^3\gamma^3e\mathcal{E}_0}{\omega^3}\right)^{1/2}Y(\psi_s)$	$2\left(\frac{\beta c\gamma^2e\mathcal{E}_0}{\omega\beta^2E}\right)^{1/2}Y(\psi_s)$

The equilibrium beam distribution must be a function of the Hamiltonian, i.e. $\rho[H(\Delta W/\omega, \Delta\psi)]$. In small bunch approximation, the Hamiltonian becomes

$$H = -\frac{\omega^3}{2mc^3\beta^3\gamma^3}(\frac{\Delta W}{\omega})^2 - \frac{1}{2}e\mathcal{E}_0\cos\psi_s\,(\Delta\psi)^2. \tag{3.334}$$

A Gaussian beam distribution with small bunch area becomes

$$\rho(\frac{\Delta W}{\omega}, \Delta\psi) = \frac{1}{2\pi\sigma_{\Delta W/\omega}\sigma_{\Delta\psi}} e^{H/H_0}, \tag{3.335}$$

$$\sigma_{\Delta W/\omega} = \sqrt{\frac{H_0 mc^3\beta^3\gamma^3}{\omega^3}} = \sqrt{\frac{\mathcal{A}_{\rm rms}}{\pi}} \left(\frac{mc^3\beta^3\gamma^3 e\mathcal{E}_0 \cos\psi_{\rm s}}{\omega^3}\right)^{1/4},$$

$$\sigma_{\Delta\psi} = \sqrt{\frac{H_0}{e\mathcal{E}_0\cos\psi_{\rm s}}} = \sqrt{\frac{\mathcal{A}_{\rm rms}}{\pi}} \left(\frac{\omega^3}{mc^3\beta^3\gamma^3 e\mathcal{E}_0\cos\psi_{\rm s}}\right)^{1/4},$$

where H_0 is related to the thermal energy of the beam and the rms energy spread and bunch width are given by where $\mathcal{A}_{\rm rms}$ is the rms phase space area in (eVs), i.e. $\mathcal{A}_{\rm rms} = \pi\sigma_{\Delta W/\omega}\sigma_{\Delta\psi}$. The bunch length in τ-coordinate is given by $\sigma_\tau = \sigma_{\Delta\psi}/\omega$. Note that, for a constant phase space area $\mathcal{A}_{\rm rms}$, we find $\sigma_{\Delta W} \sim (\omega\mathcal{E}_0)^{1/4}(\beta\gamma)^{3/4}$, and $\sigma_\tau \sim (\omega\mathcal{E}_0)^{-1/4}(\beta\gamma)^{-3/4}$. However, the fractional momentum spread will decrease when the beam energy is increased:

$$\sigma_{\Delta p/p} = \sqrt{\frac{\mathcal{A}_{\rm rms}}{\pi}} \left(\frac{\omega e\mathcal{E}_0\cos\psi_{\rm s}}{m^3c^5\beta^5\gamma}\right)^{1/4}. \tag{3.336}$$

Examples for beam properties in the Fermilab DTL linac and SNS linacs are available in Exercises 3.8.3 and 3.8.10 respectively.

VIII.5 Transverse Beam Dynamics in a Linac

Figure 3.50 shows the electric field lines between electrodes in an acceleration gap, e.g., the drift tubes of an Alvarez linac or the irises of a high-β linac. In an electrostatic accelerator, the constant field strength gives rise to a global focusing effect because the particle at the end of the gap has more energy so that the defocusing force is weaker. This has been exploited in the design of DC accelerators such as the Van de Graaff or Cockcroft-Walton accelerators.

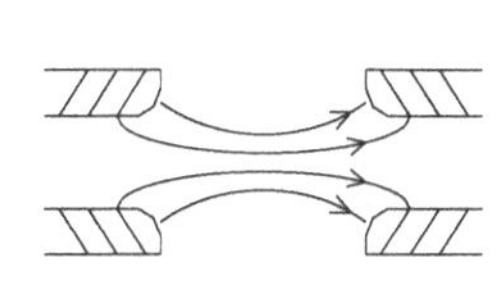

Figure 3.50: A schematic drawing of electric field lines between electrodes of acceleration cavities. Note that the converging field lines contribute to a focusing effect in electrostatic accelerators. For rf accelerators, the field at the exit end increases with time so that the defocusing effect due to the diverging field lines is larger than the focusing effect at the entrance end of the cavity gap. E.O. Lawrence placed a screen at the end of the cavity gap to straighten the electric field line. The screen produces a focusing force, but unfortunately it also causes nuclear and Coulomb scattering.

For rf linear accelerators, phase stability requires $\pi/2 > \psi_{\rm s} > 0$ (below transition energy), and field strength increases with time during the passage of a particle. Thus the defocusing force experienced by the particle at the exit end of the gap is stronger than the focusing force at the entrance of the gap.

Using Eq. (3.310), the EM field of TM_{010} mode is

$$E_{\rm s} = \mathcal{E}_0 \sin\psi, \quad E_r = +\frac{\omega r}{2v_{\rm p}}\mathcal{E}_0 \cos\psi, \quad B_\phi = +\frac{\omega r}{2c^2}\mathcal{E}_0 \cos\psi, \tag{3.337}$$

where $\psi = \left(\omega t - \omega \int ds/v_{\rm p}\right)$. The transverse force on particle motion is

$$\frac{d(\gamma m\dot{r})}{dt} = +eE_r - evB_\phi = +\frac{er\omega\epsilon_0}{2v_{\rm p}}(1 - \frac{vv_{\rm p}}{c^2})\cos\psi \overset{v=v_{\rm p}}{\longrightarrow} \frac{|e|\omega\epsilon_0\cos\psi}{2\beta\gamma^2 c}r. \tag{3.338}$$

For a relativistic particle with $\gamma \gg 1$, the transverse defocusing force is highly reduced because the transverse electric force and the magnetic force cancel each other. Assuming a zero defocusing force, Eq. (3.338) becomes

$$\frac{dp_x}{dt} = 0 \quad \text{or} \quad \frac{d}{ds}\gamma\frac{dx}{ds} = 0 \quad \Longrightarrow \quad \gamma\frac{dx}{ds} = \text{constant} = \gamma_0 x_0'. \tag{3.339}$$

Assuming $\gamma = \gamma_0 + \gamma' s$, where $\gamma' = d\gamma/ds$, we obtain

$$x - x_0 = \left(\frac{\gamma_0}{\gamma'}\ln\frac{\gamma}{\gamma_0}\right)x_0'. \tag{3.340}$$

Thus the orbit displacement increases only logarithmically with distance along a linac (Lorentz contraction), if no other external force acts on the particle. In reality quadrupoles are needed to focus the beam to achieve good transmission efficiency and emittance control in a linac.

Transverse particle motion in the presence of quadrupole elements is identical to that of betatron motion. The linear betatron equation of motion is given by

$$\frac{d^2}{ds^2}x(t,s) + K_x(s)x(t,s) = 0, \qquad \frac{d^2}{ds^2}z(t,s) + K_z(s)z(t,s) = 0, \tag{3.341}$$

where $K_x(s)$ and $K_z(s)$ are focusing functions. Since there is no repetitive focusing elements, the betatron motion in linac is an initial value problem. It should be designed from a known initial or desired betatron amplitude function and matched through the linac. A mismatched linac will produce quadrupole mode oscillations along the linac structure.

In smooth approximation, the linear betatron motion can be described by

$$\frac{d^2}{ds^2}y(t,s) + k_y^2(s)y(t,s) = 0, \tag{3.342}$$

where y is used to represent either x or z, and k_y is the wave number. Since there is no apparent periodic structure, the concept of betatron tune is not necessary. However, many linacs employ periodic focusing systems. In this case, one can *define* the betatron tune per period as $\nu_y = k_y L/2\pi$, where L is the length of a period. Betatron resonances may occur when the condition $m\nu_x + n\nu_z = \ell$ is satisfied, where m, n, and ℓ are integers. Furthermore, synchrobetatron resonances may occur when the condition $m\nu_x + n\nu_z + l\nu_{\rm syn} = \ell$ is satisfied, where l is also an integer.

Wakefield and beam break up instabilities

Applying the Panofsky-Wenzel theorem [29], we find the transverse force:

$$\nabla_\perp \int ds\, F_\| = \frac{\partial}{\partial s}\int ds\, \vec{F}_\perp\,, \qquad \vec{F}_\perp = \int ds\, \nabla_\perp F_\| = \frac{ce}{\omega}\nabla_\perp E_s\,. \tag{3.343}$$

Thus the transverse force on a charged particle is related to the transverse dependence of the longitudinal electric field; it vanishes if the longitudinal electric field is independent of the transverse positions. This is the basic driving mechanism of synchro-betatron coupling resonances.[81] Since TE modes have zero longitudinal electric field, its effect on the transverse motion vanishes as well. Thus we are most concerned with HOMs of the TM waves. These HOMs are also called wakefields. The design of cavities that minimize long range wakefields is an important task in NLC research.[82]

In the presence of a wakefield, the equation of motion is [5]

$$\frac{d^2}{ds^2}x(t,s) + k^2(t,s)x(t,s) = \frac{r_0}{\gamma(t,s)}\int_t^\infty d\tilde{t}\rho(\tilde{t})W_\perp(\tilde{t}-t)x(\tilde{t},s), \tag{3.344}$$

where t describes the longitudinal position of a particle, s is the longitudinal coordinate along the accelerator, $x(t,s)$ is the transverse coordinate of the particle, $k(t,s)$ is the betatron wave number (also called the focusing function), $\rho(t)$ is the density of particle distribution, and $W_\perp(t'-t)$ is the transverse wake function. Detailed properties of the wake function and its relation to the impedance and the transverse force can be found in Ref. [5]. We will examine its implications on particle motion in a simple macro-particle model.

We divide an intense bunch into two macro-particles separated by a distance $\ell = 2\sigma_z$. Each macro-particle represents half of the bunch charge. They travel at the speed of light c. The equation of motion in the smoothed focusing approximation is

$$x_1'' + k_1^2 x_1 = 0, \tag{3.345}$$

$$x_2'' + k_2^2 x_2 = \frac{e^2 N W_\perp(\ell)}{2E}x_1 = Gx_1, \tag{3.346}$$

where $eN/2$ is the charge of the leading macro-particle, x_1 and x_2 are transverse displacements, $W_\perp(\ell)$ is the wake function evaluated at the position of the trailing particle, and k_1 and k_2 are betatron wave numbers for these two macro-particles.

If, for some reason, the leading particle begins betatron oscillation; the trailing particle can be resonantly excited, i.e.

$$x_1 \;=\; \hat{x}_1 \sin k_1 s,$$

[81] See e.g., S.Y. Lee, *Phys. Rev.* **E 49**, 5706 (1994).

[82] See R. Ruth, p. 562 in Ref. [19].

$$\begin{aligned} x_2 &= \frac{k_1}{k_2}\hat{x}_1 \sin k_2 s + \frac{G\hat{x}_1}{k_2^2 - k_1^2}\left(\sin k_1 s - \frac{k_1}{k_2}\sin k_2 s\right). \\ &\to \hat{x}_1 \sin k_1 s + \hat{x}_1\left(\Delta k - \frac{G}{2k_1}\right) s \cos k_1 s, \end{aligned} \tag{3.347}$$

where $\Delta k = k_2 - k_1$. In the limit $\Delta k \to 0$, The amplitude of the trailing particles can grow linearly with s. This is the essence of BBU instability. If the beam bunch is subdivided into many macro-particles, one would observe nonlinear growth for trailing particles.[83]

An interesting and effective method to alleviate the beam break up instabilities is BNS damping.[84] If the betatron wave number for the trailing particle is higher than that for the leading particle by

$$\Delta k = \frac{G}{2k_1} = \frac{e^2 N W_\perp(\ell)}{4Ek_1}, \tag{3.348}$$

the linear growth term in Eq. (3.347) vanishes. This means that the dipole kick due to the wakefield is exactly canceled by the extra focusing force. The bunch will perform rigid coherent betatron oscillations without altering its shape. Note that BNS damping depends on the beam current.

The BNS damping of Eq. (3.348) can be achieved either by applying rf quadrupole field across the bunch length or by lowering the energy of trailing particles. The SLC linac uses the latter method by accelerating the bunch behind the rf crest early in the linac, and then ahead of the rf crest downstream, to restore the energy spread at the end of the linac. Since the average focusing function is related to the energy spread by the chromaticity

$$\frac{\Delta k}{k_1} = C_x \frac{\Delta E}{E}, \tag{3.349}$$

and the chromaticity $C_x \approx -1$ for FODO cells, the energy spread is equivalent to a spread in focusing strength. This method can also be used to provide BNS damping. It is also worth pointing out that the smooth focusing approximation of Eq. (3.345) provides a good approximation for the description of particle motion in a linac.

Exercise 3.8

1. Show that the phase shifts per cell for the CEBAF and SLAC linac cavities listed in Table 3.10 are $kd = \pi$ and $2\pi/3$ respectively.

---

[83]Including beam acceleration, the amplitude will grow logarithmically with energy (distance), as in Eq. (3.340) [5].

[84]V. Balakin, A. Novokhatsky, and V. Smirnov, *Proc. 12th HEACC*, p. 119 (1983).

2. Show that the peak rf magnetic flux density on the inner surface of a pillbox cylindrical cavity in TM_{010} mode is

$$\hat{B}_\phi \approx \frac{\pi\mu_0}{2Z_0}\,\mathcal{E} \quad \text{or} \quad \hat{B}_\phi\ [\mathrm{T}] \approx 50\times 10^{-4}\ \mathcal{E}\ [\mathrm{MV/m}],$$

where $Z_0 = \mu_0 c$ is the impedance of the vacuum.

3. In an Alvarez linac, the longitudinal equations of motion (3.327) and (3.328) can be expressed as mapping equations:

$$\Delta\psi_{n+1} = \Delta\psi_n - \frac{L_{\rm cell}\omega}{mc^3\beta^3\gamma^3}\,\Delta E_n,$$
$$\Delta E_{n+1} = \Delta E_n + eV\cos\psi_{\rm s}\ \Delta\psi_{n+1},$$

where $\psi_n, \Delta E_n$ are the synchrotron phase space coordinates at the nth cell, $L_{\rm cell}$ is the length of the drift tube cell, and eV is the energy gain in this cell.

(a) Using the Courant–Snyder formalism, we can derive the amplitude function for synchrotron motion similar to that for betatron motion. Show that the synchrotron phase advance per cell is

$$\Phi_{\rm syn} = 2\arcsin\left(\frac{\pi eV\cos\psi_{\rm s}}{2\beta^2\gamma^2 E}\right)^{1/2},$$

where $E = \gamma mc^2$ is the beam energy, $\mathcal{E}_{\rm av} = V/L_{\rm cell}$ is the average acceleration field, λ is the rf wave length, and $\psi_{\rm s}$ is the synchronous phase.

(b) Using the table below, calculate the synchrotron phase advance per cell for the first and last cells of cavities 1 and 2, where the synchronous phase is chosen to be $\cos\psi_{\rm s} = 1/2$. Estimate the total synchrotron phase advance in a cavity.

Fermilab Alvarez linac

Cavity Number	1	2
Proton energy in (MeV)	0.75	10.42
Proton energy out (MeV)	10.42	37.54
Cavity length (m)	7.44	19.02
Cell length (cm) (first/last)	6.04/21.8	22.2/40.8
Average field gradient (MV/m) (first/last)	1.60/2.30	2.0
Average gap field (MV/m) (first/last)	7.62/7.45	10.0/6.45
Transit time factor (first/last)	0.64/0.81	0.86/0.81
Number of cells	55	59

4. In a resonance circuit, Q is expressed as

$$\mathrm{Q} = \frac{\frac{1}{2}\omega LI^2}{\frac{1}{2}RI^2} = \frac{\omega L}{R} = 2\pi\frac{\text{stored energy}}{\text{energy dissipation per period}},$$

where $\omega = (LC)^{-1/2}$. The energy stored in the cavity volume is

$$W_{\rm st} = \frac{\mu}{2}\int_V |H_\phi|^2 dV = \frac{\epsilon}{2}\int_V |\mathcal{E}|^2 dV.$$

The power loss in the wall is obtained from the wall current,

$$P_{\rm d} = \frac{1}{2}\int_S R_{\rm s}|H|^2 dS = \frac{\mu\delta_{\rm skin}\omega}{4}\int_S |H|^2 dS,$$

where $R_{\rm s} = 1/\sigma\delta_{\rm skin}$ is the surface resistance,[85] $\delta_{\rm skin} = \sqrt{2/\mu\sigma\omega}$ is the skin depth, and σ is the conductivity. The total energy loss in one period becomes

$$\Delta W_{\rm d} = \frac{2\pi}{\omega}P_{\rm d} = \frac{\pi\mu\delta_{\rm skin}}{2}\int_S |H|^2 dS.$$

(a) Using the identity $\int_0^b J_1^2(k_r r)2\pi r dr = \pi b^2 J_1^2(k_r b)$, show that the quality factor for a pillbox cavity at TM_{010} mode is

$$\mathrm{Q} = \frac{2\int_V |H|^2 dV}{\delta_{\rm skin}\int_S |H|^2 dS} = \frac{d}{\delta_{\rm skin}}\frac{b}{d+b} = \frac{2.405 Z_0}{2R_{\rm s}(1+b/d)},$$

where b and d are the radius and length of a cavity cell, $R_{\rm s}$ is the surface resistivity, and $Z_0 = 1/\mu_0 c \approx 377\Omega$. The Q-factor depends essentially on geometry of the cavity. Since $\delta_{\rm skin} \sim \omega^{-1/2}$, we find $\mathrm{Q} \sim \omega^{+1/2}$. Find the Q-value for the SLAC copper cavity at $f = 2.856$ GHz.

(b) Show that the shunt impedance is

$$R_{\rm sh} = \frac{Z_0^2 d^2}{R_{\rm s}\pi b(b+d)J_1^2(k_r b)} \quad \text{or} \quad \frac{r_{\rm sh}}{\mathrm{Q}} = \frac{2\omega\mu}{\pi(k_r b)^2 J_1^2(k_r b)} = 0.41\omega\mu,$$

where $k_r b = 2.405$. Note here that the shunt impedance behaves like $r_{\rm sh} \sim \omega^{1/2}$. At higher frequencies, the shunt impedance is more favorable; however, the diameter of the cavity will also be smaller, which may limit the beam aperture.

5. The average power flowing through a transverse cross-section of a wave guide is

$$P = \frac{1}{2}\int E_\perp \times H_\perp dS$$

where only transverse components of the field contribute. For TM mode, The energy stored in the electric and magnetic fields are

$$\frac{\mathcal{E}_\perp}{H_\perp} = Z_0\frac{\lambda}{\lambda_{\rm g}}, \quad P = \frac{1}{2Z_0}\int \frac{k}{\beta}|\mathcal{E}_\perp|^2 dS$$

$$W_{\rm st,m} = \frac{\mu}{4}\int |H_\perp|^2 dS = \frac{\mu}{4}\frac{k^2}{Z_0^2\beta^2}\int |\mathcal{E}_\perp|^2 dS$$

$$W_{\rm st} = W_{\rm st,m} + W_{\rm st,e} = 2W_{\rm st,m}.$$

[85] In the limit that the mean free path ℓ of conduction electrons is much larger than the skin depth $\delta_{\rm skin}$, the surface resistance becomes $R_{\rm s} = (8/9)(\sqrt{3}\mu_0^2\omega^2\ell/16\pi\sigma)^{1/3}$. Since the conductivity is proportional to the mean free path ℓ, the resulting surface resistance is independent of the mean free path, and is proportional to $\omega^{2/3}$. There is little advantage to operating copper cavities at very low temperature. See G.E.H. Reuter and E.H. Sonderheimer, *Proc. Roy. Soc.* **A195**, 336 (1984).

(a) Show that the energy flow, defined by $v_e = P/W_{st}$, is $v_e = \beta c/k$.

(b) Verify that $v_g = d\omega/d\beta = v_e$.

6. The disk-loaded linac acceleration structure can be modeled by LC resonant circuits coupled with capacitors $2C_p$ shown in the figure below. The model describes only the qualitative narrowband properties of a loaded wave guide. Thus the equivalent circuit does not imply that a coupled resonator accurately represents a disk loaded structure. In the limit of large C_p, these resonators are uncoupled, which corresponds to a pillbox without holes, or equivalently, a small beam hole in a pillbox cavity corresponds to $C_p \gg C_s$.

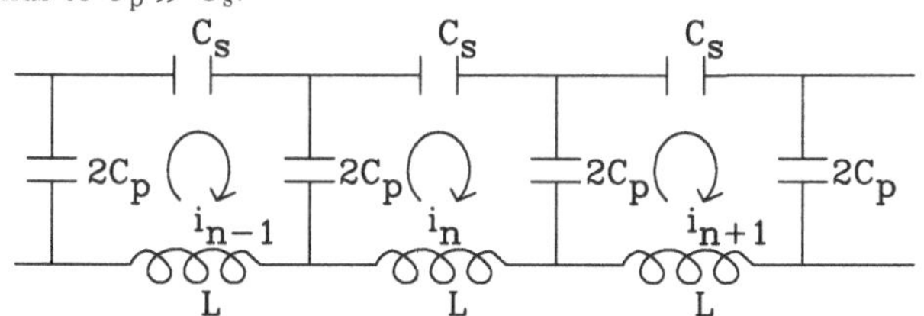

(a) Applying Kirchoff's law, show that

$$i_{n+1} - 2\cos(kd)\, i_n + i_{n-1} = 0, \quad \cos(kd) = 1 + \frac{C_p}{C_s} - \omega^2 C_p L.$$

(b) Show that the solution of the above equation is

$$i_n = e^{\pm j[nkd+\chi_0]}, \quad n = 0, 1, 2, \cdots.$$

We identify kd as the phase advance per cell, and k as the wave number. Show that the frequency is

$$\omega^2 = \omega_0^2 \left[1 + \kappa(1 - \cos kd)\right],$$

where $\omega_0 = 1/\sqrt{LC_s}$ is the natural frequency without coupling at $kd = 0$, and $\kappa = C_s/C_p$ is the coupling constant between neighboring cavities.

(c) Show that the condition for an unattenuated traveling wave is $\omega_0 \le \omega \le \omega_\pi$, where

$$\omega_\pi = \omega_0 \left(1 + 2\frac{C_s}{C_p}\right)^{1/2} \approx \omega_0 \left(1 + \frac{C_s}{C_p}\right)$$

is the resonance frequency at phase advance $kd = \pi$. Draw the dispersion curve of ω vs k. In a realistic cavity, there are higher frequency modes, which give rise to another passband (see Fig. 3.45).

(d) Find k such that the phase velocity $v_p = c$.

(e) Cavities can also be magnetically coupled. The magnetically coupled-cavity chain can be modeled by replacing $2C_p$ in the LC circuit with $L_p/2$. Show that the dispersion curve of a magnetically coupled cavity is

$$\omega_0^2 = \omega^2 \left[1 + \kappa(1 - \cos kd)\right],$$

where $\omega_0 = 1/\sqrt{LC_s}$ is the natural frequency without coupling at $kd = 0$, and $\kappa = L_p/L$ is the coupling constant between neighboring cavities. Discuss the differences between the electric and magnetic coupled cavities.

7. Using Eq. (3.318), the electric field of a standing wave rf cavity structure that consists of N cells is

$$\mathcal{E}_s = \mathcal{E}_0 \cos ks \cos\omega t,$$

where $s \in [0, Nd]$ is the longitudinal coordinate, k is the wave number, d is the cell length of one period, and ω is the frequency. The resonance condition is

$$kd = m\pi/N, \quad m = 0, 1, \cdots, N,$$

where kd is the rf phase advance per cell.

   (a) For a particle traveling at velocity v, show that the total voltage gain in passing through the cavity is

$$\Delta V = \frac{1}{2}N\, d\, \mathcal{E}_0 \left[\frac{\sin(k-(\omega/v))Nd}{(k-(\omega/v))Nd} + \frac{\sin(k+(\omega/v))Nd}{(k+(\omega/v))Nd}\right].$$

   Show that the energy gain is maximum when the phase velocity k/ω is equal to the particle velocity v. Show that the maximum voltage gain of the standing wave is $(\Delta V)_{\rm max} = Nd\,\mathcal{E}_0/2$, i.e. the energy gain of a standing wave structure is only 1/2 that of an equivalent traveling wave structure.

   (b) For a sinusoidal electric field, the power consumed in one cell is

$$|\mathcal{E}_0 d|^2/2R_{\rm sh,cell},$$

   where $R_{\rm sh,cell}$ is the shunt impedance per cell for the traveling wave. For an rf structure composed of N cells, the power is

$$P_{\rm d} = N|\mathcal{E}_0 d|^2/2R_{\rm sh,cell}.$$

   Using the definition of shunt impedance, show that the shunt impedance of a standing wave rf structure is

$$R_{\rm sh} = \frac{1}{2}NR_{\rm sh,cell}.$$

   Thus the shunt impedance for a standing wave structure is equal to 1/2 that of an equivalent traveling wave structure.[86]

8. There are two types of traveling wave structures. A constant impedance structure has a uniform multi-cell structure so that the impedance is constant and the power decays exponentially along the structure. A constant gradient structure is tapered so that the longitudinal electric field is kept constant. The electric field is related to the shunt impedance per unit length by [see Eq. (3.305)]

$$\mathcal{E}^2 = -r_{\rm sh}\frac{dP_{\rm d}}{ds} = 2\alpha\, r_{\rm sh}\, P_{\rm d}(s) \qquad \text{where} \qquad \alpha = -\frac{1}{2P_{\rm d}}\frac{dP_{\rm d}}{ds}.$$

The total energy gain for an electron in a linac of length L is

$$\Delta E = e\int_0^L \mathcal{E}ds.$$

[86]The above calculation for voltage gain in the cavity structure is not applicable for an standing wave structure with $kd = 0$ and π, where two space harmonics contribute to the electric field so that $\mathcal{E}_s = 2\mathcal{E}_0 \cos ks \cos\omega t$. This means that the voltage gain in the rf structure is $\Delta V = Nd\mathcal{E}_0$, and the shunt impedance is $R_{\rm sh} = NR_{\rm sh,cell}$.

(a) In a constant impedance structure, show that the energy gain is

$$\Delta E(L) = eL(2r_{\rm sh}P_0\alpha)^{1/2}\frac{1-e^{-\alpha L}}{\alpha L},$$

where P_0 is the power at the input point.

(b) Assuming that $r_{\rm sh}$ and Q are nearly constant in a constant gradient structure, show that

$$P_{\rm d} = P_0\left(1 - \frac{s}{L}(1-e^{-2\tau})\right),$$

where $\tau = \int_0^L \alpha(s)ds$. The group velocity is equal to the velocity of energy flow. Show that the group velocity of a constant gradient structure [see Eq. (3.307)] and the energy gain are

$$v_{\rm g} = \omega L\left(1 - \frac{s}{L}(1-e^{-2\tau})\right)\left(Q(1-e^{-2\tau})\right)^{-1},$$
$$\Delta E = e\mathcal{E}L = e\sqrt{P_0 r_{\rm sh} L\,(1-e^{-2\tau})}.$$

9. The design of the 2 MW spallation neutron source uses a chain of linacs composed of ion source, RFQ, DTL, CCL, and SCL to accelerate 2.08×10^{14} particles per pulse at 60 Hz repetition rate. An accumulator compresses the 1 ms linac pulse into a 695 ns high intensity beam pulse with 250 ns beam gap. The following table lists linac and beam parameters. Calculate the longitudinal bucket and bunch areas in (eVs). Compare the rms bunch length in (ns) and in (m) with the rms transverse beam size at exit points of linacs. Each microbunch has about $N_{\rm B} = 8.70\times10^8$ protons, what is the longitudinal brightness of the beam in number of particles per (eVs)?

	RFQ	DTL	CCL	SRFL
L (m) length of the structure	3.723	38.7	55.12	206.812
$f_{\rm rf}$ (MHz)	402.5	402.5	805	805
$\psi_{\rm s}$ (differ from linac convention by $-\pi/2$)	60°	45-65°	60-62°	20°
$\mathcal{E}_0 T$ (MV/m)		3.0	3.37	10.6
$\mathrm{KE_{inj}}$ (MeV)	0.065	2.5	86.8	185
$\mathrm{KE_{ext}}$ (MeV)	2.5	86.8	185	1001.5
$\epsilon_\parallel$ (π-MeV-deg) emittance at exit point	0.108	-	-	0.60
$\epsilon_\perp$ (π-mm-mrad) emittance at exit point	0.21	-	-	0.45
$\sigma_{\Delta W}$ (MeV)	0.0092			0.33
$\mathcal{A}_{\rm bucket}$ (eVs) at injection energy				
$\mathcal{A}_{\rm rms}$ (eVs)				
$\sigma_\mathcal{T}$ (ns)				
$k_{\rm syn}$ (m^{-1})				
$\beta_{x/z}$ at exit (m)	0.2/0.2	-	-	10.1/5.3

10. A prebuncher is usually used to prebunch the electrons from a source, which can be thermionic or rf gun. We assume a thermionic gun with a DC gun voltage V_0, which is usually about 80–150 kV. Let the electric field and the gap width of the prebuncher be $\mathcal{E}\sin(\omega t)$ and g. Electrons that arrive earlier are slowed and that arrive late are

sped up. At a drift distance away from the prebuncher, the faster electrons catch up the slower electrons. Thus electrons are prebunched into a smaller phase extension to be captured by the buncher and the main linac. Assuming a small prebuncher gap with $V_1 = \mathcal{E}g \ll V_0$, find the drift distance as a function of the V_0 and V_1. Discuss the efficiency of prebunching as a function of relevant parameters.

Chapter 4

Physics of Electron Storage Rings

Accelerated charged particles, particularly electrons in a circular orbit, radiate electromagnetic energy. As far back in 1898, Liénard derived an expression for electromagnetic radiation in a circular orbit. Modern synchrotron radiation theory was formulated by many physicists; in particular, its foundation was laid by J. Schwinger. Some of his many important results are summarized below:[1]

- The angular distribution of synchrotron radiation is sharply peaked in the direction of the electron's velocity vector within an angular width of $1/\gamma$, where γ is the relativistic energy factor. The radiation is plane polarized on the plane of the electron's orbit, and elliptically polarized outside this plane.
- The radiation spans a continuous spectrum. The power spectrum produced by a high energy electron extends to a critical frequency $\omega_c = 3\gamma^3\omega_c/2$, where $\omega_\rho = c/\rho$ is the cyclotron frequency for electron moving at the speed of light.[2]
- Quantum mechanical correction becomes important when the critical energy of the radiated photon, $\hbar\omega_c = \frac{3}{2}\hbar c\gamma^3/\rho$, is comparable to $E = \gamma mc^2$, or at the electron beam energy $mc^2(mc\rho/\hbar)^{1/2} \approx 10^3$ TeV. The beamstrahlung parameter, defined as $\Upsilon = \frac{2}{3}\hbar\omega_c/E$, is a measure of the importance of quantum mechanical effects.

Shortly after the first observation of synchrotron radiation at the General Electric 70 MeV synchrotron in 1947,[3] applications of this radiation were contemplated.[4] The

[1]J. Schwinger, *Phys. Rev.* **70**, 798 (1946); **75**, 1912 (1949); *Proc. Nat. Acad. Sci.* **40**, 132 (1954).

[2]D.H. Tomboulin and P.L. Hartman experimentally verified that electrons at high energy (70 MeV then) could emit extreme ultraviolet (XUV) photons; *Phys. Rev.* **102**, 1423 (1956).

[3]F.R. Elder *et al.*, *Phys. Rev.* **71**, 829 (1947); *ibid.* **74**, 52 (1948); *J. Appl. Phys.* **18**, 810 (1947).

[4]R.P. Madden and K. Codling at the National Bureau of Standards were the first to apply synchrotron radiation to the study of atomic physics. See *Phys. Rev. Lett.* **10**, 516 (1963); *J. Appl. Phys.* **36**, 380 (1965).

first dedicated synchrotron radiation source, Tantalus at the University of Wisconsin, was commissioned in 1968.[5] Today, nearly a hundred light sources are distributed in almost all continents. Applications of synchrotron radiation include surface physics, condensed matter physics, biochemistry, medical research, advanced manufacturing processes, etc.

A. Basic properties of synchrotron radiation from electrons

According to Larmor's theorem, the instantaneous radiated power from an accelerated electron is

$$P = \frac{1}{4\pi\epsilon_0}\frac{2e^2\dot{v}^2}{3c^3} = \frac{2r_0}{3mc}\left(\frac{d\vec{p}}{dt}\cdot\frac{d\vec{p}}{dt}\right), \tag{4.1}$$

where $\dot{v}$ is the acceleration rate and $r_0 = e^2/4\pi\epsilon_0 mc^2$ is the classical radius of the electron. The relativistic generalization of Larmor's formula (obtained by Liénard in 1898) is

$$P = \frac{2r_0}{3mc}\left(\frac{dp_\mu}{d\tau}\cdot\frac{dp_\mu}{d\tau}\right) = \frac{2r_0}{3mc}\left[\left(\frac{d\vec{p}}{d\tau}\right)^2 - \frac{1}{c^2}\left(\frac{dE}{d\tau}\right)^2\right], \tag{4.2}$$

where the proper-time element $d\tau = dt/\gamma$, and $p_\mu = (p_0, \vec{p})$ is the 4-momentum vector. In Sec. I.2 we will show that the power radiated from a circular orbit of a highly relativistic charged particle is much higher than that from a linear accelerator, i.e.

$$\left|\frac{d\vec{p}}{d\tau}\right| = \gamma\omega|p| \gg \frac{1}{c}\frac{dE}{d\tau}, \tag{4.3}$$

where ω is the angular cyclotron frequency.

The radiation power arising from circular motion is

$$P = \frac{2r_0}{3mc}\gamma^2\omega^2|\vec{p}|^2 = \frac{2r_0}{3mc}\gamma^2|F_\perp|^2 = \frac{\beta^4 c}{2\pi}C_\gamma\frac{E^4}{\rho^2} = \frac{\beta^2 e^2 c^3}{2\pi}C_\gamma E^2 B^2, \tag{4.4}$$

where $F_\perp = \omega|\vec{p}| = evB$ is the transverse force,[6] $v = \beta c$ is the speed of the particle, B is the magnetic field of the bending dipole, ρ is the local radius of curvature, and

$$C_\gamma = \frac{4\pi}{3}\frac{r_0}{(mc^2)^3} = \begin{cases} 8.846\times 10^{-5}\ \mathrm{m/(GeV)^3} & \text{for electrons} \\ 4.840\times 10^{-14}\ \mathrm{m/(GeV)^3} & \text{for muons} \\ 7.783\times 10^{-18}\ \mathrm{m/(GeV)^3} & \text{for protons.} \end{cases} \tag{4.5}$$

---

[5]See e.g. G. Margaritondo, *The evolution of a dedicated synchrotron light source*, *Phys. Today* **61**, 37 (2008); D. W. Lynch, J. Synchrotron Rad. 4, 334 (1997).

[6]In other applications, the velocity v is the component perpendicular to the magnetic field, i.e. $v = v_\perp$. Since $\langle v_\perp^2\rangle = \frac{2}{3}v^2$, the power of synchrotron radiation is $\langle P\rangle = \frac{2}{3}P = \frac{4}{3}\sigma_{\rm T} c U_{\rm rad}\beta^2\gamma^2$, where P is the power given by Eq. (4.4), $\sigma_{\rm T} = \frac{8\pi}{3}r_0^2$ is the Thomson scattering cross-section, $U_{\rm rad} = B^2/(2\mu_0)$ is the radiation energy density.

The energy radiated from the particle with nominal energy E_0 in one revolution is

$$U_0 = \oint P dt = C_\gamma \beta^3 E_0^4 R \langle \frac{1}{\rho^2} \rangle = \frac{C_\gamma \beta^3 E_0^4}{2\pi} \int_0^{2\pi R} \frac{ds}{\rho^2}, \tag{4.6}$$

where R is the average radius. For an *isomagnetic ring* with constant field strength in all dipoles, the energy loss per revolution and the average radiation power become

$$U_0 = C_\gamma \frac{\beta^3 E_0^4}{\rho} = e \frac{ecZ_0}{3\rho} \beta^3 \gamma^4, \qquad \langle P \rangle = \frac{U_0}{T_0} = \frac{c C_\gamma \beta^3 E_0^4}{2\pi R \rho}, \tag{4.7}$$

where $Z_0 = \mu_0 c = 377\Omega$ is the vacuum impedance, $e^2 c Z_0 \approx 18.1 \times 10^{-9}$ eV-m, and $T_0 = 2\pi R/\beta c$ is the orbital revolution period.

Because the power of synchrotron radiation is proportional to E^4/ρ^2, and the beam is compensated on average by the longitudinal electric field, the longitudinal motion is damped. This natural damping produces high brightness electron beams, whose applications include e^+e^- colliders for nuclear and particle physics, and electron storage rings for generating synchrotron light and free electron lasers for research in condensed matter physics, biology, medicine and material applications.

B. Synchrotron radiation sources

The brilliance of the photon beam is defined as

$$\mathcal{B} = \frac{d^4 N}{dt\, d\Omega\, dS\, (d\lambda/\lambda)} = \frac{d^4 N_{\rm ph}}{dt \sigma_x \sigma_{x'} \sigma_z \sigma_{z'} (d\lambda/\lambda)} \tag{4.8}$$

in units of photons/(s mm^2-mrad^2 0.1% of bandwidth). Neglecting the optical diffraction, the product of the solid angle and the spot size $d\Omega dS = \sigma_x \sigma_{x'} \sigma_z \sigma_{z'}$ is proportional to the product of electron beam emittances $\epsilon_x \epsilon_z$. Therefore a high brilliance photon source demands a high brightness electron beam with small electron beam emittances. Furthermore, a beam with short bunch length can also be important in time resolved experiments. Using the synchrotron radiation generated from the storage rings, one can obtain a wide frequency span tunable high brilliance monochromatic photon source.

Synchrotron radiation sources are generally classified into generations. A first generation light source parasitically utilizes synchrotron radiation in an electron storage ring built mainly for high energy physics research. Some examples are SPEAR at SLAC, and CHESS in CESR at Cornell University. A second generation light source corresponds to a storage ring dedicated to synchrotron light production, where the lattice design is optimized to achieve minimum emittance for high brightness beam operation. In Secs. II and III we will show that the natural emittance of an electron beam is $\epsilon_{\rm nat} = \mathcal{F} C_q \gamma^2 \theta^3 / \mathcal{J}_x$, where $C_q = 3.83 \times 10^{-13}$ m, $\mathcal{J}_x \approx 1$ is a damping partition number, and θ is the bending angle of one half period. The factor $\mathcal{F}$ can be optimized in different lattice designs. A third generation light source employs high brightness electron beams and insertion devices such as wigglers or undulators to optimize photon brilliance, mostly about 10^{20} photons/[s (mm-mrad)^2 0.1% bandwidth], which is about five to six orders higher than that generated in dipoles, or about ten order higher than the brilliance of X-ray tubes. Using long undulators in long straight sections of a collider ring, a few first generation light sources can provide photon beam brilliance equal to that of third generation light sources. Table 4.1 lists some machine parameters of the advanced light source (ALS) at LBNL and the advanced photon source (APS) at ANL.

The widely discussed "fourth" generation light source is dedicated to the coherent production of X-rays and free electron lasers at a brilliance at least a few orders higher than that produced in third generation light sources.[7]

Table 4.1: Properties of some electron storage rings.

	Colliders					Light Sources	
	BEPC	CESR	LER(e^+)	HER(e^-)	LEP	APS	ALS
E [GeV]	2.2	6	3.1	9	55	7	1.5
ν_x	5.8	9.38	32.28	25.28	76.2	35.22	14.28
ν_z	6.8	9.36	35.18	24.18	70.2	14.3	8.18
ρ [m]	10.35	60	30.6	165.0	3096.2	38.96	4.01
α [$\times 10^{-4}$]	400	152	14.9	24.4	3.866	2.374	14.3
C [m]	240.4	768.4	2199.3	2199.3	26658.9	1104	196.8
h	160	1281	3492	3492	31320	1296	328
f_{rf} [MHz]	199.5	499.8	476	476	352.2	352.96	499.65
ν_s	0.016	0.064	0.034	0.0522	0.085	0.006	0.0082
$\frac{\Delta E}{E_0}$ [$\times 10^{-4}$]	4.0	6.3	9.5	6.1	8.4	10	7.1
$\mathcal{A}$ [$\times 10^{-4}$eV-s]	3.5	7.2	3.1	5.7	78.	4.1	0.43
ϵ_x [nm]	450	240	64	48	51	3.1	4.8
ϵ_z [nm]	35	8	3.86	1.93	0.51	0.045	0.48

[7]M. Cornacchia and H. Winick, eds., *Proc. Fourth Generation Light Sources*, SSRL 92/02 (1992); J.L. Laclare, ed., *Proc. Fourth Generation Light Sources*, ESRF report (1996).

I Fields of a Moving Charged Particle

Let $\vec{x}'(t')$ be the position of an electron at time t' and let $\vec{x}$ be the position of the observer with $\vec{R}(t) = \vec{x} - \vec{x}'(t')$ (Fig. 4.1). The electromagnetic signal, emitted by the electron at time t' and traveling on a straight path, will arrive at the observer at time

$$t = t' + \frac{R(t')}{c} \tag{4.9}$$

where $R(t') = |\vec{x} - \vec{x}'|$; t' is called the retarded time or the emitter time; and t is the observer time. The motion of the electron is specified by $\vec{x}'(t')$ with

$$\frac{d\vec{x}'}{dt'} = \vec{\beta}c = -\frac{d\vec{R}}{dt'}.$$

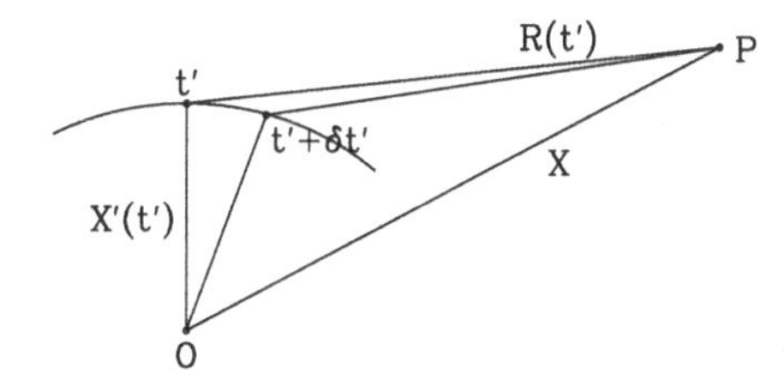

Figure 4.1: Schematic drawing of the coordinates of synchrotron radiation emitted from a moving charge. Here t' is the retarded time. The unit vector along the line joining the point of emission and the observation point P is $\hat{n} = \vec{R}(t')/R(t')$.

The retarded scalar and vector potentials (4-potential) due to a moving point charge are

$$\begin{aligned} A_\mu(\vec{x}, t) &= \frac{1}{4\pi\epsilon_0} \int\int \frac{J_\mu(\vec{x}', t')}{R}\, \delta(t' + \frac{R}{c} - t)\, d^3x'dt', \\ &= \frac{e}{4\pi\epsilon_0} \int \frac{\beta_\mu(t')}{R(t')} \delta(t' + \frac{R(t')}{c} - t)dt', \end{aligned} \tag{4.10}$$

where $J_\mu(\vec{x}', t') = ec\beta_\mu \delta(\vec{x}' - \vec{r}(t'))$ is the current density of the point charge with $\beta_\mu = (\vec{\beta}/c, 1)$, and $r(t')$ is the orbiting path of the charge particle. The delta function in Eq. (4.10) is needed to ensure the retarded condition. With the identity $\int F\delta(f(t'))dt' = F/|df/dt|$, the scalar and vector potentials become

$$\Phi(\vec{x}, t) = \frac{e}{4\pi\epsilon_0} \frac{1}{\kappa R}\bigg|_{\rm ret}, \qquad \vec{A}(\vec{x}, t) = \frac{e}{4\pi\epsilon_0 c} \frac{\vec{\beta}}{\kappa R}\bigg|_{\rm ret}, \tag{4.11}$$

with

$$\kappa = \frac{dt}{dt'} = 1 + \frac{1}{c}\frac{dR}{dt'} = 1 - \hat{n}\cdot\vec{\beta}, \tag{4.12}$$

where $\hat{n} = \vec{R}/R = \nabla R$.

The electric and magnetic fields are $\vec{E} = -\nabla\Phi - \partial\vec{A}/\partial t$ and $\vec{B} = \nabla\times\vec{A}$. Using the identity $\nabla \to \nabla R\frac{\partial}{\partial R} = \hat{n}\frac{\partial}{\partial R}$, we obtain

$$\begin{aligned}
\vec{E} &= \frac{e}{4\pi\epsilon_0}\int [\frac{\hat{n}}{R^2}\delta(t' + \frac{R}{c} - t) + \frac{\vec{\beta}-\hat{n}}{cR}\delta'(t' + \frac{R}{c} - t)]dt' \\
&= \frac{e}{4\pi\epsilon_0}\left[\frac{\hat{n}}{\kappa R^2} + \frac{1}{c\kappa}\frac{d}{dt'}\frac{\hat{n}-\vec{\beta}}{\kappa R}\right]_{\rm ret}, \qquad (4.13)\\
\vec{B} &= \frac{e}{4\pi\epsilon_0 c}\int (\hat{n}\times\vec{\beta})[-\frac{1}{R^2}\delta(t' + \frac{R}{c} - t) + \frac{1}{cR}\delta'(t' + \frac{R}{c} - t)]dt' \\
&= \frac{e}{4\pi\epsilon_0 c}\left[\frac{\vec{\beta}\times\hat{n}}{\kappa R^2} + \frac{1}{c\kappa}\frac{d}{dt'}\frac{\vec{\beta}\times\hat{n}}{\kappa R}\right]_{\rm ret}. \qquad (4.14)
\end{aligned}$$

Since the time derivative of the vector $\hat{n}$ is equal to the ratio of the vector $\vec{v}_\perp$ to R

$$\frac{d\hat{n}}{cdt'} = \frac{\hat{n}\times(\hat{n}\times\vec{\beta})}{R} = \frac{(\hat{n}\cdot\vec{\beta})\hat{n}-\vec{\beta}}{R}, \qquad (4.15)$$

we obtain

$$\vec{E}(\vec{x},t) = \frac{e}{4\pi\epsilon_0}\left[\frac{\hat{n}-\vec{\beta}}{\kappa^2R^2} + \frac{\hat{n}}{c\kappa}\frac{d}{dt'}\frac{1}{\kappa R} - \frac{1}{c\kappa}\frac{d}{dt'}\frac{\vec{\beta}}{\kappa R}\right]_{\rm ret}, \qquad (4.16)$$

$$\vec{B}(\vec{x},t) = \frac{e}{4\pi\epsilon_0 c}\left[\left(\frac{\vec{\beta}}{\kappa^2R^2} + \frac{1}{c\kappa}\frac{d}{dt'}\frac{\vec{\beta}}{\kappa R}\right)\times\hat{n}\right]_{\rm ret}. \qquad (4.17)$$

Note that the magnetic field is in fact related to the electric field by $\vec{B} = (1/c)\,\hat{n}\times\vec{E}$, a feature common to all electromagnetic radiation in free space. Thus it suffices to calculate only the electric radiation field. Using the relations

$$\frac{d\vec{\beta}}{dt'} = \dot{\vec{\beta}}, \quad \frac{dR}{cdt'} = -\hat{n}\cdot\vec{\beta}, \quad \frac{d(\kappa R)}{cdt'} = \beta^2 - \vec{\beta}\cdot\hat{n} - \frac{R}{c}\hat{n}\cdot\dot{\vec{\beta}}, \qquad (4.18)$$

we obtain the electric field as

$$\vec{E}(\vec{x},t) = \frac{e}{4\pi\epsilon_0}\left[\frac{\hat{n}-\vec{\beta}}{\gamma^2\kappa^3R^2}\right]_{\rm ret} + \frac{e}{4\pi\epsilon_0 c}\left[\frac{\hat{n}}{\kappa^3R}\times((\hat{n}-\vec{\beta})\times\dot{\vec{\beta}})\right]_{\rm ret}. \qquad (4.19)$$

The flux, defined as the energy passing through a unit area per unit time at the observer location, is the Poynting vector

$$\vec{S} = \frac{1}{\mu_0}[\vec{E}\times\vec{B}] = \frac{1}{\mu_0 c}|\vec{E}|^2\hat{n}. \qquad (4.20)$$

The total power radiated by the particle is

$$\frac{dP}{d\Omega} = (\hat{n} \cdot \vec{S}) R^2 \frac{dt}{dt'} = \kappa R^2 |\vec{E}|^2. \tag{4.21}$$

Note here that the electric field in Eq. (4.19) is composed of two terms. The first term, which is proportional to $1/R^2$, is a static field pointing away from the charge at time t. This field can be transformed into an electrostatic electric field by performing a Lorentz transformation into a frame in which the charge is at rest. The total energy from this term is zero.

The second term, related to the acceleration of the charged particle, is the radiation field, which is proportional to $1/R$. Both $\vec{E}$ and $\vec{B}$ radiation fields are transverse to $\vec{n}$ and are proportional to $1/R$.

I.1 Non-relativistic Reduction

When the velocity of the particle is small, the radiation field, the Poynting's vector (energy flux), and the power radiated per unit solid angle are

$$\vec{E}_a = \frac{e}{4\pi\epsilon_0 c}\left[\frac{\hat{n} \times (\hat{n} \times \dot{\vec{\beta}})}{R}\right]_{\rm ret}, \qquad \vec{S} = \frac{1}{\mu_0}\vec{E} \times \vec{B} = \frac{1}{\mu_0 c}|E_a|^2 \hat{n},$$
$$\frac{dP}{d\Omega} = \frac{1}{\mu_0 c}|E_a|^2 R^2 = \frac{e^2}{16\pi^2 \epsilon_0 c}|\hat{n} \times (\hat{n} \times \dot{\vec{\beta}})|^2 = \frac{e^2}{16\pi^2\epsilon_0 c^3}\dot{v}^2 \sin^2\Theta, \tag{4.22}$$

where Θ is the angle between vectors $\hat{n}$ and $\dot{\vec{\beta}}$, i.e. $\hat{n} \times \dot{\vec{\beta}} = |\dot{\vec{\beta}}| \sin\Theta$. Integration over all angles gives the same total radiated power as Larmor's formula of Eq. (4.1).

I.2 Radiation Field for Particles at Relativistic Velocities

For particles at relativistic velocity, the Poynting's vector becomes

$$\vec{S} \cdot \hat{n} = \frac{r_{\rm e} mc}{4\pi}\left[\frac{1}{\kappa^6 R^2}|\hat{n} \times ((\hat{n} - \vec{\beta}) \times \dot{\vec{\beta}})|^2\right]_{\rm ret}. \tag{4.23}$$

The total energy of radiation during the time between T_1 and T_2 is

$$W = \int_{T_1+(R_1/c)}^{T_2+(R_2/c)} (\vec{S} \cdot \hat{n}) dt = \int_{t'=T_1}^{t'=T_2} (\vec{S} \cdot \hat{n}) \frac{dt}{dt'} dt'. \tag{4.24}$$

Thus the power radiated per unit solid angle in retarded time is

$$\frac{dP(t')}{d\Omega} = R^2 (\vec{S} \cdot \hat{n}) \frac{dt}{dt'} = \kappa R^2 (\vec{S} \cdot \hat{n}) = \frac{r_{\rm e} mc}{4\pi} \frac{|\hat{n} \times ((\hat{n} - \vec{\beta}) \times \dot{\vec{\beta}})|^2}{(1 - \hat{n} \cdot \vec{\beta})^5}. \tag{4.25}$$

There are two important relativistic effects on the the electromagnetic radiation. The first arises from the denominator with $\kappa = 1 - \hat{n}\cdot\vec{\beta}$. Note that the instantaneous radiation power is proportional to $1/\kappa^5$, where $\kappa = dt/dt'$ is the ratio of the observer's time to the electron's radiation time. At relativistic energies, we have

$$\beta = (1 - \frac{1}{\gamma^2})^{1/2} \approx 1 - \frac{1}{2\gamma^2}, \qquad \frac{1}{\kappa} \approx \frac{2}{(\theta^2 + 1/\gamma^2)}\,, \tag{4.26}$$

where θ is the angle between the radiation direction $\hat{n}$ and the velocity vector $\vec{\beta}$. Since the angular distribution is proportional to $1/\kappa^5$, the radiation from a relativistic particle is sharply peaked at the forward angle within an angular cone of $\theta \approx 1/\gamma$.

The second relativistic effect is the squeeze of the observer's time: $dt = \kappa dt' \approx dt'/\gamma^2$. When the observer is in the direction of the electron's velocity vector within an angle of $1/\gamma$, the time interval of the electromagnetic radiation dt' of the electron appears to the observer squeezed into a much shorter time interval because a relativistic electron follows very closely behind the photons it emitted at an earlier time. Thus photons emitted at later times follow closely behind those emitted earlier. Therefore it appears to the observer that the time is squeezed. The resulting wavelength of the observed radiation is shortened or, equivalently, the energy of the photon is enhanced.

Example 1: linac

In a linear accelerator, $\dot{\vec{\beta}}$ is parallel to $\vec{\beta}$. The angular distribution of the electromagnetic radiation is

$$\frac{dP(t')}{d\Omega} = \frac{r_e mc\dot{v}^2}{4\pi}\frac{\sin^2\Theta}{(1-\beta\cos\Theta)^5}, \tag{4.27}$$

where Θ is the angle between $\hat{n}$ and $\dot{\vec{\beta}}$. The maximum of the angular distribution is located at

$$\Theta_{\max} = \cos^{-1}\left[\frac{1}{3\beta}(\sqrt{1+15\beta^2} - 1)\right] \to \frac{1}{2\gamma}. \tag{4.28}$$

The rms of the angular distribution is also $\langle\Theta^2\rangle^{1/2} = 1/\gamma$. The integrated power is

$$P(t) = \int \frac{dP}{d\Omega}d\Omega = \frac{1}{4\pi\epsilon_0}\frac{2e^2\gamma^6\dot{v}^2}{3c^3} = \frac{e^2}{6\pi\epsilon_0 m^2c^3}\left(\frac{dp_{\rm L}}{dt}\right)^2, \tag{4.29}$$
$$\frac{dp_{\rm L}}{dt} = m\gamma^3\dot{v} = \frac{\Delta E}{\Delta s}.$$

Here $\Delta E/\Delta s$ is the energy gain per unit length, typically, about 20 MeV/m in the SLAC linac, and 25 to 100 MV/m in future linear colliders.

Example 2: Radiation from circular motion

When the charged particle is executing circular motion due to a transverse magnetic field, $\dot{\vec{\beta}}$ is perpendicular to $\vec{\beta}$. Figure 4.2 shows the coordinate system (see Exercise 4.1.3). The power per unit solid angle is

$$\begin{aligned} \frac{dP}{d\Omega} &= \frac{e^2\dot{v}^2}{16\pi^2\epsilon_0 c^3}\frac{1}{(1-\beta\cos\Theta)^3}\left[1-\frac{\sin^2\Theta\cos^2\Phi}{\gamma^2(1-\beta\cos\Theta)^2}\right] \\ &\approx \frac{e^2\dot{v}^2}{2\pi^2\epsilon_0 c^3}\gamma^6\frac{1}{(1+\gamma^2\Theta^2)^3}\left[1-\frac{4\gamma^2\Theta^2\cos^2\Phi}{(1+\gamma^2\Theta^2)^2}\right], \end{aligned} \tag{4.30}$$

where $\dot{v}=\beta^2c^2/\rho$, and ρ is the bending radius. Therefore the radiation is also confined to a cone of angular width of $\langle\Theta^2\rangle^{1/2}\sim 1/\gamma$. The total radiated power is obtained by integrating the power over the solid angle, i.e.

$$P(t)=\int\frac{dP}{d\Omega}d\Omega=\frac{2e^2}{6\pi\epsilon_0 m^2c^3}\gamma^2\left(\frac{dp_{\rm T}}{dt}\right)^2, \tag{4.31}$$

$$\frac{dp_{\rm T}}{dt}=\gamma m\dot{v}=\gamma m\frac{\beta^2c^2}{\rho}=299.79\,\beta\,B[\mathrm{T}]\quad[\mathrm{MeV/m}].$$

Comparing Eq. (4.31) with Eq. (4.29), we find that the radiation from circular motion is at least a factor of $2\gamma^2$ larger than that from longitudinal acceleration.

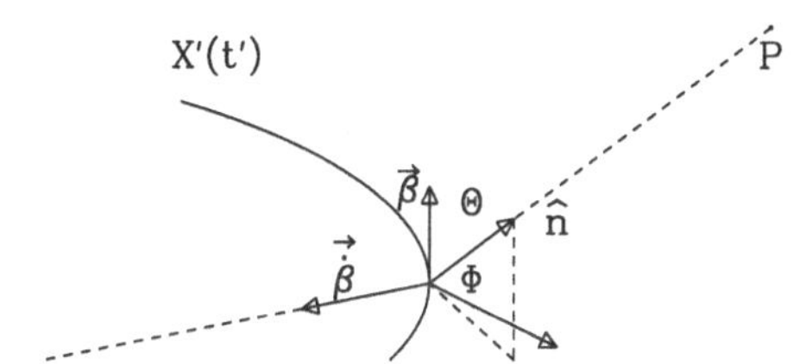

Figure 4.2: The coordinate system for synchrotron radiation from the circular motion of a charged particle.

I.3 Frequency and Angular Distribution

The synchrotron radiation from an accelerated charged particle consists of contributions from the components of acceleration parallel and perpendicular to the velocity. Since the radiation from the parallel component has been shown to be $1/\gamma^2$ smaller than that from the perpendicular component, it can be neglected. In other words, the radiation emitted by a charged particle in an arbitrary extremely relativistic motion is about the same as that emitted by a particle moving instantaneously along the arc of a circular path. In this case, the acceleration $\dot{v}_\perp$ is related to the radius of curvature ρ by $\dot{v}_\perp=v^2/\rho\approx c^2/\rho$.

The angular distribution given by Eq. (4.25) has an angular width $\langle\Theta^2\rangle^{1/2}\sim 1/\gamma$, and the charged particle illuminates the observer for a time interval $cdt'\approx\rho\Theta_{\rm rms}=$

ρ/γ. To the observer, however, the corresponding time interval Δt of the radiation and the critical frequency are

$$\Delta t \sim \frac{dt}{dt'}\Delta t' \approx \frac{1}{\gamma^2}\Delta t' = \frac{\rho}{\gamma^3 c}, \qquad \omega_c \sim \frac{1}{\Delta t} \sim \gamma^3 \frac{c}{\rho} = \gamma^3 \omega_\rho. \tag{4.32}$$

The frequency spectrum spans a broad continuous spectrum up to the critical frequency, defined as $\omega_c = \frac{3}{2}\gamma^3\omega_\rho$.

To obtain the frequency and angular distribution of the synchrotron radiation, we should study the time dependence of the angular distribution discussed in the last section. The power radiated per unit solid angle is given by Eq. (4.22), i.e.

$$\frac{dP}{d\Omega} = |\vec{G}(t)|^2, \quad \vec{G}(t) = (\frac{1}{\mu_0 c})^{1/2}[R\vec{E}]_{\rm ret} \tag{4.33}$$

with electric field $\vec{E}$ given by Eq. (4.22). Using the Fourier transform

$$\vec{G}(\omega) = \int \vec{G}(t)e^{j\omega t}dt, \quad \vec{G}(t) = \frac{1}{2\pi}\int \vec{G}(\omega)e^{-j\omega t}dt, \tag{4.34}$$

we obtain the total energy radiated per unit solid angle as

$$\frac{dW}{d\Omega} = \int_{-\infty}^{\infty} |\vec{G}(t)|^2 dt = \frac{1}{2\pi}\int_{-\infty}^{\infty} |\vec{G}(\omega)|^2 d\omega, \quad \text{(Parseval's theorem)}. \tag{4.35}$$

Since the function $\vec{G}(t)$ is real, the Fourier component has the property $\vec{G}(-\omega) = \vec{G}^*(\omega)$. Since the negative frequency is folded back to the positive frequency, we can define the energy radiation per unit solid angle per frequency interval as

$$\frac{dW}{d\Omega} = \int_0^{\infty} \frac{dI(\omega)}{d\Omega} d\omega, \tag{4.36}$$

with

$$\frac{dI(\omega)}{d\Omega} = |\vec{G}(\omega)|^2 + |\vec{G}(-\omega)|^2 = 2|\vec{G}(\omega)|^2. \tag{4.37}$$

The Fourier amplitude $\vec{G}(\omega)$ is

$$\begin{aligned} \vec{G}(\omega) &= (\frac{e^2}{32\pi^3\epsilon_0 c})^{1/2}\int_{-\infty}^{\infty} \frac{\hat{n}\times((\hat{n}-\vec{\beta})\times\dot{\vec{\beta}})}{\kappa^3} e^{j\omega t}dt \\ &= (\frac{e^2}{32\pi^3\epsilon_0 c})^{1/2}\int_{-\infty}^{\infty} \frac{\hat{n}\times((\hat{n}-\vec{\beta})\times\dot{\vec{\beta}})}{\kappa^2} e^{j\omega(t'+R/c)}dt', \end{aligned} \tag{4.38}$$

where $R = |\vec{x} - \vec{r}(t')|$ is the distance between the observer and the electron. With the observer far away from the source, we have $R = |\vec{x} - \vec{r}(t')| \approx x - \hat{n}\cdot\vec{r}(t')$, where x

is the distance from the origin to the observer. Apart from a constant phase factor, the amplitude of the frequency distribution becomes

$$\begin{aligned}\vec{G}(\omega) &= (\frac{e^2}{32\pi^3\epsilon_0 c})^{1/2}\int_{-\infty}^{\infty}\frac{\hat{n}\times((\hat{n}-\vec{\beta})\times\dot{\vec{\beta}})}{\kappa^2}e^{j\omega(t'-\hat{n}\cdot\vec{r}/c)}dt' \\ &= j\omega(\frac{e^2}{32\pi^3\epsilon_0 c})^{1/2}\int_{-\infty}^{\infty}\hat{n}\times(\hat{n}\times\vec{\beta})e^{j\omega(t'-\hat{n}\cdot\vec{r}/c)}dt', \end{aligned} \quad (4.39)$$

where we use integration by parts and the relation $\frac{\hat{n}\times((\hat{n}-\vec{\beta})\times\dot{\vec{\beta}})}{\kappa^2}=\frac{d}{dt'}\frac{\hat{n}\times(\hat{n}\times\vec{\beta})}{\kappa}$.

We now consider a group of charged particles e_j. The radiation amplitude is a linear combination of contributions from each charge, and the corresponding intensity spectrum becomes

$$\begin{aligned}&e\vec{\beta}e^{-j\omega\hat{n}\cdot\vec{r}/c}\to\sum_{k=1}e_k\vec{\beta}_k e^{-j\omega\hat{n}\cdot\vec{r}_k/c}\to\frac{1}{c}\int d^3x\vec{J}(\vec{x},t)e^{-j\omega\hat{n}\cdot\vec{x}/c}, \\ &\frac{dI(\omega)}{d\Omega}=\frac{\omega^2}{16\pi^3\epsilon_0 c^3}\left|\int dt\int d^3x\;\hat{n}\times(\hat{n}\times\vec{J})e^{-j\omega(t-\hat{n}\cdot\vec{x}/c)}\right|^2. \end{aligned} \quad (4.40)$$

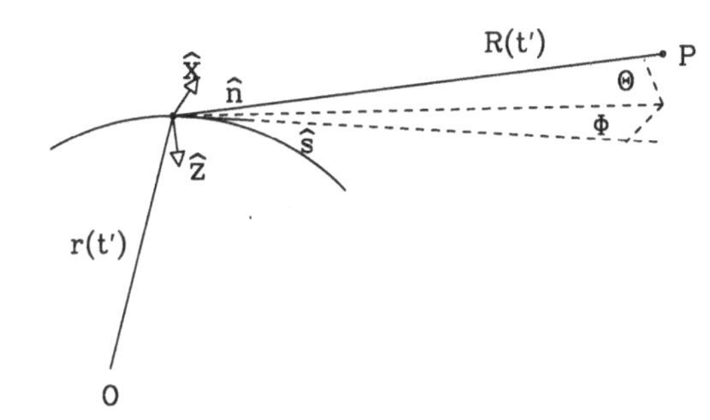

Figure 4.3: Coordinate system for a circular trajectory of electrons.

A. Frequency spectrum of synchrotron radiation

The radiation emitted by an extremely relativistic particle subject to arbitrary acceleration arises mainly from the instantaneous motion of the particle along a circular path. The radiation is beamed in a narrow cone in the forward direction of the velocity vector. The short pulse of radiation resembles a searchlight sweeping across the observer. Figure 4.3 shows the coordinate system of a particle moving along a circular orbit, where the trajectory is

$$\begin{aligned}&\vec{r}(t')=\rho(1-\cos\omega_\rho t',\ \sin\omega_\rho t',\ 0), \\ &\vec{\beta}=\beta(\sin\omega_\rho t',\ \cos\omega_\rho t',\ 0),\end{aligned}$$

where $\omega_\rho = \beta c/\rho$ is the cyclotron frequency and $\vec{\beta} c = d\vec{r}(t')/dt'$ is the velocity vector. Let

$$\hat{n} = (\cos\Theta \sin\Phi,\ \cos\Theta\cos\Phi,\ \sin\Theta) \tag{4.41}$$

be the direction of photon emission, as shown in Fig. 4.3. Because the particle is moving on a circular path, all horizontal angles are equivalent, and it is sufficient to calculate the energy flux for the case $\Phi = 0$. The vector $\hat{n}\times(\hat{n}\times\vec{\beta})$ can be decomposed into

$$\hat{n}\times(\hat{n}\times\vec{\beta}) = \beta\left[-\hat{\mathbf{e}}_{\parallel}\sin\omega_\rho t' + \hat{\mathbf{e}}_{\perp}\cos\omega_\rho t' \sin\Theta\right], \tag{4.42}$$

where $\hat{\mathbf{e}}_{\parallel}$ is the polarization vector along the plane of circular motion in the outward $\hat{x}$ direction and $\hat{\mathbf{e}}_{\perp} = \hat{n}\times\hat{\mathbf{e}}_{\parallel}$ is the orthogonal polarization vector, which is nearly perpendicular to the orbit plane.

Since the range of the t' integration is of the order of $\Delta t' \sim \rho/c\gamma$, the exponent of Eq. (4.39) can be expanded as

$$\begin{aligned}\omega(t' - \frac{\hat{n}\cdot\vec{r}}{c}) &= \omega(t' - \frac{\rho}{c}\sin\omega_\rho t'\cos\Theta) \approx \frac{\omega}{2}\left[(\frac{1}{\gamma^2}+\Theta^2)t' + \frac{c^2}{3\rho^2}t'^3\right]\left[1+\mathrm{O}(\frac{1}{\gamma^2})\right] \\ &= \frac{3}{2}\xi(x+\frac{1}{3}x^3)+\cdots,\end{aligned} \tag{4.43}$$

where

$$x = \frac{\gamma\omega_\rho t'}{(1+X^2)^{\frac{1}{2}}},\quad X = \gamma\Theta,\quad \xi = \frac{\omega}{2\omega_c}(1+X^2)^{3/2},\quad \omega_c = \frac{3}{2}\gamma^3\omega_\rho = \frac{3\gamma^3 c}{2\rho}.$$

Note that both terms in the expansion of Eq. (4.43) are of the same order of magnitude. The critical frequency ω_c has indeed the characteristic behavior of Eq. (4.32).

With the identity

$$\int_0^\infty x\sin\left[\frac{3}{2}\xi\left(x+\frac{1}{3}x^3\right)\right]dx = \frac{1}{\sqrt{3}}K_{2/3}(\xi), \tag{4.44}$$

$$\int_0^\infty \cos\left[\frac{3}{2}\xi\left(x+\frac{1}{3}x^3\right)\right]dx = \frac{1}{\sqrt{3}}K_{1/3}(\xi) \tag{4.45}$$

for the modified Bessel function, the energy and angular distribution function of synchrotron radiation becomes

$$\frac{dP(\omega)}{d\omega d\Omega} = \frac{\beta^2 e^2\omega^2}{16\pi^3\epsilon_0 c}\left|-\hat{\mathbf{e}}_{\parallel}G_{\parallel}(\omega) + \hat{\mathbf{e}}_{\perp}G_{\perp}(\omega)\right|^2 \tag{4.46}$$

where the amplitudes are

$$G_{\parallel} = \frac{1}{\omega_\rho\gamma^2}(1+X^2)\int_{-\infty}^{\infty} x e^{j\frac{3}{2}\xi(x+\frac{1}{3}x^3)}dx = \frac{2(1+X^2)}{\sqrt{3}\omega_\rho\gamma^2}K_{2/3}(\xi), \tag{4.47}$$

$$G_{\perp} = \frac{1}{\omega_\rho\gamma^2}X(1+X^2)^{\frac{1}{2}}\int_{-\infty}^{\infty} e^{j\frac{3}{2}\xi(x+\frac{1}{3}x^3)}dx = \frac{2X(1+X^2)^{1/2}}{\sqrt{3}\omega_\rho\gamma^2}K_{1/3}(\xi). \tag{4.48}$$

Thus the energy radiated per unit frequency interval per unit solid angle becomes

$$\frac{dP}{d\omega d\Omega} = \frac{3e^2}{16\pi^3\epsilon_0 c}\gamma^2(\frac{\omega}{\omega_c})^2(1+X^2)^2\left[K_{2/3}^2(\xi) + \frac{X^2}{1+X^2}K_{1/3}^2(\xi)\right], \tag{4.49}$$

where the first term in the brackets arises from the polarization vector on the plane of the orbiting electron and the second from the polarization perpendicular to the orbital plane. The angular distribution has been verified experimentally.[8] On the orbital plane, where $X = 0$, the radiation is purely plane polarized. Away from the orbital plane, the radiation is elliptically polarized.

B. Asymptotic property of the radiation

Using the asymptotic relation of the Bessel functions

$$K_\nu(\xi) \sim \begin{cases} 2^{\nu-1}\Gamma(\nu)\xi^{-\nu}, & \text{if } \xi \ll 1, \\ \sqrt{\pi/2\xi}e^{-\xi}, & \text{if } \xi \gg 1, \end{cases} \tag{4.50}$$

we find that the radiation is negligible for $\xi \geq 1$. Thus the synchrotron radiation is confined by

$$\omega \leq \frac{2\omega_c}{(1+X^2)^{3/2}} \quad \text{or} \quad \langle\Theta^2\rangle^{1/2} \leq \frac{1}{\gamma}(\frac{\omega_c}{\omega})^{1/3}. \tag{4.51}$$

The synchrotron radiation spans a continuous spectrum up to ω_c. High frequency synchrotron light is confined in an angular cone $1/\gamma$. The radiation at large angles is mostly low frequency.

C. Angular distribution in the orbital plane

In the particle orbital plane with $\Theta = 0$, the radiation contains only the parallel polarization. We find

$$\left.\frac{dI}{d\Omega}\right|_{\Theta=0} = \frac{3e^2}{16\pi^3\epsilon_0 c}\gamma^2 H_2(\frac{\omega}{\omega_c}),$$

$$H_2(y) = y^2 K_{2/3}^2(\frac{y}{2}) \sim \begin{cases} 4^{\frac{1}{3}}[\Gamma(\frac{2}{3})]^2 y^{2/3}, & \text{if } y \ll 1, \\ \pi y e^{-y}, & \text{if } y \gg 1, \end{cases} \tag{4.52}$$

with $y = \omega/\omega_c$ (Fig. 4.4). Thus the energy spectrum at $\Theta = 0$ increases with frequency as $2.91(\omega/\omega_c)^{2/3}$ for $\omega \ll \omega_c$, reaches a maximum near ω_c, and then drops to zero exponentially as $e^{-\omega/\omega_c}$ above critical frequency.

[8]F.R. Elder *et al.*, *Phys. Rev.* **71**, 829 (1947); **74**, 52 (1948); *J. Appl. Phys.* **18**, 810 (1947).

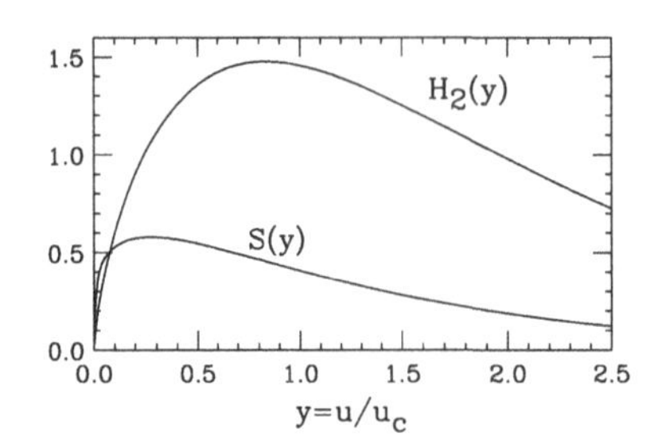

Figure 4.4: The functions $H_2(y)$ and $S(y)$ for synchrotron radiation are shown as functions of $y = u/u_c$, where $u = \hbar\omega$, $u_c = \hbar\omega_c$.

D. Angular distribution for the integrated energy spectrum

When the energy flux is integrated over all frequency (see Section 6.576 in Ref. [31]), we obtain

$$\int_0^\infty \frac{dP}{d\omega d\Omega} d\omega = \frac{7e^2}{96\pi\epsilon_0 c} \frac{\gamma^2 \omega_c}{(1+X^2)^{5/2}} \left(1 + \frac{5X^2}{7(1+X^2)}\right) \tag{4.53}$$

where the first term corresponds to the polarization vector parallel to the orbital plane, while the second term is the perpendicular component. Integrating over all angles, we find that the parallel polarization carries seven times as much energy as does the perpendicular polarization.

E. Frequency spectrum of radiated energy flux

Integrating Eq. (4.49) over the entire angular range, we obtain the energy flux[9]

$$I(\omega) = \frac{\sqrt{3}e^2}{4\pi\epsilon_0 c} \gamma \frac{\omega}{\omega_c} \int_{\omega/\omega_c}^\infty K_{5/3}(y) dy = \frac{2e^2}{9\epsilon_0 c} \gamma S(\frac{\omega}{\omega_c}), \tag{4.54}$$

$$S(y) = \frac{9\sqrt{3}}{8\pi} y \int_y^\infty K_{5/3}(y') dy', \quad \int_0^\infty S(y) = 1. \tag{4.55}$$

The total instantaneous radiation power becomes

$$P_\gamma = \frac{1}{2\pi\rho} \int_0^\infty I(\omega) d\omega = \frac{4e^2}{36\pi\epsilon_0 \rho} \gamma \omega_c = \frac{c C_\gamma}{2\pi} \frac{E_0^4}{\rho^2}, \tag{4.56}$$

where $C_\gamma = 8.85 \times 10^{-5}$ meter/(GeV)^3. This result was obtained by Liénard in 1898. The *instantaneous* power spectrum becomes

$$\tilde{I}(\omega) = \frac{1}{2\pi\rho} I(\omega) = \frac{P_\gamma}{\omega_c} S(\frac{\omega}{\omega_c}). \tag{4.57}$$

[9] J. Schwinger, *Phys. Rev.* **75**, 1912 (1949).

Since the energy of the photon is $\hbar\omega$, the photon flux density is

$$\frac{d\mathcal{F}}{d\Omega} = \left[\frac{I}{e}\right]\frac{3e^2}{16\pi^3\epsilon_0\hbar c}\frac{\delta\omega}{\omega}\left(\frac{\omega}{\omega_c}\right)^2(1+X^2)^2\left[K_{2/3}^2(\xi)+\frac{X^2}{1+X^2}K_{1/3}^2(\xi)\right], \quad (4.58)$$

$$\left.\frac{d\mathcal{F}}{d\Omega}\right|_{\Theta=0} = \left[\frac{I}{e}\right]\frac{3e^2}{16\pi^3\epsilon_0\hbar c}\frac{\delta\omega}{\omega}H_2(\frac{\omega}{\omega_c})$$

$$= 1.33\times 10^{13}E_0^2 I[A]H_2(\frac{\omega}{\omega_c})\quad\left[\frac{\text{photons}}{\text{s mr}^2\ 0.1\%\text{ bandwidth}}\right], \quad (4.59)$$

which peaks at $y=1$ or $\omega=\omega_c$. Thus the radiation due to the bending magnets has a smooth spectral distribution with a broad maximum at the critical frequency ω_c. The critical photon energy is

$$\omega_c = \frac{3c\gamma^3}{2\rho}, \qquad u_c\text{ [keV]} = \hbar\omega_c = 0.665\ E^2[\text{GeV}]\ B[\text{T}], \quad (4.60)$$

where E is the electron beam energy and B is the magnetic flux density.

Using the asymptotic properties of the modified Bessel functions of Eq. (4.50), we find that the spectral flux vanishes as $(\omega/\omega_c)^{2/3}$ for $\omega \ll \omega_c$ and as $e^{-\omega/\omega_c}$ for $\omega \gg \omega_c$. Following the traditional convention, we define $4\omega_c$ as the upper limit for useful photon frequency from bending magnet radiation.

Integrating Eq. (4.49) over the vertical angle Θ, we obtain

$$\left.\frac{d\mathcal{F}}{d\Phi}\right|_{\Theta=0} = \left[\frac{I}{e}\right]\frac{\sqrt{3}e^2}{8\pi^2\epsilon_0\hbar c}\frac{\delta\omega}{\omega}\gamma\frac{\omega}{\omega_c}\int_{\omega/\omega_c}^{\infty}K_{5/3}(y)dy$$

$$= 2.46\times 10^{13}E_0[GeV]I[A]G_1(\frac{\omega}{\omega_c})\left[\frac{\text{photons}}{\text{s mr}^2\ 0.1\%\text{ bandwidth}}\right] \quad (4.61)$$

$$G_1(y) = y\int_y^{\infty}K_{5/3}(y')dy'$$

The function $S(y)=\frac{9\sqrt{3}}{8\pi}G_1(y)$ is shown in Figure 4.4.

I.4 Quantum Fluctuation

Electromagnetic radiation is emitted in quanta of energy $u=\hbar\omega$, where $\hbar$ is Planck's constant. Let $n(u)du$ be the number of photons per unit time emitted in the frequency interval $d\omega = du/\hbar$ at frequency ω, i.e.

$$un(u)du = I(\omega)d\omega = I(\omega)\frac{du}{\hbar} \quad (4.62)$$

$$n(u) = \frac{P_\gamma}{u_c^2}F(\frac{u}{u_c}) = \frac{9\sqrt{3}}{8\pi}\frac{P_\gamma}{u_c^2}\int_{u/u_c}^{\infty}K_{5/3}(y)dy, \quad (4.63)$$

$$F(y) = \frac{1}{y}S(y), \qquad \int_0^{\infty}F(y) = \frac{15\sqrt{3}}{8}. \quad (4.64)$$

The total number of photons emitted per second, $\mathcal{N}$, and the average number of photons emitted per revolution are

$$\mathcal{N} = \int_0^\infty n(u)du = \frac{15\sqrt{3}}{8}\frac{P_\gamma}{u_c} = \frac{5\alpha c\gamma}{2\sqrt{3}\rho}, \tag{4.65}$$

$$N_\gamma = \mathcal{N} 2\pi\frac{\rho}{c} = \frac{5\pi}{\sqrt{3}}\alpha\gamma, \tag{4.66}$$

where $\alpha = e^2/4\pi\epsilon_0\hbar c$ is the fine structure constant. Table 4.2 lists synchrotron radiation properties of some storage rings. Note that the number of photons emitted per revolution is typically a few hundred to a few thousand. In Table 4.2, E is the beam energy, ρ is the bending radius, C is the circumference, T_0 is the revolution period, U_0 is the energy loss per revolution, τ_s and $\tau_\perp$ are radiation damping times of the longitudinal and transverse phase spaces (to be discussed in Sec. II), u_c is the critical photon energy, and N_γ is the average number of photons emitted per revolution.

Table 4.2: Properties of some high energy storage rings

	BEPC	CESR	LER	HER	APS	ALS	LEP	LHC
E [GeV]	2.2	6	3.2	9	7	1.5	55	7000
ρ [m]	10.35	60	30.6	165	38.96	4.01	3096.2	3096.2
C [m]	240.4	768.4	2199.3	2199.3	1104	196.8	26658.9	26658.9
T_0 [μs]	0.80	2.56	7.34	7.34	3.68	0.66	89.	89.
U_0 [MeV]	0.20	1.91	0.30	3.52	5.45	0.11	261.	0.00060
$\tau_\parallel$ [ms]	8.8	8.0	77.	19.	4.7	8.8	19.	1.0×10^9
$\tau_\perp$ [ms]	18.	16.	155.	38.	9.4	18.	38.	2.0×10^9
u_c [keV]	2.28	7.97	2.37	9.78	19.50	1.86	119.00	0.040
N_γ	285	777	415	1166	907	194	7125	494

The moments of energy distribution become

$$\langle u\rangle = \frac{1}{\mathcal{N}}\int_0^\infty un(u)du = \frac{8}{15\sqrt{3}}u_c, \qquad \langle u^2\rangle = \frac{1}{\mathcal{N}}\int_0^\infty u^2 n(u)du = \frac{11}{27}u_c^2,$$

$$\mathcal{N}\langle u^2\rangle = C_u u_c P_\gamma = \frac{3C_u C_\gamma}{4\pi}\frac{\hbar c^2}{(mc^2)^3}\frac{E_0^7}{\rho^3}, \qquad C_u = \frac{55}{24\sqrt{3}}. \tag{4.67}$$

At a fixed bending radius, the quantum fluctuation varies as the seventh power of the energy.

Exercise 4.1

1. A particle of mass m and charge e moves in a plane perpendicular to a uniform, static magnetic induction B.

   (a) Calculate the total energy radiated per unit time. Express it in terms of the constants m, e, γ and B.

   (b) Find the path of the electron.

2. Plot the angular distribution of synchrotron radiation shown in Eq. (4.49) for $\beta = 0.5$ and $\beta = 0.99$. Find the maximum angular distribution of synchrotron radiation, and show that the integrated power is given by Eq. (4.56).

3. Using the coordinate system of Fig. 4.2 with $\vec{\beta} = \beta\hat{\beta}$, $\dot{\vec{\beta}} = (\dot{v}/c)\hat{\dot{\beta}}$, and $\hat{n} = \cos\Theta\hat{\beta} + \sin\Theta\cos\Phi\hat{\dot{\beta}} + \sin\Theta\sin\Phi\hat{\jmath}$, where $\hat{\jmath} = \hat{\beta}\times\hat{\dot{\beta}}$, verify Eq. (4.30). Plot the angular distribution of synchrotron radiation shown in Eq. (4.30) for $\beta = 0.5$ and $\beta = 0.99$. Find the angle of the maximum angular distribution, and show that the integrated power is given by Eq. (4.31).

4. The synchrotron radiation generated by the circulating beam will liberate photo electrons from the chamber walls, which will desorb the surface molecules. The photon yields depend on the photon energy and the chamber wall material. Using Eq. (4.54), show that the number of primarily photons per unit energy interval in one revolution is

$$\frac{dN}{du} = \frac{9\sqrt{3}}{8\pi}\frac{U_0}{u_c^2}\int_{u/u_c}^{\infty} K_{5/3}(y)dy,$$

where $K_{5/3}$ is the Bessel function of order 5/3; $u_c = 3\hbar c\gamma^3/2\rho$ is the critical photon energy; and

$$U_0 = \frac{4\pi r_p m_p c^2\gamma^4}{3\rho} = \begin{cases} 8.85\times10^{-5}[(E\ [\mathrm{GeV}])^4/\rho\,[\mathrm{m}]]\,[\mathrm{GeV}] & \text{for electrons,} \\ 7.78\times10^{-6}[(E\ [\mathrm{TeV}])^4/\rho\,[\mathrm{m}]]\,[\mathrm{GeV}] & \text{for protons,} \end{cases}$$

is the energy loss per revolution. Show that the total number of primary photons in one revolution is given by

$$N_\gamma = \frac{15\sqrt{3}}{8}\frac{U_0}{u_c}$$

Verify N_γ of the machines in the table below.

	Proton storage rings			Electron storage rings		
	VLHC	SSC	LHC	LEP	HER(B)	APS
E [GeV]	50000	20000	8000	55	9	7
ρ [m]	15000	10108	3096.2	3096.2	165	38.96
γ	53289	21316	8526	107632	17612	13699
u_c [keV]	3.0	0.28	0.059	119	9.78	19.5
U_0 [keV]	3246	123	10.3	261495	3518	5453
N_γ	3530	1429	567	7136	1168	908

5. At 55 GeV, the magnetic flux density in a LEP dipole is $B = 592.5$ G. What happens if you design the LEP with a magnetic flux density of 0.5 T at 55 GeV beam energy? What will the energy loss per revolution be at 100 GeV? With the present LEP dipole magnets, at what energy will the beam lose all its energy in one revolution?

6. Verify Eq. (4.53), integrate the intensity over all angles, and prove that the parallel polarization carries seven times as much energy as that of the perpendicular polarization.

7. Verify Eqs. (4.59) and (4.61).

8. In designing a high energy collider, you need to take into account the problems associated with gas desorption due to synchrotron radiations.

   (a) For an accelerator with an average current I [A], show that the total synchrotron radiation power is given by

$$P_\gamma = 6.03 \times 10^{-9} \frac{\gamma^4 R\,[\mathrm{m}]\, I\,[\mathrm{A}]}{(\rho[\mathrm{m}])^2} \quad [W].$$

   (b) Show that the total number of photons per unit time (s) is given by

$$\mathcal{N} = 4.14 \times 10^{17} \frac{R}{\rho} \gamma\, I\,[\mathrm{A}].$$

   Show that the total number of photons per unit length in the dipole magnet is given by[10]

$$\frac{d\mathcal{N}}{ds} = 6.60 \times 10^{16} \frac{R}{\rho^2} \gamma\, I\,[\mathrm{A}] \quad [\mathrm{photons/m}].$$

[10]The resulting pressure increase is given by

$$\Delta P = \frac{\eta}{kS} \times \frac{d\mathcal{N}}{ds},$$

where η is the molecular desorption yield (molecules/photon), S is the pumping speed (liter/s), and $k = 3.2 \times 10^{19}$ (molecules/torr-liter) at room temperature. See O. Gröbner, p. 454 in Ref. [19].

II Radiation Damping and Excitation

The instantaneous power radiated by a relativistic electron at energy E in the magnetic field strength B is given by (4.4), i.e. $P \sim E^2B^2 \sim E^4/\rho^2$, where ρ is the local radius of curvature. The total energy radiated in one revolution and the average radiation power for an isomagnetic ring are respectively

$$U_0 = \frac{C_\gamma \beta^2 E^4}{2\pi}\oint \frac{ds}{\rho^2} \xrightarrow{\text{isomagnetic}} \frac{C_\gamma E^4}{\rho} = 26.5\ (E[\text{GeV}])^3\ B[\text{T}]\ [\text{keV}].$$
$$\langle P_\gamma\rangle = \frac{U_0}{T_0} = \frac{cC_\gamma E^4}{2\pi R\rho},$$

where $T_0 = \beta c/2\pi R$ is the revolution period, and R is the average radius of a storage ring. For example, an electron at 50 GeV in the LEP at CERN ($\rho = 3.096$ km) will lose 0.18 GeV per turn, and the energy loss per revolution at 100 GeV is 2.9 GeV, i.e. 3% of its total energy. The energy of circulating electrons is compensated by rf cavities with *longitudinal* electric field.

Since higher energy electrons lose more energy than lower energy electrons [see Eq. (4.4)] and the average beam energy is compensated by longitudinal electric field, there is radiation damping (cooling) in the longitudinal phase space. Furthermore, electrons lose energy in a cone with an angle about $1/\gamma$ of their instantaneous velocity vector, and gain energy through rf cavities in the longitudinal direction. This mechanism provides transverse phase-space damping. The damping (e-folding) time is generally equal to the time it takes for the electron to lose all of its energy.

At the same time, photon emission is discrete, random and quantized. The quantum process causes diffusion and excitation. The balance between quantum fluctuation and phase-space damping provides natural momentum spread of the beam.

The longitudinal and transverse motions are coupled through the dispersion function; there is a damping-fluctuation partition between the longitudinal and transverse radial planes. The balance between damping and excitation provides natural emittance or equilibrium beam size. The vertical emittance is determined by the residual vertical dispersion function and linear betatron coupling. In this section we discuss damping time, damping partition, quantum fluctuation, beam emittances, and methods of manipulating the damping partition number.

II.1 Damping of Synchrotron Motion

Expanding the synchrotron radiation power of Eq. (4.4) around synchronous energy, we obtain

$$U(E) = U_0 + W\Delta E, \quad W = \left.\frac{dU}{dE}\right|_{E=E_0}, \tag{4.68}$$

where E_0 is the synchronous energy. Since a particle having nonzero betatron amplitude moves through different regions of magnetic field, its rate of synchrotron radiation may differ from that of an electron with zero betatron amplitude. However, if the field is linear with respect to displacement, the radiation power averaged over a betatron cycle is independent of betatron amplitude. Thus Eq. (4.68) does not depend explicitly on betatron amplitude. We will show that the coefficient W determines the damping rate of synchrotron motion.

First, we consider the longitudinal equation of motion in the presence of energy dissipation. We assume that all particles travel at the speed of light. Let $(c(\tau + \tau_s), \Delta E)$ be the longitudinal phase-space coordinates of a particle, where ΔE is the energy deviation from the synchronous energy. Let $(c\tau_s, 0)$ be those of a synchronous particle. The path length difference between these two particles is $\Delta C = \alpha_c C \frac{\Delta E}{E}$, where α_c is the momentum compaction factor, C is the accelerator circumference. The difference of the arrival time and its time derivative become

$$\Delta\tau = \alpha_c \frac{C}{c}\frac{\Delta E}{E} = \alpha_c T_0 \frac{\Delta E}{E} \quad \Longrightarrow \quad \frac{d\tau}{dt} = \frac{\Delta\tau}{T_0} = \alpha_c \frac{\Delta E}{E}. \tag{4.69}$$

Here, the phase slip factor is $\eta = \alpha_c - 1/\gamma^2 \approx \alpha_c$ for high energy electrons.

During one revolution, the electron loses energy $U(E)$ by radiation, and gains energy $eV(\tau)$ from the rf system. Thus the net energy change is

$$\frac{d(\Delta E)}{dt} = \frac{eV(\tau) - U(E)}{T_0}, \tag{4.70}$$

where the energy loss per revolution is $U(E) = U_0 + W\Delta E$. For simplicity, we assume a sinusoidal rf voltage wave and expand the rf voltage around the synchronous phase angle $\phi_s = h\omega_0\tau_s$,

$$V(\tau) = V_0 \sin\phi = V_0 \sin\omega_{rf}(\tau + \tau_s),$$

where the rf frequency is $\omega_{rf} = h\omega_0 = 2\pi h/T_0$; ω_0 is the revolution frequency; and h is the harmonic number. Now we consider the case of a storage mode without net acceleration, where the energy gain in the rf cavity is to compensate the energy loss in synchrotron radiation, i.e.

$$U_{rf} = eV(\tau) = U_0 + e\dot{V}\tau,$$
$$U_0 = eV_0 \sin(\omega_{rf}\tau_s), \quad \dot{V} = \omega_{rf} V_0 \cos(\omega_{rf}\tau_s).$$

Thus, in small amplitude approximation, we have

$$\frac{d(\Delta E)}{dt} = \frac{1}{T_0}(e\dot{V}\tau - W\Delta E). \tag{4.71}$$

Combining with Eq. (4.69), we obtain

$$\frac{d^2\tau}{dt^2} + 2\alpha_E \frac{d\tau}{dt} + \omega_s^2 \tau = 0, \qquad \alpha_E = \frac{W}{2T_0}, \quad \omega_s^2 = -\frac{\alpha_c e\dot{V}}{T_0 E}. \tag{4.72}$$

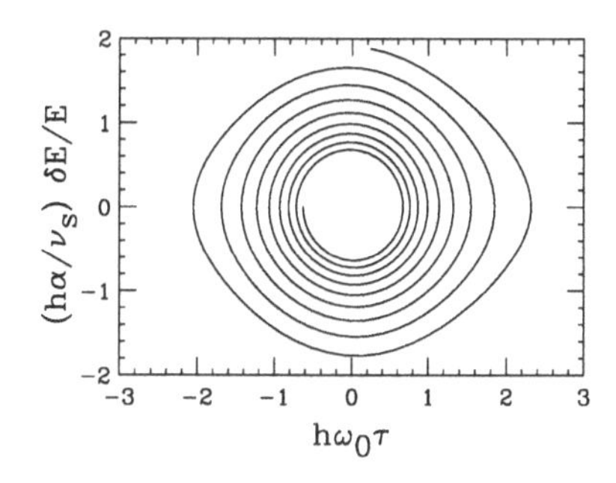

Figure 4.5: A schematic drawing of damped synchrotron motion. Particle motion is damped toward the center of the bucket.

This is the equation of a damped harmonic oscillator with synchrotron frequency ω_s and damping coefficient α_E. The longitudinal damping time is $\tau_{\parallel} = 1/\alpha_E$. Since the damping rate is normally small, i.e. $\alpha_E \ll \omega_s$, the solution can be expressed as

$$\tau(t) = Ae^{-\alpha_E t}\cos(\omega_s t - \theta_0). \tag{4.73}$$

Figure 4.5 illustrates damped synchrotron motion. Table 4.2 lists the longitudinal damping time of some storage rings. Typically, $1/e$ damping time is $10^3 - 10^4$ revolutions.

The damping partition

To evaluate the damping rate, we need to evaluate W. Since the radiation energy loss per revolution is

$$U_{\rm rad} = \oint P_\gamma dt = \oint P_\gamma \frac{dt}{ds} ds = \frac{1}{c}\oint P_\gamma (1+\frac{x}{\rho})ds = \frac{1}{c}\oint P_\gamma (1+\frac{D}{\rho}\frac{\Delta E}{E_0})ds,$$

where D is the dispersion function, ρ is the radius of curvature, and we have used $cdt/ds = (1 + x/\rho)$. The transverse displacement x is the sum of betatron displacement and off-momentum closed orbit. Since we are interested in the dependence of total radiation energy on the off-energy coordinate and $\langle x_\beta \rangle = 0$, we replace x by $D(\Delta E/E_0)$. The derivative of radiation energy with respect to particle energy is

$$W = \frac{dU_{\rm rad}}{dE} = \frac{1}{c}\oint \left\{\frac{dP_\gamma}{dE} + \frac{D}{\rho}\frac{P_\gamma}{E}\right\}_{E_0} ds.$$

Using $P_\gamma \sim E^2B^2$ of Eq. (4.4), we obtain

$$\begin{aligned}
\left.\frac{dP_\gamma}{dE}\right|_{E_0} &= 2\frac{P_\gamma}{E_0} + 2\frac{P_\gamma}{B_0}\frac{dB}{dE} = 2\frac{P_\gamma}{E_0} + 2\frac{P_\gamma}{B_0}\frac{dx}{dE}\frac{dB}{dx} = 2\frac{P_\gamma}{E_0} + 2\frac{P_\gamma}{B_0}\frac{D}{E_0}\frac{dB}{dx}, \\
\frac{dU_{\rm rad}}{dE} &= \frac{1}{c}\oint \left\{2\frac{P_\gamma}{E_0} + 2\frac{P_\gamma}{B_0}\frac{D}{E_0}\frac{dB}{dx} + \frac{P_\gamma}{E_0}\frac{D}{\rho}\right\}_{E_0} ds \\
&= \frac{U_0}{E_0}\left[2 + \frac{1}{cU_0}\oint \left\{DP_\gamma\left(\frac{1}{\rho} + \frac{2}{B}\frac{dB}{dx}\right)\right\}_{E_0} ds\right].
\end{aligned}$$

Thus the damping coefficient becomes

$$\alpha_E = \frac{1}{2T_0}\frac{dU_{\rm rad}}{dE} = \frac{U_0}{2T_0 E}(2+\mathcal{D}).$$

Here $\mathcal{D}$ is the *damping re-partition number*:

$$\mathcal{D} = \frac{1}{cU_0}\oint\left\{DP_\gamma\left(\frac{1}{\rho}+\frac{2}{B}\frac{dB}{dx}\right)\right\}_{E_0} ds = \left[\oint\frac{D}{\rho}\left(\frac{1}{\rho^2}+2K(s)\right)ds\right]\left[\oint\frac{ds}{\rho^2}\right]^{-1} \quad (4.74)$$

where $K(s) = B_1/B\rho$ is the quadrupole gradient function with $B_1 = \partial B/\partial x$. The damping re-partition number $\mathcal{D}$ is a property of lattice configuration. For isomagnetic ring, the re-partition factor becomes

$$\mathcal{D} = \frac{1}{2\pi}\oint D(s)\left(\frac{1}{\rho^2}+2K(s)\right)_{\rm dipole} ds, \quad (4.75)$$

which is to be evaluated only in dipoles.

Example 1: Damping re-partition for separate function accelerators

For an isomagnetic ring with separate function magnets, where $K(s) = 0$ in dipoles,

$$\mathcal{D} = \frac{1}{2\pi\rho}\oint\frac{D(s)}{\rho}ds = \frac{\alpha_c R}{\rho}, \quad (4.76)$$

where α_c is the momentum compaction factor. Since normally $\alpha_c \ll 1$ in synchrotrons, $\mathcal{D} \ll 1$ for separate function machines.

The damping coefficient for separate function machines becomes

$$\alpha_E = \frac{\langle P_\gamma\rangle}{2E}\left(2+\frac{\alpha_c R}{\rho}\right) \approx \frac{U_0}{ET_0} = \frac{\langle P_\gamma\rangle}{E}. \quad (4.77)$$

The damping time constant, which is the inverse of α_E, is nearly equal to the time it takes for the electron to radiate away its total energy.

Example 2: Damping re-partition for combined function accelerators

For an isomagnetic combined function accelerator, we find (see Exercise 4.2.1)

$$\mathcal{D} = 2 - \frac{\alpha_c R}{\rho} \quad (4.78)$$

and $\alpha_E \approx 2\langle P_\gamma\rangle/E$. The synchrotron motion is damped two times faster at the expense of horizontal betatron excitation, to be discussed in the next section.

II.2 Damping of Betatron Motion

A. Transverse (vertical) betatron motion

A relativistic electron emits synchrotron radiation primarily along its direction of motion within an angle $1/\gamma$. The momentum change resulting from recoil of synchrotron radiation is exactly opposite to the direction of particle motion. Figure 4.6 illustrates betatron motion with synchrotron radiation, where vertical betatron coordinate z is plotted as a function of longitudinal coordinate s. The betatron phase-space coordinates are

$$z = A\cos\phi, \quad z' = -\frac{A}{\beta}\sin\phi, \quad A^2 = z^2 + (\beta z')^2, \tag{4.79}$$

where A is betatron amplitude, ϕ is betatron phase, and β is betatron function.

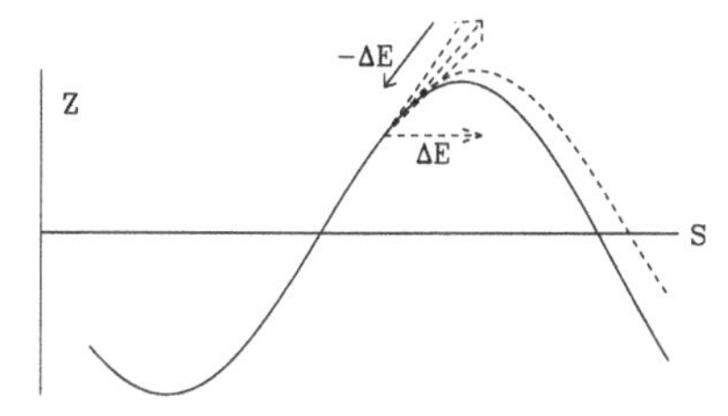

Figure 4.6: Schematic drawing of the damping of vertical betatron motion due to synchrotron radiation. The energy loss through synchrotron radiation along the particle trajectory with an opening angle of $1/\gamma$ is replenished in the rf cavity along the longitudinal direction. This process damps the vertical betatron oscillation to a very small value.

When an electron loses an amount of energy u by radiation, the momentum vector changes by $\Delta\vec{P}$, which is parallel and opposite to $\vec{P}$ with $|c\delta\vec{P}| = c\delta p = u$. Since the radiation loss changes neither slope nor position of the trajectory, the betatron amplitude is unchanged except for a small increment in effective focusing force. Putting the above statement into equation, we find

$$z'|_{\text{before}} = \frac{p_z}{p} = z'|_{\text{after}} = \frac{p_z}{p-\delta p}\left(1 - \frac{\delta p}{p}\right) \xrightarrow{\text{after cavity}} \frac{p_z}{p}\left(1 - \frac{\delta p}{p}\right),$$

where the effect of energy compensation is shown in the above equation. Thus the change of z' is

$$z' \to z'(1 - \frac{\delta p}{p}), \quad \text{or} \quad \Delta z' = -z'\frac{\delta p}{p} = -z'\frac{u}{E}. \tag{4.80}$$

The corresponding change of amplitude A in one revolution becomes

$$A\delta A = \langle \beta^2 z'\Delta z'\rangle = -\langle(\beta z')^2\rangle\frac{U_0}{E},$$

where $\langle\ldots\rangle$ averages over betatron oscillations in one revolution, and U_0 is synchrotron radiation energy per revolution. Since the betatron motion is sinusoidal, we find $\langle(\beta z')^2\rangle = A^2/2$. The time variation of the betatron amplitude function and the damping coefficient become

$$\frac{1}{A}\frac{dA}{dt} = \frac{1}{T_0}\frac{\delta A}{A} = -\frac{U_0}{2ET_0},$$
$$\alpha_z = \frac{U_0}{2ET_0} = \frac{\langle P_\gamma \rangle}{2E}. \tag{4.81}$$

The radiation loss alone does not result in betatron phase-space damping. The radiation damping arises from the combination of energy loss in the direction of betatron orbit and energy gain in the longitudinal direction from rf systems. The 1/e damping time of the vertical betatron motion is $\tau_z = 1/\alpha_z$. The damping rate of Eq. (4.81) applies also to the horizontal betatron motion.

B. Horizontal betatron motion

The horizontal motion of an electron is complicated by the off-momentum closed orbit. The horizontal displacement from the reference orbit is

$$x = x_\beta + x_{\rm e}, \quad x_{\rm e} = D(s)\frac{\Delta E}{E}, \quad x' = x'_\beta + x'_{\rm e}, \quad x'_{\rm e} = D'(s)\frac{\Delta E}{E},$$

where x_β is the betatron displacement, $x_{\rm e}$ is the off-energy closed orbit, and $D(s)$ is the dispersion function. When the energy of an electron is changed by an amount u due to photon emission, the off-energy closed orbit $x_{\rm e}$ changes by an amount $\delta x_{\rm e} = D(s)\,(u/E)$ shown schematically in Fig. 4.7. Since phase-space coordinates are not changed by any finite impulse, the resulting betatron amplitude is

$$\delta x_\beta = -\delta x_{\rm e} = -D(s)\frac{u}{E}, \qquad \delta x'_\beta = -\delta x'_{\rm e} = -D'(s)\frac{u}{E}. \tag{4.82}$$

The resulting change of betatron amplitude can be obtained from the betatron phase average along an accelerator.

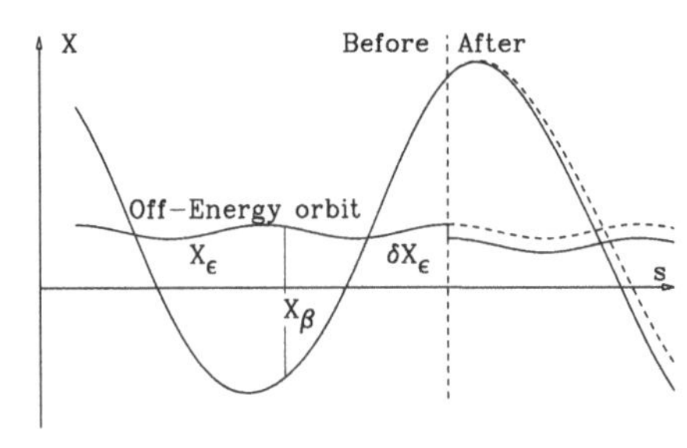

Figure 4.7: Schematic illustration, after M. Sands [2], of quantum excitation of horizontal betatron motion arising from photon emission at a location with nonzero dispersion functions. At a location marked by a vertical dashed line, the electron emits a photon, and the electron energy is changed by u, and thus the off energy closed orbit is shifted by δx_e, which perturbs the betatron motion. A small and not so important effect is a stronger focusing field for betatron motion.

We consider betatron motion with

$$x_\beta = A\cos\phi, \quad x'_\beta = -\frac{A}{\beta}\sin\phi, \quad A^2 = x_\beta^2 + (\beta x'_\beta)^2; \tag{4.83}$$

the change in betatron amplitude becomes

$$A\delta A = x_\beta \delta x_\beta + \beta^2 x'_\beta \delta x'_\beta = -(Dx_\beta + \beta^2 D' x'_\beta)\frac{u}{E}. \tag{4.84}$$

Substituting the energy loss u in an element length $\delta\ell$ with

$$u = -\frac{P_\gamma(x_\beta)}{c}\delta\ell = -\frac{1}{c}\left(P_\gamma + 2\frac{P_\gamma}{B}\frac{dB}{dx}x_\beta\right)\left(1+\frac{x_\beta}{\rho}\right)ds \tag{4.85}$$

into Eq. (4.84), we obtain the change in betatron amplitude as

$$A\delta A = x_\beta D\left(1 + \frac{2}{B}\frac{dB}{dx}x_\beta + \frac{x_\beta}{\rho}\right)\frac{P_\gamma}{cE}ds. \tag{4.86}$$

Here we use $\delta\ell = (1 + x/\rho)ds$ with $x = x_\beta$, because we are interested in the effect on betatron motion. The off-momentum closed orbit does not contribute to the change in betatron amplitude. We have also neglected all terms linear in x'_β, because their average over the betatron phase is zero. We are now looking for the time average over the betatron phase, where $\langle x_\beta\rangle = 0$ and $\langle x_\beta^2\rangle = \frac{1}{2}A^2$. The fractional betatron amplitude increment in one turn becomes

$$\frac{\langle\delta A\rangle}{A} = \frac{U_0}{2E}\left(\oint\frac{D}{\rho}\left[2K(s) + \frac{1}{\rho^2}\right]ds\right)\left(\oint\frac{1}{\rho^2}ds\right)^{-1} = \mathcal{D}\frac{U_0}{2E}, \tag{4.87}$$

where $\mathcal{D}$ is the damping re-partition number given in Eq. (4.74). In particular, we observe that the right side of Eq. (4.87) is positive, i.e. there is an increase in horizontal betatron amplitude due to synchrotron radiation. Emission of a photon excites betatron motion of the electron. This resembles the random walk problem, and the resulting betatron amplitude will increase with time. Including the phase-space damping due to rf acceleration given by Eq. (4.81), we obtain the net horizontal amplitude change per revolution and the damping (rate) coefficient:

$$\frac{\Delta A}{A} = -(1-\mathcal{D})\frac{U_0}{2E}; \qquad \alpha_x = (1-\mathcal{D})\frac{U_0}{2T_0E}, \tag{4.88}$$

where the damping re-partition $\mathcal{D}$ is given by Eq. (4.74).

In summary, radiation damping coefficients for the three degrees of freedom in a bunch are

$$\alpha_x = \mathcal{J}_x\alpha_0, \quad \alpha_z = \mathcal{J}_z\alpha_0, \quad \alpha_E = \mathcal{J}_E\alpha_0, \tag{4.89}$$

where $\alpha_0 = \langle P_\gamma\rangle/2E$, and the damping partition numbers

$$\mathcal{J}_x = 1-\mathcal{D}, \quad \mathcal{J}_z = 1, \quad \mathcal{J}_E = 2+\mathcal{D} \tag{4.90}$$

satisfy the Robinson theorem [see K. Robinson, *Phys. Rev.* **111**, 373 (1958)]:

$$\sum\mathcal{J}_i = \mathcal{J}_x + \mathcal{J}_z + \mathcal{J}_E = 4 \quad \text{or} \quad \mathcal{J}_x + \mathcal{J}_E = 3 \tag{4.91}$$

provided that all fields acting on the particle are predetermined and are not influenced by the motion of electrons. The corresponding damping time constants are

$$\begin{aligned}
\tau_x &= \frac{2E}{\mathcal{J}_x\langle P_\gamma\rangle} = \frac{4\pi R\rho}{cC_\gamma \mathcal{J}_x E^3} = \frac{2E}{\mathcal{J}_x U_0}T_0, \\
\tau_z &= \frac{2E}{\mathcal{J}_z\langle P_\gamma\rangle} = \frac{4\pi R\rho}{cC_\gamma \mathcal{J}_z E^3} = \frac{2E}{\mathcal{J}_z U_0}T_0, \\
\tau_E &= \frac{2E}{\mathcal{J}_E\langle P_\gamma\rangle} = \frac{4\pi R\rho}{cC_\gamma \mathcal{J}_E E^3} = \frac{2E}{\mathcal{J}_E U_0}T_0,
\end{aligned}$$

where T_0 is the revolution period. Note that the damping time, for constant ρ, is inversely proportional to the cubic power of energy and, for a fixed B-field, is inversely proportional to the square of energy. Some typical damping times for electron storage rings are listed in Table 4.2. The damping decrements are defined as

$$\lambda_x = T_0/\tau_x, \quad \lambda_z = T_0/\tau_z, \quad \lambda_E = T_0/\tau_E, \tag{4.92}$$

i.e. the beam phase space areas are reduced by $\exp(-\lambda_x)$, $\exp(-\lambda_z)$, and $\exp(-\lambda_E)$ per revolution respectively. The damping rate of an individual particle or a portion of a bunch can be modified if additional forces are introduced that depend on the details of particle motion. Some examples are image current on vacuum chamber wall, induced current in rf cavity, wakefields, longitudinal and transverse dampers powered by amplifiers sensing beam displacement, and electron and stochastic cooling devices.

II.3 Damping Rate Adjustment

The damping re-partition and damping times are determined by the lattice design. However, insertion devices, such as undulators and wigglers, can be used to adjust beam characteristic parameters. We discuss below some techniques for damping rate adjustment.

A. Increase U to increase damping rate (damping wiggler)

Phase-space damping rates, apart from damping partition numbers, depend on radiation energy U_0 per turn. Wiggler magnets, which consist of strings of dipole magnets with alternate polarities excited so that the net deflection is zero, can be used to increase the radiation energy and thus enhance damping rate. The resulting energy loss per revolution and the damping rate become

$$U_{\rm w} = U_0 + U_{\rm wiggler}, \qquad \alpha_{\rm w} = \frac{U_{\rm w}}{2ET_0} = \alpha_0 + \alpha_{\rm wiggler}. \tag{4.93}$$

The damping time is shortened by a factor of $(1 + U_{\rm wiggler}/U_0)^{-1}$.

B. Change $\mathcal{D}$ to re-partition the partition number

Many early synchrotrons, such as the 8 GeV synchrotron (DESY) in Hamburg, the 28 GeV PS at CERN, the 33 GeV AGS at BNL, etc., used combined function isomagnetic magnets, where $\mathcal{D} \approx 2$ (see Exercise 4.2.1). Thus the energy oscillations are strongly damped ($\mathcal{J}_E \approx 4$) and the horizontal oscillations become anti-damped ($\mathcal{J}_x \approx -1$).

At the CERN PS, in facilitating the acceleration of e^+/e^- from 0.6 to 3.5 GeV as part of the LEP injection chain, horizontal emittance is an important issue. The growth time at 3.5 GeV is about 76 ms ($\mathcal{J}_x \approx -1$, $\rho = 70$ m), which is much shorter than the cycle time of 1.2 s. Stability of the electron beam can be achieved only by having a positive damping partition number, which can be facilitated by decreasing the orbit radius R.

The reason for the change in damping partition due to orbit radius variation is as follows. The potential for betatron motion in a quadrupole is

$$V_\beta = \frac{1}{2}K(s)\ (x^2 - z^2), \tag{4.94}$$

where $K = (1/B\rho)(\partial B_z/\partial x)$ is the focusing function; $K > 0$ for a focusing quadrupole, and $K < 0$ for a defocusing quadrupole. If the rf frequency is increased without changing the dipole field, the mean radius will move inward, and the change of radius ΔR is

$$\frac{\Delta f}{f} = -\frac{\Delta R}{R} = -\alpha_c \delta_s. \tag{4.95}$$

The actual closed orbit can be expressed as $x = x_{\rm co} + x_\beta$, where $x_{\rm co} < 0$ is a new closed orbit relative to the center of a quadrupole, and x_β is the betatron coordinate. The potential for betatron motion becomes

$$V_\beta = \frac{1}{2}K(s)\ (x_\beta^2 - z^2 + 2x_\beta x_{\rm co} + x_{\rm co}^2). \tag{4.96}$$

Since $x_{\rm co} < 0$, the effective dipole field $x_{\rm co}K(s)$ in a quadrupole and the quadrupole field have opposite signs, i.e. $(1/B_z)(\partial B_z/\partial x) < 0$. This is similar to the effect of a Robinson wiggler, discussed below. The combined effect is that the damping re-partition $\mathcal{D}$ will get a negative contribution from these quadrupoles. The effective dipole field arising from the closed orbit in a quadrupole is given by $B\rho K(x_\beta + x_{\rm co})$, where $B\rho$ is the momentum rigidity. Substituting the contribution of quantum excitation from the quadrupole into Eq. (4.85) gives the additional change of betatron amplitude in Eq. (4.86) as

$$\Delta\left(\frac{\delta A}{A}\right) = \frac{C_\gamma E^3}{2\pi} \oint K^2 D^2 ds\ \delta_s, \tag{4.97}$$

where we have used $x_{\rm co} = D\delta_s$, and the fractional off-momentum shift $\delta_s = -\Delta f/\alpha_c f$. The resulting change of damping re-partition is (see also Exercise 4.2.9)

$$\frac{\Delta\mathcal{D}}{\delta_s} = \left(2\int K^2 D^2 ds\right)\left[\oint \frac{1}{\rho^2} ds\right]^{-1}. \tag{4.98}$$

The CERN PS lattice is composed of $N_{\rm cell} = 50$ nearly identical combined function FODO cells with a mean radius of $R = 100$ m. Using Eq. (4.74) or Eq. (4.85) we get the change in damping partition due to closed orbit variation (see Exercise 4.2.9),

$$\Delta\mathcal{D} = \frac{\rho}{\pi}\oint D\left(\frac{B'}{B\rho}\right)^2 ds\Delta R \approx \frac{8N_{\rm cell}^2}{\pi^2 R}\Delta R. \tag{4.99}$$

Figure 4.8 shows $\mathcal{J}_x, \mathcal{J}_E$ vs ΔR for the CERN PS.

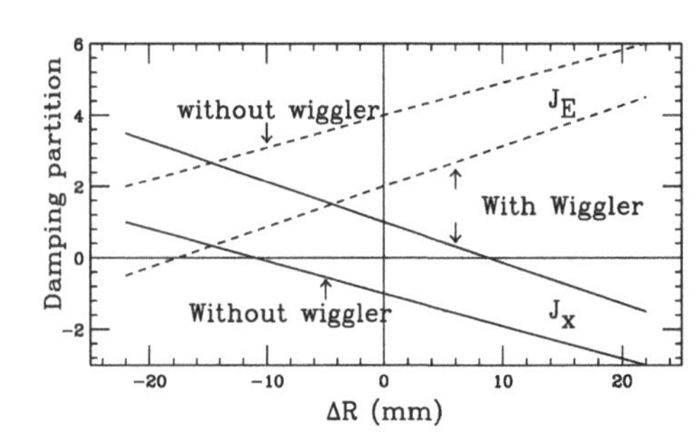

Figure 4.8: The variation of the damping partition number of the CERN PS with the strength of the Robinson wiggler. Without the Robinson wiggler, a fairly large change in ΔR is needed to attain $\mathcal{J}_x = 1$, with loss of useful aperture. From K. Hubner, CERN 85-19, p. 226 (1985).

C. Robinson wiggler

Without a Robinson wiggler, changing the damping re-partition requires a large shift of the mean orbiting radius (Fig. 4.8), and this limits the dynamical aperture of circulating beams. Thus, it is preferable to change the damping re-partition number by using the Robinson wiggler, which consists of a string of four identical magnet blocks having zero net dipole and quadrupole fields so that it will not produce global orbit and tune distortion in the machine. If the gradient and dipole field of each magnet satisfy $K\rho < 0$, as shown in Fig. 4.9, the damping re-partition of Eq. (4.74) can be made negative.

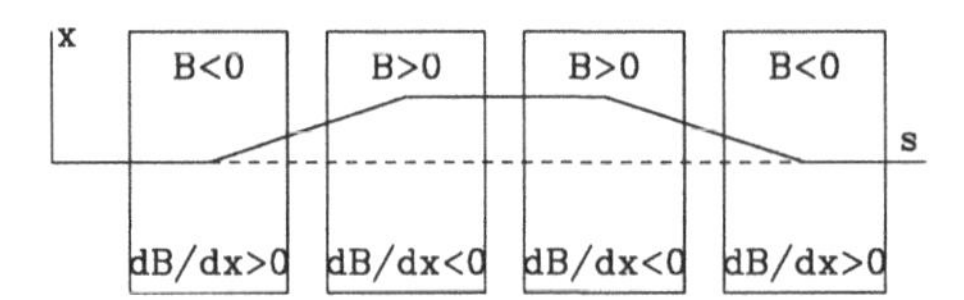

Figure 4.9: Schematic drawing of a Robinson wiggler, where gradient dipoles with $B\frac{dB}{dx} < 0$ are used to change the damping re-partition number.

Since these magnets have $K_{\rm w}\rho_{\rm w} < 0$, the wiggler contributes a negative term to the damping re-partition of Eq. (4.74). The change of damping re-partition is

$$\Delta\mathcal{D} = 2\langle D\rangle \frac{1}{B_{\rm w}}\frac{dB_{\rm w}}{dx}\frac{4L_{\rm w}\rho}{2\pi\rho_{\rm w}^2}\left(1+\frac{4L_{\rm w}\rho}{2\pi\rho_{\rm w}^2}\right)^{-1}, \tag{4.100}$$

where $\rho_{\rm w}, B_{\rm w}, dB_{\rm w}/dx$, and $L_{\rm w}$ are respectively the bending radius, the wiggler field strength, its derivative, and the length of each wiggler; and ρ is the bending radius of ring magnets and $\langle D\rangle$ the average dispersion function in wiggler locations. The Robinson wiggler has been successfully employed in the CERN PS to produce $\mathcal{J}_x \approx 2$, which enhances damping of horizontal emittance and reduces damping in energy oscillation. The resulting line density of beam bunches is likewise reduced to prevent collective instabilities.

II.4 Radiation Excitation and Equilibrium Energy Spread

Electromagnetic radiation is emitted in quanta of discrete energy. When a photon is emitted, the electron energy makes a small discontinuous jump. The emission time is short and thus the synchrotron radiation can be considered as instantaneous. This can be verified as follows. In a semi-classical picture, the time during which a quantum is emitted is about

$$\frac{\rho\Theta}{c} \approx \frac{\rho}{c\gamma} \approx \frac{6}{B[\text{Tesla}]} \times 10^{-12}\ \text{s},$$

where γ is the relativistic Lorentz factor, ρ is the radius of curvature, and B is the magnetic flux density. Since this time is very short compared with the revolution period and the periods of synchrotron and betatron oscillations, quantum emission can be considered instantaneous.

Another important feature of synchrotron radiation is that emission times of individual quanta are statistically independent. Since the energy of each photon [keV] is a very small fraction of electron energy, the emission of successive quanta is a purely random process, i.e., the probability of an electron emitting n photons per second is given by a Poisson distribution $f(n) = p^n e^{-p}/n!$. Here $p = \langle n\rangle$ is the average rate per second. The variance of Poisson distribution σ^2 is equal to p. In the limit of large p, Poisson distribution approaches Gaussian distribution, i.e. $P(n) = (1/\sqrt{2\pi p})e^{-(n-p)^2/2p}$.

Discontinuous quantized photon emission disturbs electron orbits. The cumulative effect of many such small disturbances introduces diffusion similar to random noise. The amplitude of oscillation will grow until the rates of quantum excitation and radiation damping are on the average balanced. The damping process depends only on the average rate of energy loss, whereas the quantum excitation fluctuates about its average rate.

A. Effects of quantum excitation

When a quantum of energy $\hbar\omega$ is emitted, the energy of the electron is suddenly decreased by an amount $\hbar\omega$. The impulse disturbance sets up a small energy oscillation. The cumulative effect of many such random disturbances causes energy oscillation to grow (as in a random walk). The growth is limited by damping.

In the absence of any disturbance and damping, the energy deviation ΔE from the synchronous energy, expressed in complex representation, is

$$\Delta E = A_0 e^{j\omega_s(t-t_0)}, \tag{4.101}$$

where A_0 is the amplitude of synchrotron motion, and ω_s is the synchrotron frequency. Now if the energy is suddenly decreased by an amount u at instant t_1 via quantum emission, the energy oscillation of the particle becomes

$$\Delta E = A_0 e^{j\omega_s(t-t_0)} - u e^{j\omega_s(t-t_1)} = A_1 e^{j\omega_s(t-t_1)}, \quad (t > t_1), \tag{4.102}$$

where

$$A_1^2 = A_0^2 + u^2 - 2A_0 u \cos\omega_s(t_1 - t_0). \tag{4.103}$$

The quantum emission has changed the *amplitude* of synchrotron oscillation. Since the time t_1 is unpredictable, the probable change in amplitude will be

$$\delta A^2 = \langle A^2 - A_0^2 \rangle_t = u^2, \tag{4.104}$$

where $\langle \ldots \rangle_t$ stands for time average. Qualitatively, the amplitude growth rate becomes

$$\langle \frac{dA^2}{dt} \rangle = \frac{d\langle A^2 \rangle}{dt} = \mathcal{N} u^2, \tag{4.105}$$

where $\mathcal{N}$ is the rate of photon emission.

B. Equilibrium rms energy spread

Since damping time of the amplitude A is $\tau_E = 1/\alpha_E$, as shown in Eq. (4.72), the damping time of A^2 is $\tau_E/2$. The equation for the synchrotron amplitude thus becomes

$$\frac{d\langle A^2 \rangle}{dt} = -2\frac{\langle A^2 \rangle}{\tau_E} + \mathcal{N} u^2, \tag{4.106}$$

where the stationary state solution is $\langle A^2 \rangle = \frac{1}{2}\mathcal{N} u^2 \tau_E$. A qualitative estimation of the rms beam energy spread for sinusoidal energy oscillation is

$$\sigma_E{}^2 = \frac{\langle A^2 \rangle}{2} = \frac{1}{4}\mathcal{N} u^2 \tau_E. \tag{4.107}$$

For an order of magnitude estimation, we use $u \approx \hbar\omega_c$, $\mathcal{N} \approx P_\gamma/\hbar\omega_c$, and $\tau_E \approx E/P_\gamma$ to obtain an rms energy oscillation amplitude of $\sigma_E \propto \sqrt{E\hbar\omega_c} \sim \gamma^2$. The energy fluctuation is roughly the geometric mean of electron energy and critical photon energy, and is proportional to γ^2. To attain a better calculation on the equilibrium beam momentum spread, the quantum fluctuation should be obtained from the sum of the entire frequency spectrum because the photon spectrum of synchrotron radiation is continuous.

Let $n(u)du$ be the photon density at energy between u and $u+du$. The amplitude growth rate due to quantum fluctuation becomes

$$\frac{d\langle A^2\rangle}{dt}=\int_0^\infty u^2 n(u)du=\mathcal{N}\langle u^2\rangle,\quad \mathcal{N}=\int_0^\infty n(u)du. \tag{4.108}$$

This shows that the amplitude growth rate depends on mean energy loss $\langle u^2\rangle$ of electrons, which depends on electron energy E and local radius of curvature ρ. Since the radius of curvature may vary widely along the ring, and damping time τ_E and synchrotron period $1/\omega_s$ are much longer than revolution period T_0, it is reasonable to average the excitation rate by averaging $\mathcal{N}\langle u^2\rangle$ over one revolution around the accelerator. We define the mean square energy fluctuation rate G_E as

$$G_E=\langle\mathcal{N}\langle u^2\rangle\rangle_s=\frac{1}{2\pi R}\oint\mathcal{N}\langle u^2\rangle ds, \tag{4.109}$$

where the subscript s indicates an average over the ring. The mean square equilibrium energy width becomes

$$\sigma_E^2=\frac{1}{4}G_E\tau_E. \tag{4.110}$$

On the design orbit, the radiated power is

$$P_\gamma\Big|_{\text{designed orbit}}=\frac{cC_\gamma}{2\pi}\frac{E^4}{\rho^2}=\frac{\langle P_\gamma\rangle}{\langle 1/\rho^2\rangle\rho^2}. \tag{4.111}$$

Equation (4.67) then gives

$$\mathcal{N}\langle u^2\rangle\Big|_{\text{designed orbit}}=\frac{3}{2}C_u\hbar c\frac{\gamma^3}{|\rho^3|}\frac{\langle P_\gamma\rangle}{\langle 1/\rho^2\rangle}, \tag{4.112}$$

where $C_u=55/24\sqrt{3}$. Using Eq. (4.109) and $\tau_E=2E/\mathcal{J}_E\langle P_\gamma\rangle$, we obtain

$$G_E=\frac{3}{2}C_u\hbar c\gamma^3\frac{\langle P_\gamma\rangle}{\langle 1/\rho^2\rangle}\langle 1/|\rho^3|\rangle \tag{4.113}$$

and

$$\sigma_E^2=\frac{3C_u\hbar mc^3\gamma^4}{4\mathcal{J}_E\langle 1/\rho^2\rangle}\langle 1/|\rho^3|\rangle.\qquad (\frac{\sigma_E}{E})^2=\frac{C_q\gamma^2}{\mathcal{J}_E\langle 1/\rho^2\rangle}\langle 1/|\rho^3|\rangle, \tag{4.114}$$

where

$$C_q=\frac{3C_u\hbar}{4mc}=\frac{55}{32\sqrt{3}}\frac{\hbar}{mc}=3.83\times 10^{-13}\ \text{m}. \tag{4.115}$$

For an isomagnetic ring, we obtain

$$(\frac{\sigma_E}{E})^2 = \frac{55}{48\sqrt{3}}\frac{\hbar\omega_c}{\mathcal{J}_E E} = C_q\frac{\gamma^2}{\mathcal{J}_E\rho} \quad \text{or} \quad \frac{\sigma_E}{E} \sim (0.62\times 10^{-6})\frac{\gamma}{\sqrt{\mathcal{J}_E\rho[\mathrm{m}]}}. \tag{4.116}$$

Note that the energy spread is independent of the rf voltage. For a bunch with a given momentum spread, the bunch length is shorter with higher rf voltage, and the resulting phase-space area is smaller; with lower rf voltage phase-space area is larger. In many electron storage rings, the bunch length is also affected by wakefields [5].

C. Adjustment of rms momentum spread

Insertion devices, such as undulators and wigglers, can change the rms energy spread of Eq. (4.110). Two competing effects determine the equilibrium energy spread. Insertion devices increase radiation power, which will increase quantum fluctuation G_E. Since the damping time is also shortened, the resulting equilibrium energy spread becomes

$$\sigma_E^2 = \sigma_{E0}^2\left(1+\frac{I_{3w}}{I_3}\right)\left(1+\frac{I_{2w}}{I_2}\right)^{-1}, \tag{4.117}$$

$$I_3 = \int\frac{1}{|\rho|^3}ds,\quad I_2 = \int\frac{1}{|\rho|^2}ds,\quad I_{3w} = \int\frac{1}{|\rho_{\rm w}|^3}ds,\quad I_{2w} = \int\frac{1}{|\rho_{\rm w}|^2}ds,$$

where I's are radiation integrals for ring dipoles and wigglers respectively. Because the magnetic field of insertion devices is usually larger than that of ring dipoles, i.e. $|\rho_{\rm w}| < \rho$, the rms energy spread will normally be increased by insertion devices.

D. Beam distribution function in momentum

The energy deviation ΔE at any instant t is a result of contributions from the emission of quanta at an earlier time t_i. We can write

$$\Delta E(t) = \sum u_i e^{-\alpha_E(t-t_i)}\cos\left[\omega_s(t-t_i)\right], \tag{4.118}$$

where u_i is the energy of a quantum emitted at time t_i. Since the typical value of $\Delta E(t)$ is much larger than the energy of each photon, and t_i's are randomly distributed, the sum at any time t consists of a large number of individual terms, which are positive and negative with equal probability. The central limit theorem (see Appendix A) implies that the distribution function of energy amplitude is Gaussian:

$$\Psi(\Delta E) = \frac{1}{\sqrt{2\pi}\sigma_E}e^{-\Delta E^2/2\sigma_E^2}, \tag{4.119}$$

where σ_E is the rms standard deviation. Normally the damping time is much longer than the synchrotron period, $2\pi/\omega_s$. For a particle executing synchrotron motion, we find the off-energy and the relative-time coordinate to a synchronous particle [see Eq. (4.69)] are

$$\Delta E(t) = A\cos(\omega_s t - \chi), \qquad \tau = \frac{\alpha_c A}{E\omega_s}\sin(\omega_s t - \chi), \tag{4.120}$$

where α_c is the momentum compaction factor, and χ is an arbitrary phase factor. The normalized phase-space coordinates are $(\Delta E, \theta = E\omega_s\tau/\alpha_c)$.

Since the normalized phase-space ellipse is a circle, the Gaussian distribution of a beam bunch is

$$\Psi(\Delta E, \theta) = N_B \Psi(\Delta E)\Psi(\theta), \tag{4.121}$$

where N_B is the number of particles in a bunch, and

$$\Psi(\theta) = \frac{1}{\sqrt{2\pi}\sigma_E} e^{-\theta^2/2\sigma_E^2}, \quad \Psi(\Delta E) = \frac{1}{\sqrt{2\pi}\sigma_E} e^{-(\Delta E)^2/2\sigma_E^2}.$$

The bunch length in time is $\sigma_\tau = \frac{\alpha_c}{E\omega_s}\sigma_E$, which depends on the rf voltage. We define the invariant amplitude $A^2 = \Delta E^2 + \theta^2$, and the distribution function becomes

$$g(A) = N\frac{A}{\sigma_E^2} e^{-A^2/2\sigma_E^2} = N\frac{2A}{\sigma_A^2} e^{-A^2/\sigma_A^2}. \tag{4.122}$$

where $\sigma_A^2 = \langle A^2 \rangle = 2\sigma_E^2$. Using the variable $W = A^2$ with $dW = 2AdA$, we get the probability distribution function as

$$h(W) = N\frac{1}{\langle W \rangle} e^{-W/\langle W \rangle}, \qquad \langle W \rangle = 2\sigma_E^2. \tag{4.123}$$

II.5 Radial Bunch Width and Distribution Function

Emission of discrete quanta in synchrotron radiation also excites random betatron motion. The emission of a quantum of energy u results in a change of betatron coordinates, i.e.

$$\delta x_\beta = -D\ (u/E_0), \qquad \delta x'_\beta = -D'\ (u/E_0). \tag{4.124}$$

The resulting change in the Courant-Snyder invariant is

$$\delta a^2 = \frac{2}{\beta_x}\left[Dx_\beta + (\beta_x D' - \frac{\beta'_x}{2}D)(\beta_x x' - \frac{\beta'_x}{2}x)\right]\frac{u}{E_0} + \frac{1}{\beta_x}\left[D^2 + (\beta_x D' - \frac{\beta'_x}{2}D)^2\right](\frac{u}{E_0})^2,$$

where β_x and β'_x are the horizontal betatron amplitude function and its derivative with respect to longitudinal coordinate s. Averaging betatron coordinates x_β, x'_β, the resulting amplitude growth becomes

$$\delta\langle a^2 \rangle = \mathcal{H}(\frac{u}{E_0})^2, \qquad \mathcal{H} = \frac{1}{\beta_x}\left[D^2 + (\beta_x D' - \frac{1}{2}\beta'_x D)^2\right], \tag{4.125}$$

where the $\mathcal{H}$-function depends on the lattice design. In an accelerator straight section, where there is no dipole, the $\mathcal{H}$-function is invariant; it is not invariant in regions with dipoles. The rate of change of betatron amplitude (emittance) is obtained by replacing u^2 with $\mathcal{N}\langle u^2\rangle$ and averaging over the accelerator, i.e.

$$\frac{d\langle a^2\rangle}{dt} \equiv G_x = \frac{1}{2\pi RE^2}\oint \mathcal{N}\langle u^2\rangle \mathcal{H} ds = \frac{\langle \mathcal{N}\langle u^2\rangle \mathcal{H}\rangle_s}{E^2}, \tag{4.126}$$

where $\langle\cdots\rangle_s$ stands for an average over a complete revolution. The emittance growth in a transport line is $d\epsilon/dt = \frac{1}{cE^2}\int_0^s \mathcal{N}\langle u^2\rangle \mathcal{H} ds$. Adding the damping term of Eq. (4.81), we obtain

$$\frac{d\langle a^2\rangle}{dt} = -2\frac{\langle a^2\rangle}{\tau_x} + G_x. \tag{4.127}$$

The equilibrium rms width becomes

$$\langle a^2\rangle = \frac{1}{2}\tau_x G_x \quad \text{and} \quad \sigma^2_{x\beta_x} = \frac{1}{2}\beta_x\langle a^2\rangle. \tag{4.128}$$

Using Eq. (4.67) for $\mathcal{N}\langle u^2\rangle$, we obtain

$$\begin{aligned} G_x &= \frac{3}{2}C_u\hbar c\gamma^3\frac{\langle P_\gamma\rangle\langle\mathcal{H}/|\rho|^3\rangle}{E^2\langle 1/\rho^2\rangle} = \frac{3C_q c r_0\gamma^5\langle\mathcal{H}/|\rho^3|\rangle}{3\langle\rho^2\rangle\langle 1/\rho^2\rangle}, \\ \epsilon_x &= \frac{\sigma^2_{x\beta_x}}{\beta_x} = \frac{1}{4}\tau_x G_x = C_q\frac{\gamma^2\langle\mathcal{H}/|\rho|^3\rangle}{\mathcal{J}_x\langle 1/\rho^2\rangle} \xrightarrow{\text{isomagnetic}} C_q\frac{\gamma^2\langle\mathcal{H}\rangle_{\text{mag}}}{\mathcal{J}_x\rho}, \end{aligned} \tag{4.129}$$

where $C_q = 3.83\times 10^{-13}$ m is given by Eq. (4.115) and $\langle\mathcal{H}\rangle_{\text{mag}} = \frac{1}{2\pi\rho}\int_{\text{dipole}}\mathcal{H}ds$ is the average $\mathcal{H}$-function in isomagnetic dipoles. The emittance of Eq. (4.129) is called the *natural emittance*. Since the $\mathcal{H}$-function is proportional to $L\theta^2 \sim \rho\theta^3$, where θ is the dipole angle of a half cell, the natural emittance of an electron storage ring is proportional to $\gamma^2\theta^3$. The normalized emittance is proportional to $\gamma^3\theta^3$. Unless the orbital angle of each dipole is inversely proportional to γ, the normalized natural emittance of an electron storage ring increases with energy. Comparing with the energy width for the isomagnetic ring, we find

$$\frac{\sigma^2_{x\beta_x}}{\beta_x} = \frac{\mathcal{J}_E\langle\mathcal{H}\rangle_{\text{mag}}}{\mathcal{J}_x}\left(\frac{\sigma_E}{E}\right)^2. \tag{4.130}$$

The horizontal distribution function

The distribution function for particles experiencing uncorrelated random forces with zero average in a simple harmonic potential well is Gaussian:

$$\Psi(x_\beta) = \frac{1}{\sqrt{2\pi}\sigma_{x\beta_x}}\exp\left\{-\frac{x_\beta^2}{2\sigma^2_{x\beta_x}}\right\}. \tag{4.131}$$

Since the betatron oscillation period is much shorter than the damping time, the distribution in phase-space coordinates follows the Courant-Snyder invariant

$$\Psi(x_\beta, x'_\beta) = \frac{1}{\sqrt{2\pi}\sigma_{x\beta_x}} \exp\left\{-\frac{x_\beta^2 + (\beta_x x'_\beta - (\beta'_x/2)x_\beta)^2}{2\sigma^2_{x\beta_x}}\right\}. \tag{4.132}$$

The total radial beam width has contributions from both betatron and energy oscillations. The rms beam width is Gaussian quadrature

$$\sigma_x^2 = \sigma^2_{x\beta_x} + \sigma^2_{x\epsilon} \xrightarrow{\text{isomagnetic}} C_q\frac{\gamma^2}{\rho}\left[\frac{\beta_x(s)\langle\mathcal{H}\rangle_{\rm mag}}{\mathcal{J}_x} + \frac{D^2(s)}{\mathcal{J}_E}\right]. \tag{4.133}$$

II.6 Vertical Beam Width

Synchrotron radiation is emitted in the forward direction within a cone of angular width $1/\gamma$. When the electron emits a photon at a nonzero angle with respect to its direction of motion, it experiences a small transverse impulse. Consider the emission of a photon with momentum u/c at angle θ_γ from the electron direction of motion, where we expect $\theta_\gamma \le 1/\gamma$. The transverse kick is then equal to $\theta_\gamma u/c$. The transverse angular kicks on phase-space coordinates become

$$\delta x = 0,\ \delta x' = \frac{u}{E_0}\theta_x, \quad \delta z = 0,\ \delta z' = \frac{u}{E_0}\theta_z, \tag{4.134}$$

where θ_x, θ_z are projections of θ_γ onto x, z axes respectively. Since $\delta x'$ is small compared with that of Eq. (4.124), we neglect it. We consider only the effect of random kick on vertical betatron motion. Emission of a single photon with energy u gives rise to an average change of invariant betatron emittance $\delta\langle a_z^2\rangle = (u/E_0)^2\theta_z^2\beta_z$. Including both damping and quantum fluctuation, the equilibrium beam width is

$$\sigma^2_{z\beta} = \frac{1}{4}\tau_z G_z \beta_z, \tag{4.135}$$

$$G_z = \frac{\langle\mathcal{N}\langle u^2\theta_z^2\rangle\beta_z\rangle_s}{E^2} \approx \frac{\langle\mathcal{N}\langle u^2\rangle\langle\theta_z^2\rangle\beta_z\rangle_s}{E^2} \approx \frac{\langle\mathcal{N}\langle u^2\rangle\rangle\langle\beta_z\rangle}{\gamma^2E^2}, \tag{4.136}$$

where we have used the fact that $\langle\theta_z^2\rangle \sim 1/\gamma^2$. Recalling that $\langle\mathcal{N}\langle u^2\rangle\rangle = G_E$, we obtain

$$\frac{\sigma_z^2}{\sigma_E^2} = \frac{\tau_z G_z \beta_z}{\tau_E G_E} \approx \frac{\mathcal{J}_E\langle\beta_z^2\rangle}{\mathcal{J}_z\gamma^2E^2}. \tag{4.137}$$

Using $\mathcal{J}_z = 1$, we obtain

$$\sigma_z^2 \approx C_q\langle\beta_z^2\rangle/\rho \quad \text{or} \quad \epsilon_z \approx C_q\langle\beta_z^2\rangle^{1/2}/\rho, \tag{4.138}$$

which is very small. Thus the vertical oscillation is energy independent and is less than the radial oscillation by a factor of $1/\gamma^2$. The vertical beam size is damped almost to zero.

Emittance in the presence of linear coupling

Sometimes it is desirable to introduce intentional horizontal and vertical betatron coupling. When the coupling is introduced, the quantum excitation is shared up to an equal division. Let ϵ_x and ϵ_z be the horizontal and vertical emittances with

$$\epsilon_x + \epsilon_z = \epsilon_{\rm nat}, \tag{4.139}$$

where the natural emittance $\epsilon_{\rm nat}$ is Eq. (4.129). The horizontal and vertical emittances can be redistributed with appropriate linear betatron coupling

$$\epsilon_x = \frac{1}{1+\kappa}\epsilon_{\rm nat}, \quad \epsilon_z = \frac{\kappa}{1+\kappa}\epsilon_{\rm nat}, \tag{4.140}$$

where the coupling coefficient κ is (see Exercise 4.2.8).

II.7 Beam Lifetime

We have used a Gaussian distribution function for the electron beam distribution function. Since the aperture of an accelerator is limited by accelerator components such as vacuum chambers, injection or extraction kickers, beam position monitors, etc., the Gaussian distribution, which has an infinitely long tail, is only an ideal representation when the aperture is much larger than the rms width of the beam so that particle loss is small.

A. Quantum lifetime

Even when the aperture is large, electrons, which suffer sufficient energy fluctuation through quantum emission, can produce a radial displacement as large as the aperture. If the chance of an electron being lost at the aperture limit, within its damping time, is small, then the loss probability per unit time is the same for all electrons. The loss rate becomes

$$\frac{1}{N}\frac{dN}{dt} = -\frac{1}{\tau_q}, \tag{4.141}$$

where τ_q is the quantum lifetime. We discuss quantum lifetime for radial and longitudinal motion below.

Radial oscillation

We consider radial betatron oscillation $x = a\cos\omega_\beta t$. The invariant amplitude of betatron motion is $W = a^2$. Quantum excitation and radiation damping produce an equilibrium distribution given by

$$h(W) = \frac{1}{\langle W\rangle}e^{-W/\langle W\rangle}, \quad \langle W\rangle = 2\sigma_x^2. \tag{4.142}$$

To estimate beam lifetime, we set up a diffusion equation for $h(W)$. We assume an equilibrium distribution without aperture limit and consider an electron at amplitude W_0, with $W_0 \gg \langle W \rangle$, so that the probability for the electron to have $W > W_0$ is small. Once the electron gets into the tail region ($W > W_0$) of the distribution, it is most likely to return to the main body of the distribution because of faster damping at large amplitude, i.e.

$$\frac{dW}{dt} = -\frac{2W}{\tau_x}. \tag{4.143}$$

The flux inward through W_0 due to damping is

$$Nh(W)\frac{dW}{dt}\bigg|_{W_0} = \frac{2NW_0 h(W_0)}{\tau_x}. \tag{4.144}$$

In a stationary state, an equal flux of electron passes inward and outward through W_0, i.e.

$$-\frac{dN}{dt} = \frac{2N\hat{W}h(W)}{\tau_x} = N\frac{2\hat{W}}{\tau_x\langle W\rangle}e^{-\hat{W}/\langle W\rangle}, \tag{4.145}$$

where W_0 has been replaced by $\hat{W}$. Thus the quantum lifetime is

$$\tau_q = \frac{\tau_x}{2\xi}\,e^{\xi}, \qquad \xi = \frac{\hat{W}}{\langle W\rangle}. \tag{4.146}$$

Note that the formula is valid only in a weakly damping system [See M. Bai *et al.*, *Phys. Rev.* E **55**, 3493 (1997)].

Synchrotron oscillations

For synchrotron motion, the aperture is limited by rf voltage and bucket area. The Hamiltonian of synchrotron motion is

$$H(\delta,\phi) = \frac{1}{2}h\omega\alpha_c\delta^2 + \frac{\omega eV_0}{2\pi E}\left[\cos\phi - \cos\phi_s + (\phi-\phi_s)\sin\phi_s\right], \tag{4.147}$$

where $\delta = \Delta p/p = \Delta E/E$, $\phi = hc\tau$, and h is the harmonic number. If the nonlinear term in the momentum compaction factor is negligible and the synchrotron tune differs substantially from zero, the Hamiltonian is invariant.

The Hamiltonian has two fixed points, $(0,\phi_s)$ and $(0,\pi-\phi_s)$. The value of the Hamiltonian at the separatrix is

$$H_{sx} = H(0,\pi-\phi_s) = \frac{\omega_0 eV_0}{\pi E}\left[-\cos\phi_s + \left(\frac{\pi}{2}-\phi_s\right)\sin\phi_s\right]. \tag{4.148}$$

The stable rf phase angle ϕ_s is determined by the energy loss due to synchrotron radiation with $eV_0\sin\phi_s = U_0 = C_\gamma E^4/\rho$. From Eq. (4.146), the quantum lifetime is

$$\tau_q = \frac{\tau_E}{2\xi}\,e^{\xi}, \tag{4.149}$$

where $\xi = H_{\rm sx}/\langle H\rangle$, $\langle H\rangle = h\omega_0\alpha_{\rm c}(\sigma_E/E)^2)$ is the average value of the Hamiltonian of the beam distribution, and

$$\xi = \frac{48\sqrt{3}}{55\pi h\alpha}\mathcal{J}_E\frac{eV_0}{u_c}\left[-\cos\phi_{\rm s} + \left(\frac{\pi}{2}-\phi_{\rm s}\right)\sin\phi_{\rm s}\right]. \tag{4.150}$$

B. Touschek lifetime

In the beam moving frame, the deviation of the momentum $\Delta p_{\rm b}$ of a particle from that of the synchronous particle, which has zero momentum, is related to the momentum deviation in the laboratory frame Δp by

$$\Delta p_{\rm b} = \Delta p/\gamma. \tag{4.151}$$

Thus the momentum deviation in the rest frame of the beam is reduced by the relativistic factor γ. Because of synchrotron radiation damping and quantum fluctuation in the horizontal plane, the rms beam velocity spreads in the beam moving frame satisfy the characteristic property

$$\langle (x'_\beta)^2\rangle^{1/2} \gg \langle (z'_\beta)^2\rangle^{1/2} \approx \langle (\Delta p_{\rm b}/p_0)^2\rangle^{1/2}, \tag{4.152}$$

where x_β and z_β are betatron coordinates; $x'_\beta = dx_\beta/ds$, $z'_\beta = dz_\beta/ds$ are the slopes of the horizontal and vertical betatron oscillations; and p_0 is the momentum of a synchronous particle. Since the transverse horizontal momentum spread of the beam is much larger than the momentum spread of the beam in the longitudinal plane, large angle Coulomb scattering can transfer the radial momentum to the longitudinal plane and cause beam loss. This process was first pointed out by Touschek *et al.* in the Frascati e^+e^- storage ring (AdA) [see e.g. C. Bernardini et al., *Phys. Rev. Lett.* **10**, 407 (1963)]. The Touschek effect has been found to be important in many low emittance synchrotron radiation facilities.

We consider the Coulomb scattering of two particles in their center of mass system (CMS) with momentum $\vec{p}_{1,\rm init} = (p_x, 0, 0)$ and $\vec{p}_{2,\rm init} = (-p_x, 0, 0)$, where the momenta are expressed in the $\hat{x}, \hat{s}$, and $\hat{z}$ base vectors. The velocity difference between two particles in the CMS is

$$v = 2p_x/m. \tag{4.153}$$

Since the transverse radial momentum component of the orbiting particle is much larger than the transverse vertical and longitudinal components, we assume that the initial particle momenta of scattering particles are only in the horizontal direction. In the spherical coordinate system, the differential cross-section is given by the Möller formula,

$$\frac{d\sigma}{d\Omega} = \frac{4r_0^2}{(v/c)^4}\left[\frac{4}{\sin^4\theta} - \frac{3}{\sin^2\theta}\right], \tag{4.154}$$

where r_0 is the classical electron radius. Let χ be the angle between the momentum $\vec{p}_{1,\rm scatt}$ of a scattered particle and the s-axis, and let φ be the angle between the x-axis and the projection of the momentum of the scattered particle onto the x-z plane, as shown in Fig. 4.10.

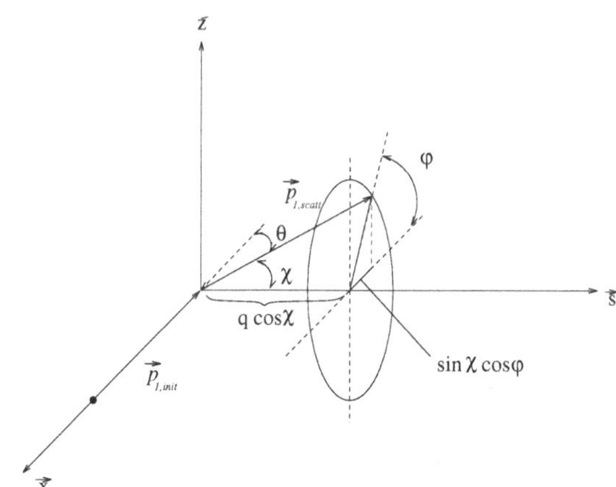

Figure 4.10: The schematic geometry of Touschek scattering, which transfers horizontal momentum into longitudinal momentum in the center of mass frame of scattering particles. We use $\vec{x}, \vec{s}$, and $\vec{z}$ as orthonormal curvilinear coordinate system. Particle loss resulting from large angle Coulomb scattering gives rise to the Touschek lifetime, which becomes a limiting factor for high brightness electron storage rings.

With the geometry shown in Fig. 4.10, the momentum of a scattered particle is

$$\vec{p}_{1,\rm scatt} = (p_x \sin\chi\cos\varphi,\ p_x\cos\chi,\ p_x\sin\chi\sin\varphi), \tag{4.155}$$

where the momentum of the other scattered particle is $-\vec{p}_{1,\rm scatt}$. The scattering angle θ is related to χ and φ by $\cos\theta = \sin\chi\cos\varphi$, and the momentum transfer to the longitudinal plane in the CMS is $\Delta p_{\rm cms} = p_x|\cos\chi|$. Now we assume that the scattered particles will be lost if the scattered longitudinal momentum is larger than the momentum aperture, i.e.

$$|\cos\chi| \ge \frac{\hat{\Delta p}}{\gamma p_x} \tag{4.156}$$

where $\hat{\Delta p} = (2\nu_{\rm s}/h\alpha_{\rm c})Y(\phi_{\rm s})$ is the rf bucket height. Thus the total cross-section leading to particle loss in the CMS is $\sigma_{\rm T} = \int_{|\cos\chi|\ge\hat{\Delta p}/\gamma p_x}(d\sigma/d\Omega)d\Omega$, i.e.

$$\begin{aligned}\sigma_{\rm T} &= \frac{4r_0^2}{(v/c)^4}\int_0^{\cos^{-1}(\hat{\Delta p}/\gamma p_x)}\sin\chi d\chi\int_0^{2\pi}d\varphi\left[\frac{4}{(1-\sin^2\chi\cos^2\varphi)^2}-\frac{3}{1-\sin^2\chi\cos^2\varphi}\right]\\ &= \frac{8\pi r_0^2}{(v/c)^4}\left[\frac{\gamma^2p_x^2}{(\hat{\Delta p})^2}-1+\ln\frac{\hat{\Delta p}}{\gamma p_x}\right].\end{aligned} \tag{4.157}$$

The number of particles lost by Touschek scattering in the CMS becomes

$$dN = 2\sigma_{\rm T}\,N\,n\,dx, \tag{4.158}$$

where n is the density of the beam bunch, ndx is the target thickness, N is the total number of particles in the bunch, and the factor 2 indicates that two particles are lost in each Touschek scattering. Thus the loss rate in the CMS is $dN/dt = 2\int \sigma_{\rm T} v n^2 dV$, where dV is the volume element, and $v = dx/dt$.

In the laboratory frame, the Touschek loss rate becomes

$$\frac{dN}{dt} = \frac{2}{\gamma^2}\int v\sigma_{\rm T} n^2 dV, \tag{4.159}$$

where the factor $1/\gamma^2$ takes into account the Lorentz transformation of $\sigma_{\rm T} v$ from the CMS to the laboratory frame. Since Touschek scattering takes place only in the horizontal plane, the vertical and longitudinal planes can be integrated easily, and the Touschek loss rate becomes

$$\frac{dN}{dt} = 2\frac{N^2}{\gamma^2}\frac{1}{4\pi\sigma_z\sigma_s}\int v\sigma_{\rm T}\rho(x_1, x_1')\rho(x_2, x_2')\delta(x_1 - x_2)dx_1 dx_1' dx_2 dx_2', \tag{4.160}$$

where σ_z and σ_s are respectively the rms bunch height and bunch length, and the function $\delta(x_1 - x_2)$ indicates that the scattering process takes place in a short range between two particles. For Gaussian longitudinal and vertical distributions, the integrals of the vertical and longitudinal planes are respectively $(2\sqrt{\pi}\sigma_z)^{-1}$ and $(2\sqrt{\pi}\sigma_s)^{-1}$. Using the Gaussian horizontal density function

$$\rho(x, x') = \frac{\beta_x}{2\pi\sigma_x^2}\exp\left[-\frac{1}{2\sigma_x^2}\left(x^2 + (\beta_x x' - \frac{\beta_x'}{2}x)^2\right)\right], \tag{4.161}$$

we easily integrate the integral of Eq. (4.160) to obtain

$$\frac{1}{N}\frac{dN}{dt} = \frac{N r_0^2 c}{8\gamma^2\pi\sigma_x\sigma_z\sigma_s}\left(\frac{\gamma mc}{\hat{\Delta p}}\right)^3 D(\xi), \tag{4.162}$$

where $\xi = (\hat{\Delta p}/\gamma\sigma_{p_x})^2 = (\beta_x\hat{\Delta p}/\gamma^2 mc\sigma_x)^2$; $\sigma_{p_x} = \gamma mc\sigma_x/\beta_x$; and

$$D(\xi) = \sqrt{\xi}\int_0^\infty \frac{1}{(u+\xi)^2}\left[u - \frac{1}{2}\xi\ \ln\frac{u+\xi}{\xi}\right]e^{-(u+\xi)}du. \tag{4.163}$$

The Touschek loss rate is inversely proportional to the 3D volume $\sigma_x\sigma_z\sigma_s$.

With typical parameters $\Delta p \approx \sigma_p$; $\sigma_p/p = \sqrt{C_q\gamma}/\sqrt{\mathcal{J}_E\rho}$ [see Eq. (4.114)]; $\sigma_x = \sqrt{\beta_x\epsilon_x}$; and $\epsilon_x = \mathcal{F}C_q\gamma^2\theta^3/\mathcal{J}_x$ [see Eq. (4.167) Sec. III]; the parameter ξ is

$$\xi \approx \frac{10}{\gamma\,\theta^{3/2}}\sqrt{\frac{\beta_x\mathcal{J}_x}{\rho\mathcal{F}\mathcal{J}_E}}, \tag{4.164}$$

where $\mathcal{J}_x$ and $\mathcal{J}_E$ are the damping partition numbers, ρ is the bending radius, $\mathcal{F}$ is the lattice dependent factor, and θ is the orbital bending angle in one half period.

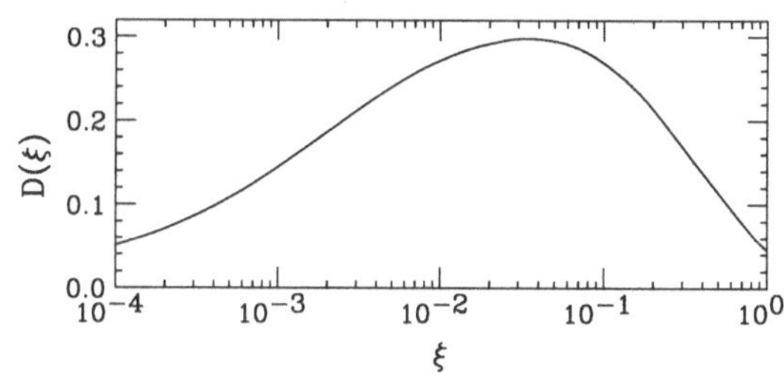

Figure 4.11: The Touschek integral $D(\xi)$ of Eq. (4.163).

Thus the typical ξ parameter for Touschek scattering is about 10^{-3} to 1. In this parameter region, $D(\xi)$ is a slow varying function of the parameter ξ (see Fig. 4.11).

In a low emittance storage ring, the betatron amplitude function can change appreciably. The actual Touschek scattering rate should be averaged over the entire ring, i.e.

$$\frac{1}{\tau_{\mathrm{T}}} = \left\langle \frac{1}{N}\frac{dN}{dt} \right\rangle_s = \frac{1}{2\pi R}\oint \frac{1}{N}\frac{dN}{dt} ds. \tag{4.165}$$

Since $D(\xi)$ a a slowly varying function, we can approximate $\langle D(\xi)\rangle = 1/6$ to obtain

$$\tau_{\mathrm{T}} \approx \frac{48\gamma^2\pi\sigma_x\sigma_z\sigma_s}{N r_0^2 c}\left(\frac{\hat{\Delta p}}{p}\right)^3. \tag{4.166}$$

The Touschek lifetime is a complicated function of machine parameters. It can be affected by linear coupling, rf parameters, peak intensity, etc. If we choose $\hat{\Delta p} \approx 10\sigma_p$, i.e. the rf voltage increases with energy with $V_{\mathrm{rf}} \propto \gamma^2$, we obtain $\tau_{\mathrm{T}} \propto \gamma^6$, and at a fixed energy the Touschek lifetime is proportional to V_{rf} because $\sigma_s \propto V_{\mathrm{rf}}^{-1/2}$. Actual calculation of Touschek lifetime should include the effect of the dispersion function. See, e.g., J. LeDuff, CERN 89-01, p. 114 (1989). Touschek lifetime calculation is available in MAD [23] and other optics codes. The beam current in many high brightness synchrotron radiation light sources is limited by the Touschek lifetime.

II.8 Summary: Radiation Integrals

To summarize the properties of electron beams, we list radiation integrals in the left column and the corresponding physical quantities in the right column. Here $\langle S\rangle$ is the spin polarization, $P_{\mathrm{ST}} = -8/5\sqrt{3}$ is the Sokolov-Ternov radiative polarization limit, C_γ and C_q are respectively given by Eqs. (4.5) and (4.115).

$I_1 = \int \frac{D}{\rho}\, ds$	$\alpha_c = \frac{I_1}{2\pi R}$
$I_2 = \int \frac{1}{\rho^2}\, ds$	$U_0 = \frac{C_\gamma}{2\pi} E^4 I_2$
$I_3 = \int \frac{1}{\lvert\rho\rvert^3}\, ds$	$\left(\frac{\sigma_E}{E}\right)^2 = \frac{C_q \gamma^2 I_3}{2I_2 + I_4} = \frac{C_q \gamma^2 I_3}{2I_2 \mathcal{J}_E}$
$I_{3a} = \int \frac{1}{\rho^3}\, ds$	$\langle S \rangle = P_{\rm ST} \frac{I_{3a}}{I_3}$
$I_4 = \int \frac{D}{\rho}\left(\frac{1}{\rho^2} + 2K\right) ds$	$\mathcal{D} = \frac{I_4}{I_2}; \quad \mathcal{J}_x = 1 - \mathcal{D}, \quad \mathcal{J}_E = 2 + \mathcal{D}, \quad \mathcal{J}_z = 1$
$K = \frac{1}{B\rho}\frac{\partial B_z}{\partial x}$	$\alpha_x = \frac{U_0}{2T_0 E}\mathcal{J}_x, \quad \tau_x = \frac{2E}{\mathcal{J}_x U_0} T_0,$
	$\alpha_z = \frac{U_0}{2T_0 E}\mathcal{J}_z, \quad \tau_z = \frac{2E}{\mathcal{J}_z U_0} T_0,$
	$\alpha_E = \frac{U_0}{2T_0 E}\mathcal{J}_E, \quad \tau_E = \frac{2E}{\mathcal{J}_E U_0} T_0,$
$I_5 = \int \frac{\mathcal{H}}{\lvert\rho\rvert^3}\, ds$	$\epsilon_x = \frac{C_q \gamma^2 I_5}{I_2 - I_4} = \frac{C_q \gamma^2 I_5}{I_2 \mathcal{J}_x}$

Exercise 4.2

1. Show that the damping partition number is $\mathcal{D} = 2 - (\alpha_c R/\rho)$ for an isomagnetic combined function lattice, and $\mathcal{D} = (\alpha_c R/\rho)$ for an isomagnetic separate function lattice with sector magnets.

   (a) In thin-lens approximation, show that the damping re-partition number for an isomagnetic combined function accelerator made of N FODO cells is given by

   $$\mathcal{D} \approx 2 - \frac{R}{\rho}\left(\frac{2\pi}{2N \sin(\Phi/2)}\right)^2,$$

   where R is the mean radius of the accelerator, ρ is the bending radius of the dipole, and Φ is the phase advance per cell.

   (b) Show that the damping re-partition number for a separated function double bend achromat with sector dipoles [see Eq. (2.195)] is

   $$\mathcal{D} = \left(1 - \frac{\sin(\pi/N)}{(\pi/N)}\right) \approx \frac{\theta^2}{6},$$

   where N is the number of DBA cells for the entire lattice, and θ is the bending angle of a half DBA cell. The damping re-partition number of DBA lattices is independent of the betatron tunes.

(c) Use the midpoint rule to evaluate the integral of the damping re-partition $\mathcal{D}$, and show that the damping re-partition number for the separate function FODO cell lattice is

$$\mathcal{D} \approx \frac{R\,\theta^2}{\rho\,\sin^2(\Phi/2)},$$

where R is the average radius of the ring, ρ is the bending radius of the dipoles, θ is the bending angle of a half FODO cell, and Φ is the phase advance per cell.

2. The damping partition number $\mathcal{D}$ for energy spread and natural emittance is given by $\mathcal{D} = (I_{4a} + 2I_{4b})/I_2$, where the radiation integrals are

$$I_{4a} = \oint \frac{D}{\rho^3} ds, \qquad I_{4b} = \oint K\frac{D}{\rho} ds, \qquad I_2 = \int \frac{1}{\rho^2} ds.$$

Here ρ is the bending radius, D is the dispersion function, and $K = (1/B\rho)\partial B_z/\partial x$ is the quadrupole gradient function.

(a) For a separate function isomagnetic machine with sector dipoles, show that $I_{4a} = 2\pi\alpha_c R/\rho^2$ and $I_{4b} = 0$, where α_c is the momentum compaction factor and R is the average radius of the synchrotron.

(b) Show that the contribution from the edge angles of a non-sector type magnet to the integral I_{4b} is [see e.g. R.H. Helm, M.J. Lee, P.L. Morton, and M. Sands, *IEEE Trans. on Nucl. Sci.* **NS-20**, 900 (1973)]

$$I_{4b} = D_1\frac{\tan\delta_1}{2\rho^2} + D_2\frac{\tan\delta_2}{2\rho^2},$$

where δ_1 and δ_2 are entrance and exit angles of the beam, D_1 and D_2 are values of the dispersion function at the entrance and exit of the dipole with $D_2 = (1 - \cos\theta)\rho + D_1\cos\theta + (\rho D_1' + D_1\tan\delta_1)\sin\theta$.

3. The beam energy spread of a collider should be of the order of the width of the resonance in the energy region of interest. For example, $\Gamma(J/\psi : 3100) = 0.063$ MeV and $\Gamma(\psi' : 3685) = 0.215$ MeV. The rms beam energy spread is given by Eq. (4.114). Show that

$$\sigma_E[\mathrm{MeV}] = 1.21\frac{E^2[\mathrm{GeV}]}{\sqrt{\mathcal{J}_E\rho[\mathrm{m}]}}.$$

For a SPEAR-like ring, with $\rho = 12$ m and $\mathcal{J}_E \approx 2$, find the energy spread at the J/ψ and ψ' energies. Note that, when the energy spread is large, the production rate is reduced by a factor of Γ/σ_E.

4. From the previous problem, we learn that the beam energy spread can reduce the effective reaction rate. Now imagine that you want to design an interaction region (IR) such that the higher energy electrons will collide with lower energy positrons or vice versa. What is the constraint of the IR design such that the total center of mass energies for all electron-positron pairs are identical? Discuss possible difficulties. Discuss your result.

5. Verify Eq. (4.125) for the change of betatron amplitude in photon emission.

6. Show that the vertical emittance resulting from residual vertical dispersion is given by
$$\epsilon_z = C_q\gamma^2 \frac{\langle \mathcal{H}_z/|\rho|^3\rangle}{\mathcal{J}_z\langle 1/\rho^2\rangle},$$
where
$$\mathcal{H}_z = \frac{1}{\beta_z}[D_z^2 + (\beta_z D_z' + \alpha_z D_z)^2],$$
β_z and α_z are vertical betatron amplitude functions, and D_z and D_z' are the residual vertical dispersion function and its derivative with respective to s. Make a realistic estimate of the magnitude of the vertical emittance arising from the residual vertical dispersion.

7. Near a betatron coupling resonance, the horizontal action of each particle can interchange with its vertical action, while the total action of the particle is conserved, as shown in Eq. (2.221). Use the following model to find emittances of electron storage rings. The equation of motion for emittance of an electron storage ring near a linear coupling resonance is
$$\begin{aligned}\frac{d\epsilon_x}{dt} &= -C(\epsilon_x - \epsilon_z) - \alpha_x(\epsilon_x - \epsilon_0),\\ \frac{d\epsilon_z}{dt} &= -C(\epsilon_z - \epsilon_x) - \alpha_z\epsilon_z,\end{aligned}$$
where α_x, α_z are damping rates, ϵ_0 is the natural emittance, and C is the linear coupling constant.

   (a) Show that the equations of motion for horizontal and vertical emittances are
$$\begin{aligned}\frac{d^2\epsilon_x}{dt^2} + (\alpha_x + \alpha_z + 2C)\frac{d\epsilon_x}{dt} + [\alpha_x\alpha_z + C(\alpha_x + \alpha_z)]\epsilon_x &= \alpha_x(\alpha_z + C)\epsilon_0,\\ \frac{d^2\epsilon_z}{dt^2} + (\alpha_z + \alpha_z + 2C)\frac{d\epsilon_z}{dt} + [\alpha_x\alpha_z + C(\alpha_x + \alpha_z)]\epsilon_z &= \alpha_x C\epsilon_0.\end{aligned}$$
   Find the equilibrium emittances.

   (b) For $\alpha_x = \alpha_z = \alpha$, show that the emittance can be expressed by Eq. (4.140) where the κ parameter is given by
$$\kappa = \frac{C}{\alpha + C}.$$

8. The damping partition $\mathcal{D}$ can be decreased by moving the particle orbit inward. Use the following steps to derive the expression for $\Delta\mathcal{D}/\Delta R$.

   (a) The synchrotron radiation power is
$$P_\gamma = \frac{c^3 C_\gamma e^2}{2\pi} E^2 B^2.$$
   If the rf frequency is altered, the average radius and the beam energy are changed by $\Delta R/R = -\Delta f/f_0$ and $E_0 + \delta_e$, and the magnetic field can be expanded as

$B = B_0 + B'x_{co} + B'x_\beta$. Using Eq. (4.86) show that the average rate of betatron amplitude diffusion per revolution is

$$\frac{\delta A}{A} = \frac{c^3 C_\gamma e^2}{4\pi c E}(E_0 + \delta_e)^2 \oint D \left[\frac{(B_0 + B'x_{co})^2}{\rho} + 2B'B_0 + 2B'^2 x_{co}\right] ds.$$

(b) Show that the change in damping partition with respect to x_{co} is

$$\frac{\Delta \mathcal{D}}{\Delta x_{co}} \approx \left(2 \oint D \left(\frac{B'}{B\rho}\right)^2 ds\right)\left(\oint \frac{1}{\rho^2} ds\right)^{-1}.$$

For an isomagnetic FODO cell combined function machine, show that

$$\frac{\Delta \mathcal{D}}{\Delta x_{co}} \approx \frac{8N_{\rm cell}^2}{\pi^2 R},$$

where $N_{\rm cell}$ is the number of FODO cells.

(c) The above analysis assumes that $x_{co} = \Delta R$. In fact, $x_{co} = D\delta_s$, where

$$\delta_s = -\frac{1}{\alpha_c}\frac{\Delta f}{f_0} = \frac{1}{\alpha_c}\frac{\Delta R}{R}$$

is the fractional momentum deviation from the momentum at frequency f_0. Show that the variation of the damping partition with respect to δ_s is

$$\frac{\Delta \mathcal{D}}{\Delta \delta_s} \approx \left(2 \oint D^2 \left(\frac{B'}{B\rho}\right)^2 ds\right)\left(\oint \frac{1}{\rho^2} ds\right)^{-1}.$$

For a FODO cell combined function lattice, show that

$$\frac{\Delta \mathcal{D}}{\Delta \delta_s} \approx \frac{8}{\sin^2(\Phi/2)} \qquad \text{and} \qquad \Delta \mathcal{D} = \frac{8N_{\rm cell}^2}{\pi^2 R}\Delta R.$$

(d) Compare your estimation with that in Fig. 4.8 for the CERN PS.

9. Consider a weak focusing synchrotron (Exercise 2.4.5) with focusing index $0 < n < 1$. Show that $\langle \mathcal{H} \rangle = \rho/(1-n)^{3/2}$; $\mathcal{D} = (1-2n)/(1-n)$; $\mathcal{J}_x = n/(1-n)$; $\mathcal{J}_E = (3-4n)/(1-n)$; and $\epsilon_x = C_q\gamma^2/n\sqrt{1-n}$, where ρ is the bending radius. Show also that the quadrature horizontal beam size of the electron beam is $\sigma_x^2 = 3\rho C_q\gamma^2/[n(3-4n)]$.

10. The displacement vector from a reference orbit for a particle is $x = x_\beta + D\delta$, and $x = x_\beta + D\delta$, where x_β, x'_β are phase space coordinates of betatron motion and δ is the off-momentum parameter. If the betatron motion and synchrotron motion are independent, show that (see Exercise 2.2.14):

$$\sigma_x^2 = \beta_x\epsilon_x + D^2\sigma_\delta^2, \quad \sigma_{x'}^2 = \gamma_x\epsilon_x + D'^2\sigma_\delta^2, \quad \sigma_{x,x'} = -\alpha_x\epsilon_x + DD'\sigma_\delta^2,$$

where $\alpha_x(s), \beta_x(s)$ and $\gamma_x(s)$ are the betatron amplitude functions, $D(s)$ and $D'(s)$ are dispersion functions, and ϵ_x and σ_δ are the rms emittance and the rms off-momentum width of the beam. Show also that the effective emittance defined as that of Eq. (2.57) is

$$\epsilon_{x,\rm eff} \equiv \sqrt{\sigma_x^2\sigma_{x'}^2 - \sigma_{xx'}^2} = \sqrt{\epsilon_x\left[\epsilon_x + \mathcal{H}(s)\sigma_\delta^2\right]},$$

where $\mathcal{H}(s) = \gamma_x D^2 + 2\alpha_x DD' + \beta_x D'^2$.

11. The horizontal beta-function and dispersion function at s=0 of a long drift space from s=?10m to 10m (straight section without any magnet elements) are measured to be $\beta_{x0} = 20.0$ m, $\alpha_{x0} = -2.00$, $D_{x0} = 2.0$ m and $D'_{x0} = +0.10$.

    (a) Find the location of the minimum horizontal beta-function, and its value. Find the value of the $\mathcal{H}$-function at the end of the straight section, i.e. $s = +10$ m.

    (b) What is the phase advance from $s = 0$ to $s = 10m$?

    (c) A Gaussian beam is measured to have $\epsilon_{rms} = 15.0$ nm (known as π-nm), and $\sigma_\delta = 3.0 \times 10^{-4}$. What is the effective rms emittance of the beam at $s = 5$ m?

12. The brilliance of the photon beam is inversely proportional to $\sigma_x\sigma_{x'}$ and $\sigma_z\sigma_{z'}$. In Exercise 2.4.19, we derived the effective emittance, defined as $\sqrt{\sigma_x^2\sigma_{x'}^2 - \sigma_{xx'}^2}$, for the horizontal plane of the electron beam. However, if the photon beam does not retain the correlation of the electron beam, show that the effective source emittance becomes

$$\epsilon_{x,\text{eff}} = \langle\sigma_x\sigma'_x\rangle = \left\langle \sqrt{(\beta_x(s)\epsilon_x + D^2\sigma_\delta^2)\,(\gamma_x(s)\epsilon_x + D'^2\sigma_\delta^2)} \right\rangle,$$

where $\langle ...\rangle$ is the average of s within the length of the photon emission. In reality, $\sigma_{x'} = \text{Maximum}(\ K_\text{W}/\gamma,\ 1/\gamma,\ \sqrt{\gamma_x\epsilon_x}\ )$. If the angular divergence of the photon beam K_W/γ or $1/\gamma$ is larger than the angular divergence of the electron beam $\sqrt{\gamma_x\epsilon_x}$, the photon brilliance will be independent of the divergence of the electron beams, and the brilliance is proportional to

$$\frac{1}{\langle\sigma_x\rangle} = \frac{1}{\left\langle\sqrt{\beta_x(s)\epsilon_x + D^2\sigma_\delta^2}\right\rangle}.$$

13. Particle loss through random and non-resonant processes in a storage ring can be described by

$$N(t) = N_0 \exp\left(-\frac{t}{\tau}\right), \qquad \frac{1}{\tau} = \sum_i \frac{1}{\tau_i},$$

where τ is called the lifetime of the beam, $1/\tau_i$ is the decay rate the i-th process. The decay rates of independent processes add. The half-life is equal to 0.693τ. Mechanisms contribute to beam loss are (1) the interaction of the beam with the target, (2) intrabeam interaction induced emittance growth resulting in beam loss due to longitudinal or transverse acceptance of the ring, (3) scattering loss due to large angle Coulomb scattering with residual gas atoms; elastic and inelastic nuclear reaction; (4) diffusion processes due to finite dynamic aperture, etc. Signal on a wall gap monitor or Pick-up electrode (PUE) is proportional to the number of particle in the beam. Show that the lifetime of the beam is related to the decreases of beam signal power (in unit of dB) in a time interval Δt via

$$\tau = 8.7\frac{\Delta t}{\text{Decrease of beam power in dB}},$$

III Emittance in Electron Storage Rings

The synchrotron light emitted from a dipole spans vertically an rms angle of $1/\gamma$ around the beam trajectory at the point of emission, where γ is the Lorentz factor. Horizontally, the synchrotron light fans out to an angle equal to the bending angle of the dipole magnet. The critical frequency or the critical photon energy of the synchrotron light spectrum is given by Eq. (4.60). Beyond the critical photon frequency, the power of the synchrotron light decreases exponentially $e^{-\omega/\omega_c}$. Because synchrotron light sources from electron storage rings are tunable, they have been widely applied in basic research areas such as atomic, molecular, condensed matter, and solid state physics, chemistry, cell biology, microbiology, electronic processing, etc. The brilliance of a photon beam, defined in Eq. (4.8), is *inversely* proportional to the product of electron beam emittances $\epsilon_x\epsilon_z$. Thus a small electron beam emittance is desirable for a high brightness synchrotron radiation storage ring.

The amplitudes of the betatron and synchrotron oscillations are determined by the equilibrium between the quantum excitation due to the emission of photons and the radiation damping due to the rf acceleration field used to compensate the energy loss of the synchrotron radiation. The horizontal (natural) emittance of Eq. (4.129) is

$$\epsilon_x = C_q\gamma^2\frac{\langle\mathcal{H}/|\rho|^3\rangle}{\mathcal{J}_x\langle 1/\rho^2\rangle} \xrightarrow{\text{isomagnetic}} C_q\gamma^2\frac{\langle\mathcal{H}\rangle_{\text{dipole}}}{\mathcal{J}_x\rho},$$

where $C_q = 3.83\times10^{-13}$ m, $\mathcal{J}_x \approx 1$ is the damping partition number, ρ is the bending radius, the $\mathcal{H}$-function is given by Eq. (2.159), and $\langle\mathcal{H}\rangle$ is the average of $\mathcal{H}$-function in an isomagnetic ring. The objective of low emittance optics is to minimize $\langle\mathcal{H}\rangle$ in dipoles. Computer codes such as MAD [23], SYNCH [24] or ELEGANT [25] can be used to optimize $\langle\mathcal{H}\rangle$. However, it would be useful to understand the theoretical limit of achievable emittance in order to determine the optimal solution for a given lattice. Since $\mathcal{H} \sim L\theta^2 = \rho\theta^3$, the $\langle\mathcal{H}\rangle$ and the resulting natural emittance obey the scaling laws:

$$\langle\mathcal{H}\rangle/\mathcal{J}_x = \mathcal{F}_{\text{lattice}}\rho\theta^3 \quad \text{and} \quad \epsilon_x = \mathcal{F}_{\text{lattice}}C_q\gamma^2\theta^3, \tag{4.167}$$

where the scaling factor $\mathcal{F}_{\text{lattice}}$ depends on the design of the storage ring lattices, and θ is the total dipole bending angle in a bend-section. The resulting normalized emittance $\epsilon_{\text{n}} = \gamma\epsilon_x = \mathcal{F}_{\text{lattice}}C_q(\gamma\theta)^3$. depends essentially only on the lattice design factor $\mathcal{F}_{\text{lattice}}$ for electron storage rings at constant $\gamma\theta$.

III.1 Emittance of Synchrotron Radiation Lattices

Storage ring lattices are designed to attain desirable electron beam properties. Electron storage rings have many different applications, and each application has its special design characteristics. For example, the lattice of a high energy collider is usually composed of arcs with many FODO cells and low β insertions for high energy particle detectors. The function of arcs is to transport beams in a complete revolution. On the other hand, lattices for synchrotron radiation sources are usually arranged such that many insertion devices can be installed to enhance coherent radiation while attaining minimum emittance for the beam. Popular arrangements include the double-bend achromat (DBA), three-bend achromat (TBA), FODO cells, etc. In this section, we review the properties of these lattices.

A. FODO cell lattice

FODO cells have been widely used as building blocks for high energy colliders and storage rings. Some high energy colliders have been converted into synchrotron light sources in parasitic operation mode. A FODO cell is usually configured as $\{\frac{1}{2}\mathrm{Q_F}\ \mathrm{B}\ \mathrm{Q_D}\ \mathrm{B}\ \frac{1}{2}\mathrm{Q_F}\}$, where $\mathrm{Q_F}$ and $\mathrm{Q_D}$ are focusing and defocusing quadrupoles and B is a dipole magnet (see Chap. 2, Sec. II). The $\mathcal{H}$-function, given in Eq. (2.160), is invariant outside the dipole region. The ratio $\mathcal{H}_{\rm F}/\mathcal{H}_{\rm D}$ is typically less than 1, as shown in the left plot of Fig. 4.12. Note that the dispersion invariant

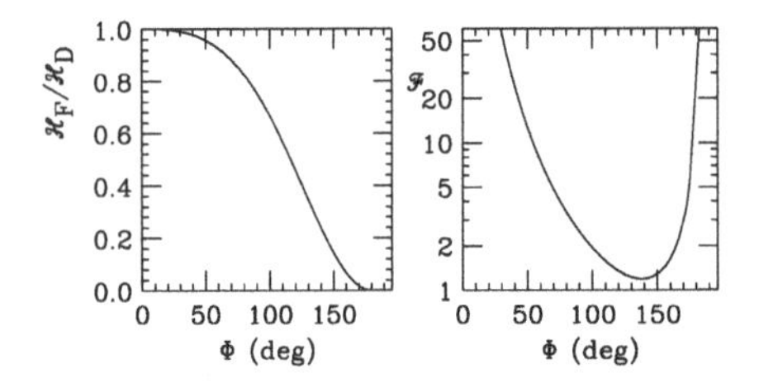

Figure 4.12: Left: the ratio $\mathcal{H}_{\rm F}/\mathcal{H}_{\rm D}$; right: the lattice factor $\mathcal{F}$ with $\mathcal{J}_x = 1$ for a FODO cell lattice.

Since the dispersion invariant does not vary much from QF to QD, The $\langle\mathcal{H}\rangle$ in the dipole can be approximated by averaging $\mathcal{H}_{\rm F}$ and $\mathcal{H}_{\rm D}$ (see Exercise 4.3.1) to obtain the lattice coefficient $\mathcal{F}$ of the FODO cell:

$$\begin{aligned}
\langle\mathcal{H}\rangle &\approx \frac{1}{2}\rho\theta^3\frac{\cos(\Phi/2)}{\sin^3(\Phi/2)}\left[\frac{(1+\frac{1}{2}\sin(\Phi/2))^2}{(1+\sin(\Phi/2))}+\frac{(1-\frac{1}{2}\sin(\Phi/2))^2}{(1-\sin(\Phi/2))}\right].\\
\mathcal{F}_{\rm FODO} &= \frac{1-\frac{3}{4}\sin^2(\Phi/2)}{\sin^3(\Phi/2)\cos(\Phi/2)}\,\mathcal{J}_x^{-1}. \qquad (4.168)
\end{aligned}$$

The right plot of Fig. 4.12 shows the coefficient $\mathcal{F}$ as a function of phase advance per cell, where we assume $\mathcal{J}_x = 1$. The coefficient decreases rapidly with phase advance of the FODO cell. The factor $\mathcal{F}$ has a minimum of about 1.3 at $\phi \approx 140°$. At this phase advance, the chromaticity and the sextupole strength needed for chromaticity correction are large. Nonlinear magnetic fields can become critical in determining the dynamical aperture.

One can employ focusing quadrupole and the defocusing combined function dipole for the FODO cells to increase packing factor and thus reduce emittance. The left plot of Fig. 4.13 shows the lattice structure (top), $\mathcal{H}/\rho\theta^3$ (middle), and betatron amplitude functions (bottom) at $\Phi_x = 99°$. The right plot shows $\mathcal{H}/\rho\theta^3$ as the betatron phase advance increases. At large betatron phase advance, the dispersion function inside the dipole is minimum at the middle of dipole, but the $\mathcal{H}$-function is larger inside dipole than that at the quadrupole, although the $\mathcal{H}$-function decreases with the betatron phase advance.

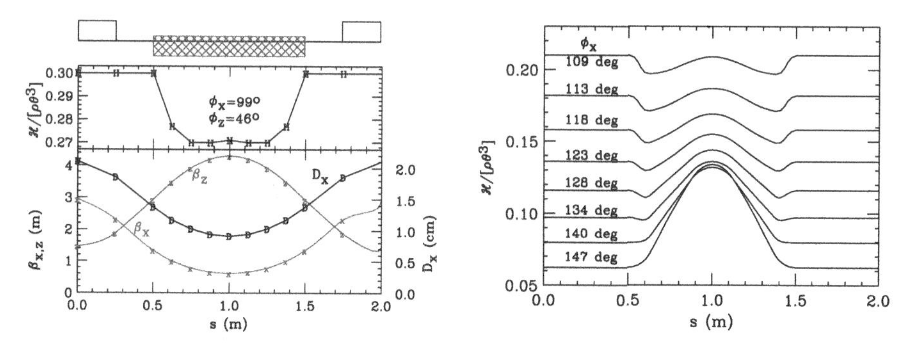

Figure 4.13: Left: the lattice function (bottom), $\mathcal{H}/\rho\theta^3$ (middle), and lattice configuration (top) with $L_q = 0.5$ m, $K_{1q} = 3.8$ m^{-2} $L_{dipole} = 1.0$ m, and $K_{1,dipole} = -1.5$. The betatron tunes and the chromaticities of the cell are $\mu_x = 0.2749$, $\mu_z = 0.1266$, $C_x = -0.311$ and $C_z = -0.215$. Right: $\mathcal{H}/\rho\theta^3$ as phase advance ϕ_x increases.

The scaling properties of $\langle\mathcal{H}\rangle$ normalized to the TME lattice of $\rho\theta^3/[12\sqrt{15}]$ of Eq. (4.180), and the normalized chromaticities, C_x/ν_x and C_z/ν_z marked with symbol "X" and "Z" respectively, are shown in Fig. 4.14.

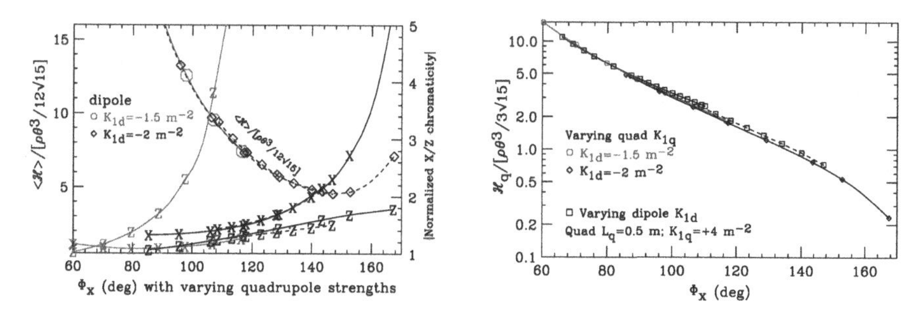

Figure 4.14: Left: the scaling property of $\langle\mathcal{H}\rangle/[\rho\theta^3/12\sqrt{15}]$ vs the horizontal phase advance for the FODO cell. The normalized chromaticities are marked "X" or "Z" respectively. Right: the scaling property of $\langle\mathcal{H}\rangle_q/[\rho\theta^3/3\sqrt{15}]$ vs the normalized phase advance.

The $\langle\mathcal{H}\rangle$ depends only on the horizontal phase advance Φ_x. If the strength of defocusing dipole is too weak (e.g. $|K_{1d}| \leq 1.5$), the vertical plane becomes unstable too quickly (see the necktie diagram in Fig. 2.10), and the corresponding vertical betatron function will be too large and the vertical chromaticity also becomes very large (see the $K_{1d} = -1.5$ m^{-2} cases in Fig. 4.14, where the $\langle\mathcal{H}\rangle$ is still large, while the vertical chromaticity is already large). If the strength of the focusing quadrupole is too large, the horizontal chromaticity also becomes large. The $\mathcal{H}$-function at the focusing quadrupole, normalized to the TME value $\rho\theta^3/[3\sqrt{15}]$ of Eq. (4.180), is shown in the right plot of Fig. 4.14. It depends only on ϕ_x. Since $\mathcal{H}_q$ of the FODO cell arc section is small, it requires a special matching this section to rest of the accelerator lattice.

It is tempting to replace the focusing quadrupole of by a combined function focusing dipole at the same strength so that we need to make only one kind of dipole. Figure 4.15 shows the lattice functions and the corresponding normalized $\mathcal{H}$-function.

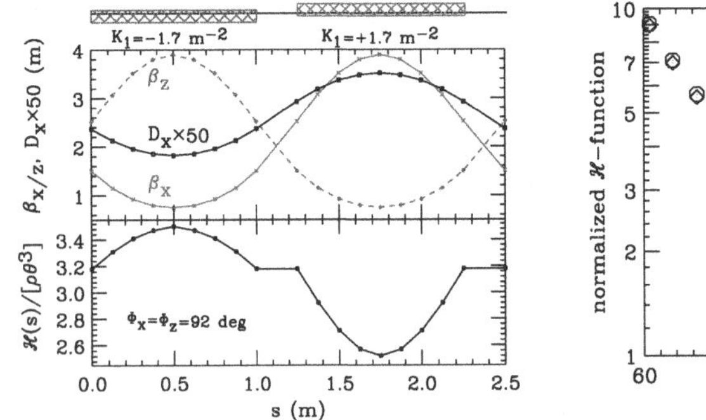

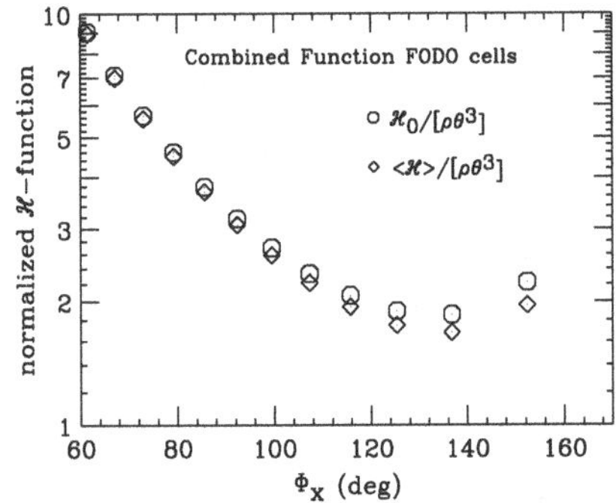

Figure 4.15: Left: the lattice configuration of a combined function FODO cell (top), the lattice function (middle), and the normalized $\mathcal{H}$-function (bottom). Right: the scaling property of $\langle\mathcal{H}\rangle_q/[\rho\theta^3]]$ vs the normalized phase advance.

Note that the $\mathcal{H}$-function in defocusing dipole is higher than that in the focusing dipole. The average of the $\mathcal{H}$-function turn out to be nearly equal to the $\mathcal{H}$ function in straight section. Although the dipole angle in this configuration is smaller by a factor of 2, i.e. its $\mathcal{H}$-function is smaller by a factor of 8, the resulting $\langle\mathcal{H}\rangle$ is still larger than that of the configuration with focusing quadrupole shown in Fig. 4.13 by about a factor of 2. If the magnet sagitta is not an issue, so that all the dipole magnets are straight, we need only one kind of dipole magnet in this FODO cell section. Possible schemes of the low emittance FODO section is schematically shown in Fig. 4.16.

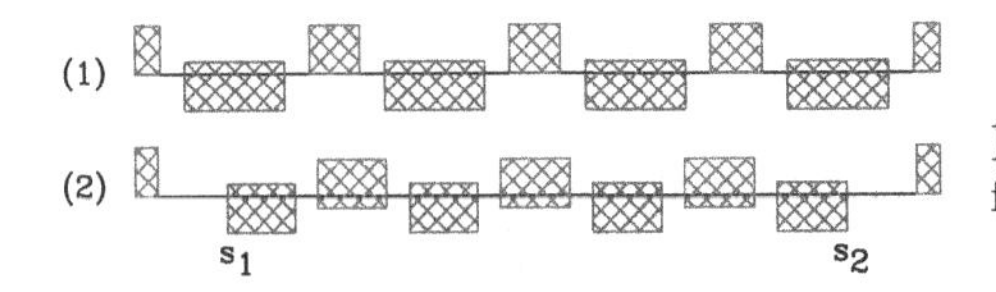

Figure 4.16: Possible FODO arc schemes for low emittance lattices.

The wholly combined function FODO cell of the scheme (2) has a problem of damping partition number. The contribution of the radiation integral I_4 to the damping partition of the combined function arc-cell section in Fig. 4.16(2) is

$$\int_{s_1}^{s_2} \frac{2K_1 D_x}{\rho} ds = \frac{2L_{\rm dipole} - 2\rho\,[D'_x(s_2) - D'_x(s_1)]}{\rho^2}, \tag{4.169}$$

where $L_{\rm dipole}$ is the total length of the arc dipoles, ρ is the bending radius of dipoles, and D'_x is the slope of the dispersion function. Detailed calculations show that the dipole length is nearly compensated by the slope of the dispersion-function term. It is worth noting that the number of FODO cells can be optimized to attain proper phase advances (both the horizontal and vertical planes) for nonlinear optics optimization.

B. Double-bend achromat (Chasman-Green lattice)

The simplest Chasman-Green lattice is made of two dipole magnets with a focusing quadrupole between them to form an achromatic cell (see Exercise 2.5.14). A possible configuration is {[OO] B {O Q$_{\rm F}$ O} B [OO]}. The betatron function matching [OO] section can be made of doublets or triplets for attaining optical properties suitable for insertion devices such as wigglers, undulators, and rf cavities. The {O Q$_{\rm F}$ O} section may consist of a single focusing quadrupole, or a pair of doublets, or triplets with reflection symmetry for dispersion matching. Since the dispersion function is nonzero only in this section, chromatic sextupoles are also located in this section. In general, the dispersion function inside the dipole is

$$D = \rho(1 - \cos\phi) + D_0 \cos\phi + \rho D'_0 \sin\phi, \tag{4.170}$$

$$D' = \left(1 - \frac{D_0}{\rho}\right) \sin\phi + D'_0 \cos\phi\ , \tag{4.171}$$

where $\phi = s/\rho$ is the bend angle at a distance s from the entrance of the dipole, ρ is the bending radius, and D_0 and D'_0 are respectively the values of the dispersion function and its derivative at $s = 0$. For the Chasman-Green lattice, we need $D_0 = 0$ and $D'_0 = 0$ to attain the achromatic condition.

The evolution of the $\mathcal{H}$-function in a dipole is (see Exercise 2.4.14)

$$\begin{aligned}\mathcal{H}(\phi) \ &= \ \mathcal{H}_0 + 2(\alpha_0 D_0 + \beta_0 D'_0)\sin\phi - 2(\gamma_0 D_0 + \alpha_0 D'_0)\rho(1-\cos\phi) \\ &\quad + \beta_0 \sin^2\phi + \gamma_0\rho^2(1-\cos\phi)^2 - 2\alpha_0\rho\sin\phi(1-\cos\phi),\end{aligned} \tag{4.172}$$

where $\mathcal{H}_0 = \gamma_0 D_0^2 + 2\alpha_0 D_0 D'_0 + \beta_0 D'^2_0$, and α_0, β_0, and γ_0 are the Courant-Snyder parameters at $s = 0$. Averaging the $\mathcal{H}$-function in the dipole, we get

$$\begin{aligned}\langle\mathcal{H}\rangle \ &= \ \mathcal{H}_0 + (\alpha_0 D_0 + \beta_0 D'_0)\theta E(\theta) - \frac{1}{3}(\gamma_0 D_0 + \alpha_0 D'_0)\rho\theta^2 F(\theta) \\ &\quad + \frac{\beta_0}{3}\theta^2 A(\theta) - \frac{\alpha_0}{4}\rho\theta^3 B(\theta) + \frac{\gamma_0}{20}\rho^2\theta^4 C(\theta),\end{aligned} \tag{4.173}$$

where L and $\theta = L/\rho$ are the length and bending angle of dipole(s) in a half DBA cell, and

$$E(\theta) = 2(1-\cos\theta)/\theta^2, \quad F(\theta) = 6(\theta - \sin\theta)/\theta^3, \quad A(\theta) = (6\theta - 3\sin 2\theta)/(4\theta^3),$$
$$B(\theta) = (6 - 8\cos\theta + 2\cos 2\theta)/\theta^4, \quad C(\theta) = (30\theta - 40\sin\theta + 5\sin 2\theta)/\theta^5.$$

In the small angle limit, we find $A \to 1,\ B \to 1,\ C \to 1,\ E \to 1,\ F \to 1$. With the normalized scaling parameters

$$d_0 = \frac{D_0}{L\theta}, \quad d_0' = \frac{D_0'}{\theta}, \quad \tilde{\beta}_0 = \frac{\beta_0}{L}, \quad \tilde{\gamma}_0 = \gamma_0 L, \quad \tilde{\alpha}_0 = \alpha_0, \tag{4.174}$$

the $avg\mathcal{H}$-function becomes

$$\begin{aligned}\langle\mathcal{H}\rangle &= \rho\theta^3 \Bigg\{ \left[\tilde{\gamma}_0 d_0^2 + 2\tilde{\alpha}_0 d_0 d_0' + \tilde{\beta}_0 d_0'^2\right] + \left[\tilde{\alpha}_0 E - \frac{\tilde{\gamma}_0}{3}F\right] d_0 \\ &\quad + \left[\tilde{\beta}_0 E - \frac{\tilde{\alpha}_0}{3}F\right] d_0' + \frac{\tilde{\beta}_0}{3}A - \frac{\tilde{\alpha}_0}{4}B + \frac{\tilde{\gamma}_0}{20}C \Bigg\}. \end{aligned} \tag{4.175}$$

B1. Minimum emittance DBA lattice

Applying the achromatic condition with $d_0 = d_0' = 0$, we get the average $\mathcal{H}$-function as

$$\langle\mathcal{H}\rangle = \rho\theta^3 \left[\frac{\tilde{\beta}_0}{3}A - \frac{\tilde{\alpha}_0}{4}B + \frac{\tilde{\gamma}_0}{20}C\right], \tag{4.176}$$

where $\tilde{\beta}_0\tilde{\gamma}_0 = (1 + \tilde{\alpha}_0^2)$. The Courant-Snyder parameters that minimize $\langle\mathcal{H}\rangle$ and its minimum value are (see Exercise 4.3.3)

$$\tilde{\beta}_0 = \frac{6C}{\sqrt{15}G}, \quad \tilde{\alpha}_0 = \frac{\sqrt{15}B}{G}, \quad \tilde{\gamma}_0 = \frac{8\sqrt{5}A}{\sqrt{3}G}, \quad \to \quad \langle\mathcal{H}\rangle_{\rm MEDBA} = \frac{G}{4\sqrt{15}}\rho\theta^3, \tag{4.177}$$

where $G = \sqrt{16AC - 15B^2}$, shown in Fig. 2.40, decreases slightly with increasing θ. The corresponding minimum β-function value and its location are $\beta^*_{\rm MEDBA} = \frac{3}{4\sqrt{60}}L$ and $s^*_{\rm MEDBA} = \frac{3}{8}L$.

The dispersion action $\mathcal{H}(\theta)$ outside the dipole is an important parameter in determining the aperture requirement. For a minimum emittance (ME) DBA lattice, we find $\mathcal{H} = 0$ at $s = 0$, and the $\mathcal{H}$-function at the end of the dipole is

$$\begin{aligned}\mathcal{H}(\theta) &= \frac{\rho\theta^3}{\sqrt{15}G}\left\{6C[\frac{\sin^2\theta}{\theta^2}] - 15B[\frac{2\sin\theta(1-\cos\theta)}{\theta^3}] + 10A[\frac{4(1-\cos\theta)^2}{\theta^4}]\right\} \\ &\to \frac{1}{\sqrt{15}}\rho\theta^3 \quad \text{(thin lens approximation).}\end{aligned} \tag{4.178}$$

One can understand the result of Eq. (4.177) as follows. Since $\mathcal{H}(\phi) \sim \phi^3$, the average of $\mathcal{H}$ is $\frac{1}{4}$ of its maximum value, i.e. $\langle\mathcal{H}\rangle = \frac{1}{4}\mathcal{H}(\theta)$, The minimum emittance DBA lattice factor of Eq. (4.167) is $\mathcal{F}_{\rm MEDBA} = 1/(4\sqrt{15}\mathcal{J}_x)$.

In zero gradient approximation, the horizontal betatron phase advance across a dipole for the MEDBA lattice is 156.7°, and the phase advance in the dispersion matching section is 122° (see Exercise 4.3.3). Thus the minimum phase advance for the MEDBA module is 435.4°, which does not include the phase advance of the zero dispersion betatron function matching section for the insertion devices. Thus each MEDBA module will contribute about 1.2 unit to the horizontal betatron tune. Since the phase advance is large, the chromatic properties of lattices should be carefully corrected. The resulting emittance is smaller than the corresponding FODO cell lattice by a factor of 20 to 30.

B2. Examples of low emittance DBA lattices

Many high brilliance synchrotron radiation light sources employ low emittance DBA lattice for the storage ring. Figure 4.17 shows the lattice functions of a nearly minimum emittance DBA lattice ELETTRA at Trieste in Italy for 2 GeV electron storage ring (left), and the low emittance DBA lattice of APS at Argonne for 7 GeV electron storage ring (right). The total phase advance of each ELETTRA DBA-period is about 429°, while the corresponding phase advance of APS DBA-period is about 319°. The ELETTRA lattice employs defocusing combined-function dipole with $q = \sqrt{|B_1|/B\rho}\ell_{\rm dipole} = 0.9439$ to increase damping partition number $\mathcal{J}_x$. The resulting horizontal emittances of these lattices are respectively $\epsilon_{\rm elettra}/\epsilon_{\rm MEDBA} \approx 1.38$ and $\epsilon_{\rm aps}/\epsilon_{\rm MEDBA} \approx 3.64$.

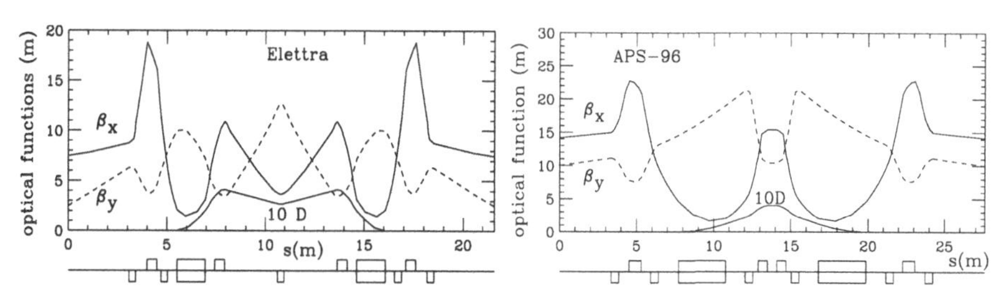

Figure 4.17: The low emittance lattice functions for a superperiod of ELETTRA (left) and APS (right). The ELETTRA lattice has 12 superperiods, and the APS lattice has 40 superperiods.

B3. Triplet DBA lattice

A variant of the DBA lattice is the triplet DBA, where a quadrupole triplet is used in the dispersion matching section for the achromat condition. Because there is no quadrupole in the [OO] section of the DBA, the lattice is very simple (see the lower plot of Fig. 2.34). The minimum emittance lattice factor is[11] $\mathcal{F}_{\rm ME,triplet} = \frac{2\beta^*}{3\ell\mathcal{J}_x}$, where β^* is the value of the betatron amplitude function at the symmetry point of the dispersion free straight section, and ℓ is the length of the dipole.

C. Theoretical Minimum Emittance (TME) lattice

Without the achromat constraint, each module of an accelerator lattice has only one dipole. The optical functions that minimize $\langle\mathcal{H}\rangle$, i.e. the dispersion and betatron amplitude functions, are symmetric with respect to the center of the dipole.

From Eq. (4.173), the minimization procedure for $\langle\mathcal{H}\rangle$ can be achieved through the following steps. First, $\langle\mathcal{H}\rangle$ can be minimized by finding the optimal dispersion functions with

$$\frac{\partial\langle\mathcal{H}\rangle}{\partial d_0} = 0, \quad \frac{\partial\langle\mathcal{H}\rangle}{\partial d_0'} = 0,$$

where we obtain

$$d_{0,\rm min} = \frac{1}{6}F, \quad d'_{0,\rm min} = -\frac{1}{2}E, \qquad \langle\mathcal{H}\rangle = \frac{1}{12}\rho\theta^3\left(\tilde\beta_0\tilde A - \tilde\alpha_0\tilde B + \frac{4\tilde\gamma_0}{15}\tilde C\right) \qquad (4.179)$$

with $\tilde A = 4A - 3E^2$, $\tilde B = 3B - 2EF$, and $\tilde C = \frac{9}{4}C - \frac{5}{4}F^2$. With the relation $\tilde\beta_0\tilde\gamma_0 = 1 + \tilde\alpha_0^2$, the minimum emittance is

$$\langle\mathcal{H}\rangle_{\rm ME} = \frac{\tilde G}{12\sqrt{15}}\rho\theta^3, \qquad (4.180)$$

where $\tilde G = \sqrt{16\tilde A\tilde C - 15\tilde B^2}$ is also shown in Fig. 2.40. The corresponding lattice properties at the entrance of the dipole are

$$\tilde\beta_0 = \frac{8\tilde C}{\sqrt{15}\tilde G}, \quad \tilde\alpha_0 = \frac{\sqrt{15}\tilde B}{\tilde G}, \quad \tilde\gamma_0 = \frac{2\sqrt{15}\tilde A}{\tilde G}. \qquad (4.181)$$

The waist of the optimal betatron amplitude function for minimum $\langle\mathcal{H}\rangle$ is located at the center of the dipole, i.e. $s^*_{\rm ME} = L/2$. The corresponding minimum betatron amplitude function and dispersion function at the waist are $\beta^*_{\rm ME} = L/\sqrt{60}$ and $D^*_{\rm ME} = L\theta/24$ respectively. The required minimum betatron amplitude function is $\beta^*_{\rm ME} = \frac{4}{3}\beta^*_{\rm MEDBA}$. The attainable theoretical minimum emittance lattice factor of Eq. (4.167) is $\mathcal{F}_{\rm ME} = 1/(12\sqrt{15}\mathcal{J}_x)$. To attain the minimum emittance, the betatron phase advance across the dipole is 151°, and the dispersion matching section is 133.4°. Thus the horizontal betatron tune of this minimum emittance single dipole module is 284.4°

[11] See Exercise 4.3.9, where we find that the stability condition is incompatible with the achromat condition. Therefore, this minimum emittance condition can not be reached.

(see Exercise 4.3.4). Each minimum emittance module with a single dipole would contribute a horizontal betatron tune of 0.79.

The values of the dispersion $\mathcal{H}$-function on both sides of the dipole are important in determining the beam size in the straight sections, where insertion devices such as undulators are located. Using Eq. (4.179) and (4.181) for the ME condition, we obtain

$$\mathcal{H}(0) = \mathcal{H}(\theta) = \frac{1}{3\sqrt{15}}\rho\theta^3 \left\{6\tilde{C}E^2 - \frac{15}{2}\tilde{B}EF + \frac{5}{2}\tilde{A}F^2\right\}\tilde{G}^{-1}. \tag{4.182}$$

In small bending angle approximation, we have $\mathcal{H}(\theta) = \frac{1}{3\sqrt{15}}\rho\theta^3 = 4\langle\mathcal{H}\rangle_{\rm ME}$, which is equal to $\frac{1}{3}\mathcal{H}(\theta)|_{\rm MEDBA}$.

The brilliance of the photon beam from an undulator depends essentially on the electron beam width. The horizontal beam width is given by the quadrature of the betatron beam width and the momentum beam width. It is appropriate to define the dispersion emittance as

$$\epsilon_d \equiv \gamma_x(D\sigma_\delta)^2 - \beta_x'(D\sigma_\delta)(D'\sigma_\delta) + \beta_x(D'\sigma_\delta)^2 = \mathcal{H}(0)\sigma_\delta^2, \tag{4.183}$$

where $\sigma_\delta^2 = (\sigma_E/E)^2 = C_q\gamma^2/\rho\mathcal{J}_E$ is the equilibrium energy spread of the beam below the microwave instability threshold. Because the $\mathcal{H}$-function is invariant in the straight section, ϵ_d is invariant in the straight section. Substituting $\mathcal{H}(0)$ of Eq. (4.182) into Eq.(4.183), we obtain

$$\epsilon_d = \frac{1}{3\sqrt{15}}\frac{C_q\gamma^2\theta^3}{\mathcal{J}_E}. \tag{4.184}$$

For a separated function lattice, we find $\mathcal{J}_E \approx 2$, $\mathcal{J}_x \approx 1$ or $\mathcal{J}_E \approx 2\mathcal{J}_x$. The effective 1D emittance for a bi-Gaussian distribution becomes

$$\epsilon_{x,\rm 1D} = \epsilon_x + \mathcal{H}_{\rm ID}\sigma_\delta^2 = \epsilon_{\rm ME} + \epsilon_d = \frac{1}{4\sqrt{15}}\frac{C_q\gamma^2\theta^3}{\mathcal{J}_x} = \epsilon_{\rm MEDBA}, \tag{4.185}$$

where $\mathcal{H}_{\rm ID}$ is the $\mathcal{H}$-function at the ID locations. The decrease in betatron beam size in minimum emittance lattice is accompanied by an equal amount of increment in the dispersion beam size, i.e. the 1D effective emittances of the TME and MEDBA lattices are equal. The brilliance of a photon beam is inversely proportional to the phase space areas $\sigma_x\sigma_{x'}\sigma_z\sigma_{z'} \sim \epsilon_{x,\rm eff}\epsilon_{z,\rm eff}$, where the effective horizontal and vertical emittances are $\epsilon_{x,\rm eff} = \sqrt{\epsilon_x\epsilon_{x,\rm 1D}}$ and $\epsilon_{z,\rm eff} = \epsilon_z$ [see Exercises 2.4.19 and 4.2.12].

D. Three-bend achromat

Now we are ready to discuss the minimum emittance for three-bend achromat (TBA) lattices, which have been used in synchrotron radiation sources such as the Advanced

Light Source (ALS) at LBNL, the Taiwan Light Source (TLS), the Pohang Light Source (PLS), etc. The TBA is a combination of DBA lattices with a single dipole cell at the center.

To simplify our discussion, we use small angle approximation, which is good approximation provided that the bending angle for each dipole is less than 30°. Since the $\mathcal{H}$-function is invariant in the optical matching section without dipoles, equating Eqs. (4.178) and (4.182), we find

$$\rho_2\theta_2^3 = 3\rho_1\theta_1^3, \quad \text{or} \quad \frac{L_2^3}{\rho_2^2} = 3\frac{L_1^3}{\rho_1^2}, \tag{4.186}$$

where ρ_1, θ_1, L_1 and ρ_2, θ_2, L_2 are respectively the bending radii, bending angles and lengths of the outer and inner dipoles. This is the necessary condition for achieving dispersion phase space matching.[12] The matching condition of Eq. (4.186) requires $L_2 = 3^{1/3}L_1$ for isomagnetic storage rings, or $\rho_1 = \sqrt{3}\rho_2$ for storage rings with equal length dipoles.

Thus we have proved a theorem stating that an isomagnetic TBA with equal length dipoles can *not* be matched to attain the advertised minimum emittance. For an isomagnetic storage ring, the center dipole for the TBA should be longer by a factor of $3^{1/3}$ than the outer dipoles in order to achieve dispersion matching. In this case, we can prove the following trivial theorem: The emittance of the matched minimum TBA (QBA, or nBA) lattice is

$$\epsilon_{\rm METBA} = \frac{C_q\gamma^2\theta_1^3}{4\sqrt{15}\mathcal{J}_x}, \tag{4.187}$$

where θ_1 is the bending angle of the outer dipoles, provided the middle dipole is longer by a factor of $3^{1/3}$ than the outer dipoles. The formula for the attainable minimum emittance is identical to that for the MEDBA.

E. Summary of Lattice Properties and QBA

The lattice factor $\mathcal{F}_{\rm lattice}$ is generally smaller for the nBA lattices. The natural emittance depends also on the number of dipoles. Figure 4.18 shows a compilation of achieved natural emittance for published synchrotron light sources scaled to 3 GeV energy.[13]

---

[12] S.Y. Lee, *Phys. Rev.* **E 54**, 1940 (1996). The necessary condition for finite angle can be obtained by equating Eq. (4.178) and Eq. (4.182).

[13] G. Mülhaupt, EPAC1990, 65 (1990); M. Böge, *et al*, EPAC1998, 623 (1998); M. Böge, EPAC2002, 39 (2002); L. Dallin, *et al*, PAC2003, 220 (2003); M.-P. Level, *et al*, PAC2003, 229 (2003); R.P. Walker, PAC2003, 232 (2003); R. Hettel *et al*, PAC2003, 235 (2003); D. Einfeld *et al*, PAC2003, 238 (2003); B. Podobedov *et al*, PAC2003, 241 (2003); H. Ohkuma *et al*, PAC2003, 883 (2003); Greg LeBlanc, *et al.*, PAC2003, 2321 (2003); M. Eriksson *et al*, EPAC2004, 2392 (2004);

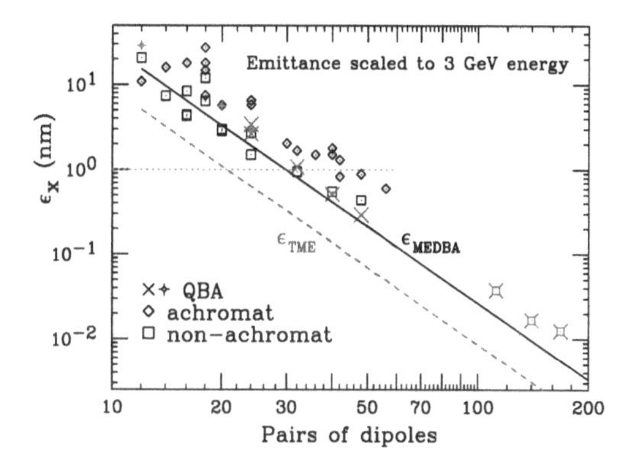

Figure 4.18: Compilation of the natural emittances in unit of nm achieved for some synchrotron light sources vs the number of pairs of dipoles, including some of the recently upgraded 6 GeV light sources. All emittance is scaled down to 3 GeV beam energy. For example, a 12 cell 7BA corresponds to 42 pairs of dipoles. The lines are the minimum emittance for DBA lattice and the theoretical minimum emittance (TME) respectively. The minimum emittance of nBA lattice is given by Eq. (4.187).

The diamond symbols are emittances of the achromatic lattices. Many storage ring designers choose non-achromatic (TME) concept in order to reduce the natural emittance by a factor of 3, shown as the square symbols. For the multiple-bend lattices, the pairs of dipoles is not equal to the number of cells. For example, a lattice with 12(7BA) is equivalent to 42 pairs of dipoles.

Although the non-achromatic lattice can reduce the natural emittance by a factor of 3, the effective emittance does not gain as much as one wishes. Since it is difficult to design a lattice reaching MEDBA or TME condition, we ask a simple question: would it be better off with the achromatic or the non-achromatic mode of operation?

We assume that we can design a lattice at the achromatic mode with $\epsilon_{x,\rm a} = f_{\rm a}\epsilon_{\rm MEDBA}$, and another non-achromatic mode with $\epsilon_{x,\rm na} = f_{\rm na}\epsilon_{\rm TME}$, where both $f_{\rm a}$ and $f_{\rm na}$ are typically about 2∼4. The effective emittance (see Exercise 4.2.10) of these lattices can be compared as follows:

$$\epsilon_{x,\rm a} = \frac{f_{\rm a}}{4\sqrt{15}\mathcal{J}_x}C_q\gamma^2\theta^3, \qquad \epsilon_{x,\rm 1D,na} = \left[\frac{f_{\rm na}}{12\sqrt{15}\mathcal{J}_x} + \frac{1}{3\sqrt{15}\mathcal{J}_E}\right]C_q\gamma^2\theta^3,$$
$$\frac{\epsilon_{x,\rm eff,na}}{\epsilon_{x,\rm a}} \approx \frac{\sqrt{f_{\rm na}(f_{\rm na}+2)}}{3f_{\rm a}}. \tag{4.188}$$

where we have used $\mathcal{H}_{\rm ID} \approx \mathcal{H}_{\rm TME}$ with $\mathcal{J}_x \approx 1$ and $\mathcal{J}_E \approx 2$. If both $f_{\rm a}$ and $f_{\rm na}$ are 2, the effective emittance of the non-achromatic lattice is about 0.47 of the achromatic one, i.e. the effective emittance is only reduced by about a factor of 2. However, all straight sections are non-achromatic, and the insertion devices in these non-achromatic straight sections can increase the natural emittance to be discussed

J.B. Murphy *et al*, EPAC2004, 2457 (2004); A. Jackson, PAC2005, 102 (2005); Z. Zhao *et al*, PAC2005, 214 (2005); R.O. Hettel, PAC2005, 505 (2005); G. Vignola *et al*, PAC2005, 587 (2005); V.M. Tsakanov, PAC2005, 629 (2005); E.S. Kim, APAC2004, 85 (2004); C.C. Kuo, *et al.*, PAC2005, 2989 (2005); D. Einfeld *et al*, PAC2005, 4203 (2005). K. Tsumaki, *et al*, PAC1989, 1358 (1989); D.Einfeld and M.Plesko, Nucl. Instru. and Methods, A **335**, 402 (1993).

in Sec. III.3. The effect of emittance increase is most important in low energy storage rings, where high power wavelength shifters are needed for hard X-ray production. These wavelength shifters in the dispersive straight section can dramatically increase the natural emittance of the beam.

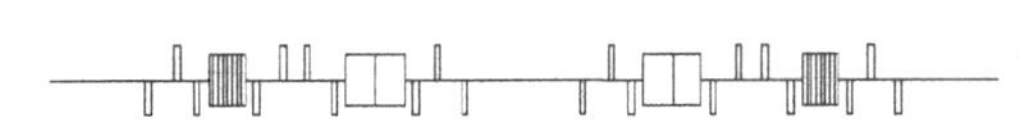

Figure 4.19: A basic QBA cell. Several QBA cells can be grouped together to form a superperiod with different optical properties in the straight sections.

This defect can be remedied by the introduction of the quadruple-bend-achromatic (QBA) cell, defined as a super cell made of two double-bend (DB) cells (see Fig. 4.19), where the two inner dipoles have larger bending angles than those of two outer dipoles required for dispersion function matching. For the same number of dipole, the natural emittance of a QBA lattice is about half of that of the DBA, while retaining 50% of achromatic straight sections. The number of straight sections is the same as that of the DB cells. In thin lens approximation, the necessary condition for a minimum emittance is $L_{\text{inner dipole}}/L_{\text{outer dipole}} = 3^{1/3}$ as shown in Eq. (4.186) for isomagnetic lattice. In actual lattice design, the ratio is found to be about 1.5~1.6 for an optimal matching. Figure 4.20 shows the optical functions of a QBA lattice with 12 QBA cells (equivalent to 24 DBA cells) with circumference 486 m.[14]

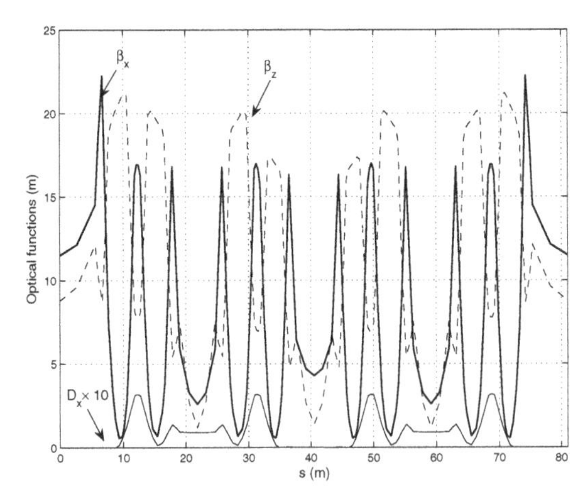

Figure 4.20: The Optical functions of two QBA cells with a 10.91 m and a 5.31 m dispersion-free and two 5.31 m dispersive straight sections. The dipole length ratio is $L_2/L_1 = 1.5$. The circumference is $C = 486$ m. At 3 GeV, the resulting natural emittance is about 3.0 nm.

The QBA lattice provides an advantage over the DBA by reducing the emittance by a factor of 2, and over the double-bend non-achromat in performance by providing 50% zero-dispersion straight-sections. The X and star symbols on Fig. 4.18 are the natural emittance obtained from the QBA lattices.

[14]M.H. Wang, *et al.* Review of Scientific Instruments, **78**, 055109 (2007).

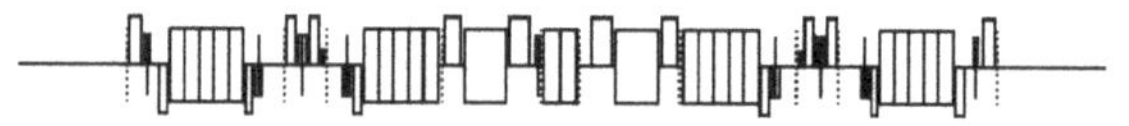

Figure 4.21: Lattice configuration of an upgrade low emittance lattice cell for the ESRF-U.

F. Design concepts of recent light source upgrades

In recent years, there are efforts toward diffraction limited light source upgrade projects at Argonne Photon Source (APS-U), European Synchrotron Radiation Facility (ESRF-U), and the Spring-8-U in Japan. An interesting idea proposed by ESRF-U design is the combination of DBA and small emittance FODO-cell arc, shown in Fig. 4.21.[15]

The optical functions of the ESRF upgrade lattice is shown in Figure 4.22. The three low emittance FODO cells in the middle of the arc is matched to dispersion free-straight sections by 2 DBA-like matching sections. The matching section of each DBA-section can create a large dispersion bump for chromatic correction. The ESRF lattice has also an interesting trick in the dipoles of the DBA-like section. The dipole field is stronger at the low $\mathcal{H}$-function location so that its $\langle\mathcal{H}/|\rho|^3\rangle$ in these DBA-like dipoles contributes about equally to the radiation integral I_5 of the low emittance FODO cells. A simple model of longitudinal gradient dipole on the dispersion function and $\langle\mathcal{H}/|\rho|^3\rangle$ is given in Exercise 4.3.7.

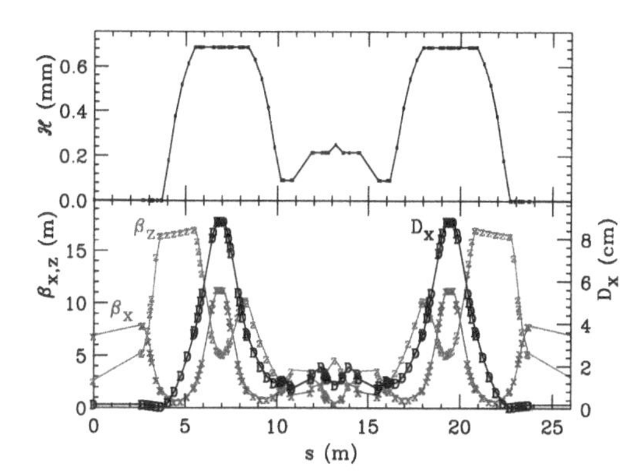

Figure 4.22: Optical functions of an upgrade low emittance lattice of ESRF. Top: The $\mathcal{H}$-function in low emittance cell. Note that the $\mathcal{H}$-function is increased to a large number for achieving a large dispersion function for non-linear chromatic correction. These two-dipoles on both sides of the cell resemble the DBA-like structure. In the middle of the superperiod, there are 3 low emittance FODO cells, with phase advance of about 101 degree per cell.

It is worth noting that the lattice of the APS upgrade project employs reverse-bend dipoles inside the DBA-like structure to control dispersion function matching without affecting the betatron amplitude functions.[16] Independent control of the dispersion function and betatron amplitude functions can provide a better matching of the low emittance lattice.

[15]L. Farvacque *et al.*, IPAC2013, 79 (2013)
[16]M. Borland *et al.*, IPAC2015, 1776 (2015).

All light source upgrade projects are limited by the existing tunnel and the optical beam-lines, the cell length can not be changed. All storage ring lattices with very low emittance have very small dynamic aperture, due to very strong sextupoles needed for chromaticity correction. Increasing the dispersion function at the matching section can minimize the strength of sextupoles. For green-field light sources, the number of the low emittance FODO cells may be optimized to create proper phase advances between sextupoles for nonlinear dynamic aperture optimization.

III.2 Insertion Devices

The straight sections in storage rings can house insertion devices (IDs), e.g. wigglers or undulators, that can greatly enhance the brilliance and wavelength of photon beams. The IDs are normally made of dipole magnets with alternating dipole fields so that the orbit outside the device is un-altered. A simple planer undulator with vertical sinusoidal fields, that satisfies Maxwell's equation, is

$$B_z = B_{\rm w}\cosh(k_{\rm w}z)\ \cos(k_{\rm w}s), \tag{4.189}$$

where $k_{\rm w} = 2\pi/\lambda_{\rm w}$ is the wiggler wave number, $\lambda_{\rm w}$ is the wiggler period, and $B_{\rm w}$ is the magnetic field at mid-plane.[17] The corresponding horizontal and longitudinal magnetic fields, and the vector potential are $B_x = 0$, $B_s = -B_{\rm w}\sinh k_{\rm w}z\sin k_{\rm w}s$, and

$$A_x = -\frac{1}{k_{\rm w}}B_{\rm w}\cosh k_{\rm w}z\sin k_{\rm w}s, \quad A_z = 0, \quad A_s = 0. \tag{4.190}$$

Thus the Hamiltonian of particle motion is

$$H = \frac{1}{2}(p_x + \frac{1}{k_{\rm w}\rho_{\rm w}}\cosh k_{\rm w}z\sin k_{\rm w}s)^2 + \frac{1}{2}p_z^2, \tag{4.191}$$

where $\rho_{\rm w}$ is the bending radius of the wiggler field, i.e. $B_{\rm w}\rho_{\rm w} = p/e$ with particle momentum p. The equation of motion is

$$\begin{cases} x'' = \dfrac{1}{\rho_{\rm w}}\cosh k_{\rm w}z\cos k_{\rm w}s, \\ z'' + \dfrac{\sin^2 k_{\rm w}s\sinh 2k_{\rm w}z}{\rho_{\rm w}^2 \quad 2k_{\rm w}} = \dfrac{p_x}{\rho_{\rm w}}\sinh k_{\rm w}z\sin k_{\rm w}s \approx 0. \end{cases} \tag{4.192}$$

The nonlinear magnetic field can be neglected if the vertical betatron motion is small with $k_{\rm w}z \ll 1$. The horizontal closed orbit becomes

$$x_{\rm co} = \frac{1}{\rho_{\rm w}k_{\rm w}^2}(1-\cos k_{\rm w}s), \quad x'_{\rm co} \equiv \beta_\perp = \frac{1}{\rho_{\rm w}k_{\rm w}}\sin k_{\rm w}s = \frac{K_{\rm w}}{\gamma}\sin k_{\rm w}s, \tag{4.193}$$

$$K_{\rm w} = \frac{eB_{\rm w}\lambda_{\rm w}}{2\pi mc} = 0.934\, B_{\rm w}\,[{\rm T}]\,\lambda_{\rm w}\,[{\rm cm}]. \tag{4.194}$$

[17] The magnetic field $B_{\rm w}$ at $z = 0$ is related to the field $\hat{B}$ at the pole-tip by $B_{\rm w} = \hat{B}/\cosh(\pi g/\lambda_{\rm w})$, where g is the wiggler gap-height. Small period wigglers demands small-gap operation with $g \le 0.5\lambda_{\rm w}$. The dynamic and physical aperture of small gap wigglers may be small.

where we use $1/\rho_{\rm w}k_{\rm w} = K_{\rm w}/\beta\gamma \approx K_{\rm w}/\gamma$ with $K_{\rm w}$ as the wiggler parameter. Table 4.3 lists some wiggler parameters of some insertion devices for the third generation light sources.

Table 4.3: parameters of some undulators and wigglers

Machine	E [GeV]	B [T]	$\lambda_{\rm w}$ [cm]	L [m]	$K_{\rm w}$
ALS	1.5	5	14	1.96	65
		1.15	9	4.8	9.7
APS	7	0.65	2.2	5	1.3

The transverse electron angular divergence inside the undulator or wiggler is equal to $K_{\rm w}/\gamma$. For $K_{\rm w} \sim 1$, the device is called an undulator; for $K_{\rm w} \gg 1$, it is called a wiggler. The velocity vector of an electron in the planar undulator is $\vec{\beta} = \beta_\perp \hat{x} + \beta_\parallel \hat{s}$, where $\beta^2 = \beta_\perp^2 + \beta_\parallel^2 = 1 - 1/\gamma^2$, or

$$\begin{aligned} \beta_\parallel &= \sqrt{\beta^2 - \beta_\perp^2} = \left(1 - \frac{1}{\gamma^2} - \frac{K_{\rm w}^2}{\gamma^2}\sin^2 k_{\rm w}s\right)^{1/2} \\ &\approx 1 - \frac{1 + K_{\rm w}^2 \sin^2 k_{\rm w}s}{2\gamma^2} = 1 - \frac{1 + K_{\rm w,rms}^2}{2\gamma^2} + \frac{K_{\rm w,rms}^2}{2\gamma^2}\cos 2k_{\rm w}s. \end{aligned} \tag{4.195}$$

Note that the magnitude of the longitudinal velocity oscillates at two times the undulator wave number. The quantity $K_{\rm w,rms} \equiv \frac{1}{\sqrt{2}}K_{\rm w}$ is called the rms undulator parameter for planer undulator. The photon emitted in each wiggler period is enhanced by a resonance condition to be discussed in the next section.

A: Ideal helical undulators or wigglers

We next consider a helical wiggler with magnetic field[18]

$$\vec{B}_{\rm w} = B_{\rm w}\,(\hat{x}\cos\frac{2\pi s}{\lambda_{\rm w}} + \hat{z}\sin\frac{2\pi s}{\lambda_{\rm w}}), \tag{4.196}$$

where $(\hat{x}, \hat{s}, \hat{z})$ are unit vectors of the curvilinear coordinate system for the transverse radial, longitudinal, and transverse vertical directions. The transverse equation of motion for electrons traveling at nearly the speed of light in the longitudinal direction inside the wiggler is

$$\gamma mc\frac{d\vec{\beta}}{dt} = e\beta c\hat{s} \times \vec{B}, \qquad \text{or} \qquad \frac{d\vec{\beta}}{ds} = \frac{eB_{\rm w}}{\gamma mc}\,(\hat{z}\cos k_{\rm w}s - \hat{x}\sin k_{\rm w}s),$$

[18]In order for the ideal helical magnetic field to obey the Maxwell's equation, we need to include higher order nonlinear terms in the magnetic field. For linear betatron motion, we neglect all higher order terms in the following discussions.

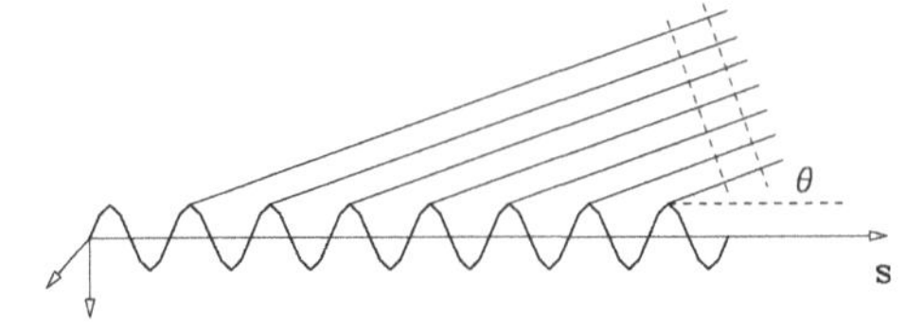

Figure 4.23: Coherent addition of radiation from electrons in wigglers or undulators. Longitudinal coherence gives rise to resonance condition of single frequency of diffraction like structure.

$$\vec{\beta} = \frac{K_{\rm w}}{\gamma}(\hat{x}\cos k_{\rm w}s + \hat{z}\sin k_{\rm w}s) + \beta_{\parallel}\hat{s}, \tag{4.197}$$

$$\beta^2 = \beta_{\perp}^2 + \beta_{\parallel}^2 = 1 - \frac{1}{\gamma^2}, \qquad \beta_{\parallel} \approx 1 - \frac{1+K_{\rm w}^2}{2\gamma^2}.$$

where the wiggler parameter $K_{\rm w}$ is defined in Eq. (4.194). Note that the magnitude of the transverse velocity vector is $\beta_{\perp} = K_{\rm w}/\gamma$. Unlike the planer undulator, the helical undulator does not produce a large tune shift in linear approximation.

The displacement vector of the electron in the wiggler is obtained by integrating Eq. (4.197):

$$\frac{\vec{r}(t')}{c} = \frac{K_{\rm w}}{\omega_{\rm w}\gamma}(\hat{x}\sin\omega_{\rm w}t' - \hat{z}\cos\omega_{\rm w}t') + \beta_{\parallel}t'\hat{s}; \tag{4.198}$$

where t' is the reference frame of the moving electrons. Let the observer be located at one end of the wiggler. The $\hat{n}$ can be written as (see Fig. 4.23),

$$\hat{n} = \phi\hat{x} + \psi\hat{z} + (1 - \frac{1}{2}\theta^2)\hat{s} \tag{4.199}$$

with $\phi^2 + \psi^2 = \theta^2$, where these angles are of the order of $\frac{1}{\gamma}$. The observer's time t is related to the electron's time t' via the retarded condition, i.e.

$$t = t' - \frac{\hat{n}\cdot\vec{r}(t')}{c} = \frac{1+K_{\rm w}^2+\gamma^2\theta^2}{2\gamma^2}t' - \frac{\phi K_{\rm w}}{\omega_{\rm w}\gamma}\sin\omega_{\rm w}t' + \frac{\psi K_{\rm w}}{\omega_{\rm w}\gamma}\cos\omega_{\rm w}t'. \tag{4.200}$$

Let $\xi = \omega_{\rm w}t'$, Eq. (4.200) can be transformed to

$$\frac{2\gamma^2}{1+K_{\rm w}^2+\gamma^2\theta^2}\omega_{\rm w}t = \xi - \frac{2\gamma\phi K_{\rm w}}{1+K_{\rm w}^2+\gamma^2\theta^2}\sin\xi + \frac{2\gamma\phi K_{\rm w}}{1+K_{\rm w}^2+\gamma^2\theta^2}\cos\xi. \tag{4.201}$$

It is apparent to see that the periodic motion of the electron in the wiggler are transformed to the observer at a frequency boosted by the factor shown in Eq. (4.201). Let us use the notation $\omega_{\rm L}$ for the laser frequency, i.e.

$$\omega_{\rm L} = \frac{2\gamma^2}{1+K_{\rm w}^2+\gamma^2\theta^2}\,\omega_{\rm w}. \tag{4.202}$$

Here $\omega_{\rm L}$ corresponds to the characteristic frequency of the device in the observer's frame. We can rewrite Eq. (4.201) as $\omega_{\rm L} t = \xi - p\sin\xi + q\cos\xi$, where

$$p = \frac{2\gamma\phi K_{\rm w}}{(1+K_{\rm w}^2+\gamma^2\theta^2)}; \quad q = \frac{2\gamma\psi K_{\rm w}}{(1+K_{\rm w}^2+\gamma^2\theta^2)}. \tag{4.203}$$

The integrand of the radiation integral in Eq. (4.39) of the classical radiation formula is

$$\hat{n}\times(\hat{n}\times\vec{\beta}) \approx \hat{x}[\phi - \frac{K_{\rm w}}{\gamma}\cos\omega_{\rm w}t'] + \hat{z}[\phi - \frac{K_{\rm w}}{\gamma}\sin\omega_{\rm w}t'], \tag{4.204}$$

where we retain only terms up to the order of $1/\gamma$. The radiation integral of Eq. (4.39) becomes

$$\vec{G} = -j\left(\frac{\omega^2 e^2}{8\pi^2\omega_{\rm w}^2 c}\right)^{1/2}\int_{-\infty}^{\infty}\left([\phi - \frac{K_{\rm w}}{\gamma}\cos\xi]\hat{x} + [\phi - \frac{K_{\rm w}}{\gamma}\sin\xi]\hat{z}\right)e^{-j\omega t}d\xi.$$

For a periodic wiggler, we obtain

$$\begin{aligned}\vec{G} &= -j\frac{2\gamma\omega_0 K_{\rm w}}{\omega_{\rm L}(0)[1+K_{\rm w}^2]}\left(\frac{e^2}{8\pi^2 c}\right)^{1/2} N_{\rm w}\tilde{S}(\omega/\omega_{\rm L})\int_{-\pi}^{\pi}\left[\left(\phi - \frac{K_{\rm w}}{\gamma}\cos\xi\right)\hat{x}\right.\\ &\left. + \left(\phi - \frac{K_{\rm w}}{\gamma}\sin\xi\right)\hat{z}\right]\times\exp\left\{-j\frac{\omega}{\omega_{\rm L}}(\xi - p\sin\xi + q\cos\xi)\right\}d\xi, \end{aligned}\tag{4.205}$$

where $N_{\rm w}$ is the number of the wiggler period, the apparent angular frequency $\omega_{\rm L}(0)$ at the forward direction is

$$\omega_{\rm L}(0) = \omega_{\rm L}(\theta=0) = \frac{2\gamma^2}{1+K_{\rm w}^2}\omega_{\rm w}, \quad \lambda_{\rm L} = \frac{\lambda_{\rm w}}{2\gamma^2}(1+K_{\rm w}^2) \tag{4.206}$$

and the spectral coherent factor $\tilde{S}(\omega/\omega_{\rm L})$ is sharply peaked at integer harmonics of $\omega_{\rm L}(\theta)$:

$$\tilde{S}(\omega/\omega_{\rm L}) = \left[\frac{\sin N_{\rm w}\frac{\omega}{\omega_{\rm L}}\pi}{N_{\rm w}\sin\frac{\omega}{\omega_{\rm L}}\pi}\right] \approx \pm\left[\frac{\sin\pi N_{\rm w}(\omega-\omega_n)/\omega_{\rm L}}{\pi N_{\rm w}(\omega-\omega_n)/\omega_{\rm L}}\right], \tag{4.207}$$

$$\omega_n = n\omega_{\rm L}(\theta) = n\omega_{\rm L}(0)\left(1+\frac{\gamma^2\theta^2}{1+K_{\rm w}^2}\right)^{-1}.$$

The corresponding photon energy at the fundamental frequency is

$$\epsilon_1[\text{keV}] = \hbar\omega(0) = \frac{0.95E_e^2[\text{GeV}]}{(1+K_{\rm w}^2)\lambda_{\rm w}[\text{cm}]}; \quad \text{or} \quad \lambda_1[\mathring{A}] = \frac{13.1(1+K_{\rm w}^2)\lambda_{\rm w}[\text{cm}]}{E_e^2[\text{GeV}]}. \tag{4.208}$$

Thus the photon energy can be adjusted by tuning the electron energy, or the wiggler parameters, $\lambda_{\rm w}$ and $K_{\rm w}$. The spectral distribution of the diffraction pattern has a full width half maximum at the n-th harmonic:

$$\frac{\Delta\omega}{\omega_n} \approx \frac{2.7}{\pi n N_{\rm w}} \approx \frac{0.85}{nN_{\rm w}}. \tag{4.209}$$

Due to the coherent interference nature, the frequency spectrum is discrete. The maximum power is proportional to $N_{\rm w}^2$. The photon flux is proportional to the number of electron due to incoherent nature. The frequency spectrum will also be broadened by the momentum spread of the electron beam.

B. Characteristics of radiation from undulators and wigglers

If the wiggler parameter is large, i.e. $K_{\rm w} \gg 1$, the spectra are similar to those of synchrotron radiation from dipoles. Synchrotron radiation has a continuous spectrum up to the critical frequency $\omega_{\rm c,w} = 3\gamma^3 c/2\rho_{\rm w}$. Since a wiggler magnet may have a stronger magnetic field, the synchrotron radiation spectrum generated in a wiggler is shifted upward in frequency. Such a wiggler is also called a wavelength shifter.

If $K_{\rm w} \leq 1$, the radiation from each undulator period can coherently add up to give rise to a series of spectral lines given by

$$\lambda_n = \frac{1 + K_{\rm w}^2 + \gamma^2\Theta^2}{2n\gamma^2}\lambda_{\rm w} = \frac{13.1\,\lambda_{\rm w}\,[\mathrm{cm}]}{nE^2\,[\mathrm{GeV}]}(1 + K_{\rm w}^2 + \gamma^2\Theta^2)\ [\mathring{A}] \quad (n = 1, 2, \ldots), \tag{4.210}$$

where Θ is the observation angle with respect to the undulator plane. For reference, the critical wavelength from a dipole is

$$\lambda_{\rm c,\,\mathit{dipole}} = \frac{4\pi mc}{3eB\gamma^2} = \frac{0.007135}{B\ [\mathrm{T}]\ \gamma^2}\ [m] = \frac{18.6}{B\ [\mathrm{T}]\ E^2\ [\mathrm{GeV}]}\ [\mathring{A}].$$

The resonance condition for constructive interference is achieved when the path length difference between the photon and electron, during the time that the electron travels one undulator period, is an integer multiple of the electromagnetic wavelength, i.e. $\lambda_{\rm w}/\beta_{\|} - \lambda_{\rm w}\cos\Theta = n\lambda_n$. Figure 4.24 shows schematically the sinusoidal electron orbit and electromagnetic radiation (vertical bars), where the electron (circle) lags behind the electromagnetic wave by one wave length in traversing one undulator period for $n = 1$.

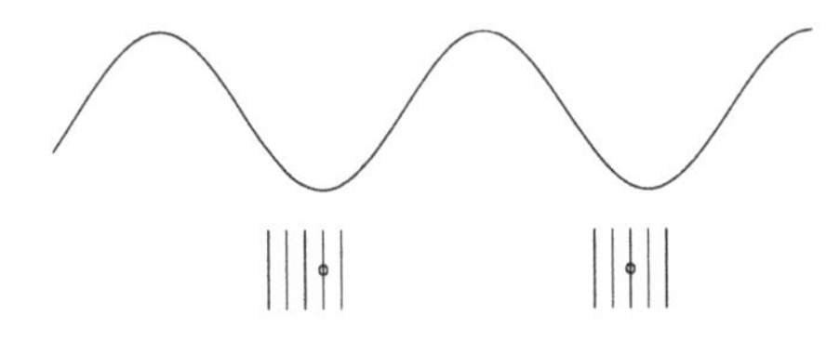

Figure 4.24: Schematic drawing of the sinusoidal orbit of an electron in a planar undulator and the electromagnetic wave (vertical bars). The resonance condition is achieved when the electron travels one undulator period, it lags behind the electromagnetic wave by one full wave length for the $n = 1$ mode.

The pulse length of a photon from a short electron bunch is

$$\Delta t = \frac{N_{\rm w}\lambda_{\rm w}}{\beta_{\|}c} - \frac{N_{\rm w}\lambda_{\rm w}}{c}\cos\Theta \approx \frac{N_{\rm w}\lambda_1}{c}. \tag{4.211}$$

Thus the frequency bandwidth is

$$\Delta\omega = \frac{\pi}{\Delta t} = \frac{\pi c}{N_{\rm w}\lambda_1} = \frac{\omega_1}{2N_{\rm w}}, \tag{4.212}$$

where λ_1 is the wavelength of the fundamental radiation. The fractional bandwidth is then $\Delta\omega/\omega_1 = 1/(2N_{\rm w})$ with the angular aperture $\langle\Theta^2\rangle^{1/2} = \sqrt{\lambda/N_{\rm w}\lambda_{\rm w}}$ and the source radius $\sqrt{\lambda N_{\rm w}\lambda_{\rm w}}/4\pi$, i.e. the emittance of the photon beam is $\lambda/4\pi$. Optical resonance cavities have been used to enhance the radiation called the free electron laser (FEL).[19] The number of photons, emitted within the solid angle $\lambda/(N_{\rm w}\lambda_{\rm w})$ and bandwidth $\Delta\omega/\omega$, is $\mathcal{N}_\gamma = \pi\xi\alpha[{\rm JJ}]^2$ per undulator period, where $\xi = K^2/(2+K^2)$, and the factor $[{\rm JJ}] = J_0(\frac{1}{2}\xi) - J_1(\frac{1}{2}\xi)$ comes from the planar undulator.

With the progress in small emittance beam sources from photocathode rf guns, emittance preservation in linacs, longitudinal bunch compression, and precise undulators, efforts are being made in many laboratories to demonstrate the self-amplified spontaneous emission (SASE) principle, to produce an infrared FEL, and to achieve single pass X-ray FELs such as the Linac Coherent Light Source (LCLS) at SLAC and DESY.

III.3 Effect of IDs on beam dynamics

The rm beam emittance ϵ_x and the fractional energy spread σ_E/E in electron storage rings are given by the the radiation integrals listed in Sec. II.8). The natural emittance and the energy spread of the beam in the presence of IDs become

$$\epsilon_x = \epsilon_{x0}\left(1+\frac{I_{5\rm w}}{I_{50}}\right)\left(1+\frac{I_{2\rm w}-I_{4\rm w}}{I_{20}-I_{40}}\right)^{-1} \approx \epsilon_{x0}\frac{1+I_{5\rm w}/I_{50}}{1+I_{2\rm w}/I_{20}}, \tag{4.213}$$

$$\left(\frac{\sigma_E}{E}\right)^2 = \left(\frac{\sigma_E}{E}\right)_0^2\left(1+\frac{I_{3\rm w}}{I_{30}}\right)\left(1+\frac{2I_{2\rm w}+I_{4\rm w}}{2I_{20}+I_{40}}\right)^{-1} \approx \left(\frac{\sigma_E}{E}\right)_0^2\frac{1+I_{3\rm w}/I_{30}}{1+I_{2\rm w}/I_{20}}, \tag{4.214}$$

where I_{20}, I_{30}, I_{40}, and I_{50} are radiation integrals evaluated in bending dipoles, and $\epsilon_{x0} = C_q\gamma^2 I_{50}/(I_{20}-I_{40})$ and $(\sigma_E/E)_0^2 = C_q\gamma^2 I_{30}/(2I_{20}+I_{40})$ is the emittance of the lattice without IDs and $I_{2\rm w}$, $I_{3\rm w}$, $I_{4\rm w}$, and $I_{5\rm w}$ are radiation integrals evaluated in wigglers with $I_{40} \ll I_{20}$, $I_{4w} \ll I_{2\rm w}$, and $I_{4w} \ll I_{20}$ for high energy storage rings with separated function magnets. In particular, we find $I_{2\rm w}/I_{20} = U_{\rm w}/U_0$, where

$$U_0 = C_\gamma E^4/\rho_0,$$
$$U_{\rm w} = \begin{cases} C_\gamma E^4 L_{\rm w}/(4\pi\rho_{\rm w}^2), & \text{planer undulator,} \\ C_\gamma E^4 L_{\rm w}/(2\pi\rho_{\rm w}^2), & \text{helical undulator,} \end{cases}$$

[19]The free electron laser was realized in 1977 by J. Madey's group [D.A.E. Deacon *et al.*, *Phys. Rev. Lett.* **38**, 892 (1977)]. Since then, this field has been very active, with many regular workshops and conferences. See, e.g. R.H. Pantell, p. 1708 in Ref. [16]; G. Dattoli, A. Torre, L. Giannessi, and S. Dopanfilis, CERN 90-03 p. 254 (1990); G. Dattoli and A. Torre, CERN 89-03 (1989).

are the energy losses in the storage ring dipoles and in an undulator or wiggler in one revolution respectively. Here ρ_0 and $\rho_{\rm w}$ are the bending radii of storage ring dipoles and wiggler magnets, $L_{\rm w}$ is the length of the undulator or wiggler, and E is the beam energy. Depending on the values of these radiation integrals, the emittance and the energy spread can increase or decrease.

A. Effect of IDs on beam emittances

Normally, the change of the $\mathcal{H}$-function due to the ID with rectangular magnets is small. The natural beam emittance depends on $\oint \mathcal{H}/|\rho|^3 ds \equiv \langle\mathcal{H}\rangle \oint (1/\rho^3) ds = 2\pi\langle\mathcal{H}\rangle/\rho_0^2$, where $\langle\mathcal{H}\rangle$ is the average of $\mathcal{H}$-function in dipoles. Let the parameter f_h be defined as $\langle\mathcal{H}\rangle = f_h\mathcal{H}_{\rm ID}$, where $\mathcal{H}_{\rm ID}$ is the value of the $\mathcal{H}$-function at an ID section. Thus the value of f_h may vary slightly in different straight sections. For example, we find $f_h = \infty$ in an achromatic straight section, $f_h = 0.25$ for a TME lattice, and a typically well designed non-achromatic lattice has $f_h \approx 0.5 \sim 0.8$. Since $\mathcal{H} = \mathcal{H}_{\rm ID}$ is constant in the entire ID region, we find

$$I_{50} = \oint \frac{\mathcal{H}}{|\rho|^3} ds \approx \langle\mathcal{H}\rangle \oint \frac{1}{|\rho|^3} ds = \frac{2\pi}{\rho_0^2} f_h \mathcal{H}_{\rm ID}, \tag{4.215}$$

$$I_{5\rm w} = \mathcal{H}_{\rm ID} \int_W \frac{1}{|\rho|^3} ds \approx \begin{cases} \dfrac{4}{3\pi}\dfrac{\mathcal{H}_{\rm ID} L_{\rm w}}{\rho_{\rm w}^3} & \text{planer,} \\[2ex] \dfrac{\mathcal{H}_{\rm ID} L_{\rm w}}{\rho_{\rm w}^3} & \text{helical,} \end{cases} \tag{4.216}$$

where we assume sinusoidal magnetic field for undulators. The ratio of the radiation integrals becomes

$$\frac{I_{5\rm w}}{I_{50}} \approx \begin{cases} \dfrac{8\rho_0}{3\pi f_h \rho_{\rm w}} \dfrac{U_{\rm w}}{U_0} = \dfrac{8B_{\rm w}}{3\pi f_h B_0} \dfrac{U_{\rm w}}{U_0}, & \text{planer} \\[2ex] \dfrac{\rho_0}{f_h \rho_{\rm w}} \dfrac{U_{\rm w}}{U_0} = \dfrac{B_{\rm w}}{f_h B_0} \dfrac{U_{\rm w}}{U_0}, & \text{helical} \end{cases} \tag{4.217}$$

where $B_{\rm w}$ and B_0 are the magnetic flux densities of the undulator and the main dipole magnet, and the emittance becomes

$$\frac{\epsilon_x}{\epsilon_{x0\rm min}} = \begin{cases} \left(1 + \sum\limits_{\rm w} \dfrac{8}{3\pi f_h} \dfrac{\rho_0}{\rho_{\rm w}} \dfrac{U_{\rm w}}{U_0}\right) \left(1 + \sum\limits_{\rm w} \dfrac{U_{\rm w}}{U_0}\right)^{-1}, & \text{planer} \\[2ex] \left(1 + \sum\limits_{\rm w} \dfrac{1}{f_h} \dfrac{\rho_0}{\rho_{\rm w}} \dfrac{U_{\rm w}}{U_0}\right) \left(1 + \sum\limits_{\rm w} \dfrac{U_{\rm w}}{U_0}\right)^{-1}, & \text{helical.} \end{cases} \tag{4.218}$$

The emittance will decrease slightly when the undulators field is $B_{\rm w} \le \frac{3\pi f_h}{8} B_0$, and the natural emittance will increase for strong field wigglers. To minimize the emittance increase, strong field IDs should be installed in straight sections with smaller

$\mathcal{H}_{\rm ID}$. For a given $\langle\mathcal{H}\rangle$, we would prefer a larger $\mathcal{H}$ in the dispersion matching section, and a smaller $\mathcal{H}$ in the ID section. To minimize emittance degradation in the non-achromatic lattice, one chooses a strong main dipole field by making the dipole length short, or ρ_0 small. The drawback is that the value of dispersion function in the dispersion matching section, which is proportional to $\rho_0\theta^2$, becomes smaller, and thus the chromatic sextupole strength becomes larger. Optimization of the dynamic aperture becomes a difficult task in the design of such a machine.

Since $U_0 \gg U_{\rm w}$ in high energy photon sources such as the ESRF, APS, and Spring-8, non-zero dispersion function in the IDs does not cause much emittance degradation, and thus there is a small advantage of non-achromatic DB mode over the DBA mode of operation as shown in Eq. (4.188).

Each ID increases radiation power in a storage ring. If an ID is located in an achromatic straight section, where $\mathcal{H}_{\rm ID} = 0$ or $f_h = \infty$, the natural beam emittance becomes

$$\epsilon_x \approx \frac{\epsilon_{x0}}{1 + U_{\rm w}/U_0}, \tag{4.219}$$

where we assume the effect of the ID on dispersion function is small so that $I_{5\rm w} \approx 0$. Since $\epsilon_{x0\rm DBA} \approx 3\epsilon_{x0\rm min}$, the energy loss $U_{\rm w}$ in the damping IDs must be at least twice that in the dipole U_0 to achieve the same emittance as the non-achromat lattice. In order to maximize emittance damping effect, one chooses a low main dipole field option, the dipole length L is increased, and the dispersion function ($\sim \rho_0\theta^2$) in the dispersion matching section is larger. This, in turn, reduces the sextupole strength needed for chromatic correction. The dynamic aperture is still a very difficult problem, but is relatively easier to handle than in the high field dipole option. However, the DBA design requires a larger circumference or potentially, a smaller fraction of available space for IDs.

B. Effect of IDs on momentum spread

Evaluating the radiation integrals in Eq. (4.214), we obtain

$$I_{30} = \frac{2\pi}{\rho_0^2}, \qquad I_{3\rm w} = \begin{cases} \dfrac{4}{3\pi}\dfrac{L_{\rm w}}{\rho_{\rm w}^3} \\[2mm] \dfrac{L_{\rm w}}{\rho_{\rm w}^3} \end{cases}, \qquad \frac{I_{3\rm w}}{I_{30}} = \begin{cases} \dfrac{8}{3\pi}\dfrac{\rho_0}{\rho_{\rm w}}\dfrac{U_{\rm w}}{U_0} \\[2mm] \dfrac{\rho_0}{\rho_{\rm w}}\dfrac{U_{\rm w}}{U_0} \end{cases},$$

$$\left(\frac{\sigma_E}{E}\right)^2 = \left(\frac{\sigma_E}{E}\right)_0^2 \times \begin{cases} \left(1 + \dfrac{8}{3\pi}\dfrac{\rho_0}{\rho_{\rm w}}\dfrac{U_{\rm w}}{U_0}\right)\left(1 + \dfrac{U_{\rm w}}{U_0}\right)^{-1}, & \text{planer} \\[2mm] \left(1 + \dfrac{\rho_0}{\rho_{\rm w}}\dfrac{U_{\rm w}}{U_0}\right)\left(1 + \dfrac{U_{\rm w}}{U_0}\right)^{-1}, & \text{helical.} \end{cases} \tag{4.220}$$

If $B_{\rm w} \le 3\pi B_0/8$ for the planer undulator, the beam momentum spread will decrease. On the other hand, the high field IDs can increase momentum spread, particularly important for the low main dipole field design.

C. Effect of ID induced dispersion functions

We consider a simple ideal vertical field wiggler (Fig. 4.25), where $\rho_{\rm w} = p/eB_{\rm w}$ is the bending radius, $\theta = \Theta_{\rm w} = L_{\rm w}/\rho_{\rm w}$ is the bending angle of each dipole, and $L_{\rm w}$ is the length of each wiggler dipole. Since the rectangular magnet wiggler is an achromat (see Exercise 2.4.20), the wiggler, located in a zero dispersion straight section, will not affect the dispersion function outside the wiggler. However, this wiggler can generate it's own dispersion (see Exercise 2.4.20), i.e.

$$D(s) \approx \begin{cases} -s^2/2\rho_{\rm w} \\ -[2L_{\rm w}^2 - (2L_{\rm w}-s)^2]/2\rho_{\rm w} \end{cases}, \quad D'(s) \approx \begin{cases} -s/\rho_{\rm w}, & 0 \le s < L_{\rm w} \\ -[2L_{\rm w}-s]/\rho_{\rm w}, & L_{\rm w} < s \le 2L_{\rm w} \end{cases}.$$

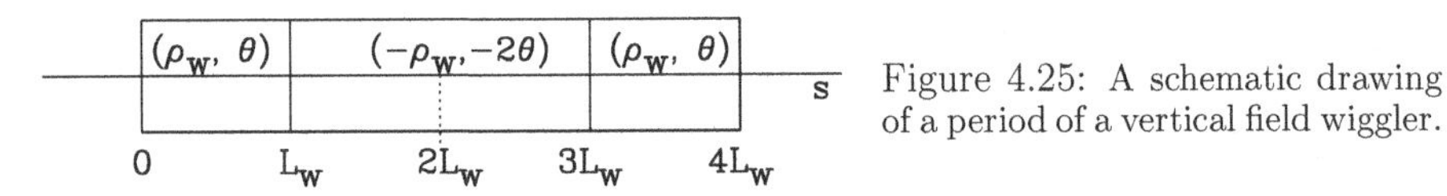

Figure 4.25: A schematic drawing of a period of a vertical field wiggler.

We assume that the center of the wiggler magnet is located at the symmetry point so that $\beta_x(s) = \beta_x^* + (s-2L_{\rm w})^2/\beta_x^*$. The radiation integral $I_{5\rm w}$ in the wiggler becomes

$$I_{5\rm w} = \frac{4}{3}\beta_x^* \frac{L_{\rm w}^3}{\rho_{\rm w}^5}\left(1 + \frac{103}{40}\frac{L_{\rm w}^2}{\beta_x^{*2}}\right) \approx \frac{4\beta_x^*}{3\rho_{\rm w}^2}\Theta_{\rm w}^3, \tag{4.221}$$

where we make an approximation that $\beta_x^* \gg L_{\rm w}$ with $\Theta_{\rm w} = L_{\rm w}/\rho_{\rm w}$. The approximation implies that each pole contributes an equal amount to the radiation integral $I_{5\rm w}$. The contribution of each wiggler period to I_2 is $I_{2\rm w} = 4L_{\rm w}/\rho_{\rm w}^2 = 4\Theta_{\rm w}/\rho_{\rm w}$. For $N_{\rm w}$ wiggler periods in the straight section, the emittance of Eq. (4.213) becomes

$$\frac{\epsilon_{x,\rm w}}{\epsilon_{x0}} \approx \left(1 + \frac{2\rho_0^2 N_{\rm w}\beta_x^*\Theta_{\rm w}^3}{3\pi\rho_{\rm w}^2\,\langle\mathcal{H}\rangle_0}\right)\left(1 + \frac{2\rho_0 N_{\rm w}\Theta_{\rm w}}{\pi\rho_{\rm w}}\right)^{-1} = \frac{1 + F_\epsilon \times (U_{\rm w}/U_0)}{1 + (U_{\rm w}/U_0)}, \tag{4.222}$$

$$F_\epsilon = \frac{\rho_0\beta_x^*\Theta_{\rm w}^2}{3\rho_{\rm w}\langle\mathcal{H}\rangle_0} \le \frac{16\sqrt{15}}{3}N_{\rm w}\left(\frac{\theta_{\rm w}}{\theta_0}\right)^3, \tag{4.223}$$

where we have used $I_{50} = 2\pi\langle\mathcal{H}\rangle_0/\rho_0^2$ and $I_{20} = 2\pi/\rho_0$ for the isomagnetic storage ring with $I_{40} \approx 0$ and $I_{4\rm w} \approx 0$, $\beta_x^* \sim 4L_{\rm w}N_{\rm w} \approx N_{\rm w}\lambda_{\rm w}$, $\langle\mathcal{H}\rangle_0 \ge \frac{1}{4\sqrt{15}}\rho_0\theta_0^3$, and θ_0 is the bending angle of the outer nBA or DBA dipole. For a sinusoidal wiggler, $\theta_{\rm w} = K_{\rm w}/\gamma$. For most high field wigglers, we find $F_\epsilon < 0.01$ and the wigglers in dispersion-free regions will reduce the beam emittance. The emittance reduction is approximately given by Eq. (4.219). At a constant field strength, i.e. constant $\rho_{\rm w}$, the factor F_ϵ is smaller if $\Theta_{\rm w}$ or $L_{\rm w}$ is reduced while maintaining a constant $N_{\rm w}\Theta_{\rm w}$, i.e. increasing the number of poles by reducing the length of each pole.

Similarly, the dispersion functions in the insertion region induced by a sinusoidal vertical wiggler field of Eq. (4.189) is

$$D = -\frac{1}{\rho_{\rm w} k_{\rm w}^2}\left(1 - \cos\left[k_{\rm w}\left(s + \frac{L_{\rm w}}{2}\right)\right]\right), \quad D' = -\frac{1}{\rho_{\rm w} k_{\rm w}} \sin\left[k_{\rm w}\left(s + \frac{L_{\rm w}}{2}\right)\right], \tag{4.224}$$

where we have assumed $D_0 = D_0' = 0$ at $s = -L_{\rm w}/2$, and $L_{\rm w}$ is the wiggler length. The sinusoidal wiggler is achromatic if $k_{\rm w} L_{\rm w}$ is an integer multiple of 2π. With $\beta_x = \beta^* + s^2/\beta^*$, the radiation integrals of the sinusoidal wiggler are

$$I_{2\rm w} = \frac{L_{\rm w}}{2\rho_{\rm w}^2}, \quad I_{3\rm w} = \frac{4L_{\rm w}}{3\pi\rho_{\rm w}^3}, \quad I_{5\rm w} \approx \frac{8N_{\rm w}\langle\beta_x\rangle}{15\rho_{\rm w}^5 k_{\rm w}^3} = \frac{8N_{\rm w}\langle\beta_x\rangle}{15\rho_{\rm w}^2}\theta_{\rm w}^3, \tag{4.225}$$

where $\theta_{\rm w} = 1/(\rho_{\rm w} k_{\rm w}) = K_{\rm w}/\beta\gamma$ is the maximum orbit angle of the electron beam in the wiggler. Using $I_{2\rm w}/I_{20} = U_{\rm w}/U_0$, we find

$$\epsilon_x = \epsilon_{x0}\frac{1 + F_\epsilon \times (U_{\rm w}/U_0)}{1 + (U_{\rm w}/U_0)}, \qquad \sigma_E^2 = \sigma_{E0}^2 \frac{1 + F_{\Delta E} \times (U_{\rm w}/U_0)}{1 + (U_{\rm w}/U_0)}, \tag{4.226}$$

$$F_\epsilon \le \frac{64\langle\beta_x\rangle}{\sqrt{15}L_{\rm w}} N_{\rm w}\left(\frac{\theta_{\rm w}}{\theta_0}\right)^3 \approx \frac{64}{\sqrt{15}} N_{\rm w}\left(\frac{\theta_{\rm w}}{\theta_0}\right)^3, \qquad F_{\Delta E} = \frac{8\rho_0}{3\pi\rho_{\rm w}}, \tag{4.227}$$

where we use $\langle\mathcal{H}\rangle_0 \ge \frac{1}{4\sqrt{15}}\rho_0\theta_0^3$ in main-dipoles and $\langle\beta_x\rangle \sim L_{\rm w}$. Since $F_\epsilon \sim 0.01$, the damping wigglers will generally reduce the emittance, given approximately by Eq. (4.219). For a given $U_{\rm w}/U_0 = (\rho_0/2\rho_{\rm w})N_{\rm w}\theta_{\rm w}$ with a constant $N_{\rm w}\theta_{\rm w}$, the factor F is smaller for a smaller $\theta_{\rm w}$. However, wigglers with very large $K_{\rm w}$ values may increase the beam emittance. The momentum spread of the beam will increase or decrease depending on whether the magnetic field $B_{\rm w}$ of the planer wiggler is larger or smaller than $(3\pi/8)B_0$ of the main dipole field.

D. Effect of IDs on the betatron tunes

The IDs with rectangular magnets is achromatic, and the edge defocusing in the rectangular vertical field magnets cancels the dipole focusing gradient of $1/\rho^2$. Thus there is no net focusing in the horizontal plane. The focal length of the vertical betatron motion and the tune shift resulting from the rectangular wiggler dipole are respectively

$$f \approx \frac{\rho_{\rm w}}{4\tan\Theta_{\rm w}} \approx \frac{\rho_{\rm w}^2}{4L_{\rm w}}, \qquad \Delta\nu_z = \frac{1}{4\pi}\int\frac{\beta_z}{f}ds \approx \frac{L_{\rm w,total}\langle\beta_z\rangle}{4\pi\rho_{\rm w}^2}, \tag{4.228}$$

where $L_{\rm w,total} = 4N_{\rm w}L_{\rm w}$ is the total length of the undulator, and $\langle\beta_z\rangle$ is the average betatron amplitude function in the wiggler region.

Similarly, the vertical sinusoidal field undulator of Eq. (4.189) generates average vertical focusing strength and vertical betatron tune shift:

$$\langle\frac{\sin^2 k_{\rm w}s}{\rho_{\rm w}^2}\rangle = \frac{1}{2\rho_{\rm w}^2}, \qquad \Delta\nu_z = \frac{1}{4\pi}\int\frac{\beta_z(s)ds}{2\rho_{\rm w}^2} = +\frac{\langle\beta_z\rangle L_{\rm w,total}}{8\pi\rho_{\rm w}^2}, \tag{4.229}$$

where $L_{\rm w}$ is the total length of the undulator. The tune shift induced by the sinusoidal undulator is a factor of 2 smaller than the dipole undulator because the effective (rms) dipole field is $1/\sqrt{2}$ of the peak dipole field. The nonlinear field in the wiggler can also affect the dynamical aperture. The betatron tunes should avoid all low order nonlinear resonances.

III.4 Beam Physics of High Brightness Storage Rings

High brilliance photon beams are generally produced by the synchrotron radiation of high brightness electron beams, which can be attained by high quality linacs with high brightness rf-gun electron sources or by high brightness storage rings. Here, we discuss only the physics issues relevant to high brightness electron storage rings. Some of these issues are the emittance, dynamical aperture, beam lifetime, beam intensity limitation, beam brightness limitation, etc.

A. Low emittance lattices and the dynamical aperture

In Sec. III.1, we have studied methods of attaining a small natural emittance. At the same time, the vertical emittance is determined mainly by the residual vertical dispersion and the linear betatron coupling. The beam brightness is proportional to $N_{\rm B}/\epsilon_x\epsilon_z$, where $N_{\rm B}$ is number of electrons per bunch. If we neglect the effects of the residual vertical dispersion function, the vertical emittance is arrived from the linear betatron coupling with $\epsilon_x + \epsilon_z = \epsilon_{\rm nat}$. Thus, minimizing the natural emittance in an accelerator lattice and minimizing the vertical emittance by correcting the linear coupling will provide higher beam brightness.

The natural emittance obeys the scaling law $\epsilon_{\rm nat} = \mathcal{F}C_q\gamma^2\theta_1^3$, where θ_1 is the total bending angle of dipoles in a half-cell or of dipoles in the outer half-cell of nBA lattice. The lattice factor $\mathcal{F}$ is

	ME	MEnBA	Triplet DBA	FODO
$\mathcal{F}$	$\frac{1}{12\sqrt{15}\mathcal{J}_x}$	$\frac{1}{4\sqrt{15}\mathcal{J}_x}$	$\frac{2\beta^*}{3L\mathcal{J}_x}$	$\frac{5+3\cos\Phi}{2(1-\cos\Phi)\sin\Phi\,\mathcal{J}_x}$

Here β^* is the betatron amplitude function at the symmetry point of the dispersion free straight section, L is the length of the dipole, and Φ is the phase advance of a FODO cell. To maximize beam brightness for synchrotron radiation with insertion devices, lattices with zero dispersion-function straight sections are favorable. Thus DBA, TBA, or nBA lattices are often used in the design of synchrotron radiation sources (see Sec. III.1).

Low emittance lattices require strong focusing optics. The correction of large chromaticities in these lattices requires strong chromaticity sextupoles. Dynamical aperture can be limited by strong nonlinear resonances and systematic chromatic stop-bands. Multiple-families of sextupoles are needed to correct geometric and chromatic aberrations. Since a strong focusing machine is much more sensitive to the dipole and quadrupole errors, the lifetime and brightness of the beam can be considerably reduced by power supply ripple, ground motion, and other error sources.

B. Diffraction limit

The conjugate phase space coordinates of a wave packet obey the uncertainty relation $\sigma_x \sigma_{kx} \sim \frac{1}{2}$, where σ_x and σ_{kx} are the rms beam width and the rms value of the conjugate wave number. The equality is satisfied for a Gaussian wave packet. Thus $\sigma_x \sigma_{x'} = \sigma_x(\sigma_{kx}/k) \sim 1/(2k) = \lambda/(4\pi)$, where $k = 2\pi/\lambda$ is the wave number in the longitudinal direction. The emittance of photon with wavelength λ is $\epsilon_{\rm photon} = \Delta x_{\rm r} \Delta x'_{\rm r} = \Delta z_{\rm r} \Delta z'_{\rm r} = \sigma_{\rm r} \sigma'_{\rm r} \sim \lambda/4\pi$,where the subscript r stands for radiation. The electron beam emittance that reaches the diffraction limit is $\epsilon_{\rm diff} \sim \epsilon_{\rm photon}$.

The electron beam emittance for producing hard X-ray at energies 10 keV is about $\epsilon_{\rm diff} \approx 10^{-11}$ m. High energy linacs can reach such a small emittance. For a storage ring, the emittance is given by Eq. (4.187). The photon wave length in undulator radiation is given by Eq. (4.210). To reach hard X-ray energy of 10 keV or wavelength of 0.1 nm for undulator wavelength of $\lambda_{\rm w} = 15$ mm and $K_{\rm w} = 1$, the electron energy must be larger than 6 GeV. We consider the 11BA lattice as a candidate of this storage ring. Each 11BA cell is made of 11 dipoles with middle dipole $\theta_2 = 1.5\theta_1$ for dispersion function matching. We find that the bending angle is $\theta_1 = 0.77°$, and the required number of superperiods is $N = 2\pi/](2 + 9 \times 1.5)\theta_1] \approx 30$. The dipole bending radius is determined energy spread of Eq.(4.116). With the energy spread about $\sigma_\delta = 5 \times 10^{-4}$, the bending radius is $\rho = 115$ m. the circumference is about 2400 m, and the compaction factor of this 30-11BA storage ring will be about 9×10^{-6}. With broadband impedance of $|Z_\parallel/n| \sim 0.5\Omega$, the single bunch peak current for the microwave threshold is about 180 mA. This is about a thousand times smaller than that can be achieved in linacs.

Now we consider vacuum ultra-violet (VUV) photons of 100 eV. The photon emittance is about 1 nm. A number of storage ring at 1 GeV can produce respectable emittance that reaches the diffraction limit. As an example, we consider a storage ring with $\rho = 3.5$ m, $\theta_1 = 20°$, $C = 120$m, the emittance of TME lattice is 1.34 nm. The momentum spread and compaction factor of such a lattice are $\sigma_\delta = 4.6 \times 10^{-4}$ and $\alpha_{\rm c} = 1.9 \times 10^{-3}$. The microwave threshold peak current is about 5 A.

C. Beam lifetime

Since high energy photons can desorb gases in a vacuum chamber, vacuum pressure is particularly important to beam lifetime in synchrotron radiation sources (see Exercise 4.1.8). The beam gas scattering processes include elastic and inelastic scattering with electrons and nuclei of the gases,[20] bremsstrahlung, ionization, ion trapping, etc. The beam-gas scattering lifetime is

$$\tau_{\rm g} = -\frac{1}{N}\frac{dN}{dt} = \sigma_{\rm tot}\beta c n, \tag{4.230}$$

where $\sigma_{\rm tot}$ is the total cross-section, $n = 3.22 \times 10^{22} P\,[\text{Torr}]\ \text{m}^{-3}$ is the density of the gas, and βc is the speed of the particle.

Because of these problems, many high brightness storage rings employ positrons with full energy injection. An effect associated with beam gas scattering is the multiple small angle Coulomb scattering, which results in emittance dilution. The small angle multiple Coulomb scattering between beam particles within a bunch is called the intrabeam scattering. This is particularly important for high-charge density low-energy beams (≤ 1 GeV). The beam emittances in low energy storage rings are usually determined by the intrabeam scattering.

The quantum lifetime can be controlled by the rf cavity voltage. The Touschek scattering discussed in Sec. II.7 arises from the Coulomb scattering that transfers transverse horizontal momentum into longitudinal momentum. If the longitudinal momentum of the scattered particle is outside the rf bucket, the particle will be lost. The Touschek lifetime depends on a high power of γ, and it is usually alleviated by increasing the beam energy. Another solution is to increase the rf voltage. However, the corresponding bunch length will be decreased and the peak beam current may be limited by collective beam instabilities.

D. Collective beam instabilities

Collective instabilities are important to high intensity electron beams. The single beam instabilities are usually driven by broadband impedance. The turbulent bunch lengthening or microwave instability leads to increase in bunch length and momentum spread (see Sec. VII.4; Chap. 3). The broadband impedance can be reduced by minimizing the discontinuities in the vacuum chamber. The transverse microwave instability has usually a larger threshold provided that the chromaticities are properly corrected.

In storage ring, there are high-Q components such as the rf cavities, un-shielded beam position monitors, etc. These accelerator components can lead to coupled

[20]See e.g., E. Weihreter, CERN 90-03, p. 427 (1990) for analysis on the vacuum requirement for compact synchrotron radiation sources.

bunch oscillations. The results are emittances dilution, fluctuation, lifetime degradation, intensity limitation, etc. Methods to combat these collective instabilities are minimizing the impedance by careful design of vacuum chamber, de-Qing HOMs of rf cavities, enlarging the tune spread with Landau cavities, and active feedback systems to damp the collective motion [5]. Besides collective beam instabilities, stability of the beam orbit is also an important issue. Power supply ripple, ground motion, mechanical vibration, and/or ground motion can perturb the closed orbit of the beam. Bunch-by-bunch feedback system is normally implemented for the control of the beam orbit.

Exercise 4.3

1. Dividing the dipole into two pieces, we can express the dispersion function transfer matrix of the half dipole by

$$M_{\frac{1}{2}B} = \begin{pmatrix} 1 & \frac{1}{2}L & L\theta/8 \\ 0 & 1 & \theta/2 \\ 0 & 0 & 1 \end{pmatrix},$$

where L and θ are the length and the bending angle of the dipole in the half cell.

(a) Use thin lens approximation and show that the dispersion function at the center of the dipole of a separate function FODO cell is

$$D = \frac{L\theta(1 - \frac{1}{8}\sin^2\frac{\Phi}{2})}{\sin^2(\Phi/2)}, \qquad D' = -\frac{\theta}{\sin(\Phi/2)},$$

where Φ is the phase advance per cell and L is the half cell length; and show that the dispersion invariant at the center of the dipole is

$$\mathcal{H} = \frac{\rho\theta^3}{\sin^3(\Phi/2)\cos(\Phi/2)}(1 - \frac{3}{4}\sin^2\frac{\Phi}{2} + \frac{1}{32}\sin^4\frac{\Phi}{2}).$$

(b) Apply 3-point Simpson's rule and show that the average of $\mathcal{H}$-function in dipole of a separate-function FODO cell is

$$\langle\mathcal{H}\rangle = \rho\theta^3\left[\frac{1 - \frac{3}{4}\sin^2(\Phi/2) + \frac{1}{48}\sin^4(\Phi/2)}{\sin^3(\Phi/2)\cos(\Phi/2)}\right].$$

The number in brackets is the $\mathcal{F}$ factor of Eq. (4.167), where the numerator depends slightly on the dipole configuration.

2. In a zero gradient dipole, the dispersion transfer matrix is

$$M = \begin{pmatrix} \cos\varphi & \rho\sin\varphi & \rho(1-\cos\varphi) \\ -(1/\rho)\sin\varphi & \cos\varphi & \sin\varphi \\ 0 & 0 & 1 \end{pmatrix}$$

where ρ is the bending radius, and $\varphi = s/\rho$ is the beam bending angle along the dipole. Using Exercise 2.4.11, show that the average of the $\mathcal{H}$-function in the dipole is

$$\begin{aligned}\langle\mathcal{H}\rangle &= \mathcal{H}_0 + 2(\alpha_0 D_0 + \beta_0 D_0')\frac{(1-\cos\theta)}{\theta} - 2(\gamma_0 D_0 + \alpha_0 D_0')\rho\,(1 - \frac{\sin\theta}{\theta}) \\ & -2\alpha_0\frac{\rho}{\theta}(\frac{3}{4} - \cos\theta + \frac{\cos 2\theta}{4}) + \frac{\beta_0}{2}(1 - \frac{\sin 2\theta}{2\theta}) + \gamma_0\rho^2(\frac{3}{2} - \frac{2\sin\theta}{\theta} + \frac{\sin 2\theta}{4\theta}),\end{aligned}$$

where θ is the bending angle of dipoles in a half cell and ρ is the bending radius. Express the $\langle\mathcal{H}\rangle$ in Eq. (4.173).

3. Show that average $\mathcal{H}$ in a dipole for a double-bend achromat (DBA) is

$$\langle\mathcal{H}\rangle = \rho\theta^3\left(\frac{\beta_0}{3\ell} - \frac{\alpha_0}{4} + \frac{\gamma_0\ell}{20}\right),$$

in small angle approximation, where $\ell = \rho\theta$, ρ and θ are respectively the length, the bending radius and bending angle of the dipole. If the betatron amplitude functions are normalized with $\tilde\alpha_0 = \alpha_0$, $\tilde\beta_0 = \beta_0/\ell$ and $\tilde\gamma_0 = \gamma_0\ell$, the average of the $\mathcal{H}$-function in the dipole becomes $\langle\mathcal{H}\rangle = \rho\theta^3\left(a\tilde\beta_0 - b\tilde\alpha_0 + c\tilde\gamma_0\right)$, with $a = \frac{1}{3}$, $b = \frac{1}{4}$, $c = \frac{1}{20}$.

   (a) Consider the function $F(\tilde\alpha_0, \tilde\beta_0, \tilde\gamma_0) = a\tilde\beta_0 - b\tilde\alpha_0 + c\tilde\gamma_0$ with $\tilde\beta_0\tilde\gamma_0 = 1 + \tilde\alpha_0^2$. Show that the minimum of the function F is $F = \sqrt{4ac - b^2}$ with $\tilde\alpha_0 = b/\sqrt{4ac - b^2}$, $\beta_0 = 2c/\sqrt{4ac - b^2}$, and $\gamma_0 = 2a/\sqrt{4ac - b^2}$. Using the result to show that the minimum of $\langle\mathcal{H}\rangle$ is

$$\langle\mathcal{H}\rangle_{\rm MEDBA} = \frac{1}{4\sqrt{15}}\rho\theta^3 \quad \text{with} \quad \alpha_0 = \sqrt{15}, \quad \beta_0 = \frac{6}{\sqrt{15}}\ell, \quad \gamma_0 = \frac{8\sqrt{15}}{3\ell},$$

   where the minimum of the $\langle\mathcal{H}\rangle$ function occurs at $s^* = \frac{3}{8}\ell$ with $\beta^* = \frac{3}{4\sqrt{60}}\ell$.

   (b) Show that the horizontal betatron phase advance across the dipole for a MEDBA lattice is

$$(\tan^{-1}\sqrt{15} + \tan^{-1}5\sqrt{15}/3).$$

   (c) Evaluate $(\alpha D + \beta D')/\sqrt{\beta_x}$ and $D/\sqrt{\beta_x}$ at the exit points of the dipole magnet for the MEDBA condition and show that

$$\frac{1}{\sqrt{\beta_x}}(\alpha D + \beta D') = \pm\frac{7\ell^{3/2}}{8(15)^{1/4}\rho}, \qquad \frac{1}{\sqrt{\beta_x}}D = \frac{(15)^{1/4}\ell^{3/2}}{8\rho}.$$

   Use the result to show that the phase advance of the dispersion function matching section of the MEDBA is $2\tan^{-1}(7/\sqrt{15})$.

4. When the $\langle\mathcal{H}\rangle$ of the minimum emittance (ME) lattice in Eq. (4.179) is minimized, show that the dispersion and the betatron amplitude functions inside the dipole are

$$D(s) = \frac{\ell^2}{24\rho} + \frac{1}{2\rho}(s - \frac{\ell}{2})^2, \qquad \beta = \beta^* + \frac{1}{\beta^*}(s - \frac{\ell}{2})^2,$$

where $s = 0$ corresponds to the entrance edge of the dipole, ℓ is the length of the dipole magnet, and $\beta^* = \ell/\sqrt{60}$. Show that the horizontal betatron phase advance in the dipole is $2\tan^{-1}\sqrt{15}$. Evaluate $(\alpha D + \beta D')/\sqrt{\beta_x}$ and $D/\sqrt{\beta_x}$ at the exit point of the dipole magnet for the ME condition and show that

$$\frac{1}{\sqrt{\beta_x}}(\alpha D + \beta D') = \pm\frac{3\ell^{3/2}}{4\sqrt{2}(15)^{1/4}\rho}, \qquad \frac{1}{\sqrt{\beta_x}}D = \frac{(15)^{1/4}\ell^{3/2}}{12\sqrt{2}\rho}.$$

Use the result to show that the phase advance of the matching section is $2\tan^{-1}(9/\sqrt{15})$.

5. A minimum emittance n-bend achromat (MEnBA) module is composed of $n-2$ ME modules inside a MEDBA module. Show that the necessary condition for matching ME modules to the MEDBA module in small angle approximation is

$$\left.\frac{\ell^3}{\rho^2}\right|_{\rm ME} = 3\left.\frac{\ell^3}{\rho^2}\right|_{\rm MEDBA}.$$

Thus an isomagnetic nBA can achieve optical matching for the minimum emittance only if the middle dipole is longer than the outer dipole by a factor of $3^{1/3}$. Find the minimum emittance. Extend your result to find a formula for the condition necessary for a matched MEnBA without using small angle approximation.

6. The dispersion function in the combined function dipole is

$$\begin{aligned} D &= \frac{1}{\rho K}(\cosh\phi - 1) + D_0\cosh\phi + \frac{1}{\sqrt{K}}D_0'\sinh\phi, \\ D' &= (D_0\sqrt{K} + \frac{1}{\rho\sqrt{K}})\sinh\phi + D_0'\cosh\phi, \end{aligned}$$

where $K = -K_x = |+B_1/B\rho + 1/\rho^2|$ is the defocusing strength with $B_1 = \partial B_z/\partial x$, $\phi = \sqrt{K}s$ is the betatron phase, $s = 0$ corresponds to the entrance of the dipole, and D_0 and D_0' are respectively the values of the dispersion function and its derivative at $s = 0$. The evolution of the $\mathcal{H}$-function in a dipole is

$$\begin{aligned} \mathcal{H}(\phi) &= \mathcal{H}_0 + \frac{2}{\rho\sqrt{K}}(\alpha_0 D_0 + \beta_0 D_0')\sinh\phi - \frac{2}{\rho K}(\gamma_0 D_0 + \alpha_0 D_0')(\cosh\phi - 1) \\ &\quad + \frac{\beta_0}{\rho^2 K}\sinh^2\phi + \frac{\gamma_0}{\rho^2 K^2}(\cosh\phi - 1)^2 - \frac{2\alpha_0}{\rho^2 K^{3/2}}\sinh\phi\,(\cosh\phi - 1), \end{aligned}$$

where $\mathcal{H}_0 = \gamma_0 D_0^2 + 2\alpha_0 D_0 D_0' + \beta_0 D_0'^2$, and α_0, β_0, and γ_0 are the Courant-Snyder parameters at $s = 0$.

(a) Averaging the $\mathcal{H}$-function in the dipole, show that

$$\begin{aligned} \langle\mathcal{H}\rangle &= \mathcal{H}_0 + \rho\theta^3\left[(\tilde\alpha_0 d_0 + \tilde\beta_0 d_0')E(q) - \frac{1}{3}(\tilde\gamma_0 d_0 + \tilde\alpha_0 d_0')F(q)\right. \\ &\quad \left. + \frac{\tilde\beta_0}{3}A(q) - \frac{\tilde\alpha_0}{4}B(q) + \frac{\tilde\gamma_0}{20}C(q)\right], \end{aligned}$$

where $\theta = L/\rho$ is the bending angle of the dipole, L is the length of the dipole, $q = \sqrt{KL}$ is the defocusing strength of the dipole, the normalized betatron amplitude functions are $\tilde{\alpha}_0 = \alpha_0$, $\tilde{\beta}_0 = \beta_0/L$, $\tilde{\gamma}_0 = \gamma_0 L$, $d_0 = D_0/L\theta$, $d'_0 = D'_0/\theta$, and

$$E(q) = \frac{2(\cosh q - 1)}{q^2}, \quad F(q) = \frac{6(\sinh q - q)}{q^3}, \quad A(q) = \frac{3\sinh 2q - 6q}{4q^3},$$
$$B(q) = \frac{6 - 8\cosh q + 2\cosh 2q}{q^4}, \quad C(q) = \frac{30q - 40\sinh q + 5\sinh 2q}{q^5}.$$

(b) The minimization procedure can be achieved through the following steps. First, $\langle\mathcal{H}\rangle$ can be minimized by finding the optimal dispersion functions with

$$\frac{\partial\langle\mathcal{H}\rangle}{\partial d_0} = 0, \quad \frac{\partial\langle\mathcal{H}\rangle}{\partial d'_0} = 0.$$

Show that the solution is $d_{0,min} = \frac{1}{6}F(q)$, $d'_{0,min} = -\frac{1}{2}E(q)$, with

$$\langle\mathcal{H}\rangle = \frac{1}{12}\rho\theta^3(\tilde{\beta}_0\tilde{A} - \tilde{\alpha}_0\tilde{B} + \frac{4\tilde{\gamma}_0}{15}\tilde{C}),$$

where $\tilde{A} = 4A - 3E^2, \tilde{B} = 3B - 2EF, \tilde{C} = \frac{9}{4}C - \frac{5}{4}F^2$.

(c) Using the relation $\tilde{\beta}_0\tilde{\gamma}_0 = 1 + \tilde{\alpha}_0^2$, show that the minimum $\langle\mathcal{H}\rangle$ is

$$\langle\mathcal{H}\rangle_{\rm ME} = \frac{\tilde{G}}{12\sqrt{15}}\rho\theta^3, \quad \text{with} \quad \tilde{\beta}_0 = \frac{8\tilde{C}}{\sqrt{15}\tilde{G}}, \ \tilde{\alpha}_0 = \frac{\sqrt{15}\tilde{B}}{\tilde{G}}, \ \tilde{\gamma}_0 = \frac{2\sqrt{15}\tilde{A}}{\tilde{G}},$$

where $\tilde{G} = \sqrt{16\tilde{A}\tilde{C} - 15\tilde{B}^2}$. Plot $\tilde{G}$ vs q. Show also that the value of the dispersion $\mathcal{H}$-function at both ends of the dipole for the ME condition is

$$\mathcal{H}(0) = \mathcal{H}(q) = \frac{1}{3\sqrt{15}\tilde{G}}\rho\theta^3\{6\tilde{C}E^2 - \frac{15}{2}\tilde{B}EF + \frac{5}{2}\tilde{A}F^2\}.$$

(d) Show that the damping partition number $\mathcal{J}_x$ for horizontal motion is

$$\mathcal{J}_x = 1 - \alpha_c\frac{R}{\rho} + \frac{4}{q^2}\left(\cosh q - 1 - \frac{1}{2}q^2\right).$$

Discuss the effect of damping partition number for the combined function ME lattice.

7. Consider a dipole in an achromat cell, shown schematically as follows, where (a) is a uniform field dipole, and (b) is divided into two sections having different dipole field strength. These two dipole schemes have the same bending angle, i.e.

$$\theta = \frac{L}{\rho} = \theta_1 + \theta_2 = \frac{L}{2\rho_1} + \frac{L}{2\rho_2}.$$

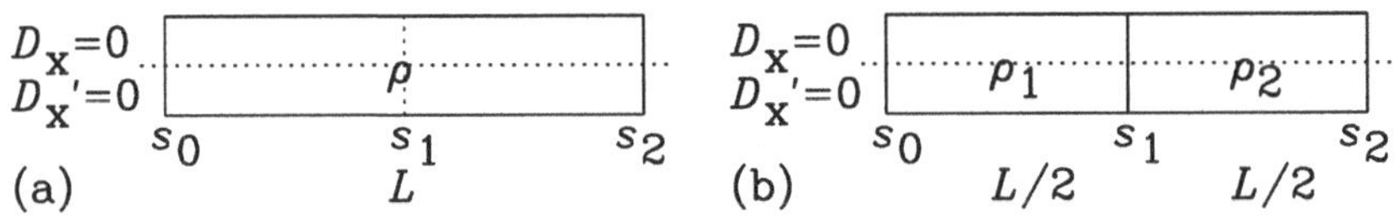

(a) Show that the dispersion function at the non-achromat end of the uniform field dipole in small angle approximation is $D_x(s_2) = \frac{1}{2}L\theta$ and $D'_x(s_2) = \theta$.

(b) Show that the dispersion function at the end of different strength dipole in small angle approximation is $D_x(s_2) = \frac{1}{2}L\theta + \frac{1}{4}L(\theta_1 - \theta_2)$ and $D'_x(s_2) = \theta$.

The dispersion function is larger than that of the uniform dipole field if $\theta_1 > \theta_2$. This means that the dispersion function at the non-achromatic end of a dipole with higher field at the achromatic end will be larger. The theoretical $\langle\mathcal{H}/\rho^3\rangle_{\text{minimum}}$ is not smaller for the longitudinal gradient dipole than that of the uniform field dipole. However, it appears easier to minimize $\langle\mathcal{H}/\rho^3\rangle$ in realistic lattice design.

8. Show that the dispersion function generated by a helical wiggler in a dispersion-free straight section is

$$D = \frac{K_{\text{w}}\lambda_{\text{w}}}{2\pi\gamma}\left[\hat{x}\sin\frac{2\pi s}{\lambda_{\text{w}}} + \hat{z}\left(1 - \cos\frac{2\pi s}{\lambda_{\text{w}}}\right)\right].$$

Find the ratio of the transverse beam sizes arising from the dispersion function and betatron motion, where we assume $K_{\text{w}} = 5$, $\gamma = 800$, $\lambda_{\text{w}} = 8.0$ cm, $\sigma_\delta = 0.1$ %, $\beta_x = \beta_z = 10$ m, and $\epsilon_{\text{N}} = 2\pi$ mm-mrad.

9. A variant of the double bend achromat is to replace the focusing quadrupole by a triplet. The configuration of the basic cell is[21]

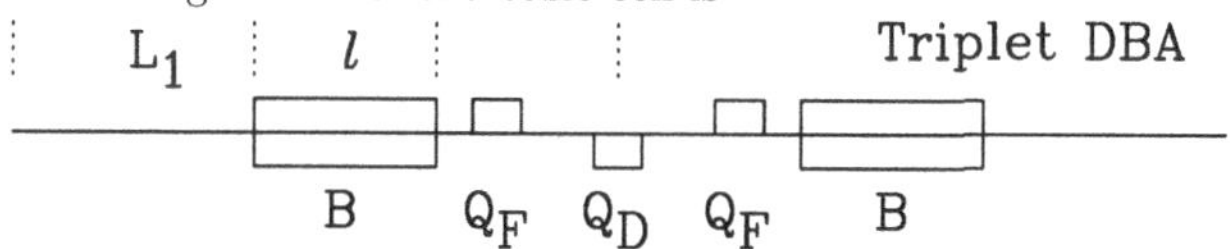

Here $2L_1$ is the length of the zero dispersion straight section, and ℓ is the length of the dipole. Since the mid-point of the straight section is the symmetry point for the lattice function, the betatron amplitude function inside the dipole is

$$\beta = \beta^* + \frac{(s + L_1)^2}{\beta^*},$$

where $s = 0$ corresponds to the entrance of the dipole.

(a) In small angle approximation, show that the average of the $\mathcal{H}$ function in the dipole is

$$\langle\mathcal{H}\rangle = \rho\theta^3\left[\frac{\beta^*}{3\ell} + \frac{1}{\beta^*}\left(\frac{L_1^2}{3\ell} + \frac{L_1}{4} + \frac{\ell}{20}\right)\right],$$

where $2L_1$ is the length of the zero dispersion straight section, and ℓ is the length of the dipole. Show that the minimum emittance of the triplet DBA is

$$\langle\mathcal{H}\rangle_{\min} = \rho\theta^3\frac{2\beta^*}{3\ell} \quad \text{with} \quad \frac{\beta^*}{\ell} = \sqrt{\frac{L_1^2}{\ell^2} + \frac{3L_1}{4\ell} + \frac{3}{20}}.$$

[21] Because there is no quadrupole in the straight section, such a configuration has the advantage of a very compact storage ring for synchrotron light with dispersion free insertions. This configuration appears in the SOR Ring in Tokyo and the ACO Ring in Orsay, where combined function dipoles are used.

Since the emittance is proportional to the betatron amplitude function at the insertion region, the emittance will be altered by insertion devices that alter the betatron amplitude function.

(b) Show that the betatron phase advance of the dispersion matching section for a minimum emittance triplet DBA is

$$\psi = 2\arctan\left(\frac{4\xi^2 + \frac{9}{2}\xi + \frac{13}{10}}{\sqrt{\xi^2 + \frac{3}{4}\xi + \frac{3}{20}}}\right),$$

where $\xi = L_1/\ell$. Plot the betatron phase advance of the matching section and β^*/ℓ as a function of ξ.

(c) What happens to $\langle\mathcal{H}\rangle$ and the natural emittance if the dipole is replaced by a combined function magnet?

(d) Study the linear stability of the triplet DBA lattice.

Chapter 5

Special Topics in Beam Physics

In preceding chapters, we have focused on particle dynamics of betatron and synchrotron motions, nonlinear beam dynamics, the effects of space-charge force, linac, impedance and collective beam instabilities, radiation damping and quantum fluctuation in electron storage rings, and synchrotron radiation. However, this introductory textbook does not address advanced topics including free-electron laser, laser-particle interaction, beam-beam interaction, beam cooling, advanced nonlinear beam dynamics, and collective beam instabilities. There are many textbooks and workshop proceedings on these advanced topics [14, 15, 16, 17, 18]. Nevertheless, this chapter provides introduction to two topics: free electron laser and beam beam interaction.

In Chapter 4, we discussed incoherent spontaneous synchrotron radiation of each individual electron in dipole or wiggler fields. The radiated electromagnetic wave plays no role on the motion of electrons. The radiation is incoherent and the power or intensity of the radiation is proportional to the number of electrons in a bunch. To produce coherent photons, it is necessary to induce laser oscillation in a laser cavity consisting of undulator and mirrors.[1] The idea has been extended to vacuum ultraviolet (VUV) and X-ray production by a process called Self-Amplified Spontaneous Emission (SASE), a collective instability induced microbunching in electron beam through interaction with the EM fields.

The center of mass energy available in fixed target experiments is limited by the kinematic transformation. Since 1960's, there are great efforts in developing colliders, where two counter-traveling beams collide at interaction points. The space charge force between two counter-traveling bunches produces large impulse on each other. The force is highly non-linear. It perturbs the beam distribution, degrades beam lifetime, causes beam instability, induces noises in the detector area, and plays a major role in limiting the luminosity of all high energy colliders.

[1]J.M.J. Madey, *J. Appl. Phys.* 42 (1971) 1906; R.L. Elias *et al.*, *Phys. Rev. Lett.* 36 (1976) 717; see also C.W. Robinson and P. Sprangle, *a Review of FELs*, p. 914 in Ref. [14].

I Free Electron Laser (FEL)

Lasers are coherent and high power light (radiation) sources. The radiation is generated by coherent transition from population-inverted states to a low-lying state of a lasing medium made of atomic or molecular systems. A free electron laser employs relativistic electron-beams in undulators to generate tunable, coherent, high power radiation. Its optical properties possess characteristic of conventional lasers: high spatial coherence and near the diffraction limit. Its wavelengths are tunable from millimeter to visible and potentially ultraviolet to x-ray.[2] Figure 5.1 summarizes the existing laboratories with free electron laser research facilities. For a complete updated list, see the World Wide Web Virtual Library of the Free Electron Laser at http://sbfel3.ucsb.edu/www/vl_fel.html.

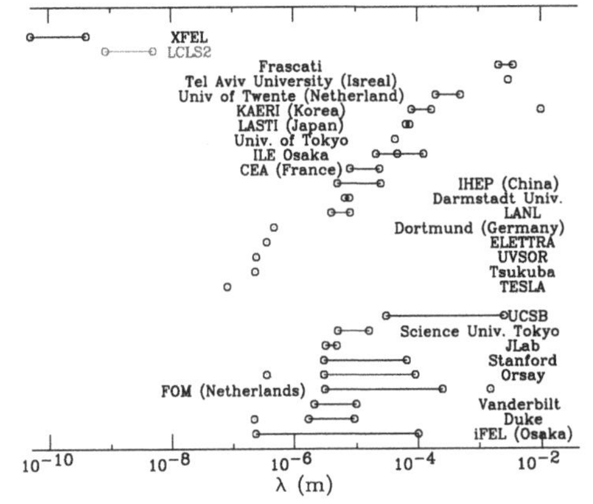

Figure 5.1: A compilation of the existing FEL laboratories with associated FEL wavelength. High gain X-ray FEL projects, such as the LCLS and XFEL to be completed around 2007, are not listed on this graph. The wavelengths of these projects are of the order of 0.1 nm. Besides these operational facilities listed in the graph, there are about 10 FEL development centers in universities and National Laboratories.

The circularly polarized plane electromagnetic (EM) wave, produced by the relativistic electron-beam in a helical wiggler magnetic field $\vec{B}_{\rm w}$ of Eq. (4.196)[3] is

$$\vec{E} = E_0[\hat{x}\sin(k_0 s - \omega_0 t + \phi_0) + \hat{z}\cos(k_0 s - \omega_0 t + \phi_0)], \quad \vec{B} = \frac{1}{c}\hat{s}\times\vec{E} \tag{5.1}$$

propagating along the wiggler axis $\hat{s}$. Here $k_0 = 2\pi/\lambda$, $\omega_0 = 2\pi c/\lambda$, ϕ_0 is an arbitrary initial phase of the EM wave, s is the longitudinal distance, and t is the time coordinate. In the presence of the electromagnetic fields, the equation of motion for electron is

$$\frac{d\vec{p}}{dt} = e\vec{E} + ec\vec{\beta}\times(\vec{B}+\vec{B}_{\rm w}) + F_{\rm s.c.} + F_{\rm radiation}. \tag{5.2}$$

[2]See Ref. [32] and *FEL Physics*, in SLAC-R-521 Chapter 4, (2001) and reference there in.

[3]For radiation from planar undulator, see Exercise 5.1.1. For simplicity, we limit our discussion to 1D FEL-theory, where the amplitude E_0 is independent of the transverse coordinates, and is a slowly varying function of t and s.

where βc is the speed of the electron. The space charge force $F_{\rm s.c.}$ is proportional to $1/\gamma^2$ (see Exercise 2.3.2), it is negligible at the energy of our consideration. The radiation reaction force $F_{\rm radiation}$ is also small in the energy E of electrons of our interest because

$$\frac{F_{\rm radiation}}{ecB_{\rm w}} = \frac{P_\gamma}{ec^2B_{\rm w}} = \frac{2r_0c}{3mc^2}\gamma^2 B_{\rm w} = 4.22\times 10^{-6}\,(E[\rm GeV])^2\, B_{\rm w}[\rm T] \ll 1.$$

Here P_γ is the instantaneous radiation power in the wiggler and r_0 is the electron classical-radius.

The wiggler field provides electron trajectory while the EM-fields interact with electrons for energy exchange. Using Eqs. (5.1) and (4.197), we obtain the energy-exchange between the electron and electromagnetic wave: $mc^2\dot\gamma = -e\vec{E}\cdot\vec{\beta}c$, or

$$\gamma' \equiv \frac{d\gamma}{ds} = -\frac{eE_0\beta_\perp}{mc^2}\sin\phi = -\frac{eE_0K_{\rm w}}{\gamma mc^2}\sin\phi, \tag{5.3}$$

$$\phi = (k_{\rm w}+k_0)s - \omega_0 t + \phi_0. \tag{5.4}$$

With $ds/dt = c\beta_\parallel$ of Eq. (4.197), the stationary phase (or **resonance**) condition to maximize the energy exchange is

$$0 = \phi' \equiv \frac{d\phi}{ds} = \frac{2\pi}{\lambda_{\rm w}\lambda}\left[\lambda + \lambda_{\rm w}\left(1-\frac{1}{\beta_\parallel}\right)\right] \approx k_{\rm w}\left(1 - \frac{\lambda_{\rm w}(1+K_{\rm w}^2)}{2\gamma^2\lambda}\right), \tag{5.5}$$

For a planar undulator, the longitudinal velocity vector is given by Eq. (4.195), and thus the wiggler parameter in the resonance condition should be replaced by the rms undulator parameter $K_{\rm w,rms} = K_{\rm w}/\sqrt{2}$ (See also Exercise 5.1.1). The resonance condition

$$\gamma_r^2 = \frac{\lambda_{\rm w}(1+K_{\rm w}^2)}{2\lambda} \quad \text{or} \quad \lambda_r = \lambda_{\rm w}\frac{1+K_{\rm w}^2}{2\gamma^2}. \tag{5.6}$$

agrees with Eq. (4.210). When this resonance condition is satisfied, electrons lag behind the EM wave by one wavelength as electrons advance one wiggler period, i.e. $\omega_0\Delta t = \omega_0(\frac{\lambda_{\rm w}}{\beta_\parallel c} - \frac{\lambda_{\rm w}}{c}) = 2\pi$, graphically represented in Fig. 4.24.

The equation of motion for the phase angle ϕ becomes

$$\phi' = k_{\rm w}(1 - \frac{\gamma_r^2}{\gamma^2}). \tag{5.7}$$

The coupled equations (5.3) and (5.7) can be derived from the Hamiltonian

$$H = k_{\rm w}(1+\frac{\gamma_r^2}{\gamma^2})\gamma - \frac{eE_0K_{\rm w}}{\gamma_r mc^2}\cos\phi, \tag{5.8}$$

where (γ,ϕ) are conjugate phase-space coordinates and longitudinal distance s serves as the independent "time" coordinate. The energy exchange between the electron and the external electromagnetic fields can be obtained by solving Hamilton's equation.

I.1 Small Signal Regime

In a small radiation loss regime, $\gamma \simeq \gamma_r$, we define a small parameter

$$\eta = \frac{\gamma - \gamma_r}{\gamma_r}. \tag{5.9}$$

The Hamiltonian and Hamilton's equations of motion are

$$H = k_\mathrm{w}\eta^2 - \frac{eE_0K_\mathrm{w}}{mc^2\gamma_r^2}\cos\phi; \qquad \eta' = -\frac{eE_0K_\mathrm{w}}{mc^2\gamma_r^2}\sin\phi, \quad \phi' = 2k_\mathrm{w}\eta, \tag{5.10}$$

where (η, ϕ) are conjugate phase-space coordinates and the longitudinal coordinate s is the independent variable. The Hamiltonian resembles that of the synchrotron motion. The small amplitude wave-number κ is

$$\kappa^2 = \frac{2k_\mathrm{w}eE_0K_\mathrm{w}}{mc^2\gamma_r^2} = \frac{eE_0\lambda_\mathrm{w}K_\mathrm{w}}{\pi\gamma_r^2mc^2}k_\mathrm{w}^2. \tag{5.11}$$

$$\frac{\kappa}{k_\mathrm{w}} = \sqrt{\frac{eE_0\lambda_\mathrm{w}K_\mathrm{w}}{\pi\gamma_r^2mc^2}} = 4.03\times10^{-5}\frac{\sqrt{K_\mathrm{w}\lambda_\mathrm{w}[\mathrm{cm}]E_0[\mathrm{MV/m}]}}{E[\mathrm{GeV}]}. \tag{5.12}$$

Here, the quantities k_w and κ/k_w play the roles of the orbital wave number and "synchrotron" tune respectively. The Hamiltonian and Hamilton's equations of motion become

$$H = k_\mathrm{w}\eta^2 - \frac{1}{2}k_\mathrm{w}\left(\frac{\kappa}{k_\mathrm{w}}\right)^2\cos\phi = \frac{\kappa^2}{2k_\mathrm{w}}\left[\frac{1}{2}\left(\frac{\phi'}{\kappa}\right)^2 - \cos\phi\right], \tag{5.13}$$

$$\phi' = 2k_\mathrm{w}\eta, \quad \eta' = -\frac{k_\mathrm{w}}{2}(\frac{\kappa}{k_\mathrm{w}})^2\sin\phi, \tag{5.14}$$

where $(\phi, \frac{\phi'}{\kappa})$ forms a set of normal phase-space coordinates. Figure 5.2 shows tori of Hamiltonian flow in phase-space $(\phi'/\kappa = 2k_\mathrm{w}\eta/\kappa, \phi)$. The separatrix corresponds to a Hamiltonian torus that passes through $(\phi = \pm\pi, \frac{\phi'}{\kappa} = 0)$ and $(\phi = 0, \frac{\phi'}{\kappa} = \pm 2)$:

$$H_\mathrm{sx} = k_\mathrm{w}\eta_\mathrm{sx}^2 - \frac{eE_0K_\mathrm{w}}{mc^2\gamma_r^2}\cos\phi = \frac{eE_0K_\mathrm{w}}{mc^2\gamma_r^2}. \tag{5.15}$$

The energy exchange between the electron and EM-fields depends on the electron trajectory in phase-space. The electron can lose or gain energy. To calculate the energy transfer, we have to integrate the electron paths and average over the initial condition of all electrons in the beam bunch. The time evolution of the electron beam distribution function $f(\phi, \eta, s)$ is governed by the Vlasov equation.

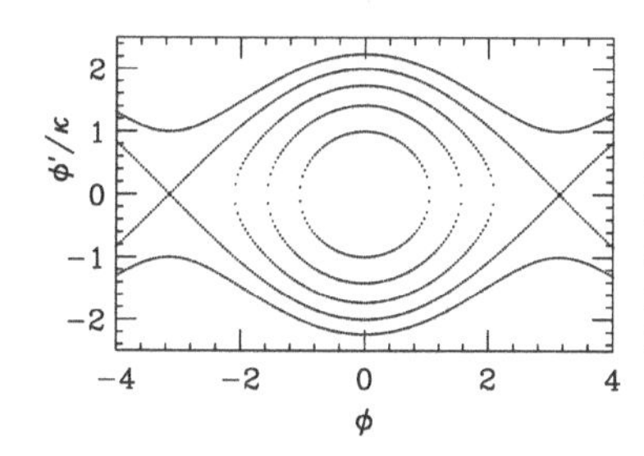

Figure 5.2: The phase-space ellipses of the energy loss ϕ'/κ vs. ϕ for electrons in an undulator. The energy exchange between the electron and the EM fields resembles the synchrotron motion for a particle in the rf system.

A. Vlasov equation in longitudinal phase-space coordinates

When the transverse and longitudinal oscillations are decoupled, the longitudinal distribution function obeys the Vlasov equation:

$$\frac{\partial f}{\partial s} + \phi' \frac{\partial f}{\partial \phi} + \eta' \frac{\partial f}{\partial \eta} = 0, \tag{5.16}$$

where the conjugate coordinates (ϕ, η) satisfy Eq. (5.13). A steady-state solution of the Vlasov equation is a function of the Hamiltonian: $f_{eq} = f_{eq}(H(\phi, \eta))$, which is an even function of η. If an initial beam distribution is *not* a function of the Hamiltonian, the distribution function evolves with "time", s. Depending on the initial condition, the time evolution of the bunch can be obtained by solving the Vlasov equation. Once the distribution is obtained, the energy exchange can be calculated by averaging the variable η over distribution function in phase-space coordinates, i.e.

$$\langle \eta \rangle = \int \eta f(\phi, \eta) d\phi d\eta.$$

Any equilibrium distribution function that is an even function of η gives $\langle \eta \rangle = 0$. The equilibrium distribution function has no net energy-exchange between electrons and EM-fields. For example, we consider a distribution that is initially uniform in ϕ, i.e.

$$f_0(\phi, \eta) = f(\phi, \eta, s = 0) = \frac{n_0}{2\pi} g(\eta). \tag{5.17}$$

where n_0 is the number of particle per unit volume. The maximum change in the energy-deviation coordinate in one wiggler period is $\Delta\eta|_{\max} \approx \pi(\frac{\kappa}{k_{\rm w}})^2$ as shown in Eq. (5.14). Since $\frac{\kappa}{k_{\rm w}}$ is a small number, we expand the distribution function in power series of $(\frac{\kappa}{k_{\rm w}})^2$ and solve the distribution function iteratively:

$$f = \sum_n f_n(\phi, \eta, s) (\frac{\kappa}{k_{\rm w}})^{2n},$$

$$\frac{\partial f_n}{\partial s} + 2k_{\rm w}\eta \frac{\partial f_n}{\partial \phi} = \frac{k_{\rm w}}{2} \sin\phi \frac{\partial f_{n-1}}{\partial \eta}, \tag{5.18}$$

$$f_1 = -\frac{n_0}{8\pi\eta} \frac{dg}{d\eta} (\cos\phi - \cos(\phi - 2\eta k_{\rm w} s)), \tag{5.19}$$

where we use f_0 of Eq. (5.17) to obtain f_1, which is not a uniform function of ϕ, i.e. the electron beam becomes bunched.

The exchange of energy between the electron and the EM-fields is

$$\langle\eta\rangle = \int \eta f_0 d\eta d\phi + (\frac{\kappa}{k_{\rm w}})^2 \int \eta f_1 d\eta d\phi \equiv \langle\eta\rangle_0 + \left(\frac{\kappa}{k_{\rm w}}\right)^2 \langle\eta\rangle_1, \tag{5.20}$$

where $\Delta\langle\eta\rangle = \langle\eta\rangle - \langle\eta\rangle_0 = \langle\eta\rangle_1 = 0$ in the ϕ-integral with the distribution function (5.19). The beam is bunched in the first order perturbation expansion without producing any energy exchange.

To obtain an energy exchange, we need to perform a second order perturbation calculation by expanding f_2 as

$$f_2 = \sum_{m=-2}^{m=2} f_{2m} e^{im\phi}. \tag{5.21}$$

The term that contributes to the energy exchange is f_{20} with

$$\begin{aligned} \frac{\partial f_{20}}{\partial s} &= \frac{n_0 k_{\rm w}}{32\pi}\left[\left(\frac{\partial}{\partial\eta}\left[\frac{1}{\eta}\frac{\partial g}{\partial\eta}\right]\right)\sin(2\eta k_{\rm w}s) + 2k_{\rm w}s\left(\frac{1}{\eta}\frac{\partial g}{\partial\eta}\right)\cos(2\eta k_{\rm w}s)\right] \\ f_{20} &= \frac{n_0}{64\pi}\frac{\partial}{\partial\eta}\left[\frac{1}{\eta^2}\frac{\partial g}{\partial\eta}[1-\cos(2\eta k_{\rm w}s)]\right]. \end{aligned} \tag{5.22}$$

Averaging η over the distribution function f_{20}, we obtain

$$\Delta\langle\eta\rangle = \langle\eta\rangle - \langle\eta\rangle_0 = \frac{n_0}{16}\left(\frac{\kappa}{k_{\rm w}}\right)^4 (2k_{\rm w}s)^3 \int g(\eta)F(2\eta k_{\rm w}s)d\eta, \tag{5.23}$$

with

$$F(\tau) = \frac{1}{\tau^3}\left[\cos\tau - 1 + \frac{\tau}{2}\sin\tau\right] = \frac{1}{4}\frac{d}{d\tau}[S(\tau)]^2, \qquad S(\tau) = \frac{\sin(\tau/2)}{(\tau/2)}, \tag{5.24}$$

where $S(\tau)$ is the line-shape function of the spontaneous emission in an undulator. The line shape function is equal to the square of the diffraction coherent function in Eq. (4.207). The factor $\frac{n_0}{16}\left(\frac{\kappa}{k_{\rm w}}\right)^4$ in Eq. (5.23) can be expressed as

$$\frac{n_0}{16}\left(\frac{\kappa}{k_{\rm w}}\right)^4 = \frac{1}{2}(2\rho_{\rm fel})^3 \frac{U_{\rm em}}{\gamma_r mc^2}, \tag{5.25}$$

where $U_{\rm em} = \langle U_{\rm e} + U_{\rm m}\rangle = \epsilon_0 E_0^2$ is the average energy-density of electromagnetic fields, $\rho_{\rm fel}$ is the FEL (or the Pierce) parameter:

$$(2\rho_{\rm fel})^3 = \frac{\mu_0 n_0 e^2 \lambda_{\rm w}^2 K_{\rm w}^2}{8\pi^2\gamma_r^3 m} = 2\left(\frac{K_{\rm w}\omega_{\rm pl}}{\gamma\omega_{\rm w}}\right)^2 = \frac{\hat{I}}{I_{\rm A}}\frac{\lambda_{\rm w}^2 K_{\rm w}^2}{2\pi\Sigma_e\gamma^3}. \tag{5.26}$$

Here, $\omega_{\rm pl} = \sqrt{n_0 e^2/\epsilon_0 \gamma m}$ is the plasma frequency of the electron beam, $\omega_{\rm w} = k_{\rm w}c$, $\hat{I} = n_0 \Sigma_e ec$ is the peak current, $I_{\rm A} = ec/r_0 = 17.0$ kA is the Alfvén current, and Σ_e is the electron beam cross-sectional area. Equation (5.23) allows us to evaluate the average electron energy exchange at the exit of the wiggler: $s = N_{\rm w}\lambda_{\rm w}$ and $\tau = \eta 2k_{\rm w}s = \eta 4\pi N_{\rm w}$. The gain function $F(\tau)$ is plotted in Fig. 5.3. Electrons lose energy if $\tau \geq 0$ and gain energy if $\tau < 0$. The gain is proportional to the slope of the spontaneous emission line-shape function.

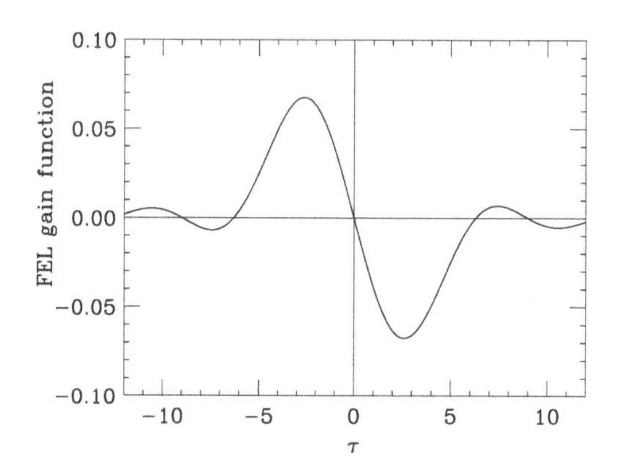

Figure 5.3: The gain function $F(\tau) = [\cos\tau - 1 + (\tau/2)\sin\tau]/\tau^3$ of the FEL is plotted as a function of the parameter $\tau = 2\eta k_{\rm w}s = 4\pi N_{\rm w}\eta$.

If the initial energy distribution is narrow compared with the width of the function $F(\tau)$, e.g. $4\pi N_{\rm w}\Delta\eta < 2$ or $\Delta\eta < \frac{1}{2\pi N_{\rm w}}$, we can approximate $g(\eta) = \delta(\eta - \eta_0)$. The energy exchange becomes

$$\Delta\langle\eta\rangle = \frac{n_0}{16}\left(\frac{\kappa}{k_{\rm w}}\right)^4 (4\pi N_{\rm w})^3 F(\tau_0) = \left\{4(4\pi\rho_{\rm fel}N_{\rm w})^3 F(\tau_0)\right\}\frac{U_{\rm em}}{\gamma_r mc^2}, \tag{5.27}$$

where $\tau_0 = 4\pi N_{\rm w}\eta_0$. The maximum energy loss is obtained with $\tau_0 = 2.6$ or $\eta_0 = \frac{0.2}{N_{\rm w}}$, which corresponds to $F(\tau_0) = 0.0675$.

B. The free electron laser gain

The energy loss or gain of the electron beam bunch transforms energy to or extracts energy from the electromagnetic fields. The gain of the free electron laser, defined as the fractional increase of electromagnetic wave intensity of the spontaneous emission in a single pass, is

$$G_0 = \frac{N_{\rm w}\lambda_{\rm w}\Sigma_e\gamma_r mc^2\langle\Delta\eta\rangle}{N_{\rm w}\lambda_{\rm w}\Sigma_\gamma U_{\rm em}} = 4(4\pi\rho_{\rm fel}N_{\rm w})^3 F(\tau_0), \tag{5.28}$$

where Σ_e and Σ_γ are the cross-sectional areas of the electron and photon beams and $\Sigma_e = \Sigma_\gamma = N_{\rm w}\lambda_{\rm w}\lambda/(16\pi)$ is assumed at the undulator.

Electrons lose energy if $\gamma > \gamma_r$ and gain energy if $\gamma < \gamma_r$. There is no energy exchange if $\gamma = \gamma_r$. The maximum energy gain occurs at the condition $\tau_0 = 2.6$:

$$\Delta\gamma\Big|_{\max} \approx \frac{5.2}{4\pi N_{\rm w}}\gamma_r \approx \frac{\gamma_r}{2N_{\rm w}}. \tag{5.29}$$

The efficiency for spontaneous emission in an undulator is $\leq \frac{1}{2N_{\rm w}}$. For a beam with finite momentum spread, the gain function is reduced by folding integral of Eq. (5.23). If the fractional momentum width is larger than $1/2N_{\rm w}$, i.e. $\sigma_\gamma/\gamma \geq 1/2N_{\rm w}$, the effective gain becomes nearly zero. The natural momentum width of the beam, given by Eq. (4.114), is normally well within the limit.

The FEL gain G_0 in Eq. (5.28) is proportional to the peak current $\hat{I}$. It is important to increase the peak current in wiggler region in order to enhance the FEL gain. Since the FEL gain is only a few percent, an optical cavity with two mirrors in the simplest configuration can be added to induce FEL oscillation. One of these mirrors is assumed to be a perfect reflector, while the other is assumed to transmit a fraction g_0 of the incident light (see Fig. 5.4). Neglecting the possible loss of light in the cavity, the system can be a laser oscillator if the gain is larger than the loss, i.e. $G \geq g_0$. When $G = g_0$, the system is in steady state operation. At the steady state, the laser output power is

$$P_{\rm L} = \text{efficiency} \ \times EI_{\rm av}, \tag{5.30}$$

where $I_{\rm av}$ is the average electron particle current and E is the energy of the electron. The efficiency of the device, i.e. the fraction of energy transfer, is about $\frac{1}{2N_{\rm w}}$.

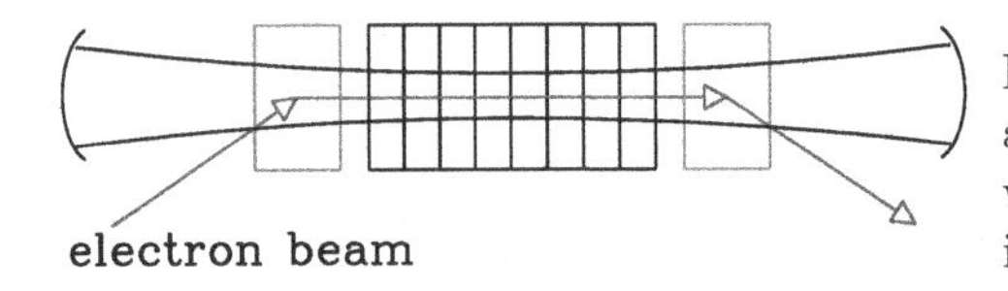

Figure 5.4: A schematic drawing of an optical cavity for FEL resonator with mirrors, while the electron beam is guided by the deflectors.

Because the laser gain is proportional to the peak electron beam current, the space time structure of the laser beam reflects the electron beam structure. For a bunch beam operation, the laser pulse length is equal to the electron pulse length σ_s. The time structure of the laser line width becomes

$$\sigma_\omega = \frac{2\pi c}{\sigma_s}; \quad \text{or} \quad \frac{\sigma_\omega}{\omega} = \frac{\lambda}{\sigma_s} = \frac{L_{\rm w}(1+K_{\rm w}^2)}{2\gamma^2\sigma_s N_{\rm w}}, \tag{5.31}$$

which is smaller than the diffraction limit of $1/2N_{\rm w}$. To sustain amplification, the synchronization of the laser pulses with the electron beam bunches is important as well. The synchronization procedure can be achieved by adjusting the length of the optical cavity.

I.2 Interaction of the Radiation Field with the Beam

We obtained the gain of EM field-intensity through the electron motion. When the gain is large, the system is coupled. The evolution of the electromagnetic fields is obtained through Maxwell's equation:

$$\nabla \times \vec{B} - \frac{1}{c^2}\frac{\partial \vec{E}}{\partial t} = \mu_0 \vec{J}, \qquad \vec{J} = ec\sum \vec{\beta}_{\perp i}\delta(\vec{r}-\vec{r}_i). \tag{5.32}$$

where J is the transverse electric-current: The electric and magnetic fields are (see Eq. (5.1))

$$\begin{aligned}\vec{E} &= E_0(s,t)[\hat{x}\sin(k_0 s - \omega_0 t + \phi_0) + \hat{z}\cos(k_0 s - \omega_0 t + \phi_0)],\\ \vec{B} &= \frac{1}{c}E_0(s,t)[\hat{x}\cos(k_0 s - \omega_0 t + \phi_0) - \hat{z}\sin(k_0 s - \omega_0 t + \phi_0)].\end{aligned}$$

The amplitude and phase, $E_0(s,t)$ and $\phi_0(s,t)$, are slowly varying functions in coordinates s and t within one optical wave length, i.e.

$$\frac{\partial E_0}{\partial t} \ll \omega_0 E_0; \quad \frac{\partial E_0}{\partial s} \ll \frac{2\pi}{\lambda}E_0; \quad \frac{\partial \phi_0}{\partial t} \ll \omega_0\phi_0; \quad \frac{\partial \phi_0}{\partial s} \ll \frac{2\pi}{\lambda}\phi_0.$$

Carrying out some algebraic manipulation, we obtain

$$\begin{aligned}&\frac{\partial E_0}{\partial s} + \frac{1}{c}\frac{\partial E_0}{\partial t} = -\mu_0 c J_a, \qquad E_0\left(\frac{\partial \phi_0}{\partial s} + \frac{1}{c}\frac{\partial \phi_0}{\partial t}\right) = -\mu_0 c J_b, \\ &J_a = J_x \sin(k_0 s - \omega_0 t + \phi_0) + J_z\cos(k_0 s - \omega_0 t + \phi_0) = \frac{ecK_{\rm w}}{\gamma_r} n_0\left\langle\frac{\sin\phi}{1+\eta}\right\rangle, \\ &J_b = J_x \cos(k_0 s - \omega_0 t + \phi_0) - J_z\sin(k_0 s - \omega_0 t + \phi_0) = \frac{ecK_{\rm w}}{\gamma_r} n_0\left\langle\frac{\cos\phi}{1+\eta}\right\rangle.\end{aligned} \tag{5.33}$$

Here $\langle ...\rangle$ is the ensemble average over the beam distribution in phase space coordinates ϕ and $\eta = \frac{\gamma-\gamma_r}{\gamma_r}$ given by Eqs. (5.3) and (5.9), and we use Eq. (5.32) to obtain J_a and J_b. The ensemble average of any function g is defined as

$$\langle g\rangle = \int d\phi \int d\eta g(\phi,\eta) f(\phi,\eta,s) \tag{5.34}$$

where the electron-beam distribution-function $f(\phi,\eta)$ satisfies the Vlasov equation Eq. (5.16). The system of coupled equations can be solved numerically at a given initial condition to obtain the time and space evolution of the electron beam and of the EM fields.

A. Perturbation solution of the Maxwell-Vlasov equations

The perturbed distribution function of Eq. (5.19), up to the first order of $(\frac{\kappa}{k_{\rm w}})^2$, is

$$f = \frac{n_0}{2\pi}g(\eta) + \frac{n_0}{8\pi}\frac{1}{\eta}\frac{\partial g}{\partial \eta}[-\cos\phi + \cos(\phi - 2\eta k_{\rm w} s)]\left(\frac{\kappa}{k_{\rm w}}\right)^2. \tag{5.35}$$

This solution was obtained by assuming a constant electric field-amplitude E_0. Using the zeroth order distribution function, we find $\langle\cos\phi\rangle = 0$ and $\langle\sin\phi\rangle = 0$, and this $J_a = 0$ and $J_b = 0$.

Using the first order perturbation distribution function, which carries the information of beam bunching, we obtain J_a and J_b as

$$\begin{aligned} J_a &= \frac{ecK_{\rm w}}{16\gamma_r} n_0 (\frac{\kappa}{k_{\rm w}})^2 \frac{1}{\eta^2} [\sin(2\eta k_{\rm w} s) - (2\eta k_{\rm w} s)\cos(2\eta k_{\rm w} s)], \\ J_b &= \frac{ecK_{\rm w}}{16\gamma_r} n_0 (\frac{\kappa}{k_{\rm w}})^2 \frac{1}{\eta^2} [\cos(2\eta k_{\rm w} s) - 1 + (2\eta k_{\rm w} s)\sin(2\eta k_{\rm w} s)]. \end{aligned}$$

The change in the electric field-amplitude E_0, up to first order in $(\frac{\kappa}{k_{\rm w}})^2$, and the gain of electromagnetic-field energy can be obtained by integrating Eq. (5.33):

$$\Delta E_0 = \frac{\mu_0 n_0 e c^2 \lambda_{\rm w} K_{\rm w}}{32\pi\gamma_r} (\frac{\kappa}{k_{\rm w}})^2 (2k_{\rm w} s)^3 F(\tau_0). \tag{5.36}$$

$$G_0 = \frac{2\Delta E_0}{E_0} = 4(4\pi\rho_{\rm fel} N_{\rm w})^3 F(\tau_0),$$

which is identical to that of Eq. (5.28). This verifies the fact that energy loss or gain in electron beam is equal to the energy gain or loss in the electromagnetic radiation.

B. High gain regime

The wavelength of the electromagnetic wave radiated by electrons in an undulator is determined by the resonance condition Eq. (5.5) with a line-width of (4.209). The radiation is not transversely coherent and its power is proportional to the peak beam current, or the number of electrons. Under certain conditions, the EM-wave can cause the electron-beam to bunch itself into microbunches with bunch-length equal to the wavelength of the EM-wave. All electrons in each microbunch radiate coherently in-phase. The radiation-intensity is proportional to the square of the total charge in a microbunch; and its power is greatly enhanced.

The population inversion in the free-electron coherent-lasing process is produced by the electron beam microbunching. The electromagnetic radiation occurs through coherent transition of the microbunched electron beam as a single identity. The process can be described by the coupled Vlasov-Maxwell equations. We consider a simple one-dimensional (1D) approximation:

$$\frac{\partial f}{\partial s} + \phi' \frac{\partial f}{\partial \phi} + \eta' \frac{\partial f}{\partial \eta} = 0, \tag{5.37}$$

$$\frac{\partial E_0}{\partial s} + \frac{1}{c}\frac{\partial E_0}{\partial t} = -\mu_0 c J_a, \tag{5.38}$$

$$\phi' = 2k_{\rm w}\eta, \quad \eta' = -\frac{k_{\rm w}}{2}(\frac{\kappa}{k_{\rm w}})^2 \sin\phi, \qquad J_a = \frac{n_0 e c K_{\rm w}}{\gamma_r} \langle \frac{\sin\phi}{1+\eta} \rangle.$$

With the coordinate-transformation from (s,t) to (s,ϕ) of Eq. (5.4), Maxwell's equation (5.38) becomes (see Exercise 5.1.2)

$$\frac{\partial E_0}{\partial s} \approx \left(\frac{\mu_0 n_0 c^2 e K_{\rm w}}{\gamma_r}\right)\left[\frac{1}{2\pi}\int d\eta d\phi\ f(\eta,\phi)\sin\phi\right], \tag{5.39}$$

where we assume E_0 is independent of ϕ or the detuning of the electric field amplitude E_0 is neglected. The amplitude of the EM-field can be enhanced by the electron-beam distribution function.

Now we consider perturbation to the electron beam distribution function with $f = f_0(\eta) + f_1(\eta,\phi)$, where f_0 is the unperturbed distribution, and f_1 is the perturbed distribution function resulting from the interaction with the EM-field. With $\kappa/k_{\rm w}$ from Eq. (5.12), the linearized Vlasov equation is

$$\frac{\partial f_1}{\partial s} + 2k_{\rm w}\eta\frac{\partial f_1}{\partial \phi} = +j\left(\frac{eK_{\rm w}}{2\gamma_r^2 mc^2}\frac{\partial f_0}{\partial \eta}\right)\left(e^{-j\phi} - e^{+j\phi}\right)E_0. \tag{5.40}$$

The perturbed electron beam distribution f_1 is coupled to the ponderomotive force of the electric field.

Self-consistent solution of the electric field and the perturbed distribution function can be expressed as

$$E_0 = \tilde{E}_0 \exp[j(\Omega s] + \text{c.c.}, \tag{5.41}$$

$$f_1 = \tilde{f}_{1+}\exp[j(\Omega s - \phi)] + \tilde{f}_{1-}\exp[j(\Omega s + \phi)] + \text{c.c.}. \tag{5.42}$$

Substituting the above equation to Eqs (5.39) and (5.40), we obtain the equation for eigen-wave-number Ω:[4]

$$\Omega + (2\rho_{\rm fel}k_{\rm w})^3\int\frac{\partial f_0/\partial\eta}{\Omega - 2k_{\rm w}\eta}d\eta = \Omega - (2\rho_{\rm fel}k_{\rm w})^3\int\frac{f_0(\eta)}{(\Omega - 2k_{\rm w}\eta)^2}d\eta = 0. \tag{5.43}$$

For an initial delta-function distribution function with $f_0 = \delta(\eta)$, the eigenvalues are

$$\Omega_1 = 2\rho_{\rm fel}k_{\rm w},\quad \Omega_2 = 2\rho_{\rm fel}k_{\rm w}\exp\left(j\frac{2\pi}{3}\right),\quad \Omega_3 = 2\rho_{\rm fel}k_{\rm w}\exp\left(-j\frac{2\pi}{3}\right), \tag{5.44}$$

where $\rho_{\rm fel}$ is the FEL parameter given by Eq. (5.26). There is a mode that grows exponentially. The exponential growth-factor of the electric field $|\tilde{E}_0|$ and the microbunching in the electron beam distribution function is determined by the imaginary part of the eigenvalues: $|{\rm Im}(\Omega_2)| = \sqrt{3}\rho_{\rm fel}k_{\rm w}$. The evolution of the magnitude of electric field and power is

$$|E_0| \sim e^{|{\rm Im}(\Omega)|s} = e^{\sqrt{3}\rho_{\rm fel}k_{\rm w}s},\qquad \text{Power} \sim |E_0|^2 \sim e^{2\sqrt{3}\rho_{\rm fel}k_{\rm w}s} = e^{s/L_{\rm g}}.$$

$$L_{\rm g,1D} = \frac{1}{2|Im(\Omega)|} = \frac{\lambda_{\rm w}}{4\sqrt{3}\ \pi\rho_{\rm fel}}. \tag{5.45}$$

[4] R. Bonifacio, C. Pellegrini, and L. Narducci, *Opt. Commun.*, **50**, 373 (1984); K.-J. Kim, *Nucl. Instru. and Methods*, **A250**, 396 (1986); J.M. Wang and L.H. Yu, *Nucl. Instru. and Methods*, **A250**, 484 (1986).

where the power gain-length is defined as the e-folding distance of the electromagnetic-field energy. The electric field gain-length is twice the power gain-length. The exponential growth will eventually saturate at a saturation length about $20L_g$.

The fact that the beam microbunching arises from the shot-noise and its effect is amplified by the beam-laser interaction. This instability is called *self-amplified spontaneous emission* (SASE) process.

I.3 High Gain FEL Facilities

Since 1980, experiments using high gain FEL as an amplifier have been successfully carried out at LLNL.[5] The high peak beam-current of the induction linac accelerators is used as the amplifier from microwave to CO_2 laser attaining up to 35% efficiency with tapered undulators.

In 2000, there were many successful SASE-FEL experiments.[6] These experiments verified the exponential growth of FEL power, statistical nature of the SASE process, and the transverse coherence of the photon beam in diffraction pattern. The high-gain harmonic generation (HGHG) concept uses the high-gain amplifier by dividing the undulator into a modulator, a dispersive section for electron beam-bunching, and a radiator section for harmonic generation. The resulting coherent radiation can be greatly enhanced at a narrower bandwidth and shorter wavelength.[7]

Upon the verification of the SASE-FEL and HGHG principles, many proposals aimed to produce single pass high gain FEL from vacuum-ultra-violet (VUV) to X-ray. In 2009, the LCLS at SLAC were successfully commissioned.[8] Development of coherent light sources is an important topic in accelerator physics research.

Exercise 5.1

1. In a planar undulator, the closed orbit and the velocity vector are given by Eqs. (4.193) and (4.195). The interaction of the electron with the EM field becomes

$$\frac{d\gamma}{ds} = -\frac{eE_0(s,t)K_{\rm w}}{2mc^2\gamma}\sin k_{\rm w}s\left(e^{j(k_0s-\omega_0t)}+c.c.\right)$$

---

[5] See e.g. T.J. Orzechowski, p. 1840 in [14] and references therein.

[6] See e.g. J. Rossbach, in *Proc. of Linac Conference 2002*, p.582 (2002) and references therein; M. Hogan *et al.*, *Phys. Rev. Lett.*, **81**, 4867 (1998); S. Milton *et al.*, *Phys. Rev. Lett.*, **85**, 988 (2000), Science **292**, 2037 (2001); J. Andruszkow *et al.*, *Phys. Rev. Lett.* **85**, 3825 (2000); V. Ayvazyan *et al.*, *Phys. Rev. Lett.* **88**, 104802 (2002); see also G.T. More, *Nucl. Instrum. Methods* **239**, 19 (1985); K.J. Kim, *Phys. Rev. Lett.*, **57**, 1871 (1986); S. Krinsky and L.H. Yu, *Phys. Rev.* A **35**, 3406 (1987); L.H. Yu, S. Krinsky and R. Gluckstern, *Phys. Rev. Lett.*, **64**, 3011 (1990); Z. Huang and K.J. Kim, *Nucl. Instrum. Methods* A **475**, 59 (2001);

[7] L.H. Yu *et al.*, Science **289**, 932 (2000); see also S.G. Biedron, Ph.D. thesis, University of Lund (2001).

[8] J.N. Galayda, *Proc. of IPAC 2010*, p. 11.

$$\approx \quad +\frac{eE_0(s,t)K_\mathrm{w}}{4mc^2\gamma}\left(je^{j\psi} - je^{j\psi - 2k_\mathrm{w}s} + c.c.\right),$$

where $E_0(s,t)$ is a slow-varying amplitude of the electric field and the phase factor ψ is defined as $\psi = (k_\mathrm{w} + k_0)s - \omega_0 t$.

(a) The energy exchange is maximum at the stationary phase condition: $\langle d\psi/ds \rangle = 0$. Show that the resonance condition and the equation for the phase factor ψ are

$$\gamma_r^2 = \frac{\lambda_\mathrm{w}(1 + \frac{1}{2}K_\mathrm{w}^2)}{2\lambda}, \quad \text{or} \quad \lambda_r = \lambda_\mathrm{w}\frac{1 + \frac{1}{2}K_\mathrm{w}^2}{2\gamma^2}.$$

$$\frac{d\psi}{ds} = 2k_\mathrm{w}\eta + 2k_\mathrm{w}b\cos 2k_\mathrm{w}s,$$

where $\eta = \Delta\gamma/\gamma_r$ as defined in Eq. (5.9) and the constant $b = \frac{1}{4}K_\mathrm{w}^2/(1 + \frac{1}{2}K_\mathrm{w}^2)$. The phase factor ψ is not monotonic, it oscillates at twice the undulator wave number. Defining the phase factor $\phi = \psi + b\sin 2k_\mathrm{w}s$, show that the equation for the phase factor ϕ becomes

$$\frac{d\phi}{ds} = 2k_\mathrm{w}\eta.$$

(b) Expanding the $\exp(-jb\sin 2k_\mathrm{w}s)$ in Bessel functions (see Sec III in Appendix B), show that the energy equation of the electron becomes

$$\frac{d\gamma}{ds} = -\frac{eE_0K_\mathrm{w}[\mathrm{JJ}]}{2mc^2\gamma_r}\sin\phi, \quad \text{or} \quad \frac{d\eta}{ds} = -\frac{eE_0K_\mathrm{w}[\mathrm{JJ}]}{2mc^2\gamma_r^2}\sin\phi,$$

where the factor [JJ] is defined as $[\mathrm{JJ}] \equiv J_0(b) - J_1(b)$ and $J_0(b)$ and $J_1(b)$ are the Bessel functions of order 0 and 1. Note that (ϕ, η) forms a set of conjugate phase-space coordinates. The spontaneous emission of electrons in an undulator is identical to that of a helical undulator with the FEL parameter of Eq. (5.26) replaced by

$$(2\rho_\mathrm{fel})^3 = \frac{\mu_0 n_0 e^2 \lambda_\mathrm{w}^2 K_\mathrm{w}^2 [\mathrm{JJ}]^2}{16\pi^2\gamma_r^3 m}.$$

2. With the coordinate-transformation from (s,t) to $(s, \phi = [k_\mathrm{w} + k_0]s - \omega_0 t + \phi_0)$:

$$\left(\frac{\partial}{\partial s}\right)_t = \left(\frac{\partial}{\partial s}\right)_\phi + \left(\frac{\partial\phi}{\partial s}\right)_t\left(\frac{\partial}{\partial\phi}\right)_s = \left(\frac{\partial}{\partial s}\right)_\phi + (k_\mathrm{w} + k_0)\left(\frac{\partial}{\partial\phi}\right)_s,$$

$$\left(\frac{\partial}{\partial t}\right)_s = \left(\frac{\partial\phi}{\partial t}\right)_s\left(\frac{\partial}{\partial\phi}\right)_s = -\omega_0\left(\frac{\partial}{\partial\phi}\right)_s,$$

show that Maxwell's equation (5.38) becomes

$$\frac{\partial E_0}{\partial s} + \frac{1}{c}\frac{\partial E_0}{\partial t} = \frac{\partial E_0}{\partial s} + k_\mathrm{w}\frac{\partial E_0}{\partial\phi} \approx \frac{\partial E_0}{\partial s} = \left(\frac{\mu_0 n_0 c^2 e K_\mathrm{w}}{\gamma_r}\right)\left[\frac{1}{2\pi}\int d\eta d\phi\ f(\eta,\phi)\sin\phi\right].$$

II Beam-Beam Interaction

Coulomb force plays an important role in high brightness beams. The small-angle *intrabeam scattering* causes emittance dilution. The Touschek scattering is a process of momentum-transfer from the horizontal to longitudinal planes. The *mean field* of Coulomb force is called the space charge force, which is proportional to γ^{-2} due to cancellation between the electric and magnetic fields. Thus the beam brightness at low energy is normally set by the space charge limit.

For high energy colliders, the beam-beam interaction describes the force between two oppositely moving colliding beam bunches. The force is enhanced by the addition of both the electric and magnetic fields. The luminosity of all colliders are limited by the beam-beam force. This section addresses its effect on particle motion. The electric potential Φ of a bunch can be obtained from the Poisson equation, i.e. $\nabla^2\Phi = -\rho/\epsilon_0$, where ρ is the beam charge distribution. However, for simplicity, we discuss only the Gaussian charge distribution where the electric potential can easily be obtained. Although a real charge-distribution may not be Gaussian, results based on Gaussian distribution quantitatively agree with most beam-beam phenomena.

II.1 The Beam-Beam Force in Round Beam Geometry

We consider head-on collisions between two Gaussian round beams of length L with transverse charge distribution:

$$\rho(r) = \frac{Ne}{2\pi\sigma^2}\exp\left[-\frac{r^2}{2\sigma^2}\right], \tag{5.46}$$

where $Ne = \int \rho(r)2\pi r dr$ is the charge per unit length, $\sigma^2 = \frac{1}{2}\langle r^2\rangle$ is the rms beam width. The Lorentz force on a test particle due to the opposing bunch at a radius r is $\vec{F}_\perp = e(\vec{E} + \vec{v}\times\vec{B}) = e(E_r + \beta c B_\phi)\hat{r}$:

$$\vec{F}_\perp = \pm\frac{\gamma Ne^2(1+\beta^2)}{4\pi\epsilon_0 r}\left(1-\exp\left[-\frac{r^2}{2\sigma^2}\right]\right)\hat{r} \quad \stackrel{\beta\to 1}{\longrightarrow} \quad \pm\frac{\gamma Ne^2}{4\pi\epsilon_0\sigma^2}(x\hat{x} + z\hat{z} + \cdots),$$

where $\beta = v/c$, $\gamma = \sqrt{1-\beta^2}$, and $\pm$ signs correspond to the force seen by the like/unlike charges. The kick-angle $\Delta x'$ is (in linear approximation)

$$\Delta x' = \frac{\int F_x\,(ds/\gamma)}{\gamma mc^2\beta^2} \quad \stackrel{\beta\to 1}{\longrightarrow} \quad \pm\frac{N_{\rm B}r_0}{\gamma\sigma^2}x + \cdots \tag{5.47}$$

where $N_{\rm B} = \int N ds$ is the number of particles in the opposing bunch, $r_0 = \frac{e^2}{4\pi\epsilon_0 mc^2}$ is the classical radius. The focal length of the beam-beam kick is $\frac{1}{f} = -\frac{\Delta x'}{x} = \mp\frac{N_{\rm B}r_0}{\gamma\sigma^2}$. The linear beam-beam parameter $\xi_{\rm bb}$ is defined as the magnitude of the linear beam-beam tune shift:

$$\xi_{\rm bb} = |\Delta\nu| = \left|\frac{1}{4\pi}\frac{\beta^*}{f}\right| = \frac{N_{\rm B}r_0\beta^*}{4\pi\gamma\sigma^2} = \frac{N_{\rm B}r_0}{4\pi\epsilon_N}, \tag{5.48}$$

where β^* is the betatron amplitude function at the interaction point and ϵ_N is the normalized emittance of the bunch. The beam-beam force is highly nonlinear, and $\xi_{\rm bb}$ serves as the scaling factor for the nonlinear beam-beam force.

A. The beam-beam potential

The beam-beam kick can be derived from a beam-beam potential $V_{\rm bb}(x,z)$, which is obtained by solving Poisson's equation for a beam distribution function. For a Gaussian beam distribution, the beam-beam potential is (see Exercise 5.2.1):

$$\begin{aligned} V_{\rm bb}(x,z) &= \mp\frac{Nr_0}{\gamma}\int_0^\infty \frac{1-e^{-\frac{x^2}{2\sigma_x^2+t}-\frac{z^2}{2\sigma_z^2+t}}}{\sqrt{(2\sigma_x^2+t)(2\sigma_z^2+t)}}dt \\ &= \mp\frac{Nr_0}{\gamma}\left(\frac{x^2}{\sigma_x(\sigma_x+\sigma_z)}+\frac{z^2}{\sigma_z(\sigma_x+\sigma_z)}\right) \\ &\quad \pm\frac{Nr_0}{4\gamma\sigma_x^2(\sigma_x+\sigma_z)^2}\left(\frac{2+R}{3}x^4+\frac{2}{R}x^2z^2+\frac{1+2R}{3R^3}z^4\right)+\cdots \end{aligned} \tag{5.49}$$

where σ_x and σ_z are the rms radii of the beam, $R=\sigma_z/\sigma_x$ is called the round-beam parameter. In linear approximation with $x \ll \sigma_x$ and $z \ll \sigma_z$, the beam-beam kicks, the focal lengths, and the beam-beam parameters are (see Exercise 5.2.1)

$$\Delta x' = \pm\frac{2N_{\rm B}r_0}{\gamma\sigma_x(\sigma_x+\sigma_z)}x, \qquad \Delta z' = \pm\frac{2N_{\rm B}r_0}{\gamma\sigma_z(\sigma_x+\sigma_z)}z. \tag{5.50}$$

$$\frac{1}{f_x} = \mp\frac{2N_{\rm B}r_0}{\gamma\sigma_x(\sigma_x+\sigma_z)}, \qquad \frac{1}{f_z} = \mp\frac{2N_{\rm B}r_0}{\gamma\sigma_z(\sigma_x+\sigma_z)}. \tag{5.51}$$

$$\xi_x = \frac{N_{\rm B}r_0\beta_x^*}{2\pi\gamma\sigma_x(\sigma_x+\sigma_z)}, \qquad \xi_z = \frac{N_{\rm B}r_0\beta_z^*}{2\pi\gamma\sigma_z(\sigma_x+\sigma_z)}, \tag{5.52}$$

The focal length is positive (focusing) for the collision of unlike-charges, and negative (defocusing) for the collision of like-charges.

B. Dynamics betatron amplitude functions

At $\xi_{\rm bb} \sim 0.05$, the focal length $|f_{\rm bb}| \sim \beta^*$ is usually small and the beam-beam force is strong. Thus the linear lattice is also strongly perturbed by the beam-beam interaction. The one-turn map, including a thin-lens beam-beam kick, is

$$\mathbf{M} = \begin{pmatrix} \cos\Phi_0+\alpha_0^*\sin\Phi_0 & \beta_0^*\sin\Phi_0 \\ -\gamma_0^*\sin\Phi_0 & \cos\Phi_0-\alpha_0^*\sin\Phi_0 \end{pmatrix}\begin{pmatrix} 1 & 0 \\ -\frac{1}{f} & 1 \end{pmatrix}, \tag{5.53}$$

where β_0^*, α_0^*, and γ_0^* are the values of the unperturbed betatron amplitude function at IP, Φ_0 is the unperturbed betatron phase advance in one revolution. Identifying the one-turn map $\mathbf{M}$ with Courant-Snyder parametrization, we obtain

$$\cos\Phi = \cos\Phi_0 - \frac{\beta_0^*}{2f}\sin\Phi_0 = \cos\Phi_0 \mp 2\pi\xi_{\rm bb}\sin\Phi_0, \tag{5.54}$$

$$\frac{\beta^*}{\beta_0^*} = \frac{\sin\Phi_0}{\sin\Phi} \approx 1 \mp 2\pi\xi_{\rm bb}\cot\Phi_0, \tag{5.55}$$

where β^* is the value of the perturbed betatron amplitude function at IP, Φ is the perturbed betatron phase advance in one turn. The betatron tune-shift due to the beam-beam interaction is $\Delta Q = (\Phi - \Phi_0)/(2\pi) \approx \pm\xi_{\rm bb}$, and the betatron amplitude function is *dynamically* modified. The tune-shift can cause betatron tunes to overlap with betatron resonances. This can result in emittance blow-up, beam halo, and beam loss. The mismatched betatron amplitude-function can induce lattice-function modulation (β-beat) if the resulting betatron tune is near a half-integer stopband.

The linear stability condition for the betatron motion is

$$|\cos\Phi_0 \mp 2\pi\xi_{\rm bb}\sin\Phi_0| \le 1 \quad \Longrightarrow \quad \xi_{\rm bb} \le \frac{1}{2\pi}\cot\frac{\Phi_0}{2}, \quad \text{or} \quad \xi_{\rm bb} \le \frac{1}{2\pi}\tan\frac{\Phi_0}{2}. \tag{5.56}$$

The shaded area in Fig. 5.5 shows the stable region of the beam-beam parameter. Experimentally, the measured beam-beam tune parameters are about 0.05 for e^+e^- colliders and 0.03 for hadron colliders. The actual beam-beam parameter may be limited by the effects of nonlinearity in the beam-beam interaction, noises, time-dependent tune modulation, and nonlinear magnetic fields in the storage ring.

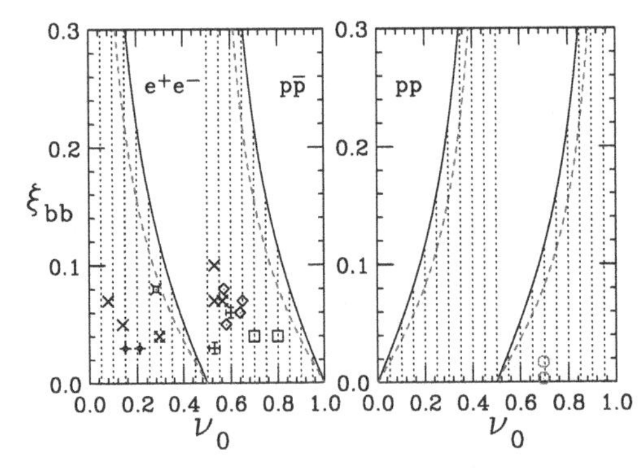

Figure 5.5: The shaded area corresponds to stable condition of Eq. (5.56), where ν_0 is the fractional part of the bare betatron tune $\Phi_0/(2\pi)$. The left plot is for the e^+e^- or $p\bar{p}$ colliders, and the right plot is for colliders with like-charges. The actual achieved beam-beam parameters, shown as various symbols, are smaller than that of the linear stability limit. The coherent beam-beam limit is shown as the dashed line.

C. Disruption factor

The disruption factor, defined as the ratio of the bunch length to the focal length of the beam-beam interaction,

$$\mathcal{D} = \frac{\sigma_s}{f} = \frac{Nr_0\sigma_s}{\gamma\sigma^2} = 4\pi\xi_{\rm bb}\frac{\sigma_s}{\beta^*}, \tag{5.57}$$

is commonly used to gauge the strength of beam-beam interaction in linear colliders. If the disruption factor is larger than 1, the beam particles are focused toward or defocused away from each other within the bunch length. For colliders with $\beta^* \approx \sigma_s$ and $\xi_{\rm bb} \approx 0.05$ at a single IP, we find $\mathcal{D} \approx 0.63$, If this beam-beam tune shift is produced by a single interaction point, the focal length of beam-beam interaction is about $f \approx 2\beta^*$. For e^+e^- linear colliders, $\mathcal{D}$ can be much larger than 1 in order to achieve luminosity enhancement (pinch effect).

II.2 The Coherent Beam-Beam Effects

For two beams with similar intensity, if one beam is slightly displaced with respect to the other, coherent oscillations can be induced, which may lead to unstable motion.[9] We consider two counter circulating bunches, specified by indices 1 and 2 respectively. The center of mass motion relative to each other is

$$y_1' = -\frac{g}{f_1}(y_1 - y_2), \quad y_2' = \frac{g}{f_2}(y_1 - y_2), \tag{5.58}$$

where g is the geometric overlapping factor that depends on the transverse distributions of two beams. In rigid beam approximation, the c.m. linear betatron motion (closed orbit) of the two beams is obtained from the one turn transfer matrix. The transfer matrix in the normalized coordinates becomes (see Exercise 5.2.4)

$$\begin{aligned}\mathbf{M} &= \begin{pmatrix} \cos\Phi_1 & \sin\Phi_1 & 0 & 0 \\ -\sin\Phi_1 & \cos\Phi_1 & 0 & 0 \\ 0 & 0 & \cos\Phi_2 & \sin\Phi_2 \\ 0 & 0 & -\sin\Phi_2 & \cos\Phi_2 \end{pmatrix} \begin{pmatrix} 1 & 0 & 0 & 0 \\ -4\pi g\xi_1 & 1 & 4\pi g\xi_1 & 0 \\ 0 & 0 & 1 & 0 \\ 4\pi g\xi_2 & 0 & -4\pi g\xi_2 & 1 \end{pmatrix} \\ &= \begin{pmatrix} \cos\Phi_1 - 4\pi g\xi_1\sin\Phi_1 & \sin\Phi_1 & 4\pi g\xi_1\sin\Phi_1 & 0 \\ -\sin\Phi_1 - 4\pi g\xi_1\cos\Phi_1 & \cos\Phi_1 & 4\pi g\xi_1\cos\Phi_1 & 0 \\ 4\pi g\xi_2\sin\Phi_2 & 0 & \cos\Phi_2 - 4\pi g\xi_2\sin\Phi_2 & \sin\Phi_2 \\ 4\pi g\xi_2\cos\Phi_2 & 0 & -\sin\Phi_2 - 4\pi g\xi_2\cos\Phi_2 & \cos\Phi_2 \end{pmatrix}.\end{aligned}$$

The stability of the system is determined by the eigenvalue of the transfer matrix $\mathbf{M}$, i.e. $|\lambda - \mathbf{M}| = 0$. Consider a simple example of two beams with identical intensity and betatron amplitude functions with $\xi_{\rm bb} = \xi_1 = \xi_2$ and $\Phi_0 = \Phi_1 = \Phi_2$. Two of four eigenvalues are given by

$$\lambda_\sigma = \cos\Phi_0 \pm j\sin\Phi_0, \tag{5.59}$$

i.e., the same eigenvalues as the original unperturbed system. This is identified as the σ-mode, where two beams oscillate in-phase with each other, and produce no coherent beam-beam effect on betatron motion.

The eigenvalues of π-mode are solutions of the equation:

$$\lambda_\pi^2 - 2(\cos\Phi_0 - 4\pi g\xi_{\rm bb}\sin\Phi_0)\lambda_\pi + 1 = 0, \tag{5.60}$$

[9] A. Piwinski, *Proc. 8th Int. Conf. on High Energy Accelerators*, p. 357 (1971).

where two beams oscillate out of phase. In small linear beam-beam parameter approximation, the coherent tune shift is $\Delta Q = 2g\xi_{\rm bb}$. For rigid Gaussian distribution, the coherent beam-beam tune-shift factor is $2g = \sqrt{2}$. Yokoya *et al.* carried out careful analysis of coherent motion using Vlasov equation and found that the Yokoya-factor is about $2g \approx 1.21 \sim 1.33$. Experimental observations show that the Chao-Yokoya factor agreed well with Yokoya's analysis.[10] The stability condition is

$$|\cos\Phi_0 - 4\pi g\xi_{\rm bb}\sin\Phi_0| \le 1, \quad \text{or} \quad \xi_{\rm bb} \le \frac{1}{4\pi g}\cot\frac{\Phi_0}{2}. \tag{5.61}$$

This condition is more stringent than that of Eq. (5.56) by a factor $1/(2g)$, shown as dashed line in Fig. 5.5.

II.3 Nonlinear Beam-Beam Effects

With beam beam interaction, the effective particle Hamiltonian in impulse approximation is

$$H = \frac{1}{2}(x'^2 + K_x x^2) + \frac{1}{2}(z'^2 + K_z z^2) + V_{\rm bb}(x,z)\delta(s), \tag{5.62}$$

which is highly nonlinear. The integrated beam-beam potential $V_{\rm bb}(x,z)$ of Eq. (5.49) is represented by point interaction, which is rich in perturbing harmonics. As an example, we consider the nonlinear octupole-like beam-beam force for a round beam:

$$\Delta x' = -\frac{\pi\xi_{\rm bb}}{\beta^*\sigma^2}(x^3 + xz^2). \tag{5.63}$$

This octupole-like nonlinear beam-beam force differs from a regular octupole pole magnetic force shown in Eq. (2.19), and is difficult to be compensated by octupole magnets. Even if we were using an octupole to compensate the x^3 nonlinearity, the required octupole strength, $B_3\ell = \frac{\pi\xi_{\rm bb}}{\beta^*\sigma^2}B\rho$, is very very large.

Using the Floquet transformation of Eq. (2.55) and the orbital angle $\theta = s/R$ as the time coordinate, we obtain

$$\begin{aligned}
\tilde{H} &= \nu_x J_x + \nu_z J_z + U(J_x, J_z, \psi_x, \psi_z; \theta),\\
U &= \frac{1}{2\pi}\sum K_{m,n}(J_x, J_z)e^{-j(m\psi_x + n\psi_z - \ell\theta)}\\
K_{m,n} &= \frac{Nr_0}{(2\pi)^2\gamma}\int\int d\phi_x d\phi_z dt \frac{1-\exp\{-\frac{\beta_x J_x\cos^2\phi_x}{2\sigma_x^2+t} - \frac{\beta_z J_z\cos^2\phi_z}{2\sigma_z^2+t}\}}{\sqrt{(2\sigma_x^2+t)(2\sigma_z^2+t)}}e^{j(m\phi_x+n\phi_z)},
\end{aligned}$$

[10] A. Chao, *Beam-beam Instability*, in AIP Conference Proceedings #127, *Physics of High Energy Accelerators*, p. 202 (AIP, NY, 1983); K. Yokoya *et al.*, KEK Preprint 89-14 (1989); K. Yokoya and H. Koiso, Particle Accelerators **27**, 181 (1990); W. Fischer *et al.*, Proceedings of PAC2003, p. 135 (2003); J.T. Seeman, *Luminosity and beam-beam interactions*, in AIP Proceedings #592, p. 163 (AIP, NY, 2001).

where we have assumed transverse Gaussian beam distribution. If the beam distribution is symmetric in x and z, we find only even order resonances, i.e., m and n must be even. Using the generating functions for the Bessel functions listed in Sec. III Appendix B, we obtain the detuning term:

$$K_{0,0} = \frac{Nr_0}{\gamma}\int dt \frac{Z_0(\frac{\beta_x J_x}{2\sigma_x^2+t})Z_0(\frac{\beta_z J_z}{2\sigma_z^2+t})}{\sqrt{(2\sigma_x^2+t)(2\sigma_z^2+t)}}; \qquad Z_0(a) = e^{-a}I_0(a). \tag{5.64}$$

The Hamiltonian becomes

$$\tilde{H} = \nu_x J_x + \nu_z J_z + K_{0,0}(J_x, J_z) + \frac{1}{2\pi}\sum_{m,n\neq 0} K_{m,n}(J_x, J_z)e^{-j(m\psi_x+n\psi_z-\ell\theta)}. \tag{5.65}$$

The beam-beam interaction creates an amplitude dependent betatron tune through $K_{0,0}$, and produces nonlinear resonances at $m\nu_x + n\nu_z = \ell$, where m and n are even-integers for head on collisions. If two beams are colliding off-axis, or with an angle, the odd order resonances appear. The strengths of these resonances are usually strong. Similar to what we have discussed in Sec. VII in Chapter 2, nonlinear resonances can profoundly influence the beam distribution function in phase-space, reduce beam lifetime, and cause beam loss. When higher order resonances are important, the available resonance-free-tune space becomes very small as shown in Fig. 2.57 in Chap. 2.

II.4 Experimental Observations and Numerical Simulations

In 1960-1990, the e^+e^- colliders have played an important role in the discovery and exploration of elementary particle physics, such as J/Ψ, Υ, etc. Because e^+ and e^- particles have opposite charges, e^+e^- beams can sometimes be confined in a single storage ring. Some of e^+e^- colliders were the SPEAR, PEP, and SLC at SLAC, DORIS and PETRA at DESY, CESR at Cornell, BEPC in China, VEPP-4M in Novosibirsk, TRISTAN at KEK, and LEP at CERN.

In the 1990's, the LEP and SLC have provided careful tests of the electro-weak theory of the standard model. In 2000's, particle factories (the B-factories: PEP-II and KEKB the DAΦNE at Frascati, and the tau-charm factory at BEPC in China) further our understanding of the fundamental symmetry in the force of nature. The luminosity of all high energy colliders is limited by the beam-beam interaction. There are many experiments, numerical simulations, methods of compensation, and workshops conducted on this subject.

We consider two interacting beams with identical intensity. The luminosity for a head-on collision is

$$\mathcal{L} = \frac{BN_1N_2f_0}{2\pi\Sigma_x\Sigma_z} \xrightarrow{(N_1=N_2,\ \sigma_{y1}=\sigma_{y2})} \frac{N_{\rm B}^2 f_{\rm coll}}{4\pi\sigma_x\sigma_z} = \frac{(I/e)^2}{4\pi\sigma_x\sigma_z f_{\rm coll}}, \tag{5.66}$$

where N_1 and N_2 are numbers of particles in counter-moving bunches, B is the number of bunches, f_0 is the revolution frequency, $f_{\rm coll} = Bf_0$ is the bunch collision frequency,

Σ_x and Σ_z are effective transverse rms beam widths of these colliding bunches at the interaction point (IP) with $\Sigma_x^2 = \sigma_{x1}^2 + \sigma_{x2}^2$ and $\Sigma_z^2 = \sigma_{z1}^2 + \sigma_{z2}^2$. Here, σ_{x1}, σ_{z1} and σ_{x2}, σ_{z2} are the rms horizontal and vertical beam widths of two interacting beam bunches. When the colliding bunches have equal number of particles with $N_B = N_1 = N_2$ and equal transverse beam widths ($\sigma_{x1} = \sigma_{x2}$, $\sigma_{z1} = \sigma_{z2}$), the luminosity formula is further reduced as shown in the right-side of Eq. (5.66), where $I/e = N_B f_{coll}$ is the particle current, $4\pi\sigma_x\sigma_z$ is the effective cross-section area of the beam at the IP:

$$\sigma_x^2 = \beta_x^*\epsilon_x + \left(D_x^*\frac{\Delta p}{p_0}\right)^2, \qquad \sigma_z^2 = \beta_z^*\epsilon_z + \left(D_z^*\frac{\Delta p}{p_0}\right)^2. \tag{5.67}$$

Here $\epsilon_{x,z}$ are the horizontal and the vertical rms emittances, $\beta_{x,z}^*$ and $D_{x,z}^*$ are the values of the horizontal and vertical betatron and dispersion functions at IP. Normally, the collider lattice is designed such that $D_z^* = 0$ and $D_x^* = 0$ in order to maximize the luminosity. Table 5.1 lists parameters of some high luminosity e^+e^- colliders.

Table 5.1: Parameter-list of high luminosity e^+e^- colliders

	KEKB		PEP2		BEPC	DAΦNE	LEP
E (GeV)	3.5	8	3.12	8.97	1.55	0.511	98
C (m)	3016.3		2199.318		240.4	97.69	26659
ρ (m)	16.3	104.5	13.75	165	10.35	1.4	3096.2
$\tau_{damping}$ (ms)	43	46	63	37	44	36	6.5
ϵ_x (nm)	18	24	40	49	390	160	45
ϵ_z (nm)	0.36	0.36	4	2	3.9	1.4	0.1
θ_x (mrad)	11	11	0	0	0	25	0
β_x^* (cm)	59	63	50	50	120	25	150
β_z^* (cm)	0.7	0.7	1.25	1.25	5	0.9	5
σ_ℓ (mm)	7	7	13	13	45	30	13
N_b (10^{10})	7.3	5.4	9.9	5.3	21.6	8.9	43
B	1284	1284	1658	1658	1	45	4
ξ_x	0.097	0.074	0.065	0.075	0.04	0.03	0.021
ξ_z	0.066	0.05	0.048	0.06	0.04	0.042	0.083
$\mathcal{L}$ (10^{30})	19000		9200		15	436	100

Because of the importance of the beam-beam interaction, many numerical simulations and experimental studies have been conducted to understand the underlying beam-beam physics. Two theoretical models are (1) weak-strong (incoherent) beam-beam model and (2) strong-strong (coherent) beam-beam model. In the incoherent or the weak-strong model, a test particle interacts with the mean field produced by the counter rotating beam bunch. The stability of the test particle depends essentially on the single particle dynamics. In the coherent beam-beam model, beam stability and bunch shape deformation are dynamically excited by coherent mode interactions. In both models, the *linear beam-beam parameters* ξ_x and ξ_z of Eq. (5.52) serve as scaling strength-parameters for beam-beam interaction.

Past experiments show that the luminosity of colliders is determined mainly by the beam-beam interaction. Although the horizontal emittance of electron storage rings is much larger than the vertical emittance, the beam-beam parameters ξ_z and ξ_x can be made equal by setting $\beta_x^*/\epsilon_x \simeq \beta_z^*/\epsilon_z$. Figure 5.6 shows the typical beam-beam parameter ξ_z achieved for some e^+e^--colliders: DAΦNE, VEPP2M, DCI, ADONE, SPEAR, BEPC, CESR, PEP, KEKB, PEP2, PETRA, HERA, TRISTAN, and LEP.[11] The energy dependence of the beam-beam parameter is not obvious. Typical value achieved is about 0.05-0.08 for e^+e^--colliders and about 0.005-0.025 for hadron-colliders. The curves on the right plot correspond to $\xi_0 + (\xi_{\rm LEP} - \xi_0)(\lambda_{\rm damping}/\lambda_{\rm LEP})^a$, where $\lambda_{\rm damping}$ is the damping decrement, $\lambda_{\rm LEP}$ is the damping decrement for LEP at 102.7 GeV, $\xi_0 = 0.025$, $\xi_{\rm LEP} = 0.115$ and $a = 0.175$ (solid) and 0.35 (dashes) respectively. The dependence of the beam-beam parameters on the damping decrement has not been fully established. There is no theoretical basis for these curves.

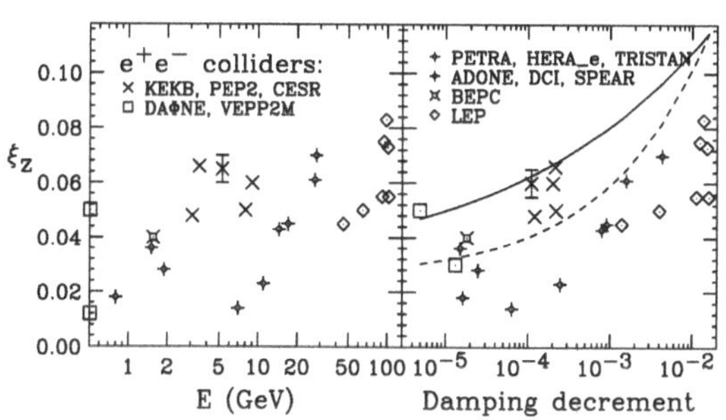

Figure 5.6: Left: A compilation of the achieved linear beam-beam parameter ξ_z vs. the beam-energy of e^+e^- colliders. Right: The same data plotted vs the transverse damping-decrement.

In e^+e^- colliders, we have $\sigma_x \gg \sigma_z$ and thus $\xi_z > \xi_x$. One expresses the luminosity in terms of ξ_z as

$$\mathcal{L} = \frac{\gamma(I/e)\xi_z}{2r_e\beta_z^*}(1+\frac{\sigma_z}{\sigma_x}) = \frac{\gamma^2 Bf\pi}{r_e^2}\xi_z^2\frac{\sigma_z\sigma_x}{\beta_z^{*2}}(1+\frac{\sigma_z}{\sigma_x})^2. \tag{5.68}$$

If ξ_z is the limiting factor for colliders, the luminosity is increased by increasing the beam intensity and beam emittance while maintaining a constant ξ_z. For a constant ξ_z, the luminosity can be increased by decreasing β_z^* value at the IP, increasing σ_z, or increasing the number of bunches B.[12] However, the maximum luminosity occurs

[11] See e.g. J. Seeman, *Observation of the Beam-Beam Interaction*, in *Lecture Notes in Physics* #247, p. 121 (Springer-Verlag, Berlin, 1985); R. Assmann and K. Cornelis, Proceedings of EPAC2000, p. 1187 (EPAC, 2000).

[12] If the counter rotating e^+e^- beams are stored in the same storage ring, these two counter-rotating beams should be separated by orbit separation schemes to minimize the long-range beam-beam tune shift.

when $\beta_z^* \simeq \sigma_s$.

If $\beta_z^* < \sigma_s$, the hourglass effect becomes important and the effective luminosity decreases. This arises from the fact that the betatron amplitude function in the interaction region is given by $\beta_z(s) = \beta_z^* + s^2/\beta_z^*$, where s is the distance away from the IP (see Exercise 2.2.17). Since the betatron amplitude function within the bunch length increases, the effective cross-sectional area of the beam increase and the effective emittance is reduced. The luminosity reduction factor for the flat beam and the round beam are respectively

$$\mathcal{L} = \begin{cases} \mathcal{L}_0\sqrt{\dfrac{A_z}{\pi}}\exp(\dfrac{A_z^2}{2})K_0(\dfrac{A_z^2}{2}) & \text{flat beam with } A_z = \beta_z^*/\sigma_s \\ \mathcal{L}_0 \times \sqrt{\pi}\exp(A^2)\ \mathrm{erfc}(A) & \text{round beam with } A = \beta^*/\sigma_s \end{cases}$$

where $\mathcal{L}_0$ is the luminosity for zero bunch length, $K_0(A_z^2/2)$ is the zeroth order modified Bessel function, erfc(A)=1-erf(A) is the error function, and A_z and A are lattice variation parameters.

If e^+ and e^- beams are of the same energy, the counter-circulating e^+ and e^- beams can be stored in a single storage ring. The number of bunches is limited by avoiding un-wanted beam-beam interactions other than the interaction points for physics experiments. Electrostatic separators are installed in these colliders to separate these counter-circulating beams, called *pretzel scheme* for achieving separate closed orbits for these two beams. In order to produce high luminosity in B-meson (KEKB and PEPII), Φ-meson (DAΦNE), and τ-charm (BEPC) factories, two storage rings crossing only at an interaction area are used for multi-bunch operation. Since the luminosity of e^+e^- colliders is usually limited by the beam-beam parameter, the design strategy differs from that of synchrotron radiation sources, where the emittance is minimized to maximize the beam brightness.

For hadron colliders, the $p\bar{p}$ colliders (TEVATRON or S$p\bar{p}$S) have used one ring strategy to minimize construction cost. These colliders have also been limited by the beam-beam interaction. In order to avoid un-wanted beam-beam interaction, electrostatic beam separators are installed in the storage ring for multi-bunch operation. Proton-proton and heavy ion colliders require two independent rings for counter-circulating beams to collide at a few interaction points. In this case, the parasitic long-range beam-beam interactions at the interaction area can also set limit on the number of bunches in the ring.

Because of the importance of this subject, there have been many workshops devoted to this subjects. For related and current topics on this subject, the proceedings of these workshops are handy.[13]

[13]See e.g. Lecture Notes in Physics **247**, *Nonlinear Dynamics Aspects of Particle Accelerators* (Springer-Verlag, NY, 1985); Proceedings of the *ICFA Beam Dynamics Workshop on Beam-Beam Effects in Circular Colliders*, (Novosibirsk, 1989); AIP Conf. Proc. **693**, (AIP, New York, 2003).

II.5 Beam-Beam Interaction in Linear Colliders

The beam-beam interactions in linear collider at TeV energies are usually characterized by disruption factors $\mathcal{D}_x$ and $\mathcal{D}_z$, disruption deflection angle θ_0 for colliding particles, and the beamstrahlung parameter Υ for pair production through beamstrahlung.

In linear approximation, the effect of beam-beam interaction is characterized by the focal length f_x and f_z in Eq. (5.51). The disruption deflection angle for a particle at 1σ amplitude is

$$\theta_0 = \frac{\sigma_x}{f_x} = \frac{\sigma_x}{f_z} = \frac{2N_{\rm B}r_0}{\gamma(\sigma_x+\sigma_z)}, \tag{5.69}$$

and the disruption factors $\mathcal{D}_x$ and $\mathcal{D}_z$ are defined as

$$\mathcal{D}_{x,z} = \frac{2N_{\rm B}r_0\sigma_s}{\gamma\sigma_{x,z}(\sigma_x+\sigma_z)} = \frac{\sigma_s}{f_{x,z}} = \frac{\xi_{x,z}}{A_{x,z}}, \tag{5.70}$$

where $\xi_{x,z}$ are the linear beam-beam tune shift parameter, $A_{x,z} = \frac{\beta_{x,z}}{\sigma_s}$ is the lattice variation parameter at the interaction point.

Defining the normalized coordinates:[14]

$$X = \frac{x}{\sigma_x}; \quad Z = \frac{z}{\sigma_z}; \quad S = \frac{s}{\sigma_s},$$

we find linearized equations of motion within the bunch crossing as

$$\frac{d^2X}{dS^2} = -\mathcal{D}_x X; \quad \frac{d^2Z}{dS^2} = -\mathcal{D}_z Z.$$

The solution is sinusoidal. In the range of opposing bunch, $S \in (-1,1)$, the number of oscillation is $n = \sqrt{\mathcal{D}}/\pi$ or $\mathcal{D} = \pi^2 n^2$. These oscillations can also be viewed as plasma oscillation with

$$\omega_p \Delta t = \omega_p \frac{\sqrt{2\pi}\sigma_s}{c} = 2\pi n = 2\sqrt{\mathcal{D}}, \quad \text{or} \quad \mathcal{D} = \sqrt{\frac{\pi}{2}}\frac{\sigma_s^2\omega_p^2}{c^2} = \sqrt{\frac{\pi}{2}}(\frac{2\pi\sigma_s}{\lambda_p})^2.$$

Note that the disruption parameter is proportional to the number of plasma oscillations within the bunch length. Another quantity of interest is the Debye length, λ_D, defined as the transverse amplitude of the plasma oscillation. The time for the maximum amplitude is about $\frac{1}{4}T_p$. Define the emittance as $\epsilon = \pi\sigma_x\sqrt{\langle v_\perp^2\rangle}/c$, one obtains

$$\lambda_D = \frac{1}{4}\sqrt{\langle v_\perp^2\rangle}T_p = \frac{2\pi}{4}\frac{\sqrt{\langle v_\perp^2\rangle}}{\omega_p}, \quad \text{or} \quad \frac{\lambda_D}{\sigma_x} = (\frac{\pi}{2})^{1/4}\frac{\sigma_s}{2\beta^*\sqrt{\mathcal{D}}} = (\frac{\pi}{2})^{1/4}\frac{1}{2A\sqrt{\mathcal{D}}}.$$

[14] R. Hollebeck, NIM **184**, 333 (1981); AIP Proc. **184**, 680 (1988).

The Debye length is normally less than the bunch width for a reasonable machine parameter A. Particle motion in the e^+e^- linear collider is trapped in the bunch during the collision.

When the disruption factor is large, the number of particle oscillation within the opposing bunch may be large. As the disruption factor increases, the luminosity can be enhanced by the pinch effect, i.e. $\mathcal{L} = \mathcal{L}_0 \times H_D$ where H_D is a function of $\mathcal{D}_x$, $\mathcal{D}_z$, A_x, and A_z. The luminosity enhancement for a nearly round SLC beams was found to be about 2.[15]

The strong beam-beam interaction in e^+e^- colliders at TeV energies can cause beam particles to lose substantial amount of their energy to synchrotron radiation. The quantity is characterized by the beamstrahlung parameter:

$$\Upsilon = \frac{2}{3}\frac{\hbar\omega_c}{E} = \frac{\hbar\gamma^2}{mc\rho} = \gamma\frac{2B}{B_c}, \tag{5.71}$$

where B is the magnetic field produced by the colliding bunches, the factor 2 takes into account the contributions both the electric and magnetic fields, and $B_c = m^2c^2/(e\hbar) \approx 4.4 \times 10^9$ T is the Schwinger critical field. Averaging the induced magnetic field over the beam distribution, the average beamstrahlung parameter is

$$\langle\Upsilon\rangle \approx \frac{5}{12\pi}\frac{N_{\rm B}r_0\lambda_c\gamma}{\sigma_s(\sigma_x+\sigma_z)} = \frac{5}{24\pi}\frac{\gamma^2\mathcal{D}_z\lambda_c\sigma_z}{\sigma_s^2}, \tag{5.72}$$

where $\lambda_c = h/mc$ is the Compton wavelength. When the beamstrahlung parameter $\langle\Upsilon\rangle$ becomes large, quantum-electrodynamics processes are important in the beam-beam interaction. The beamstrahlung parameter is about 10^{-3} for the SLC, and about 0.1 for the ILC design. These topics are actively researched in the quest of linear collider design studies.[16]

Exercise 5.2

1. Follow the following steps to derive the beam-beam interaction potential for a beam with N particles per unit length with Gaussian charge density:

$$\rho(x,z) = \frac{Ne}{2\pi\sigma_x\sigma_z}\exp\{-\frac{x^2}{2\sigma_x^2} - \frac{z^2}{2\sigma_z^2}\},$$

   (a) Show that the Fourier transform of the Poisson equation of the $\nabla^2\Phi = -\frac{\rho}{\epsilon_0}$, where ρ is the charge distribution, and ϵ_0 is the vacuum permittivity, for the

[15] T. Barklow *et al.*, *Proc. of PAC1999*, p. 307 (IEEE, 1999).

[16] See K. Yokoya, in *High Quality Beams*, AIP Proceedings **592**, p. 185 (AIP, N.Y. 2001); D. Schulte, *Proc. of PAC1999*, p. 1688 (IEEE, 1999); P. Chen, in *Handbook of Accelerator Physics and Engineering*, edited by A. Chao and M. Tigner, p. 140 (World Scientific, Singapore, 1999).

electrostatic potential Φ of a beam is

$$(k_x^2+k_z^2)\tilde{\Phi}(k_x,k_z)=\frac{1}{\epsilon_0}\tilde{\rho}(k_x,k_z),$$
$$\tilde{\Phi}=\frac{1}{4\pi^2}\int \Phi(x,z)e^{ik_xx+ik_zz}dxdz,\quad \tilde{\rho}=\frac{1}{4\pi^2}\int \rho(x,z)e^{ik_xx+ik_zz}dxdz.$$

For a Gaussian beam distribution, show that

$$\tilde{\rho}=\frac{Ne}{4\pi^2}e^{-\frac{1}{2}\sigma_x^2k_x^2-\frac{1}{2}\sigma_z^2k_z^2},\quad \tilde{\Phi}=\frac{Ne}{4\pi^2\epsilon_0}\frac{1}{k_x^2+k_z^2}e^{-\frac{1}{2}\sigma_x^2k_x^2-\frac{1}{2}\sigma_z^2k_z^2}.$$

(b) Show that the electrostatic potential Φ is[17]

$$\Phi(x,z)=\frac{Ne}{4\pi\epsilon_0}\int_0^\infty \frac{-1+\exp\{-\frac{x^2}{2\sigma_x^2+t}-\frac{z^2}{2\sigma_z^2+t}\}}{\sqrt{(2\sigma_x^2+t)(2\sigma_z^2+t)}}dt,$$

where the singularity at $x=z=0$ is removed by the addition of the -1 term in the numerator of the integrand.

(c) Show that the beam-beam kick at $\beta\approx 1$ is (see Sec. II.1)

$$\Delta x'=-\frac{\partial V}{\partial x},\ \Delta z'=-\frac{\partial V}{\partial z},\quad V(x,z)=\frac{Nr_0}{\gamma}\int_0^\infty \frac{1-\exp\{-\frac{x^2}{2\sigma_x^2+t}-\frac{z^2}{2\sigma_z^2+t}\}}{\sqrt{(2\sigma_x^2+t)(2\sigma_z^2+t)}}dt$$

(d) For small amplitude particle motion with $x\ll\sigma_x, z\ll\sigma_z$, show that[18]

$$\Delta x'=\frac{2Nr_0}{\gamma\sigma_x(\sigma_x+\sigma_z)}x,\quad \Delta z'=\frac{2Nr_0}{\gamma\sigma_z(\sigma_x+\sigma_z)}z$$

(e) In many electron storage rings, we have $\sigma_x\gg\sigma_z$. Now we define

$$r=\frac{\sigma_z}{\sigma_x},\quad a=\frac{x}{\sqrt{2(\sigma_x^2-\sigma_z^2)}},\quad b=\frac{z}{\sqrt{2(\sigma_x^2-\sigma_z^2)}},\quad s^2=\frac{2\sigma_z^2+t}{2\sigma_x^2+t},$$

show that

$$V=\frac{2Nr_0}{\gamma}\int_r^1\frac{1}{1-s^2}\left(1-\exp\left[-a^2(1-s^2)-b^2\left(\frac{1}{s^2}-1\right)\right]\right)ds.$$

(f) Show that the beam-beam kick is[19]

$$\Delta x'-j\Delta z'=-\frac{4Nr_0}{\gamma\sqrt{2(\sigma_x^2-\sigma_z^2)}}e^{-(a+jb)^2}\int_{ar+j\frac{b}{r}}^{a+jb}e^{\zeta^2}d\zeta$$

[17] S. Kheifeit, "Potential of a three dimensional Gauss-bunch", PETRA Note 119 (1976). Note that

$$\frac{1}{k_x^2+k_z^2}=\frac{1}{4}\int_0^\infty e^{-\frac{1}{4}t(k_x^2+k_z^2)}dt.$$

[18] Change the dummy variable to $s^2=(2\sigma_z^2+t)/(2\sigma_x^2+t)$.

[19] Hint: change the dummy variable to $\zeta=as+ib/s$ with $(r<s<1)$.

(g) Using the complex error function [30] of Sec. III.3 in Appendix B, show that[20]

$$\Delta x' - j\Delta z' = j\frac{2Nr_0\sqrt{\pi}}{\gamma\sqrt{2(\sigma_x^2-\sigma_z^2)}}\left[w(a+jb) - e^{-(a+jb)^2+(ar+j\frac{b}{r})^2}w(ar+j\frac{b}{r})\right]$$

or

$$\Delta x' = -\frac{2Nr_0\sqrt{\pi}}{\gamma\sqrt{2(\sigma_x^2-\sigma_z^2)}}Im\left[w(\frac{x+iz}{\sqrt{2(\sigma_x^2-\sigma_z^2)}}) - e^{-\frac{x^2}{2\sigma_x^2}-\frac{z^2}{2\sigma_z^2}}w(\frac{x\frac{\sigma_z}{\sigma_x}+iz\frac{\sigma_x}{\sigma_z}}{\sqrt{2(\sigma_x^2-\sigma_z^2)}})\right]$$

$$\Delta z' = -\frac{2Nr_0\sqrt{\pi}}{\gamma\sqrt{2(\sigma_x^2-\sigma_z^2)}}Re\left[w(\frac{x+iz}{\sqrt{2(\sigma_x^2-\sigma_z^2)}}) - e^{-\frac{x^2}{2\sigma_x^2}-\frac{z^2}{2\sigma_z^2}}w(\frac{x\frac{\sigma_z}{\sigma_x}+iz\frac{\sigma_x}{\sigma_z}}{\sqrt{2(\sigma_x^2-\sigma_z^2)}})\right]$$

2. Using Eq. (5.65), show that [See A. Chao, AIP127, 202 (1979)]

$$\Delta\nu_x = \frac{\xi(1+\frac{1}{r})}{2}\int_0^\infty \frac{du}{(1+u)^{3/2}(1+\frac{u}{r^2})^{1/2}}Z_1(\frac{\beta_x J_x/\sigma_x^2}{1+u})Z_0(\frac{\beta_z J_z/\sigma_z^2}{1+\frac{u}{r^2}})$$

$$\Delta\nu_z = \frac{\xi(1+r)}{2}\int_0^\infty \frac{du}{(1+u)^{3/2}(1+r^2u)^{1/2}}Z_1(\frac{\beta_x J_x/\sigma_x^2}{1+ur^2})Z_0(\frac{\beta_z J_z/\sigma_z^2}{1+u})$$

with $r=\frac{\sigma_z}{\sigma_x}$, $Z_0(x)=e^{-x}I_0(x)$ and $Z_1(x)=e^{-x}[I_0(x)-I_1(x)]$.

3. Using the normalized phase-space coordinates $Y=y/\sqrt{\beta_0^*}$ and $\mathcal{P}_Y=(\alpha_0^* y+\beta_0^* y')/\sqrt{\beta_0^*}$, show that the one-turn transfer matrix $\mathbf{M}$ of Eq. (5.53) becomes

$$\mathbf{M} = \begin{pmatrix}\cos\Phi_0 & \sin\Phi_0\\ -\sin\Phi_0 & \cos\Phi_0\end{pmatrix}\begin{pmatrix}1 & 0\\ -4\pi\xi & 1\end{pmatrix},$$

where $\xi=\beta_0^*/(4\pi f)$ is the linear beam-beam tune-shift parameter.

[20]Numerical calculation of beam-beam interaction by using the complex error functions in CERN library is considerably more accurate and faster than that obtained from the numerical integration. The derivation of this homework problem was due to M. Bassetti, and G.A. Erskine, CERN-ISR-TH/80-06 (1980).

Appendix A

Classical Mechanics and Analysis

I Hamiltonian Dynamics

I.1 Canonical Transformations

Based on the variational principle of the Lagrangian, $\delta \int L dt = 0$, particle motion obeys Lagrange's equation:

$$\frac{dp_i}{dt} - \frac{\partial L}{\partial q_i} = 0, \quad p_i \equiv \frac{\partial L}{\partial \dot{q}_i}, \tag{A.1}$$

where the Lagrangian is a function of the coordinate $(q_i, \dot{q}_i)$. Hereafter the subscripts of the phase-space coordinates are omitted when there is no ambiguity. The Hamiltonian and Hamilton's equation of motion are

$$H(q, p, t) = \sum_i \dot{q}_i p_i - L(q, \dot{q}, t), \tag{A.2}$$

$$\frac{dp}{dt} = -\frac{\partial H}{\partial q}, \quad \frac{dq}{dt} = \frac{\partial H}{\partial p}; \quad \frac{\partial H}{\partial t} = -\frac{\partial L}{\partial t}. \tag{A.3}$$

Hamilton's equation of motion is derived from the variational principle of the Lagrangian. The conjugate variables (q, p) of the coordinates and momenta can be transformed to another set (Q, P) by a total differential (*contact transformation*) with a generating function. Four forms of generating functions are $F_1(q, Q, t)$, $F_2(q, P, t)$, $F_3(p, Q, t)$ and $F_4(p, P, t)$. The corresponding canonical transformations are

$$G = F_1(q, Q, t): \; p = \frac{\partial F_1}{\partial q}, \; P = -\frac{\partial F_1}{\partial Q}; \quad \mathcal{H}(Q, P, t) = H(q, p, t) + \frac{\partial F_1}{\partial t}; \tag{A.4}$$

$$G = F_2(q, P, t): \; p = \frac{\partial F_2}{\partial q}, \; Q = \frac{\partial F_2}{\partial P}; \quad \mathcal{H}(Q, P, t) = H(q, p, t) + \frac{\partial F_2}{\partial t}; \tag{A.5}$$

$$G = F_3(p, Q, t): \; q = -\frac{\partial F_3}{\partial p}, \; P = -\frac{\partial F_3}{\partial Q}; \quad \mathcal{H}(Q, P, t) = H(q, p, t) + \frac{\partial F_3}{\partial t}; \tag{A.6}$$

$$G = F_4(p, P, t): \; q = -\frac{\partial F_4}{\partial p}, \; Q = \frac{\partial F_4}{\partial P}; \quad \mathcal{H}(Q, P, t) = H(q, p, t) + \frac{\partial F_4}{\partial t}. \tag{A.7}$$

I.2 Fixed Points

Fixed points of Hamiltonian flow are phase space points where both $\dot{q} = 0$ and $\dot{p} = 0$. Thus the velocity field at fixed point is zero. Fixed points are classified into stable fixed points (SFPs) and unstable fixed points (UFPs). Near the SFP, the Hamiltonian flow resembles elliptical motion, and it is also called elliptical fixed point. Near the UFP, the Hamiltonian flow is hyperbolic, thus it is also called hyperbolic fixed point.

The fixed points are important in the Hamiltonian dynamics, because they determine the topology of Hamiltonian flow in the phase space. The Hamiltonian flow (torus) that pass through the UFP is called the separatrix, which separates the Hamiltonian flow into stable and unstable regions.

I.3 Poisson Bracket

The Poisson bracket of two functions, $u(q,p), v(q,p)$ of the phase-space coordinates is defined as

$$[u,v] = \sum_i \left(\frac{\partial u}{\partial q_i}\frac{\partial v}{\partial p_i} - \frac{\partial v}{\partial q_i}\frac{\partial u}{\partial p_i}\right). \tag{A.8}$$

By definition, we have $[q_i, q_j] = 0$; $[q_i, p_j] = \delta_{ij}$; $[p_i, p_j] = 0$. From the definition, the Poisson brackets satisfy the following properties:

- anti-commutative : $[u,v] = -[v,u]$;
- Jacobi's identity : $[[u,v],w] + [[v,w],u] + [[w,u],v] = 0$.

Using the Poisson bracket, we can express Hamilton's equation and the time derivative of an arbitrary function $F(q,p,t)$ as

$$\frac{dq}{dt} = [q,H], \quad \frac{dp}{dt} = [p,H]; \qquad \frac{dF}{dt} = [F,H] + \frac{\partial F}{\partial t}. \tag{A.9}$$

If the Poisson bracket $[F,H] = 0$ and F is not an explicit function of time, then F is a constant of motion. Clearly, if the Hamiltonian H is not an explicit function of time, then H is a constant of motion. If H is independent of coordinate q_i then the conjugate momentum p_i is a constant of motion. This can be observed easily from the Hamilton's equation. If a canonical transformation can be found that transforms all momenta to constants, the complete solution can be obtained through inverse transformation. Some examples of Hamiltonian systems are given in this section.

I.4 Liouville Theorem

Let $H(t, q_1, \cdots, q_N, p_1, \cdots, p_N)$ be the Hamiltonian of an isolated dynamical system, where t is the time coordinate, and $(q_1, \cdots, q_N, p_1, \cdots, p_N)$ are the generalized phase-space coordinates with

$$\dot{q}_i = \frac{\partial H}{\partial p_i}, \quad \dot{p}_i = -\frac{\partial H}{\partial q_i}. \tag{A.10}$$

Let $\rho(q_1, \cdots, q_N, p_1, \cdots, p_N)$ be the *density function*, and $\rho d\tau$ be the number of the system within phase-space volume $d\tau = \prod dq_i \prod dp_i$. The rate of increasing phase-space points inside volume V and the rate of phase-space points flowing out of the volume are respectively

$$\frac{\partial}{\partial t}\int_V \rho d\tau \quad \text{and} \quad \int \rho(\vec{v}\cdot\vec{n})d\sigma,$$

where $\vec{v}$ is the vector field of the Hamiltonian flow, $\vec{n}$ is the normal vector on the surface of the volume V, and $d\sigma$ is the surface integral differential. For a non-dissipative Hamiltonian system with no source and sink, we obtain the continuity equation with Gauss's theorem:

$$\frac{\partial\rho}{\partial t} + \nabla\cdot(\rho\vec{v}) = 0, \quad \Longleftarrow \quad \int \rho(\vec{v}\cdot\vec{n})d\sigma = \int_V \nabla\cdot(\rho\vec{v})d\tau. \tag{A.11}$$

Using Hamilton's equation, Eq. (A.10), we get the equation of continuity

$$\frac{d\rho}{dt} = \frac{\partial\rho}{\partial t} + \sum_i \dot{q}_i\frac{\partial\rho}{\partial q_i} + \sum_i \dot{p}_i\frac{\partial\rho}{\partial p_i} = 0. \tag{A.12}$$

This is called the Liouville theorem.

I.5 Floquet Theorem

We consider the linear Hill's equation of motion $y'' + K(s)y = 0$,where y and y' are conjugate phase space coordinates, $K(s)$ is the focusing function, and the prime is the derivative with respect to the independent variable s. In particle accelerators, $K(s)$ is a periodic function of s with period L, i.e. $K(s+L) = K(s)$. It is advantageous to make the Floquet transformation and express the solution in amplitude and phase functions. The Floquet theorem states that the amplitude and phase functions satisfy a periodic periodic boundary condition similar to that of the potential function $K(s)$, i.e.

$$y(s) = w(s)e^{j\psi(s)}, \qquad w(s) = w(s+L), \quad \psi(s+L) - \psi(s) = 2\pi\mu, \tag{A.13}$$

where the phase advance μ in one period is independent of s. Although the periodic boundary condition is not necessary, it would simplify the solution of the differential condition. Using the Floquet transformation, we get the differential equation

$$2w'\psi' + w\psi'' = 0, \quad \psi' = \frac{1}{w^2}, \quad \psi = \int_{s_0}^{s}\frac{dt}{w^2}, \tag{A.14}$$

$$w'' + K(s)w - w\psi'^2 = 0\ , \quad w'' + K(s)w - \frac{1}{w^3} = 0, \tag{A.15}$$

where we have chosen a normalization for the amplitude function in Eq. (A.14). By defining $Y = w^2$, we obtain

$$\frac{d^3Y}{ds^3} + 4K\frac{dY}{ds} + 2\frac{dK}{ds}Y = 0.$$

The amplitude function can be solved easily for some special function K. The second order differential equation has two independent solutions $y_1 = we^{j\psi}$ and $y_2 = we^{-j\psi}$. It is easy to verify that the Wronskian $W = y_1y_2' - y_1'y_2$ is invariant. The solution y of Hill's equation is a linear combination of y_1 and y_2; it satisfies the Courant-Snyder invariant,

$$\epsilon = w^{-2}y^2 + (w'y - y'w)^2, \tag{A.16}$$

where $\pi\epsilon$ is the phase-space area enclosed by the ellipse of particle motion.

II Stochastic Beam Dynamics

Electrons in storage rings emit synchrotron radiation, which is a quantum process. Since the photon emission is discrete and random, the quantum process causes also diffusion and excitation. The balance between damping and excitation provides a natural emittance or beam size for the electron beam bunch in a storage ring. Because the synchrotron radiation spectrum depends weakly on the energy of photons up to the critical frequency, the emission of photons can be approximated by white noise, i.e. electrons are acted on by a Langevin force. For random noise, an important theorem is the central limit theorem discussed below.

II.1 Central Limit Theorem

If the probability $P(u)$ of each quantum emission is statistically independent, and the probability function falls off rapidly as $|u| \to \infty$, then the probability distribution function for the emission of n photons is a Gaussian,

$$\mathcal{P}_n(w) = \frac{1}{\sqrt{2\pi}\sigma_n} e^{-(w-w_n)^2/2\sigma_n^2}, \tag{A.17}$$

$$w_n = n\langle u\rangle, \quad \langle u\rangle = \int uP(u)du, \quad \sigma_n^2 = n\sigma_u^2, \quad \sigma_u^2 = \int (u - \langle u\rangle)^2 P(u)du.$$

The theorem is important in all branches of information science. We provide a mathematically non-rigorous proof as follows. Since the quantum emission is statistically independent, the probability of n photons being emitted is

$$P(u_1)du_1P(u_2)du_2\cdots P(u_n)du_n.$$

Thus we have

$$\mathcal{P}_n(w) = \int\int\int P(u_1)P(u_2)\cdots P(u_n)\delta(w - \sum_{i=1}^{n} u_i)du_1du_2\cdots du_n. \quad \text{(A.18)}$$

Using the identity

$$\delta(w - \sum_{i=1}^{n} u_i) = \frac{1}{2\pi}\int_{-\infty}^{\infty} e^{jk(w-\sum u_i)}, \quad \text{(A.19)}$$

we obtain

$$\mathcal{P}_n(w) = \frac{1}{2\pi}\int_{-\infty}^{\infty} e^{-jkw}\,[Q(k)]^n, \quad \text{(A.20)}$$

$$Q(k) = \int_{-\infty}^{\infty} due^{jku}P(u) \approx 1 + jk\langle u\rangle - \frac{1}{2}\langle u^2\rangle k^2 + \cdots. \quad \text{(A.21)}$$

Since $Q(k)$ is small for large k, we can expand it in power series shown in Eq. (A.21). Substituting into Eq. (A.20) and using the formula $\ln(1+y) = y - \frac{1}{2}y^2 + \cdots$, we obtain

$$\mathcal{P}_n(w) = \frac{1}{2\pi}\int_{-\infty}^{\infty} e^{-j(kw+n\ln[Q(k)])}dk = \frac{1}{\sqrt{2\pi}\sigma_n}e^{-(w-w_n)^2/2\sigma_n^2}, \quad \text{(A.22)}$$

where ω_n and σ_n are given in Eq. (A.17), which is called the Einstein relation in the random walk problem. This result indicates that the distribution function is Gaussian, and the square of the rms width increases linearly with the number of photons emitted. The balance between diffusion and phase-space damping gives rise to an equilibrium beam width.

II.2 Langevin Equation of Motion

We consider a 1D *noisy damping* dynamical system: $H_0 = \frac{1}{2}p^2 + U(x)$. In the presence of damping and white noise, the unperturbed Hamiltonian and equations of motion are

$$x' = p, \quad p' = -\frac{dU}{dx} - Ap + D\xi(t), \qquad \langle\xi(t)\rangle = 0, \quad \langle\xi(t)\xi(t')\rangle = \delta(t-t')\,. \quad \text{(A.23)}$$

Here (x, p) are conjugate phase-space coordinates, and $U(x)$ is the potential energy, D is the diffusion coefficient, A is the phase-space damping coefficient, and $\xi(t)$ is the white noise function. The stochastic differential equation becomes

$$x'' + Ax' + \frac{dU}{dx} = D\xi(t), \quad \text{(A.24)}$$

To solve the stochastic differential equation numerically, we can use several numerical algorithms of *stochastic integration methods* listed as follows.

A. Random walk method

Including the quantum emission of photons, the difference *tracking* equation for the normalized synchrotron phase-space coordinates is

$$\begin{cases} x_{i+1} = x_i + 2\pi\nu_s(-Ax_i + p_i) \\ p_{i+1} = p_i - 2\pi\nu_s \left.\dfrac{dU}{dx}\right|_{i+1} + (2\pi\nu_s)^{1/2} DW(t) \,, \end{cases} \qquad W(t) = \frac{1}{T_0^{1/2}} \int_t^{t+T_0} \xi(t')dt' \,,$$

where the subscript indicates the revolution number, and ν_s is the (synchrotron) tune, the $W(t)$ is Wiener process function, T_0 is the time for one revolution in the ring, and $\xi(t)$ is the white-noise function. Thus the variance of the Wiener process function becomes $\langle W(t)W(t)\rangle = 1$ .In the tracking equations, a Wiener process $W(t)$ can be imitated by a random walk of ± 1 per revolution. In the smooth approximation, the above tracking equation is equivalent to the differential equations of motion:

$$\begin{cases} \dfrac{dx}{dt} = -\left(\dfrac{2\pi\nu_s}{T_0}\right) Ax + \left(\dfrac{2\pi\nu_s}{T_0}\right) p \\ \dfrac{dp}{dt} = -\left(\dfrac{2\pi\nu_s}{T_0}\right) \dfrac{dU}{dx} + \left(\dfrac{2\pi\nu_s}{T_0}\right)^{1/2} D\xi(t) \,. \end{cases} \tag{A.25}$$

Here t is the real time for particle motion in a storage ring.

B. Other stochastic integration methods

For one stochastic variable x, the general Langevin equation has the form

$$\dot{x}(t) = f(x) + g(x)\xi(t). \tag{A.26}$$

The Langevin force $\xi(t)$ is assumed to be a Gaussian random variable with zero mean and δ-function correlation shown in Eq. (A.23). The integration of Eq. (A.26) is

$$x(t+h) = x(t) + f(x)h + g(x)\sqrt{h}W(h),$$
$$W(h) = \frac{1}{\sqrt{h}} \int_t^{t+h} ds\; \xi(s).$$

Two widely used methods for solving stochastic differential equations numerically are Euler's and Heun's.

B.1 Euler's scheme

Euler's integration scheme includes terms up to order h for additive noise. To integrate stochastic differential equations from $t = 0$ to $t = T$, we first divide the time interval T into N small finite steps of length h

$$t_n = nh, \qquad h = T/N, \quad n = 1, 2, ..., N.$$

The stochastic variable at a later time t_{n+1}, $x_{n+1} = x(t_{n+1}) = x((n+1)h)$, is calculated according to $x_{n+1} = x_n + f(x_n)h + g(x_n)\sqrt{h}W_n(h)$, where $W_1(h)$, $W_2(h)$, ..., $W_N(h)$ are independent Gaussian-distributed random variables with zero mean and variance 1, i.e. $\langle W_n \rangle = 0$, $\langle W_n W_m \rangle = \delta_{nm}$. A possible choice of the set of Gaussian random variable W_n is

$$W_n(h) = \sum_{i=1}^{M} \sqrt{\frac{12}{M}}(r_i - 0.5), \tag{A.27}$$

where r_i is a random number with $0 \le r_i < 1$, and M is an arbitrary non-zero large integer, e.g. $M \ge 10$.

B.2 Heun's scheme

Heun's scheme is second order in h. The difference from Euler's scheme is an additional predictor step,

$$x_{n+1} = x_n + \frac{1}{2}(f(x_n) + f(y_n))h + g(x_n)\sqrt{h}W_{2n-1}(h) \tag{A.28}$$

with $y_n = x_n + f(x_n)h + g(x_n)\sqrt{h}W_{2n}(h)$. In this case we need $2N$ independent random variables $\{W_n(h)\}$. The equilibrium distribution function does not depend on the method of stochastic integration used in numerical simulations. It is, however, worth pointing out that a non-symplectic integration method can lead to a slightly different $E_{\rm th}$ due essentially to the change in the effective A parameter.

II.3 Fokker-Planck Equation

The equilibrium distribution function of a stochastic differential equation (A.24) satisfies the Fokker-Planck equation

$$\frac{\partial \Psi}{\partial t} = \left[-p\frac{\partial}{\partial x} + A\frac{\partial}{\partial p}p + \frac{dU}{dx}\frac{\partial}{\partial p} + \frac{D^2}{2}\frac{\partial^2}{\partial p^2}\right]\Psi. \tag{A.29}$$

The solution of the Fokker-Planck equation is

$$\Psi = \frac{1}{\mathcal{N}}\exp\{-\frac{H}{E_{\rm th}}\}, \qquad E_{\rm th} = \frac{D^2}{2A}, \tag{A.30}$$

where $\mathcal{N}$ is the normalization, and $E_{\rm th}$ is the "thermal" energy. In the small bunch approximation, the normalization constant becomes $\mathcal{N} = E_{\rm th}$. This is the Einstein relation, where the diffusion coefficient is proportional to the thermal energy $E_{\rm th} = kT$, where k is the Boltzmann constant and T is the temperature.

If the potential is nearly quadratic, i.e., the restoring force is simple harmonic, the distribution is bi-Gaussian. Thus the central limit theorem of white noise gives rise to a Gaussian distribution. In reality, if the potential is nonlinear, the distribution may not be Gaussian in coordinate space x.

The rms phase-space area $\mathcal{A}$ of the beam distribution is

$$\frac{\mathcal{A}}{\pi} = \sqrt{\mathrm{var}(x)\mathrm{var}(p) - (\mathrm{covar}(x,p))^2}, \tag{A.31}$$

where $\mathrm{var}(x) = \langle (x - \langle x \rangle)^2 \rangle$, $\mathrm{var}(p) = \langle (p - \langle p \rangle)^2 \rangle$, $\mathrm{covar}(x,p) = \langle xp \rangle - \langle x \rangle \langle p \rangle$. Here $\langle \cdots \rangle$ denotes an average over the beam distribution. In a small bunch nearly Gaussian approximation, the rms phase-space area is equal to $\pi E_{\rm th}$, i.e. the emittance $\mathcal{A}/\pi$ is equal to the thermal energy $E_{\rm th}$.

III Methods of Data Analysis in Beam Physics

The linear response of a dynamical system is represented by the relation between the N_b-dimensional observation vector $\mathbf{y}(t)$, i.e. the number of BPMs, and the N_s-dimensional source-signal vector $\mathbf{s}(t)$ by

$$\mathbf{y}(t) = \mathbf{A}\mathbf{s}(t) + \mathcal{N}(t) \tag{A.32}$$

where $N_b \geq N_s$, N_s is unknown a priori, $\mathbf{A} \in \Re^{N_b \times N_s}$ is the mixing matrix, and $\mathcal{N}(t)$ is the noise vector assumed to be stationary, zero mean, temporally white and statistically independent of source signal $\mathbf{s}(t)$. The task is to determine the mixing matrix $\mathbf{A}$ and the source signals $\mathbf{s}(t)$ from the measured sample signal $\mathbf{y}(t)$.

The source signals $\mathbf{s}(t)$ in most physical processes are independent and temporally un-correlated, i.e.

$$\langle s_i(t) s_j(t-\tau) \rangle = \int s_i(t) s_j(t-\tau) dt = C_{ii}\delta_{ij} \tag{A.33}$$

for an arbitrary non-zero time-lag constant τ. Here $\langle \cdots \rangle$ stands for mathematical expectation value or the ensemble average of the source signal. In particle accelerators, sources are betatron motion, synchrotron motion, power supply ripple, collective beam instabilities due to wake fields, ground motion, high frequency noises, etc. The data sampled by BPMs around the ring are put into a data matrix

$$\mathbf{y} = \begin{pmatrix} y_1(1) & y_1(2) & \dots & y_1(N) \\ y_2(1) & y_2(2) & \dots & y_2(N) \\ \vdots & \vdots & \ddots & \vdots \\ y_m(1) & y_m(2) & \dots & y_m(N) \end{pmatrix} \tag{A.34}$$

where N is the total number of turns, $m = N_b$ is the number of BPMs. The element $y_i(j)$ is the reading of the i'th BPM on the j'th turn. BPM gains may be applied to correct the BPM calibration error if necessary and available.

Traditionally, the model independent analysis (MIA) method analyzes data by making SVD decomposition to the data matrix, i.e.

$$\mathbf{y} = \mathbf{U}\mathbf{\Lambda}\mathbf{V}^T, \tag{A.35}$$

where $\mathbf{U}$ and $\mathbf{V}$ are unitary real matrices with $\mathbf{U}^T\mathbf{U} = \mathbf{I}$ and $\mathbf{V}^T\mathbf{V} = \mathbf{I}$, and $\mathbf{\Lambda}$ is a diagonal matrix. The MIA procedure is equivalent to $\mathbf{y}\mathbf{y}^T = \langle \mathbf{y}\mathbf{y}^T \rangle = \mathbf{U}\mathbf{\Lambda}^2\mathbf{U}^T$, i.e. making *equal-time correlation* to the data matrix. The eigenvalues in $\mathbf{\Lambda}$ of equal time correlation may sometimes become degenerate, and cause mode-mixing.

For example, we consider the betatron motions: $\mathbf{y} = \sqrt{2\beta_y J}\sin(\nu_y\phi)$, where $\beta_y(s)$ is the betatron amplitude function and J is the action. The SVD of the M-BPMs, N-turn data-matrix becomes

$$\mathbf{U} = \begin{pmatrix} P\sqrt{\frac{2\beta_y}{M}}\sin(\nu_y\phi_1) & P\sqrt{\frac{2\beta_y}{M}}\cos(\nu_y\phi_1) & 0 & \dots \\ P\sqrt{\frac{2\beta_y}{M}}\sin(\nu_y\phi_2) & P\sqrt{\frac{2\beta_y}{M}}\cos(\nu_y\phi_2) & 0 & \dots \\ \vdots & \vdots & \vdots & \dots \end{pmatrix},$$

$$\mathbf{V}^T = \begin{pmatrix} \sqrt{\frac{2}{N}}\cos(2\pi\nu_y\cdot 0) & \sqrt{\frac{2}{N}}\cos(2\pi\nu_y\cdot 1) & \dots & \sqrt{\frac{2}{N}}\cos(2\pi\nu_y\cdot (N-1)) \\ \sqrt{\frac{2}{N}}\sin(2\pi\nu_y\cdot 0) & \sqrt{\frac{2}{N}}\sin(2\pi\nu_y\cdot 1) & \dots & \sqrt{\frac{2}{N}}\sin(2\pi\nu_y\cdot (N-1)) \\ 0 & 0 & \dots & 0 \\ \vdots & \vdots & \ddots & \vdots \end{pmatrix}$$

where P is the normalization factor. The diagonal $\mathbf{\Lambda}$ matrix is $\mathbf{\Lambda}_{11} = \sqrt{2JMN}/(2P)$ and $\mathbf{\Lambda}_{22} = \sqrt{2JMN}/(2P)$. The spatial wave function of $\mathbf{U}$ matrix describes the betatron amplitude function, and the temporal wave function of $\mathbf{V}$ describes the sinusoidal betatron amplitude function. The SVD decomposition usually works well in identifying modes in the data matrix $\mathbf{y}$, except when the eigenvalues are degenerate, or weak modes that can be mixed by random noise.

The independent component analysis (ICA) extends the data analysis of MIA into *unequal-time correlation*, and thus it has the potential to separate these independent-modes. The source signal separation is to jointly diagonalize the covariance matrices with selected time-lag constants τ with data whitening procedure listed as follows.[1]

1. Compute the $N_b \times N_b$ sample covariance matrix $\mathbf{C}_\mathbf{y}(0) \equiv \langle \mathbf{y}(t)\mathbf{y}(t)^T \rangle$. Perform eigenvalue decomposition to $\mathbf{C}_\mathbf{y}(0)$ to obtain

$$\mathbf{C}_\mathbf{y}(0) = (\mathbf{U}_1, \mathbf{U}_2)\begin{pmatrix} \mathbf{D}_1 & 0 \\ 0 & \mathbf{D}_2 \end{pmatrix}\begin{pmatrix} \mathbf{U}_1 \\ \mathbf{U}_2 \end{pmatrix} \tag{A.36}$$

[1] see X. Huang, *et al.*,PRSTAB, **8**, 064001 (2005); X. Huang, Ph.D. Thesis, Indiana University (2005); F. Wang, and S.Y. Lee, PRSTAB **11**, 050701 (2008); F. Wang, Ph.D. Thesis, Indiana University (2008); X. Pang, S.Y. Lee, Journal of Applied Physics, **106**, 074902 (2009); X. Pang, Ph.D. Thesis, Indiana University (2009); J. Kolski, Ph.D. Thesis, Indiana University (2010).

where $\mathbf{D}_1$,$\mathbf{D}_2$ are diagonal matrices with $\min(\mathbf{D_1}_{ii}) > \lambda_c > \max(\mathbf{D_2}_{ii}) \geq 0$, λ_c is a cut-off threshold set to remove the singularity of the data matrix, or equivalently removing the noise background. Defining the matrix $\mathbf{V}$ as $\mathbf{V} \equiv \mathbf{D_1}^{-1/2}\mathbf{U_1}^T$, we construct an N_s-component vector as $\mathbf{Y} = \mathbf{Vy}$. The Vector $\mathbf{Y}$ is called white because $\langle \mathbf{YY}^T \rangle = \mathbf{I}$, where $\mathbf{I}$ is the $N_s \times N_s$ identity matrix. This MIA procedure reduces the dimension of the data space, separates the noise from the original data, and de-correlates and normalizes the data.

2. For a selected set of time-lag constants $\{\tau_k | k = 1, 2, \ldots, K\}$, compute the time-lagged covariance matrix $\{\mathbf{C_Y}(\tau_k) = \langle \mathbf{Y}(t)\mathbf{Y}(t-\tau_k)^T \rangle | k = 1, 2, \ldots, K\}$ and form symmetric matrix $\overline{\mathbf{C}}_\mathbf{Y}(\tau_k) = (\mathbf{C_Y}(\tau_k) + \mathbf{C_Y}(\tau_k)^T)/2$, and find a unitary matrix $\mathbf{W}$ that diagonalize all matrices $\overline{\mathbf{C}}_\mathbf{Y}(\tau_k)$ of this set, i.e. $\overline{\mathbf{C}}_\mathbf{Y}(\tau_k) = \mathbf{W}^T\mathbf{D_k}\mathbf{W}$, where $\mathbf{D_k}$ is diagonal.[2]

3. The source signals are found by $\mathbf{s} = \mathbf{WVy}$, i.e. the mixing matrix is $\mathbf{A} = \mathbf{V}^{-1}\mathbf{W}^T$.

For digitized sample data $y_i(t)$, constants τ_k have to be integers. The expectation functional $\langle \cdots \rangle$ are replaced with sample average in practice.[3] The application of ICA to synchrotron beam diagnosis involves three phases: data acquisition and pre-processing, source signal separation and beam motion identification. A pinger or rf resonant kicker is used excite the beam, the coherent transverse motion of turn-by-turn data is digitized. ICA algorithm is then applied to data matrix $\mathbf{y}$ to extract the mixing matrix $\mathbf{A}$ and source signals $\mathbf{s}$. Each source signal $\mathbf{s}_i$ and its spatial distribution $\mathbf{A_i}$, where $\mathbf{A_i}$ is the i'th column of $\mathbf{A}$, is called a mode. The physical origin of a mode can be identified by its spatial and temporal pattern. For example, the betatron amplitude and phase functions of a betatron mode are

$$\beta_i = a^2(A_{b1,i}^2 + A_{b2,i}^2), \quad \psi_i = \tan^{-1}\frac{A_{b1,i}}{A_{b2,i}}, \tag{A.37}$$

where a is a scaling factor depending on the kick amplitude and BPM calibrations. The dispersion function is $\mathbf{D} = b\mathbf{A}_s$. Here, b is also the scaling factor, that depends on the magnitude of synchrotron motion.

[2]The time-lagged covariance matrix is in fact symmetric if there is no error in noise and finite sampling. The symmetrization is used to guarantee a real solution in matrix diagonalization. Since the source signals are independent as shown in Eq. (A.33), the matrices $\mathbf{C_Y}(\tau_k)$ can be jointly diagonalized with an identical eigenvector matrix $\mathbf{W}$ and the eigenvalue-matrix $\mathbf{C_s}(\tau_k)$ for each τ_k. In practice, joint diagonalization can be achieved only approximately due to finite sampling error and noise. See J. F. Cardoso and A. Souloumiac, SIAM J. Mat. Anal. Appl., **17**, 161, (1996).

[3]The algorithm could be improved by robust whitening or by combining the non-stationary and time-correlation. See e.g. Aapo Hyvarinen, Juha Karhunen, Erkki Oja, *Independent Component Analysis*, (John Wiley & Sons, New York, 2001).

Appendix B

Numerical Methods and Physical Constants

I Fourier Transform

Spectral analysis of beam properties has many applications in beam physics. When a detecting device picks up a beam current or position signal as the beam passes by, the time structure and its frequency can be analyzed to uncover characteristic properties of the beam in the accelerator.

Let $y(t)$ be a physical quantity of the beam, e.g. a transverse betatron coordinate, a transverse sum signal, or a longitudinal phase coordinate of the beam. The Fourier spectrum function of $y(t)$ and its inverse Fourier transforms are

$$Y(\omega) = \int_{-\infty}^{\infty} y(t)\, e^{-j\omega t}\, dt,$$

$$y(t) = \frac{1}{2\pi}\int_{-\infty}^{\infty} Y(\omega)\, e^{j\omega t}\, d\omega$$

The variable t is time and the conjugate variable ω is the angular frequency.

In a synchrotron, the beam passes through the detector in a discrete sampling time at revolution period T_0. The measured physical quantities are

$$y_n = y(nT_0), \quad n = 1, 2, 3, \ldots,$$

where the sampling rate for beam motion in a synchrotron is usually equal to the revolution period.

I.1 Nyquist Sampling Theorem

The Nyquist theorem of discrete sampling states that, if the data are taken in time interval T_0, then their spectral content is limited by the Nyquist critical frequency

$$f_c = \frac{1}{2T_0}. \tag{B.1}$$

In terms of betatron tune, the Nyquist critical tune is $q_c = f_c T_0 = \frac{1}{2}$. The discrete sampling of beam motion can provide power spectrum only within the frequency range $(-f_c, f_c)$. The power spectrum of all outside frequencies is folded into the range $(-f_c, f_c)$. This is called aliasing. Frequency components outside the critical frequency range are aliased into the range by discrete sampling. In other words, we can assume that the Fourier component is nonzero only inside the frequency range $-f_c$ to f_c for our discrete sampling data.

I.2 Discrete Fourier Transform

Now we would like to find the Fourier transform $Y(\omega)$ of the phase space coordinate $y(t)$, where we have collected N consecutive data samples,

$$y_k = y(kT_0), \qquad k = 0, 1, 2, \cdots, N-1. \tag{B.2}$$

The data can provide the Fourier amplitude for all frequencies within the range $-f_c$ to f_c. For the N data points, we can estimate the Fourier amplitude at discrete frequencies

$$\omega_n = \frac{n}{N}\omega_0, \quad n = -\frac{N}{2}, \ldots, \frac{N}{2}, \tag{B.3}$$

where $\omega_0 = 2\pi/T_0$ is the angular revolution frequency. The discrete Fourier transform and its inverse transform are[1]

$$Y_n = Y(\omega_n) = \int_{-\infty}^{\infty} y(t)\, e^{-j\omega_n t} dt \approx T_0 \sum_{k=0}^{N-1} y_k\, e^{-j2\pi kn/N}. \tag{B.4}$$

$$y_k = \frac{1}{NT_0} \sum_{n=0}^{N-1} Y_n\, e^{j2\pi kn/N}.$$

Here n varies from $-N/2$ to $N/2$. We note that Y_n of Eq. (B.4) has a period N with $Y_{-n} = Y_{N-n}$, $n = 1, 2, \cdots$. Thus we can let n in Y_n vary from 0 to $N-1$. Here the frequency range $0 \le f < f_c$ corresponds to $0 \le n \le N/2 - 1$, the frequency range $-f_c < f < 0$ corresponds to $N/2 + 1 \le n \le N - 1$, and $f = f_c$ and $f = -f_c$ give rise to $n = N/2$. The discrete Fourier transform has properties similar to those of the Fourier transform of continuous functions, e.g.

y_n real	$Y_{-n} = Y_{+n}$
y_n imaginary	$Y_{-n} = -Y^*_{+n}$
y_n even/odd	Y_n even/odd

[1]The discrete Fourier transform can be optimized by an algorithm called the Fast Fourier Transform (FFT), which uses clever numerical algorithms to minimize the number of operations for the calculation of Y_n in Eq. (B.4). See, e.g., W.H. Press, B.P. Flannery, S.A. Tukolsky, and W.T. Vetterling *Numerical Recipes* (Cambridge Press, New York, 1990).

Since the tune is equal to the number of (betatron or synchrotron) oscillations per orbital revolution, the discrete Fourier transform gives a tune within the range $0 \le q \le 1/2$. Equation (B.3) of Nyquist's theorem implies also that the spectrum resolution from N sampling data points is

$$\Delta\omega = \frac{1}{N}\omega_0, \quad \text{or} \quad \Delta q = \frac{1}{N}. \tag{B.5}$$

For example, the betatron tune resolution is 0.001 from 1000 digitized data points. Figure 2.22 shows the FFT spectrum of horizontal betatron oscillations excited by a magnetic kicker. Since 385 data points are used in obtaining the FFT spectrum, the betatron tune resolution is about 0.003.

Discrete sampling of the phase space coordinate also gives rise to aliasing. Figure B.1 shows that the discrete data points can be fitted by sinusoidal functions with tunes $Q = m \pm q$, where m is an integer, and q is the fractional part of the betatron tune.

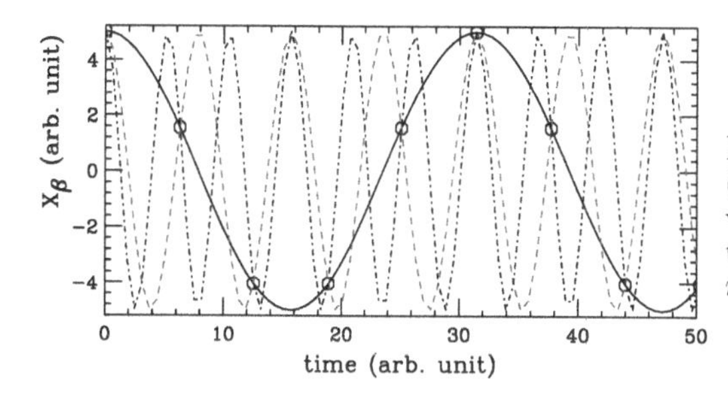

Figure B.1: Discrete data points (circles) fitted by sinusoidal functions with $Q = m \pm q$, where q is the fractional part of the betatron tune, and m is an integer.

I.3 Digital Filtering

The digitized data for a low intensity beam can be contaminated by many sources of noise, e.g. cable noise, cable attenuation, amplifier noise, power supply ripple, ground motion, etc., and the resulting betatron and synchrotron phase space coordinate data can be noisy.[2] A possible way to enhance the signal to noise (S/N) ratio is to filter the FFT spectrum by multiplying it by a filtering function $F(q)$. Performing the inverse FFT transformation with the filtered data can provide a much clearer beam signal. The filtering function can be a low pass filter, a high pass filter, a band pass filter, or a notch filter to remove only narrow bands of unwanted frequencies. A DC offset such as a closed orbit or BPM offset can be filtered by removing the running average of the BPM signals.

[2]See e.g., R.W. Hamming, *Digital filters* (Prentice Hall, Englewood Cliffs, NJ, 1977).

I.4 Some Simple Fourier Transforms

$\delta(\theta-\theta') = \frac{1}{2\pi}\sum_{n=-\infty}^{\infty} e^{jn(\theta-\theta')}$ $\qquad$ $\sum_{\ell=-\infty}^{\infty}\delta(t-\ell T_0) = \frac{1}{T_0}\sum_{n=-\infty}^{\infty} e^{j2\pi nt/T_0}$

$\delta(t) = \frac{1}{2\pi}\int_{-\infty}^{\infty} e^{j\omega t}d\omega$ $\qquad$ $\delta(\omega) = \frac{1}{2\pi}\int_{-\infty}^{\infty} e^{-j\omega t}dt$

$y(t) = \begin{cases} 1 & -T < t < T \\ 0 & \text{otherwise} \end{cases}$ $\qquad$ $Y(\omega) = \frac{1}{\pi}\frac{\sin\omega T}{\omega}.$

$Y(\omega, T)\Big|_{T\to\infty} = \delta(\omega).$

$\theta(t) = \begin{cases} -\frac{1}{2} & t < 0 \\ \frac{1}{2} & t > 0 \end{cases}$ $\qquad$ $\theta(\omega) = -\frac{1}{2\pi j\omega}.$

$\Theta(t) = \begin{cases} 0 & t < 0 \\ 1 & t > 0 \end{cases}$ $\qquad$ $\Theta(\omega) = -\frac{1}{2\pi j\omega} + \frac{1}{2}\delta(\omega).$

$y(t) = e^{-t/\tau}, \quad (t \geq 0),$ $\qquad$ $Y(\omega) = \frac{1}{2\pi j}\frac{1}{\omega - j/\tau}.$

$y(t) = \frac{1}{\pi}\frac{\tau}{t^2+\tau^2}$ $\qquad$ $Y(\omega) = \frac{1}{2\pi}e^{-\omega|\tau|}.$

II Cauchy Theorem and the Dispersion Relation

II.1 Cauchy Integral Formula

If $f(z)$ is an analytic function within and on a contour C, then $\oint_C f(z)dz = 0$. Let $f(z)$ be an analytic function within a closed contour C, and continuous on the contour C, then

$$f(a) = \frac{1}{2\pi j}\oint \frac{f(z)}{z-a}dz, \tag{B.6}$$

where a is any arbitrary point within C. The denominator of a Cauchy integral can usually be represented by

$$\frac{1}{z-a\mp j\epsilon} = \frac{1}{z-a}\Big|_{\text{P.V.}} \pm j\pi\delta(z-a), \tag{B.7}$$

where P.V. stands for the principle value of the integral, i.e.

$$\int_{\text{P.V.}} \frac{f(z)}{z-z_1}dz = \int_{\text{P.V.}} \frac{f(z)-f(z_1)}{z-z_1}dz.$$

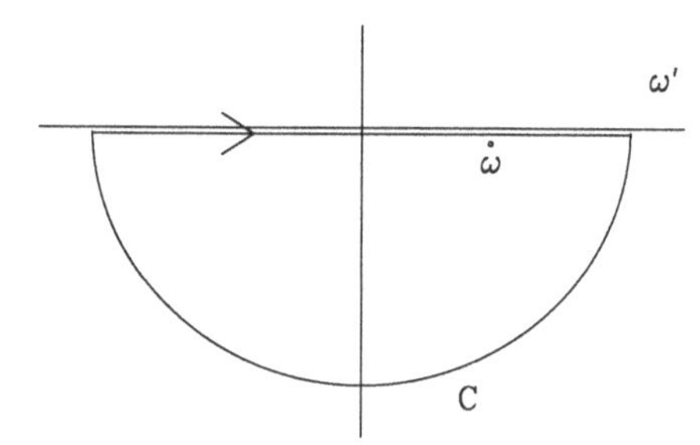

Figure B.2: The contour integral of the impedance in the complex ω' plane. Because the impedance is analytic in the lower complex ω' plane, the Cauchy integral formula can be used to obtain the dispersion relation.

II.2 Dispersion Relation

Since the impedance must be analytic in the lower complex plane, we obtain

$$Z(\omega) = \frac{-1}{2\pi j} \oint_C \frac{Z(\omega')}{\omega' - \omega} d\omega', \tag{B.8}$$

where the contour integral C is shown in Fig. B.2. Assuming that $Z(\omega) \to 0$ as $\omega \to \infty$, we obtain

$$Z(\omega) = \frac{-1}{2\pi j} \oint_C \frac{Z(\omega')}{\omega' - \omega + j\epsilon} d\omega' = \frac{-1}{2\pi j} \left[\oint_{\text{P.V.}} \frac{Z(\omega')}{\omega' - \omega} d\omega' - j\pi Z(\omega) \right], \tag{B.9}$$

where P.V. means taking the principal value of the integral. Thus the real and the imaginary impedance are related by the Hilbert transform

$$\operatorname{Re} Z_{\parallel}(\omega) = -\frac{1}{\pi} \int_{\text{P.V.}} d\omega' \frac{\operatorname{Im} Z_{\parallel}(\omega')}{\omega' - \omega}, \qquad \operatorname{Im} Z_{\parallel}(\omega) = +\frac{1}{\pi} \int_{\text{P.V.}} d\omega' \frac{\operatorname{Re} Z_{\parallel}(\omega')}{\omega' - \omega}.$$

III Useful Handy Formulas

III.1 Generating Functions for Bessel Functions

$$e^{jz\cos\theta} = \sum_{k=-\infty}^{\infty} j^k J_k(z) e^{jk\theta} = \sum_{k=-\infty}^{\infty} j^k J_k(z) \cos(k\theta)$$

$$\cos(z\cos\theta) = J_0(z) + 2\sum_{k=1}^{\infty} (-1)^k J_{2k}(z) \cos(2k\theta)$$

$$\sin(z\cos\theta) = 2\sum_{k=0}^{\infty} (-1)^k J_{2k+1}(z) \cos((2k+1)\theta)$$

$$e^{jz\sin\theta} = \sum_{k=-\infty}^{\infty} J_k(z) e^{jk\theta}$$

$$\cos(z\sin\theta) = J_0(z) + 2\sum_{k=1}^{\infty} J_{2k}(z)\cos(2k\theta)$$

$$\sin(z\sin\theta) = 2\sum_{k=0}^{\infty} J_{2k+1}(z)\sin((2k+1)\theta)$$

$$e^{\frac{1}{2}z(t+1/t)} = \sum_{k=-\infty}^{\infty} t^k I_k(z)$$

$$e^{z\cos\theta} = I_0(z) + 2\sum_{k=1}^{\infty} I_k(z)\cos k\theta$$

$$e^{z\sin\theta} = I_0(z) + 2\sum_{k=1}^{\infty}(-1)^k I_{2k}(z)\cos 2k\theta + 2\sum_{k=0}^{\infty}(-1)^k I_{2k+1}\sin(2k+1)\theta$$

III.2 The Hankel Transform

$$f(r) = \int_0^{\infty} g(k)J_n(kr)kdk \quad g(k) = \int_0^{\infty} f(r)J_n(kr)rdr$$

$$\delta(r-r') = r\int_0^{\infty} J_n(kr)J_n(kr')kdk \qquad \delta(k-k') = k\int_0^{\infty} J_n(kr)J_n(k'r)rdr$$

III.3 The Complex Error Function [30]

$$w(z) = e^{-z^2}\mathrm{erfc}(-jz) = \frac{j}{\pi}\int_{-\infty}^{\infty}\frac{e^{-t^2}}{z-t}dt = e^{-z^2}\left[1 + \frac{2j}{\sqrt{\pi}}\int_0^z e^{\zeta^2}d\zeta\right]$$

III.4 A Multipole Expansion Formula

$$\frac{b^2 - r^2}{b^2 + r^2 - 2br\cos\chi} = 1 + 2\sum_{n=1}^{\infty}\left(\frac{r}{b}\right)^n \cos n\chi$$

III.5 Cylindrical Coordinates

A point in rectangular space coordinates is represented in cylindrical coordinates by $x = \rho\cos\phi$, $z = \rho\sin\phi$, $s = s$, so that the unit vectors $\hat{\rho}$, $\hat{\phi}$ and $\hat{s}$ form the orthonormal basis with

$$\frac{d\hat{\phi}}{d\phi} = -\hat{\rho}, \qquad \frac{d\hat{\rho}}{d\phi} = \hat{\phi}.$$

The Jacobian is ρ, i.e. the volume element is $dV = \rho\, d\rho\, d\phi\, ds$. Any vector in the space can be expanded in the coordinate system by $\vec{A} = A_\rho\hat{\rho} + A_\phi\hat{\phi} + A_s\hat{s}$, where

A_ρ, A_ϕ and A_s are components. The position vector is $\vec{r} = \rho\hat{\rho} + s\hat{s}$, and the gradient operator and the Laplacian are respectively,

$$
\begin{aligned}
\nabla &= \hat{\rho}\frac{\partial}{\partial\rho} + \hat{\phi}\frac{1}{\rho}\frac{\partial}{\partial\phi} + \hat{s}\frac{\partial}{\partial s}, \\
\nabla^2 &= \frac{1}{\rho}\frac{\partial}{\partial\rho}\left(\rho\frac{\partial}{\partial\rho}\right) + \frac{1}{\rho^2}\frac{\partial^2}{\partial\phi^2} + \frac{\partial^2}{\partial s^2}, \\
\nabla\,\Phi &= \hat{\rho}\frac{\partial\Phi}{\partial\rho} + \hat{\phi}\frac{1}{\rho}\frac{\partial\Phi}{\partial\phi} + \hat{s}\frac{\partial\Phi}{\partial s}, \\
\nabla\cdot\vec{A} &= \frac{1}{\rho}\frac{\partial}{\partial\rho}(\rho A_\rho) + \frac{1}{\rho}\frac{\partial A_\phi}{\partial\phi} + \frac{\partial A_s}{\partial s}, \\
\nabla\times\vec{A} &= \hat{\rho}\left(\frac{1}{\rho}\frac{\partial A_s}{\partial\phi} - \frac{\partial A_\phi}{\partial s}\right) + \hat{\phi}\left(\frac{\partial A_\rho}{\partial s} - \frac{\partial A_s}{\partial\rho}\right) + \hat{s}\left(\frac{1}{\rho}\frac{\partial}{\partial\rho}(\rho A_\phi) - \frac{1}{\rho}\frac{\partial A_\rho}{\partial\phi}\right), \\
\nabla^2\Phi &= \frac{1}{\rho}\frac{\partial}{\partial\rho}\left(\rho\frac{\partial\Phi}{\partial\rho}\right) + \frac{1}{\rho^2}\frac{\partial^2\Phi}{\partial\phi^2} + \frac{\partial^2\Phi}{\partial s^2}.
\end{aligned}
$$

III.6 Gauss' and Stokes' Theorems

$$
\oint_S \vec{A}\cdot d\vec{S} = \int_V (\nabla\cdot\vec{A})\;dV \qquad\qquad \oint_C \vec{A}\cdot d\vec{s} = \int_S \nabla\times A\cdot d\vec{S}
$$

Here S is the surface area that encloses the volume V, $d\vec{S}$ is the differential for the surface integral, $d\vec{s}$ is the differential for the line integral, and C is the line enclosing the surface area S.

III.7 Vector Operation

$$
\begin{aligned}
&\vec{A}\cdot(\vec{B}\times\vec{C}) = (\vec{A}\times\vec{B})\cdot\vec{C} \\
&\vec{A}\times(\vec{B}\times\vec{C}) = \vec{B}\;(\vec{A}\times\vec{C}) + \vec{C}\;(\vec{A}\times\vec{B}) \\
&\nabla(\vec{A}\cdot\vec{B}) = \vec{B}\times(\nabla\times\vec{A}) + \vec{A}\times(\nabla\times\vec{B}) + (\vec{B}\cdot\nabla)\vec{A} + (\vec{A}\cdot\nabla)\vec{B} \\
&\nabla\cdot(\vec{A}+\vec{B}) = \nabla\cdot\vec{A} + \nabla\cdot\vec{B} \\
&\nabla\cdot(f\vec{A}) = (\nabla f)\cdot\vec{A} + f\;(\nabla\cdot\vec{A}) \\
&\nabla\cdot(\vec{A}\times\vec{B}) = (\nabla\times\vec{A})\cdot\vec{B} - \vec{A}\cdot(\nabla\times\vec{B}) \\
&\nabla\times(\vec{A}+\vec{B}) = \nabla\times\vec{A} + \nabla\times\vec{B} \\
&\nabla\times(f\vec{A}) = (\nabla f)\times\vec{A} + f\;(\nabla\times\vec{A}) \\
&\nabla\times(\nabla\times\vec{A}) = \nabla(\nabla\cdot\vec{A}) - \nabla^2\vec{A} \\
&\nabla\times(\vec{A}\times\vec{B}) = \vec{A}(\nabla\cdot\vec{B}) - \vec{B}(\nabla\cdot\vec{A}) + (\vec{B}\cdot\nabla)\vec{A} - (\vec{A}\cdot\nabla)\vec{B}
\end{aligned}
\tag{B.10}
$$

The term $(\vec{B}\cdot\nabla)\vec{A}$ obtains from the operation of the differential operator $(\vec{B}\cdot\nabla)$ on the vector function $\vec{A}$.

III.8 2D Magnetic Field in Multipole Expansion

order	Normal multipole ($B_0 b_n$)	
n	B_x	B_z
0		$B_0 b_0$
1	$B_0 b_1 z$	$B_0 b_1 x$
2	$B_0 b_2 (2xz)$	$B_0 b_2 (x^2 - z^2)$
3	$B_0 b_3 (3x^2 z - z^3)$	$B_0 b_3 (x^3 - 3xz^2)$
4	$B_0 b_4 (x^3 z - xz^3)$	$B_0 b_4 (x^4 - 6x^2 z^2 + z^4)$
5	$B_0 b_5 (5x^4 z - 10x3z^3 + z^5)$	$B_0 b_5 (x^5 - 10x3z^2 + 5xz^4)$

order	Skew multipole ($B_0 a_n$)	
n	B_x	B_z
0	$B_0 a_0$	
1	$B_0 a_1 x$	$-B_0 a_1 z$
2	$B_0 a_2 (x^2 - z^2)$	$-B_0 a_2 (2xz)$
3	$B_0 a_3 (x^3 - 3xz^2)$	$-B_0 a_3 (3x^2 z - z^3)$
4	$B_0 a_4 (x^4 - 6x^2 z^2 + z^4)$	$-B_0 a_4 (x^3 z - xz^3)$
5	$B_0 a_5 (x^5 - 10x3z^2 + 5xz^4)$	$-B_0 a_5 (5x^4 z - 10x3z^3 + z^5)$

IV Maxwell's Equations

The following table lists Maxwell's equations in a homogeneous medium, and the scalar and vector potentials:

$$\nabla \cdot \vec{E} = \frac{\rho}{\epsilon} \qquad \oint_S \vec{E} \cdot d\vec{S} = Q_{\text{encl.}} / \epsilon$$

$$\nabla \cdot \vec{B} = 0 \qquad \vec{B} = \nabla \times \vec{A},$$

$$\nabla \times \vec{E} = -\frac{\partial \vec{B}}{\partial t} \qquad \vec{E} = -\nabla \Phi - \frac{\partial \vec{A}}{\partial t}$$

$$\nabla \times \vec{B} = \mu \vec{J} + \mu \epsilon \frac{\partial \vec{E}}{\partial t} \qquad \oint_C \vec{H} \cdot d\vec{s} = I_{\text{encl.}}$$

Here ϵ and μ are the permittivity and permeability of the medium, $\vec{B} = \mu \vec{H}$ is the magnetic flux density, J is the current density, $Q_{\text{encl.}}$ is the charge inside the enclosed volume, and $I_{\text{encl.}}$ is the total current enclosed by a contour C.

The equation of continuity $\nabla \cdot \vec{J} + \partial \rho / \partial t = 0$ is a consequence of Maxwell's equation. The boundary condition at the interface of two material is $(\epsilon_2 \vec{E}_2 - \epsilon_1 \vec{E}_1) \cdot \hat{n} = \sigma$, where σ is the surface charge density on the boundary, $(\vec{E}_2 - \vec{E}_1)_{\parallel} = 0$, $(\vec{B}_2 - \vec{B}_1) \cdot \hat{n} = 0$, and $(\vec{H}_2 - \vec{H}_1)_{\parallel} = K_s$, where K_s is the surface current density per unit length along the boundary.

IV.1 Lorentz Transformation of EM Fields

The electromagnetic fields $(\vec{E}, \vec{B})$ at a rest frame (x, y, z) are transformed to another inertial reference frame (x', y', z') moving at a velocity $\vec{v}$ relative to the rest frame with velocity in the $+\hat{x}$ direction by

$$E'_x = E_x, \qquad E'_y = \gamma(E_y - vB_z), \qquad E'_z = \gamma(E_z + vB_y),$$
$$B'_x = B_x, \qquad B'_y = \gamma(B_y + \frac{v}{c^2}E_z), \qquad B'_z = \gamma(B_z - \frac{v}{c^2}E_y),$$

where $\gamma = 1/\sqrt{1-\beta^2}$, $\beta = v/c$.

IV.2 Cylindrical Waveguides

The electromagnetic fields in a cylindrical waveguide can be expressed in cylindrical coordinate system as $\vec{E} = E_\rho\hat{\rho} + E_\phi\hat{\phi} + E_s\hat{s}$ and $\vec{H} = H_\rho\hat{\rho} + H_\phi\hat{\phi} + H_s\hat{s}$. For the propagation of electromagnetic fields in a uniform sourceless medium (e.g. the free space), only two components of the EM fields, e.g., E_s and H_s, are sufficient to determine $\vec{E}$ and $\vec{H}$.

Without loss of generality, we consider the forward propagation mode with $e^{j[\omega t - k_s s]}$. The resulting field equations are

$$\left(\frac{\partial^2}{\partial\rho^2} + \frac{1}{\rho}\frac{\partial}{\partial\rho} + \frac{1}{\rho^2}\frac{\partial^2}{\partial\phi^2} + k_\rho^2\right) E_s = 0, \tag{B.11}$$
$$\left(\frac{\partial^2}{\partial\rho^2} + \frac{1}{\rho}\frac{\partial}{\partial\rho} + \frac{1}{\rho^2}\frac{\partial^2}{\partial\phi^2} + k_\rho^2\right) H_s = 0, \tag{B.12}$$

where $k_\rho^2 = (\omega/c)^2 - k_s^2$ with $c = 1/\sqrt{\epsilon\mu}$. The longitudinal components of the EM fields can be solved by the method of separation of variables, and all other components of the EM fields are

$$E_\rho = -\frac{j}{k_\rho^2}\left(k_s\frac{\partial E_s}{\partial\rho} + \frac{\omega\mu}{\rho}\frac{\partial H_s}{\partial\phi}\right), \qquad H_\rho = +\frac{j}{k_\rho^2}\left(\frac{\omega\epsilon}{\rho}\frac{\partial E_s}{\partial\phi} - k_s\frac{\partial H_s}{\partial\rho}\right),$$
$$E_\phi = -\frac{j}{k_\rho^2}\left(\frac{k_s}{\rho}\frac{\partial E_s}{\partial\phi} - \omega\mu\frac{\partial H_s}{\partial\rho}\right), \qquad H_\phi = -\frac{j}{k_\rho^2}\left(\omega\epsilon\frac{\partial E_s}{\partial\rho} + \frac{k_s}{\rho}\frac{\partial H_s}{\partial\phi}\right).$$

One can replace k_s by $-k_s$ for backward traveling waves.

The EM fields are conveniently classified into transverse magnetic (TM) modes, where $H_s = 0$ or $H_\parallel = 0$, and transverse electric (TE) modes, where $E_s = 0$ or $E_\parallel = 0$.

A. TM modes: $H_s = 0$

The solution of the longitudinal electric field is

$$E_s = A\cos(m\phi + \chi)J_m(k_\rho\rho), \tag{B.13}$$

where m = integer is the azimuthal mode number, $J_m(k_\rho \rho)$ is Bessel's function of order m, $k_\rho = \sqrt{\omega^2/c^2 - k_s^2}$ is the radial wave number, and k_s is the longitudinal wave number. In a perfectly conducting wave guide with radius b, the longitudinal component of the electric field must vanish on the wave guide wall, i.e. $J_m(k_\rho b) = 0$, or $k_{\rho,mn} = j_{mn}/b$, where j_{mn}, listed in Table B.1, are zeros of $J_m(z)$. The corresponding wave propagation mode is called TM_{mn} mode. Other components of the electromagnetic field are

$$E_\rho = -\frac{jk_s}{k_\rho} A\cos(m\phi + \chi) J'_m(k_\rho \rho), \qquad H_\phi = \frac{\omega\epsilon}{k_s} E_\rho,$$

$$E_\phi = +\frac{jk_s m}{k_\rho^2 \rho} A\sin(m\phi + \chi) J_m(k_\rho \rho), \qquad H_\rho = -\frac{\omega\epsilon}{k_s} E_\phi.$$

The wave impedance is

$$Z_{\text{TM}} = \frac{E_\rho}{H_\phi} = -\frac{E_\phi}{H_\rho} = \sqrt{\frac{\mu}{\epsilon}} \frac{k_s}{\omega/c}. \tag{B.14}$$

Table B.1: Zeros of Bessel function for TM and TE modes

	j_{mn} (TM)				j'_{mn} (TE)			
m\n	1	2	3	4	1	2	3	4
0	2.405	5.520	8.654	11.79	3.832	7.016	10.174	13.32
1	3.832	7.016	10.17	13.32	1.841	5.331	8.536	11.71
2	5.136	8.417	11.62	14.80	3.054	6.706	9.969	13.17
3	6.380	9.761	13.02	16.22	4.201	8.015	11.35	14.59

B. TE modes: $E_s = 0$

The solution of the longitudinal electric field is

$$H_s = A\cos(m\phi + \chi) J_m(k_\rho \rho), \tag{B.15}$$

where m = integer is the azimuthal mode number, $J_m(k_\rho \rho)$ is Bessel's function of order m, $k_\rho = \sqrt{\omega^2/c^2 - k_s^2}$ is the radial wave number, and k_s is the longitudinal wave number. Other components of the electromagnetic field are

$$E_\rho = +\tfrac{jm\omega\mu}{k_\rho^2 \rho} A\sin(m\phi + \chi) J_m(k_\rho \rho), \qquad H_\phi = \frac{k_s}{\omega\mu} E_\rho,$$

$$E_\phi = +\tfrac{j\omega\mu}{k_\rho} A\cos(m\phi + \chi) J'_m(k_\rho \rho), \qquad H_\rho = -\frac{k_s}{\omega\mu} E_\phi.$$

In a perfectly conducting wave guide with radius b, the ϕ-component of the electric field must vanish on the wave guide wall, i.e. $J'_m(k_\rho b) = 0$, or $k_{\rho,mn} = j'_{mn}/b$, where j'_{mn}, shown in Table B.1, are zeros of $J'_m(z)$. The corresponding wave propagation mode is called TE_{mn} mode. The wave impedance is

$$Z_{\text{TM}} = \frac{E_\rho}{H_\phi} = -\frac{E_\phi}{H_\rho} = \sqrt{\frac{\mu}{\epsilon}} \frac{\omega/c}{k_s}. \tag{B.16}$$

IV.3 Voltage Standing Wave Ratio

The voltage standing wave ratio (VSWR) measures the amount of reflection in a transmission line. Let the forward wave be $e^{j(\omega t-ks)}$ and the reflected wave $R\,e^{j(\omega t+ks)}$. Then the wave amplitude along the transmission line is $(e^{-jks}+R\,e^{+jks})e^{j\omega t}$.

The wave in the transmission line appears to oscillate in phase with respect to time but has spatial modulations due to interference. When $R=0$ the spatial modulation disappears, and when $R=1$ (100% reflection) the modulation looks like a standing wave on a string with nodes and peaks. The VSWR is defined as

$$\mathrm{VSWR}=\frac{\mathrm{Max}|e^{-jks}+R*e^{+jks}|}{\mathrm{Min}|e^{-jks}+R*e^{+jks}|},\qquad \text{or}\qquad R=\frac{\mathrm{VSWR}-1}{\mathrm{VSWR}+1}.$$

V Physical Properties and Constants

Microwave transmission in wave guide

We list some useful electromagnetic wave transmission properties of some media. Here ρ is resistivity; $\delta_{\rm skin}$, skin depth; ω, microwave frequency; ϵ, permittivity; μ, permeability; c, capacitance per unit length; and ℓ, inductance per unit length.

$\delta_{\rm skin}=\sqrt{2\rho/\omega\mu}$	skin depth
$Z=(1+j)\rho/\omega\delta_{\rm skin}$	resistive wall impedance
$c=2\pi\epsilon/\ln(r_2/r_1)$	capacitance per unit length in a coaxial cable
$\ell=(\mu/2\pi)\ln(r_2/r_1)$	inductance per unit length in a coaxial cable
$Z_c=\sqrt{L/C}$	characteristic impedance of a transmission line
$v=1/\sqrt{LC}=1/\sqrt{\mu\epsilon}$	speed of a wave in a transmission line

Thermodynamic law of dilute gases

The ideal gas law $PV=NkT=nRT$ is often used in the calculation of molecules in vacuum chamber, where P,V,N,k,T,n and R are respectively the pressure, volume, number of molecules, Boltzmann's constant, temperature, number of moles, and the ideal gas constant. Since there are a composition of gases in the vacuum chamber, partial pressure is usually used with $P=\sum P_i$, $N=\sum N_i$, and $n=\sum n_i$, where the ideal gas law becomes $P_iV=N_ikT=n_iRT$, where P_i,N_i and n_i, partial pressure, number of molecules, and number of moles for the ith gas species.

The target thickness t, defined as the number of molecules per m^2, is given by $t=CN/V=CP/kT$, where C is the circumference of the accelerator.

Critical temperature T_c of some superconducting materials

	Nb_3Sn	Nb	Pb	Hg	$YaBa_2Cu_3O_7$	BiSrCaCuO	TlBaCaCuO
T_c (K)	18.05	9.46	7.18	4.15	90	105	125

Resistivity and density of some materials

Resistivity	Ag	Cu	Au	Al	SS304	W
ρ_{c0} [$10^{-8}\Omega$m] (at 20°C)	1.59	1.7	2.44	2.82	7.3	5.5
α_c [10^{-3} /° C]	3.8	3.9	3.4	3.9	5.0	4.5
Resistivity at temperature T	$\rho_c(T) = \rho_{c0}[1 + \alpha_c(T - T_0)]$					
Density [g/cm³]	10.5	8.92	19.3	2.70	7.87	19.3

Units of physical quantities

Quantity	unit	SI unit	SI derived unit
Capacitance	F (farad)	m^{-2} $kg^{-1}s^4A^2$	C/V
Electric charge	C (coulomb)	As	
Electric potential	V (volt)	m^2 kg $s^{-3}A^{-1}$	W/A
Energy	J (joule)	m^2 kg s^{-2}	Nm
Force	N (newton)	m kg s^{-2}	N
Frequency	Hz (hertz)	s^{-1}	
Inductance	H (henry)	m^2 kg $s^{-2}A^{-2}$	Wb/A
Magnetic flux	Wb (weber)	m^2 kg $s^{-2}A^{-1}$	Vs
Magnetic flux density	T (tesla)	kg $s^{-2}A^{-1}$	Wb/m^2
Power	W (watt)	m^2 kg s^{-3}	J/s
Pressure	Pa (pascal)	m^{-1} kg s^{-2}	N/m^2
Resistance	Ω (ohm)	m^2 kg $s^{-3}A^{-2}$	V/A

Magnetic flux density is in Tesla [T], where 1 T $=10^4$ G. Magnetic field in the SI unit is (A/m). However, The cgs unit of Oe (in honor of Oersted) is also commonly used. The unit conversion is 1 Oe $= 1000/4\pi$ A/m, or 1 A/m $=4\pi \times 10^{-3}$Oe.

Radiation dose units and EPA limit

Activity	Ci	3.70×10^{10} disintegrations/s
	Bq	1 disintegration/s
Energy deposit	rad	amount of radiation that deposits energy 1.00×10^{-2} J/kg
	Gy	1 Gy = 100 rad = 1 J/kg in absorber
quality factor	QF	RBE (relative biological effect)
effective dose	rem	dose (in rad × QF)
	Sv	1 sievert = dose (in Gy × QF) = 100 rem

The RBE (QF) factors are 1.0 for X-rays and γ-rays, 1.0–1.7 for β-particles, 10–20 for α-particles, 4–5 for slow neutrons, 10 for fast neutrons, 10 for protons, and 20 for heavy ions.

Low level radiation dosage from natural sources accounts for about 130 mrems/year. The upper limit of radiation dose recommended by the U.S. government is 500 mrems/year apart from background radiation and exposure related to medical procedures. The upper limit of radiation dosage for radiation worker is 5 rems/year (or 50 mSv/year) for the entire body.

Quantities associated with nuclear collisions

1. Target thickness: $t = \rho\ell_t$ in (mass/area), where ρ is the density of the target material and ℓ_t is the thickness of the target.
2. Target thickness: $N_t = N_{\rm A}\rho\delta s/A$, usually in (number of atoms/cm^2), where $N_{\rm A}$ is the Avogadro number and A is the atomic mass in one mole.
3. Luminosity for fixed target: $\mathcal{L} = \mathcal{F}\times N_t$, where the incident flux is the number of incident beam particle per unit time, i.e. $\mathcal{F} = dN_{\rm beam\ particles}/dt$. The dimension of the luminosity is usually cm^{-2}s^{-1}.
4. Counting rate: $R = \mathcal{L}\sigma$, where σ is the collision cross-section.
5. Absorption length: $\lambda_{\rm abs} = A/(N_{\rm A}\sigma_{\rm inel}\rho)$, where A is the atomic mass, $N_{\rm A}$ is the Avogadro number, $\sigma_{\rm inel}$ is the inelastic cross-section, and ρ is the density of the material.

Unit definition and conversion often used in beam physics:

$$1\ {\rm cal} = 4.186\ {\rm J};\qquad 1\ {\rm J} = 10^7\ {\rm erg};\qquad 1\ {\rm eV} = 1.60217733\times 10^{-19}\ {\rm J}$$

$$E_{\rm photon}[{\rm eV}] = \frac{1239.84\ [{\rm eV-nm}]}{\lambda\ [{\rm nm}]}$$

$$1\ {\rm in.} = 0.0254\ {\rm m};\qquad 1\ {\rm Angstrom}\ [{\rm \AA}] = 10^{-10}\ {\rm m}$$

$$1\ {\rm barn} = 10^{-28}\ {\rm m}^2 = 10^{-24}\ {\rm cm}^2;$$

$$1\ {\rm atm} = 760\ {\rm torr} = 1.01325\times 10^5\ {\rm Pa} = 1.01325\ {\rm bar}$$

Power units:

$$1\ {\rm dBm} = 10\log({\rm P}/1.0\,{\rm mW})$$

$${\rm dB\ gain\ of\ power\ amplifier} = 10\log({\rm P_{out}/P_{in}}) = 20\log({\rm V_{out}/V_{in}})$$

Momentum rigidity of a beam:

$$p = mv = qB\rho,\qquad B\rho\ [{\rm T-m}] = \frac{p}{q} = \frac{A}{Z}\times 3.33564\times p\ [{\rm GeV/c/u}],$$

Longitudinal action in mm-mrad vs longitudinal phase space area in eV-s

The relation between the longitudinal action I_s (in mm-mrad) and the longitudinal phase space area $\mathcal{A}$ (in eV-s):

$$I_s = (c/\beta E)\mathcal{A} = \begin{cases} 1.60\times 10^5(\mathcal{A}\ [{\rm eV\,s}]/\beta\gamma)\ [\pi\mu{\rm m}] & \text{proton synchrotron} \\ 2.93\times 10^8(\mathcal{A}\ [{\rm eV\,s}]/\beta\gamma)\ [\pi\mu{\rm m}] & \text{electron synchrotron} \end{cases}$$

Fundamental physical constants

Physical constant	symbol	value	unit
Avogadro's number	N_A	6.022141×10^{23}	/mol
atomic mass unit ($\frac{1}{12}m(C^{12})$)	m_u or u	1.660539×10^{-27}	kg
Boltzmann's constant	k	1.38065×10^{-23}	J/K
Bohr magneton	$\mu_B = e\hbar/2m_e$	9.274009×10^{-24}	J/T
		5.788382×10^{-5}	eV/T
Bohr radius	$a_0 = 4\pi\epsilon_0\hbar^2/m_e c^2$	0.529177×10^{-10}	m
classical radius of electron	$r_e = e^2/4\pi\epsilon_0 m_e c^2$	2.81794×10^{-15}	m
classical radius of proton	$r_p = e^2/4\pi\epsilon_0 m_p c^2$	$1.5346986 \times 10^{-18}$	m
elementary charge	e	1.602176×10^{-19}	C
fine structure constant	$\alpha = e^2/2\epsilon_0 hc$	1/137.036	
$m_u c^2$		931.494	MeV
mass of electron	m_e	9.10938×10^{-31}	kg
$m_e c^2$		0.5109989	MeV
mass of proton	m_p	$1.6726216 \times 10^{-27}$	kg
$m_p c^2$		938.272	MeV
mass of neutron	m_n	1.674927×10^{-27}	kg
$m_p c^2$		939.5655	MeV
molar gas constant	$R = N_A k$	8.314	J/mol K
neutron magnetic moment	μ_n	$-0.966236 \times 10^{-26}$	J/T
proton g factor	$g_n = \mu_n/\mu_N = G_n/2$	-1.913427	
nuclear magneton	$\mu_p = e\hbar/2m_u$	5.05073×10^{-27}	J/T
Planck's constant	h	6.626069×10^{-34}	J s
permeability of vacuum	μ_0	$4\pi \times 10^{-7}$	N/A^2
permittivity of vacuum	ϵ_0	8.854188×10^{-12}	F/m
proton magnetic moment	μ_p	1.410607×10^{-26}	J/T
proton g factor	$g_p = \mu_p/\mu_N = G_p/2$	2.7928473	
speed of light (exact)	c	299792458	m/s
vacuum impedance	$Z_0 = 1/\epsilon_0 c = \mu_0 c$	376.73	Ω

Bibliography

[1] M. Stanley Livingston and John P. Blewett, *Particle Accelerators* (McGraw-Hill, New York, 1962)

[2] M. Sands, in *Physics with Intersecting Storage Rings*, edited by B. Touschek (Academic Press, N.Y. 1971), see also SLAC report SLAC-r-121 (1971).

[3] D. Edwards and M. Syphers, *An Introduction to the Physics of High Energy Accelerators* (Wiley, N.Y. 1993)

[4] T. Wangler, *Principle of RF Linear Accelerators*, (Wiley, N.Y. 1998).

[5] A. Chao, *Physics of Collective Beam Instabilities in High Energy Accelerators* (Wiley, N.Y. 1993)

[6] Bruno W. Zotter and Semyon Kheifets, *Impedance and Wakes in High-Energy Particle Accelerators* (World Scientific, 1998).

[7] K.Y. Ng, *Physics of Intensity Dependent Beam Instabilities* (World Scientific, 2005).

[8] M. Reiser, *Theory and Design of Charged Particle Beams* (Wiley, N.Y. 1994).

[9] S. Humphries, *Principle of Charge Particle Acceleration* (Wiley, N.Y. 2012).

[10] L. Michelotti, *Intermediate Classical Mechanics with Applications to Beam Physics* (Wiley, N.Y. 1995)

[11] H. Wiedemann, *Particle Accelerator Physics: Basic Principles and Linear Beam Dynamics* (Springer-Verlag, 1993)

[12] H. Wiedemann, *Particle Accelerator Physics II: Nonlinear and Higher-Order Beam Dynamics* (Springer-Verlag, 1995).

[13] N.S. Dikanskii and D. Pestrikov*The Physics of Intense Beams and Storage Rings* (AIP, N.Y. 1994).

[14] M. Month (ed.), AIP Conference Proceedings No. 249: The Physics of Particle Accelerators (Upton, N.Y. 1989, 1990)

[15] M. Month, (ed.), AIP Conference Proceedings No. 184: The Physics of Particle Accelerators (Ithaca, N.Y. 1988)

[16] M. Month, (ed.), AIP Conference Proceedings No. 153: The Physics of Particle Accelerators (Fermilab, 1984, SLAC, 1985)

[17] M. Month, (ed.), AIP Conference Proceedings No. 127: The Physics of Particle Accelerators (BNL, 1983)

[18] M. Month, S. Turner (eds.), Lecture Notes in Physics, *Frontiers of Particle Beams* (Springer-Verlag, Heidelberg, 1988).

[19] M. Dienes, M. Month, S. Turner (eds.), Lecture Notes in Physics, No. 400, *Frontiers of Particle Beams: Intensity Limitations* (Springer-Verlag, Heidelberg, 1990).

[20] E.D. Courant, H. Snyder, Theory of the Alternating Gradient Synchrotron, *Ann. Phys.* **3**,1 (1958). E.D. Courant, M.S. Livingston, and H.S. Snyder, *Phys. Rev.* **88**, 1190 (1952); E.D. Courant, M.S. Livingston, H.S. Snyder, and J.P. Blewett, *Phys. Rev.* **91**, 202 (1953).

[21] E.M. McMillan, Phys. Rev., **68**, 143 (1945); V.I. Veksler, *Compt. Rend. Acad. Sci. U.S.S.R.*, **43**, 329 (1944); **44**, 365 (1944).

[22] J.D. Jackson, *Classical Electrodynamics*, (John Wiley & Sons, New York, 1963).

[23] H. Grote and F.C. Iselin, The MAD Program, Version 8.1, User's Reference Manual, CERN/SL/90-13(AP) (1991); see http://mad.web.cern.ch/mad/

[24] A. Garren, The SYNCH Program.

[25] M. Borland, THE ELEGANT program; https://www3.aps.anl.gov/forums/elegant/

[26] L. Sanchez *et al.*, COMFORT (Control of Machine Functions, Orbits, and Trajectories) Version 4.0, unpublished.

[27] S.Y. Lee, *Spin Dynamics and Snakes in Synchrotrons*, (World Scientific, Singapore, 1997).

[28] W.H. Press *et al.*, *Numerical Recipes in Fortran, The Art of Scientific Computing*, 2nd ed., Cambridge University Press (1992).

[29] W.K.H. Panofsky and W.A. Wenzel, *Rev. Sci. Inst.* **27**, 967 (1956).

[30] M. Abramowitz and I.A. Stegun, eds, *Handbook of Mathematical Functions*, National Bureau of Standards, Applied Mathematics Series **55**, 9th printing (1970).

[31] I.S. Gradshteyn and I.M. Ryzlik, *Table of Integrals, Series, and Products* (Academic Press, New York, 1980); E.T. Whittaker and G.N. Watson, *A Course of Modern Analysis*, 4th edition, pp.404–427 (Cambridge Univ. Press, 1962).

[32] S. Turner, Editor, *Synchrotron Radiation and Free Electron Laser*, CERN-90-03 (CERN, Geneva, 1990); H. Winick and S. Doniach, *Synchrotron Radiation Research* (Plenum press, N.Y. 1980); C. Kunz, Editor, *Synchrotron Radiation*, in Topics in Current Research No.**10** (Springer-Verlag, Berlin, 1979).

Index

A

accelerators and laboratories
  Argonne National Laboratory (ANL)
    APS, 148, 239, 400, 412, 449, 465
  Brookhaven National Laboratory (BNL)
    AGS, 52, 127, 151, 239, 252, 286, 357
    AGS booster, 81, 88, 301
    RHIC, 151, 239, 252, 286, 301, 357
  BEPC, 239, 400, 412, 494
  CERN,
    ISR, 17, 310, 361
    LEP, 239, 400, 412, 413, 494
    LHC, 151, 412, 413
    PS (CPS), 139, 292, 424
    PSB, 311
    SPS, 324
  CESR, 400, 412, 493
  ELETTRA, 449
  Fermilab,
    booster, 44, 114, 170, 234, 286, 290
    MI, 252, 286, 301
    Tevatron, 151, 496
  IUCF (now IU CEEM)
    CIS, 19, 79, 303, 337
    Cooler, 90, 195, 239, 249, 251, 283
  JLAB (TJNAF), 23, 143
    CEBAF, 378, 391
  KEK
    PS, 201, 286, 357
    KEKB, 494
    TRISTAN, 239
  KURRI, 199
  LBNL
    ALS, 239, 400, 412
  LLNL,
    induction linac, 7
  LANL,
    CCL, CCDTL, 382, 395
    PSR, 94, 113, 364
  NSCL
    superconducting cyclotrons, 12
  SLAC,
    NLC damping ring, 239, 254
  SSC, 24, 151, 239, 413
accelerator applications, 22
accelerator components,
  acceleration cavities (rf system), 329
  dipoles (see dipole)
  quadrupole (see quadrupole)
  miscellaneous components, 22
accelerator lattice (see betatron motion)
achromats (see dispersion function)
action-angle variables,
  transverse (see betatron motion)
  longitudinal (see synchrotron motion)
adiabatic damping (see betatron motion)
adiabatic time (see synchrotron motion)
adiabaticity condition (see synchrotron motion)
admittance, 59
Alfvén current, 481
α-bucket (see synchrotron motion)
atomic-beam polarized ion source, 78
attractor (see resonances)

B

barrier rf bucket (see rf system)
beam-beam effects, 488
  beam-beam parameter, 488

coherent beam-beam effects, 491
dynamic beta, 489
in linear collider, 497
beamstrahlung (see beamstrahlung)
disruption deflection angle, 498
disruption factor, 490, 497
hour-glass effect (see luminosity)
nonlinear effects, 492
pretzel scheme, 496
tune shift, 488
beam-gas scattering, 121, 468
beam loading, 339
fundamental theorem, 340
phasor, 340
Robinson instability, 343
steady state, 341
beam manipulation, 292, 300
bunch compression (rotation), 305
(pre)buncher, 385, 395
capture, 302, 384
debunching, debuncher, 252, 308
extraction, 108
fast, 87, 109
slow, 110
injection, 108
charge exchange, 108
multi-turn injection, 119
strip (see charge exchange)
mismatch, 76, 254
phase space stacking, 109, 308
rf knock-out, 89
tune jump, (see tune)
beam position monitor (BPM), 32, 99
Δ-signal, 100
Σ-signal, 100
beam transfer function, 91, 275, 348
beamstrahlung parameter, Υ, 397, 497
betatron, betatron principle, 7
betatron motion, 44
accelerator lattice
Chasman-Green (see DBA)
combined function DBA, 445
DBA, 134, 149, 440
doublet cell, 52
Fixed-Field-Alternating-Gradient (FFA), 146
FMC, 141
FODO cell, 50, 124
missing dipole, 125
FOFO cell, 72
insertions, 151, 166
nBA, 452
QBA, 452
TBA, 452
triplet DBA, 473
action-angle, 53
adiabatic damping, 60, 236, 245
betatron tune (see tune)
betabeat (beta-beat), 96, 167
betatron amplitude matrix, 49
chromatic aberration, 158
correction, 163
chromaticity, 159
chromaticity corrections, 159
chromaticity measurement, 159
specific chromaticity, 159
Courant-Snyder invariant, 55
Courant-Snyder parametrization, 47
closed orbit, 80
integer stopband (see stopbands)
correction, 84
closed orbit bump, 84, 118
golden orbit, 122
off-momentum, 122
envelope equation (see envelope equation)
errors (dipole, quadrupole, see errors)
geometric aberration, 161, 186
momentum compaction, 129, 232
DBA and TME lattices, 150
flexible γ_T lattice, 141
FODO cell lattice, 128
γ_T manipulation, 138
necktie diagram, 61
normalized phase space coordinates, 55, 256
path length, 70, 85, 129, 154, 233, 267, 385
stability, 61

symplecticity, 61, 235, 239, 507
Beth representation (see magnetic field)
bifurcation (see resonance)
bucket area (see synchrotron motion)
bunched beams, 130, 242
bunch area (see synchrotron motion)
bunch height (see synchrotron motion)
bunching factor, 117, 245, 362

C

canonical perturbation, 208, 246, 254, 328
canonical transformation, 501
cavity (see rf systems)
central limit theorem, 504
Child's law, 31
chromatic aberration (see betatron motion)
chromaticity (see betatron motion)
Chasman-Green lattice (see lattice DBA)
circumference, 50
closed orbit (see betatron motion)
coasting (DC) beam, 107, 354, 360
Cockcroft-Walton, 6
collective instabilities, 210, 348
 beam break up (BBU), 383, 389
 head-tail, 159, 221, 255
 Keil-Schnell criterion, 360
 Landau damping, 210
 longitudinal, 347
 microwave, 353
 Robinson instability (see beam loading)
 transverse, 210
collider (colliding beam facility), 16, 400, 494
 e^+e^- colliders, 16, 494
compaction factor (see betatron motion)
cooling, beam cooling, 17, 109, 415
 stochastic cooling, 3, 88
 electron cooling, 3
 laser cooling, 3
 ionization cooling, 3
 synchrotron radiation cooling, 3, 415
Courant-Snyder invariant (see betatron motion)
Courant-Snyder parametrization (see betatron motion)
critical frequency ω_c, 411
cyclotron, 11
 AVF, sector-focused, 13
 isochronous, 13
 K-value, 12
 superconducting, 13
 synchrocyclotron, 14
 separate-sector (ring), 13
cyclotron frequency ω_{cyc}, 11

D

damping decrement, 265, 298, 422
Debuncher (see beam manipulation)
diffraction function $\zeta_P(u)$, 165
dipole (magnet), 20, 28
 sector dipole, 40, 45, 69, 124
 edge focusing, 69
dipole mode, 217, 260
dispersion function, 122
 achromat, 133
 dispersion action, 126
  $\mathcal{H}$-function, 126
  $\langle\mathcal{H}\rangle$ minimization, 146
 dispersion suppression, 133
 integral representation, 128
 momentum compaction, 129, 232
 phase slip factor, 129, 233
 transition energy, 129, 138, 153, 233
distribution functions,
 transverse, 59, 63, 204, 217, 430
 longitudinal, 106, 244, 255, 271, 289, 350
double bend achromat (see betatron motion)
double rf system (see rf system)
dynamic aperture, 198

E

electrostatic accelerators, 5
 Cockcroft-Walton, 6
 tandem, Van de Graaff, 6

X-ray tube, 6
EM fields (see Maxwell's equation)
emittance, 55
effective emittance, 441, 453
electron storage rings, 397, 443
IDs on emittances, (see undulator)
growth (see beam gas scattering)
growth (see injection mismatch)
longitudinal emittance (see phase space area)
normalized emittance, 60
measurement, 57
quadrupole tuning method, 58
moving screen method, 58
momentum spread, 244
momentum spread (quantum fluctuation), 428
statistical definition, 56, 73
transverse emittance, 56
envelope equation
of betatron motion, 48, 62, 75
KV envelope equation, 65
errors,
dipole field error, 80
quadrupole field error, 95
extraction (see beam manipulation)

F

feed-down, 95
FEL (see synchrotron radiation)
ferrite, 332
fixed point (see resonances)
fixed-field-alternating-gradient (FFA), 23, 198
Floquet theorem, 497
Floquet transformation, 48, 53, 154, 497
FMC (see betatron motion, lattice)
focal length, 45
focusing of atomic beams, 78
focusing: lithium lens, 78
focusing: solenoid, 42, 72
focusing index, 21, 39, 70, 153, 441
Fokker-Planck equation (see Langevin force)
Fourier transform, 104, 188, 339, 406, 511
discrete Fourier transform, 512
digital filtering, 513
Nyquist theorem, 511
free electron laser (see synchrotron radiation)
Frenet-Serret coordinate system, 35
frequency map (see tune)

G

gain (see FEL)
gain-length (see FEL)
geometric aberration (see betatron motion)
Green's function, 80, 92
gradient error, (see errors)
half-integer stopband (see stopbands)
group velocity (see velocity)

H

$\mathcal{H}$-function (see dispersion function)
Hamilton's equation, 501
Hill's equation, 39, 44, 503
HOM, 332, 383, 389
hour-glass effect (see luminosity)

I

impedance, 210, 355
characteristic, 333
longitudinal, 355
RLC, 335
shunt impedance, 335, 371
transverse, 210
vacuum impedance Z_0, 211, 357, 373
independent component analysis (ICA), 509
inductance of accelerator magnets
solenoid, 27
dipole, 28
quadrupole, 29
injection (see beam manipulation)
insertions (see betatron motion, lattice)

insertion devices (IDs) (see undulator)
instabilities,
  see resonances
  see collective instabilities
interaction point (IP), 77, 169
isochronous
  cyclotron (see cyclotron)
  quasi-isochronous (see cyclotron)
isomagnetic, 399

J

Jacobian, 53, 59, 235
Jacobian elliptical function, 258, 296, 315
[JJ] factor, 461

K

KAM theorem, 198, 273
K-value (see cyclotron)
Keil-Schnell (see collective instabilities)
kicker, 88
kicker lever arm, 71, 88
Kilpatrick limit, 305
klystron, 10, 369
KV distribution, 63
KV equation (see envelope equation)

L

Lambertson septum, 89
Landau damping (see collective instabilities)
Landé g-factor, 78
Langevin force, 505
  Fokker–Planck equation, 507
Larmor theorem, 398
Laslett tune shift (see space charge)
lattice (see betatron motion)
linac, 9, 367
  Alvarez, DTL, 9, 369, 377
  beam breakup instability, 390
  BNS damping, 390
  CCL, 382
  RFQ, 10
  standing wave, 330, 370, 381
  superconducting, 372, 524
  traveling wave, 381
    constant gradient, 394
    constant impedance, 395
  Wideröe, 9
linear coupling, 171
  solenoid, 171
  skew quadrupole, 171
  coupling coefficient, 172
lithium lens (see focusing)
loss factor (see rf system)
luminosity, 26, 77, 493
  hour-glass effect, 77
Lyapunov exponent, 200

M

magnetic moment (dipole), 78
magnetic field,
  Beth representation, 38
  multipole expansion, 38
  pressure, 79
Mathieu instability, 113, 276
Maxwell's equations, 330, 357, 483, 518
  Electromagnetic fields
    TE mode, 519
    TM mode, 520
microtron, 14
microwave instability (see instabilities)
model independent analysis (MIA), 94, 509
momentum compaction (see betatron motion)
momentum rigidity, 4, 20, 523
momentum spread (see emittance)

N

necktie diagram (see betatron motion)
negative mass, 233, 358
normalized emittance (see emittances)
nonlinear resonances (see resonances)

nonlinear time (see synchrotron motion)

O

OFHC (oxygen free high conductivity), 21
off-momentum closed orbit (see closed orbit)
optically pumped polarized ion source, 78
orbit response matrix (ORM), 91, 132

P

Panofsky-Wenzel theorem, 211, 227, 389
paraxial ray equation, 31, 66, 78, 111
path length (see betatron motion)
perveance (see space charge)
phase detector, 251
phase displacement acceleration, 309
phase focusing, 130, 230, 383
phase slip factor (see dispersion function)
phase space area
    longitudinal (see emittance)
    transverse (see emittance)
phase space stacking (beam manipulation)
phase stability (see phase focusing)
phase velocity (see velocity)
Pierce parameter (see synchrotron radiation)
phasor (see beam loading)
Poisson bracket, 502
Poisson equation, 31, 488, 498
Poisson distribution, 425
Poincaré surface of section, 54, 174, 266
pretzel scheme (see beam-beam interaction)
proper time, 25, 398
PUE (Pick Up Electrode), (see BPM)

Q

Q-factor (see rf systems)
quadrupole, 21, 28
quadrupole mode,
    longitudinal, 278, 366, 388
    transverse, 75

R

radiation length, 121
reaction length, 79
response matrix (see orbit response matrix)
RFQ linac (see linac)
Robinson instability (see beam loading)
Robinson theorem, 421
Robinson wiggler (see wiggler)
resonances,
    attractor, 268
    bifurcation, 195, 261
    bifurcation tune, 260
    devil's stair case, 195
    difference resonance 172, 188, 203
    fixed point (FP), 190, 241, 256, 278, 502
        elliptical, stable (SFP), 190, 241
        hyperbolic, unstable (UFP), 190, 241
        separatrix, 190, 241, 297, 433, 502
    hysteretic phenomena, 272
    island tune (see tune)
    linear coupling, 171
    nonlinear resonance, 186
    parametric resonance, 259
    sum resonances, 185, 188, 205
    sum rule (see sum rule)
    synchro-betatron resonance (SBR), 227
    torus, 53, 190, 236, 247, 262
retarded
    scalar and vector potential, 36, 401, 520
    retarded time, 401
rf systems, 329
    barrier bucket, 316
    beam loading (see beam loading)
    coaxial cavities, 332
    double rf system, 310
    ferrite loaded cavities, 333
    filling time, 336, 372
    loss factor, 339
    pillbox, 330
    Q-factor, 334, 372
    rf accelerators, 8
    rf cavities, 330
    RLC equivalent circuit, 335

shunt impedance (see impedance)
transit time factor, 231, 238, 370
wake function, 339
rf knockout (see beam manipulation)
rotation harmonics, 104, 214

S

σ–matrix, 57
Schottky noise, 106, 349
sensitivity factor, 83
separatrix (see resonance)
septum, 88
shunt impedance (see impedance)
skin depth, 211, 333, 345, 357, 392, 521
slow extraction (see beam manipulation)
solenoid, 27, 42, 72, 172, 182
Sokolov-Ternov radiative polarization, 437
space charge, 62
tune shift, 68, 112, 117
perveance, 64, 112, 117
ion source, 31
potential (longitudinal), 356
potential (transverse), 68
spectrum, beam spectrum
longitudinal, 104, 250, 349
roll-off frequency, 106, 349
spectrum analyzer (SA), 350
transverse, 104, 181, 214
spin polarization, 103, 437
intrinsic spin resonances, 103
imperfection spin resonances, 103
Stern-Gerlach effect, 78
stochastic integration methods, 506
stopbands,
integer stopband, 82
half-integer stopband, 96
systematic (chromatic) stopband, 165
storage rings, 16, 397
sum rule (resonances), 258, 283, 317
superconductor (type II), 17, 521
superperiod, 46
surface resistivity, surface resistance, 334
symplecticity (see betatron motion)
synchronous particle, 130, 231
synchrotron, 14
synchrotron motion, 229
action-angle variables, 246, 256
adiabaticity coefficient $\alpha_{\rm ad}$, 302
adiabatic synchrotron motion, 241
adiabatic time, 286
α-bucket, 295
barrier bucket (see rf systems)
bucket area, 242, 298
bucket height, 242
bunch area, 244
bunch rotation (see beam manipulations)
double rf system (see rf systems)
mapping equation, 235
non-adiabatic, 285
nonlinear time, 290
rf phase modulation, 259
rf voltage modulation, 275
synchrotron sidebands, 260, 349
synchrotron tune (see tune)
torus (see nonlinear resonances)
synchrotron radiation, 397
FEL, 476
efficiency, 482
FEL (Pierce) parameter, 485
gain, 481
gain-length, 485
critical frequency (see critical frequency)
damping, 415
longitudinal, 417, 422
partition number, 418
re-partition number, 418
transverse (horizontal), 420, 422
transverse (vertical), 420, 422
emittance (see emittance)
radiation excitation, 425
lifetime
quantum, 432
Touschek, 434
measurement, 442
momentum spread (see emittance)
photon flux, 411

power, 410
quantum excitation (see radiation)
quantum fluctuation, 411
radiation integrals, 438
undulator, (see undulator)
wigglers (see undulator)

T

tandem (see electrostatic accelerators)
target (foil) thickness, 26, 121, 524
three bend achromat (see lattice)
torus (see resonances)
Touschek lifetime (see synchrotron radiation)
transport notation, 136
tune,
betatron tune, 50
frequency map, 195
instantaneous, or turn-by-turn, 195
island tune, 207, 261, 278
nonlinear detuning, 196, 260
synchrotron tune, 234
tune diffusion rate, 200
tune jump, 102
tune shift,
beam-beam (see beam-beam effect)
of quadrupole error, 96
space charge (see space charge)
transfer matrix, 44
transit time factor (see rf systems)
transition energy (see dispersion function)

U

U parameter (impedance), 215, 359
undulator, 456
on beam dynamics, 461
on emittances, 461
on momentum spread, 461
resonance wavelength, 460
Robinson wiggler, 424
wavelength shifter, 460
undulator or wiggler parameter, 456

V

V parameter (impedance), 215, 359
vector and scalar potential, 36, 401, 520
velocity,
group velocity, 374
phase velocity, 374
vacuum impedance (see impedance)
Van de Graaff, 6
VSWR, 338, 521

W

wavelength shifter (see wiggler)
wigglers (see undulator)
wall gap monitor, 250
wake function (see rf system)
wakefield, 214, 339
Weierstrass function, 296
Wronskian, 46, 504

XYZ

X-ray tube, Coolidge, 5
zero gradient synchrotron (ZGS), 69